Developmental Biology

A COMPREHENSIVE SYNTHESIS

Volume 2

The Cellular Basis of Morphogenesis

Developmental Biology
A COMPREHENSIVE SYNTHESIS

General Editor
LEON W. BROWDER
University of Calgary
Calgary, Alberta, Canada

Developmental Biology

A COMPREHENSIVE SYNTHESIS

Volume 2

The Cellular Basis of Morphogenesis

Edited by
LEON W. BROWDER

University of Calgary
Calgary, Alberta, Canada

PLENUM PRESS • NEW YORK AND LONDON

Library of Congress Cataloging in Publication Data

Developmental biology.

Includes bibliographies and index.
Contents: v. 1. Oogenesis—v. 2. The cellular basis of morphogenesis.
1. Developmental biology—Collected works. I. Browder, Leon W.

QH491.D426 1985   574.3   85-3406
ISBN 0-306-42164-X

Cover illustration: Low power scanning electron micrograph of a landscape of deep cells migrating on the internal yolk syncytial layer of a *Fundulus* mid-gastrula–stage embryo. This image is visible by removing the overlying blastoderm. (From J. P. Trinkaus and C. A. Erickson, 1983, Protrusive activity, mode and rate of locomotion, and pattern of adhesion of *Fundulus* deep cells during gastrulation, *J. Exp. Zool.* **228**: 45.)

© 1986 Plenum Press, New York
A Division of Plenum Publishing Corporation
233 Spring Street, New York, N.Y. 10013

Printed in the United States of America

To Johannes Holtfreter

Through the elegant experiments he performed, Dr. Holtfreter demonstrated that our understanding of morphogenesis can be improved by studying it at the cellular level. His example and inspiration have convinced developmental biologists that complex morphogenic processes can be resolved into more simple ones, which can be separately analyzed. The progress made by successive generations of scientists in understanding morphogenesis confirms his faith in this approach to analyzing the embryo.

Ray Keller
Leon Browder

Contributors

Patricia Calarco-Gillam   Department of Anatomy, School of Medicine, University of California, San Francisco, California 94143

P. J. Donovan   Department of Anatomy, St. George's Hospital Medical School, London SW17 ORE, England

Carol A. Erickson   Department of Zoology, University of California—Davis, Davis, California 95616

Mark Harris Hankin   Department of Anatomy and Cell Biology, Center for Neuroscience, University of Pittsburgh, School of Medicine, Pittsburgh, Pennsylvania 15261

S. Robert Hilfer   Department of Biology, Temple University, Philadelphia, Pennsylvania 19122

Ray E. Keller   Department of Zoology, University of California—Berkeley, Berkeley, California 94720

John Kolega   Department of Biology, Yale University, New Haven, Connecticut 06511.   *Present address:* Department of Dermatology, New York University Medical Center, New York, New York 10016

James W. Lash   Department of Anatomy, School of Medicine, University of Pennsylvania, Philadelphia, Pennsylvania 19104

Clive Lloyd   Department of Cell Biology, John Innes Institute, Norwich NR4 7UH, England

Nadine C. Milos   Department of Anatomy, University of Alberta, Edmonton, Alberta, T6G 2H7, Canada

Frances Moody-Corbett   Department of Physiology, Tufts University School of Medicine, Boston, Massachusetts 02111.   *Present address:* Division of Basic Medical Sciences, Faculty of Medicine, Memorial University of Newfoundland, Saint John's, Newfoundland A1B 3V6, Canada

David Ostrovsky Department of Biology, Millersville University of Pennsylvania, Millersville, Pennsylvania 17551

Esmond J. Sanders Department of Physiology, University of Alberta, Edmonton, Alberta, T6G 2H7, Canada

Thomas E. Schroeder Friday Harbor Laboratories, University of Washington, Friday Harbor, Washington 98250

Robert L. Searls Department of Biology, Temple University, Philadelphia, Pennsylvania 19122

Jerry Silver Department of Developmental Genetics and Anatomy, School of Medicine, Case Western Reserve University, Cleveland, Ohio 44106

Michael Solursh Department of Biology, University of Iowa, Iowa City, Iowa 52242

Evelyn Spiegel Department of Biological Sciences, Dartmouth College, Hanover, New Hampshire 03755

Melvin Spiegel Department of Biological Sciences, Dartmouth College, Hanover, New Hampshire 03755

D. Stott Department of Anatomy, St. George's Hospital Medical School, London SW17 ORE, England

Masatoshi Takeichi Department of Biophysics, Faculty of Science, Kyoto University, Kyoto 606, Japan

C. C. Wylie Department of Anatomy, St. George's Hospital Medical School, London SW17 ORE, England

Sara E. Zalik Department of Zoology, University of Alberta, Edmonton, Alberta T6G 2E9, Canada

Preface

This series was established to create comprehensive treatises on specialized topics in developmental biology. Such volumes are especially vital in developmental biology, since it is a very diverse field that receives contributions from a wide variety of disciplines. This series is a meeting-ground for the various practitioners of this science, facilitating an integration of heterogeneous information on specific topics.

Each volume is intended to provide the conceptual basis for a comprehensive understanding of its topic as well as an analysis of the key experiments upon which that understanding is based. The specialist in any aspect of developmental biology should understand the experimental background of the field and be able to place that body of information in context to ascertain where additional research would be fruitful. At that point, the creative process takes over, and new experiments are designed. This series is intended to be a vital link in that ongoing process of learning and discovery. If it facilitates scholarship, it will serve an important function.

In this volume, the various mechanisms that operate at the cellular level to assemble multicellular embryos are discussed. Our current understanding of the roles of the cytoskeleton, cell surface, and extracellular matrix molecules in cellular shape changes, cellular interactions, and cell motility in development is presented in detail. Topics covered run the gamut from a consideration of generalized cell dynamics to the acquisition of specific cell shapes and the establishment of specific cell–cell interactions that are necessary to produce functional entities. Additional information on the role of the cell surface in development is presented in Volume 3 of this series. That book (*The Cell Surface in Development and Cancer*, edited by Malcolm Steinberg) also chronicles the properties of the cell surface in transformed cells. An understanding of the surfaces of malignant cells is important not only in determining the basis for malignant transformation but also in devising therapeutic means to counteract the deadly consequences of proliferation and spread of cancer cells.

Johannes Holtfreter is in large part responsible for developing the idea that the complex processes involved in formation of a multicellular embryo are explainable by the behaviors of individual cells. This reductionist approach to morphogenesis has been highly successful and accounts for virtually all of our progress in understanding the basis for morphogenesis in the past 40-odd years. There remain great challenges in understanding how the information in the genome of a zygote can be utilized differentially in individual cells of the embryo to result in massive and highly specific cell rearrangements, establishment of functional interactions, cell shape changes, and acquisition of the ultimate form of the adult organism. The molecular biology of morphogenesis is an exciting challenge for the future. Not only must we be able to account for the expression of individual genes encoding proteins that perform significant morphogenic functions, but we must explain how the expression of those genes is controlled and coordinated with the expression of other morphogenic-significant genes. With our current knowledge of cellular function in morphogenesis as a launching pad, progress in molecular morphogenesis is inevitable and will be exciting to study.

Leon Browder

Contents

II. Cell–Cell Interactions in Development

Chapter 8 • Cell–Cell Interactions in Mammalian Preimplantation Development

Patricia Calarco-Gillam

Chapter 9 • Molecular Basis for Teratocarcinoma Cell–Cell Adhesion

Masatoshi Takeichi

III. Cell Migration

Chapter 10 • Migration of Sea Urchin Primary Mesenchyme Cells

Michael Solursh

IV.   Cellular Dynamics in Morphogenesis

Chapter 15 • Mechanisms of Axonal Guidance: The Problem of
            Intersecting Fiber Systems

Mark Harris Hankin and Jerry Silver

Chapter 16 • Formation of the Vertebrate Neuromuscular Junction

Frances Moody-Corbett

I

The Cytoskeleton in Morphogenesis

Chapter 1

Cytoskeletal Dynamics in Animal Morphogenesis

S. ROBERT HILFER and ROBERT L. SEARLS

1. Introduction

Morphogenesis of animals involves dramatic changes in cell shape and massive rearrangements of cells, including migration of cells over relatively long distances. Both cellular shape change and motility are mediated by elements of the cytoskeleton. This chapter introduces the cytoskeleton of animal cells and discusses its roles in cellular dynamics. Chapter 2 discusses the function of the cytoskeleton in plant cell morphogenesis. Chapter 3 then discusses the egg cortex, which is rich in cytoskeletal elements, in the context of its role in early echinoderm development. Because of its central role in morphogenesis, virtually every aspect of the acquisition of adult form, shape, and function is affected by the cytoskeleton. Thus, the importance of the cytoskeleton is evident throughout this volume.

1.1. Definition of Cytoskeleton

The term *cytoskeleton* is somewhat misleading, because it covers not only those cytoplasmic elements that give rigidity or stability to cell shape, but also those elements that are involved in change of cell shape and in motility. The cytoskeleton consists of all those intracellular elements that remain in place after membranes have been removed with detergent and the cells have been thoroughly washed with physiological saline. Upon fractionation on sodium dodecyl sulfate (SDS)–polyacrylamide gels, the cytoskeleton is found to consist of a very large number of proteins. Most of these have not yet been further investigated and have no known function. When cytoskeletons are prepared by treating cells with detergent and washing, scanning or transmission electron microscopy demonstrates that they contain filaments, granules, and amor-

S. ROBERT HILFER and ROBERT L. SEARLS • Department of Biology, Temple University, Philadelphia, Pennsylvania 19122.

phous material. Not all these materials can be discerned in electron micrographs of intact cells. Two types of filaments, distinguished on the basis of size, are seen in electron micrographs of intact cells. Thinner filaments, with a diameter of 6–8 nm, are called **microfilaments.** Thicker filaments, with a diameter of 8–11 nm, are called **intermediate filaments.** Occasionally, short thicker filaments are encountered that resemble myosin rods in structure. Thick tubular structures in the cytoplasm, with a diameter of 24 nm, are called **microtubules.** A number of other proteins that copurify with, and that are considered part of, the filaments or tubules are known. Other proteins have been purified and identified by immunohistochemistry as components of the cytoskeleton.

1.2. Historical Summary

The concept of an elastic or contractile framework in the cytoplasm is not new. During the late 1800s, it was proposed by investigators such as Flemming, DuJardin, and Heidenhain to explain the structures they saw in living and fixed cells. However, much of this speculation was based on fixation artifacts (see Wilson, 1924, for review). Holtfreter (1943) recognized densities at the free surfaces of cell layers that were undergoing morphogenetic changes. He interpreted these dense layers as elastic surface coats. These were especially prominent during formation of the dorsal lip during amphibian gastrulation. These regions now are known to have well-organized microfilament bundles. Warren Lewis (1947) also proposed that a contractile mechanism causes neurulation. The pioneering work of Hoffman-Berling demonstrated contractility of glycerinated fibroblast cytoplasm in the presence of ATP (e.g., Hoffman-Berling and Weber, 1953).

However, cytoskeletal elements were not clearly demonstrated until the 1960s. Microtubules were recognized in stable structures such as cilia and flagella during the early days of electron microscopy. They were not recognized as a normal cytoplasmic component, however, until refinements in technique, such as adoption of aldehydes as fixatives, fixation at room temperature, and the use of epoxy resins, became available during the early 1960s. The isolation of contractile proteins by biochemical techniques from nonmuscle cells (e.g., Bettex-Galland et al., 1962), the morphological identification of actin in electron microscopic preparations (Ishikawa et al., 1969), and the recognition of a correlation between microfilament organization and cellular shape changes (Cloney, 1966) led to studies of the functional organization of microfilament meshworks and bundles. Investigation of the microfilaments led to the observation of filaments too wide to be microfilaments and too narrow to be myosin: intermediate filaments (Ishikawa et al., 1969). The technique of immunohistochemistry emerged during this same period. Numerous proteins were isolated as components of the cytoskeleton in cells of one type, antibodies prepared, and the same or similar proteins were then discovered in numerous other cell types by immunohistochemistry. Finally, certain drugs have been found to act specifically, or almost specifically, on individual components of the cytoskeleton such as the microtubules (Weisenberg et al., 1968) and micro-filaments (Schroeder, 1969). These agents permitted the manipulation of cytoskeleton organization to assess the role of specific elements.

2. Organization of the Cytoskeleton

2.1. Microfilaments

The microfilaments consist of cytoplasmic actin. Bundles of microfila-ments may also contain cytoplasmic myosin and various regulatory proteins, or they may contain one or more of an assortment of actin-binding proteins. The composition of this part of the cytoskeleton has been the subject of several reviews (e.g., Pollard and Weihing, 1974; Clarke and Spudich, 1977; Pollard, 1981).

2.1.1. Actin

Cytoplasmic actin consists of two gene products, **β-actin** and **γ-actin.** Cytoplasmic β- and γ-actins differ from sarcomeric α-actin by the absence of the N-terminal amino acid and by 25- and 24-amino acid replacements respec-tively out of a total of 374 amino acids (Vandekerckhoeve and Weber, 1978). All of the actins have a molecular weight of approximately 42,000 M_r and cannot be separated from one another on SDS–polyacrylamide gels. The β- and γ-actins differ from each other by four amino acid replacements, all in the N-terminal 18-amino acids. They are present in different cell types in different ratios but are present in different fractions of a single cell type in the same ratio. It is not known whether they have the same function.

The globular monomers (G-actin) assemble into a double helix with a diameter of 6–7 nm (F-actin). Polarity of the actin filaments can be demon-strated by binding to them the "head" portions, heavy meromyosin (HMM), of myosin molecules (see Section 2.1.2). The "head" portions bind at an angle of approximately 45°, forming arrowheads. All the arrowheads point in the same direction—the direction in which the actin filaments would move if contrac-tion of actomyosin complexes were to occur. The direction in which the ar-rowheads point is called the pointed end; the other end of the filament is the barbed end. Polymerization of G-actin in vitro proceeds in three steps. Mono-mers link to form small oligomers. These short chains act as nucleation sites for additional monomer. Finally, oligomers join end to end to form long filaments. Formation of F-actin proceeds until a critical concentration of monomer re-mains in solution. At this point, a steady state is reached with loss of monomer at one end, while addition occurs at the same rate at the other end. This process, called **treadmilling,** depends on assembly and disassembly occurring at different rates at the two ends (Neuhaus et al., 1983). Treadmilling and the difference in critical concentrations at the two ends of actin filaments have been demonstrated using actin bundles from Limulus sperm (Bonder et al.,

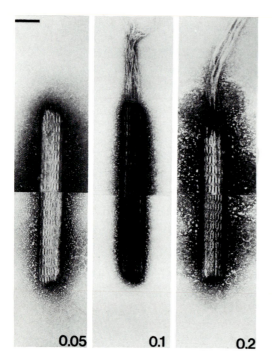

Figure 1. Composite micrographs prepared from the two ends of actin filament bundles isolated from *Limulus* sperm. Bundles were incubated in 0.05-, 0.1-, or 0.2-µM actin monomers. The upper half of each bundle shows continued nucleation of the barbed ends of the filaments down to the 0.1-µM monomer, whereas the pointed ends do not nucleate even at the 0.2-µM monomer. The nucleated assembly no longer occurs at the barbed ends at the 0.05-µM monomer. Scale bar: 0.1 µm. (Reprinted with permission from Bonder *et al.*, 1983. © Massachusetts Institute of Technology, published by the MIT Press.)

1983). When F-actin decorated with myosin heads is added to a high concentration of G-actin, undecorated regions extend the chain at both ends. At a concentration of 0.5 mM actin, the pointed ends no longer elongate, but the concentration must be lowered to 0.05 mM actin to prevent elongation of the barbed ends (Fig. 1). This differential critical concentration exists only in the presence of micromolar Mg^{2+} concentrations. When Ca^{2+} is added, the critical concentration of G-actin is the same at both ends.

2.1.2. Myosin

Cytoplasmic myosin does not cross-react with antibodies to sarcomeric myosin. However, it has the same structure: two identical heavy chains of about 200,000 M_r, two identical light chains of 20,000 M_r, and two identical light chains of 15,000–17,000 M_r. When lightly digested with papain, the molecules split into a rod portion, light meromyosin (LMM), possessing the ability to assemble into short filaments, and a head portion, HMM, that contains the light chains. Further protease treatment will split the head portion into two fragments with (S_1) and without (S_2) the light chains. The HMM and S_1 fragments possess actin-binding activity (HMM fragments are used to decorate actin filaments) and myosin ATPase activity. The intrinsic ATPase activity is greatly increased in the presence of actin and Ca^{2+}. The Ca^{2+} is bound by the calcium regulatory protein, **calmodulin.** This serves to activate myosin light-chain kinase, a protein that transfers a phosphate from ATP to the 20,000-M_r

light chain of the myosin head (Yerna *et al.*, 1979). Both the fragment of the heavy chain remaining in the myosin head and the 20,000-M_r light chain interact with actin (Sutoh, 1982). Monomers of cytoplasmic myosin assemble into short bipolar rods that have the heads pointing outward in a spiral fashion. Thus, the polymer is a symmetrical bipolar unit (Stossel and Pollard, 1973). When assembled *in vitro*, myosin polymers from nonmuscle cells have a diameter of 8–12 nm and consist of relatively few myosin molecules compared with sarcomeric myosin.

2.1.3. Other Proteins in Microfilaments

Accessory proteins have been reviewed by Schliwa (1981) and Geiger (1983). These molecules are involved in actin–myosin interactions, in regulating the length and state of polymerization of actin, and in attaching actin to the plasmalemma.

Nonmuscle cells contain a nonmuscle **tropomyosin** but not troponins (Adelstein, 1982). There have been reports that the Ca^{2+} modulation of the actin–myosin interaction in smooth muscle involves proteins called leiotonins, which serve much the same purpose as the troponins (Ebashi *et al.*, 1982). Leiotonins have not been reported in nonmuscle cells. Some nonmuscle cells also contain α-actinin, one component of the Z disc in sarcomeric muscle.

A number of other proteins interact with actin. These actin-modifying proteins fall into two groups: those that are calcium independent and those whose action is calcium dependent. The first group includes **filamin** (250,000 M_r), **actin-binding protein** (250,000 M_r), **fimbrin** (68,000 M_r), and **fascin** (58,000 M_r), all of which cause actin filaments to form bundles or meshworks *in vitro*. The second group, the calcium-dependent proteins, include **profilin** (16,000 M_r), **villin** (95,000 M_r), and **gelsolin** (91,000 M_r). Villin and gelsolin act in the presence of Ca^{2+} to break actin chains and to prevent elongation of actin filaments. Profilin sequesters G-actin in the absence of Ca^{2+} and releases the actin to form F-actin in the presence of Ca^{2+}.

The principal membrane-attachment protein seems to be **vinculin** (130,000 M_r) (Geiger, 1979; Geiger *et al.*, 1981). Attachment to the cell membrane seems to be through spectrinlike molecules (Davis and Bennett, 1983; Hirokawa *et al.*, 1983a). **Spectrin** is a dimer (260,000 and 225,000 M_r) that provides mechanical stability to erythrocyte cell membranes. Erythrocyte spectrin also binds actin filaments through protein band 4.1. Avian erythrocyte spectrin also binds calmodulin in the presence of Ca^{2+}, although mammalian spectrin does not. Similar proteins have been found in brain (**fodrin**, 265,000 and 260,000 M_r) and in intestinal brush border terminal web (**TW 260/240**); both proteins bind actin and calmodulin in the presence of Ca^{2+}. Molecules of similar size that cross-react immunologically can be detected in lens, liver, kidney, testes, lung, and heart (Davis and Bennett, 1983). Spectrin, fodrin, and TW 260/240 proteins have different peptide maps and different molecular weights, although they cross-react immunologically. It may be that other tissues contain additional proteins of this family.

2.1.4. Drugs That Interfere with Microfilaments

The mushroom poison **phalloidin** and its derivatives cap the pointed end of actin filaments and prevent depolymerization (Wieland and Faultish, 1978). **DNAse I** in contrast prevents polymerization (Blikstadt *et al.*, 1978). These compounds have been useful in studying the role of treadmilling in cell motility. Another family of drugs that has been useful in studying shape changes is **cytochalasin.** Although the specificity of their action was questioned at one time, it now appears that cytochalasins bind with high affinity to the barbed end of actin filaments and cause depolymerization of F-actin in a fashion similar to that of villin and gelsolin (Schliwa, 1982). The action of some of the cytochalasins, especially cytochalasins B and D, are reversible by washing. Thus, the cell can be returned to normal function after a period of inhibition.

The role of calcium in regulating motility has been studied with agonists and antagonists of calcium transport. The ionophore **A23187** inserts into cell membranes and provides a specific channel for movement of Ca^{2+} into cells. **Verapamil** and **papaverine** prevent transport of Ca^{2+} into cells by stopping the calcium pump. The **phenothiazines** interfere with Ca^{2+}-dependent processes by binding to calmodulin in a Ca^{2+}-dependent way and inhibiting it from acting (Weiss and Levin, 1978).

2.2. Intermediate Filaments

Intermediate filaments are found in virtually all cell types. All filaments in this category are morphologically similar, are insoluble in a broad range of salt concentrations, and seem to share some antigenic determinants. They consist of a large number of different proteins transcribed from different genes. They have been categorized into five main classes by biochemical and immunological criteria (reviewed by Lazarides, 1981; Osborn and Weber, 1982; Fuchs and Hanukoglu, 1983). Four of these classes contain only a single protein. **Desmin** (52,000 M_r) is found in skeletal, visceral, and certain smooth muscle cells. **Vimentin** (55,000 M_r) is found in cell types such as muscle, fibroblasts, neural tube, lens, and erythroid cells. Astrocytes, support cells of the central nervous system (CNS), contain **glial filaments** (53,000 M_r). Most neurons contain **neurofilaments** (68,000, 145,000, and 200,000 M_r). Of these neurofilament proteins, only the 68,000-M_r proteins can form filaments by themselves *in vitro*; the other proteins copolymerize if present (Geisler and Weber, 1981). The fifth class consists of filaments similar to the filaments of hair, wool, and feather and have been called **cytokeratins.** This class consists of 20–30 distinct proteins (40,000–70,000 M_r). They have been divided into two approximately equal classes on the basis of nucleotide and amino acid sequence homology, **type I** (40,000–55,000 M_r) and **type II** (55,000–70,000 M_r) (Fuchs *et al.*, 1981). Cytokeratins are found in epithelial and mesothelial cells. Every epithelial cell contains one or more cytokeratins of each class, but each epithelial cell type contains a unique catalog of cytokeratins (Moll *et al.*, 1982). The mesothelial cytokeratins are largely different from the epithelial cytokeratins (Wu *et al.*, 1982). All these proteins (desmin, vimentin, glial filaments, neurofilaments, and cytokeratins) show 30–90% amino acid sequence homology insofar as either the gene or the protein has been sequenced, and all the sequences would appear to fold with the same tertiary structure (Fuchs and Hanukoglu, 1983; Hanukoglu and Fuchs, 1983).

2.3. Microtubules

Microtubules have been the subject of several recent reviews (e.g., De Brabander and May, 1980; McKeithan and Rosenbaum, 1984). In sections of embedded material, microtubules have a diameter of 24 nm and appear as hollow tubules. When assembled *in vitro*, the diameter approaches 30 nm. Microtubules dissociate into dimers consisting of two subunits called **α-tubulin** and **β-tubulin.** Each has a molecular weight of 50,000. *In vitro*, α- and β-tubulin assemble into rows of alternating subunits to make protofilaments. In most cells, free microtubules consist of 13 rows of protofilaments. Two high-molecular-weight proteins are associated with microtubules and copurify with extracted tubulin. These **microtubule-associated proteins (MAPs)** have molecular weights of 345,000 (**MAP$_1$**) and 271,000 or 286,000 (**MAP$_2$**) and change the assembly characteristics of tubulin solutions. A smaller set of proteins, having molecular weights of 55,000–70,000, the τ **proteins,** stimulate the rate and extent of polymerization. MAPs and τ proteins appear to form projections along the protofilaments with a 32-nm periodicity.

Assembly of microtubules *in vitro* occurs in the presence of GTP and Mg^{2+}. ATP and Ca^{2+} inhibit assembly and promote disassembly of preformed microtubules. Assembly continues until a critical concentration is reached, when the rate of loss of monomer equals the rate of addition and a steady state is established. The addition of MAPs to purified tubulin solutions lowers the critical concentration and increases the rate of assembly; tubulin assembly in the absence of MAPs occurs only under nonphysiological conditions. The critical concentration differs at the two ends of a microtubule. In a flagellum the distal end is the site of assembly, and the proximal end is the site of disassembly. At subunit concentrations between the two critical concentrations, treadmilling occurs, a process first described for actin. At these concentrations, monomer loss from the proximal end equals monomer addition at the distal end. In intact cells, most microtubules appear to be anchored. Centrioles or basal bodies serve as anchor points for flagella and cilia and centrioles for the mitotic spindle. Cells with large numbers of oriented microtubules possess **microtubule organizing centers (MTOCs),** which seem to serve the same function. These points represent the proximal end of the tubules, and assembly occurs at the free end. These regions contain proteins (visualized as amorphous material by electron microscopic examination) that serve as nucleation sites for tubulin assembly. An initial lag in assembly rate *in vitro* is interpreted as the time necessary to form small obligomers that act as sites for additional more rapid assembly (nucleation). The MTOC may serve to block disassembly at the proximal end and thus promote elongation.

Several drugs have been invaluable in the study of microtubule assembly. The first to be discovered were **colchicine** and its derivative **colcemid.** These have been shown to bind specifically to tubulin monomers and to prevent assembly (Weisenberg, 1972). Other drugs with the same action are **vinblastine, vincristine,** and **podophyllotoxin.** Recently, side effects of these drugs have been demonstrated, especially inhibition of osmosis (Beebe *et al.,* 1979). **Nocodazole** is a new drug that lacks this side effect and that binds specifically to tubulin monomers (Hoebeke *et al.,* 1976). **Taxol** has the opposite effect; it induces assembly by preventing depolymerization (Schiff *et al.,* 1979).

3. The Cytoskeleton of Adult Cells

3.1. General Distribution

The various components of the cytoskeleton were first seen in characteristic places in particular adult cell types. Microtubules were first detected in cilia and flagella and in mitotic spindles. These locations indicated that microtubules are involved in some way with movement. Later, they were found in many other cell types, frequently radiating from the region of the centrioles but extending throughout the cytoplasm. In neurons, they were not only in the soma but extended down the dendrites and axons.

The cytokeratins were observed very early as filaments extending from desmosomes. It was not obvious at first that the filaments at desmosomes in the spinous layer of the epidermis are chemically similar to the filaments in the corneum. Desmin was first described as a skeleton that was left in muscle after the actin and myosin had been extracted, i.e., skeletin (Small and Sobieszek, 1977) and desmin (Lazarides and Hubbard, 1978). Vimentin was first identified in intermediate filament protein of mesenchyme-derived cells (Hynes and Destree, 1978; Franke *et al.,* 1978). It was observed very early that axons and dendrites and the processes extending from astrocytes contain many 10-nm solid core filaments that lie in parallel with the microtubules.

The microfilaments were first observed at the apices of several epithelial cell types. These apical accumulations were first called terminal webs on the basis of light microscopic observations. Electron microscopic examination demonstrated that the microfilaments form dense bundles at the level of the zonula adherens.

Cells that are spread in culture or are motile in the organism have microfilament bundles called **stress fibers**. These are especially well organized in fibroblastic cell types, although they are also present in epithelial and ameboid cells. Immunocytochemical techniques (Fig. 2) have shown that stress fibers contain actin along with myosin and regulatory proteins. The actin filaments are anchored at the plasma membrane in regions that bind antivinculin anti-

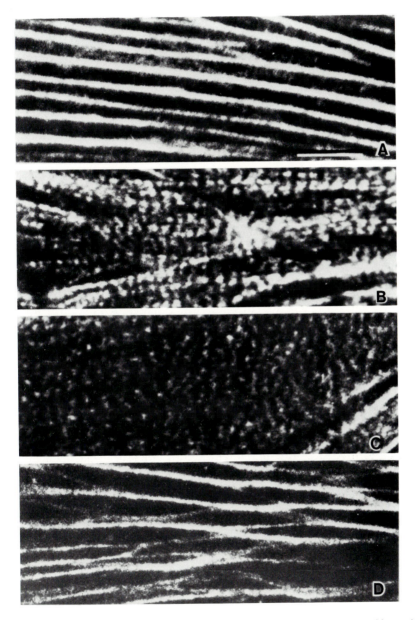

Figure 2. Indirect immunofluroescent localization of contractile proteins in stress fibers of gerbil fibroma cells. Equivalent regions of cells are shown after incubation with (A) antiactin, (B) antimyosin, (C) anti-α-actinin, and (D) antimyosin and α-actinin double staining. Note the complementary localization of the two components in (D). Scale bar: 5 μm. (Reprinted with permission from Gordon, 1978.)

body (Geiger, 1983). The actin filaments adjacent to the membrane are polarized with the barbed ends at the membrane. Within the bundles farther from the membrane, the polarity of the actin fibers is reversed with the pointed end toward the membranes. Since stress fibers stretch from the membrane into the cortical cytoplasm, it seems likely that the actin filaments in a fiber are relatively short as compared with the length of the entire stress fiber. In many spread cells myosin, tropomyosin, and α-actinin appear to be arranged in a periodic fashion along the actin bundles. In some cells, intermediate filaments also are part of the stress fibers. In these cells α-actinin may act to anchor actin bundles to each other or to intermediate filaments. During movement, stress fibers are lost and reformed. Actin is also the principal component of microvilli and microspikes.

3.2. The Intestinal Brush Border

The organization of microfilament bundles has been studied most extensively in the brush border of the intestinal epithelium (Burgess, 1982; Hirokawa *et al.*, 1983*b*; Mooseker, 1983, 1984). Each microvillus has a core of actin filaments polarized with the barbed ends at the tip of the microvillus (Fig. 3). The actin filaments in the microvillus are associated with villin, the actin-splitting protein, and with fimbrin, the actin-bundling protein. A 110,000-M_r protein attaches the core filaments to the membrane, and calmodulin is bound to both the 110,000-M_r protein and to villin. Beneath the microvilli in the apical cytoplasm, the core of actin filaments extends into a meshwork of fine filaments. The rootlets of the core actin filaments, where they enter the meshwork of fine filaments, contain myosin and tropomyosin. The major constituent of the meshwork of fine filaments is a spectrinlike dimer (TW 260/240). The 240,000-M_r subunit cross-reacts with antibody to spectrin and to fodrin of brain. The meshwork of fine filaments extends apically to the plasmalemma of the microvillar surface and laterally to the vicinity of the adherent junctions, where bundles of microfilaments form a circumferential ring. The circumferential ring contains actin filaments in antiparallel arrangements, attached to the plasmalemma by vinculin. The circumferential ring also contains myosin, tropomyosin, and α-actinin. The myosin appears to be in the form of small rods or as monomers. Beneath the rootlets of the core filaments and the meshwork of fine filaments is a meshwork of intermediate filaments that radiate from the desmosomes of the junctional complex.

Addition of ATP and Ca^{2+} to epithelial sheets or isolated brush borders results in constriction of the cell apices to half their former diameter (Burgess, 1982; Hirokawa *et al.*, 1983*b*). Concurrently, most of the TW 260/240 links become soluble, and the ends of the rootlets move closer together. Constriction is calmodulin dependent and results in phosphorylation of the 20,000-M_r myosin light chain in a calcium-dependent manner (Keller and Mooseker, 1982). No change is seen in the microvillar structure during constriction; higher cal-

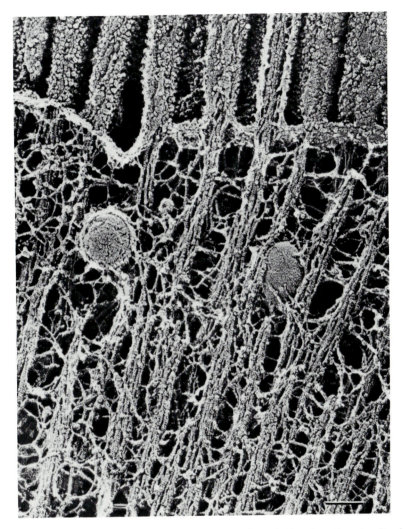

Figure 3. Terminal web region of an isolated brush border from chicken intestine visualized by the quick-freeze, deep-etch technique. Actin bundles from the microvilli at the top of the figure pass through the apical cytoplasm. TW 260/240 filaments in the superficial cytoplasm link adjacent actin rootlets, vesicles, and the plasmalemma. Myosin rods link the rootlets at deeper levels. Scale bar: 0.1 μm. (Micrograph courtesy of N. Hirokawa, University of Tokyo, and M. S. Mooseker, Yale University. Reproduced from *The Journal of Cell Biology*, 1984, vol. 99, pp. 104S–112S by copyright permission of The Rockefeller University Press.)

cium concentrations result in solubilization of the core filaments, however. Since microvilli do not contain myosin, it has been proposed that calmodulin acts as a calcium sink to preserve microvillar structure rather than in a contractile mechanism in the microvillar cores. A number of minor proteins copurify with the terminal web constituents and have not yet been characterized.

3.3. Teleost Photoreceptors

Another model system that has been useful in studying cell motility is the photoreceptor layer of teleostean fish (Burnside, 1978, 1981; Burnside *et al.*, 1982). The rods and cones move from a position apposed to the pigmented epithelium to a position several microns from their surface (Fig. 4). The lengthening is linear and is triggered by light. Cone photoreceptors shorten when lights are turned on and lengthen when lights are turned off; rod photoreceptors undergo lengthening or shortening by the reverse stimuli. The change in cell height involves a change in the length of the inner segment, the portion of the cell distal to the cell body and proximal to the outer segment—the photoreceptor. In cone photoreceptors, the inner segments contain longitudinally oriented bundles of actin filaments polarized with their barbed ends at the base of the outer segment. These bundles continue into the cell body, where they interdigitate with actin filaments that are polarized in the opposite direction and are anchored at the cell base. Between these two sets of actin filaments are thicker filaments that have the dimensions and appearance of myosin filaments. Large numbers of central microtubules are oriented parallel to the long axes of the cells. Shortening of the cones is inhibited by cytochalasins, and elongation is inhibited by colchicine. However, colchicine does not affect shortening, and cytochalasin does not affect elongation. When the cells are extracted with nonionic detergent and treated with physiological concentrations of calcium and ATP, they shorten in the normal fashion. In the absence of either calcium or ATP, shortening does not occur. Shortening does not occur if the detergent-extracted cells are treated with HMM that has been rendered incapable of releasing actin in the presence of ATP (N-ethylmaleimide-treated HMM) before Ca^{2+} and ATP are added (Porrello *et al.*, 1983). Regulatory proteins have not been studied in this system. The system provides an exceptionally good model for studies on cell motility because of its similarities to muscle: Actin and myosin are arranged in a sarcomerelike structure, shortening is ATP and calcium dependent, and the response is proportional to the stimulus intensity.

3.4. Motile Cells

Microfilament meshworks are found at the leading edges of motile cells in the organism and in cell culture. The actin filaments appear to be attached to the plasma membrane through vinculin, with a polarity indicating assembly at the membrane. There is conflicting evidence as to whether these meshworks contain myosin. Tropomyosin and α-actinin are not found in meshworks at the leading edge. When the cells are treated with cytochalasin, these meshworks collapse, and motility ceases. Motility resumes and the meshworks reappear when the drug is washed out. Motility involving microfilament meshworks could occur by distortion of the entire meshwork through its association with stress fibers, through a direct interaction of actin and myosin within the mesh-

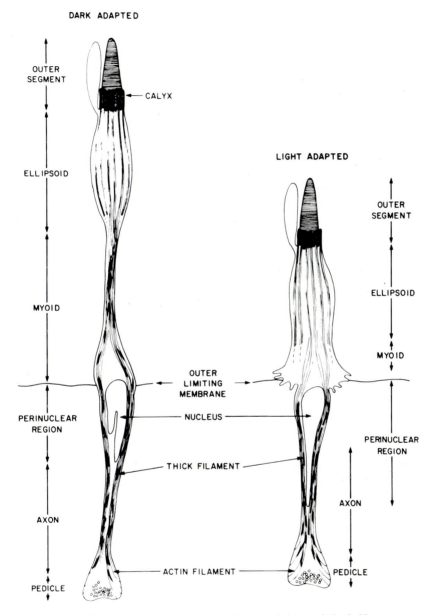

Figure 4. Schematic illustration showing the distribution of thin and thick filaments in retinal cones after shortening and elongation. Thin filaments are attached at the tips of the microvilluslike calyces, extend in bundles around the ellipsoid mitochondria, and course down the myoid to the nuclear region. Thin filaments also are present in the axon. Thick filaments are found in axon, perinuclear, and myoid regions in elongated cones, but accumulate in the axon in shortened cones. (Courtesy of Dr. B. Burnside. Reproduced from *The Journal of Cell Biology*, 1978, vol. 78, pp. 227–246 by copyright permission of The Rockefeller University Press.)

work, or through cycles of assembly and disassembly controlled by actin-binding proteins.

Intermediate filaments seem to be quite stable in all cell types. They connect with organelles such as nuclei, muscle cell Z discs and M bands, and with the cell periphery—particularly at desmosomes. However, the intermediate filaments are capable of reorganization, changing, and, in some cases, disassembling at mitosis. Intermediate filaments seem to collapse when a cell is treated with colchicine, indicating a connection with microtubules.

Microtubles are the most obvious component of cilia, flagella, and mitotic spindles, suggesting that microtubules are in some way involved in motility. Microtubules are also clearly capable of rapid assembly and disassembly, as when cells undergo mitosis. Parallel microtubules are a major constituent of axons and are thought to be involved in transport of synaptic vesicles from the soma to axonal endings. Colchicine has been found to inhibit the release of many sorts of vesicles (exocytosis) and to inhibit movements in the cell membrane such as capping. Microtubules are found in large numbers and in parallel orientation in cells undergoing elongation. In polarized migrating cells, treatment with drugs that depolymerize microtubules results in rounding. Microtubules are associated with actin through links that involve MAPs. In the brush border, the TW 240/260 also seems to be involved. In cilia and flagella the microtubules extend the length of the cilia or flagella. Movement seems to be by sliding of the microtubules with respect to each other. This sliding is propelled by **dynein,** a protein with three ATPase-containing heads (Shimizu and Johnson, 1983). Dynein has not been detected in other locations. Movement of microtubules elsewhere has been attributed to treadmilling (see Section 2.3), net movement being due to assembly at one end of the microtubule and disassembly at the other. Translocation of cytoplasm as in axonal flow and saltatory motion are examples.

4. Models of Function

4.1. Types of Morphogenetic Movements

Morphogenetic changes that might involve the cytoskeleton can be classified as those unique to sheets (epithelia), those characteristic of single cells (mesenchyme), and those found in both cell types (Fig. 5). Epithelial sheets undergo three major forms of deformation: folding, pouching, and branching. These morphogenetic changes are described in detail in Chapter 4. Folding occurs in organs such as the neural plate and intestine (Fig. 5A1). In both cases, few cells may actually be involved in the change in shape—those at the midline in formation of the V of the chick spinal cord, for instance. Formation of a pouch as an evagination or invagination is characteristic of a variety of early primordia, including olfactory, otic, optic, lens, and thyroid placodes (Fig.

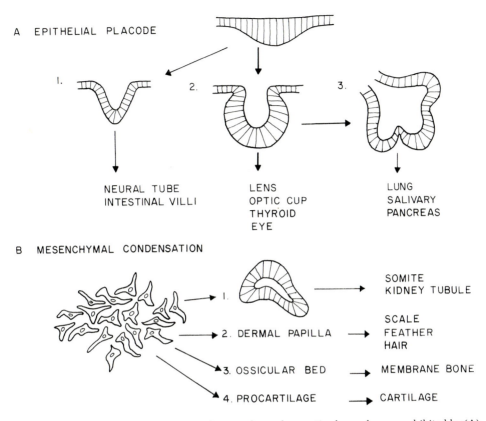

Figure 5. Diagrammatic representation of types of morphogenetic shape changes exhibited by (A) epithelial and (B) mesenchymal primordia. See text for explanation. (Drawing by Eva K. Hilfer.)

5A2). In these cases, more cells change shape, and the primordium has a rounded surface rather than the V shape of a fold. Branching is a type of folding that occurs in tubular organs (Fig. 5A3). It is characteristic of glands such as the salivary and pancreatic primordia and the lung. A fold, or cleft, appears in the surface of terminal bulbous extensions. The result is subdivision into two new bulbous end buds. Another characteristic of epithelial primordia is increase in cell height, or **palisading.** Palisading often precedes other morphogenetic events.

Individual cells of the mesenchyme or neural crest undergo **condensation** (compaction) or **dispersal** (Fig. 5B). Sometimes dispersal originates from an epithelial structure, such as in the formation of sclerotome from the somites (Fig. 5B1) (see Chapter 14). Condensation can result in a connective tissue such as bone or cartilage (Fig. 5B2,3,4), or it can be the first step in the formation of an epithelium, as in shaping of kidney tubules (Fig. 5B1). Dispersal clearly involves cell migration. It is not clear how much condensation depends on migration and how much is contributed by cell division.

4.2. Constriction

The most obvious way in which the cytoskeleton could be involved in morphogenesis would be through a contractile mechanism involving actin. Contraction of actin and myosin has been proposed as the cause of constriction of cell apices in a number of primordia (for review, see Wessells et al., 1971; Schroeder, 1973; Spooner, 1975). Many epithelial primordia have circumferentially arranged microfilament bundles in the appropriate locations to account for changes in cell shape that accompany folding, invagination/evagination, and branching. Just before the epithelial sheet begins to bend, circumferentially arranged microfilament bundles frequently develop at the cell apices, where the bending is to occur. While the cell sheet is bending, the apices of the cells in the bend narrow, and the microfilament bundles become shorter and thicker. Circumferential microfilament bundles appear in the neural plate at the onset of neurulation (Burnside, 1973a; Karfunkel, 1972; Baker and Schroeder, 1967), during growth of uterine glands (Wrenn and Wessells, 1970), and during formation of the pancreas (Wessells and Evans, 1968), salivary gland (Spooner, 1973), and lens (Wrenn and Wessells, 1969). Filament bundles are already present in the thyroid (Shain et al., 1972) and eye primordia (Hilfer et al., 1981) before pouching begins.

It is assumed that all apical or basal microfilament bundles contain actin on the basis of their morphology. Filaments within the bundles in salivary and lung primordia have been identified as actin both morphologically, by their ability to bind HMM (Spooner et al., 1973), as well as biochemically (Spooner et al., 1978). In the neural plate, the cell apices bind antibody to myosin (Lee et al., 1983) as well as antibody to actin (Lee et al., 1983; Sadler et al., 1982). Further evidence that circumferential bundles contain actin comes from treatment of primordia with cytochalasin B. Morphogenesis is stopped and is often reversed after treatment, and electron micrographs show disorganization or collapse of the filament bundles in salivary primordia (Spooner and Wessells, 1972), oviduct (Wrenn and Wessells, 1970), and neural plate (Karfunkel, 1972).

Functional studies in thyroid (Hilfer et al., 1977), nasal placode (Smuts, 1981), and eye (Hilfer et al., 1981; Maloney and Wakely, 1982) have shown that invagination is an ATP-dependent process. When primordia are permeabilized with detergent and incubated in millimolar ATP in the presence of physiological Ca^{2+} concentrations, they undergo a rapid organotypic shape change (Fig. 6). The cells also increase in height during this process. Substitution of other nucleotides, pyrophosphate, or ATP analogues in the presence of calcium ions does not cause precocious organogenesis. Invagination in the neural plate (Moran, 1976), salivary primordium (Ash et al., 1973), eye vesicle (Brady and Hilfer, 1982), and thyroid primordium (S. R. Hilfer, unpublished data) is calcium dependent, as is the formation of epithelial somites (Chernoff & Hilfer, 1982). The ionophore A23187 promotes precocious invagination, whereas antagonists of Ca^{2+} transport inhibit invagination. The source of Ca^{2+} appears to be external to the cells, because precocious development is not promoted by ionophore in calcium-free medium.

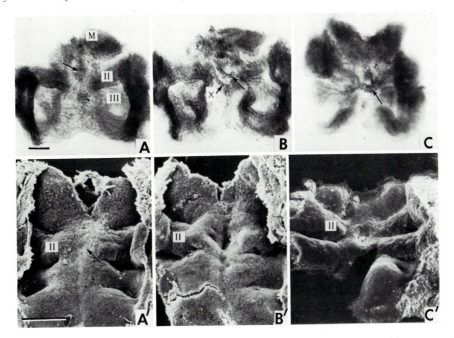

Figure 6. Changes in thyroid shape during normal development and after artificial formation. (A, B, C) Living pharynxes isolated from chick embryos. (A', B', and C') Scanning electron micrographs of isolated pharyngeal regions. (A) Normal pharynx of 22-somite embryo. The left half of the mandibular arch (M) was removed. Arches II and III lie to either side of the pharyngeal floor. The thyroid primordium is visible as a dense patch (arrow) between the second pair of arches. (B) The same pharynx after 10 min incubation in 0.05% Triton X-100 and 5 min in 1.0 mM ATP. A ridge is visible in the thyroid region (arrow). The mandible has rotated downward, bringing a remnant of the bulbus arteriosus into focus (x). (C) The ventral pharynx of a 28-somite embryo. The thyroid forms a small pouch (arrow). (A') Ventral pharynx of a 22-somite embryo fixed after 20 min in Triton X-100. Arch II and the thyroid placode (arrow) are indicated. (B') Ventral pharynx of another 22-somite embryo after treatment with Triton X and ATP. The region between the second arches (II) is deeply indented. (C') Ventral pharynx of a normal 28-somite embryo. The specimen was tilted to show the ridge surrounding the thyroid pouch, between the second arches (II). (A, B, C) Same magnification. Scale bar: 100 μm. (A', B', C') Same magnification, scale bar: 100 μm. (Reprinted from Hilfer *et al.*, 1977.)

The effect of Ca^{2+} appears to be mediated through calmodulin in optic cup formation. Treatment with trifluoperazine or chlorpromazine, which bind specifically to calmodulin when it is complexed with Ca^{2+} (Weiss and Levin, 1978), prevents invagination. The effect of these phenothiazines appears to be on calmodulin rather than on some nonspecific process, because the sulfated derivatives of the drugs do not block optic cup formation, the K_D for the reaction of the phenothiazines with the optic cup is the same as the K_D for the reaction of phenothiazines with calmodulin, and A23187 does not cause precocious invagination after treatment with phenothiazine. Thus, although the cells of organ primordia have microfilament bundles in appropriate locations (and in those primordia that have been tested the process is Ca^{2+} and ATP dependent), further work must be done to establish actin–myosin contraction as a common mechanism for shape change. It would be interesting to know the distribution and types of regulatory proteins present in the cells as well as the relationship between actin and myosin both before and after a morphogenetic event.

4.3. Anchorage

Microfilament bundles could be under tension generated by actin–myosin interaction without actually shortening, similar to the tension exerted by some smooth muscles. Such bundles might act to resist external forces that otherwise would stretch the cells out of shape. In some primordia, such as thyroid (Shain et al., 1972; Hilfer, 1973), filament bundles traverse the cell apices instead of being circumferentially arranged. Since bundles with this orientation are found in cells that do not constrict, it has been suggested that they function to maintain shape rather than to cause a shape change. Bundles oriented from the apex to base of cells may serve a similar function. During invagination of the thyroid, a series of concentric grooves forms on the basal surface of the primordium. The cells that form the grooves contain longitudinally oriented bundles of filaments (Fig. 7A). These cells do not elongate during invagination, whereas the rest of the cells become approximately 50% taller. It is interesting that these cells with longitudinal filament bundles also do not incorporate thymidine into DNA and presumably are not dividing (Smuts et al., 1978). Since the thyroid occupies a confined space in the floor of the pharynx, its shape changes may result partly from folding as the cell number increases. The grooves may serve as points around which the evaginating primordium folds (Fig. 7B).

The optic primordium also contains a band of cells with longitudinally oriented filament bundles during early development. The initial optic diverticulum forms a vesicle and stalk (Hilfer et al., 1981). As the vesicle expands, it grows dorsally to lie parallel to the wall of the dorsally expanding brain. At the juncture of the optic stalk and optic vesicle, a series of grooves develop on the dorsal surface where the vesicle changes its direction of outgrowth. The cells that form the groove contain the longitudinal bundles and occupy an analogous position to the cells in the thyroid with similarly oriented bundles.

Maintenance of shape could be a passive function of microfilament bundles. Bundles of filaments that do not contain myosin could act as struts. Yet they would be dynamic structures if they underwent polymerization and depolymerization. Microfilaments polymerized at random rather than in bundles could cause a region of the cytoplasm to form a gel. A number of proteins (see Section 2.1.3 or the more detailed review by Schliwa, 1981) cause actin to gel, and in some cases the gel–sol transition is under control of Ca^{2+} ions. Isolation of the appropriate actin-binding proteins would be the only evidence that either of these mechanisms is actually operating.

Microtubules that orient parallel to the long axis of cells could serve a similar function in maintaining cell height. Once cells elongate, microtubules

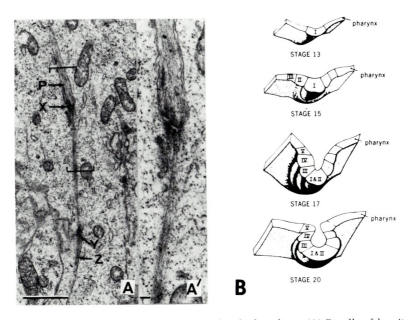

Figure 7. Evagination of the thyroid placode in the chick embryo. (A) Bundle of longitudinally oriented microfilaments in cells forming a groove in the basal surface of a primordium in a 25-somite embryo. A cell process (P) filled with microfilaments terminates at a dense region on another cell (X). A microfilament bundle traverses that cell to end at a density on a tangentially cut membrane (Y). A third bundle (Z) lies along the lateral membrane of another cell basal to the other two. Serial sections of this preparation show that each cell stretches from base to apex of the primordium and contains longitudinal bundles at all levels. Scale bar: 1 μm. (A′) Enlargement of area bracketed in A to show microfilaments. Scale bar: 0.1 μm. (Reprinted from Hilfer, 1973.) (B) Diagrammatic representation of the events occurring during thyroid evagination. See text for explanation. (Reprinted from Smuts et al., 1978.)

may maintain the shape change. Large numbers of longitudinally oriented microtubules exist in cells of the neural and optic primordia, the lens, oviduct, pancreas, and salivary bud. The increased height of the primordia may act as an amplifier of laterally directed shape changes. Thus, a small change in diameter of cell apex or base will have a greater effect on curvature of a sheet of tall cells than it will on a sheet of short cells.

4.4. Spreading

A third type of morphogenetic movement that might involve mirofilaments is spreading of epithelial sheets. Spreading results in increased diameter of the cells and in decreased thickness. Spreading occurs in the adjacent ectoderm during formation of (1) the neural plate and the primitive streak, (2) the future retinal pigmented epithelium of the eye cup, and (3) the respiratory epithelium from the terminal buds of the lung. In many cases, it has been assumed that spreading results from tension, causing the same number of cells to cover a greater surface. Spreading might also require a relaxation or disassembly of microfilaments.

During formation of the optic cup, spreading occurs only at one cell surface (Hilfer *et al.*, 1981). The bases of optic vesicle and lens placode are attached to each other by fusion of their basal laminae (Johnston *et al.*, 1979; Hendrix and Zwaan, 1975). The apposed cell bases do not change in area as invagination proceeds. As the lens cells invaginate, their apices constrict on the inside of the curve. On the outside of the curve, the apices of the future neural retinal cells must enlarge. This expansion of the retinal cell apices is accompanied by a change in the organization of the cytoskeleton. Before invagination, all of the optic vesicle cells contain bundles of circumferentially arranged microfilaments in the apical cytoplasm. During invagination, the filament bundles become more pronounced, and the apices constrict around the margin of the optic cup. At the center of the neural retina, however, the cell apices no longer contain microfilament bundles (Fig. 8). Instead, few filaments form a loose meshwork in the apical cytoplasm. It appears that spreading of the apical cell surfaces may require disorganization of the filament bundles.

4.5. Elongation

Cell elongation, or palisading, is one of the first signs of organ formation in epithelial primordia. In thyroid, eye, lens, nasal, otic, and neural primordia, palisading results in formation of a placode that is taller than the neighboring epithelium. Often elongation is accompanied by loss of extracellular space between the epithelial cells and tighter lateral cell contacts, suggesting changes in the properties of the lateral plasmalemma. In the early lens (Zwaan and Hendrix, 1973) and neural plate (Burnside, 1973a) where the cell volume has been measured, there is little change in cell volume during elongation. Elongation of cells with no increase in cell volume and a decrease in extracellular space must involve increase in cell density, either by cell division or by cells moving closer together. Elongation is accepted frequently as being caused by microtubules. Elongation is generally inhibited by colchicine or its derivatives, and the presence of larger numbers of longitudinally oriented microtubules correlates with cell elongation in formation of oviduct (Wrenn and Wessells, 1970), pancreatic diverticulum (Wessells and Evans, 1968) as well as in the above primordia. Colchicine has side effects, however. It blocks not only cell division but also change in cell volume, presumably by inhibiting water flow into the cells. During later stages of lens development, the cells at the periphery of the lens vesicle elongate to form lens fibers. During elongation, the cells increase in volume. Elongation is blocked by colchicine but not by nocodazole, a drug that depolymerizes microtubules but does not affect the volume change (Beebe *et al.*, 1981). Thus, it seems important to reexamine the other primordia in which volume of individual cells has not been measured.

In neural plate formation, Burnside (1973a) concluded that elongation cannot be caused by either elongation of microtubules or sliding of micro-

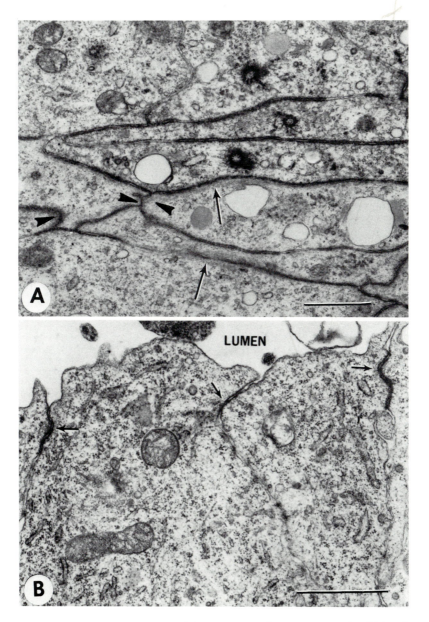

Figure 8. Cell apices during invagination of the optic primordium in a 22-somite chick embryo. (A) Section parallel to the cell surface passing through the level of the junctional complexes, near the margin of the optic cup. The cell apices have become narrowed tangential to the rim of the cup. Microfilament bundles lie parallel to the plasmalemma (arrows) and traverse ends of the flattened apices (arrowheads). (B) Cell apices at the center of the retinal disc, in a section cut perpendicular to the cell surfaces. Microtubules and microfilaments are scattered throughout the cytoplasm. Tangled, short filaments are visible near cell junctions (arrows), but bundles of filaments are not seen. Scale bar: 1 μm. (Reprinted with permission from Dr. S. R. Hilfer, Dr. R. C. Brady, and Dr. J.-J. W. Yang. Reproduced from *Ocular Size and Shape: Regulation during Development*, 1981, (S. R. Hilfer and J. B. Sheffield, eds.), pp. 47–78, Springer Verlag, New York.

tubules past one another. This was based on measurements that showed little increase in total microtubule length and no change in the number of paired microtubules before and after cell elongation. She concluded that elongation might occur by flow of cytoplasmic organelles along the oriented microtubules.

It is possible that microtubules are not responsible for cell elongation but rather for stabilization of the elongation once it has occurred. Many elongating cells have a meshwork of fine filaments in their lateral cortical cytoplasm. Constriction of this meshwork could produce an elongation, which would be unstable without the longitudinally oriented microtubules. A preliminary report by Burnside (1973b) suggested that cytochalasin B interfered with elongation in isolated neural ectoderm.

It is also possible that elongation is mediated in some way by interkinetic nuclear migration. Most placodes become pseudostratified columnar epithelia, and the nuclei migrate from the cell base during the S phase of the division cycle to the apex during the M phase. It is not known what causes nuclear movement up and down in the cytoplasm, but it is suspected that microtubules are involved. Cell elongation may be an extension of the forces that move the nuclei.

Elongation must involve either insertion of membrane laterally and loss of membrane from apical and basal surface or a shift in position of junctional complexes so that apical and basal membrane become lateral. These changes must involve at least a rearrangement of microfilament and intermediate filament meshworks.

4.6. Migration

Movement of single cells or cell clusters plays a major role in development. Migration occurs during dispersal of neural crest cells, formation of somites, their dispersal to form sclerotome and dermis, and formation of secondary corneal stroma, to name only a few places. It is not clear whether migration occurs during condensation to form such structures as dermal papillae and kidney tubules, but the cells form extensions during the process. Axon elongation is another process that involves outgrowth of cell extensions, not only in the form of the axon itself but also in the microspikes that project from the growth cone and seem to determine the direction of growth. Cell migration is inhibited by cytochalasins but not by colchicine. The latter does, however, destroy polarity and interfere with the direction of migration. The leading edge of migrating cells contains a meshwork of actin filaments and may contain some myosin. Microspikes and other projections of the leading edge contain microfilament bundles that in some cases have been shown to consist of polarized actin filaments. Nerve cell motility appears to be a calcium-dependent process (Letourneau and Wessells, 1974), and neural crest cell migration correlates with elevated calcium levels at the leading edge (Moran, 1983). Cell migration during somite formation also is cytochalasin D sensitive, calcium dependent, and blocked by inhibitors of calmodulin (Chernoff and Hilfer, 1982).

During cell migration there is some evidence for interaction of the cytoskeleton with extracellular matrix. For instance, the direction of neural crest migration and outgrowth of axons have been reported to depend on environmental cues acting at the level of the cell membrane. Neural crest cells follow trails of fibronectin (Newgreen and Thiery, 1980; Newgreen et al., 1982) and may be oriented along fibronectin-coated collagen bundles in the embryo (see Chapter 13). Nerve axons, in the grasshopper at least (Bentley and Keshishian, 1982), project microspikes from their growth cones. When the microspikes touch specific neuron cell bodies, called guidepost cells, they orient the growth cone to grow along the preexisting axon of the guidepost cells.

5. Conclusions

The cytoskeleton consists of many different polypeptides, which are organized into filaments, bundles, meshworks, and tubules. Not all these polypeptides have been fully characterized. In addition, much remains to be learned about the control of assembly and disassembly of the cytoskeletal elements. Very little is known about the molecular basis of interactions of the different cytoskeletal proteins with each other. Even less is known about their interaction with intramembranous and transmembranous proteins. Although there has been evidence that the cytoskeleton plays a role in capping of membrane proteins, the mechanism has remained a matter of speculation. More recently it has become apparent that movement of other cytoplasmic membranous structures involve links with the cytoskeleton. These phenomena include exocytosis and endocytosis; they may include interkinetic nuclear migration as well. Until the molecular bases for these movements are known, it will be difficult to assess the role of coexisting cellular structures, such as circumferential microfilament bundles, in shape changes.

Much of our knowledge of the causes of organotypic shape changes is derived from morphological studies. Although this information on distribution of cytoskeletal polypeptides and the effect of drugs on their integrity is important, correlations of shape changes with alterations of cytoskeletal elements provide only circumstantial evidence of causality.

It has been recognized that the cytoskeleton is involved in a variety of cellular shape changes that occur during organogenesis. It is clear that all shape changes are not caused by a single cytoskeletal element. Rather, several cytoskeletal elements may act sequentially or simultaneously to produce a specific change. These sequential actions could occur within the same responding cell or within different groups of cells within the same organ. The same elements might produce one shape change in one group of cells and a different shape change in another group of cells. Alternatively, different cytoskeletal elements may produce the same shape change through different mechanisms in different organ primordia. It is reasonable to suspect that this diversity may arise from differential gene expression in different primordia. However, a direct relationship between a particular cytoskeletal structure and a particular shape change is difficult to demonstrate, because so little is known about the molecular changes in the cytoskeleton under different conditions. A great deal more research both on the control of the cytoskeletal elements and on the control of organotypic shape change is clearly required.

References

Adelstein, R. S., 1982, Calmodulin and the regulation of the actin–myosin interaction in smooth muscle and nonmuscle cells, *Cell* **30**:349–350.

Ash, J. F., Spooner, B. S., and Wessells, N. K., 1973, Effects of papaverine and calcium-free medium on salivary gland morphogenesis, *Dev. Biol.* **33**:463–469.

Baker, P. C., and Schroeder, T. E., 1967, Cytoplasmic filaments and morphogenetic movements in the amphibian neural tube, *Dev. Biol.* **15**:432–450.

Beebe, D. C., Feagans, D. E., Blanchette-Mackie, E. J., and Nau, M., 1979, Lens epithelial cell elongation in the absence of microtubules: Evidence for a new effect of colchicine, *Science* **206**:836–838.

Beebe, D. C., Johnson, M. C., Feagans, D. E., and Compart, P. J., 1981, The mechanism of cell elongation during lens fiber cell differentiation, in: *Ocular Size and Shape: Regulation During Development* (S. R. Hilfer and J. B. Sheffield, eds.), pp. 79–98, Springer-Verlag, New York.

Bentley, D., and Keshishian, H., 1982, Pathfinding by peripheral pioneer neurons in grasshoppers, *Science* **218**:1082–1088.

Bettex-Galland, M., Portzehl, H., and Lüscher, E. F., 1962, Dissociation of thrombosthenin into two components comparable with actin and myosin, *Nature (Lond.)* **193**:777.

Blikstad, I., Markey, F., Carlsson, L., Persson, T., and Lindberg, U., 1978, Selective assay of monomeric and filamentous actin in cell extracts, using inhibition of deoxyribonuclease I, *Cell* **15**:935–943.

Bonder, E. M., Fishkind, D. J., and Mooseker, M. S., 1983, Direct measurement of critical concentrations and assembly rate constants at the two ends of an actin filament, *Cell* **34**:491–501.

Brady, R. C. and Hilfer, S. R., 1982, Optic cup formation—A calcium-regulated process, *Proc. Natl. Acad. Sci. USA* **79**:5587–5591.

Burgess, D. R., 1982, Reactivation of intestinal epithelial cell brush border motility: ATP-dependent contraction via a terminal web contractile ring, *J. Cell Biol.* **95**:853–863.

Burnside, B., 1973a, Microtubules and microfilaments in amphibian neurulation, *Am. Zool.* **13**:989–1006.

Burnside, B., 1973b, In vitro elongation of isolated neural plate cells: Possible roles of microtubules and contractility, *J. Cell Biol.* **59**:41a.

Burnside, B., 1978, Thin (actin) and thick (myosin like) filaments in cone contraction in the teleost retina, *J. Cell Biol.* **78**:227–246.

Burnside, B., 1981, Mechanism of cell shape determination in teleost retinal cones, in: *Ocular Size and Shape: Regulation During Development* (S. R. Hilfer and J. B. Sheffield, eds.), pp. 25–45, Springer-Verlag, New York.

Burnside, B., Smith, B., Nagata, M., and Porrello, K., 1982, Reactivation of contraction in detergent-lysed retinal cones, *J. Cell Biol.* **92**:199–206.

Chernoff, E. G., and Hilfer, S. R., 1982, Calcium dependence and contraction in somite formation, *Tissue Cell* **14**:435–449.

Clarke, M., and Spudich, J. A., 1977, Nonmuscle contractile proteins: The role of actin and myosin and shape determination, *Annu. Rev. Biochem.* **46**:797–822.

Cloney, R. A., 1966, Cytoplasmic filaments and cell movements: Epidermal cells during ascidian metamorphosis, *J. Ultrastruct. Res.* **14**:300–328.

Davis, J., and Bennett, V., 1983, Brain spectrin: Isolation of subunits and formation of hybrids with erythrocyte spectrin subunits, *J. Biol. Chem.* **258**:7757–7766.

DeBrabander, M. and DeMey, J. (eds.), 1980, *Microtubules and Microtubule Inhibitors*, Elsevier/ North-Holland Biomedical Press, Amsterdam.

Ebashi, S., Nonomura, Y., Nakamura, S., Nakasone, H., and Kohama, K., 1982, Regulatory mechanism in smooth muscle: Actin-linked regulation, *Fed. Proc.* **41**:2863–2867.

Franke, W. W., Schmid, E., Osborn, M., and Weber, K., 1978, Different intermediate-sized filaments distinguished by immunofluorescence microscopy, *Proc. Natl. Acad. Sci. USA* **75**:5034–5038.

Fuchs, E., and Hanukoglu, I., 1983, Unraveling the structure of the intermediate filaments, *Cell* **34**:332–334.

Fuchs, E. V., Coppock, S. M., Green, H., and Cleveland, D. W., 1981, Two distinct classes of keratin genes and their evolutionary significance, *Cell* **27**:75–84.

Geiger, B., 1979, A 130K protein from chicken gizzard: Its localization at the termini of microfilament bundles in cultured chicken cells, *Cell* **18**:193–205.

Geiger, B., 1983, Membrane–cytoskeleton interaction, *Biochim. Biophys. Acta* **737**:305–341.

Geiger, B., Dutton, A. H., Tokuyasu, K. T., and Singer, S. J., 1981, Immunoelectron microscopic studies of membrane–microfilament interactions: The distributions of α-actinin, tropomyosin, and vinculin in intestinal epithelial brush border and in chicken gizzard smooth muscle cells, *J. Cell Biol.* **91**:614–628.

Geisler, N., and Weber, K., 1981, Comparison of the proteins of two immunologically distinct intermediate-sized filaments by amino acid sequence analysis: Desmin and vimentin, *Proc. Natl. Acad. Sci. USA* **78**:4120–4123.

Gordon, W. E., III, 1978, Immunofluroescent and ultrastructural studies of "sarcomeric" units in stress fibers of cultured non-muscle cells, *Exp. Cell Res.* **117**:253–260.

Hanukolu, I., and Fuchs, E., 1983, The cDNA sequence of a type II cytoskeletal keratin reveals constant and variable structural domains among keratins, *Cell* **33**:915–924.

Hendrix, R. W., and Zwaan, J., 1975, The matrix of the optic vesicle-presumptive lens interface during induction of the lens in the chicken embryo, *J. Embryol. Exp. Morphol.* **33**:1023–1049.

Hilfer, S. R., 1973, Extracellular and intracellular correlates of organ initiation in the embryonic chick thyroid, *Am. Zool.* **13**:1023–1038.

Hilfer, S. R., Palmatier, B. Y., and Fithian, E. M., 1977, Precocious evagination of the embryonic chick thyroid in ATP-containing medium, *J. Embryol. Exp. Morphol.* **42**:163–175.

Hilfer, S. R., Brady, R. C., and Yang, J-J. W., 1981, Intracellular and extracellular changes during early ocular development in the chick embryo, in: *Ocular Size and Shape: Regulation During Development* (S. R. Hilfer and J. B. Sheffield, eds.), pp. 47–78, Springer-Verlag, New York.

Hirokawa, N., Cheney, R. E., and Willard, M. B., 1983a, Location of a protein of the fodrin-spectrin TW 260/240 family in the mouse intestinal brush borders, *Cell* **32**:953–965.

Hirokawa, N., Keller T., III, Chasan, R., and Mooseker, M., 1983b, Mechanism of brush border contractility studied by the quick-freeze, deep etch method, *J. Cell Biol.* **96**:1325–1336.

Hoebeke, J., van Nijen, G., and DeBrabander, M., 1976, Interaction of nocodazole (R17934), a new antitumoral drug, with rat brain tubulin, *Biochem. Biophys. Res. Commun.* **69**:319–324.

Hoffman-Berling, H., and Weber, H. H., 1953, Vergleich der Motilität von Zellmodellen und Muskelmodellen, *Biochim. Biophys. Acta* **10**:629–630.

Holtfreter, J., 1943, A study of the mechanics of gastrulation, *J. Exp. Zool.* **94**:261–318.

Hynes, R. O., and Destree, A. T., 1978, Relationship between fibronectin (LETS protein) and actin, *Cell* **15**:875–886.

Ishikawa, H., Bischoff, R., and Holtzer, H., 1969, Formation of arrowhead complexes with heavy meromyosin in a variety of cell types, *J. Cell. Biol.* **43**:312–328.

Johnston, M. C., Noden, D. M., Hazelton, R. D., Coulombre, J. L., and Coulombre, A. J., 1979, Origins of avian ocular and periocular tissues, *Exp. Eye Res.* **29**:27–43.

Karfunkel, P., 1972, The activity of microtubules and microfilaments in neurulation in the chick, *J. Exp. Zool.* **181**:289–302.

Keller, T. C. S., III, and Mooseker, M. S., 1982, Ca^{2+}-Calmodulin-dependent phosphorylation of myosin and its role in brush border contraction in vitro, *J. Cell Biol.* **95**:943–959.

Lazarides, E., 1981, Intermediate filaments—Chemical heterogeneity in development, *Cell* **23**:649–650.

Lazarides, E., and Hubbard, B. D., 1976, Immunological characterization of the subunit of the 100 A filaments from muscle cells, *Proc. Natl. Acad. Sci. USA* **73**:4344–4348.

Lee, H. Y., Kosciuk, M. C., Nagele, R. G., and Roisen, F. J., 1983, Studies on the mechanisms of neurulation in the chick: Possible involvement of myosin in elevation of neural folds, *J. Exp. Zool.* **225**:449–457.

Letourneau, P. C., and Wessells, N. K., 1974, Migratory cell locomotion versus nerve axon elongation. Differences based on the effects of lanthanum ion, *J. Cell Biol.* **61**:56–69.

Lewis, W. H., 1947, Mechanics of invagination, *Anat. Rec.* **97**:139–156.

Maloney, C., and Wakely, J., 1982, Microfilament patterns in the developing chick eye: Their role in invaginations, *Exp. Eye Res.* **34**:877–886.

McKeithan, T. W., and Rosenbaum, J. L., 1984, The biochemistry of microtubules, in: *Cell and Muscle Motility*, Vol 5 (J. W. Shay, ed.), pp. 255–288, Plenum Press, New York.

Moll, R., Franke, W. W., Schiller, D. L., Geiger, B., and Krepler, R., 1982, The catalog of human cytokeratins: Patterns of expression in normal epithelia, tumors and cultured cells, *Cell* **31**:11–24.

Mooseker, M. S., 1983, Actin binding proteins of the brush border, *Cell* **35**:11–13.

Mooseker, M. S., Bonder, E. M., Conzelman, K. A., Fishkind, D. J., Howe, C. L., and Keller, T. C. S., III, 1984, Brush border cytoskeleton and integration of cellular functions, *J. Cell Biol.* **99**:104s–112s.

Moran, D. J., 1976, A scanning electron microscopic and flame spectrometry study on the role of Ca^{2+} in amphibian neurulation using papaverine inhibition and ionophore induction of morphogenetic movement, *J. Exp. Zool.* **198**:409–416.

Moran, D., 1983, Fluorescent localization of calcium at sites of cell attachment and spreading, *J. Exp. Zool.* **229**:81–89.

Neuhaus, J. M., Wanger, M., Keiser, T., and Wegner, A., 1983, Treadmilling of actin, *J. Muscle Res. Cell Motil.* **4**:507–527.

Newgreen, D., and Thiery, J.-P., 1980, Fibronectin in early avian embryos: synthesis and distribution along the migration pathways of neural crest cells, *Cell Tissue Res.* **211**:269–291.

Newgreen, D. F., Gibbins, I. L., Sauter, J., Wallenfels, B., and Wütz, R., 1982, Ultrastructural and tissue culture studies on the role of fibronectin, collagen and glycosaminoglycans in the migration of fowl embryo neural crest cells, *Cell Tissue Res.* **21**:521–549.

Osborn, M., and Weber, K., 1982, Intermediate filaments: Cell-type-specific markers in differentiation and pathology, *Cell* **31**:303–306.

Pollard, T. D., 1981, Cytoplasmic contractile proteins, *J. Cell Biol.* **91**:156S–165S.

Pollard, T. D., and Weihing, R. R., 1974, Actin and myosin and cell movement, *CRC Crit. Rev. Biochem.* **2**:1–65.

Porrello, K., Cande, W. Z., and Burnside, B., 1983, N-Ethylmaleimide-modified subfragment-1 and heavy meromyosin inhibit reactivated contraction in motile models of retinal cones, *J. Cell Biol.* **96**:449–454.

Sadler, T. W., Greenberg, D., Coughlin, P., and Lessard, J. L., 1982, Actin distribution patterns in the mouse neural tube during neurulation, *Science* **215**:172–174.

Schiff, P. B., Fant, J., and Horwitz, S. B., 1979, Promotion of microtubule assembly in vitro by taxol, *Nature (Lond.)* **277**:665–667.

Schliwa, M., 1981, Proteins associated with cytoplasmic actin, *Cell* **26**:587–590.

Schliwa, M., 1982, Action of cytochalasin D on cytoskeletal networks, *J. Cell Biol.* **92**:79–91.

Schroeder, T. E., 1969, The role of "contractile ring" filaments in dividing *Arbacia* eggs, *Biol. Bull.* **137**:413.

Schroeder, T. E., 1973, Cell constriction: Contractile role of microfilaments in division and development, *Am. Zool.* **13**:949–960.

Shain, W. G., Hilfer, S. R., and Fonte, V. G., 1972, Early organogenesis of the embryonic chick thyroid. I. Morphology and biochemistry, *Dev. Biol.* **28**:202–218.

Shimizu, T., and Johnson, K. A., 1983, Kinetic evidence for multiple dynein ATPase sites, *J. Biol. Chem.* **258**:13841–13846.

Small, J. V., and Sobieszek, A., 1977, Ca-regulation of mammalian smooth muscle actomyosin via a kinase-phosphatase-dependent phosphorylation and dephosphorylation of the 20,000-M_r light chain of myosin, *Eur. J. Biochem.* **76**:521–530.

Smuts, M. S., 1981, Rapid nasal pit formation in mouse embryos stimulated by ATP-containing medium, *J. Exp. Zool.* **216**:409–414.

Smuts, M. S., Hilfer, S. R., and Searls, R. L., 1978, Patterns of cellular proliferation during thyroid organogenesis, *J. Emb. Exp. Morphol.* **48**:269–286.

Spooner, B. S., 1973, Microfilaments, cell shape changes, and morphogenesis of salivary epithelium, *Am. Zool.* **13**:1007–1022.

Spooner, B. S., 1975, Microfilaments, microtubules and extracellular materials in morphogenesis, *BioScience* **25**:440–451.

Spooner, B. S., and Wessells, N. K., 1972, An analysis of salivary gland morphogenesis: Role of cytoplasmic microfilaments and microtubules, *Dev. Biol.* **27**:38–54.

Spooner, B. S., Ash, J. F., Wrenn, J. T., Frater, R. B., and Wessells, N. K., 1973, Heavy meromyosin binding to microfilaments involved in cell and morphogenetic movement, *Tissue Cell* **5**:37–46.

Spooner, B. S., Ash, J. F., and Wessells, N. K., 1978, Actin in embryonic organ epithelia, *Exp. Cell Res.* **114**:381–387.

Stossel, T. P., and Pollard, T. D., 1973, Myosin in polymorphonuclear leukocytes, *J. Biol. Chem.* **248**:8288–8294.

Sutoh, K., 1982, Identification of myosin-binding sites on the actin sequence, *Biochemistry* **27**:3654–3661.

Vandekerckhove, J., and Weber, K., 1978, Mammalian cytoplasmic actins are the products of at least two genes and differ in primary structure in at least 25 identified positions from skeletal muscle actins, *Proc. Natl. Acad. Sci. USA* **75**:1106–1110.

Weisenberg, R. C., 1972, Microtubule formation *in vitro* in solutions containing low calcium concentration, *Science* **177**:1104–1105.

Weisenberg, R. C., Borisy, G. G., and Taylor, E. W., 1968, The colchicine binding protein of mammalian brain and its relationship to microtubules, *Biochemistry* **7**:4466–4479.

Weiss, B., and Levin, R. M., 1978, Mechanism for selectively inhibiting the activation of cyclic nucleotide phosphodiesterase and adenylate cyclase by antipsychotic agents, *Adv. Cyclic Nucleotide Res.* **9**:285–303.

Wessells, N. K., and Evans, J., 1968, Ultrastructural studies of early morphogenesis and cytodifferentiation in the embryonic mammalian pancreas, *Dev. Biol.* **17**:413–446.

Wessells, N. K., Spooner, B. S., Ash, J. F., Bradley, M. O., Luduena, M. A., Taylor, E. L., Wrenn, J. T., and Yamada, K. M., 1971, Microfilaments in cellular and developmental processes, *Science* **171**:135–143.

Wieland, T., and Faultish, M., 1978, Amatoxin, phallolysin, and antaminide: The biologically active components of poisonous Amantia Mushrooms, *CRC Crit. Rev. Biochem.* **5**:185–260.

Wilson, E. B., 1924, *The Cell in Development and Heredity*, 3rd ed., Macmillan, New York.

Wrenn, J. T., and Wessells, N. K., 1969, An ultrastructural study of lens invagination in the mouse, *J. Exp. Zool.* **171**:359–368.

Wrenn, J. T., and Wessells, N. K., 1970, Cytochalasin B: effects upon microfilaments involved in morphogenesis of estrogen-induced glands of oviduct, *Proc. Natl. Acad. Sci. USA* **66**:904–908.

Wu, Y.-J., Parker, L. M., Binder, N. E., Beckett, M. A. Sinard, J. H., Griffiths, C. T., and Rheinwald, J. G., 1982, The mesothelial keratins: A new family of cytoskeletal proteins identified in cultured mesothelial cells and nonkeratinizing epithelia, *Cell* **31**:693–703.

Yerna, M-J., Dabrowska, R., Hartshorne, D. J., and Goldman, R. D., 1979, Calcium-sensitive regulation of actin–myosin interactions in baby hamster kidney (BHK-21) cells, *Proc. Natl. Acad. Sci. USA* **76**:184–188.

Zwaan, J., and Hendrix, R. W., 1973, Changes in cell and organ shape during early development of the ocular lens, *Am. Zool.* **13**:1039–1049.

Chapter 2

Microtubules and the Cellular Morphogenesis of Plants

CLIVE W. LLOYD

1. Introduction

A major distinction between the morphogenesis of animals and plants is that in the former the migration of cells can play a large part in shaping tissues, whereas in plants the relative positions of cells are mapped out in ways that exclude cell locomotion. Another difference is that animal development may involve a progressive loss of totipotency, but mature plant cells can retain their capacity to divide, not only to repair local wounds, but in order to initiate new organs and—as illustrated by the regeneration of plants from tissue culture cells—entire organisms. These different traits reflect the different ways in which growth contributes to form in the animal and plant kingdoms. Organ formation is largely a once-and-for-all event in animals, but plants construct their organs (i.e., leaves, lateral roots) repetitively and retain the capacity to do so in mature organisms—a feature exploited in recovery from injury and predation and during asexual propagation.

The pattern formed by a group of plant cells depends, at least initially, on the way in which the new cross wall is oriented across dividing cells. Pattern formation and cell division are therefore interlinked, but—once established in meristematic regions—pattern can be modified according to the way in which cells yield to turgor pressure. To yield in a particular direction (as occurs, for instance, when cells elongate) depends on the nonrandom deposition of components within the wall and—in cells with polylamellate walls—on the way in which complex wall patterns combine to resist expansion in one plane but to succumb to it in another. Still further changes in tissue morphology can occur when growth factors—as part of a developmental program—plasticize and permit the reorientation of wall components. Similar but unprogrammed changes may occur when the tissue is wounded. It can therefore be seen that control over the plane of cell division, cell size, and cell expansion all combine

CLIVE W. LLOYD • Department of Cell Biology, John Innes Institute, Norwich NR4 7UH, England.

to affect the final shape achieved by a group of walled plant cells (although shape can be further sculptured by specialized patternings of wall components as occur in xylem tracheids).

Looked at as a problem of wall placement, it would therefore appear that plants achieved morphological diversity by modeling and remodeling the strain-resisting components of their extracellular matrices—either during interphase (when the direction of cell expansion is consolidated or modified) or during division (when new cross walls are deposited). These two separate opportunities for tissue modeling are mirrored, within the cell, by the appearance of distinctive sets of microtubules—the interphase array being implicated in the shaping of side walls, whereas three other arrays, seen only in dividing cells, are involved in positioning the cross wall. Analysis of the role of plant microtubules is inextricably linked with wall fibrils. After outlining the way in which these two elements influence the shape of individual cells, the way in which this might be coordinated at the supracellular level is discussed.

2. Patterns of Cell Growth and Division

Two systems that introduce some of the principles of how plant growth is patterned are described. An example of how plant form develops from programmed placement of walls comes from the work of Gunning and colleagues (1978a), who reconstructed the fate map of every single cell in the root of the water fern, *Azolla*. All cells can be traced, in a lineage, to their formation from an apical cell (Fig. 1). The apical cell in the root, for instance, is like a wedge with three "cutting" faces that divide cells into the root. These faces divide in turn, either clockwise or counterclockwise, to form **merophytes.** These merophytes are cells that—by virtue of subsequent set patterns of cell division, expansion, and differentiation—can be considered developmental packages. Within these packages are two classes of division that we shall see occurring in other developmental contexts: **formative** and **proliferative divisions.** Formative divisions are those generally longitudinal divisions that tend to increase the girth of the root by positioning cells side by side. This generates, in the transverse plane, precursors of new cell types—precursors that then undergo proliferative transverse divisions that multiply the number of a particular cell type within a cell file. The extremely miniaturized root of *Azolla* exhibits predictable patterns of growth that raise issues, not only concerning division planes and cell shape, but of coordination at the supracellular level also. For instance, roots can be right-handed or left-handed according to whether the apical cell cuts off merophytes in a clockwise or counterclockwise direction. This patterning is amplified by subsequent divisions and illustrates for this small organism that the sequence of wall placements is rigidly programmed from the outset (Gunning, 1982). However, for more complex and larger organs, it is conceivable that the developmental program is modified or activated by external cues.

Another exemplary system in which microtubules are well characterized is

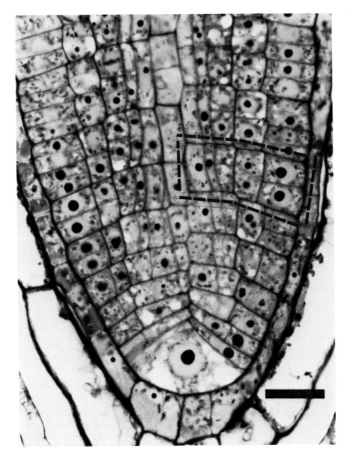

Figure 1. Longitudinal section of the water fern *Azolla pinnata* illustrating the way in which patterns of division affect root morphology. The apical cell divides at one of its three inner surfaces (two are visible). The cells thus produced are merophytes (one is outlined), which overlap at the midline. These basic units of construction first undergo a precise sequence of longitudinal, or formative, divisions, which produce intial cells of cortex, pericycle, endodermis, and so forth. Initials then undergo transverse proliferative divisions, so-called because they increase the number of cells within a file. Formative and proliferative divisions are therefore divisions of quality and quantity, respectively. Scale bar: 20 μm. (Reproduced with permission from Gunning *et al.*, 1978a.)

provided by the development of the stomatal complex (Pickett-Heaps and Northcote, 1966) (Fig. 2). By a sequence of **symmetrical** and **asymmetrical divisions,** epidermal cells in adjacent files cooperate to form a stomatal complex in which guard cells flank the stomatal pore. A smaller guard mother cell (GMC) is formed by the asymmetrical division of an epidermal cell, and the GMC so produced is flanked by other epidermal cells that also divide asymmetrically to yield small subsidiary cells that cup the GMC. After these two asymmetrical divisions, the GMC itself divides longitudinally, but incompletely, in a symmetrical division to produce guard cells. The walls of the cells, which thicken at their opposing faces, split apart to form the pore.

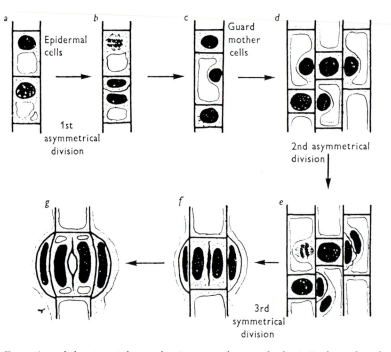

Figure 2. Formation of the stomatal complex in young leaves of wheat. Each nucleus becomes situated at the same end (a) of epidermal cells which divide asymmetrically to form small and compact guard mother cells and larger vacuolated cells (b). Each cell elongates slightly (c) and at the next preprophase stage, the nuclei of the large vacuolated cells, flanking the GMC in adjacent files, come to rest at the wall common to them and the GMC (d). These flanking epidermal cells then undergo a second asymmetrical division to form small subsidiary cells that cup the GMC (e). Finally, the guard mother cell divides longitudinally and symmetrically to form guard cells, with the faces abutting the pore representing the site of increased thickening of wall (f, g). In this system, the division plane (whether symmetrical or asymmetrical) is anticipated by the preprophase band and, as in the shaping of the stomatal pore (see Fig. 4), indicates the close involvement of microtubules in key stages of histogenesis. (From Pickett-Heaps and Northcote, 1966.)

It will be seen that the sequence, symmetry, and plane of division all affect the morphology of the stomatal complex, as does the uneven thickening of cell walls. The involvement of microtubules in these processes is examined further—first in nondividing cells and then during cell division.

3. Microtubules during Interphase

3.1. Microtubules and Cell Polarity

When prepared from cells by removing wall material with cellulases, protoplasts are spherical, even though the cortical microtubules are still in place (albeit in an altered arrangement). This is to be contrasted with the removal of animal tissue cells from their matrix, for in such cells the microtubules are generally depolymerized. Clearly, the cortical microtubules in plant cells, although more stable, do not by themselves maintain polarized cell shape; this, instead, would seem to be a property of wall components that were deposited in an oriented manner in the presence of microtubules (Lloyd *et al.*, 1980). The establishment and maintenance of cell polarity therefore has a cytoplasmic as well as an extracellular aspect.

Roelofsen and Houwink noted in 1953 that the innermost walls of cells such as root hairs possess fibrils in a mainly transverse orientation. We shall discuss this observation and the way in which the orientation of fibrils in successive wall lamellae progressively changes, in Section 3.3. However, it was studies such as these that introduced the idea that turgor pressure would be channeled in favor of cell elongation if wall fibrils were arranged transversely to discourage lateral bulging. A corollary of this notion of wall organization was that new wall layers were actually apposed from the plasma membrane.

In testing this collection of hypotheses on the long, cylindrical cells of the filamentous alga, *Nitella*, Green (1963) performed experiments that pointed to a cytoplasmic explanation for cell polarity. In the presence of the microtubule depolymerizing drug colchicine, *Nitella* internodes grew into spheres, and this loss in polarity was accompanied by loss of organized wall texture as judged with polarized light. After washing away the drug, transverse order was restored and was better than in cells recovering in an artificial restraining jacket of dialysis tubing. It was concluded that (1) transverse orientation does not depend on existing wall texture, (2) transverseness is enhanced by strain (i.e., less ordered in the presence of a restraining jacket), and (3) there is a cytoplasmic colchicine-sensitive template responsible for determining and restoring transverse deposition. These hypothetical elements of the cytoplasmic framework were analogized to spindle fibre protein. It is against this background that ideas concerning wall–cytoplasm interactions have developed.

3.1.1. Microtubules Parallel Inner Wall Fibrils

In an electron microscopic analysis of glutaraldehyde-fixed plant cells, Ledbetter and Porter (1963) described the then-novel cytotubules as being circumferentially arranged, i.e., in a plane normal to the long axis of the cell and parallel to one another. These workers also stressed the parallelism between these microtubules (MTs) and microfibrils (MFs) in the wall, suggesting that the tubules be considered as agents for orienting the inner wall fibrils. This study represents a turning point between the already extensive work on the orientation of wall fibrils and ideas of Green and of others that led to the hypothesis that control of cell polarity originates in the cytoplasm.

Since that time, parallelism between cortical microtubules and the innermost fibrils of the cell wall has been recorded and reviewed many times. Thus, we shall look only at selected cases that exemplify the important features.

One of the difficulties of studying the cell wall is that cellulose microfibrils may represent only a fraction of total wall components and are not easy to delineate amongst the matrix materials. Fortunately, the unicellular alga

Oocystis has large amounts of cellulose; here, the thick cellulose microfibrils parallel one another, but the collective orientation switches by 90° from layer to layer (Fig. 3). MTs parallel the innermost MFs; thus, the orientation of MTs must switch as each new layer of microfibrils switches by 90° relative to the preceding layer (see Robinson and Quader, 1982).

This illustrates the point that MT/MF parallelism is not static but dynamic. It is another theme that we shall return to, for this coupling of MT and MF orientation through time can also be seen in higher plants during successive shifts in the orientation of wall fibrils during the formation of a polylamellate wall.

In *Allium cepa*, GMCs in the epidermis divide, and the cell plate eventually assumes a longitudinal orientation (Palevitz and Hepler, 1974). Microtubules are not uniformly arranged along this wall but are clustered in the middle, at the future site of the stomatal pore. The MTs radiate outward toward their apposing longitudinal walls and are later matched by microfibrils of the innermost wall layer. Wall becomes accumulated preferentially at these central zones, and—as the cells grow in volume—they assume a kidney shape where the concave, thickened walls of the apposing cell faces pull apart and define the pore (Fig. 4). This suggests that MTs together with oriented wall fibrils play a part in achieving the differentiated shape of these cells. In timothy grass (*Phleum pratense*), final shape of guard mother cells is affected by further events (Palevitz, 1982). At a stage equivalent to the swelling of *Allium* guard mother cells, those of timothy grass undergo a transient swelling but then reconstrict and elongate. This results in formation of a long slitlike pore, surrounded by two dumbbell-shaped guard mother cells in contrast to the kidney shape in the stomatal complex of *Allium*. Electron microscopic analysis shows this extra transition to be marked by a change from radial alignment of microtubules and fibrils to a near-axial (steep criss-cross) parallelism achieved within the elongated cells.

Twenty years after Ledbetter and Porter's original description, it can be concluded that microtubule–wall fibril parallelism is widely based, but it is now important to acknowledge that parallelism is not restricted to a transverse girdling of elongated cells and that it persists throughout a variety of programmed and unprogrammed changes in cell shape.

3.1.2. A Dynamic Parallelism

That the initial alignment of fibrils at the plasma membrane is influenced by the orientation of microtubules, and not vice versa, is usually derived from the following sorts of argument:

3.1.2a. Microtubule Depolymerizing Drugs Alter the Texture of the Cell Wall.
In the stomatal complex of *Allium*, colchicine causes disappearance of microtubules, the cells swell, wall material is deposited abnormally, and pore development is inhibited. Wall fibrils therefore continue to be deposited in the

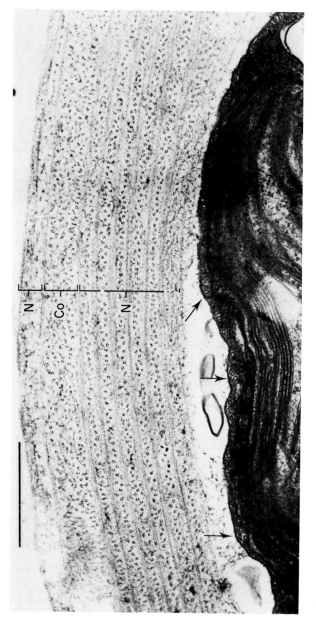

Figure 3. In the alga *Oocystis solitaria*, cellulose microfibrils occur parallel to each other, but their direction alternates by 90° from lamella to lamella, building up a cross-ply structure. Microtubules are always parallel to one of these directions; when MTs are depolymerized with colchicine, the regular alternation is inhibited. After washing out the colchicine, microtubules reappear (arrows), and the criss-cross pattern is resumed, yielding a sandwich wall in which the thick colchicine layer (Co) lies between normally deposited lamellae (N). This demonstrates two important points: (1) Depolymerization of microtubules need not necessarily result in randomization of cellulose microfibrils, but it does hinder the switching of angle from lamella to lamella; and (2) each new lamella is inserted by apposition at the plasma membrane and moves outward as newer lamellae displace it. Scale bar: 0.5 μm. (Reproduced by permission from Robinson and Quader, 1982.)

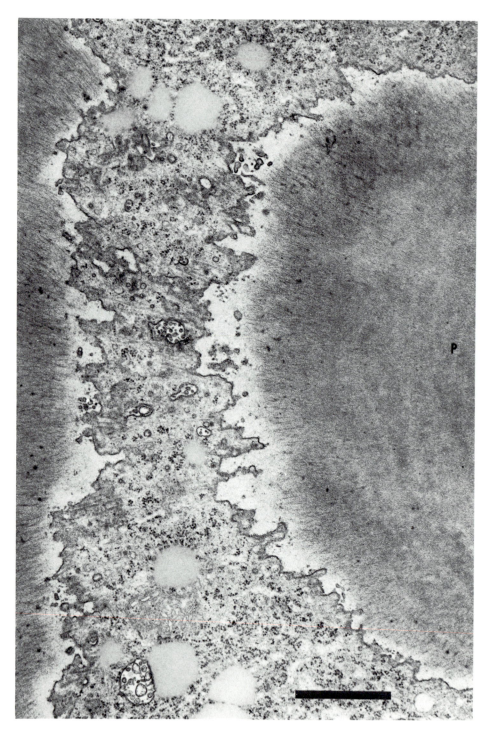

Figure 4. Cortical microtubules in the cytoplasm run parallel to inner fibrils of the cell wall. This grazing section through a kidney-shaped guard cell of *Allium cepa* L. shows that both microtubules and wall fibrils radiate out from the heavily thickened site of the pore (P), which is seen on the right. Scale bar: 1 μm. (Reproduced with permission from Palevitz and Hepler, 1976.)

absence of microtubules, but not in a manner that leads to the attainment of normal, differentiated shape (see Palevitz, 1982).

This is also the case in the alga *Oocystis* (see Robinson and Quader, 1982). As discussed above, cellulose is deposited in layers of parallel microfibrils, in which each layer normally alternates its orientation by 90° from left-hand to right-hand helices. Microtubules remain parallel to the innermost cellulose microfibrils throughout. Colchicine depolymerizes microtubules, but, as before, the wall continues to be deposited, except that the layers do not alternate in sign, with the microfibrils continuing to be deposited in the direction prevailing before drug treatment (see Fig. 3).

The more complex issue of wall patterning in higher plants will be discussed later, but it is noted here that their polylamellate walls also contain layers with alternating helical sign. In some (such as epidermal cells) the alternation is not as stark as in *Oocystis*, for there are layers of wall with intermediate angles; rotation of sign is therefore gradual. Again, colchicine does not lead to an irregular deposition of wall fibrils, but the rhythm of wall lamellation—and, hence, of switching from left-hand to right-hand lamellae—is inhibited (Takeda and Shibaoka, 1981; Vian *et al.*, 1982).

It can be concluded that colchicine treatment does not necessarily lead to a randomization of fibrillar deposition; fibrils may still pack side to side within a layer to form parallel arrays, but the ability of each layer to rotate its angle of deposition is hampered.

3.1.2b. Parallelism Survives Unprogrammed and Radical Changes in Direction of Growth.

We have seen that wall fibrils can change their orientation from layer to layer and that microtubules are required in order to anticipate changes in microfibrillar orientation. However, this switching of microtubular orientation occurs not just during programs of rhythmic changes but where experimental treatments induce unprogrammed and more dramatic alternations.

Ethylene causes epidermal cells to expand radially rather than to elongate. This is accompanied by a tendency for microtubules to be arranged longitudinally rather than transversely, and this is matched by a change in the orientation of the most recently deposited fibrils. Where MTs and fibrils are not parallel, the MTs are generally displaced toward the predicted future direction of fibril deposition (Lang *et al.*, 1982).

Whereas ethylene permits radial expansion of cells, gibberellic acid increases cell length (Sawhney and Srivastava, 1974), and this can be inhibited by colchicine. Gibberellic acid encourages transverse deposition in epidermal cell walls, which normally tend toward the crossed polylamellate pattern in which longitudinal, transverse, and intermediate orientations occur. Colchicine negates this effect of GA_3: The fibrils of the inner lamella are still parallel to one another, but the angle at which the fibrils are deposited (and oblique, longitudinal, and transverse fibrils occurred with equal frequency) fails to rotate between lamellae as occurs in crossed-polylamellate controls (Takeda and Shibaoka, 1981).

When leaves are detached from the succulent plant, *Graptopetalum para-guayense*, dormant meristems develop into shoots, and in the epidermis MTs and fibrils undergo a coordinated change in orientation by as much as 90°. In these epidermal cells, microtubules could be seen to be rearranged before the emergence of the new primordium (Hardham *et al.*, 1980).

This and related evidence combines to suggest that parallel packing of wall fibrils does not require MTs but that they are required during changes in orientation that routinely occur in polylamellate walls, during *ad hoc* initiation of a new lateral outgrowth, and in order to respond to experimental treatments that alter the axis of cell expansion.

3.2. Cellulose Synthesis at the Plasma Membrane

New material is incorporated into the wall as cells expand. In principle, this could be performed in two ways: It could be incorporated at the plasma membrane in a process of **apposition** while the older layers somehow reorganize as the cell stretches, or the expansion of wall could be accommodated by intercalation of new material at various points throughout the thickness of the wall, by a process of **intussusception.** This argument has a long history, but a study by Ray in 1967 is especially relevant. Feeding oat coleoptiles and pea stems with tritiated glucose in the presence or absence of the growth hormone indoleacetic acid, Ray was able to demonstrate that radioactivity (as judged by EM autoradiography) was incorporated throughout the thickness of the wall. But then, by selectively extracting different classes of polysaccharides from the wall, it was shown that cellulose was formed on the surface of the protoplast, whereas other components, such as hemicelluloses, were inserted throughout the wall, i.e., by apposition and by intussusception, respectively.

Another demonstration of growth by apposition can be derived from Robinson and Quader's studies on *Oocystis,* an alga whose wall has a pronounced pattern of cellulose microfibrils. Colchicine treatment prevents the alternation of wall plies, but regular alternation is restored after washing away the drug. In this way, a sandwich wall is constructed where the layers marked by absence of alternation are swept outward as new and alternating layers are deposited internally to them, at the plasma membrane (see Fig. 3). How, then, is a large crystalline polymer such as cellulose assembled from precursors at the outer face of the plasma membrane? Roelofsen (1958) proposed that the appropriate enzymes would be found at the growing ends of the microfibrils, and Preston (1961) visualized ordered granules in the membrane that were believed to be the appropriate synthetic complexes.

The intramembranous nature of certain proteins provided a keystone in Singer and Nicolson's (1972) hypothesis that launched the idea that membranes are fluid seas of lipid in which mobile proteins—which were either peripheral or integral to the membrane—were free to cluster. A corollary of this **fluid mosaic model** was that cytoskeletal proteins could be attached to membrane proteins, thereby restricting but also guiding their movement along the bilayer.

In an attempt to explain how intracellular microtubules could mirror extracellular wall fibrils, Heath (1974) combined these principles; he hypothesized that the membrane-bound cellulose synthases move along the plasma membrane in a direction somehow dictated by the closely apposed cortical microtubules. In this way, cellulose could be deposited parallel to underlying microtubules. It is not known what would propel cellulose synthases, but subspecies of Heath's original hypothesis now countenance various possibilities (see Heath and Seagull, 1982). Synthases might be attached directly to moveable microtubules by radial crossbridges; they might be propelled by contractile actomyosin, whereas microtubules provide tracks, or the synthases might be propelled and directed by the very extrusion of the linear cellulose crystal within channels delineated by underlying, membrane-associated microtubules (Fig. 5). A key element in these hypotheses is that cellulose synthases be located within the membrane, and there are now several studies that support this.

In the algae *Oocystis* and *Glaucocystis*, freeze fracture of the cell surface reveals linear complexes of particles, whereas in *Spirogyra*, *Closterium*, the fern *Adiantum*, and the desmid *Micrasterias* the complexes occur as rosettes of particles (see Herth, 1983, and Montezinos, 1982, and references cited therein). Similar rosettes are seen in higher plants (i.e., in cells of corn, pine, and mung

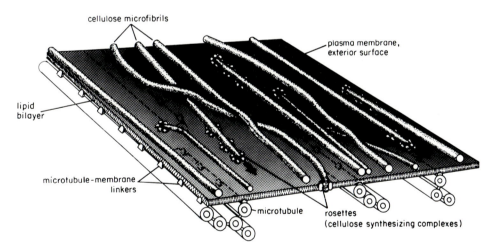

Figure 5. Various putative membrane-bound assemblies have been proposed to account for parallelism between cortical cytoplasmic microtubles and cellulose microfibrils of the inner cell wall. Some models envisage microtubules as providing tracks upon which the membrane-bound cellulose synthetases move; others hypothesize that motile force is generated by actomyosin. In the model illustrated here (reproduced with permission from Staehelin and Giddings, 1982), submembranous microtubules are shown forming channels that limit the freedom of movement of membrane synthases. Once oriented in this way, the continued protrusion of a linear and rigid cellulose microfibril would help maintain the direction in which the synthesizing enzymes move within the lipid bilayer.

bean seedlings), and microfibrils have been found associated with terminal particles embedded in the bilayer. If these multiparticulate complexes should prove to be the sites of cellulose biosynthesis *in vivo*, their complexity could account for the difficulty in inducing isolated plasma membranes to synthesize this polymer *in vitro*.

So far, we have considered microfibril–microtubule coalignment as perhaps mediated by intramembranous cellulose-synthesizing complexes. Even though this embodies the idea of mobile membrane particles, the concept is still essentially two-dimensional and static, for it takes no account of how coalignment is preserved throughout those longer periods of time in which the orientation of wall fibrils changes from layer to layer.

3.3. Patterns in the Wall

The way in which each new layer of wall is apposed to its predecessor affects both the texture of the wall and the way in which the wall is able to yield to turgor pressure. As we have seen, layers of wall fibrils are deposited in different ways in different cells and can be changed by agents such as ethylene and gibberellic acid. We have also seen that MT–MF parallelism can be preserved throughout quite radical reorientations. This requires a dynamic attitude to MT orientation that is missing from the concept of MT–MF parallelism at its simplest. Clearly, MTs do reorient, but it is an open question whether this is achieved by depolymerization of the old array with repolymerization of MT subunits in a new configuration or by a remodeling of the existing array without depolymerization.

What, then, is the overall structure of the cortical microtubule array of interphase cells, and how can it change with time while still paralleling wall fibrils? Ledbetter and Porter (1963) described the circumferential microtubules as hoops but also entertained the possibility that they could form a low-pitched spiral (helix). It is difficult to construct a picture of the overall MT array from electron microscope thin sections, and this presupposes that the array is unperturbed by fixation and embedding. By immunofluorescence, however, entire MT arrays can be seen, and it is now clear that helical arrays exist in various cells (see Fig. 6) (Lloyd, 1983). This supports a **dynamic helical model** (Lloyd, 1984), which has three premises, i.e., (1) the interphase microtubular array is integral, not fragmented, (2) it is based on the ability of microtubules to form helices, and (3) the helices are capable of adopting different conformations—from flat to steeply pitched—which are interconvertible. We shall return to this idea at the end of the present section.

As for the wall, its fibrils have also been described as forming helices, but there are separate schools of thought on how these patterns should be deciphered. By metal-shadowing the inner and outer walls of a variety of growing plant cells, Roelofsen and Houwink (1953; see also Roelofsen, 1965) observed that inner wall fibrils have a net transverse orientation, appearing as a flat spiral and regarded as the mechanical basis for spiral growth of sporangio-

Figure 6. Onion root hair, stained by indirect immunofluorescence with antibodies to tubulin, in which microtubules are arranged helically. By contrast, other root hairs, such as radish, have net axial microtubules. It has therefore been proposed (Lloyd, 1983, 1984) that the helix is the conformation underlying various microtubule arrays—from net transverse to net axial. This is consistent with the idea that wall fibrils are also deposited in helices of varying angles. Scale bar: 10 μm.

phores and hair cells. According to their **multinet growth hypothesis (MGH)**, fibrils are intertwined in bundles and are to be envisaged as a sheaf of fishing nets. As a flat helix both weft and warp have an average transverse arrangement, but as that lamella progresses outwards, due to deposition of newer lamellae, it comes under extension strain, which gradually realigns the fibrils so that they assume a preferred longitudinal orientation (rather like closing the gates of a scissors) (Fig. 7a). Both weft and warp were envisaged as being deposited within a single lamella, but subsequent work has generally shown that adjacent fibrils within a lamella are more or less parallel to each other. Any angular dispersion of neighboring fibrils over the common direction is—within a lamella—generally slight and—like interweaving—is accommodated within some hypotheses for microfibril orientation (see Fig. 5). It is difficult to see how angular dispersion of fibrils within a lamella could be exaggerated by strain realignment to yield a more distinct cross-ply texture. It is more likely that a lamella containing fibrils of more or less similar helical sign is succeeded by another lamella whose helical sign or angle is different. This requires that we should not seek the explanation for, say, a crossed-helical wall texture within a lamella, but we should ask what causes the angle to shift between lamellae.

Preston (1982) recommends that the multinet part of the MGH be abandoned, since it is no longer descriptively accurate, but he advocates that the passive reorientation component be retained (Fig. 7b), i.e., fibrils are deposited in helices and realigned by extension strain as the lamella progresses outward. Passive realignment may well be a viable concept; still, it is difficult to reconcile various observations with even a modified MGH. For instance, using freeze etch–freeze fracture to exposure wall lamellae of elongating cortical parenchyma cells of poplar, Itoh and Shimaji (1976) found that the innermost wall could contain axially arranged fibrils, or they could be transverse or oblique. This is at odds with the MGH, which proposes that fibrils only achieve a net

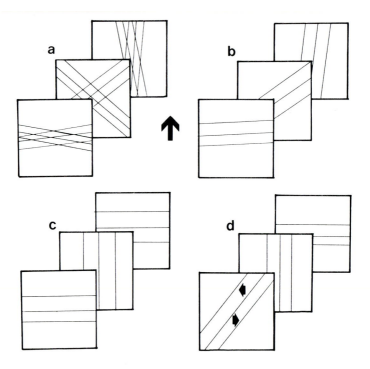

Figure 7. Comparison of models for wall lamellation. (a) Multinet growth hypothesis (MGH) (Roelofsen and Houwink, 1953). Cellulose microfibrils were conceived as being deposited more or less transversely in a mesh—like "a sheaf of fishing nets." This lamella, as each new lamella distances it from the plasma membrane, becomes subject to longitudinal extension strain (large arrow), which reorients the fibrils. From inward to outward direction, the growing wall therefore displays a gradient of fibrillar alignments from net transverse to net axial (or even diffuse). (b) Modified MGH (see, e.g., Preston, 1974). According to various modifications of the MGH, wall fibrils do not constitute an intertwined meshwork but, within one lamella, parallel each other. To account for his observations on the walls of lower plants, Preston suggested that fibrils were deposited in a flat helix; i.e., more or less transversely. However, passive reorientation of fibrils was still thought to occur, as in the original MGH, except that extension strain is envisaged as increasing the pitch of the initial flat helix. (c) Ordered subunit hypothesis (Roland and Vian, 1979). Fibrils are described as being deposited in alternating longitudinal and transverse layers, which build up a plywoodlike structure. Fibrils are therefore envisaged as sometimes being deposited at the plasma membrane in longitudinal arrays; thus, this hypothesis is contrary to the MGH, where such alignment is only anticipated in outer layers in which extension strain has had its effect. Transitional strata with intermediate degrees of alignment are also hypothesized to occur between longitudinal and transverse lamellae. These orderly patterns are thought to permit cell expansion and only become dissipated in the outer wall, with inner order persisting throughout a significant proportion of the still-expanding inner wall. (d) MGH–OSH (Sargent 1978). This solution puts forward ideas to reconcile the MGH and the OSH. As in the OSH, the angle of fibrillar deposition is thought to change regularly from layer to layer. However, this deposited pattern is not immutable, for it is envisaged that longitudinal extension strain (reintroduced from the MGH) will reorient those fibrils that do not lie on a dominant strain vector. Model d shows that inner fibrils deposited obliquely will be reoriented (arrows) toward the long axis, but the middle and outer layers depicted would have been unperturbed, since their alignment already coincided with dominant strain vectors. This concept therefore sees wall lamellation as being dynamically, rather than statically, accumulated.

axial arrangement as a result of passive reorientation in their journey away from the plasma membrane. Itoh and Shimaji (1976) were able to confirm that lamellae could be deposited in a crossed-helical pattern, since pit fields provided openings through which oblique fibrils of opposite helical sign could be seen. Furthermore, lamellae containing oblique or axial or transverse fibrils were all observed to occur throughout the thickness of the wall and this—as other reports also conclude—is counter to the idea embodied in the MGH that the gradient of orientations from in to out runs from net transverse to net axial.

Wall patterns such as these, with the helical sign progressively and repeatedly rotating from lamella to lamella to build up a complex rhythmic plywood, form the basis of Roland and Vian's (1979) **ordered subunit hypothesis (OSH)** (Fig. 7c); the term subunit is used here in a neutral sense and covers wall fibrils that are noncellulosic as well as cellulosic. The crux of the OSH is that the complex wall texture is regarded as influencing the direction of expansion rather than as being influenced by expansion. However, outer wall layers are, as in the MGH, described as disordered, and this is seen as the dissipation of formerly ordered, inner layers, which move outward during growth (Vian et al., 1982). The OSH therefore differs from the MGH in suggesting that inner walls need not be net transverse, that the lamellar pattern of the inner-ordered layers is actually constructed in that way and not evolved through passive reorientation, and that this order extends well into the thickness of the wall before being dissipated. A variety of orientations—not just transverse, as conceived in the MGH—are therefore possible for inner wall fibrils.

Perhaps the final solution to understanding these complex wall patterns will involve elements of both the MGH and the OSH, as Sargent (1978) has suggested. According to this compromisory interpretation (Fig. 7d), the angle of fibrillar deposition alters rhythmically, as in the OSH, but then becomes passively reoriented toward dominant strain vectors, as in the MGH. But in a sense, what happens to the orientation of fibrils away from the plasma membrane is irrelevant to the argument of MT–MF parallelism, and the question remains as to what influences the alignment of the newly deposited fibrils. Itoh and Shimaji (1976) admit a role for microtubules in orienting wall fibrils, for they observe oblique and axial as well as transverse orientations of MTs and inner wall fibrils. So, even in cells with complex polylamellate walls, microtubules still parallel the innermost wall fibrils and continue to do so when radial cell expansion (and deposition of longitudinal fibrils) is encouraged by ethylene (Lang et al., 1982) or when cell elongation (and deposition of transverse fibrils) is encouraged by gibberellic acid (Takeda and Shibaoka, 1981). In the latter study, colchicine treatment prevented gibberellic acid (GA$_3$) from exerting its effect: Inner walls were either transversely or obliquely or longitudinally arrayed instead of exhibiting the predominantly transverse orientation seen with GA$_3$ alone. On the basis of the dynamic helical model (Lloyd, 1984), Roberts et al. (1985) suggested that these three orientations represent helical microtubular arrays exhibiting varied pitches. Ethylene is therefore seen as unwinding the helix (i.e., converting from transverse to longitudinal alignment), much as a bedspring unwinds as it is stretched out. Immunofluores-

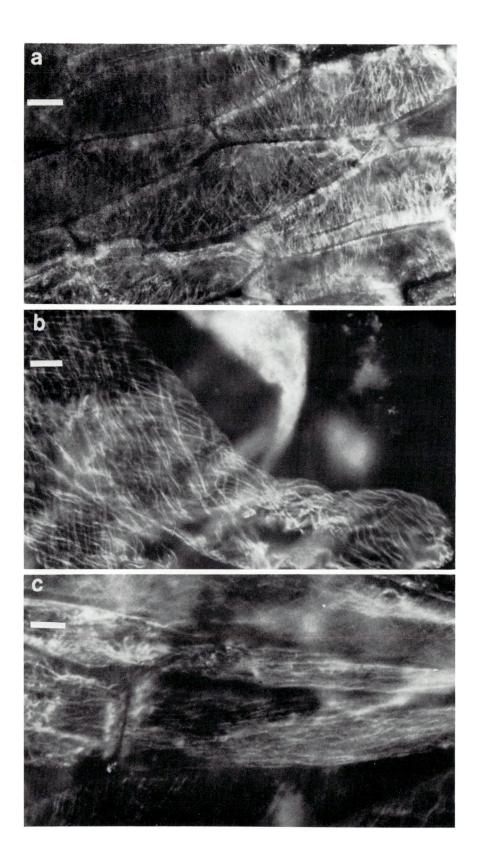

cence observations of ethylene-treated pea and mung bean epidermal cells support this interpretation. The intermediate form (which would probably be glimpsed in electron microscopic sections as oblique MTs) clearly exists as multistart 45° helices, similar to those previously seen in onion root hairs (Lloyd, 1983). The fact that cells devoid of microtubules are not encountered suggests strongly that the three arrays interconvert by a wholesale reorientation of MTs rather than by bouts of de- and repolymerization (Fig. 8). Obviously, this forms part of a continuing story, but the idea that microtubules form arrays of variable helical pitch seems to offer a way of explaining how wall lamellae, themselves of variable helical pitch, can be deposited as part of a rhythmic shifting program in which growth regulators stabilize or labilize the flat-pitched (transverse) helical MT array.

4. Microtubules in Dividing Cells

4.1. The Preprophase Band of Microtubules and the Plane of Cell Division

The cortical microtubule arrays of the preceding section influence cell shape during interphase in cycling cells and also underlie specialized wall thickenings—as in guard cells and xylem tracheids—in terminally differentiated cells. However, Pickett-Heaps and Northcote (1966) noted, during the formation of stomata in wheat leaves, that the more or less even distribution of MTs along the cell length was replaced by an equatorial concentration of tubules. This cortical band of MTs was most prominent before prophase and was less obvious once the chromatin was condensed. The intriguing thing was that during each of the three divisions (see Fig. 2) that formed the stomatal complex, this **preprophase band** (PPB) was detected in the zone where the future cell plate would meet the walls of the mother cell to complete cell division.

This ability of the PPB to predict the future division plane is not restricted to stomatal cells. In a remarkable analysis of the ontogeny of *Azolla* roots, Gunning *et al.* (1978b) demonstrated that the PPB correctly predicts the cytokinetic plane in every single division. By the time mitosis is complete and cytokinesis is under way, the PPB microtubules have long disappeared, but

Figure 8. Epidermal cells of pea, stained with antitubulin antibodies. Microtubules are shown to be wound more or less transversely around control cells (a). Ethylene inhibits cell elongation and causes these cells to bulge out sideways, effecting a shift in polarity of 90°. Antitubulin staining demonstrates that microtubules become longitudinally (c), rather than transversely, aligned after ethylene treatment and that 45° helices represent an intermediate stage (b) in this transition. Occurring without gross depolymerization, it appears that microtubules reorient by "stretching" the helix much in the way a bedspring unwinds as it is pulled out. This ability of microtubules to alter their helical angle is suggested to be involved in the formation of helicoidal walls in which the fibrils of successive lamellae shift their helical angle (Roberts, Lloyd, and Roberts, 1985). Other physiological effects (e.g., the response to gibberellic acid) might also be transduced by the microtubular cytoskeleton. Scale bars: 20 μm, 10 μm, and 10 μm, respectively.

developmental patterns in *Azolla* are so precisely understood that it can be concluded that the plate bisects the site once occupied by the PPB with an accuracy of 0.1–0.25 μm.

Whether the alignment of the PPB is the cause of cell polarization or is itself the result of some other polarizing influences is still unknown. The precision with which the PPB anticipates the position of the new cell plate might imply that this band is an expression of positional cues operating in a morphogenetic program. However, it is now clear that the PPB can predict division planes, which—because those divisions are experimentally induced—could not be part of a rigid preprogram. For example, in epidermal strips peeled from *Nautilocalyx* (Venverloo *et al.*, 1980) and then grown *in vitro*, the vacuolated epidermal cells are stimulated to divide parallel to the surface of the strip; these cells contain a band of MTs that defines where the cell plate will meet older side walls. This is also the case when normally quiescent cortical cells of pea root are reprogrammed to divide after wounding (Hardham and McCully, 1982).

Furthermore, Clayton and Lloyd (1984) demonstrated that onion meristematic cells can be induced to form tripolar or multipolar mitotic spindles in the presence of CIPC or griseofulvin. These are followed in the cell cycle by phragmoplasts that successfully form extra limbs between the extra restitution nuclei. Clearly, these are unprogrammed division planes, but in this case the PPBs are apparently normal and do not reflect the tortuous division planes that are laid down at cytokinesis. This cannot, however, be taken to imply that the PPB has no forecasting ability, since the drugs might have exerted their effects after PPB formation, at the time when the spindle poles are becoming active. What these observations do show is that the phragmoplast can grow out to cortical sites that could not have been imprinted by a PPB.

Concerning the PPB, there is still much to speculate about. For instance, in three-dimensional tissue, neighboring cells might reasonably be expected to influence the plane in which an internal cell divides. Yet, PPBs can develop in the absence of lateral influences, i.e., normal to the growth axis of cells in the single files that constitute developing trichomes (Busby and Gunning, 1980). How, then, is the division plane constructed in these and in single cells, and what larger forces are at work in tissues?

4.2. Mitosis, Cytokinesis, and the Division Plane

The PPB begins to disappear at prophase as the mitotic spindle develops from ill-defined poles (Fig. 9). The spindle equator does not necessarily coincide with the division plane. The guard mother cell in *Allium*, for instance, forms an obliquely aligned spindle apparatus such that daughter chromosomes segregate to opposite corners of the cell. Then the developing cell plate reorientates from an oblique angle to one in which the cell eventually becomes divided longitudinally by the continued centrifugal growth of the plate, which meets side walls perpendicularly at the zone previously demarcated by a PPB (Palevitz and Hepler, 1976).

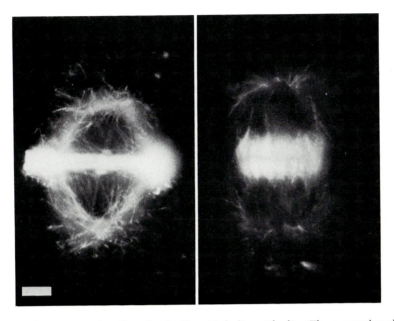

Figure 9. Fixed onion root tip cells stained with antitubulin antibodies. The preprophase band of microtubules is seen (left) as an equatorial cortical band. In many types of cell it anticipates the future division plane but disappears at about prophase. After the chromatids have been separated on the mitotic spindle, the phragmoplast (right) forms and helps construct the new cell plate across the cell. The phragmoplast consists of short overlapping microtubules arranged normal to the plate. This circular arrangement seems to help guide plate-containing vesicles to the midline; it develops centrifugally, growing out like a ripple in a pond, to meet the side walls of the mother cell at the location previously indicated by the preprophase band. Scale bar: 10 μm. (Courtesy of Dr. Lesley Clayton.)

This suggests that the PPB site is a better predictor of the division plane than of spindle orientation, but the forces involved in this are unclear. In some *Triticum* species two successive GMCs in one file of epidermal cells may be flanked by a single, long epidermal cell in the adjacent file. This long cell may be simultaneously induced to form two PPBs in the expected position, except that the single nucleus of the cell occupies a position between them. At cytokinesis, the new cell plate may curve from one PPB zone to the other within the same long cell. Such behavior does not negate the predictive ability of the PPB so much as underline the point that the direction taken by the cytokinetic phragmoplast need not stringently adhere to the plane of the PPB and implies other controlling influences (Galatis et al., 1983).

The major problem is that we lack knowledge of how the nucleus is positioned and of how the central, early cell plate grows out centrifugally to the side walls. The PPB marks the perimeter of this zone, but we might ask if there are other structures that guide the initially central cell plate out toward the cell periphery. Recently, F-actin has been detected in the phragmoplast of onion root tip cells using rhodaminyl lysine phallotoxin. Staining codistributes with the microtubules. The preprophase band does not, however, stain with this agent, and so it would seem that F-actin is involved in cytokinesis (perhaps in

propelling wall-containing vesicles to the new cell plate) but not in preposi-
tioning or predicting the site where the plate fuses with the side walls (Clayton
and Lloyd, 1985). Future work will tell if there are yet other cytoskeletal elements involved in imprinting the PPB zone, for polarization evidently endures
from before prophase until cytokinesis—a period in which microtubules display dynamic, rather than persistent, qualities.

5. Cells into Tissues

Individual cells modulate their shape according to the way in which their
side walls yield to turgor pressure. In dividing cells an important component of
tissue patterning is the way in which the orientation of the cross wall defines
the relationships of sister cells in space. But how are such events involving one
or two cells coordinated in a manner consistent with the production of organized tissue?

5.1. Conveyance of Spatial Information by Microtubule Organizing Centers

Arising from a detailed ultrastructural study of the development of *Azolla*
roots, Gunning *et al.* (1978c) put forward ideas of how spatial information—as
expressed by microtubule organizing centers (MTOCs)—could be transferred
from cell to cell. These ideas flow from two basic premises: (1) MTOCs residing
along the edges of polyhedral cells control the development of interphase microtubule arrays, and (2) bisection of the former preprophase band zone by the
cell plate represents an opportunity for the transfer of MTOCs between generations. Clusters of osmiophilic material (presumptive MTOCs) are seen along
cell edges in this water fern, and it is postulated that if they become active in
initiating MT assembly along, say, longitudinal edges (by the overlap of microtubules initiated from opposing edges) transverse arrays will be produced,
whereas a more restricted activation of the MTs along a cell edge at preprophase would give rise to the PPB. How the osmiophilic material becomes concentrated along cell edges—and there are obvious problems presented by tubular cells, which have no edges—is not known, but there are two possibilities:
Either MTOCs are deposited at preprophase in a cortical band that is later
bisected by the cell plate, or phragmoplast MTOCs are deposited at the junction
of cell plate and older side walls as the cytokinetic apparatus completes its
centrifugal development. Either way, supracellular space is envisaged as consisting of a three-dimensional gridwork of cell edges, along which MTOCs are
selectively activated to help control cell polarity during interphase or the plane
of division at mitosis/cytokinesis. It is not known how these possible tactics are
coordinated as part of the larger strategy of tissue formation, but Gunning
(1982a) presents a thoughtful analysis.

5.2. Modulation of Cell Polarity by Dynamic, Helical Microtubule Arrays

It now seems, however, that the interphase microtubule array of some higher plant cells is constructed along different lines to those envisaged for *Azolla*. This means that different paradigms will be required to account for the ways in which microtubule arrays are oriented and reoriented as part of a developmental program.

Conventional electron microscopy may have underestimated the length of microtubules for they can measure well in excess of 10 μm (Lloyd, 1984). The direct implication is that microtubules can encircle the cell so that the array need not have been formed by short microtubules overlapping in relays, between cell edges. Whole cell immunofluorescence tends to confirm this picture for although it does not measure microtubules it shows that the interphase cortical array is unitary rather than fragmentary and exists as variable helices that wind around the cell without reference to edges (Lloyd, 1983, 1984; Roberts et al., 1985).

At the transition between cytokinesis and interphase, immunofluorescence studies show that the new interphase microtubules—at least in onion root tip cells—radiate between nucleus and cortex. There appears to be no preferred alignment of MTs at this initial stage for only later do they wind around the cortex in a particular orientation. At the radiating stage it is therefore important to know whether MTs grow from MTOCs positioned at the nucleus or at the cortex. A human autoimmune serum appears to stain amorphous MTOCs in plants (as it does in animal cells) and has been used to show that MTOCs (or, better, nucleation sites) are located around the nucleus (Clayton et al., 1985). No nucleation sites were located in the PPB zone and so it would appear that the interphase array in onion cells is not pre-patterned by MTOCs. Instead, it seems that MTs growing out from the nucleus hunt out a particular axis only when they reach the cortex. There are at least two ideas concerning this: (1) MT associated proteins (attachment sites) may be pre-aligned along the plasma membrane (either deposited along "stress" lines or inherited from the previous interphase array), or (2) long plasma membrane-associated MTs could self-assemble into an array whose orientation depends upon either cell dimensions and/or growth regulators to which the orientation of MTs is known to be sensitive (see Lloyd and Barlow, 1982).

5.2.1. Persistence of Microtubule Attachment Sites

According to the first hypothesis, Green and Lang (1981) proposed that membrane attachment sites persist throughout division so that newly regenerating interphase arrays in the daughter cells return to the previous orientation. Longitudinal divisions are more problematic for these T divisions insert new cross walls at right angles to the transversely oriented, preexisting cross walls. Such divisions provide watersheds in development—generating, for instance, new cell files or initiating lateral organs. In their description of organogenesis,

Green and Lang (1981) further postulated that newly forming arrays could align upon attachment sites provided by the PPB in addition to those inherited from the mother cell's interphase array. T divisions therefore present instances in which two sets of hypothetical attachment sites are normal to one another, with the influence of the PPB zone (which anticipated the novel division plane, 90° to its predecessors) depending upon its size relative to the dimensions of the cell. Some such latter clause is important because there are cases where, following some T divisions, the interphase arrays in daughters are parallel to the previous PPB whereas in others they develop at right angles to it.

5.2.2. The Integral Microtubule Array and Its Response to Growth Substances

An alternative hypothesis places emphasis on the ability of the cortical microtubule array to alter its orientation in response to morphogenetic signals, rather than on the pre-alignment of microtubule attachment or initiation sites. It was concluded from immunofluorescence observations on the effects of ethylene on sheets of epidermal cells (Roberts et al., 1985) that the entire microtubule array reorientated from a flat pitched transverse helix, through an intermediate stage in which the helical nature of the array is seen unambiguously, to a steeply pitched axial conformation. This not only supports the view of the array as an integral and dynamic structure but illustrates the way in which the axis of expansion can swing through 90° (see Fig. 8). Adapting Green's (1963) original contracting ring hypothesis to the helical device, maximization of crossbridging between adjacent microtubules would flatten the pitch of the array (encouraging cell elongation) whereas minimization of interaction would steepen the pitch, leading to lateral cell expansion. Ethylene is one physiological factor that modulates the orientation of the array; gibberellic acid is another and produces the opposite effect of stabilizing the array in the "transverse", flat-pitched conformation (Takeda and Shibaoka, 1981). At the tissue level, therefore, cell polarity might be interpreted by responsive cortical microtubule arrays according to the sensitivity of the cell to, or the programmed production of, appropriate growth factors. Demonstrably, these responses occur in mature tissues, but they also suggest ways in which polarity may be controlled during ontogeny.

Returning to the initiation of cortical arrays, other growth factors have been suggested to have an indirect influence on orientation. The basic mapping-out of cell patterns can be looked at as an interaction between cell expansion and cell division such that relatively long cells are favored by a low rate of cell partitioning but short cells are produced in regions of high mitotic activity. In this way, the gradient of cell sizes, from short to long, observed along the growth axis of an apical meristem can be described in terms of a gradient of a division promoting factor(s) (see Lloyd and Barlow, 1982). An analogous case would be the increased series of lengths obtained when carrot suspension cells are subcultured into a series of decreasing concentrations of the division-promoting auxin 2,4-dichlorophenoxyacetic acid (Lloyd et al., 1980). Allowing for the fact that cortical microtubules form helices rather than hoops, Green's

(1963) contracting hoop hypothesis (see also Lloyd and Barlow, 1982) suggests that maximal crossbridging between adjacent microtubules would tend to tighten the array, thereby seeking the least circumference. In cells that are longer than wide, the newly forming cortical array would define the shorter, transverse axis whereas in cells that are broader than long, the cortical array would tend to re-form at 90°. By determining the size at which an expanding cell is partitioned, mitogenic substances therefore control the axial ratio of the daughters and, according to this interpretation, the initial orientation of new microtubule arrays.

In concluding this section on the influences that bear upon the orientation of cortical arrays, not only within established arrays but between generations, it is apparent that there are two basic concepts: one emphasizes inheritance of cortical alignment sites whereas the other relies upon autoregulation of the array in response to growth factors and cellular dimensions. It should be possible to decide whether either or a combination of these ideas is valid when more becomes known about the crossbridges that unite cortical microtubules to each other and to the plasma membrane.

5.3. Influence of Supracellular Forces on the Alignment of the Cell Plate

Long standing rules state that the new plate aligns at right angles to long walls, or at right angles to the direction of maximal growth of the parent cell, or as being of minimal surface area for halving the cell's volume. One particular analogy likens cross walls to partitions in soap bubbles that are oriented by surface tension forces (see Gunning, 1982b; Lloyd and Barlow, 1982). But in a perceptive analysis of the positioning of new cell walls, Sinnott and Bloch (1940) observed that no rule was inviolate and that local conditions were frequently overriding. An important outcome of this study for the practical understanding of plate alignment was that they described a fusion of cytoplasmic strands— quite unlike a weightless liquid film—that forms across the vacuole. The nucleus of a cell stimulated to divide by tissue wounding migrates from a lateral location into the centre of this **phragmosome** where mitosis takes place. At prophase, the phragmosome anticipates the plane of cell division, and in epidermal cells of *Nautilocalyx* it has subsequently been discovered that a band of microtubules, resembling a preprophase band, also exists where the phragmosome connects with existing side walls (Venverloo et al., 1980). Although an unprogrammed PPB is present, the formation of this fused network of cytoplasmic strands between the central nucleus and the cortex is at least as predictive of the division plane as the PPB. It also has a more direct influence on nuclear positioning and represents the path followed by the cytokinetic phragmoplast to the cortex. The phragmosome is not composed of structurally amorphous cytoplasm but, in the long fusiform cambial cells of ash, contains both microtubules and bundles of 7-nm filaments (Goosen-de Roo et al., 1984). Cytoskeletal elements therefore span the area between the nucleus and the imprinted cortical cytoplasm during pre- and post-mitotic stages. They are suggested to have a role in guiding the centrifugally growing phragmoplast to the cortex but in considering forces that orientate the division plane before mitosis, the microfilaments (should they prove to be composed of contractile actomyosin) should be suspected of being involved in strain alignment of the prospective division plane.

Concerning the biophysical forces that impinge upon the dividing cell, Lintilhac (1974) argued that the precursor of the cell plate is sensitive to shear and that the new cross wall is aligned along a shear-free plane that exists perpendicular to the direction of the largest principal stress exerted on the cell. As cells grow, they are subjected to compressive stresses, and Lintilhac's view of morphogenetic space consists of trajectories of stress that pass through tissues and align cell plates. The shape of cells, and of the tissues they form, therefore "feed-forward" on the growth of subsequent generations—stress representing the epigenetic continuity within developing structures.

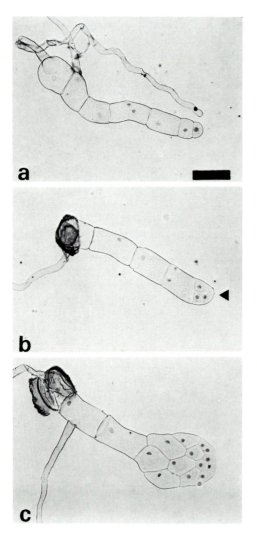

Figure 10. Young gametophytes of the fern *Onoclea sensibilis* L. illustrate the possible role of cell proportions in influencing the occurrence of T divisions that alter the growth axis. (a) The cells continue to grow as a file as long as the extending apical cell is partitioned transversely. (b) When the apical cell becomes broader than long (as the division rate slows relative to the elongation rate) longitudinal cross walls are encouraged (arrow). (c) This marks the transition from one-dimensional filamentous growth to two-dimensional growth of the flat prothallus. Miller (1980) suggests that the cell plate aligns in the shear-free plane that exists perpendicular to the direction of the largest principal stress exerted on the cell. Scale bar: 100 μm. (Reproduced with permission from Miller, 1980.)

Surrounding cells probably exert compressive forces on internal, dividing cells, and for this reason Miller's (1980) study of the apical cell of filamentous fern gametophytes provides a valuable insight into the internal factors that affect the division plane, since this cell is free from all but its subapical neighbor (Fig. 10). With continued transverse division, this file of cells grows unidirectionally, but if the apical cell divides longitudinally, as it does when an increased rate of division has made it broader than long, this initiates a transition to two-dimensional growth as a flat plate of cells—the **prothallus.** Miller argues that in a long apical cell the elastic protoplast expands more in one direction than another, thereby generating shear stress on the cytoplasm. The dimensions of the cell are therefore an important cueing device. Broad apical cells were seen to divide longitudinally, whereas long cells often divided transversely. This was taken to support the idea that plate alignment is determined by stresses operating at the time of cell division—stresses influenced by the shape of the mother cell.

From studies such as these it is reasoned that the relative rates of cell division and cell elongation represent an in-built element for determining the plane of division (see Lloyd and Barlow, 1982). In directionally enlarging cells, a slow rate of division with rapid elongation would encourage transverse cell partitioning, whereas critical T divisions would occur when a broad cell is formed as a result of rapid division and slow elongation.

Coordination of the division pattern over a field of cells would, in terms of stresses, be more difficult to analyze, but if cell proportions (in polarized cells, where diameter is regulated by circumferential microtubules) exert influence on the plane of division, then it can be envisaged that gradients of division or elongation-promoting factors could play a controlling role. According to this way of looking at morphogenesis, genetic control is exerted *via* growth factors that regulate cell volume across fields of cells, but the fine-tuning control over division planes and direction of expansion is an epigenetic one determined by biomechanical principles that influence the appropriate cytoskeletal assemblies.

References

Busby, C. H., and Gunning, B. E. S., 1980, Observations on pre-prophase bands of microtubules in uniseriate hairs, stomatal complexes of sugar cane, and Cyperus root meristems, *Eur. J. Cell Biol.* **21**:214–223.

Clayton, L. C., and Lloyd, C. W., 1984, The relationship between the division plane and spindle geometry in Allium cells treated with CIPC and griseofulvin: An anti-tubulin study, *Eur. J. Cell. Biol.* **34**:248–253.

Clayton, L. C., and Lloyd, C. W., 1985, Actin organization during the cell cycle in meristematic plant cells. Actin is present in the cytokinetic phragmoplast, *Exp. Cell Res.* **156**:231–238.

Clayton, L. C., Black, C. M., and Lloyd, C. W., 1985, Microtubule nucleating sites in higher plant cells identified by an auto-antibody against pericentriolar material, *J. Cell Biol.* **101**:319–324.

Galatis, B., Apostolanos, P., and Katsaros, C., 1983, Synchronous organization of two preprophase microtubule bands and final cell plate arrangement in subsidiary cell mother cells of some *Triticum* species, *Protoplasma* **117**:24–39.

Goosen-De Roo, L., Bakhuizen, R., Van Spronsen, P. C., and Libbenga, K. R., 1984, The presence of extended phragmosomes containing cytoskeletal elements in fusiform cambial cells of *Fraxinus excelsior* L., *Protoplasma* **122**:145–152.

Green, P. B., 1963, On mechanisms of elongation, in: *Cyto-differentiation and Macromolecular Synthesis* (M. Locke, ed.), pp. 203–234, Academic Press, New York.

Green, P. B., and Lang, J. M., 1981, Toward a biophysical explanation of organogenesis: Birefringence observations on regenerating leaves in the succulent *Graptopetalum paraguayense* E. Walther, *Planta* **151**:413–426.

Gunning, B. E. S., 1982a, The root of the water fern *Azolla*. Cellular basis of development and multiple roles for cortical microtubules, in: *Developmental Order: Its Origin and Regulation* (S. Subtelny and P. B. Green, eds.), pp. 379–421, Alan R. Liss, New York.

Gunning, B. E. S., 1982b, The cytokinetic apparatus: Its development and spatial regulation. in: *The Cytoskeleton in Plant Growth and Development* (C. W. Lloyd, ed.), pp. 229–292, Academic Press, New York.

Gunning, B. E. S., Hughes, J. E., and Hardham, A. R., 1978a, Formative and proliferative cell divisions, cell differentiation and developmental changes in the meristem of *Azolla* roots, *Planta* **143**:121–144.

Gunning, B. E. S., Hardham, A. R., and Hughes, J. E., 1978b, Preprophase bands of microtubules in all categories of formative and proliferative cell division in *Azolla* roots, *Planta* **143**:145–160.

Gunning, B. E. S., Hardham, A. R., and Hughes, J. E., 1978c, Evidence for initiation of microtubules in discrete regions of the cell cortex in *Azolla* root tip cells, and an hypothesis on the development of cortical arrays of microtubules, *Planta* **143**:161–179.

Hardham, A. R., and McCully, M. E., 1982, Reprogramming of cells following wounding in pea (*Pisum sativum* L.) roots. I. Cell division and differentiation of new vascular elements, *Protoplasma* **112**:143–151.

Hardham, A. R., Green, P. B., and Lang, J. M., 1980, Reorganization of cortical microtubules and cellulose deposition during leaf formation in *Graptopetalum paraguayense*, *Planta* **149**:181–195.

Heath, I. B., 1974, A unified hypothesis for the role of membrane bound enzyme complexes and microtubules in plant cell wall synthesis, *J. Theor. Biol.* **48**:445–449.

Heath, I. B., and Seagull, R. W., 1982, Oriented cellulose fibrils and the cytoskeleton: A critical comparison of models, in: *The Cytoskeleton in Plant Growth and Development* (C. W. Lloyd, ed.), pp. 163–182, Academic Press, New York.

Herth, W., 1983, Arrays of plasma membrane "rosettes" involved in cellulose microfibril formation of *Spirogyra, Planta* **159**:347–356.

Itoh, T., and Shimaji, K., 1976, Orientation of microtubules and microtubules in cortical parenchyma cells of poplar during elongation growth, *Bot. Mag. Tokyo* **89**:291–308.

Lang, J. M., Eisinger, W. R., and Green, P. B., 1982, Effects of ethylene on the orientation of microtubules and cellulose microfibrils of pea epicotyl cells with polylamellate cell walls, *Protoplasma* **110**:5–14.

Ledbetter, M. C., and Porter, K. R., 1963, A "microtubule" in plant cell fine structure, *J. Cell Biol.* **19**:239–250.

Lintilhac, P. M., 1974, Differentiation, organogenesis, and the tectonics of cell wall orientation. III. Theoretical considerations of cell wall mechanics, *Am. J. Bot.* **61**:230–237.

Lloyd, C. W., 1983, Helical microtubular arrays in onion root hairs, *Nature (Lond.)* **305**:311–313.

Lloyd, C. W., 1984, Toward a dynamic helical model for the influence of microtubules on wall patterns in plants, *Int. Rev. Cytol.* **86**:1–51.

Lloyd, C. W., and Barlow, P. W., 1982, The co-ordination of cell division and elongation: The role of the cytoskeleton, in: *The Cytoskeleton in Plant Growth and Development* (C. W. Lloyd, ed.), pp. 203–228, Academic Press, New York.

Lloyd, C. W., Lowe, S. B., and Peace, G. W., 1980, The mode of action of 2,4-D in counteracting the elongation of carrot cells grown in culture, *J. Cell Sci.* **45**:257–268.

Lloyd, C. W., Slabas, A. R., Powell, A. J., and Lowe, S. B., 1980, Microtubules, protoplasts and plant cell shape. An immunofluorescence study, *Planta* **147**:500–506.

Miller, J. H., 1980, Orientation of the plane of cell division in fern gametophytes: The roles of cell shape and stress, *Am. J. Bot.* **67**:534–542.

Montezinos, D., 1982, The role of the plasma membrane in cellulose microfibril assembly, in: *The Cytoskeleton in Plant Growth and Development* (C. W. Lloyd, ed.), pp. 147–162, Academic Press, New York.

Palevitz, B. A., 1982, The stomatal complex as a model of cytoskeletal participation in cell differentiation, in: *The Cytoskeleton in Plant Growth and Development* (C. W. Lloyd, ed.), pp. 346–376, Academic Press, New York.

Palevitz, B. A., and Hepler, P. K., 1976, Cellulose microfibril orientation and cell shaping in developing guard cells of *Allium*: the role of microtubules and ion accumulation, *Planta* **132**:71–93.

Pickett-Heaps, J. D., and Northcote, D. H., 1966, Cell division in the formation of the stomatal complex of the young leaves of wheat, *J. Cell Sci.* **1**:121–128.

Preston, R. D., 1961, Cellulose–protein complexes in plant cell walls, in: *Macromolecular Complexes* (M. V. Edds, ed.), pp. 229–253, Ronald Press, New York.

Preston, R. D., 1974, *Physical Biology of Plant Cell Walls*, Chapman and Hall, London.

Preston, R. D., 1982, The case for multinet growth in growing walls of plant cells, *Planta* **155**:356–363.

Ray, P. M., 1967, Radioautographic study of cell wall deposition in growing plant cells, *J. Cell Biol.* **35**:659–674.

Roberts, I. N., Lloyd, C. W., and Roberts, K., 1985, Ethylene-induced microtubule reorientations: Mediation by helical arrays, *Planta* **164**:439–447.

Robinson, D. G., and Quader, H., 1982, The microtubule–microfibril syndrome, in: *The Cytoskeleton in Plant Growth and Development* (C. W. Lloyd, ed.), pp. 109–126, Academic Press, New York.

Roelofsen, A., 1958, Cell wall structure as related to surface growth, *Acta Bot. Neerl.* **7**:77–89.

Roelofsen, P. A., 1965, Ultrastructure of the wall in growing cells and its relation to the direction of growth, *Adv. Bot. Res.* **2**:68–149.

Roelofsen, P. A., and Houwink, A. L., 1953, Architecture and growth of the primary wall in some plant hairs and in the *Phycomyces* sporangiophore, *Acta Bot. Neerl.* **2**:218–225.

Roland, J-C., and Vian, B., 1979, The wall of the growing plant cell: Its three-dimensional organization, *Int. Rev. Cytol.* **61**:129–166.

Sargent, C., 1978, Differentiation of the crossed-fibrillar outer epidermal wall during extension growth in *Hordium vulgare* L., *Protoplasma* **95**:309–320.

Sawhney, V. K., and Srivastava, L. M., 1974, Gibberellic acid induced elongation of lettuce hypocotyls and its inhibition by colchicine, *Can. J. Bot.* **52**:259–264.

Singer, S. J., and Nicolson, G. L., 1972, The fluid mosaic model of the structure of cell membranes, *Science* **175**:720–731.

Sinnott, E. W., and Bloch, R., 1940, Cytoplasmic behaviour during division of vacuolate plant cells, *Proc. Natl. Acad. Sci. USA* **26**:223–227.

Staehelin, L. A., and Giddings, T. H., 1982, Membrane-mediated control of cell wall microfibrillar order, in: *Developmental Order: Its Origin and Regulation* (S. Subtelny and P. B. Green, eds.), pp. 133–147, Alan R. Liss, New York.

Takeda, K., and Shibaoka, H., 1981, Effects of gibberellin and colchicine on microfibril arrangement in epidermal cell walls of *Vigna angularis* Ohwi et Ohashi epicotyls, *Planta* **151**:393–398.

Venverloo, C. J., Hovenkamp, P. H., Weeda, A. J., and Libbenga, K. R., 1980, Cell division in *Nautilocalyx* explants. I. Phragmosome, preprophase band and plane of division, *Z. Pflanzenphysiol.* **100**:161–174.

Vian, B., Mosniak, M., Reis, D., and Roland, J-C., 1982, Dissipative process and experimental retardation of the twisting in the growing plant cell wall. Effect of ethylene-generating agent and colchicine: A morphogenetic reevaluation, *Biol. Cell.* **46**:301–310.

Chapter 3

The Egg Cortex in Early Development of Sea Urchins and Starfish

THOMAS E. SCHROEDER

The cytoplasm can no longer be looked upon as so much passive material which the nucleus elaborates during development, but it has a complex organization and a development which may be synchronous with certain changes in the nucleus and the division mechanism, and yet is independent of these.

(Painter, 1916, p. 522)

1. Introduction

The outer few microns of cytoplasm in eggs and embryos exert a powerful organizing influence during development. The cell cortex, as this "layer" is called, is a cellular domain whose dynamic properties account for much of the appearance and behavior of eggs and early embryos. In an abstract sense, the cortex can be considered a separate component of an egg, but in actuality it interacts with other domains such as the deeper cytoplasm and the plasma membrane. There is no distinct demarcation between the egg cortex and other parts.

In modern terms, the essential component of the cell cortex is the **cortical matrix.** Its ultrastructural organization, composition, and behavior are only partially understood. It is composed in part of a meshwork of macromolecules comprising several families variously referred to as fibrillar, cytoskeletal, or contractile proteins. By interacting with other cytoplasmic constituents, these fibrillar proteins exhibit a bewildering variety of properties; indeed, the organization of the egg cortex is elusive, especially at early stages of development, probably because these multifunctional macromolecules are so labile.

In addition to the cortical matrix, inclusions reside within the cortex (e.g., cortical granules and vacuoles of various kinds). They also account for certain

THOMAS E. SCHROEDER • Friday Harbor Laboratories, University of Washington, Friday Harbor, Washington 98250.

59

cortical properties, but usually only passively. It is the intervening cortical matrix that determines the dynamics of the cortex during development.

Besides its ultrastructural organization and biochemical composition, the cortical matrix imparts well-defined mechanical and physical properties to the egg or embryo. The cortex can be contractile (i.e., able to generate mechanical forces), passively pliable, or actively stiff. Contractility in one region, combined with localized relaxation elsewhere, can cause cytoplasmic movements and changes in cell shape. When the cortical matrix is cytoskeletal (i.e., organized for statically reinforcing a cell or for resisting counterforces that tend to deform or distort it), it does not initiate movements but can still be important in an architectural sense. Contractile and cytoskeletal properties of the cortex can be separated conceptually, but they can also coexist in time and space within the cortex, thereby creating complex physical characteristics that can change dynamically.

This chapter discusses two model systems in which the cell cortex functions as an important organizer of developmental change. Sea urchin eggs and starfish oocytes have become well-known subjects for analyzing such events. Even though sea urchins and starfish are taxonomically related echinoderms, their gametes and patterns of embryonic development are actually quite different so that the two examples complement one another. For example, sea urchin eggs are fully mature (haploid, pronucleus stage) as stored in the ovary; they can be artificially fertilized without further preparation. This condition is virtually unprecedented among animal eggs, which are more commonly stored at an earlier stage of maturation. In this sense starfish oocytes are more representative; fully grown ovarian "eggs" of the starfish are immature oocytes (tetravalent diploid, germinal vesicle stage of meiotic prophase). Starfish oocytes require hormonal stimulation in order to resume meiosis; only then, about midway through the process of meiotic maturation, can they be fertilized. In general, sea urchin development has been more thoroughly examined than starfish development.

These two "model eggs" differ in other ways. Eggs and embryos of both, sea urchins and starfish are surrounded by jelly and fertilization envelopes, presumably for protection and as aids to fertilization, but a distinct hyaline layer occurs only in sea urchin embryos. During early development, there are other differences. For example, sea urchins exhibit a stereotypical developmental pattern (the familiar three-tiered array of mesomeres, macromeres and micromeres at the 16-cell stage), but starfish embryos are somewhat less definitely organized at early stages.

The purpose of this chapter is to discuss the structure and function of the cortex of sea urchin eggs and starfish oocytes and to explore its roles in early embryonic development. The element of comparison between two echinoderm groups elevates the analysis above mere description and suggests probable functions. It is concluded that the cortex performs essential roles in the control and maintenance of cell shape, surface topography, blastomere position, embryonic polarity, and overall embryonic architecture.

2. Surface Topography in Sea Urchin Embryogenesis

2.1. Early Postfertilization Changes

In a strictly formal sense, the plasma membrane defines a cell surface, but the plasma membrane alone cannot explain the development of surface topography in eggs and embryos because it is dimensionally too thin and mechanically too inconsequential. It is the cell cortex, lying immediately beneath the plasma membrane and somehow attached to it, that is thought to generate the fine details of surface contours on cells. Grosser features of cell shape and certain large protrusions may be attributed to the combined behavior of the cortex, endoplasm, and even extracellular coats.

The surfaces of most eggs and the external surfaces of blastomeres of most embryos are covered with microvilli (Figs. 1 and 14). These cylindrical projections are about 0.15 μm in diameter and may vary considerably in length and density. In most cases, microvilli are about 0.5 μm in length, yet in special circumstances they may be several micrometers long.

In living cells, microvilli are typically unbranched, solitary, and nearly straight. When preserved for ultrastructural examination they are frequently distorted and damaged in ways that make them appear to be branched or tortuous. As a generalization, microvilli are held erect by an inner core of fibrous material composed of polymerized actin microfilaments cross-linked into bundles by specific actin-binding molecules (Fig. 2). These bundles of microfilaments may also be bonded to the plasma membrane along the microvillus shaft. Sometimes, the core of microfilaments is not visible ultrastructurally, a loss that is likely due to suboptimal preservation.

The unfertilized sea urchin egg is metabolically rather quiescent. The surface topography before fertilization is quite simple, although it is not smooth. Short microvilli cover the outer surface uniformly (Fig. 1); they are only slightly longer than they are wide. Biochemical analysis reveals that a considerable quantity of unpolymerized actin is bound into the cortex before fertilization (see Section 5.2), yet microfilaments are essentially never detected at this time, either within the short microvilli or elsewhere within the cortex. It is believed that a bound form of unpolymerized actin awaits a cellular message delivered at fertilization and that polymerization then takes place.

The **cortical reaction** resulting from fertilization brings about a radical reorganization of membranes, of the cortical cytoskeleton, and of the granular inclusions that occupy the cortex (Schuel, 1978; Vacquier, 1981). Dramatic remodeling proceeds for several minutes before a new equilibrium state is reached. During this time, several thousand membrane-bound cortical granules fuse with the plasma membrane and exocytose their contents into the perivitelline space. The merged membranes together form a composite mosaic membrane at the surface, causing a transient doubling of the surface area of the eggs. A competing process of membrane reclamation by endocytosis begins almost immediately after fertilization (Schroeder, 1979; Fischer and Rebhun,

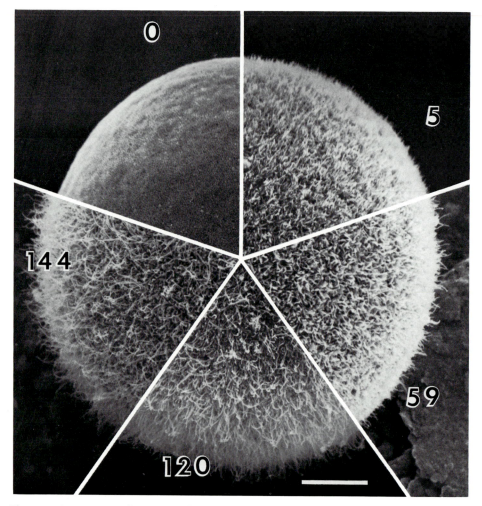

Figure 1. A composite of scanning electron micrographs of sea urchin eggs (*Strongylocentrotus purpuratus*) showing the changing length of microvilli at the surface at various times (in minutes) after fertilization. The vitelline coat is visible at 0 min, but the plasma membrane is directly visible thereafter, because the vitelline layer and the forming hyaline layer were removed soon after fertilization. The "second burst" is visible as the change in length between 59 min and 120 min. First cleavage begins immediately after the 144-min stage. Scale bar: 10 μm.

1983; Carron and Longo, 1984). Microvilli that are originally short and evenly spaced are temporarily displaced and deformed by the dynamics of exocytosis and the insertion of large quantities of new membrane from the cortical granules. Simultaneously, a somewhat disorganized core of microfilaments appears within these microvilli, which become stiff and transiently somewhat longer. This period, as it affects microvilli, has been designated the **first burst of elongation** (see Fig. 1).

Understandably, the three-dimensional topography of the sea urchin egg is

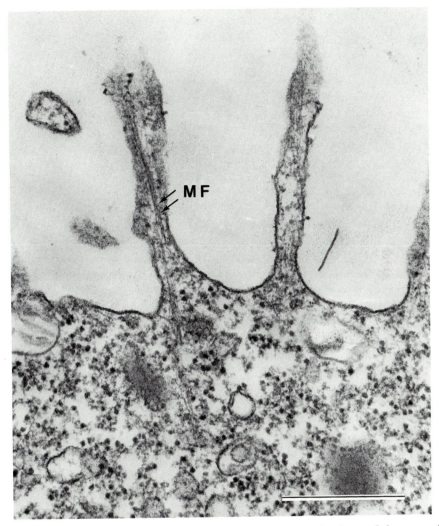

Figure 2. Electron micrograph of microvilli projecting from a fertilized egg of the sea urchin *Arbacia punctulata* after the "second burst" of elongation. The hyaline layer was removed soon after fertilization and is not present. The microvilli contain core bundles of microfilaments (MF) that function to hold the microvilli erect. Core bundles extend a short distance into the submicrovillar region of the cortex. Additional polymerization of actin into the microfilaments causes the microvilli to elongate. The cortical matrix beneath the microvilli presumably also contains actin microfilaments that are able to contract, but preservation of this component of the cell is nearly always disappointing. Scale bar: 0.5 μm.

extremely complex for the first few minutes after fertilization; various transient cavities, pits, ridges, bumps, and folds are created by exocytosis of cortical granules. Nevertheless, within 10–15 min after fertilization, the dramatic changes have subsided. Microvilli are slightly longer than before fertilization, are once again distributed evenly over the surface of an egg, and are self-supported by core bundles of microfilaments. The force for microvillus erection after the vitelline layer lifts away is thought to arise from polymerization of cortical actin as it assembles into the core bundles of microfilaments.

Increases in intracellular pH and cytosol calcium ion concentration are thought to be instrumental in triggering actin polymerization in microvilli of sea urchin eggs (Carron and Longo, 1982; Begg et al., 1982). Experimental manipulations in which these two factors are separately controlled suggest that increased free calcium ion stimulates both polymerization of microfilaments and the elongation of "floppy" microvilli. Increased intracellular pH permits the microfilaments to be crosslinked into core bundles, as a consequence of which the microvilli become erect. Crosslinking of actin has been attributed to actin-binding proteins (see Section 5.2), but many additional details of the regulatory mechanism and organization of microvilli remain to be discovered.

At the site of sperm entry, a protrusion or cluster of protrusions appears transiently during fertilization (Hamaguchi and Hiramoto, 1980; Schatten, 1982). The **fertilization cone,** as this structure is called, is obvious in some species of sea urchins and starfish but is virtually undetectable in others.

Much has been made of the fact that the fertilization cone contains a large number of actin microfilaments (Tilney and Jaffe, 1980). Accordingly, some workers believe that the fertilization cone serves a motile function in drawing a sperm into the egg. Considering that a fertilization cone is seen in only some organisms, however, it seems unlikely that it is essential for sperm incorporation. The fertilization cone may simply represent a local exaggeration of the same cortical changes that cause microvilli to elongate—namely a localized explosion of actin polymerization and bundle formation. In any case, the fertilization cone is a transient event at the site of sperm entry; several minutes after sperm entry, no long-lasting trace of it remains.

2.2. Later Changes in Microvilli

When first observed by Herbst (1900), microvilli on sea urchin eggs at the time of first cleavage were visible as long, thin lines radiating from the surface at the time of first cleavage (Figs. 1–4). Their visibility may now be explained by two facts: (1) the eggs had been treated with artificial calcium-free seawater, which Herbst first developed, and therefore lacked the hyaline layer that usually obscures them; and (2) by the time of first cleavage, microvilli on sea urchin eggs are considerably longer than at times soon after fertilization, due to a phenomenon recently named the **second burst of elongation.** The "second burst" of elongation of microvilli has been detected in a number of species of sea urchins by ordinary light microscopy, polarization microscopy, light-scattering studies, and by scanning EM (Schroeder, 1978, 1981a).

Normal embryos in natural seawater exhibit a "thickening" of the hyaline layer shortly before cleavage. Dan (1960) originally noticed that the hyaline layer is actually separated from and lifted above the egg surface and that it is held in place by microvilli that bridge the narrow space (Fig. 3). By comparing the behavior of microvilli in denuded eggs and of the hyaline layer in intact eggs, the close temporal correlation suggests that the mechanism for elevating the hyaline layer rests with the autonomous elongation of the microvilli known as the second burst.

The fact that microvilli are rather straight when the hyaline layer is lifted indicates that they behave as mechanically rigid rods. In this sense, microvilli resemble other stiff cellular projections such as acrosomal processes on activated sperm, "spikes" of starfish oocytes (see Section 3.2), fertilization cones, stereocilia of the cochlear epithelium, and the stiff twisted arrays of microvilli on sea anemone eggs (Schroeder, 1982). Rodlike rigidity is undoubtedly attributable to crosslinking in the tightly bundled microfilaments forming the cores of microvilli (Burgess and Schroeder, 1977; Spudich *et al.*, 1979; Otto *et al.*, 1980).

It is assumed that the molecular factors responsible for the second burst are similar to those during the "first burst," i.e., an increase in intracellular pH, availability of cytosol calcium, and additional cytoplasmic factors controlled by these changes, perhaps including actin-binding proteins. On the other hand, microvillus length is apparently readjusted within each cell cycle, judging from the evidence by Yoneda *et al.* (1978) on the cyclic behavior of the hyaline layer. Therefore, regulation of microvillus length may be quite complex.

2.3. Functions of Microvilli

Several functions can be suggested for microvilli, including (1) facilitation of absorption and other transport phenomena by increasing the cell surface area, (2) providing a storage depot for plasma membrane, (3) as motile or contractile structures, (4) mediating intercellular adhesion between blastomeres, and (5) mediating adhesion between the egg cortex and the hyaline layer.

The first of these functions for microvilli is likely quite real and significant, but it may be relatively incidental to morphogenetic mechanisms in embryogenesis. Likewise, microvilli as storage depots for membrane is plausible theoretically but may be superfluous, since membrane turnover or synthesis is known to occur quite rapidly. The recently fashionable notion that microvilli are motile structures probably arises because they are known to contain actin, but it is easily refuted by direct observations that microvilli neither wag back and forth nor pulsate lengthwise. Similarly, the suggestion that microvilli are important for intercellular adhesion is refuted by the fact that microvilli do not occur between blastomeres (Figs. 4 and 12).

The fifth suggested function for microvilli as mediators of cortical adhesion to the hyaline layer introduces a set of functional propositions not ordinarily addressed, namely, ideas concerning the role of the hyaline layer in embryogenesis. Considering that sea urchin embryos exhibit an association between microvilli and the hyaline layer, whereas starfish embryos do not (because they lack both components), it is plausible that the association is somehow linked to the stereotypic developmental pattern of sea urchins. If this is correct (see Section 7.1), the ability of microvilli to adhere to the hyaline layer may be one of their vital functions.

Adhesiveness between microvilli and the hyaline layer is easily verified by direct observation, as in Figures 3 and 4. Specifically, at times after the second

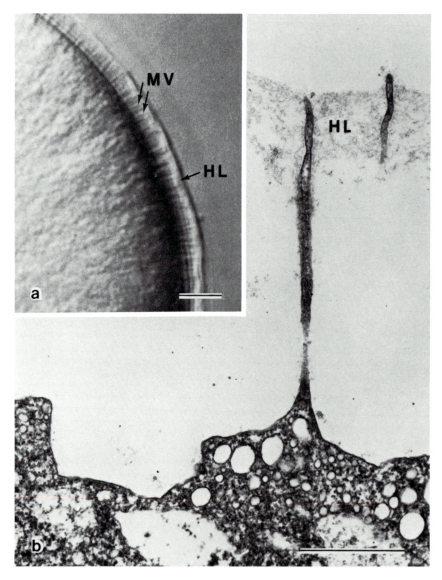

Figure 3. Microvilli (MV) on sea urchin embryos are adherent to the hyaline layer (HL). (a) Nomarski micrograph showing a portion of the surface of a fertilized egg (*Strongylocentrotus droebachiensis*) after the "second burst" of microvillus elongation, which causes the hyaline layer to lift off the egg surface. (b) Electron micrograph of the same showing the intimate association of the outer portions of elongated microvilli with the lightly fibrillar hyaline layer. Scale bars: (a) 10 μm and (b) 1 μm.

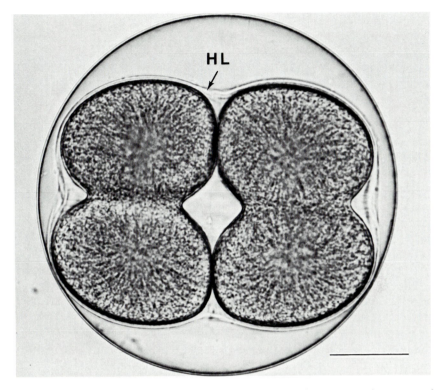

Figure 4. Mid-cleavage of the second division in a sea urchin embryo (*Strongylocentrotus droe-bachiensis*). The entire embryo is surrounded by a spherical fertilization envelope. The hyaline layer (HL) is secreted at fertilization and remains associated with the external surfaces of the embryo throughout early development. Microvilli on the surfaces of the blastomeres, even at this early stage, remain embedded in the hyaline layer. Other surfaces, including those between blastomeres, are smooth and lack microvilli. The cleavage furrow forms normally on both microvillus-covered and smooth surfaces. Scale bar: 50 μm.

burst of elongation, it is obvious that the hyaline layer adheres to the distal half of each microvillus. Perhaps the plasma membrane of this segment corresponds to the membrane of the original microvillus soon after fertilization, when the hyaline layer is initially consolidated; subsequent elongation may generate a new nonadhesive segment of microvillus plasma membrane. Additional indications of this adhesive bond between microvilli and the hyaline layer are seen during cell cleavage when microvilli pull the hyaline layer partway into the cleavage furrow and become somewhat stretched out in the process (Fig. 4).

The molecular basis for the adhesive association between microvilli and the hyaline remains to be determined, but it likely depends on calcium ions and is established soon after fertilization when the contents of the cortical granules are released and encounter calcium ions in the medium. Presumably all microvilli on a sea urchin egg become embedded in the hyaline layer at this time. Within a few minutes, the hyaline layer congeals into a relatively inelastic layer, or shell, that persists on the outer aspect of the embryo.

Since microvilli very often show a coordinate behavior (or distribution) with the pigmented submicrovillar cortex (see Section 4.2), one can postulate a cortex–hyaline layer complex that has a functional significance far beyond mere topographical complexity. The developmental consequences of such a cortex–hyaline layer complex in sea urchin embryogenesis are discussed in Section 7.1.

3. Topography of Maturing Starfish Oocytes

3.1. Hormone-Induced Maturation

Unlike sea urchin eggs, the oocytes of starfish are stored in the ovary at the germinal vesicle stage arrested at meiotic prophase. Meiosis is reinitiated by the action of the hormone 1-methyladenine (1-MA), which originates in the follicle cells surrounding each stored oocyte (Kanatani, 1973; Meijer and Guerrier, 1984). In this general scheme, starfish oocytes resemble the oocytes of many other animals, including vertebrates in which progestins arising from follicle cells stimulate oocyte maturation. As such, and because 1-MA can be applied *in vitro*, starfish oocytes are an excellent model system for studying the events of meiotic maturation. These events are not readily accessible for study in sea urchins.

1-MA apparently acts at the oocyte cell surface. Transduction of this hormonal stimulation of surface receptors transfers a signal of unknown nature to the cytoplasm. It is likely that the initial steps of the transduction events occur within the cortex. Although the transduction events are unknown, the early effects of 1-MA on the cortex are clearly of interest.

In addition to changes elicited in the cortex and cytoplasm, the germinal vesicle envelope eventually breaks down, meiosis proceeds directly through two polar body divisions, and a fully mature haploid oocyte is formed. This natural sequence of events can be accurately followed and studied *in vitro* when exogenous 1-MA is applied to isolated oocytes. Under controlled circumstances normal fertilization can only occur in the brief period between germinal vesicle breakdown and the formation of the first polar body.

3.2. Surface Changes Mediated by Actin

One of the earliest effects of 1-MA on starfish oocytes, at least *in vitro*, is to trigger a massive reorganization of cortical actin. Oocytes treated with 1-MA rapidly develop long actin-filled surface protrusions that have been called spikes (Schroeder, 1981b). Soon after they form, spikes again withdraw and disappear from the surface. Spike formation is more exaggerated after enzymatic removal of the vitelline envelope, so it is not yet totally clear what it represents *in situ*. When the vitelline envelope is present, spikes sometimes force it away from the oocyte surface in a fashion analogous to the effect of the "second burst" of microvillus elongation on the hyaline layer in sea urchin embryos.

The unstimulated immature oocyte is covered with short microvilli very much like the unfertilized sea urchin egg. Within 1 min after 1-MA treatment, the pattern of microvilli is disrupted by emerging spikes that continue to elongate up to about 10 min after treatment (Schroeder and Stricker, 1983). At this stage spikes are visible in the light microscope (Fig. 5) as straight projections 5–15 μm in length. They then gradually diminish in length and, by about 30 min, are no longer detectable even by scanning EM except as occasional thin flaccid strands of plasma membrane.

During the formation of surface spikes, microvilli are rapidly eliminated, apparently because their membrane is borrowed for use by the spikes. When spikes withdraw and the membranous remnants slough off, very few microvilli ever reemerge during maturation. This transient episode of spike formation and withdrawal causes a net loss of about 40% of the original plasma membrane as detected by scanning EM. A comparable net loss is also seen electrophysiologically as a decline of membrane capacitance (Moody and Lansman, 1983). Since spike formation is only prominent after removal of the vitelline coat, this electrophysiological finding indicates that something comparable to spike formation occurs in intact oocytes.

In addition to its effect on the plasma membrane and cell topography, the process of spike formation involves a prominent reorganization of actin and, thus, is another example of topographical change that is initiated by changes in the cytoskeleton. At 10 min after 1-MA, when spikes are most prominent, they are filled by massive bundles of polymerized actin microfilaments. This can be seen in thin sections by transmission EM, *in situ* by staining with the actin-specific dye NBD-phallacidin (Fig. 5), or in isolated cortices (see Fig. 9b) by negative staining or immunocytochemistry (see Section 5.3). When spikes withdraw, all signs of polymerized actin also vanish (Schroeder and Stricker, 1983; Otto and Schroeder, 1984a).

The process of spike formation continues to be explored. Because it is specifically triggered by 1-MA, the process can potentially help to elucidate the physiological factors of the early transductional events of 1-MA reception, as well as the assembly of actin bundles. One suggested function for spikes emerges from the observation that a starfish oocyte usually develops about the same number of spikes (~6500) as there are cellular processes that extend from the surrounding follicle cells to make junctional contact with each oocyte in the ovary (Schroeder, 1981b). If spikes actually form during natural hormone stimulation in the ovary, they might help to disconnect the intercellular junctions between oocyte and follicle cells as a first step in ovulation. The junctional sites may also represent specialized regions of the oocyte surface (and cortex) where 1-MA receptors and transduction molecules are concentrated. Instead of actually being useful or beneficial, it is conceivable that spikes could form merely as a consequence of localized stimulating factors, analogous to the nonutilitarian interpretation of fertilization cones postulated in Section 2.1.

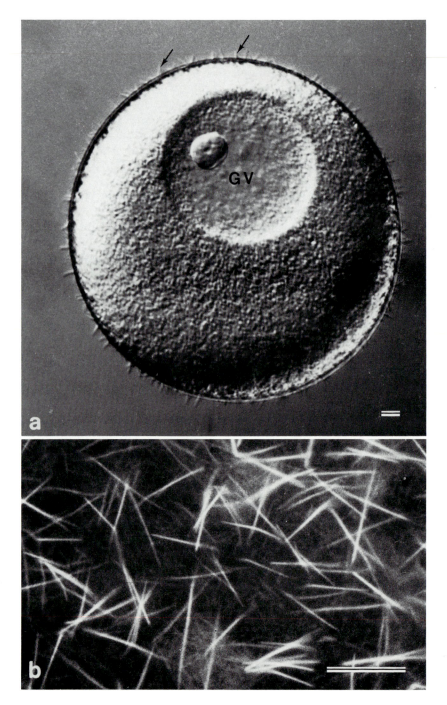

Figure 5. A few minutes after an immature oocyte of the starfish *Pisaster ochraceus* is treated with the maturation-inducing hormone (1-methyladenine), long actin-filled "spikes" (arrows) protrude from the cell surface. (a) Nomarski micrograph showing the spikes all over the oocyte surface about 10 min after hormone treatment; in an additional 10 min these spikes will be retracted. GV, germinal vesicle. (b) Surface view of an oocyte at the same stage stained with the actin-specific stain NBD-phallacidin and viewed in the fluorescence microscope. The white lines represent the stained bundles of actin microfilaments within the spikes. Scale bars: 10 μm.

The formation of spikes may reflect a step in the hormone-induced re-organization of the cortical cytoskeleton essential for the acquisition of fertilizability and competence for development. Actin polymerization during spike formation depletes the cortex of a large amount of bound unpolymerized actin (Otto and Schroeder, 1984a) (see Section 5.3). When spikes withdraw and their bundles of actin microfilaments disassemble, the actin does not revert to its formerly bound state. Thus, spike formation may represent a transient stage in the net "unbinding" of cortical actin from the cortex, and this may be a necessary early step toward later developmental change.

4. Submicrovillar Cortex *in Situ*

4.1. Nebulous Structure of the Cortical Matrix

The idea that the cortex is a distinct part of an egg has deep roots in the history of cell biology. The early micromanipulation studies carried out on developing starfish oocytes and embryos demonstrated that the peripheral cytoplasm and egg surface exhibited a cohesiveness that was not usually found in the endoplasm (reviewed by Chambers and Chambers, 1961). Some of the older empirical evidence for the existence of the cortex was probably based on the behavior of the extracellular coats attached to the egg surface *via* microvilli; therefore, it is not entirely relevant in the modern discussion of the submicrovillar cortex as such. Another aspect of that evidence was based on the obvious contractility of the cortex, as expressed during changes in shape, and this is still highly relevant.

Although the properties and structural organization of the egg cortex have proved very difficult to analyze with precision, it is now believed that cortical contractility arises in a region of the cortex situated just beneath the microvilli, i.e., the **submicrovillar cortex.** Indeed, many investigators are inclined to consider this region the real cortex because of its putatively more dynamic role in morphogenesis compared with the more static microvilli. In reality, however, both the microvillar and the submicrovillar components of the cortex work together during embryogenesis.

Because of the persistent evidence of its contractility, the submicrovillar cortex is usually expected to exhibit a prominent fibrillar organization something like the terminal web of differentiated epithelial cells. Unfortunately, this expectation is nearly always met with disappointment (see Figs. 2 and 3), presumably due to improper methods of detection and preservation. Whereas the microfilaments in microvilli of sea urchin eggs sometimes preserve well for electron microscopy (especially in *Arbacia* and much less satisfactorily in other species), microfilaments in the submicrovillar cortex are only rarely observed (Chandler and Heuser, 1981) except in isolated cortices (see Section 5.2) and certain special cases such as when the contractile ring is present during the obvious contractile event of cell cleavage (see Section 6.2). Recent findings by Usui and Yoneda (1982) provide a refreshing exception; by carefully sectioning tangentially to the egg surface, they have established the presence of a loose meshwork of microfilaments coincident with a period of overall egg contraction called the "global contraction" that procedes cell cleavage.

Selective staining of actin microfilaments in the submicrovillar cortex for light microscopic detection has not yet been successful either, although the application of improved immunocytological localizations to whole mounts or sections should prove informative in the future. Recent analytical polarization microscopy, however, has shown the existence of molecular anisotropy in the tangential plane beneath the surface of sea urchin eggs (Shoji et al., 1981). This represents an improvement over earlier studies, in which radial birefringence was attributed to microvilli (and their contents), and the only detectable tangential birefringence was attributed to the hyaline layer. Nevertheless, the level of the recently found birefringence in the submicrovillar cortex, and therefore the degree of organization of oriented cortical material, is quite low.

Thus, with a few exceptions, knowledge of the fine structural organization of the "real" cortex is still very primitive, considering the abundant empirical evidence favoring its participation in contraction. Of course, it is conceivable that a nebulous organization is sufficient for its subtle contractile activities and that even improved methods of analysis will still show that the contractile cortical matrix exhibits a low level of organization.

4.2. Functional Clues from the Behavior of Inclusions

Cellular inclusions and organelles that reside in the cortex of an egg sometimes represent useful markers of activity states attributable to the cortical cytoskeleton. This is especially true in the case of echinochrome pigment vacuoles, but even cortical granules have the potential for revealing underlying qualities of the cortical matrix.

Cortical granules in sea urchin eggs and starfish oocytes are closely applied to the plasma membrane. In ripe sea urchin eggs they are rapidly exocytosed soon after fertilization (Schuel, 1978). Exocytosis of cortical granules in starfish oocytes, however, will only occur if hormone-stimulated maturation is already under way at the time of fertilization; cortical granules are not completely released if immature oocytes are inseminated. In starfish oocytes, the acquisition of fertilizability and of a normal cortical reaction depends on breakdown of the germinal vesicle. It is not yet known what cortical changes occur during this event.

The motivating mechanism of cortical granules exocytosis is also still uncertain, but some studies have suggested that a slingshot mechanism of actin-based contractility is involved (Vacquier, 1975). The microfilament inhibitor cytochalasin B sometimes prevents cortical granule release (Morton and Nishioka, 1983), as would be expected from such a mechanism. However, the drug is also known to have a generally disrupting effect on the egg cortex, so other explanations of its action are possible. Exocytosis is not prevented by

DNase I, which also depolymerizes microfilaments (Sasaki, 1984). Another important argument against the slingshot mechanism is that at the time of fertilization the sea urchin egg has virtually no polymerized actin (undoubtedly the form necessary for contraction).

So-called pigment granules can be very prominent and exhibit more interesting behavior than cortical granules. In a few sea urchin species they are pigmented, whereas in many others the homologous vacuoles are not pigmented, despite their name. Pigment granules are apparently universally present in eggs of regular sea urchins, where there are about 5000 per egg, but they have not been reported in sand dollars. These granules are about 0.5–1.5 μm in diameter. These membrane-bounded vacuoles are known to concentrate certain fluorescent compounds such as Acridine Orange or 9-aminoacridine, indicating that their contents are acidic (Christen, 1983; Lee and Epel, 1983). The echinochrome pigments, when present, are naphthoquinone derivatives exhibiting shades of yellow, red, or brown, as well as fluorescent colors. Low-speed centrifugation experiments by Morgan and Spooner (1909) clearly established that pigment granules could be displaced into unnatural parts of an egg without affecting development. Thus, neither the actual echinochrome pigment nor the vacuoles are essential for development, so their function remains a mystery.

The normal behavior of pigment granules as cortical inclusions appears to provide information about the functional capability of the cortical matrix in which they are embedded (Schroeder, 1975, 1980b, 1981a). For example, pigment granules undergo a variety of cytochalasin B-sensitive and colchicine-insensitive activities, including (1) migration out to the cortex soon after fertilization, except in *Paracentrotus* eggs, where they are precociously localized in the cortex before fertilization (see Fig. 6b) (Sawada and Osanai, 1980), indicating a postfertilization recruitment of actin microfilaments into the cortex from the endoplasm [interestingly, if the cortical reaction is restricted to only part of an egg, pigment granules only migrate to the cortex in that portion (Allen and Rowe, 1958)]; (2) saltatory motions within the cortex (Belanger and Rustad, 1972; Schroeder, 1980b; Tanaka, 1981) (see Fig. 7); and (3) mass migration away from the vegetal pole region during development (Schroeder, 1980b). All these motions suggest that actin microfilaments in the cortex are intimately involved in a variety of subtle contractile and motile events that would probably pass unnoticed without the presence of pigment granules as inclusions. They further indicate that the pigment granules are either caught up in a web of actin microfilaments or are actually attached to them.

In comparison with sea urchin eggs, starfish oocytes also contain apparently homologous pigment granules that never contain actual echinochrome pigment. Vacuoles, generally larger than their counterparts in sea urchin eggs, are lined up under the plasma membrane even in immature oocytes, whereas they are endoplasmic in sea urchin oocytes. The vacuoles concentrate Acridine Orange and 9-aminoacridine, as they do in sea urchin eggs, but there is no evidence that they ever undergo saltatory or migratory motion. Nevertheless, they are excluded from a small circular zone around the animal pole, just above the germinal vesicle (Schroeder, 1985). The behavior of these pigment granules later in starfish development has not been studied.

4.3. Expressions of Polarity

In echinoderms, the primary embryonic axis—the animal–vegetal (A–V) axis—is established in the egg before fertilization. It is not usually apparent by simple observation until early embryonic stages, but in some circumstances the A–V axis can be detected by either extracellular or intracellular markers. Indeed, it can even be detected in the immature ovarian oocytes of both sea urchins and starfish, suggesting that the genesis of the axis is in the ovary before meiotic maturation.

Information on the A–V axis is primarily restricted to expressions of polarity, i.e., manifestations that conform to the A–V axis despite the lack of any known causal relationship. The root causes or "determinants" of polarity are totally unknown. Final establishment of the A–V axis is likely to be the cumulative effect of many events. It is logical to assume that organizational features in an egg underlie the determination and maintenance of the axis and that these organizational features develop progressively. Thus, although the cortex sometimes expresses the existence of the A–V axis, it cannot be stated with certainty that the egg cortex has any role in axis determination.

Aspects of embryonic polarity in echinoderms can be traced back to the ovarian stages of oogenesis. For example, in immature oocytes of sea urchins and starfish, the germinal vesicle is situated eccentrically next to the animal pole. In the regular sea urchins, but not in sand dollars or starfish, the jelly coat surrounding the egg exhibits a conical channel or jelly canal that can be detected by immersing unfertilized eggs in India or "sumi" ink (Schroeder, 1980a) (see Fig. 6a). This jelly canal already exists in immature oocytes directly above the germinal vesicle. When polar bodies can be seen, they lie within the jelly canal, having been pinched off from the animal pole during meiotic maturation. After fertilization, the developmental pattern consistently confirms that the jelly canal identifies the site of the animal pole. Thus, the jelly canal represents an extracellular marker of the A–V axis that arises during oogenesis. Certain intracellular markers can also be seen before meiotic maturation.

In immature starfish oocytes there is a premeiotic aster located in the narrow cytoplasmic space at the animal pole between the germinal vesicle and the cell surface (Otto and Schroeder, 1984b). Microtubules of the astral rays of this premeiotic aster appear to project into and interact with the cortex of the animal pole (Schroeder and Otto, 1984), in which additional cortical microtubules occur (see Section 5.3) (Fig. 10). This zone of aster–cortex interaction corresponds to the zone in which pigment granules (see Section 4.2) are excluded. Furthermore, actin-filled spikes (see Section 3.2) do not form in this zone when an oocyte is stimulated with 1-MA, although the causal aspects of these correlations have not been determined. More important, it remains undemonstrated that the premeiotic aster plays any role in generating and maintaining the eccentric location of the germinal vesicle or that it has any other

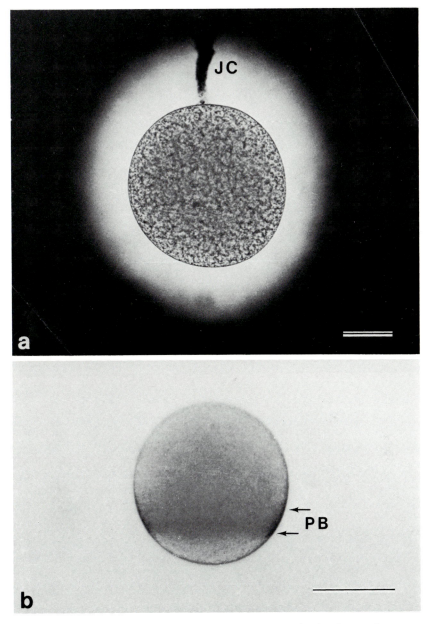

Figure 6. Two markers of the animal–vegetal axis of polarity in unfertilized sea urchin eggs. (a) The transparent jelly layer surrounding a sea urchin egg (*Strongylocentrotus droebachiensis*) exhibits a conical cleft or defect that has been filled with ink particles (black). This jelly canal (JC) is situated directly over the animal pole. (b) In some eggs of *Paracentrotus lividus* the cortically-situated red pigment granules are especially concentrated in a pigment band (PB) located subequatorially in the vegetal hemisphere. A zone surrounding the vegetal pole is devoid of pigment granules, whereas the entire animal hemisphere contains an intermediate density of pigment granules. The pigment band is an intracellular marker of the animal–vegetal axis, whereas the jelly canal is an extra-cellular marker. Scale bars: 50 μm.

consequence upon axis determination. Nevertheless, each of these intracellular morphological features indicate a specialization of the cortex at the animal pole, even if none convincingly identifies a causal or determining role in setting up the embryonic axis (Schroeder, 1985).

An interesting intracellular marker of the A–V axis in sea urchin eggs occurs in *Paracentrotus* eggs as the pigment band (Boveri, 1901; Schroeder, 1980b). The pigment band was originally described as a belt of cortical pigment located subequatorially between an unpigmented animal hemisphere and a smaller clear zone around the vegetal pole; this pattern was originally interpreted as evidence that separate pigment-excluding events occurred at both the animal and vegetal pole. These two unpigmented zones became cornerstones of the double-gradient hypothesis (Horstadius, 1973). When *Paracentrotus* eggs were recently reexamined (Fig. 6b), the animal hemisphere of "banded" eggs was found to contain uniformly distributed pigment granules of an intermediate density equivalent to the distribution of unbanded eggs; thus, the pigment band was redefined and can now be attributed to a single event of pigment granule migration away from the vegetal pole. By itself, this new finding does not invalidate the double gradient hypothesis.

Selective emigration of pigment granules away from the vegetal pole is a universal feature of sea urchin development. It occurs precociously in "banded" *Paracentrotus* eggs, but an analogous emigration occurs in all species at some time before the fourth cleavage division (Fig. 7). This timely expression of the A–V axis guarantees that the four micromeres of the 16-cell embryo lack virtually all pigment granules in all species (Figs. 7 and 12b). The process of pigment granule migration does not depend upon microtubules or cell division itself, since it persists in colchicine-inhibited embryos. Rather, the mechanism of emigration is thought to depend upon a meshwork of actin microflaments within the cortical matrix. Despite the normal correlation between the axial migration of pigment granules and the location and timing of micromere formation, no causal role of the pigment granules in micromere formation has been found (Morgan and Spooner, 1909). Something else in the region of the vegetal pole, perhaps modulated by the cortical matrix, specifies the necessary precondition whereby the vegetal fourth division mitotic apparatuses stimulate the unequal cell divisions (Horstadius, 1973; Dan et al., 1983). The activation or unmasking of this undefined feature of the vegetal polar cytoplasm seems to occur simultaneously with the emigration of pigment granules.

The participation of the cortex in the determination and maintenance of polarity in eggs is poorly understood. Although *Paracentrotus* eggs, for example, express the A–V axis in the cortex, it is intriguing to note that the endoplasm is also structured according to the same axis; the first mitotic apparatus forms (from its inception before breakdown of the nuclear envelope) with its axis perpendicular to the A–V axis, as expressed in the cortex. Does this mean that the cortex communicates polarity cues to the mitotic apparatus? If so, perhaps a variation of this communication explains the asymmetry and eccentricity of the mitotic apparatus of the micromere division at fourth cleavage. Communication in the opposite direction, namely stimulation of the cortex by the mitotic apparatus, is discussed in Section 6.4.

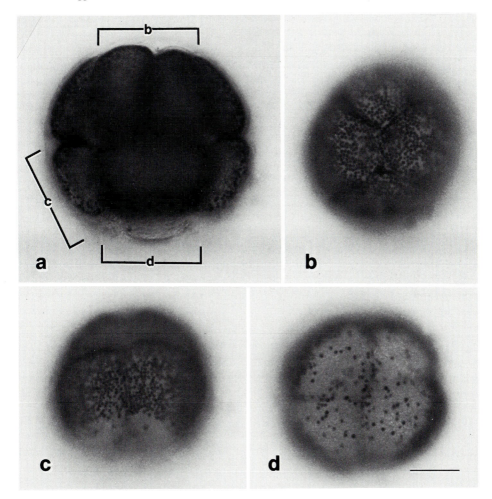

Figure 7. In all species of sea urchins whose eggs contain pigment granules, the granules are nearly all situated in the cortex after fertilization and within a few cleavages segregate in a pattern related to the animal–vegetal axis. In this *Arbacia lixula* embryo the pattern ultimately resembles the pigment band pattern of the unfertilized *Paracentrotus lividus* egg (see Fig. 6); pigment granules are relatively concentrated in the region of the macromeres and virtually eliminated from the micromeres (a). This pattern is already set up before the fourth division actually occurs (b–d). Scale bar: 20 μm. (From Schroeder, 1980b; reproduced with permission from Academic Press.)

5. The Isolated Cortex

5.1. Methods of Preparation and Limits on Interpretation

Our understanding of cortical organization and composition has been considerably advanced by techniques for isolating the cortices of eggs, although the techniques entail some important pitfalls that affect the interpretation of results. Allen (1955) was the first to take advantage of the stickiness of sea urchin

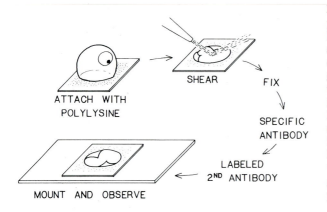

Figure 8. Diagram showing the principal steps in preparing isolated cortices of eggs for microscopic examination. Only those components that are bound together as a complex attached to the coverslip survive the shearing step and remain to be analyzed; everything else is discarded by the procedure. Even important components may be lost in this way, so the actual "isolated cortex" should be regarded as a residue of the complete cortex *in situ*.

eggs to glass surfaces for ripping away cortical egg fragments. This method of isolating cortices has since been refined (Vacquier, 1975) by pretreating glass or other surfaces with polycations such as protamine sulfate or polylysine to which denuded eggs are allowed to adhere. As illustrated in Figure 8, adherent eggs are then disrupted by a jet of defined solution administered by a syringe, pipette, or squeeze bottle. Shearing stresses from the jet rapidly break open the eggs and remove most of the endoplasm and unadhered portions. By this procedure, a lawn of attached cortical fragments remains; one fragment of cortex remains where each egg was adhered (Fig. 9). Eggs can be attached to glass slides or coverslips for microscopic examination, to carbon-coated electron microscope grids for negative staining, or to plastic or glass dishes from which isolated cortices can be scraped away and pooled for compositional analysis by gel electrophoresis or other methods. Thus, a single method permits morphological, cytochemical, and biochemical studies of the isolated cortex.

Another method for preparing cortex-rich fractions depends on breaking open eggs while they are still in suspension. Cell lysis can be achieved with a loose-fitting glass homogenizer (Sakai, 1960) or by osmotic shock in an appropriate buffer (Vacquier and Moy, 1980; Kidd and Mazia, 1980). These methods yield self-sealing vesicular fragments that can be reclaimed by centrifugation.

By whatever method, isolating the cortex from underlying endoplasmic components offers an opportunity to study the distinctive biochemical composition and organization. The advantages are the same as for any kind of cell fractionation, namely enrichment of selected cell constituents. On the other hand, contamination can be a serious problem. Contaminating endoplasmic or other noncortical cellular material can adhere to polycationized substrates or be trapped within self-sealing vesicles. If the preparations are observed microscopically, contaminants can often be distinguished and avoided, but if the preparations are analyzed only biochemically, contaminants can be misinterpreted as meaningful constituents of the cortex.

Implicit in the techniques for isolating cortical fragments from eggs is the idea that only those constituents that are bound together or integrated into the fabric of the cortex will be preserved. Unattached constituents, even if they are

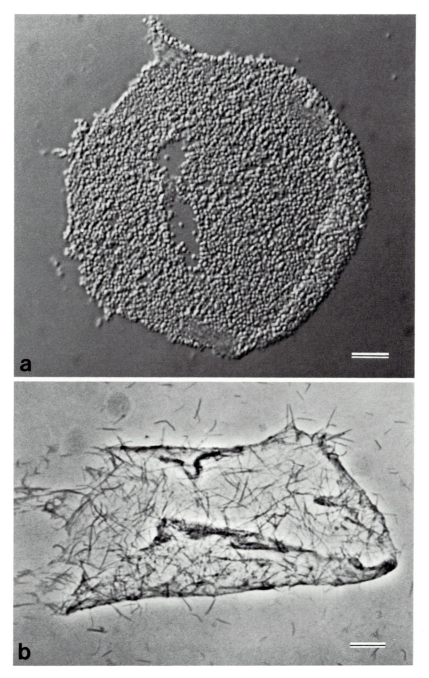

Figure 9. Examples of isolated cortices as seen in the light microscope. (a) Nomarski image of a cortex isolated from an unfertilized sea urchin egg. Numerous undischarged cortical granules dominate the field. (b) Phase-contrast image of an isolated cortex from a starfish oocyte a few minutes after hormone treatment (see Fig. 5), when actin-filled spikes are present and visible as short, dense rods. Scale bars: 10 μm.

important, will be flushed away by the shearing process or washed out by the lysis procedure. Thus, isolated cortices are actually residues of an inevitably more complex assemblage of structures in the living egg. If a distinctive structure is found, it may have real significance to the actual egg cortex; on the other hand, negative evidence is uninterpretable; i.e., if a constituent is absent from isolated cortices, its importance even to the cortex cannot be definitively denied.

5.2. The Isolated Cortex of Sea Urchin Eggs

The isolated cortex of unfertilized sea urchin eggs contains an essentially intact array of cortical granules (Fig. 9a). These structures dominate the morphological image as well as the biochemical analysis unless they are specifically removed. Likewise, the vitelline layer is also present in cortices from unfertilized sea urchin eggs unless it is removed before preparation. After fertilization, cortical granules are exocytosed and the fertilization envelope is conventionally removed before an attempt is made to isolate cortices.

Although actin and tubulin constitute the two major soluble proteins of the cytoplasm of sea urchin eggs, and comparable concentrations *in vitro* are sufficient to polymerize spontaneously, unfertilized sea urchin eggs contain no overtly polymerized actin microfilaments or microtubules. Obviously, polymerization *in situ* is inhibited in some way. Recent biochemical investigations have demonstrated the existence of inhibitory actin-binding proteins in sea urchin eggs (Mabuchi and Hosoya, 1982), although it is not known how these proteins regulate polymerization.

Despite the lack of polymerized microfilaments before fertilization, about 20% of the total amount of egg actin is retained in the isolated cortical fragment, signifying that it is bound in some way (Spudich *et al.*, 1979). The nature of the physical state of this actin is unclear.

Immediately after fertilization, microfilaments composed of polymerized actin appear in the elongating microvilli as well as in the submicrovillar cortex (Begg and Rebhun, 1979; Mabuchi *et al.*, 1980). Myosin has also been identified in these preparations (Mabuchi, 1973), providing a possibility for the necessary actin–myosin interactions involved in contractile activities. By biochemical determination, however, the amount of bound cortical actin is about the same before and after fertilization; moreover, the amount of bound cortical actin present remains constant through the first cell cleavage. Thus, for stage-dependent changes in cortical behavior, qualitative *organizational* changes in cortical actin may be much more important than quantitative changes. This idea is not consistent with the interpretation that actin is recruited into the cortex during pigment granule migration (see Section 4.2).

After the second burst of microvillus elongation, individual microvilli project radially from the perimeter of cortices isolated on polycation-coated EM grids (Burgess and Schroeder, 1977). Microfilaments in the core bundles of these microvilli have been identified as composed of actin by the technique of heavy meromyosin decoration. Each bundle consists of about 20 microfilaments with periodic cross-bridges (Schroeder, 1981a). These cross-bridge structures have been analyzed ultrastructurally (Kane, 1979; Spudich et al., 1979) and identified as the actin-binding protein fascin (see Chapter 1) by immunocytological localization (Otto et al., 1980).

Because microvilli are trapped beneath these kinds of preparations during the isolation procedure, observations by negative-staining have not unambiguously resolved the condition of the submicrovillar cortex. Certainly a very large quantity of polymerized actin microfilaments occurs within the perimeters of these preparations; much of it is probably attributable to the submicrovillar cortical matrix rather than to underlying microvilli. Indeed, thin sections of cortices isolated in suspension have confirmed the existence of a meshwork of actin microfilaments in the submicrovillar cortex.

The isolated cortex of sea urchin eggs undoubtedly contains many important constituents that have not yet been discovered. For example, one can assume that substances involved in the active endocytosis that follows fertilization (Fischer and Rebhun, 1983; Carron and Longo, 1984) can be found there. It is also likely that additional actin-binding proteins involved in nucleation of polymerization, fragmentation of microfilaments, and enzymatically coupled motility will be found.

5.3. The Isolated Cortex of Starfish Oocytes

Ovarian oocytes of starfish are arrested before meiotic maturation, unlike sea urchin eggs just mentioned. Accordingly, some of the differences encountered in the composition and morphology of the cortical matrix reflect this fundamental distinction. Unlike sea urchin eggs, starfish oocytes must be receptive and responsive to the maturation-inducing hormone 1-MA; the cell surface and cortex of a starfish oocyte are undoubtedly specialized for these functions. In addition, the immature oocyte is not yet fertilizable.

Nevertheless, as in unfertilized sea urchin eggs, there is a large amount of actin that is "bound" into cortices isolated from immature starfish oocytes—before exposure to 1-MA. This has been detected by immunofluorescence microscopy using antiactin antibodies (Otto and Schroeder, 1984a). The level of staining is quite high, indicating a large amount of total actin, despite the fact that parallel staining with NBD-phallacidin (which is selective for polymerized actin) is quite low and actual microfilaments have not been observed in negatively stained cortices by EM.

The hormone 1-MA triggers actin-filled spikes to project from the egg surface (see Section 3.2). Cortices isolated at such a time reveal that essentially all cortical actin is now distinctly polymerized and confined to the cores of the spikes. No "bound" actin in any form occurs between spikes. In other words, 1-MA stimulates a massive reorganization of cortical actin within a very few minutes. Using other antibodies, fascin and a 220,000-M_r protein are also localized within spikes, suggesting that they may be necessary for actin polymerization or the formation of bundles of actin microfilaments.

Many years before spikes were discovered, Mabuchi (1974) successfully extracted myosin from cortices of cleaving starfish oocytes. In isolated cortices analyzed by immunofluorescence, myosin is bound into the cortex of immature oocytes; it is never incorporated into the spikes of oocytes treated with 1-MA (Otto and Schroeder, 1984a). Although myosin is an actin-associating protein, it is typically involved in contractile activity rather than in the kind of static projections that spikes represent. It is likely that cortical myosin is important for cleavage furrowing.

By 30 min after 1-MA treatment, when spikes have formed and withdrawn, the isolated cortex of the maturing starfish oocyte exhibits virtually no actin, fascin, or the 220,000-M_r protein. Of the proteins studied, only myosin remains bound into the cortex. Thus, the net effect of the transient appearance of actin-filled spikes is the "unbinding" or release of unpolymerized actin from the

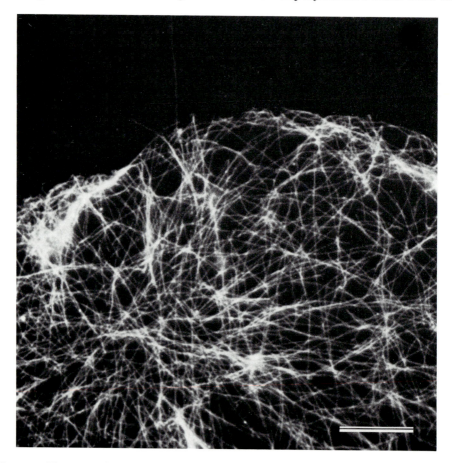

Figure 10. The cortex of immature starfish oocytes contains a meshwork of long microtubules, seen here as thin, bright lines in an isolated cortex stained with antitubulin antibody and viewed by indirect immunofluorescence microscopy. These microtubules disassemble and reassemble on a schedule that correlates with the cell cycle of maturing and developing starfish eggs (see Fig. 11). Scale bar: 10 μm. (From Otto and Schroeder, 1984b; reproduced with permission from Academic Press.)

cortex. The functional significance of this change, like its counterpart in sea urchin eggs at fertilization, is unknown. One possibility, not yet very much explored, is that the bound actin in isolated cortices of immature oocytes is a manifestation of a generalized high degree of cytoplasmic cohesiveness that accounts for the high level of overall mechanical stiffness at this stage (see Section 6.3). Furthermore, the changing organization of the actin-based cortical matrix may represent a real biological example of the interconvertible cytoskeletal and contractile assemblies that have been demonstrated for actin, myosin, fascin, and the 220,000-M_r protein *in vitro* (Kane, 1983). It may be a cortical manifestation of massive changes in cytoplasmic cohesiveness described for activated sea urchin eggs (Coffe *et al.*, 1982).

The presence of microtubules represents a striking difference between the cortices of starfish oocytes and starfish eggs that was discovered by analyzing isolated cortices. A large number of cortical microtubules forms a criss-crossing array within the cortex of immature starfish oocytes (Fig. 10). Although discovered by immunofluorescence microscopy using antitubulin antibody (Otto and Schroeder, 1984b), this fundamental observation has been confirmed by electron microscopic evaluation of isolated cortices and of sectioned whole oocytes.

In response to 1-MA, this array of microtubules repeatedly disassembles and reassembles in synchrony with the cell cycles of meiotic maturation and mitotic cell divisions (Fig. 11). Specifically, the array of cortical microtubules is assembled during phases when meiotic or mitotic microtubules are absent

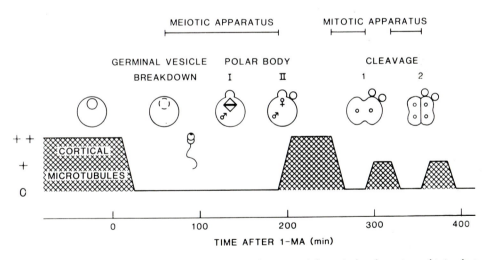

Figure 11. The cycle of cortical microtubules as determined from isolated cortices obtained at times after application of the maturation-inducing hormone 1-methyladenine (1-MA). The presence of a full array of cortical microtubules (see Fig. 10) is indicated by (++); a considerably diminished array is indicated by (+); and the absence of cortical microtubules is indicated by (0). The cortical microtubules occur whenever a meiotic or mitotic apparatus is *not* present (i.e., whenever the nucleus is intact) and vice-versa. The time of fertilization is indicated by the sperm symbol. (From Schroeder and Otto, 1984; reproduced with permission from Academic Press.)

and is disassembled when spindles are present. Thus, the cortical and spindle microtubules experience an inverse and alternate regulation. The importance of this unprecedented array of microtubules is still being explored; it is suspected to play a role in maintaining the polarized eccentricity of the germinal vesicle (Schroeder and Otto, 1984; Schroeder, 1985).

In addition to actin-based and tubulin-based cytoskeletal elements in these isolated cortices, there undoubtedly exist a large variety of other interesting constituents. As summarized by Meijer and Guerrier (1984), protein phosphorylation has been recorded in the cortex of maturing starfish oocytes, but the nature of the kinases or substrates are not yet known. Some may be involved in the process of transducing the hormonal signal from 1-MA into a cytoplasmic message. Another possibility is that the light-chain kinase of the cortical myosin needs to be activated by phosphorylation for contractility.

6. Cortical Contractility

6.1. Quantitative Measurements

The measurement of mechanical properties in sea urchin eggs and starfish oocytes is a sophisticated way to elucidate the dynamics of egg organization. Nevertheless, the mechanical properties of living cells are among the most complex in the physical world and do not lend themselves to simple physical schemes or interpretations.

The cytoplasm or cortex of living eggs is not well-behaved, even in the context of the nearly intractable physical principles of nonNewtonian viscoelasticity. Instead of displaying fixed isotropic properties, the cortex continually reorganizes itself, and its properties tend to be in dynamic flux, either as a consequence of developmental change or in response to the experimental protocol. These characteristics are not compatible with investigations that strive to overformalize the physical properties of the cortex using simple physical measurements.

To the extent that the properties of the cortex can be isolated from those of the rest of an egg, the cortex behaves like a viscoelastic structure that deforms under slowly applied stress, resists rapid deformations with unsuspected force, and recoils to a resting shape when a short-term force is released. When stressed, the cortex resists with a measurable stiffness; another attribute of this stiffness is the tensile force expressed along a tangent at the cell surface, also known as the "surface force" (to be distinguished from surface tension, which is strictly an interfacial force that develops between different substances). Cortical stiffness can be caused by a static condition of the cortex, or it can arise through contractility.

It is an essential feature in the mechanical analysis of the cortex to realize that it has a definite thickness, even when that thickness is not precisely known. Different physical theories that treat either "thick" or "thin" spherical shells have been applied to the egg cortex, based on arbitrary determinations of the ratio of the thickness of the cortex to the diameter of the egg (Wolpert, 1960; Hiramoto, 1970; Yoneda, 1973).

In order to make the physical analysis reasonably tractable, the cortex is often treated as an isotropic structure, even though the cortical cytoskeleton is clearly anisotropic at the macromolecular level of organization. As mentioned, both the structure and mechanics of the cortex are changeable. When the physical properties of the cortex undergo transitional change, there is undoubtedly an accompanying, if not causal, structural change in the cortex, even if it cannot be specifically identified. The relationship between these mechanical properties and structural changes in living cells is complex.

Contractility is a particularly knotty physical concept involving metabolically driven, force-generating mechanical properties amid dimensional and organizational changes. Confusion sometimes arises because of the complicated relationship between contractility and other more passive physical properties. Contractility inevitably implies a local increase in egg stiffness—a property that is relatively easy to measure—but stiffness itself can also increase without any force generation if the thickness or cohesiveness of the cortex rises.

Several ingenious experimental techniques have been developed to probe the mechanical behavior of the cortex. They all depend upon the application of a known imposed force and the microscopic measurement of its effect on egg geometry. Assuming that the resting shape of an egg or oocyte is spherical, deformations can be related back to this original shape; it is generally assumed that no change in volume accompanies moderate deformation. As long as the magnitude of deformation is small relative to the dimensions of the egg, stiffness measurements are probably fair representations of cortical properties; when deformations are large, however, the mechanical properties of the endoplasm begin to interfere or even dominate. Also, as long as the duration of force application is small, measurements tend accurately to indicate "instantaneous" stiffness, rather than some property generated in response to the technique. Thus, practical measurements on the egg cortex require some detailed understanding of the behavior of the eggs.

The best-known methods of measuring the mechanics of eggs include (1) compressing eggs between parallel plates, usually made of glass, which can be done on several eggs at a time with an unsupported plate of known weight acting as a passive load (Danielli, 1952; Schroeder, 1981c) or on individual eggs using a plate attached to a bending beam of known compliance (Yoneda and Dan, 1972); (2) the sessile drop method, in which gravity (or centrifugal force, when gravity is insufficient) deforms an egg against a flat surface while it is being observed from the side [for sea urchin eggs, the use of a centrifuge microscope may be necessary (Hiramoto, 1970)]; (3) sucking a localized outward bulge on an egg with a micropipette device known as a cell elastimeter, in which the negative pressure is calibrated (Mitchison and Swann, 1954; Hiramoto, 1979); and (4) the magnetic particle method, in which a smaller outward deformation is produced when an intracellularly injected iron bead is forced against the inside of the cortex by adjusting an external electromagnetic field (Hiramoto, 1974).

Without substantial knowledge of the cell being examined, even these sophisticated techniques cannot independently differentiate the mechanical contributions of extracellular materials, the cortical cytoskeleton, or the endoplasm beneath the cortex. Furthermore, these methods of quantifying mechanical properties of the cortex necessarily introduce stresses, so it is relevant to be aware that they may induce local alterations or even local breakdown of the cortical cytoskeleton, thereby potentially producing spurious results. Finally, special attention must be given to the resting shape of the cell at the start of the measurement. If it is spherical, the calculations for deriving mechanical properties probably apply, but as soon as an egg begins to change its own shape (e.g., during cleavage), the interpretation of the measurements becomes significantly more complicated.

Localized contractions of the cortex present special problems of interpretation. It is often impossible to measure the magnitudes of localized mechanical properties. In addition, it is frequently difficult to distinguish by direct observation whether a locally wrinkled, dimpled, or dented surface results from a localized contraction (at that site) or from a relative relaxation.

6.2. The Cleavage Furrow and Contractile Ring

In a symmetrically and equilaterally dividing egg, the cleavage furrow arises in accordance with regional dimensional changes: The equatorial cortex regionally shrinks in the circumferential direction. Early in cleavage the equatorial cortex also shrinks slightly in the longitudinal direction, but later it expands longitudinally. Meanwhile, the polar regions progressively expand isotropically. These geometrical changes involve displacements of the egg surface and of the cortex, usually in a coordinated fashion (reviewed in Schroeder, 1981a). In asymmetric divisions (e.g., polar body formation and micromere divisions) and unilateral cleavage divisions of certain species, the location and configuration of the equator and of the poles need to be redefined relative to the stimulating meiotic or mitotic apparatus, but the regional displacements are still basically the same.

Independently of structural considerations and in exclusively mechanical terms, the cleavage furrow can be described as a manifestation of an imbalance in the force-generating and force-resisting capabilities of regions of the cortex. As reviewed by Conrad and Rappaport (1981) and Hiramoto (1979), the cleavage furrow of a sand dollar or sea urchin egg actively exerts more force ($1-5 \times 10^{-8}$ N or $1-5 \times 10^{-3}$ dyn) than any other part of the cell. At first (and perhaps only in some species), the tension at the poles rises, as if they become passively stretched by events at the equator. As cleavage proceeds, however, resistance to stretching sometimes seems to decline actively over the polar regions. In other words, there may be progressive and opposite structural changes occurring at the poles and the equator: As the furrow cortex actively organizes itself to generate contractile force, the polar cortex may actively relax—probably by some active disorganizational change. Regardless of how the imbalance is achieved, the greater contractile force at the equator guarantees the continuing contraction of the cleavage furrow and dimension expansion at the poles.

From the start of cleavage furrowing to the finish, contraction lasts about 10 min. Ultrastructurally, during this time (but not before or after) the cortex beneath the cleavage furrow is specialized by an array of about 5000 closely packed and aligned microfilaments that make up the so-called **contractile ring** (Schroeder, 1972, 1975). This transient organelle has gained some acceptance as a model of contractile cortical organization in general, although it is perhaps an extreme and special example.

The microfilaments of the contractile ring form a thin layer within the submicrovillar cortex in a fairly close association with the plasma membrane. They are oriented in the direction of greatest contraction (i.e., circumferential) and are disrupted by cytochalasin B coincidentally with the arrested advance of the furrow. In certain animal cells (although not specifically in sea urchin eggs or starfish oocytes), contractile ring microfilaments have been shown to be composed of actin by the heavy meromyosin decoration procedure (Schroeder, 1975). Additional immunocytochemical evidence indicates that contractile ring microfilaments are normally associated with myosin, consistent with the postulated musclelike role of the contractile ring in generating the driving-force of furrowing.

The contractile role of the contractile ring in furrowing of sea urchin eggs is supported by experiments with cytochalasin B, which simultaneously disrupts the microfilaments and arrests furrowing. In addition, antimyosin antibodies injected into starfish blastomeres arrest or prevent furrowing, presumably by blocking the enzymatic or mechanical functions of endogenous myosin in the contractile ring (Mabuchi and Okuno, 1977; Kiehart et al., 1982).

Both the origin and precise mode of action of the contractile ring are largely unknown. The best candidate for the precursor of the contractile ring is the loose meshwork of cortical microfilaments visualized by Usui and Yoneda (1982). Curiously, the volume of the microfilament array, once formed, appears to decline as furrowing progresses (Schroeder, 1972). The contractile ring is believed to emerge by reorganization of actin microfilaments preexisting within the submicrovillar cortex and to gradually disassemble self-destructively as it functions. It is not yet clear whether the contractile ring represents an extreme consolidation of contractile elements of the same general type involved in other cortical phenomena such as pigment granule movements.

6.3. Cytoplasmic Regulation of Cortical Contractility

Cells commonly pass through a phase of "rounding up" before cell division. This behavior is consistent with the idea that a phase of general cell contraction precedes cell division. Several classic observations of experimentally derived fragments from eggs of various animals demonstrated that even non-nucleated portions of cells rhythmically rounded up and relaxed in approximate synchrony with the cell cycle of the parental cycle. These reports have stimulated the proposition that an important rhythm of cell contractility can exist independently of the nucleus. The idea that non-nucleated egg fragments or colchicine-treated eggs have an autonomous capability to contract and relax has received considerable support from recent experiments with amphibian eggs (Sawai, 1979; Hara *et al.*, 1980; Sakai and Kubota, 1981), oligochaete eggs (Shimizu, 1981), sea urchin eggs (Yoneda *et al.*, 1978; Schroeder, 1981c; Yoneda and Schroeder, 1984), and starfish oocytes (Yamamoto and Yoneda, 1983).

Non-nucleated fragments of sea urchin eggs, when parthenogenetically activated, exhibit cycles of tension increase and decrease that are coordinated with stepwise episodes of elevation of the hyaline layer (Yoneda *et al.*, 1978). Elevation of the hyaline layer can be attributed to the active elongation of microvilli (see Section 2.2). Apparently, cyclic elongation of microvilli and a rise of cortical tension are coupled events and are both aspects of an autonomous rhythm. The period of this rhythm is similar to the cell cycle, and the two aspects can continue coherently without a nucleus or microtubules.

Starfish oocytes also undergo rhythms of cortical tension during maturation, but the details of their independence from the nucleus is somewhat different. If non-nucleated fragments are prepared from oocytes treated with the maturation-inducing hormone before breakdown of the germinal vesicle, no cyclic rise of tension occurs (Nemoto *et al.*, 1980). On the other hand, if oocytes are bisected just after breakdown of the germinal vesicle, two—and only two—cycles of tension increase will occur, even in a vegetal half, which presumably lacks all components of the meiotic apparatus (Yamamoto and Yoneda, 1983). The peaks of tension coincide with the two polar body divisions in control oocytes. Thus, compared to activated sea urchin eggs, the regulation of the cycles of cortical contractions in maturing starfish oocytes is somewhat different. Nevertheless, for both meiotic and mitotic cell divisions, this autonomous rhythm of contractility appears to have intriguing implications for understanding the regulation and control of cell division (see Section 6.4).

It is still too early to know precisely what factors might underlie the autonomous rhythm of cortical contractility. The rhythms represent global contractions (and also global relaxations), since the entire egg cortex is involved simultaneously. Contractility presumably occurs within the submicrovillar cortex, where microfilaments have been seen at times of high contractile tension (Usui and Yoneda, 1982), and is inhibited by cytochalasin B. The coincidence of microvillus elongation and this contractility suggests that a common periodic modulation of actin behavior throughout the cortex affects both the cytoskeletal and contractile actin-based assemblies.

Before stimulation by 1-MA, immature starfish oocytes are extremely rigid, exhibiting about 3×10^{-3} N/m (3 dyn/cm) of stiffness (Nemoto *et al.*, 1980). This value is about 10-fold greater than the maximum tension generated by contraction in either sea urchin eggs or starfish oocytes during cell division (Yoneda and Dan, 1972; Hiramoto, 1979; Yamamoto and Yoneda, 1983). The rigidity or stiffness of the immature oocytes, however, does not appear to be due to cortical contractility. Rather, the rigidity extends throughout the endoplasm and is due to a high level of noncontractile consistency of the cytoplasm (Shoji et al., 1978). This highly stabilized state is sensitive to cytochalasin B (Nemoto et al., 1980) but not to colchicine (see Schroeder and Otto, 1984), suggesting that it is mediated by actin microfilaments rather than microtubules. The state of the cytoplasm may be analogous to a state of high cohesiveness observed in parthenogenetically activated sea urchin eggs (Coffe et al., 1982). In starfish oocytes, it conceivably plays a role in maintaining the eccentricity of the germinal vesicle (Schroeder and Otto, 1984), thus playing a role in the maintenance of oocyte polarity.

The high consistency of immature starfish oocytes rapidly declines under stimulation by 1-MA, beginning before germinal vesicle breakdown (Shoji et al., 1978; Nemoto et al., 1980). The decrease also occurs when non-nucleated fragments from immature oocytes are treated with 1-MA; it will be recalled that these fragments will not undergo rhythmic contractions, since they lack an essential germinal vesicle factor. Therefore, release of the actin-mediated high stiffness is independent of germinal vesicle contents, as is the assembly of actin bundles for projecting spikes (see Section 3.2). The highly stabilized mechanical state of the entire cytoplasm of immature starfish oocytes is reminiscent of the condition of the structural gel obtained from extracts of sea urchin eggs under conditions that favor associations between actin and actin-binding proteins (Kane, 1979, 1983).

6.4. Regulation of Cleavage Initiation

It is widely accepted that furrowing during cell division is accomplished by a musclelike contraction of the contractile ring at the equator. The entire process of cleavage furrow formation, constriction, and completion appears to result from structural and functional changes within the submicrovillar cortex.

There is no substantial reason to think that microvilli provide plasma membrane or actin microfilaments to this process (reviewed in Schroeder, 1981a). The processes of microvillus elongation and cleavage furrowing have been dissected by treatments with phalloidin (Hamaguchi and Mabuchi, 1982). Although it has recently been proposed that microvilli may serve an accessory role in stabilizing the contractile ring (Begg et al., 1983), microvilli are obviously not essential, since a cleavage furrow can form and proceed on surfaces that lack microvilli, as on the basolateral surfaces of blastomeres, for example (see Fig. 4).

The general scheme for actual furrowing, therefore, is somewhat clear. Nevertheless, the mechanism of cleavage initiation is poorly understood. Cleavage initiation includes (1) a so-called cleavage stimulus derived from the mitotic asters at about anaphase, and (2) the subsequent regional differentiation of cortical activities that accounts for a contracting equator and relaxing, stretching polar regions. For decades, proposed mechanisms of cleavage initiation have alternated between stimulated contraction at the equator (assuming a prior state of cortical relaxation) and a stimulated relaxation of the polar regions (sometimes viewed as a mechanically inconceivable "active expansion" in older theories). Many of the relevant ideas for this controversy have been reviewed by Rappaport (1975), Conrad and Rappaport (1981), and Schroeder (1981c). Even today, the debate continues amid an impressive variety of observational, experimental, and comparative biological data.

Many ingenious experimental manipulations of living eggs seem to demonstrate that the polar cortex cannot be implicated in cleavage initiation. They seem to disprove any hypothesis that relies upon any stimulated change over the polar regions and support the proposition that mitotic asters actively stimulate a change in the equatorial cortex so that it locally contracts. Accordingly, cortical contractile tension is portrayed as developing (i.e., increasing) in response to an influence from the asters after anaphase.

Such a mechanism as **aster-mediated stimulation of equatorial contraction** is eminently appealing, is simple to understand, and has not been ruled out by any single experiment. On the other hand, this hypothetical mechanism fails to make sense of the observations that (1) the microtubules, which dominate the structure of mitotic asters and are strong candidates for mediators of the cleavage stimulus, do not radiate out to the equatorial cortex, although they do tend to reach the polar regions (Asnes and Schroeder, 1979; Harris et al., 1980); (2) an autonomous global contraction of the cortex significantly precedes the time of cleavage stimulation and is totally independent of microtubules (Schroeder, 1981c; Yoneda and Schroeder, 1984); (3) individual asters reportedly have the capacity to mediate cortical relaxation (reviewed by Schroeder, 1981c); and (4) overall tension begins to decline at the onset of visible furrowing.

From these data, an alternate mechanism of cleavage initiation has been constructed as an updated version of one proposed originally by Wolpert (1960): It suggests that the cortical contractile capability responsible for the global contraction is utilized directly for furrow formation when the polar regions are differentially inhibited. According to this **global contraction–polar relaxation** mechanism (Schroeder, 1981c), unaltered and unresisted contractility at the equator allows a furrow to form, while the polar cortex progressively relaxes and stretches out. This mechanism suggests that asters mediate polar relaxation *via* aster microtubules, which exert an inhibitory influence over the nearby polar cortex.

These opposing propositions (aster-mediated stimulation of equatorial contraction versus global contractility–polar relaxation) summarize a conundrum of cleavage stimulation and cleavage initiation. Each scheme faithfully treats a subset of the available observational and experimental evidence. Neither scheme has been proven correct. Because they are so diametrically opposite, one begins to wonder whether we have yet asked the correct questions about cleavage initiation. Although the general tendency has been to expect a single unifying mechanism for such events, it may be that asters mediate stimulated changes in the cortex at *both* the equator and the poles. Redundancy in such developmental mechanisms may provide a special guarantee of success.

7. The Cortex and the Integration of Embryogenesis

7.1. The Cortex–Hyaline Layer Complex

The association between the cortex and an inelastic extracellular shell (the hyaline layer) is a principal difference between embryos of the two groups of echinoderms discussed here. In sea urchins embryos, the hyaline layer congeals soon after fertilization and remains a prominent coat investing the entire embryo. The egg cortex becomes fixed to this coat by long adhering microvilli and is thereby stabilized in a special way. Starfish embryos do not have such a coat and therefore are not reinforced by it.

In sea urchin embryos, new cell surfaces are defined by specifically oriented cell divisions, and there is an early segregation of distinctive cell surfaces. Each blastomere develops an external (or apical) surface that remains attached to the hyaline layer by microvilli (see Section 2.3); internal (or basolateral) surfaces are quite smooth and face either adjacent blastomeres or the embryonic cavity (blastocoel precursor) (see Figs. 4 and 12).

The distributions of the microvilli and cortical pigment granules are closely correlated so that external surfaces of blastomeres have both features, and internal surfaces have neither. An obvious axis of radial polarity is thereby set up in each blastomere (Fig. 12). Since the hyaline layer is exclusively associated with the external surfaces, a novel mechanistic perspective on embryogenesis emerges if one postulates that the association is actually utilitarian.

Persistence and stability of the cortex–hyaline layer association and the general tendency of the hyaline layer to maintain a roughly spherical shape guarantee that the arrangement of blastomeres will remain orderly. Indeed, the cortex-hyaline layer association may provide the only means for holding sea urchin blastomeres together in a well-ordered array until true intercellular adhesions and junctions are established later in development. Accordingly, the cortex–hyaline layer association seems to explain the stereotypic three-tiered architecture of the 16-cell sea urchin embryo as well as the radially polarized organization of the blastomeres (Fig. 12).

Even the presence of an obvious embryonic cavity (Fig. 12) inside the early embryo of some species can be explained in large part by reference to the cortex–hyaline layer complex. One will recall that the hyaline layer becomes inflated or elevated a few micrometers by rounds of microvillus elongation (see Sections 2.2 and 6.3). A simple geometric calculation shows that the total volume of the space between a hyaline layer and the egg surface is equivalent to the seemingly large volume of the embyronic cavity. As Dan (1960) postulated, an embryonic cavity could arise if developing blastomeres progressively pressed against the elevated hyaline layer while minimizing interblastomeric adhesion. The suggestion assumes that blastomere attachment to the hyaline layer is stronger than the binding between adjacent blastomeres, as is supported by observation. Fluid originally between the egg surface and the elevated hyaline layer is conceivably transferred to spaces between blastomeres.

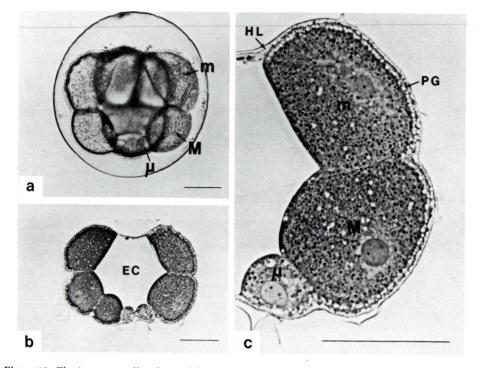

Figure 12. The intact 16-cell embryo of the sea urchin *Strongylocentrotus droebachiensis* is com-posed of three tiers: eight mesomeres (m) in the animal hemisphere, four macromeres (M), and four micromeres (μ) at the vegetal pole. (a) Light micrograph of a lateral view of a living embryo surrounded by a spherical fertilization envelope. (b) Section (1 μm) in a paraxial plane in the same orientation as (a), showing the arrangement of blastomeres around the embryonic cavity (EC)—a precursor of the blastocoel. (Two images of macromeres appear at the lower left as an artifact of the plane of sectioning.) The fertilization envelope is shrunken down around the embryo and appears folded up in some places. (c) An enlargement of the right half of (b). External surfaces of the mesomere and macromere are attached via microvilli to the hyaline layer (HL), and pigment granules (PG) underlie these surfaces. In the case of the micromere, there are fewer microvilli and pigment granules. Internal surfaces (between blastomeres and facing the embryonic cavity) are devoid of both pigment granules and microvilli in all cells. Scale bars: 50 μm.

The arrangement of blastomeres shown in Figure 12 is compatible with this idea. The essential elements for forming an embryonic cavity, therefore, include an inflatable hyaline layer, a strong cortex–hyaline layer association, cell division to partition the egg volume, nonadhesiveness between blastomeres, and relative inelasticity of the hyaline layer.

The contrasting development of sea urchin and starfish embryos correlates well with the existence or absence of a cortex–hyaline layer complex. Because starfish embryos have very few short microvilli and there is no hyaline layer, a distinct cortex–hyaline layer complex cannot be a factor in embryogenesis. Accordingly, early starfish embryos show a diversity of architectural plans from fairly well ordered to quite disarrayed aggregates of blastomeres (Kominami, 1983). In starfish there is apparently little or nothing to maintain a fixed relationship between blastomeres until intercellular junctions are developed at a later stage (Dan-Sohkawa and Fujisawa, 1980). Furthermore, as a direct result of not possessing a cortex–hyaline layer complex, early starfish embryos do not have embryonic cavities. These contrasting patterns of development are reaffirmed by additional comparisons with the developmental pattern of primitive cidaroid sea urchins, which also lack a hyaline layer and are relatively disorganized as early embryos (Schroeder, 1981d).

7.2. Autonomous Blastomere Polarization in Sea Urchins

Physical constraint by the hyaline layer may be fundamentally important for certain architectural features of sea urchin embryos, as mentioned above. Nevertheless, it is important to determine whether the presence of the hyaline layer is actually causally related to a developmental pattern or if it is merely complementary. In this context, it is interesting to compare the behavior of the cell surface and the submicrovillar cortex when the hyaline layer is intact and after it has been removed and blastomeres are cultured separately.

The hyaline layer of sea urchin eggs can be thoroughly removed soon after fertilization. When the eggs are then cultured in calcium-free seawater, cell divisions continue to occur on a normal schedule, and the blastomeres progressively separate from each other due to the lack of common association. After the fourth division, eight mesomeres, four macromeres and four micromeres routinely form, indicating that aspects of the characteristic division program are preserved even when blastomeres develop in isolation.

In the absence of the hyaline layer, microvillus elongation takes place normally (see Fig. 1); in fact, microvillus elongation is typically exaggerated, as if a mechanical restraint has been removed. Later, without a restraining hyaline layer, cleavage results in more rounded and separate daughter cells (Fig. 13).

Without the hyaline layer, pigment granules and the overlying microvilli remain evenly distributed well into the first cleavage process (Fig. 13a). This is quite unlike events in the intact egg. Ordinarily, the hyaline layer is drawn into a cleavage furrow only a small distance before it recoils (see Fig. 4). The recoiling hyaline layer appears to stretch the attached microvilli. In fact, beyond about the halfway point in first cleavage, microvilli do not appear to occur within the deepening cleavage furrow, as if the actions of the hyaline layer actually drag microvilli out of the furrow. One could therefore argue that the cortex–hyaline layer complex is so strong that the behavior of the hyaline layer causes microvilli (and thereby the submicrovillar cortex) to remain on the developing external surfaces. This tenacious attachment could conceivably explain the radial polarity of cortical features of blastomeres of intact embryos, but in fact it does not.

In this context, it is quite surprising to find that cellular polarization of the cortex of isolated blastomeres occurs autonomously and totally independently of the presence of the hyaline layer. Even in isolated blastomeres, a pattern comparable to the radial polarization of intact embryos occurs; that is, microvilli and pigment granules become confined to one surface (corresponding to

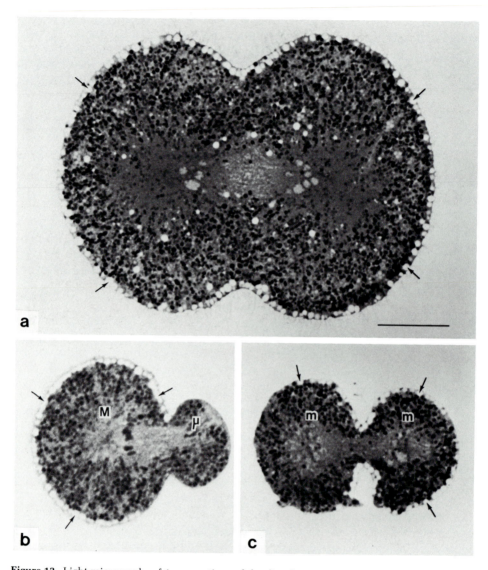

Figure 13. Light micrographs of 1-μm sections of the distribution of pigment granules (arrows) at first cleavage (a) and fourth cleavage of blastomeres of *Strongylocentrotus purpuratus* grown in continuous isolation (b,c). Pigment granules migrate out of the cortex soon after fertilization and remain uniformly distributed during most of first cleavage (a). They then undergo cyclic mass migrations that progressively evacuate zones in isolated blastomeres that appear to correspond to the basolateral surfaces of blastomeres of intact embryos. Another migration evacuates pigment granules from the precursor sites for micromeres. Thus, even though blastomeres are totally isolated from intercellular associations and the hyaline layer, the pattern of cortical pigment granules typical of the intact 16-cell stage embryo (see Fig. 12) is established by an autonomous process. That is, macromeres (M) are fairly heavily pigmented, micromeres (μ) contain few or no pigment granules, and mesomeres (m) are pigmented on one "side" only. Scale bar: 25 μm.

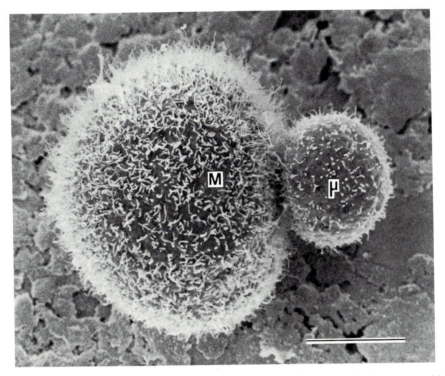

Figure 14. Scanning electron micrograph of a macromere (left cell) and attached micromere (right cell) derived from the fourth division of isolated blastomeres of a *Strongylocentrotus purpuratus* embryo. The microvillus-bearing surface automatically segregates so that nearly all the microvilli occur on the macromere (M), and relatively few occur on the micromere (μ). In the case of mesomeres, microvillus-bearing and smooth surfaces are divided equally during cell division. These distributions mirror events in the intact embryo (see Fig. 12) when microvilli are embedded in the hyaline layer and show that segregation of the microvillus-bearing surface (and underlying pigmented cortical matrix) occurs by a process of autonomous cellular polarization that does not depend on the hyaline layer or intercellular associations. Scale bar: 10 μm.

the external surface), and a counterpart of the internal surface is smooth and unpigmented (Figs. 13b, c and 14).

As correctly pointed out by Dan (1960), the first evidence of autonomous cell polarization of microvilli and pigment granules is manifested soon after the completion of first cleavage. A field of microvilli and pigment granules in each blastomere retracts away from the region of the intercellular bridge between sister cells after the end of cleavage. Scanning electron micrographs (Schroeder, 1981a) confirm the existence of such a zone lacking microvilli. This zone corresponds to the internal surface in intact embryos, and the opposite pigmented and microvillus-covered surface corresponds to the external surface. This kind of autonomous cellular polarization recurs with successive cell divisions of isolated blastomeres. The phenomenon thereby disproves the primary or causal importance of the hyaline layer in cell polarization. According to Dan (1960, p. 325), there seems to be an inherent redundancy of mechanisms in this particular aspect of sea urchin morphogenesis, whereby the apical and basolateral surfaces are segregated "by a double insuring system, consisting of: (1) a static factor, mechanically binding the cell surface to the overlying hyaline layer . . . , and (2) a more dynamic factor, which by an intrinsic activity of the blastomeres puts the cortex where it belongs after each division."

By the fourth division, mesomeres in cultures of isolated blastomeres are pigmented over one-half their surface and unpigmented over the other half (Fig. 13c). The cleavage furrow bisects these two zones perpendicularly, so that daughter mesomeres are equivalent with regard to their cortical and surface specialization. This result supports the additional idea that the programmed cleavage plane is still precisely coordinated with the axis of cellular polarization in the same way it is in an intact embryo.

Furthermore, at the same stage, the macromere–micromere division of isolated blastomeres (Figs. 13b and 14) occurs after the parent blastomere has undergone a normal segregation of its pigment granules and microvilli in conformity with both the radial cellular polarization and the A–V polarity, i.e., micromeres form with virtually no pigment granules (Fig. 13b) or microvilli (Fig. 14), as compared with the macromeres, just as they do when the hyaline layer is present. In addition, one side of the macromere is usually devoid of pigment granules and microvilli, presumably corresponding to its internal (basolateral) surface as if it were actually still part of an intact embryo.

These observations vividly demonstrate that a remarkable degree of autonomy exists in the behavior of the cortex in terms of microvillus elongation, the segregation mechanisms that generate radial cell polarization, and A–V axial differentiation. They show that the hyaline layer is not causally important in these processes.

On the other hand, when the hyaline layer is present, it clearly seems to reinforce and stabilize the cortically mediated patterns and the spatial arrangement of the blastomeres. The hyaline layer may also be important for allowing the development of an embryonic cavity. Thus, there are separate factors with overlapping actions that are responsible for major events in sea urchin embryogenesis.

7.3. Formative Factors in Embryogenesis

Distinctions between microvilli and the submicrovillar cortex, between cytoskeletal and contractile functional properties, and between behavioral patterns in the cortex of sea urchin eggs and starfish oocytes may constitute some of the important architectural elements that effect the embryogenesis of these animals and may be summarized as follows:

1. An underlying animal–vegetal polarity of the egg that is heritable by the blastomeres and is at least partially expressed in the behavior of the cortex
2. Temporal patterns of topographical change involving stiffness and length changes of microvilli

3. Regional and temporal patterns of autonomous, global cortical contractility and related motile events in the submicrovillar cortex
4. Bidirectional (or at least coordinated) influences between the asters of mitotic apparatuses and the cortex
5. Partitioning of the original egg volume by cell division to permit cell diversification
6. Radial segregation of cell surface and cortical components as "new" surfaces are generated during the early cell divisions
7. A program of oriented cleavages that are geometrically related through unknown mechanisms to both axial and radial cellular polarities of the blastomeres
8. Coordinated (if not coupled) behavior and regulation of microvilli and the submicrovillar cortex
9. Adhesion between microvilli and the hyaline layer for stabilizing the arrangement of blastomeres
10. Nonexistence of intercellular adhesion at early embryonic stages

These factors appear to be important in defining the architectural plan of an echinoderm embryo at early stages, although the details and real significance of each need to be studied further. In general, the list emphasizes the prominent role of the egg cortex in echinoderm development.

References

Allen, R. D., 1955, The fertilization reaction in isolated cortical material from sea urchin eggs, *Exp. Cell Res.* **8**:397–399.

Allen, R. D., and Rowe, E. C., 1958, The dependence of pigment granule migration on the cortical reaction in eggs of *Arbacia punctulata*, *Biol. Bull.* **114**:113–117.

Asnes, C. F., and Schroeder, T. E., 1979, Cell cleavage: Ultrastructural evidence against equatorial stimulation by aster microtubules, *Exp. Cell Res.* **122**:327–338.

Begg, D. A., and Rebhun, L. I., 1979, pH regulates the polymerization of actin in the sea urchin egg cortex, *J. Cell Biol.* **83**:241–248.

Begg, D. A., Rebhun, L. I., and Hyatt, H., 1982, Structural organization of actin in the sea urchin egg cortex: Microvillar elongation in the absence of actin filament bundle formation, *J. Cell Biol.* **93**:24–32.

Begg, D. A., Salmon, E. D., and Hyatt, H. A., 1983, Changes in the structural organization of actin in the sea urchin egg cortex in response to hydrostatic pressure, *J. Cell Biol.* **97**:1795–1805.

Belanger, A. M., and Rustad, R. C., 1972, Movements of echinochrome granules during the early development of sea urchin eggs, *Nature New Biol.* **239**:81–83.

Boveri, T., 1901, Die Polarität von Ovocyte, Ei und Larve des *Strongylocentrotus lividus*, *Zool. Jahrb. (Abt. Anat.)* **14**:630–653.

Burgess, D. R., and Schroeder, T. E., 1977, Polarized bundles of actin filaments within microvilli of fertilized sea urchin eggs, *J. Cell Biol.* **74**:1032–1037.

Carron, C. P., and Longo, F. J., 1982, Relation of cytoplasmic alkalinization to microvillar elongation and microfilament formation in the sea urchin egg, *J. Cell Biol.* **89**:128–137.

Carron, C. P., and Longo, F. J., 1984, Pinocytosis in fertilized sea urchin (*Arbacia punctulata*) eggs, *J. Exp. Zool.* **231**:413–422.

Chambers, R., and Chambers, E. L., 1961, *Explorations into the Nature of the Living Cell*, Harvard University Press, Cambridge, Mass.

Chandler, D. E., and Heuser, J., 1981, Postfertilization growth of microvilli in the sea urchin egg:

New views from eggs that have been quick-frozen, freeze-fractured, and deep etched, *Dev. Biol.* **82**:393–400.

Christen, R., 1983, Acidic vesicles and the uptake of amines by sea urchin eggs, *Exp. Cell Res.* **143**:319–325.

Coffe, G., Foucault, G., Soyer, M. O., De Billy, F., and Pudles, J., 1982, State of actin during the cycle of cohesiveness of the cytoplasm in parthenogenetically activated sea urchin egg, *Exp. Cell Res.* **142**:365–372.

Conrad, G. W., and Rappaport, R., 1981, Mechanisms of cytokinesis in animal cells, in: *Mitosis/ Cytokinesis* (A. M. Zimmerman and A. Forer, eds.), pp. 365–396, Academic Press, New York.

Dan, K., 1960, Cytoembryology of echinoderms and amphibia, *Int. Rev. Cytol.* **9**:321–367.

Dan, K., Endo, S., and Uemura, I., 1983, Studies on unequal cleavage in sea urchins. II. Surface differentiation and direction of nuclear migration, *Dev. Growth Diff.* **25**:227–237.

Dan-Sohkawa, M., and Fujisawa, H., 1980, Cell dynamics of the blastulation process in the starfish, *Asterina pectinifera, Dev. Biol.* **77**:328–339.

Danielli, J. F., 1952, Division of the flattened egg, *Nature (Lond.)* **170**:496.

Fischer, G. W., and Rebhun, L. I., 1983, Sea urchin egg cortical granule exocytosis is followed by a burst of membrane retrieval via uptake into coated vesicles, *Dev. Biol.* **99**:456–472.

Hamaguchi, M. S., and Hiramoto, Y., 1980, Fertilization process in the heart-urchin, *Clypeaster japonicus* observed with a differential interference microscope, *Dev. Growth Diff.* **22**:517–530.

Hamaguchi, Y., and Mabuchi, I., 1982, Effects of phalloidin microinjection and localization of fluorescein-labeled phalloidin in living sanddollar eggs, *Cell Motil.* **2**:103–113.

Hara, K., Tydeman, P., and Kirschner, M., 1980, A cytoplasmic clock with the same period as the division cycle in *Xenopus* eggs, *Proc. Natl. Acad. Sci. USA* **77**:462–466.

Harris, P., Osborn, M., and Weber, K., 1980, Distribution of tubulin-containing structures in the egg of the sea urchin *Stronglyocentrotus purpuratus* from fertilization through first cleavage, *J. Cell Biol.* **84**:668–679.

Herbst, C., 1900, Über das Auseinadergehen von Furchungs- und Gewebezellen in Kalkfrei Medium, *Arch. Entwicklungsmech.* **9**:424–463.

Hiramoto, Y., 1970, Rheological properties of sea urchin eggs, *Biorheology* **6**:201–234.

Hiramoto, Y., 1974, Mechanical properties of the surface of sea urchin eggs at fertilization and during cleavage, *Exp. Cell Res.* **89**:320–326.

Hiramoto, Y., 1979, Mechanical properties of the dividing sea urchin egg, in: *Cell Motility: Molecules and Organization* (S. Hatano, H. Ishikawa, and H. Sato, eds.), pp. 653–663, University Park Press, Baltimore.

Horstadius, S., 1973, *Experimental Embryology of Echinoderms*, Clarendon Press, Oxford.

Kanatani, H., 1973, Maturation-inducing substance in starfishes, *Int. Rev. Cytol.* **35**:253–298.

Kane, R. E., 1979, Actin polymerization and function in the sea urchin egg, in: *Cell Motility: Molecules and Organization* (S. Hatano, H. Ishikawa, and H. Sato, eds.), pp. 639–651, University Park Press, Baltimore.

Kane, R. E., 1983, Interconversion of structural and contractile actin gels by insertion of myosin during assembly, *J. Cell Biol.* **97**:1745–1761.

Kiehart, D. P., Mabuchi, I., and Inoue, S., 1982, Evidence that myosin does not contribute to force production in chromosome movement, *J. Cell Biol.* **94**:165–178.

Kidd, P., and Mazia, D., 1980, The ultrastructure of surface layers isolated from fertilized and chemically stimulated sea urchin eggs, *J. Ultrastruct. Res.* **70**:58–69.

Kominami, T., 1983, Establishment of embryonic axes in larvae of the starfish, *Asterina pectinifera, J. Embryol. Exp. Morphol.* **75**:87–100.

Lee, H. C., and Epel, D., 1983, Changes in intracellular acidic compartments in sea urchin eggs after activation, *Dev. Biol.* **98**:446–454.

Longo, F. J., 1980, Organization of microfilaments in sea urchin (*Arbacia punctulata*) eggs at fertilization: Effects of cytochalasin B, *Dev. Biol.* **74**:422–433.

Mabuchi, I., 1973, A myosin-like protein in the cortical layer of the sea urchin egg, *J. Cell Biol.* **59**:542–547.

Mabuchi, I., 1974, A myosin-like protein in the cortical layer of cleaving starfish eggs, *J. Biochem.* **76**:47–55.

Mabuchi, I., 1981, Purification from starfish eggs of a protein that depolymerizes actin, *J. Biochem.* **89:**1341–1344.

Mabuchi, I., and Hosoya, H., 1982, Actin-modulating proteins in the sea urchin egg. II. Sea urchin egg profilin, *Biomed. Res.* **3:**465–476.

Mabuchi, I., Hosoya, H., and Sakai, H., 1980, Actin in the cortical layer of the sea urchin egg. Changes in its content during and after fertilization, *Biomed. Res.* **1:**417–426.

Mabuchi, I., and Okuno, M., 1977, The effect of myosin antibody on the division of starfish blastomeres, *J. Cell Biol.* **74:**251–263.

Meijer, L., and Guerrier, P., 1984, Maturation and fertilization in starfish oocytes, *Int. Rev. Cytol.* **86:**130–195.

Mitchison, J. M., and Swann, M. M., 1954, The mechanical properties of the cell surface. I. The cell elastimeter, *J. Exp. Biol.* **31:**443–460.

Moody, W. J., and Lansman, J. B., 1983, Developmental regulation of Ca^{2+} and K^+ currents during hormone-induced maturation of starfish, *Proc. Natl. Acad. Sci. USA* **80:**3096–3100.

Morgan, T. H., and Spooner, G. B., 1909, The polarity of the centrifuged egg, *Arch. Entwicklungsmech.* **28:**104–117.

Morton, R. W., and Nishioka, D., 1983, Effects of cytochalasin B on the cortex of the unfertilized sea urchin egg, *Cell Biol. Int. Rep.* **7:**835–842.

Nemoto, S., Yoneda, M., and Uemura, I., 1980, Marked decrease in the rigidity of starfish induced by 1-methyladenine, *Dev. Growth Diff.* **22:**315–325.

Otto, J. J., and Schroeder, T. E., 1984a, Assembly–disassembly of actin bundles in starfish oocytes: An analysis of actin-associated proteins in the isolated cortex, *Dev. Biol.* **10:**263–273.

Otto, J. J., and Schroeder, T. E., 1984b, Microtubule arrays in the cortex and near the germinal vesicle of immature starfish oocytes, *Dev. Biol.* **10:**274–281.

Otto, J. J., Kane, R. E., and Bryan, J., 1980, Redistribution of actin and fascin in sea urchin eggs after fertilization. *Cell Motil.* **1:**31–40.

Painter, T. S., 1916, Contributions to the study of cell mechanics. I. Spiral asters, *J. Exp. Zool* **20:**509–527.

Rappaport, R., 1975, The biophysics of cleavage and cleavage of geometrically altered cells, in: *The Sea Urchin Embryo: Biochemistry and Morphogenesis* (G. Czihak, ed.), pp. 308–332, Springer-Verlag, Berlin.

Sakai, H., 1960, Studies on sulfhydryl groups during cell division of sea urchin egg: The mass isolation of the egg cortex and change in its -SH groups during cell division, *J. Biophys. Biochem. Cytol.* **8:**603–607.

Sakai, M., and Kubota, H. Y., 1981, Cyclic surface changes in the non-nucleate egg fragment of *Xenopus laevis*, *Dev. Growth Diff.* **23:**41–49.

Sasaki, H., 1984, Modulation of calcium sensitivity by a specific cortical protein during sea urchin egg cortical vesicle exocytosis, *Dev. Biol.* **101:**125–135.

Sawada, T., and Osanai, K., 1980, Movements of the pigment granules in the egg of *Temnopleurus hardwicki*, *Bull. Mar. Biol. Sta. Asamushi Tohoku Univ.* **16:**213–219.

Sawai, T., 1979, Cyclic changes in the cortical layer of non-nucleated fragments of the newt's egg, *J. Embryol. Exp. Morphol.* **51:**183–193.

Schatten, G., 1982, Motility during fertilization, *Int. Rev. Cytol.* **79:**59–163.

Schroeder, T. E., 1972, The contractile ring. II. Determining its brief existence, volumetric changes, and vital role in cleaving *Arbacia* eggs, *J. Cell Biol.* **53:**419–434.

Schroeder, T. E., 1975, Dynamics of the contractile ring, in: *Molecules and Cell Movement* (S. Inoue and R. E. Stephens, eds.), pp. 305–334, Raven Press, New York.

Schroeder, T. E., 1978, Microvilli of sea urchin eggs: A second burst of elongation, *Dev. Biol.* **64:**342–346.

Schroeder, T. E., 1979, Surface area change at fertilization: Resorption of the mosaic membrane, *Dev. Biol.* **70:**306–326.

Schroeder, T. E., 1980a, The jelly canal: Marker of polarity for sea urchin oocytes, eggs, and embryos, *Exp. Cell Res.* **128:**490–494.

Schroeder, T. E., 1980b, Expressions of the prefertilization polar axis in sea urchin eggs, *Dev. Biol.* **79:**428–443.

Schroeder, T. E., 1981a, Interrelations between the cell surface and the cytoskeleton in cleaving sea urchin eggs, in: *Cytoskeletal Elements and Plasma Membrane Organization* (G. Poste and G. L. Nicolson, eds.), pp. 170–216, Elsevier/North-Holland Press, Amsterdam.

Schroeder, T. E., 1981b, Microfilament-mediated surface change in starfish oocytes in response to 1-methyladenine: Implications for identifying the pathway and receptor sites for maturation-inducing hormones, *J. Cell Biol.* **90**:362–371.

Schroeder, T. E., 1981c, The origin of cleavage forces in dividing eggs: A mechanism in two steps, *Exp. Cell Res.* **134**:231–240.

Schroeder, T. E., 1981d, Development of a "primitive" sea urchin (*Eucidaris tribuloides*): Irregularities in the hyaline layer, micromeres, and primary mesenchyme, *Biol. Bull.* **161**:141–151.

Schroeder, T. E., 1982, Novel surface specialization on a sea anemone egg: "spires" of actin-filled microvilli, *J. Morphol.* **174**:207–216.

Schroeder, T. E., 1985, Cortical expressions of polarity in the starfish oocyte, *Dev. Growth Differ.* **27**:311–321.

Schroeder, T. E., and Otto, J. J., 1984, Cyclic assembly-disassembly of cortical microtubules during maturation and early development of starfish oocytes, *Dev. Biol.*, **103**:493–503.

Schroeder, T. E., and Stricker, S. A., 1983, Morphological changes during maturation of starfish oocytes: Surface ultrastructure and cortica actin, *Dev. Biol.* **98**:373–384.

Schuel, H., 1978, Secretory functions of egg cortical granules in fertilization and development: A critical review, *Gamete Res.* **1**:299–382.

Shimizu, T., 1981, Cyclic changes in shape of a non-nucleate egg fragment of *Tubifex*, *Dev. Growth Diff.* **23**:101–109.

Shoji, Y., Hamaguchi, M. S., and Hiramoto, Y., 1978, Mechanical properties of the endoplasm in starfish oocytes, *Exp. Cell Res.* **117**:79–87.

Shoji, Y., Hamaguchi, Y., and Hiramoto, Y., 1981, Quantitative studies on polarization optical properties of living cells. III. Cortical birefringence of the dividing sea urchin egg, *Cell Motil.* **1**:387–397.

Spudich, J. A., Spudich, A., and Amos, L., 1979, Actin from the cortical layer of sea urchin eggs before and after fertilization, in: *Cell Motility: Molecules and Organization* (S. Hatano, H. Ishikawa, and H. Sato, eds.), pp. 165–187, University Park Press, Baltimore.

Tanaka, Y., 1981, Distribution and redistribution of pigment granules in the development of sea urchin embryos, *Roux Arch. Dev. Biol.* **190**:267–273.

Tilney, L. G., and Jaffe, L. A., 1980, Actin, microvilli, and the fertilization cone of sea urchin eggs, *J. Cell Biol.* **87**:771–782.

Usui, N., and Yoneda, M., 1982, Ultrastructural basis of the tension increase in sea-urchin eggs prior to cytokinesis, *Dev. Growth Diff.* **24**:453–465.

Vacquier, V. D., 1975, The isolation of intact cortical granules from sea urchin eggs: Calcium ions trigger granule discharge, *Dev. Biol.* **43**:62–74.

Vacquier, V. D., 1981, Dynamic changes of the egg cortex, *Devel. Biol.* **84**:1–26.

Vacquier, V. D., and Moy, G. W., 1980, The cytolytic isolation of the cortex of the sea urchin egg, *Dev. Biol.* **77**:178–190.

Wolpert, L., 1960, The mechanics and mechanism of cleavage, *Int. Rev. Cytol.* **10**:163–216.

Yamamoto, K., and Yoneda, M., 1983, Cytoplasmic cycle in meiotic division of starfish oocytes, *Dev. Biol.* **96**:166–172.

Yoneda, M., 1973, Tension at the surface of sea urchin eggs on the basis of "liquid drop" concept, *Adv. Biophys.* **4**:153—190.

Yoneda, M., and Dan, K., 1972, Tension at the surface of the dividing sea-urchin egg, *J. Exp. Biol.* **57**:575–587.

Yoneda, M., and Schroeder, T. E., 1984, Cell cycle timing in colchicine-treated sea urchin eggs: Persistent coordination between the nuclear cycles and the rhythm of cortical stiffness, *J. Exp. Zool.* **231**:367–378.

Yoneda, M., Ikeda, M., and Washitani, S., 1978, Periodic change in the tension at the surface of activated non-nucleate fragments of sea-urchin eggs, *Dev. Growth Diff.* **20**:329–336.

II

Cell–Cell Interactions in Development

Chapter 4

The Cellular Basis of Epithelial Morphogenesis

JOHN KOLEGA

1. Introduction

The definitive trait of an epithelium is that its cells form cohesive cell sheets with little acellular material interposed between the cells. Nature has employed this relatively simple arrangement to fashion tissues of enormous functional and morphological diversity. In adult organisms, epithelia cover and protect the body, line the surfaces that absorb nutrients and metabolites from the environment (as in the lung and intestine), and form the tubes, ducts, and acini of glandular organs. They engage in transport, filtration, endocrine and exocrine synthetic functions, and the maintenance of chemical and electrical differentials between body compartments. Epithelia also play the lead roles in morphogenesis. During development, epithelial movement in gastrulation is responsible for much of the reorganization that lays down the overall body plan, and the folding, spreading, budding, cavitation, and delamination of epithelia produce the basic structure of most organs. In the course of morphogenesis, epithelia can form clusters, cords, or tongues of cells. They can also disperse, giving rise to mesenchymal cells, which in turn either form nonepithelial structures or reaggregate to form new epithelia. The means by which epithelial cells produce this wealth of form and function is the subject of this chapter.

2. Maintenance and Development of Epithelial Continuity

If cells are to form an epithelium, they must bind to one another along their adjoining edges. This contiguity is crucial to many epithelial functions. If the sheet is to act as a protective covering, any lack of continuity exposes the

JOHN KOLEGA • Department of Biology, Yale University, New Haven, Connecticut 06511. *Present address*: Department of Dermatology, New York University Medical Center, New York, New York 10016.

underlying tissue. Epithelia engaged in transport functions or acting as permeability barriers must also be continuous. It is probable that cell–cell attachments are also important in epithelial translocations such as occur during gastrulation, epiboly, and wound healing, where a large part of the epithelial sheet may be moved by the locomotive activity of a small subpopulation of its cells.

2.1. Intercellular Adhesions

In mature epithelia, the juxtaposition of cells can involve a number of different forms of cell–cell contact, all of which have at least some adhesive character. The most rudimentary of these is a simple, unspecialized apposition between cell surfaces, which I shall refer to as a **general close contact**. Adjacent cells in most epithelia are in general close contact over most of their lateral surfaces. Electron micrographs show that the surfaces at these contacts are separated by a very uniform gap, usually 15–20 nm wide, which is filled with an amorphous, fairly electron-lucent material. Little is known about the nature of this material, except that it contains carbohydrates and glycoproteins (Rambourg and Leblond, 1967). It is ideally located to act as the intercellular cement that, as long ago as 1862 (von Recklinghausen), was purported to hold epithelia together. However, because general close contact is found between weakly adhesive cells, such as cells of pre-blastula sea urchin embryos (Wolpert and Mercer, 1963), and in regions of highly transitory contact (e.g., beneath rapidly moving cells), it has been assumed that this type of contact is not important for long-term adhesions, and cell biologists' attention has been focused on more visible intercellular junctions. This attention is hardly unwarranted but, as discussed repeatedly in this volume, it should be noted that a number of carbohydrate-containing "adhesive factors" have been isolated from cell surfaces and shown to affect cell aggregation, presumably by altering their adhesive behavior (see also reviews by Garrod and Nicol, 1981; Roth, 1983). The ultrastructural manifestations of these adhesive changes are largely unknown, but in at least one case—the aggregation of cellular slime molds—specialized junctions are not present (Garrod and Nicol, 1981), whereas a specific glycoprotein aggregation factor has been shown to be preferentially distributed in the regions of cell–cell contact (Getolsky et al., 1980). Thus, one must be careful not to overlook the role of general close contacts in cell adhesion.

With this as warning, we shall now consider the role of cell surface specializations in holding epithelia together. Mature epithelia usually have two or more kinds of specialized intercellular junctions, and they always have at least one. In vertebrates, three different kinds of junction are frequently observed together in a distinctive arrangement known as a **junctional complex** (Farquhar and Palade, 1963). Junctional complexes may be universal constituents of adult vertebrate epithelia; they are found in epithelia of the stomach, intestine, gallbladder, uterus, oviduct, liver, pancreas, salivary gland, thyroid, parotid, and kidney (Farquhar and Palade, 1963), as well as in skin (Farquhar and Palade, 1967) and endothelia (Muir and Peters, 1962). The complex consists of an apical **zonula occludens,** or **tight junction;** a **zonula adherens,** or **intermediate junction,** which is located just subapical to the tight junction; and a **maculae adherens,** or **desmosome,** basal to that.*

The desmosomes are almost indisputably adhesive structures. They are in great abundance in highly stressed tissues, such as the uterine cervix (McNutt et al., 1971) and skin (Odland, 1958; Wilgram et al., 1964). In the latter, they are the only points of attachment between cells of the stratum spinosum (McNutt et al., 1971), and their absence has been correlated with the loss of adhesion between cells in certain skin diseases (Wilgram et al., 1964). In addition, the resistance of whole chick blastoderm (Overton, 1962) and of cultured chick heart cells (Wiseman and Strickler, 1981) to mechanical disaggregation is proportional to the number of desmosomes between cells. Ultrastructurally, the accumulation of matrix material between the cell surfaces at desmosomal appositions and the intracellular association of desmosomes with tonofilaments and electron-dense plaques further support the assumption that desmosomes are important elements in maintaining the epithelium's physical integrity.

Zonulae and fascia adherentes, as their names imply, are also believed to function in holding cells together. In addition to being part of the junctional complex, they are found between the end-to-end appositions of muscle cells (Fawcett, 1966) and at the apices of cells undergoing apical constriction during epithelial morphogenesis (see Section 5.4.2), where it is likely that particularly strong cell–cell adhesion is necessary. Like desmosomes, they are associated with filamentous elements of the cytoskeleton, but with microfilaments, rather than tonofilaments.

Although adherens junctions (including desmosomes) are extremely important in maintaining the structural integrity of epithelia, particularly where epithelia are subject to mechanical stress, other specialized junctions also contribute to cell–cell adhesion. For example, epithelial cells of the duodenal mucosa will remain joined by tight junctions when desmosomes are completely disrupted by trypsinization (Overton and Shoup, 1964). Tight junctions also remain between epithelial cells after all adherens junctions are disrupted by chelation of divalent cations (Sedar and Forte, 1964). Similarly, Muir (1967) showed that when the adherens junctions in heart muscle are disrupted by low calcium, gap junctions remain intact and adhere with such tenacity that, when cells are pulled apart, the junctions will tear from the cell surface rather than septate. In invertebrate epithelia, yet another surface specialization, the septate junction, probably serves an adhesive function (Wood, 1959). Thus, even when one considers only specialized junctions, there is still a multiplicity of ways in which epithelial cells can adhere to one another.

2.2. Transepithelial Permeability

Cell adhesion alone does not ensure that materials on one side of an epithelium will not pass to the other side *via* the intercellular space. In any

---

*The morphology of cell junctions is relatively well established and will not be discussed here. Extensive details of junction structure and function may be obtained from a variety of sources (e.g., Staehelin, 1974; Overton, 1974).

epithelium whose functions include separating fluid compartments, an occlusive seal must form between adjacent cells. This function is performed by
zonulae occludens (tight junctions), which form a beltlike region of intimate
contact between the plasma membranes of adjacent cells. That the zonula occludens acts as a barrier to diffusion was first demonstrated by Farquhar and
Palade (1963), who showed that hemoglobin infiltrated the cell–cell appositions in kidney epithelia but could not pass the tight junctions. Subsequently,
similar demonstrations using smaller tracer molecules were made in epithelia
of the intestine, pancreas, epididymis, and liver (Overton, 1962; Goodenough
and Revel, 1970; Friend and Gilula, 1972), and in endothelial cells of the brain
(Reese and Karnovsky, 1967). In many epithelia, the zonula occludens will
exclude colloidal lanthanum, demonstrating its capacity to seal out even very
small molecules. However, Macher et al. (1972) demonstrated that some
zonulae occludens are penetrated by lanthanum ions; these workers suggested
that differences in the tightness of tight junctions are responsible for differences in the electrical resistances observed across different epithelia. Claude
and Goodenough (1973) then correlated epithelial leakiness as measured by
transepithelial resistance with the morphology of tight junctions as visualized
by freeze-fracture scanning electron microscopy. Epithelia with high transepithelial resistance and, presumably, low paracellular permeability have
many interconnecting strands in a relatively broad band around the cell,
whereas leaky epithelia may have only a single tight junctional strand. Thus,
the role of tight junctions in determining paracellular permeability seems well
established.

In invertebrates, in which tight junctions are rare, occlusion of the intercellular space may be a function of septate junctions. Although septate junctions have a latticelike structure with relatively large interseptal spaces, Hand
and Gobel (1972) found that 12,000-M_r tracers did not penetrate beyond three
or four septa. They hypothesized that the interseptal space is filled with
mucopolysaccharides, which, over the great depth of most septate junctions,
would constitute a substantial barrier to paracellular diffusion.

2.3. The Development of Junctions

In animal development, blastulation constitutes the first appearance of
epithelial tissue. In the sea urchin, the cells of the morula are poorly adhesive,
being held together primarily by the hyaline layer (Dan, 1960). Only close
contacts are found between cells. Wolpert and Mercer (1963) reported that, as
the blastocoel forms, septate junctions appear between the cells of the blastocoel wall, the first epithelium of the urchin. Millonig and Giudice (1967)
showed that epithelial morphology and septate junctions also develop together
during the reaggregation of dissociated sea urchin embryos in vitro. Dan-
Sohkawa and Fujisawa (1980) observed the same coincidence between blastulation and the appearance of septate junctions in sand dollar embryos. They also
found that, in denuded embryos, cells display stage-dependent adhesive be-

havior in culture. That is, in early cleavage, cells show little tendency to associate, and they appear to be as adhesive to the glass or plastic substratum as to each other. But as they reach the age of blastulating embryos, they form distinctly epithelial aggregates, which close up and form normally developing blastulae. Septate junctions form between these cultured cells at precisely the same time as this change in adhesive behavior occurs. A detailed discussion of junction formation in sea urchin development can be found in Chapter 6.

There has not been a systematic study of the development of junctions at correspondingly early stages in vertebrate embryos. However, considerable information may be gleaned from other, more general, morphological studies of early vertebrate development. It appears that the development of epithelial morphology in vertebrates is coincident with the appearance of tight junctions. In mouse (Ducibella et al., 1975), rat (Schlafke and Enders, 1967), rabbit (Van Blerkom et al., 1973), and frog (Kalt, 1971; Sanders and Zalik, 1972), tight junctions are first observed at the onset of blastocoel formation. In teleost and avian embryos, the precise timing of tight junction development has not been reported, but tight junctions are present in the earliest epithelial stages that have been examined (Trelstad et al., 1967; Lentz and Trinkaus, 1971).

These observations suggest that the formation of tight or septate junctions is an important factor in the genesis of an epithelial morphology. It almost certainly contributes to the developing lateral adhesions between cells. However, whether the basic changes in tissue form can occur in the absence of specialized junctions is not at all clear. It is possible that the coincidence between the development of junctions and the appearance of an epithelial morphology merely reflects the importance of the junctions in the developing function of the epithelium.

In contrast to tight and septate junctions, desmosomes appear relatively late in developing epithelia, indicating that they are not necessary for the initial formation of the sheet. In teleost (Trinkaus and Lentz, 1967) and avian (Bellairs, 1963) embryos, desmosomes begin to appear after the blastocoel forms, and they continue to increase in number and maturity through gastrulation. This corresponds to a period in which the blastoderm is subject to great tensile stress as the small epithelial patch spreads over the yolk. Similarly, desmosomes develop between cells of the trophectoderm of mouse embryos at the same time that the blastocoel swells with increasing internal hydrostatic pressure (Ducibella et al., 1975). They also appear between cells of the neural plate ectoderm just before the extensive deformations of neurulation (Burnside, 1971). These observations suggest that desmosomes may be important in maintaining the structural contiguity of the epithelium, particularly when the epithelium is subject to mechanical stress. It would be interesting to determine if the number of desmosomes in an epithelium can be changed by varying the magnitude or duration of the mechanical stress to which it is exposed.

The development of junctions is also an integral part of organogenesis. Junctions appear de novo whenever a new epithelial surface is formed. For example, in the embryonic thyroid the development of follicular spaces is accompanied by the formation of junctional complexes between cells lining the lumina (Feldman *et al.*, 1961). Similarly, junctional complexes appear between cells bounding incipient bile canaliculi in the developing liver (Wood, 1963) and between cells of the nephrogenic mesenchyme as they form kidney tubules (Wartiovaara, 1966). Yet another example is the formation of intestinal villi, where junctional complexes appear in large numbers in particular regions of the multilayered intestinal epithelium. Gaps appear within regions of cell surface surrounded by complexes. The gaps expand and fuse with other gaps to form the clefts between villi (Mathan *et al.*, 1976).

Examination of junction ultrastructure during epithelial organogenesis has provided detailed descriptions of the subcellular events associated with the formation of tight junctions *in vivo*. Gilula (1973) used freeze-fracture scanning electron microscopy to examine tight junctions developing between Sertoli cells during formation of the blood–testis barrier. The immature permeable epithelium contains tight junction particles arranged in short, linear aggregates in random, discontinuous orientation. But in the mature epithelium, particles are found in continuous parallel ridges circumscribing the cells. The development of tight junctions has been further characterized in studies of the formation of the bile canaliculi (Montesano *et al.*, 1975), development of the chick chorioid epithelium (Dermietzel *et al.*, 1977), lumina development in thyroid (Luciano *et al.*, 1979), and genesis of intestinal villi (Madara *et al.*, 1981). In all cases, the same basic sequence is observed: Tight junction particles appear, then form short linear aggregates, which, in turn, fuse into smooth, anastomosing tight junctional strands (see Fig. 1). Remarkably, these events are closely paralleled in septate junction formation in invertebrates, in which an organized array of multiple rows of junctional particles develops through the same intermediates of individual particles and short linear segments (Gilula, 1973; Lane and Swales, 1982).

For desmosomes, the structure visible at the cell surface is much less elaborate. But desmosomes do have a distinctive, well-developed morphology that is readily discernible in thin sections. A probable sequence of subcellular events in the assembly of desmosomes has been constructed based on the morphology of incomplete desmosomelike structures observed in thin sections of embryonic epithelia and healing epithelial wounds *in situ* and of dissociated cells reaggregating *in vitro* (reviewed by Overton, 1974). As illustrated in Figure 2, electron-dense material appears between cells and subjacent to the plasma membrane at the location of the incipient desmosome. The electron-dense material then condenses to form the characteristic desmosomal plaques, and tonofibrils appear in association with the cytoplasmic densities.

2.4. The Control of Junction Formation

Although the morphology of junction formation is fairly well established, the dissection of the mechanisms controlling it remains far from complete. A better understanding of the intracellular events involved in junction development could tell us a great deal about how the cell manipulates its intercellular

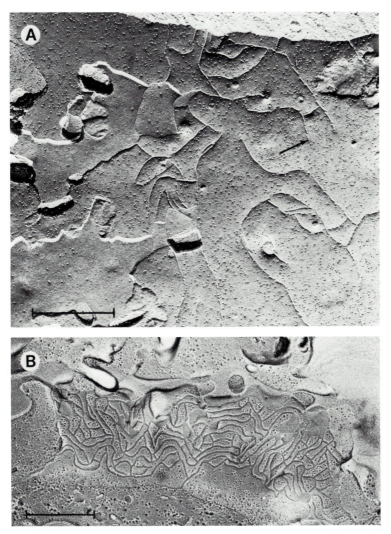

Figure 1. Development of tight junctions. Scanning electron micrographs of freeze-fracture replicas. (A) Chick skin undergoing vitamin A-induced mucous metaplasia, during which extensive proliferation of tight junctions occurs, and depicts an intermediate stage of tight junction formation. Some relatively long, continuous tight junctional strands have already formed, but many free ends and short, unconnected linear segments are present. The long strands appear to form by the fusion of shorter segments and by the addition of individual particles (arrow). Scale bar: 0.5 μm. (From Elias and Friend, 1976. Reproduced from *The Journal of Cell Biology*, 1976, vol. 68, 173–188 by copyright permission of The Rockefeller University Press.) (B) A complete tight junction in frog skin. The tight junctional strands form a smooth, anastomosing network without free, unconnected ends. Scale bar: 0.5 μm. (Courtesy of D. S. Friend.)

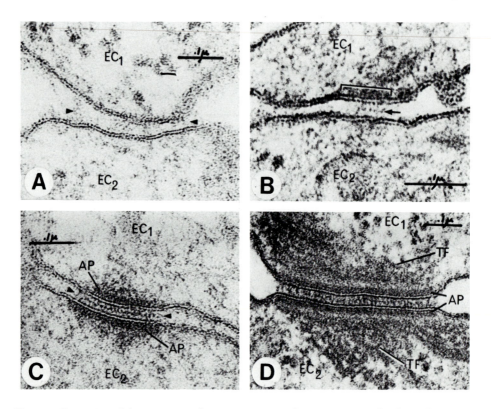

Figure 2. Formation of desmosomes. These transmission electron micrographs depict four stages in the development of desmosomes during the healing of epidermal blisters in mice. (A) The plasma membrane of two adjacent epidermal cells (EC_1 and EC_2) become closely aligned. (B) Electron-dense material appears intracellularly (bracket), and extracellular fibers accumulate between the two cells (arrow). (C) Electron-dense material continues to accumulate both intracellularly (forming the characteristic desmosomal adhesion plaques, AP) and extracellularly (arrowheads). (D) The desmosome is completed by the association of tonofilaments (TF) with the intracellular plaques. The tonofilaments generally extend deep into the cytoplasm and intermingle with other tonofilaments of the cytoskeleton. Scale bar: 0.1 μm. (From Krawczyk and Wilgram, 1973.)

attachments. The initial formation of epithelia during blastulation and organogenesis, the dissociation of epithelia into mesenchymal tissue, the aggregation of mesenchymal cells into epithelial sheets (as occurs during somite formation and formation of the coelomic lining), and the separation of epithelial cells during the invasion and metastasis of carcinomata all entail critical changes in cell–cell attachments. In addition, it has been reported that superficial cells of *Xenopus* gastrulae (Keller, 1978) and the teleost blastoderm (Kageyama, 1981; Keller and Trinkaus, 1982) rearrange and exchange nearest neighbors during epiboly, despite being all the while joined by tight junctions. Similar rearrangements also occur in the presence of septate junctions during the transformation of *Drosophila* imaginal discs (Fristrom, 1976, 1982). Thus, although junctions are usually regarded as highly stable structural elements, there can be considerable fluidity in their disposition when necessary.

Some progress toward understanding the mechanism by which cells control junction formation has been made using monolayers or suspension aggregates of cultured epithelial cells as model systems. Junctions are disrupted by the removal of calcium from the culture medium; their reformation is followed, under various conditions, when calcium is restored. When tight junctions are opened without dissociating the epithelium (e.g., by treating the sheet with calcium chelators), then allowed to close (by restoring clacium to the system), the junctions can reseal. Resealing can occur in the absence of protein synthesis (Meldolesi et al., 1978; Martinez-Palomo et al., 1980), since cycloheximide has no effect on the process. But cycloheximide does inhibit the formation of tight junctions between completely dissociated and replated cells, indicating that protein synthesis is required for the de novo formation of tight junctions (Hoi Sang et al., 1980). Colchicine also inhibits de novo junction synthesis. It may act by disrupting intracellular transport of the junctional elements to the cell surface, since it does not affect the resealing of tight junctions already present on the cell surface. Only cytochalasin has been found to inhibit the actual sealing process, suggesting that an interaction with microfilaments of the cortical cytoplasm may be involved (but see Section 4.4.2). It must be kept in mind that these experiments involve highly artificial conditions. Morphologically, the reformation of tight junction between dissociated cells does resemble closely the formation of tight junctions in vivo, lending credence to the use of these systems as models. But there is no other basis on which to assume that dissociated cells are very much like the cells of the just-forming epithelium. Of particular concern is our ignorance of what changes might occur in the composition of the cell surface during the differentiation of the epithelium and during the dissociation process. Any conclusions drawn about how the development of junctions is regulated must therefore be viewed with caution.

A question that has not been addressed in studies of tight junction formation is how the two halves of a junction come to lie in register in apposing cells. This question has been studied during desmosome development. The two halves of a desmosome are always in perfect alignment, and single, unpaired halves are very rare, suggesting that perhaps a desmosome (or the beginning of a desmosome) induces the formation of its matching half in the apposed cell. Consistent with this, Overton (1977) found that cells characteristically having a small number of desmosomes will form desmosomes with greatly increased frequency where their surfaces are placed in apposition to cells that normally form many desmosomes. This process, like the initial appearance of tight junctional particles on the cell surface, is energy dependent and sensitive to inhibitors of protein synthesis (Overton and DeSalle, 1980). However, it is not known if a similar phenomenon occurs during the formation of tight junctions.

It appears, then, that the construction of junctions can be controlled to a considerable degree on a very local basis, whether it is at the aggregation step of tight junctional particles moving in the cell surface, or the induction of the apposing half of a desmosome. Meanwhile, the cell may also regulate the formation of junctions on a more general level, e.g., by varying the number of junctional elements present on the cell surface, the mobility of surface components, or the synthesis of junctional subunits. The recent production of antibodies to specific desmosomal proteins (Franke *et al.*, 1981) has provided a means by which at least part of the intracellular pathway leading to desmosome formation might be detected. This should greatly facilitate the investigation of how junction formation is regulated and should also encourage the application of a similar approach to other junctions as well.

3. Intercellular Communication

The preceding section dealt with the mechanical integrity of epithelial sheets, i.e., how connections between cells can create barriers or give physical strength to a tissue. The continuity of epithelia may have still deeper significance in that cell–cell contacts can also permit direct communication between cells. Loewenstein and Kanno (1964) showed that the epithelial cells of *Drosophila* salivary glands are electrically coupled and that fluorescent dyes can pass from cell to cell without reaching the extracellular space. This was the first demonstration of cell coupling in nonexcitable tissue. Reports of coupling in a wide variety of epithelia [e.g., toad urinary bladder, *Chironomus* renal tubule, elasmobranch sensory epithelium (Loewenstein *et al.*, 1965), mammalian liver (Penn, 1966), and amphibian skin (Loewenstein and Penn, 1967)] soon followed. It is now apparent that electrical communication probably occurs within most epithelia (see reviews by Loewenstein, 1981; Gilula, 1980).

3.1. The Role of Cell–Cell Communication in Morphogenesis

In early embryos, electrical coupling occurs not only within tissues, but between tissues. Virtually everything is coupled to everything else. Potter *et al.* (1966) showed that the epidermis, deep cells, retina, blood vessels, and heart of the embryonic squid are each coupled to at least one other tissue. Electrical coupling is also found between most, perhaps all, cells of the morula of the newt (Ito and Hori, 1966) and at all stages from cleavage to mid-gastrula in teleost embryos (Bennett and Trinkaus, 1970). Extensive intercellular and intertissue coupling is also observed in tunicate (Miyazaki *et al.*, 1973), starfish (Tupper *et al.*, 1970), chick (Sheridan, 1968), and frog (Palmer and Slack, 1970) embryos.

Speculation has arisen that progressive uncoupling of embryonic tissues might be a way of separating cells with different developmental fates. However, the available evidence is unclear. On the one hand, the advanced differentiation of the tissues in the squid embryo just mentioned indicate that coupling persists after considerable determination has taken place. In addition, Sheridan (1968) found neural plate to be coupled to notochord, and neural tube to paraxial mesoderm—tissues that have also undergone significant determination and have acquired quite different morphologies and behaviors. On the other hand, it remains possible that the passage of larger molecules, bearing morphogenetic determinants or some other form of developmental information, is more tightly regulated than the passage of small ions, which is what is detected by measuring electrical coupling. For example, Warner and Lawrence (1973) showed that electrical coupling occurs across the sharply defined boundaries that occur between developmental compartments in insects. However, study of the same system using fluorescent dyes to assay coupling revealed that the intercellular passage of Lucifer Yellow (450 M_r) is restricted at segmental boundaries, despite the free diffusion of the dye between cells within a segment (Warner and Lawrence, 1982). Unfortunately, this is the only developmental system in which such detailed information on the degree of cell–cell communication has been reported.

Cell–cell coupling may also be important in coordinating changes within a group of cells. The sharing of small regulatory molecules (e.g., calcium or cyclic nucleotides) could coordinate such morphogenetic activities as intracellular contractions, the synthesis of extracellular matrix material, and locomotive activity. This would seem particularly useful during the formation of various organ placodes and invaginations, in which very precise movements and shape changes occur simultaneously within a discrete group of cells. In addition, the absence of coupling or a change in permeability of the coupling channels at the limits of the form change would explain the very sharp boundaries that are usually observed in such instances. Unfortunately, too few embryonic epithelia have been examined with sufficiently high spatial resolution to demonstrate whether or not this occurs *in situ*. But it has been reported that, in the human apical epidermal ridge during limb bud formation, the cells of the ridge are joined by numerous, large gap junctions, whereas few gap junctions are detected within the surrounding epidermis or between the surrounding epidermis and the cells of the ridge (Kelley and Fallon, 1976).

Coupling between cells may also be responsible for the transmission of signals controlling their growth and division. Loewenstein (1979), who gives a detailed analysis of this hypothesis, reports a strong correlation between communication incompetence and loss of contact inhibition of growth in a variety of cells *in vitro*. *In vivo*, normal epithelial cells are coupled to their neighbors, and the epithelium has well-defined growth limits. But malignant tissues often are uncoupled. Loewenstein and co-workers found no coupling between cells in several different solid tumors taken from liver (Loewenstein and Kanno, 1966), thyroid (Jamakosmanovic and Loewenstein, 1968), and stomach (Kanno and Matsui, 1968). Benign tumors are predominantly coupled (Jamakosmanovic and Loewenstein, 1968). Changes in coupling can also be observed when normal tissues are caused to undergo rapid growth. During regeneration after hepatectomy, cells of the liver display a dramatic increase in their rate of growth and division. At the same time, electrical coupling, the transmission of fluorescent dyes, and the number of gap junctions (which are believed to mediate cell–cell coupling) decline sharply (Yee and Revel, 1978; Meyer *et al.*, 1981). Thus, intercellular coupling may allow a cell to detect that it is part of a larger whole, with growth being turned on when the signal from the whole is lost or diminished. However, the nature of this proposed signal is entirely unknown.

A similar phenomenon is observed in the control of cell movement. When cuts are made in urodele skin, there is a transient loss of electrical coupling between the cells at the edge of the cut and the cells submarginal to the wound (Loewenstein and Penn, 1967). It has been suggested that this loss of electrical communication might be involved in signaling the migration of cells over the wound surface. Consistent with this hypothesis, Loewenstein and Penn also found that coupling is rapidly reestablished across the wound when margins of the wound meet. Again, the control signal is unknown. A particularly interesting possibility is that electrical gradients across the cell surface could be involved. Wounding necessarily creates a drop in electrical potential across the cell surfaces facing the wound because the cells no longer abut cells having an equal electrical potential. Recent observations in a number of laboratories indicate that potential differences can influence motile behavior (Orida and Feldman, 1982; Cooper and Keller, 1984; Nuccitelli and Erickson, 1983; Stump and Robinson, 1982), and Barker *et al.* (1982) report the presence of electrical fields in epidermal wounds in guinea pigs. These observations have also generated interest in the possibility that the electrical fields and currents observed during embryogenesis (Jaffe and Stern, 1979) and tissue regeneration (Borgens *et al.*, 1977) play an instructional role in morphogenesis.

3.2. Mechanisms of Cell–Cell Communication

There is very strong evidence that, in most cases, cell–cell coupling is mediated by gap junctions. This evidence, along with the structure and development of gap junctions and the mechanisms by which the permeability through the junction's channels might be regulated, is discussed in recent reviews (Loewenstein, 1979, 1981; Bennett *et al.*, 1981; Gilula, 1980). We shall therefore limit this discussion to a few ways, other than cytoplasmic coupling, by which direct intercellular communication might be mediated.

First, it is possible for electrical coupling to occur between cells in the absence of cytoplasmic continuity. Politoff (1977) pointed out that the ability of proteins to act as semiconductors could mediate communication across closely apposed plasma membranes. This could explain junctional rectification (the passage of current more easily in one direction than the other) and the presence of electrical coupling in the absence of dye coupling—phenomena that have been observed in a number of different systems (see Politoff, 1977; Warner and Lawrence, 1982). It is not proposed that protein semiconduction is responsible for all electrical communication, since cytoplasmic continuity is clearly demonstrated by dye coupling in a great many systems. Rather, protein semiconductance may act when cytoplasmic channels are absent or closed, perhaps even using the proteins of the gap junction.

Second, it is also possible that cell communication occurs *via* the extracellular space. The presence of occlusive seals (in the form of tight junctions) between epithelial cells can create intercellular compartments in which small, diffusable molecules might be used for relatively private communication. It is perhaps significant that the outer surface of many embryos is relatively impermeable (see Bennett and Trinkaus, 1970), creating an electrically tight compartment in which such signals would be confined.

Another model for communication between cells in intimate contact has been discussed by Glaser (1980) and Harris (1974) in attempting to explain contact inhibition of growth and movement. A cell surface may bear receptor molecules that interact with ligands on neighboring cell surfaces. Contact with another cell could then trigger internal signals in much the same fashion as the binding of a hormone at the cell surface. This concept could be extended to account for communication between cells already in contact merely by hypothesizing the addition or removal of specific surface-bound molecules.

Finally, signals may be transmitted through the mechanical connections between epithelial cells. Curtis and Seehar (1978) showed that mechanical stress causes fibroblasts to increase their rate of division. Similarly, Folkman and Moscona (1978) observed that well-spread endothelial cells, which are presumably under tension from the centripetal pull of their margins, divide faster than less flattened cells of the same type. Mechanical stress also affects the orientation and motile behavior of epithelia (see Section 4). Intercellular attachments could even coordinate epithelial shape changes. Odell *et al.* (1981) constructed a model of epithelial folding postulating that the stretch-induced contraction of microfilament bundles brings about shape changes. Computer simulations show that the contraction of a single cell, through its connections to other cells, could produce many of the epithelial shape changes that result in morphogenesis *in vivo*.

Thus, the coordinate behavior of epithelial cells could be mediated by a variety of factors, although cell coupling is perhaps the most intimate, controllable, and well-documented form of intercellular communication that has been proposed. From this plethora of putative communicative devices, we must sort out the ones that make significant contributions to morphogenesis. A major limiting factor in this task is an inadequate understanding of the processes that intercellular communication is purported to control. In other words, we do not know what must be communicated during the intercellular regulation of such phenomena as growth and motility.

4. Epithelial Polarity

The two sides of an epithelium are never the same. The free surface and the surface apposed to the substratum invariably differ in structure, biochemistry, and behavior. This sidedness is essential to many epithelial functions. In the intestine, the luminal surface is covered with microvilli presumably involved in absorption; in exocrine glands, the intracellular secretory apparatus is directed toward the free surface; and in the lining of the bladder (and in transporting epithelia, in general), the epithelium engages in directional transport, taking in ions at one surface and discharging them at the other. Polarity is manifested in cell shape, cell surface composition, and the distribution of intracellular organelles. Polarity can be observed in all epithelia and, in any given epithelium, is the same for all cells making up the sheet. How this remarkable universal organization is brought about is a question of great interest, but one for which we have few answers.

4.1. The Origins of Epithelial Polarity

During blastulation, somite formation, and the genesis of epithelial organs from mesenchyme or from solid cords of cells, polarized epithelia develop *de novo* from apolar tissue masses. For example, kidney tubules begin as a mesenchyme of loosely associated, apolar cells. These cells then aggregate into cords, which develop lumina, creating the tubule epithelium. Wartiovaara (1966) reported that the polarity of this epithelium develops in the aggregated cell mass. Nuclei assume a basal location, the Golgi apparatus and centrioles become apically oriented, and junctional complexes begin to form at a site in the interior of the aggregate where the lumen will appear. Because tight junctions appear to separate the apical and basal domains of the cell surfaces (see Section 3.2), the latter may be a critical event in the polarization of the epithelium. In developing rudiments of the liver (Wood, 1963), thyroid (Luciano et al., 1979), and intestine (Mathan et al., 1976), the appearance of a new epithelial surface is preceded by the appearance of tight junctions. But how the location of tight junctions is determined is entirely unknown. Perhaps the formation of an aggregate produces polarity by creating an inside and an outside. The development of junctions on the inside might be signalled by the increase in cell–cell contacts or by some as yet undefined diffusible substance the concentration of which would be highest in the center of the aggregate.

Another possible cue could come from the orientation of cells within electrical fields. Electrical fields and currents are known to exist in developing chick embryos (Jaffe and Stern, 1979) and regenerating limbs (Borgens et al., 1977). When fields are applied to cells *in vitro*, they can electrophorese molecules on the cell surface so that they come to lie on the side of the cell nearest the appropriate electrical pole (Orida and Feldman, 1982; Poo and Robinson, 1977). Stern (1982) has recently demonstrated that the dorsal–ventral polarity of the chick epiblast can be at least partially reversed by electrical fields of physiological strengths: Tight junctions, nuclei, and cell surface components all move from basal to apical locations when the epiblast is subjected to an electrical field of the same strength as is found across the epiblast *in vivo* but with reversed polarity. In addition, Wiley (1982) reported that Na^+,K^+-ATPase, a current-generating membrane-bound enzyme, is present only on apposed cell surfaces in murine morulae, and that its inhibition disturbs cytoplasmic polarity. He proposes that transmembrane currents generate cell polarity within the morulae by electrophoresis of the cytoplasm. This possibility warrants considerable attention in that it could provide a very tidy explanation for the development of epithelial polarity from apolar cell masses and could possibly be involved in determining the location of lumen formation during the genesis of various epithelial organs.

Still another cue for epithelial polarity might come from the surrounding extracellular matrix. All mature epithelia rest on a basal lamina *in vivo*, giving them an inherent polarity; i.e., one side is attached and the other free. The attachment of an epithelium to a fixed substratum is a very simple and direct means by which its orientation might be established. For this reason, the hypothesis that the extracellular matrix dictates epithelial polarity has been broadly championed. Epithelia that have been separated from their basal laminae (usually by enzymatically degrading the lamina) tend to lose their polar characteristics (Dodson, 1967; Banerjee *et al.*, 1977; Sugrue and Hay, 1981). Columnar cells lose their elongate shape, microtubules become randomly oriented, the basal surface extends cytoplasmic protrusions, and so on. When such stripped epithelia are placed on collagen gels (Dodson, 1967; Sugrue and Hay, 1981) or are permitted to resynthesize their own basal laminae (Banerjee *et al.*, 1977), normal polar morphology is restored.

Perhaps the most striking demonstrations of an effect of the extracellular matrix on cell polarity are the experiments of Chambard *et al.* (1981). They found that thryoid epithelial cells embedded in collagen gels form epithelial follicles with flat basal surfaces apposed to the collagen and microvillar apical surfaces facing into the lumina, as *in vivo*. But thyroid cells plated on glass or plastic form floating vesicles in which the cells have the reverse polarity, i.e., with the microvilli on the outside. If these inside-out vesicles are embedded in collagen, the epithelia reverse their polarity, restoring normal cellular orientation. Furthermore, cells plated on top of collagen gels (on a planar surface) also form epithelia with the appropriate polarity but lose their uniform orientation when overlayed with a second layer of collagen. At first they become disorganized, as some cells maintain their initial morphology while others attach to the new substratum and begin to invert. Eventually the epithelia in these collagen sandwiches reorganize to form follicles. Chambard *et al.* (1981) suggested that the extracellular matrix interacts with specific elements of the cell surface to stabilize internal elements responsible for the basal structure of the cell. The overall cell polarity then develops relative to this. That extracellular matrix materials interacting with the cell surface can indeed modulate the internal cytoskeleton has been shown by Sugrue and Hay (1981). They stripped corneal epithelium of its basal lamina and cultured it on Millipore filters. Where the epithelium is suspended across the holes in the filter, the denuded basal surface extends cytoplasmic protrusions, rather than remaining flat as it does on a complete basal lamina. But when the culture medium contains collagen, laminin, or fibronectin, the basal surface flattens, and a cortical mat of microfilaments is observed subjacent to the plasma membrane.

Although these experiments show that matrix materials can strongly influence the polarity of epithelia, it remains to be seen if changes in cell–matrix relationships coincide spatially and temporally with the appearance of asymmetry in epithelia *in vivo*. Considering the major role ascribed to the extra-cellular matrix in maintaining polarity in adult tissue, far too little attention has been given to the disposition of the matrix during the initial development of polarity. When mesenchymal cells form tight aggregates, are their basal surfaces closely apposed to surrounding extracellular materials? When does a basal lamina first appear? Does the outer surface of an aggregate take on the organization of the basal region of the epithelium before or after lumen development? During formation of intestinal villi, mesenchymal tissue protrudes into the developing villi. Do the mesencymal cells produce matrix material that might cue the development of polarity in the lumen-forming cells? Does the invasion of the mesenchyme change the position of the potential lumen-forming cells relative to the basal lamina? When do changes in matrix distribution occur during the development of thyroid follicles? The answer to these relatively simple questions should give us a better idea of how important (or unimportant) the extracellular matrix is in the determination of cell polarity.

4.2. The Development and Maintenance of Cellular Asymmetry

Whatever the signal responsible for determining the polarity of an epithelium, it must ultimately lead to the production of asymmetry within individual cells. Our knowledge of the mechanism responsible for producing and maintaining that asymmetry is sketchy at best. Some evidence suggests that the cytoskeleton plays an important role, at least in the genesis of cellular asymmetry. The preceding section noted how extracellular matrix materials can cause reorganization of intracellular microfilaments. This may, in turn, influence the distribution of cellular components elsewhere in the cell. It has been observed that the attachment of leukocytes to nylon fibers results in cell polarization in that cap formation occurs directly opposite the point of attachment (Rutishauser *et al.*, 1974). This effect appears to be mediated by the cytoskeleton, since treatment with cytochalasin or colchicine results in randomization of cap location. Cytochalasin and colchicine also inhibit the *de novo* formation of tight junctions (Hoi Sang *et al.*, 1980; Meza *et al.*, 1980), further implicating cytoskeletal elements in establishing the distribution of specific cell structures. In addition, antimicrotubule treatments have been found to cause a loss of polarity of the Golgi apparatus in the wall of the sea urchin blastocoel (Tilney and Gibbins, 1969), in the pancreatic epithelium (Moskalewski *et al.*, 1976; Meldolesi *et al.*, 1978), and in the epithelial hepatocytes lining bile canaliculi (DeBrabander *et al.*, 1978). It should be noted, however, that disruption of neither microfilaments nor of microtubules has any effect on the location of intact junctions or on morphological differences between the apical and basolateral surfaces of confluent epithelia (Meldolesi *et al.*, 1978; Hoi Sang *et al.*, 1980). It may be that the cytoskeleton is less important in maintaining these manifestations of epithelial polarity than it is in establishing them.

The tight junction itself appears to be the element primarily responsible for maintaining the asymmetric distribution of cell-surface components. The apical surface of an epithelial cell can generally be distinguished from its basolateral surface by its membrane morphology and by many biochemical criteria (e.g., the presence of specific enzymes and carbohydrate moieties, differences in lipid composition). Tight junctions form the boundary between these two distinct regions of the cell surface. When tight junctions are disrupted, intramembrane particles (Galli *et al.*, 1976; Hoi Sang *et al.*, 1979), surface proteins (Pisam and Ripoche, 1976), carbohydrates (Pisam and Ripoche, 1976; Dragsten *et al.*, 1981), and lipids (Dragsten *et al.*, 1981) that are normally confined to either the apical or basolateral surface lose their asymmetric distribution and diffuse evenly over the entire cell surface. Furthermore, Hoi Sang *et al.* (1980) found that, as tight junctions reform, random distribution of intramembrane particles is observed as long as tight junctional strands remain fragmented, but asymmetric distribution of intramembrane particles is restored as soon as a single complete tight junctional strand appears.

The tight junctions probably act as a barrier to lateral diffusion in the surface membrane (Dragsten *et al.*, 1981). Because the cell surface is constantly turning over, this requires either that the intracellular machinery for producing new surface is polarized or that the cell has some means of selectively removing materials that arrive in the wrong region of the cell surface. Current evidence indicates that at least the former mechanism is operative. When monolayers of MDCK cells are treated with antibodies to aminopeptidase, an apical surface protein of intestinal and kidney epithelia, the protein is removed from the cell surface by endocytosis. When new protein reappears on the cell surface, it does so only on the apical surface (Louvard, 1980). Similarly, viruses will bud asymmetrically from epithelial cells; i.e., certain viruses bud specifically from the apical surface, whereas others prefer the basal surface (Rodriguez Boulan and Sabatini, 1978). Because basally budding viruses and a major basal surface protein (Na^+,K^+-ATPase) have in common a large sialyl residue that apically budding viruses and apical enzymes lack, Rodriguez Boulan and Sabatini (1978) have suggested that glycosyl moieties may play an important role in determining the destinations of surface components. Further study of this system holds considerable promise for elucidating the mechanism by which epithelial cells sort their surfaces.

5. Reshaping the Epithelium

As we have already noted, epithelia assume a great many different shapes. The simple sheets, tubes, spheres, and pockets that form in early development subsequently undergo varied contortions to produce the complex morphologies of adult epithelia. It may seem that there are as many different mechanisms for producing these morphologies as there are epithelia, but all these mechanisms can be considered variations on, or combinations of, a relatively small number of fundamental morphogenetic processes. In the course of this discussion, keep in mind that none of the processes mentioned can account for all epithelial shape changes and that few, if any, epithelial shape changes involve only one of these processes.

5.1. Cellular Translocation during Epithelial Shape Changes

There are many instances in which an epithelium takes on a new form entirely through the movement of its constituent cells. Perhaps the foremost example is the spreading of epithelial sheets. Trinkaus (1976) distinguishes two types of epithelial spreading: **movement of sheets with free edges,** as in wound closure and in teleost and avian epiboly, and **movement without a free edge,** as in much of the tissue translocation that occurs during amphibian gastrulation. As gastrulation is covered in detail elsewhere in this volume (Chapter 7), we shall omit it from consideration here and focus on the other examples of epithelial movement.

The movement of an epithelium with a free edge is almost universally characterized by an adhesive, locomotive margin that pulls a less adhesive, less active sheet over a two-dimensional substratum. During wound closure in urodele epidermis (Lash, 1955) and rabbit corneal epithelium (Khodadoust et al., 1968), the migrating epithelium is firmly attached at its leading edge, but submarginally it is easily lifted off the wound surface. Broad lamellar protrusions, like those found on the leading edges of fibroblasts translocating in vitro, extend from the margins of these epithelia and from the edge of mammalian (Odland and Ross, 1968; Krawczyk, 1971; Buck, 1979) and amphibian (Radice, 1980) epidermis migrating during wound closure. Similarly, micromanipulation of the spreading blastoderm during avian epiboly reveals that it adheres almost exclusively at its margin (New, 1959), and spreading and ruffling lamellar protrusions at its margins can be observed (Bellairs et al., 1969). The enveloping layer of the teleost blastoderm also adheres exclusively at its margin during epiboly (Trinkaus, 1951; Betchaku and Trinkaus, 1978), although its movement appears to be caused by spreading of the yolk syncytial layer to which it is attached, rather than by active spreading of the margin of the enveloping layer itself (Betchaku and Trinkaus, 1978; Trinkaus, 1984a).

The fundamental locomotive machinery of epithelial cells appears to be no different from that of fibroblasts moving on planar substrata (see DiPasquale, 1975a,b). Unfortunately, in neither system is the mechanism of movement entirely understood. So, rather than dwelling on how epithelial cells move, which is really the problem of how tissue cells move (i.e., cells lacking cilia and flagella), we shall discuss those aspects of cell movement that are peculiar to epithelia. Unique to epithelia are the constraints imposed on the movement of each cell by its attachment to other cells. We have already seen one example in that most locomotive activity is restricted to the marginal cells of spreading sheets. This is probably a consequence of contact inhibition (Abercrombie, 1964): Only marginal cells are free from continuous circumferential contact with other cells. Since intercellular adhesions prevent the marginal cells of epithelia from separating from the sheet when they spread, submarginal cells rarely encounter any free space and consequently display little locomotive activity. Some submarginal locomotive activity has been observed, however, in the immediately submarginal cells of the avian blastoderm during epiboly (Downie and Pegrum, 1971), just behind the wound margin during wound closure in amphibian epidermis (Radice, 1980), and at scattered locations beneath epithelial sheets cultured on glass (DiPasquale, 1975a). The protrusive activity in all these cases appears to coincide with gaps between and beneath cells of the sheet.

As an epithelium spreads, tension must necessarily develop between the advancing marginal cells and the less motile cells to which they are attached. Virtually nothing is known about how the force generated at an epithelial cell's leading edge is transmitted through the cell to points of cell–cell adhesion. The same question, asked from a different perspective, is: How do the intercellular attachments restrain the motile margin? This restraint is evident in the continued activity of the lamellipodia at the margin of epithelia that are not spreading or that are being pulled along only very slowly. It is also apparent in the tendency of the margins of taut epithelial sheets to undergo sporadic retraction and respreading. Are the cytoskeletal elements associated with adherens junctions responsible for connecting cell–cell attachments to the motile machinery of the cell? Is the restricted availability of cell surface (due to its attachments to other cells) a limiting factor? I have observed that microfilaments in the lamellipodia of translocating epithelia are randomly oriented or loosely aligned perpendicular to the direction of motion. However, in taut, nontranslocating epithelia and in epithelia that are retracting, the microfilaments are aligned strictly parallel to the presumed lines of tension, parallel to the direction in which spreading would occur (see Fig. 3) (Kolega, 1982, 1986). Intermediate filaments and microtubules in these cells do not display a particular orientation with respect to motile behavior. Furthermore, the same reorganization of microfilaments (again, without substantial changes in the arrangement of microtubules and intermediate filaments) is also observed when epithelia are stretched with a microneedle (Kolega, 1986). This suggests that the microfilamentous cytoskeleton might play an important role in the intra- and intercellular transmission of tension. This has already been proposed to account for the alignment of microfilaments in the stress fibers observed in many cultured cells (Lewis and Lewis, 1924; Buckley and Porter, 1967; Burridge, 1981).

Tension within epithelia can have substantial effects on their behavior. Takeuchi (1979) found that when cultured corneal epithelium is stretched, spreading of its margin in the direction of tension is enhanced. At the same time, it appears that spreading perpendicular to the lines of tension is suppressed (Kolega, 1981). Thus, intraepithelial stress can focus locomotive activity along lines of tension. This may have important consequences in development. The presence of tensile fields in amphibian embryos has been clearly demonstrated (Beloussov et al., 1975). Furthermore, the polarity of much of the embryo may be strongly influenced by such fields, as suggested by Beloussov (1980). He transplanted dorsal ectodermal tissues of frog gastrulae into the field

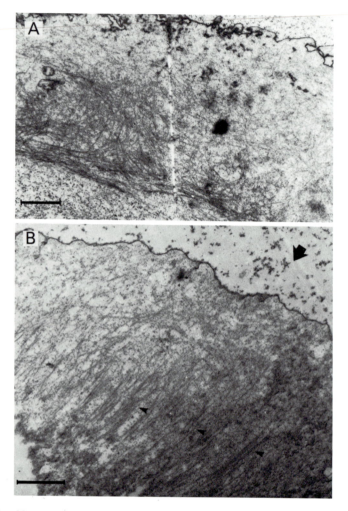

Figure 3. Microfilament organization in lamellar protrusions of translocating epithelia. (A) Electron micrograph of a thin section taken in the plane of the substratum through a lamellar protrusion extending from the leading margin of a small, translocating cluster of epidermal cells. Movement is toward the top of the micrograph. The microfilaments display no distinct orientation relative to the direction of movement. Scale bar: 2 μm. (B) Similar section taken through the *rear* margin of a translocating cluster of epidermal cells, where the lamellar protrusion is essentially being dragged backwards across the substratum (movement is in the direction of the arrow). Microfilaments (arrowheads) are oriented predominantly perpendicular to the cell's free edge; that is, parallel to the direction in which the cell is being pulled. Scale bar: 0.5 μm.

of tension lateral to the neural folds or into the tissue's original tensile field after rotating it 90°. In both cases, he observed dramatic reorientation of axial rudiments to lie along the axis of tension. In addition, when ventral tissues no longer sensitive to inductive influences were transplanted into the stretched dorsal epithelium of the gastrula, they were found to mimic the morphogenetic movements of their new location. Cells elongated parallel to cells of the host tissue, and the grafts even imitated the shape of the neural groove. The concept that physical forces play an important part in directing morphogenesis has been with us, at least in abstraction, for some time (see D'Arcy Thompson's *On Growth and Form*, particularly Chapters 6 and 16). Now that techniques are available to probe the ultrastructural manifestations of these forces, the idea merits renewed attention.

The locomotive tendencies of the margins of epithelia, coupled with the interconnections between cells, create the possibility of a novel form of movement. Epithelial cells may move as small groups, or clusters, in which the sheet does not merely spread but translocates *en masse* from one position to another. Examples of cells moving as clusters have been observed *in vivo:* Precardiac mesodermal cells in the developing chick move directionally in small, contiguous aggregates (DeHaan, 1963), and pigmented epithelial cells of the teleost *Blennius pholis* also move directionally as clusters during their embryonic migrations (Trinkaus, 1982). Moving epithelial clusters are also likely intermediates in carcinoma invasion, the disaggregation of somite cells during sclerotome formation, and other morphogenetic movements involving the dispersion of epithelia.

The directional movement of epithelial clusters poses the problem of how such directed movement is achieved by an aggregate of cells that, acting as a normal epithelium, should spread equally in all directions. This question has been addressed in two studies of epithelial clusters *in vitro*. The first showed that cluster speed is independent of cluster size (Albrecht-Buehler, 1979), indicating that cluster translocation is probably not just a two-dimensional tug-of-war. The second showed that clusters move with considerable persistence; i.e., they tend to continue moving in one direction (Kolega, 1981). Interestingly, cells on the rear margin of persistent clusters are often highly elongate perpendicular to the direction of movement (Fig. 4). On the basis of this observation, it was proposed that tension among the divergently pulling edges of the epithelium elongates these cells, suppressing their locomotive activity. Tension within the sheet (due to the continued centripetal pull of other marginal cells) maintains this condition, resulting in directional cluster movement. This is another example of how simple physical forces might direct morphogenesis. It is also an example of how the locomotive mechanism itself can reinforce movement in a particular direction.

Another situation in which cell movement alone reshapes the epithelium is during the evagination of imaginal disks in *Drosophila*. The epithelium of the disc undergoes a total change in form, going from a compact, convoluted disk to an extended adult structure such as a leg, wing, or antenna. Fristrom (1976) has shown that evagination occurs without any change in the total number of cells present. Moreover, except for a slight flattening of the cells, there is almost no change in cell shape. What does change is the distribution of cells. The number of cells along particular axes of the structure increases or decreases in accordance with the lengthening or shortening of that region. Thus, the transformation appears to be brought about by relocation of cells relative to one another. Fristrom and Fristrom (1975) point out that the shapes

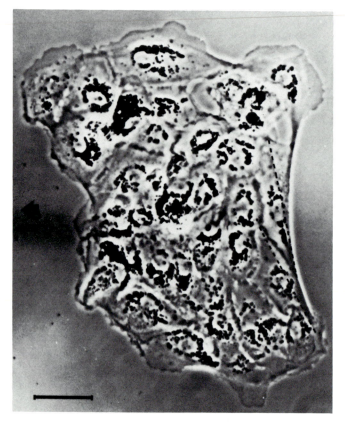

Figure 4. Directional movement of clusters of epidermal cells *in vitro*. A cluster of approximately 30 epidermal cells is shown as it moves in the direction of the arrow. The cells on the rear margin of the cluster, whose long axes are marked by the broken lines, are markedly elongate in an orientation nearly perpendicular to the direction of movement. Scale bar: 50 μm. (From Kolega, 1981.)

of clonal patches in genetic mosaics are consistent with this explanation. For example, patches are long and narrow in legs but are compact wedges in eyes undergoing little shape change during transformation. The closeness of cells in clonal patches also indicates that cells do not move very far relative to their nearest neighbors during this rearrangement, although some small displacements would account for the presence of small discontinuities occasionally observed in patches. The rearrangement is energy dependent and is inhibited by cytochalasin B, but the cytoplasmic protrusions associated with motility in other systems are absent. The rearrangement also occurs despite the continuous presence of zonulae adherentes, gap junctions, and septate junctions between cells. However, mild trypsinization, which disrupts nonjunctional contact but not the specialized junctions, accelerates evagination (Poodry and Schneiderman, 1971). This has been taken to indicate that loosening of adhesions facilitates rearrangement, perhaps by allowing easier exchange of cell–cell connections.

It is noteworthy that considerable cell rearrangement also takes place during amphibian and teleost epiboly (Keller, 1978; Kageyama, 1981; Keller and Trinkaus, 1982). As the margins of the superficial layer converge in the latter half of epiboly, cells must leave the margin (and intercalate into the submarginal pavement) as the yolk plug closes. Extensive changes in the nearest-neighbor relationships among cells occur during this process.

Although it is not known how such reorganization is brought about in epithelia, it is clear from examining the evagination of imaginal discs that small movements of individual cells can produce enormous changes in the overall structure of an epithelium, and that they do so with minimal exchange of cell adhesions and very small movements of any given cell. The possibility that similar rearrangements play a major role in other epithelial shape changes emphasizes the need for more information on how contact relationships between cells are maintained and altered. How do cells change their intercellular attachments? How are attachment sites localized to particular aspects of the cell surface? What is the relationship of contacts to the underlying cytoskeleton? Can cell–cell adhesions be moved about on the cell surface? If so, how? The subtlety of the changes involved in cell rearrangement may necessitate that we attack these more rudimentary questions before we can hope to perceive the true extent and significance of this phenomenon *in vivo*.

5.2. Cell Proliferation

Cell proliferation is fundamentally important in morphogenesis as a means of supplying the cellular materials from which tissues are sculpted. It also plays a major role in determining the thickness of multilayered epithelia. For example, the thickness of mammalian epidermis is dictated by the equilibrium between cells gained through the mitotic activity of basal cells and cells lost through the sloughing of cornified cells (Bullough, 1972; see also Houck, 1976, and Goss, 1972, for more on the regulation of tissue growth). But can cell growth and division cause morphogenetic movements, such as the spreading and folding of epithelia? This possibility receives little attention in the recent literature, being overshadowed by "more modern" mechanisms focusing on the cytoskeleton and cell movements. Nonetheless, cell proliferation is so extensive and widespread during morphogenesis that closer consideration of its potential contributions to epithelial morphogenesis is warranted.

One of the earliest propositions for growth-driven morphogenesis in epithelia was that wound closure might be driven by the overgrowth of cells in the undamaged tissue surrounding the wound (Fraisse, 1885). Although increased mitotic activity is commonly observed near the site of an epithelial wound, this theory was eventually discarded after it was shown that increased cell division does not occur until well after the start of closure (see review by Arey, 1932). More recently, a similar mechanism was proposed to explain the epithelial spreading that takes place during avian epiboly. Spratt (1963) theorized that massive cell proliferation in the center of the chick blastoderm is the driving force behind the spreading of the blastoderm. However, spreading of the blastoderm is dependent on attachment of the edge (New, 1959); also, if a cut is made behind the spreading edge, the edge continues to move while the rest of the sheet retracts (Schlesinger, 1958). In other words, the edge pulls the epithelium, the epithelium does not push the edge. This is generally the case for any epithelium with a free edge. Spreading of the sheet is a function of locomotive activity at or near its margin. Cell growth and division may limit how far the sheet can spread, but there is no evidence that animal epithelia ever spread by growing.

The most famous hypothesis for the involvement of cell proliferation in epithelial morphogenesis was put forth by His (1874). He suggested that amphibian neurulation might be caused by growth pressure; i.e., cell growth and division within the epithelium produces crowding, which in turn causes the sheet to buckle, or fold. Consistent with this notion, Derrick (1937) found an increased frequency of mitotic indices in the floor of the developing neural tube of the chick. On the other hand, Bragg (1938) failed to observe this localized increase in cell division in toad neurulae but did report a high rate of activity in the ectoderm immediately lateral to the folds. This could drive the folding by the same mechanism, but the force would be exerted from the opposite side of the folds. Thus, the distribution of mitotic activity is consistent with the His hypothesis. However, the theory fails to explain a number of other observations. For example, Roux (1885) showed that neural ectoderm removed from the embryo folds of its own accord. In the absence of the lateral restrictions imposed by the presence of the rest of the embryo, mitotic pressure would merely cause spreading; folding would not occur in the isolated epithelium. Similarly, when the neural folds are separated from the lateral ectoderm by a small incision, folding not only continues, but accelerates (Lewis, 1947; Karfunkel, 1974). Furthermore, Gillette (1944) found that there is, if anything, a slight decrease in the total volume of the ectoderm during neurulation. Unless there is a dramatic loss of tissue volume elsewhere in the neurula, this is inconsistent with the notion that the neural tube forms under growth pressure. Given these inadequacies and the presence of more attractive models, such as the apical constriction model (see Section 5.4.2), the proposal that cell proliferation drives neurulation has fallen quite justifiably by the wayside.

Mitotic pressure could still provide the forces behind epithelial shape changes in other instances, especially in smaller folds or bends, where the buckling of the sheet would be highly localized. For example, during the evagination of the thyroid from the pharyngeal epithelium (Smuts et al., 1978) and the invagination of the lens placode from the head ectoderm (Zwaan and Hendrix, 1973), there is a high rate of cell division in the region that will divert from the original plane of the primordium. Because the outside diameter of the primordium remains constant, an increasingly large number of cells must be crowded into the plane of the epithelium, unless the sheet bends outward (or inward) to form a pocket. It is important to note that, at least in the lens rudiment, the average cell volume remains constant, so the increase in cell number is accompanied by a corresponding increase in total volume of the epithelium (Zwaan and Hendrix, 1973). Although these observations are consistent with the model, the evidence extends no further. It has not been demonstrated that mitosis is necessary for folding to occur. In fact, a partial evagination of the thyroid rudiment can be induced by the addition of ATP to detergent-extracted pharyngeal epithelium of the correct developmental stage (Hilfer et al., 1977) (see Chapter 1). This does not necessarily mean that cell proliferation makes no contribution to the forces that eventually produce the full evagination, but it does indicate that the epithelium possesses a mechanism for folding that produces the basic morphological changes in the absence of growth. No such experiments have been reported for the lens epithelium or for other invaginating sheets, so it is still possible that mitotic pressure drives invagination in other instances. But, in truth, this possibility stands more on the absence of counterevidence than on the presence of rigorous testing.

Growth pressure is a more likely candidate as the force causing an epithelium to form accordion folds. For example, after formation of the pancreatic diverticulum in the mouse embryo, the pancreatic epithelium undergoes extensive growth, ultimately forming many branched lobules. The lobules first develop as irregular, unevenly distributed folds, as would be expected if the sheet were buckling to relieve mitotic pressure (Pictet et al., 1972). Richman et al. (1975) suggested that a similar phenomenon produces the convolutions of the cerebral cortex in the human brain. They reported excessive folding in microgyric brain, in which growth of the cortex is disproportionately large, and reduced folding in lissencephalic brains, which display diminished cortical growth. In addition, the shape and spacing of the folds in these cases and in normal brains is consistent with that predicted by a mathematical analysis of the stresses to which the sheet would be subject during growth. Similarly, during formation of the ciliary body in the chick eye, the neural retina epithelium is thrown into folds whose random configuration suggests that the sheet is buckling under a tangential force. This folding is accompanied by a sudden increase in the total area of the surface of the epithelium, although the sheet remains confined in the same space in the eye (Bard and Ross, 1982). Furthermore, folding may be induced simply by swelling the tissue with 50% ethanol, which causes the sheet to expand faster than the limitations on its boundaries. This observation indicates that folds can be produced in the absence of fancy cytoskeletal machinations. It also suggests that an increase in the area of the sheet, such as occurs during growth, is sufficient to drive the observed morphogenesis.

In all three cases of accordion folding, it is very plausible that growth expansion of the sheet causes the folding, but again the evidence is largely circumstantial, and there have been no further tests of the hypothesis. In particular, we lack a demonstration of the dependence of morphogenesis on growth or division. There is also a dearth of information concerning other possible mechanisms. For example, are there microfilament bundles in the furrows of the folds? Further examination of these systems should be undertaken, if only to obtain a cleaner demonstration that cell proliferation can or cannot contribute to the forces that produce this type of epithelial shape change.

Finally, there is strong evidence that cell proliferation plays a major role in

pithelia. Examples of budding may be found in the development
gland and the lung. In both cases, the budding epithelium
h rate of cell division. In the salivary gland, mitotic activity is
in the tips of the developing buds (Bernfield *et al.*, 1972). There is
aɪᵤ   d proliferation among the cells budding from tracheal epithelium *in
vitro* (Goɪᵤ.n and Wessells, 1979), although earlier studies using less sensitive
methods failed to demonstrate differential mitotic activity *in vivo* (Wessells,
1970). The tracheal bud system has also been used to demonstrate a strong
correlation between the overall rate of division within the epithelium and the
total amount of budding that occurs: X-irradiation or treatment with antimitotic
drugs inhibits both epithelial growth and budding (Alescio, 1966; Alescio and
DiMichele, 1968), and treatments that stimulate mitosis (e.g., coculture with
mesenchyme) increase bud formation (Alescio and DiMichele, 1968). Furthermore, localized budding may be induced by apposition of tracheal epithelium
with agarose pellets impregnated with epidermal growth factor (Goldin and
Opperman, 1980). Here, budding is presumed to be caused by the growth-
stimulating property of epidermal growth factor, but the possibility that the
factor is acting through a more complex inductive mechanism has not been ruled
out.

It should be noted that, at least in the case of epithelial budding, cell
division is necessary because of the expansive nature of the process; i.e., buds
are not merely bends in the sheet but also outgrowths of it. Buds enlarge and
can themselves produce branches and buds, which may, in turn, branch and
bud still further. The requirement for growth and division during budding may
simply be to provide the cells necessary for such an extensive process. The
actual shaping of the epithelium may be more a function of constraints on
where the additional material can go.

This brings up an important point regarding growth pressure as a morphogenetic mechanism. Growth alone, even highly localized growth, merely
expands the sheet. In order for bending to occur, there must be structural
limitations on that expansion. In branching morphogenesis, the extracellular
matrix is likely to play a major role in imposing such limitations (see Section
4.5). Another interesting possibility is that a supracellular network of microfilament bundles defines the boundaries of folds and outpocketings—not by acting
as contractile elements, but by providing passive resistance to deformation
(Hilfer, 1973).

5.3. Changes in Cell–Cell Adhesion

The importance of cell–cell adhesion in the formation of epithelia has
already been discussed (see Section 2.1). It has also been proposed that cell–
cell and cell–substratum adhesions play a major role in the reshaping of epithelial sheets. The first invocation of such a mechanism came from Moore
(1930) to account for the invagination of the vegetal plate at the beginning of sea
urchin gastrulation. He pointed out that an epithelium should thicken if there

is an increase in the lateral attachments between cells, and, if the attachments are concentrated toward one side of the sheet, the sheet should bend. Implicit in this argument is that cell–substratum adhesions or the internal structure of the cells would cause them to spread or at least remain cuboidal in the absence of lateral adhesions. The sheet would then curve as the apices of adjacent cells are "zippered" together (Gustafson and Wolpert, 1967). Brown *et al.* (1941) used the same model to explain the formation of the amphibian neural plate and its folding to form the neural tube. Gustafson and Wolpert (1963) added the notion that decreasing adhesion can also lead to folding. Reduced lateral adhesion would permit the cells to flatten and spread, increasing the area of the sheet. If the epithelium is constrained to lie within fixed boundaries, it would be forced to buckle to accomodate the increased area.

Despite the recurrent consideration of this model for more than 50 years, there is vanishingly little experimental evidence that changes in adhesion accompany epithelial folding. Gustafson and Wolpert (1963) catalogue various epithelial bendings and thickenings that occur during sea urchin morphogenesis and delineate how each could be elicited by changes in cell–cell and cell–substratum adhesion. However, they provide no demonstration that such changes actually occur, other than a single instance in which they note that the thickened plate of the anal arm bud is more resistant to dissociation than other epithelia of the embryo. Unfortunately, this very crude observation may be about the best assay available for differences in adhesion *in situ*. The dearth of evidence for adhesive changes may therefore be largely due to the technical difficulties associated with determining adhesiveness in living material (see Chapter 5 in Trinkaus, 1984*b*).

In the absence of a more direct and sensitive method for measuring adhesiveness, we must rely heavily on inferences from morphology. Ettensohn (1984) measured the extent of contact between cells of the invaginating vegetal plate of *Lytechinus pictus* and found a decrease in contact during invagination. This is inconsistent with the idea that increasing cell–cell adhesion toward the inside of the bend causes folding of the sheet. On the other hand, reduced lateral adhesion could permit flattening of the epithelium. The resulting increase in area could, in turn, cause the sheet to buckle. But Ettensohn also calculated the increase in the area of the sheet that would be permitted by the decrease in cell height and showed that it was too small to account for the depth of the invagination. This is not to say that changes in adhesion do not occur; clearly they do, since surface contacts must be made and broken as cells become apposed over more or less of their surfaces. But it appears that, at least in this instance, adhesive changes alone are inadequate to account for the observed morphogenesis. More studies like this one, involving a careful quantitative examination of the epithelial geometry at both the cell and tissue levels, are needed. If nothing else, they will provide stricter limits for hypothetical morphogenetic mechanisms.

Another approach to the question of whether adhesive changes are involved in epithelial morphogenesis is to alter artificially the adhesive relationships between cells and see if the predicted shape changes occur. Nardi

(1981) used the tendency of epithelial explants to circularize as an indication of the adhesiveness among epidermal cells in lepidopteran wings. He then showed that epidermis transplanted into a region of the epidermis with higher adhesiveness evaginated, and transplants to less adhesive epidermis caused invagination of the sheet. It must be cautioned that wing epidermis does not normally fold at all. However, the close resemblance of these artificial foldings to invaginations that do occur *in vivo* suggests that adhesive changes could drive some of epithelial morphogenesis. Although this type of approach does not lend itself to broad application in testing this hypothesis in other systems, we may be able to manipulate cell adhesions through other means. For example, the ongoing attempts at biochemical dissection of adhesive behavior have defined a number of factors (mostly glycoproteins) that appear to be involved in adhesion. The use of antibodies to localize these materials *in situ* may provide clues as to when and where adhesiveness changes during morphogenesis. Such antibodies may also be useful as a tool for interfering with adhesive changes *in vivo;* some form of experimental manipulation of adhesion will ultimately be necessary if we are to determine its true role in morphogenesis.

5.4. The Cytoskeleton

5.4.1. Microtubules

Epithelial foldings and invaginations are very often presaged by the formation of a thickened placode. Witness the optic, nasal and otic placodes, the neural plate, the vegetal plate of the beginning echinoderm gastrula, the thickening of the pharyngeal wall preceding thyroid formation, and so on. The thickening of these epithelia is not brought about by cell division, but by the elongation of its constituent cells; that is, the individual cells of the epithelia change shape from flat or cuboidal to columnar. Current dogma attributes this cell elongation to microtubules. Microtubule involvement was first proposed by Byers and Porter (1964) after they observed marked changes in the number and orientation of microtubules in elongating cells of the developing chick lens. During formation of the lens placode, the number of microtubules increases sharply, and their orientation changes from predominantly parallel to the free surface of the cells to parallel to the axis of elongation. After lens invagination, an increase in microtubule abundance and a distinct orientation of microtubules parallel to the long axes of the cells is also apparent in the elongating cells of the posterior lens epithelium. Microtubules parallel to the long axis of columnar cells were subsequently reported in the neural plates of salamanders (Waddington and Perry, 1966), frogs (Baker and Schroeder, 1967), and chicks (Messier, 1969), as well as in the elongate cells of the developing mouse pancreas (Wessells and Evans, 1968). In addition, a number of laboratories have demonstrated that treatments causing disruption of microtubules— cold, hydrostatic pressure, the antimicrotubule drugs vinblastin, colchicine, and colcemid—all prevent cell elongation (Pearce and Zwaan, 1970; Karfunkel, 1971, 1972; Piatigorski *et al.*, 1972).

As discussed in Chapter 1, there are a number of mechanisms by which microtubule reorganization might produce cell elongation. However, it is also quite possible that the appearance of oriented microtubules is an epiphenomenon of elongation; i.e., microtubules may merely move and reorient to fill the space created by some other mechanism of elongation. In the chick lens, if microtubules are disrupted after elongation has already taken place, the elongation of cells is not reversed (Pearce and Zwaan, 1970; Piatigorsky et al., 1972). Although this is in contrast to the elongation of cells in the neural plate, which is reversed by antimicrotubule treatments (Karfunkel, 1971, 1972), it indicates that, at least in some instances, an additional element—perhaps lateral adhesion—is required for the maintenance of the elongate form. This additional element could be the actual cause of elongation in the first place. The inhibition and reversal of elongation in tissues treated to disrupt microtubules may be due to more general debilitation of the tissue. Cold and hydrostatic pressure clearly affect other aspects of a cell. Even the reasonably specific anti-microtubule drugs have broad systemic effects, owing to the involvement of microtubules in cell division and intracellular transport. Since microtubules are also believed to play an important role in the de novo formation of junctions (see Section 1.4), their disruption could prevent essential changes in cell adhesion. Thus, although the hypothesis that microtubules bring about cell elongation is supported by the variety of treatments that disrupt both microtubules and elongation, it seems less proven than is often implied.

5.4.2. Microfilaments

In 1902, Rhumbler pointed out that small changes in the shapes of individual cells could produce dramatic changes in the form of a tissue. In particular, he illustrated how a slight narrowing of the apices of the cells in an epithelium would cause bending of the sheet. Such a change in shape could be brought about by a contraction of one surface of the epithelium, as was proposed by Gillette (1944) as the mechanism for urodele neurulation. He theorized that the contraction came from a superficial gel layer at the surface of the neurula. Lewis (1947) further championed this hypothesis, using rubber bands and brass bars to construct a mechanical model that demonstrated how different tensions at the epithelial surface could produce various shape changes. He also reported that, in Ambystoma embryos, individual cells change shape in a variety of epithelial bends: folding of the neural plate, evagination of the optic vesicle, and invagination of the optic cup, ear vesicle, nasal pit, and blastopore. Cells in these bending epithelia are generally much narrower at their ends bordering the inside of the bend.

How this constriction might take place became illuminated in the electron microscope. Balinsky (1961) observed a dense fibrous material immediately below the apical surfaces of neural ectoderm cells in frog neurulae and suggested that its contraction brings about the folding of the neural plate. When Baker (1965) reported that similar material in the invaginating cells of the frog blastopore is made up of microfilaments and Cloney (1966) reported that microfilaments are probably engaged in the contraction of the epidermis during ascidian tail resorption, interest in microfilaments as the possible force generator in epithelial folding burgeoned. Microfilament bundles were subsequently found in the apices of cells in the neural folds of frog (Baker and Schroeder, 1967), chick (Karfunkel, 1972), and rat (Freeman, 1972). They were also observed during lens invagination (Wrenn and Wessells, 1969), tubular gland formation (Wrenn and Wessells, 1970), elevation of the pancreatic diverticulum (Wessells and Evans, 1968), the budding of salivary glands (Spooner and Wessells, 1970), and thyroid evagination (Shain et al., 1972).

In most cases, microfilament bundles are located around the apex of the cell in a circular or polygonal array, which becomes smaller and thicker as folding proceeds. During urodele neurulation, microfilament length has been estimated from the size and number of microfilament bundles in thin sections, and it appears that there is little change in the total microfilament length per cell during constriction (Burnside, 1971), This suggests that the bundles contract by antiparallel sliding of microfilaments, as in a muscle fiber. In addition, the cell apex narrows during constriction, and the apical surface is thrown into folds, consistent with an intracellular constriction of the end of the cell.

A number of methods have been used to demonstrate more directly the involvement of microfilaments in epithelial folding. The most common approach is to attempt to inhibit morphogenesis with cytochalasin. Cytochalasin inhibits salivary gland morphogenesis (Spooner and Wessells, 1970), tubular gland formation (Wrenn and Wessells, 1970), and neurulation (Karfunkel, 1972), while simultaneously disrupting circumferential bands of apical microfilaments. Removal of the drug permits folding to resume with concommittant reappearance of the bundles. Although microfilament organization is the element of cell morphology most visibly altered by cytochalasin, it must be kept in mind that cytochalasin also disturbs the cell surface, affecting membrane fluidity, transport properties, and cell adhesion (Schaeffer et al., 1973; Plagemann et al., 1978; Tsuchiya et al., 1981). Therefore, drug studies alone are only suggestive of microfilament involvement, in that disruption of morphogenesis could arise from disruption of other cell functions. However, evidence from several other lines of experimentation (see Chapter 1) is consistent with the interpretation that an actin–myosin system capable of contracting to produce folding or invagination is present in epithelia and may be the basis for much of epithelial morphogenesis.

5.4.3. Intermediate Filaments

Intermediate filaments are a major component of the cytoplasm of most epithelial cells. In some squamous epithelia, they may make up as much as 30% of the total cell protein (Sun et al., 1979; see also reviews by Osborn et al., 1982, and Zackroff et al., 1981). In thin sections, a dense matrix of curving intermediate filaments is usually observed surrounding the nucleus and filling much of the perinuclear cytoplasm. Bundles of intermediate filaments loop through desmosomal plaques and so are believed to give tensile strength to

epithelia and to disperse sheer forces (Kelly, 1966). This is undoubtedly important in maintaining the integrity of a sheet during the more dramatic contortions of embryogenesis, such as neurulation, gastrulation, and epiboly. In the
mouse embryo, intermediate filaments first appear in the trophectoderm of the
blastocyst, at the same time as the first appearance of desmosomes in the
embryo (Jackson et al., 1980). The development of the trophectoderm also
coincides with the swelling of the blastocoel, presumably by hydrostatic pressure, which would necessitate a mechanically strong epithelium. Unfortunately, the distribution of intermediate filaments has not been systematically examined in other instances of morphogenesis in vivo.

There has been much speculation that intermediate filaments are involved
in maintaining cell shape. There is good reason for this. When cultured epithelia are heavily extracted with detergent, little is left but intermediate filaments and remnants of nuclei. These intermediate filament skeletons retain, to
a very good approximation, the original shape of the cells (Jones et al., 1982). In
addition, the nuclei remain suspended above the substratum within the intermediate filament matrix, suggesting that the intermediate filaments might be
involved in nuclear centration (Lehto et al., 1978). In further pursuit of this
question, I have examined the orientation of intermediate filaments in thin
sections of cultured epidermis and found that they tend to lie parallel to the
nearest cell surface (Fig. 5). They roughly follow the contours of the cell while
showing little or no specific orientation relative to the motile behavior of a cell
and changing orientation only slightly when the sheet is stretched (Kolega,
1982; 1986).

These observations suggest that the intermediate filament matrix might
indeed give epithelial cells their shape, as well as their tensile strength. However, it has recently been reported that the injection of antibodies against intermediate filament proteins into cultured epithelial cells does not effect cell

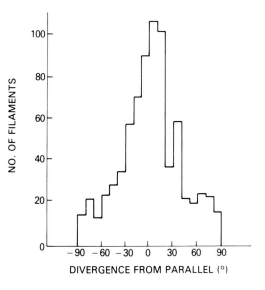

Figure 5. Intermediate filament orientation and cell shape. Intermediate filaments
were selected at random on electron micrographs of cultured epidermal cells. The
orientation of each filament relative to the
nearest cell surface cut by the plane of section was then determined. Orientation was
quantitated as the angle between a line
drawn tangential to the cell surface and a
line drawn tangential to the intermediate
filament. Thus, an angle of 0° indicates
that the filament is parallel to the cell surface at the point of measurement. The frequency at which intermediate filaments
are found with a given orientation is represented in this histogram. The clustering of
values around 0° demonstrates a strong
tendency for intermediate filaments to follow the contours of the cell.

срupting the organization of the intermediate filament network ...mkowsky, 1982). On the other hand, treatment of cells with ...β, which does not disrupt intermediate filaments, can cause dra- ...ges in cell shape. Thus, the orientation of the intermediate filaments ...o the cell contours may be a result of their movement in response to th... ...ε cell-shaping forces. In other words, intermediate filaments may be merely cellular clay, sculpted by other elements, such as the adhesive relationships of the cell or its microfilamentous cytoskeleton. Their function may be to give the cell body bulk and increased resistance to deformation by extracellular forces.

5.5. The Role of the Extracellular Matrix

The putative morphogenetic mechanisms discussed thus far all involve changes in form or behavior that are entirely intrinsic to the cells of the epithelium. We should also be aware that epithelial shape may be dictated, at least in part, by the extracellular scaffolding on which it lies. When the basal lamina is enzymatically removed from developing salivary glands, the epithelium loses its lobular morphology (Bernfield et al., 1972). When the epithelium is allowed to resynthesize a new basal lamina, its lobular form is restored (Banerjee et al., 1977). Moreover, glycosaminoglycans of the basal lamina are rapidly degraded on the growing, distal portion of the developing lobes, and the basal lamina displays frequent discontinuities. But, in the clefts between lobes, glycosaminoglycans are relatively stable, and the basal lamina has no gaps (Bernfield and Banerjee, 1982). A similar distribution of glycosaminoglycans is observed in the developing seminal vesicle (Cunha and Lung, 1979). Thus, the extracellular matrix may play an instructive, or at least a permissive, role in determining where an epithelium will change shape.

6. Summary

Epithelial tissues are ubiquitous in metazoan organisms, performing many different functions and assuming a variety of shapes. This diversity of form and function is ultimately dependent on the behavior of the cells within the epithelia. For example, it is intercellular adhesion and the control of paracellular permeability by cell junctions that permit epithelia to form barriers and act as selective filters. It is cellular polarity that enables absorptive epithelia to extract materials from a particular side of the sheet; it is the collective contributions of cell proliferation, cellular translocation, and changes in cell shape that sculpt epithelia from simple sheets into folds, pouches and tubes. Clearly, a complete understanding of epithelial morphogenesis is inextricably entwined with questions of cell behavior in general, such as how any cell adheres, moves, and maintains its shape. The study of epithelial systems has lent considerable insight into these problems and should continue to do so, just as examination of the behavior and architecture of nonepithelial cells will undoubtedly clarify many aspects of the cellular events underlying epithelial morphogenesis.

Although the action of individual cells ultimately shapes epithelia, coordination of that action is necessary for the development of a coherent tissue. Attention must therefore be given to integrative mechanisms in epithelial morphogenesis. How do the many cells in an epithelial sheet act in virtual unison during folding? What defines the boundaries of epithelial invaginations? How does an individual cell detect its position within, and thereby know its role in the morphogenesis of, the epithelial whole of which it is a part? At the most elementary level, epithelial cells interact via their physical attachments to one other. Even such rudimentary communication affects cell shape, movement, and possibly proliferation and also plays a part in the maintenance of epithelial polarity. Additional signals pass among epithelial cells by a number of other mechanisms as well, most notably electrical coupling. However, many questions remain regarding the quality and quantity of what is communicated between epithelial cells. Accordingly, elucidating the means by which supracellular order is maintained in epithelial tissues may still be regarded as the major problem in the study of epithelial morphogenesis.

ACKNOWLEDGMENTS. I heartily thank C. A. Ettensohn and J. P. Trinkaus for many helpful discussions and for their critical reading of this manuscript. My own recent research, including the writing of this chapter, has been supported by funds from Public Health Service grant number USPHS-CA-22451, awarded by the National Cancer Institute, Department of Health and Human Services, to J. P. Trinkaus.

References

Abercrombie, M. A., 1964, Behavior of cells toward one another, Adv. Biol. Skin **5**:95–112.

Albrecht-Buehler, G., 1979, Group locomotion of PtK1 cells, Exp. Cell Res. **122**:402–407.

Alescio, T., 1966, Response to X-irradiation of mouse embryonic lung cultured in vitro. Radiation effect on the epithelium growth rate, Exp. Cell Res. **43**:459–473.

Alescio, T., and DiMichele, M., 1968, Relationship of epithelial growth to mitotic rate in mouse embryonic lung developing in vitro, J. Embryol. Exp. Morphol. **19**:227–237.

Arey, L. B., 1936, Wound Healing, Physiol. Rev. **16**:327–406.

Baker, P. C., 1965, Fine structure and morphogenetic movements in the gastrula of the treefrog, Hyla regilla, J. Cell Biol. **24**:95–116.

Baker, P. C., and Schroeder, T. E., 1967, Cytoplasmic filaments and morphogenetic movements in the amphibian neural tube, Dev. Biol. **15**:432–450.

Balinsky, B. I., 1961, Ultrastructural mechanism of gastrulation and neurulation, in: Symposium on Germ Cells and Development, pp. 550–563, Institute International d'Embryologie and Fondazione A. Baselli, Pallanza.

Banerjee, S. D., Cohn, R. H., and Bernfield, M. R., 1977, Basal lamina of embryonic salivary epithelia. Production by the epithelium and role in maintaining lobular morphology, J. Cell Biol. **73**:445–463.

Bard, J. B. L., and Ross, A. S. A., 1982, The morphogenesis of the ciliary body of the avian eye. II. Differential enlargement causes an epithelium to form radial folds, Dev. Biol. **92**:87–96.

Barker, A. T., Jaffe, L. F., and Vanable, J. W., Jr., 1982, The glabrous epidermis of cavies contains a powerful battery, Am. J. Physiol. **242**:R358–366.

Bellairs, R., 1963, Differentiation of the yolk sac of the chick studied by electron microscopy, *J. Embryol. Exp. Morphol.* **11**:201–225.

Bellairs, R., Boyde, A., and Heaysman, J. E. M., 1969, The relationship between the edge of the chick blastoderm and the vitelline membrane, *Arch. Entwicklungsmech. Org.* **163**:113–121.

Beloussov, L. V., 1980, The role of tensile fields and contact polarization in the morphogenesis of amphibian axial rudiments, *Arch. Entwicklungsmech. Org.* **188**:1–7.

Beloussov, L. V., Dorfman, J. G., and Cherdantzev, V. G., 1975, Mechanical stresses and morphological patterns in amphibian embryos, *J. Embryol. Exp. Morphol.* **34**:559–574.

Bennett, M. V. L., and Trinkaus, J. P., 1970, Electrical coupling between embryonic cells by way of extracellular space and specialized junctions, *J. Cell Biol.* **44**:592–610.

Bennett, M. V. L., Spray, D. C., and Harris, A. L., 1981, Electrical coupling in development, *Am. Zool.* **21**:413–427.

Bernfield, M. R., and Banerjee, S. B., 1982, The turnover of basal lamina glycosaminoglycan correlates with epithelial morphogenesis, *Dev. Biol.* **90**:291–305.

Bernfield, M. R., Banerjee, S. B., and Cohn, R. H., 1972, Dependence of salivary epithelial morphology and branching morphogenesis upon acid mucopolysaccharide-protein (proteoglycan) at the epithelial surface, *J. Cell Biol.* **52**:674–689.

Betchaku, T., and Trinkaus, J. P., 1978, Contact relations, surface activity, and cortical microfilaments of marginal cells of the enveloping layer of the yolk syncytial and yolk cytoplasmic layers of *Fundulus* before and during epiboly, *J. Exp. Zool.* **206**:381–426.

Borgens, R. B., Vanable, J. W., Jr., and Jaffe, L. F., 1977, Bioelectricity and regeneration. I. Initiation of frog limb regeneration by minute currents, *J. Exp. Zool.* **200**:403–416.

Bragg, A. N., 1938, The organization of the early embryo of *Bufo cognatus* as revealed especially by the mitotic index, *Z. Zellforsch. Mikrosk. Anat.* **28**:154–178.

Brown, M. G., Hamburger, V., and Schmitt, F. O., 1941, Density studies on amphibian embryos with special reference to the mechanism of organizer action, *J. Exp. Zool.* **88**:353–372.

Buck, R. C., 1979, *In vitro* migration of epidermal cells in suction blisters of rat skin, *In vitro* **15**:258–262.

Buckley, I. K., and Porter, K. R., 1967, Cytoplasmic fibrils in living cultured cells. A light and electron microscope study, *Protoplasma* **64**:349–380.

Bullough, W. S., 1972, The control of epidermal thickness, *Br. J. Dermatol.* **87**:187–199.

Burnside, B., 1971, Microtubules and microfilaments in newt neurulation, *Dev. Biol.* **26**:416–441.

Burridge, K., 1981, Are stress fibres contractile?, *Nature (Lond.)* **294**:691–692.

Byers, B., and Porter, K. R., 1964, Oriented microtubules in elongating cells of the developing lens rudiment after induction, *Proc. Natl. Acad. Sci. USA* **52**:1091–1099.

Chambard, M., Gabrion, J., and Mauchamp, J., 1981, Influence of collagen gel on the orientation of epithelial cell polarity: Follicle formation from isolated thyroid cells and from preformed monolayers, *J. Cell Biol.* **91**:157–166.

Claude, P., and Goodenough, D. A., 1973, Fracture faces of zonulae occludentes from "tight" and "leaky" epithelia, *J. Cell Biol.* **58**:390–400.

Cloney, R. A., 1966, Cytoplasmic microfilaments and cell movements: Epidermal cells during ascidian metamorphosis, *J. Ultrastruct. Res.* **14**:300–328.

Cooper, M. S., and Keller, R. E., 1984, Perpendicular orientation and directional migration of amphibian neural crest cells in dc electrical fields, *Proc. Natl. Acad. Sci. USA* **81**:160–164.

Cunha, G. R., and Lung, B., 1979, The importance of stroma in morphogenesis and functional activity of urogenital epithelium, *In Vitro* **15**:50–71.

Curtis, A. S. G., and Seehar, G. M., 1978, The control of cell division by tension or diffusion, *Nature (Lond.)* **274**:52–53.

Dan, K., 1960, Cyto-embryology of echinoderms and amphibia, *Int. Rev. Cytol.* **9**:321–367.

Dan-Sohkawa, M., and Fujisawa, H., 1980, Cell dynamics of the blastulation process in the starfish, *Asterina pectinifera*, *Dev. Biol.* **77**:328–339.

DeBrabander, M., Wanson, J.-C., Mosselmans, R., Geuens, G., and Drochmans, P., 1978, Effects of antimicrotubular compounds on monolayer cultures of adult rat hepatocytes, *Biol. Cell.* **31**:127–140.

DeHaan, R. L., 1963, Migration patterns of precardiac mesoderm in the early chick embryo, *Exp. Cell Res.* **29**:544–560.

Dermietzel, R., Meller, K., Tetzloff, W., and Waelsch, M., 1977, *In vivo* and *in vitro* formation of the junctional complex in choroid epithelium. A freeze-etching study, *Cell Tissue Res.* **181**:427–441.

Derrick, G. E., 1937, An analysis of the early development of the chick by means of the mitotic index, *J. Morphol.* **61**:257–284.

DiPasquale, A., 1975a, Locomotory activity of epithelial cells in culture, *Exp. Cell Res.* **94**:191–215.

DiPasquale, A., 1975b, Locomotion of epithelial cells. Factors involved in extension of the leading edge, *Exp. Cell Res.* **95**:425–439.

Dodson, J. W., 1967, The differentiation of epidermis. I. The interrelationship of epidermis and dermis in embryonic chicken skin, *J. Embryol. Exp. Morphol.* **17**:83–105.

Downie, J. R., and Pegrum, S. M., 1971, Organization of the chick blastoderm edge, *J. Embryol. Exp. Morphol.* **26**:623–635..

Dragsten, P. R., Blumenthal, R., and Handler, J. S., 1981, Membrane asymmetry in epithelia: Is the tight junction a barrier to diffusion in the plasma membrane?, *Nature (Lond.)* **294**:718–722.

Ducibella, T., Albertini, D. F., Anderson, E., and Biggers, J. D., 1975, The pre-implantation mammalian embryo: Characterization of intercellular junctions and their appearance during development, *Dev. Biol.* **45**:231–250.

Elias, P. M., and Friend, D. S., 1976, Vitamin-A-induced mucous metaplasia. An *in vitro* system for modulating tight and gap junction differentiation, *J. Cell Biol.* **68**:173–188.

Ettensohn, C. A., 1984, Primary invagination of the vegetal plate during sea urchin gastrulation, *Am. Zool.* **24**:571–588.

Farquhar, M. G., and Palade, G. E., 1963, Junctional complexes in various epithelia, *J. Cell Biol.* **17**:375–412.

Farquhar, M. G., and Palade, G. E., 1967, Cell junctions in amphibian skin, *J. Cell Biol.* **26**:263–291.

Fawcett, D., 1966, *The Cell: An Atlas of Fine Structure*, W. B. Saunders, Philadelphia.

Feldman, J. D., Vazquez, J. J., and Kurtz, S. M., 1961, Maturation of the rat fetal thyroid, *J. Biophys. Biochem. Cytol.* **11**:365–383.

Folkman, J., and Moscona, A., 1978, Role of cell shape in growth control, *Nature (Lond.)* **273**:345–349.

Fraisse, P., 1885, *Die Regeneration von Geweben und Organen bei den Webeltieren, besonders Amphibien und Reptilien*, Theodore Fischer, Berlin.

Franke, W. W., Schmid, E., Grund, C., Müller, H., Engelbrecht, I., Moll, R., Stadler, J., and Jarasch, E.-D., 1981, Antibodies to high molecular weight polypeptides of desmosomes: Specific localization of a class of junctional proteins in cells and tissues, *Differentiation* **20**:217–241.

Freeman, B. G., 1972, Surface modifications of neural epithelial cells during formation of the neural tube in the rat embryo, *J. Embryol. Exp. Morphol.* **28**:437–448.

Friend, D. S., and Gilula, N. B., 1972, Variations in tight and gap junctions in mammalian tissues, *J. Cell Biol.* **53**:758–776.

Fristrom, D., 1976, The mechanism of evagination of imaginal discs of *Drosophila melanogaster*. III. Evidence for cell rearrangement, *Dev. Biol.* **54**:163–171.

Fristrom, D. K., 1982, Septate junctions in imaginal disks of *Drosophila*: A model for the redistribution of septa during cell rearrangement, *J. Cell Biol.* **94**:77–87.

Fristrom, D., and Fristrom, J. W., 1975, The mechanism of evagination of the imaginal disc of *Drosophila melanogaster*. I. General considerations, *Dev. Biol.* **43**:1–23.

Galli, P., Brenna, A., DeCamilli, P., and Meldolesi, J., 1976, Extracellular calcium and the organization of tight junctions in pancreatic acinar cells, *Exp. Cell Res.* **99**:178–183.

Garrod, D. R., and Nicol, A., 1981, Cell behaviour and molecular mechanisms of adhesion, *Biol. Rev.* **56**:199–242.

Getolsky, J. E., Birdwell, C. R., Weseman, J., and Lerner, R. A., 1980, A glycoprotein involved in aggregation of *D. discoidem* is distributed on the cell surface in a nonrandom fashion favoring cell junctions, *Cell* **21**:339–345.

Gillette, R., 1944, Cell number and cell size in the ectoderm during neurulation (*Ambystoma maculatum*), *J. Exp. Zool.* **96**:201–222.

Gilula, N. B., 1973, Development of cellular junctions, *Am. Zool.* **13:**1109–1117.

Gilula, N. B., 1980, Cell-to-cell communication and development, in: *The Cell Surface: Mediation of Developmental Processes* (S. Subtelny and N. K. Wessells, eds.), pp. 23–41, Academic Press, New York.

Glaser, L., 1980, From cell adhesion to growth control, in: *The Cell Surface: Mediation of Developmental Processes* (S. Subtelny and N. K. Wessells, eds.), pp. 79–97, Academic Press, New York.

Goldin, G. V., and Opperman, L. A., 1980, Induction of supernumerary tracheal buds and stimulation of DNA synthesis in the embryonic chick lung and trachea by epidermal growth factor, *J. Embryol. Exp. Morphol.* **60:**235–243.

Goldin, G. V., and Wessells, N. K., 1979, Mammalian lung development: The possible role of cell proliferation in the formation of supernumerary tracheal buds and in branching morphogenesis, *J. Exp. Zool.* **208:**337–346.

Goodenough, D. A., and Revel, J. P., 1970, A fine structural analysis of intercellular junctions in the mouse liver, *J. Cell Biol.* **45:**272–290.

Goss, R. J., 1972, Theories of growth regulation, in: *Regulation of Organ and Tissue Growth* (R. J. Goss, ed.), pp. 1–11, Academic Press, New York.

Gustafson, T., and Wolpert, L., 1963, The cellular basis of morphogenesis and sea urchin development, *Int. Rev. Cytol.* **15:**139–214.

Gustafson, T., and Wolpert, L., 1967, Cellular movement and contact in sea urchin morphogenesis, *Biol. Rev.* **42:**442–498.

Hand, A. R., and Gobel, S., 1972, The structural organization of the septate and gap junctions of *Hydra, J. Cell Biol.* **52:**397–408.

Harris, A., 1974, Contact inhibition of cell locomotion, in: *Cell Communication* (R. P. Cox, ed.), pp. 147–185, Wiley, New York.

Hilfer, S. R., 1973, Extracellular and intracellular correlates of organ initiation in the embryonic chick thyroid, *Am. Zool.* **13:**1023–1038.

Hilfer, S. R., Palmatier, B. Y., and Fithian, E. M., 1977, Precocious evagination of the embryonic chick thyroid in ATP-containing medium, *J. Embryol. Exp. Morphol.* **42:**163–175.

His, W., 1874, *Unsere Köpferform und das physiologische Probleme ihrere Entstehung,* Vogel, Leibzig.

Hoi Sang, U., Saier, M. H., Jr., and Ellison, M. H., 1979, Tight junction formation is closely linked to the polar redistribution of intramembranous particles in aggregating MDCK epithelia, *Exp. Cell Res.* **122:**384–391.

Hoi Sang, U., Saier, M. H., Jr., and Ellison, M. H., 1980, Tight junction formation in the establishment of intramembranous particle polarity in aggregating MDCK cells. Effects of drug treatment, *Exp. Cell Res.* **128:**223–235.

Houck, J. C. (ed.), 1976, *Chalones,* American Elsevier, New York.

Ito, S., and Hori, N., 1966, Electrical characterization of *Triturus* egg cells during cleavage, *J. Gen. Physiol.* **49:**1019–1027.

Jackson, B. W., Grund, C., Schmid, E., Bürki, K., Franke, W. W., and Illmensee, K., 1980, Formation of cytoskeletal elements during mouse embryogenesis. Intermediate filaments of the cytokeratin type and desmosomes in preimplantation embryos, *Differentiation* **17:**161–179.

Jaffe, L. F., and Stern, C. D., 1979, Strong electrical currents leave the primitive streak of chick embryos, *Science* **206:**569–571.

Jamakosmanovic, A., and Lawrence, W. R., 1968, Intercellular communication and tissue growth. III. Thyroid cancer, *J. Cell Biol.* **38:**556–561.

Jones, J. C. R., Goldman, A. E., Steinert, P. M., Yuspa, S., and Goldman, R. D., 1982, Dynamic aspects of the supramolecular organization of intermediate filament networks in cultured epidermal cells, *Cell Motil.* **2:**197–213.

Kageyama, T., 1981, Cellular basis of epiboly of the enveloping layer in the embryo of the medaka, *Oryzias latipes.* II. Evidence for cell rearrangement, *J. Exp. Zool.* **219:**241–256.

Kalt, M. R., 1971, The relationship between cleavage and blastocoel formation in *Xenopus laevis.* II. Electron microscopic observations, *J. Embryol. Exp. Morphol.* **26:**51–66.

Kanno, Y., and Matsui, Y., 1968, Cellular uncoupling in cancerous stomach epithelium, *Nature (Lond.)* **218**:775–776.

Karfunkel, P., 1971, The role of microtubules and microfilaments in neurulation in *Xenopus, Dev. Biol.* **25**:30–56.

Karfunkel, P., 1972, The activity of microtubules and microfilaments in neurulation in chick, *J. Exp. Zool.* **181**:289–301.

Karfunkel, P., 1974, The mechanisms of neural tube formation, *Int. Rev. Cytol.* **38**:245–271.

Keller, R. E., 1978, Time-lapse cinemicrographic analysis of superficial cell behavior during and prior to gastrulation in *Xenopus laevis, J. Morphol.* **157**:223–247.

Keller, R. E., and Trinkaus, J. P., 1982, Cell rearrangement in a tightly-joined epithelial layer during *Fundulus* epiboly, *J. Cell Biol.* **95**:325a.

Kelley, R. O., and Fallon, J. F., 1976, Ultrastructural analysis of the apical ectodermal ridge during vertebrate limb morphogenesis. I. The human forelimb with special reference to gap junctions, *Dev. Biol.* **51**:241–256.

Kelly, D. E., 1966, Fine structure of desmosomes, hemidesmosomes, and an adepidermal globular layer in developing newt epidermis, *J. Cell Biol.* **28**:51–72.

Khodadoust, A. A., Silverstein, A. M., Kenyon, K. R., and Dowling, J. E., 1968, Adhesion of regenerating corneal epithelium. The role of basement membrane, *Amer. J. Ophthalmol.* **65**:339–348.

Kolega, J., 1981, The movement of cell clusters *in vitro:* Morphology and directionality, *J. Cell Sci.* **49**:15–32.

Kolega, J., 1982, Organization of cytoskeletal filaments in relation to motility of epidermal cell clusters, *J. Cell Biol.* **95**:325a.

Kolega, J., 1986, Effects of mechanical tension on protrusive activity and microfilament and intermediate filament organization in an epidermal epithelium moving in culture, *J. Cell Biol.* (in press).

Krawczyk, W. S., 1971, Pattern of epidermal cell migration during wound healing, *J. Cell Biol.* **49**:247–263.

Krawczyk, W. S., and Wilgram, G. F., 1973, Hemidesmosome and desmosome morphogenesis during wound healing, *J. Ultrastruct. Res.* **45**:93–101.

Lane, E. B., and Klymkowsky, M. W., 1982, Epithelial tonofilaments: Investigating their form and function using monoclonal antibodies, *Cold Spring Harbor Symp. Quant. Biol.* **46**:387–402.

Lane, N. J., and Swales, L. C., 1982, Stages in the assembly of pleated and smooth septate junctions in developing insect embryos, *J. Cell Sci.* **56**:245–262.

Lash, J. W., 1955, Studies on wound closure in urodeles, *J. Exp. Zool.* **128**:13–28.

Lehto, V.-P., Virtanen, I., and Kurki, P., 1978, Intermediate filaments anchor the nuclei in nuclear monolayers of cultured human fibroblasts, *Nature (Lond.)* **272**:175–177.

Lentz, T. L., and Trinkaus, J. P., 1971, Differentiation of the junctional complex of surface cells in the developing *Fundulus* blastoderm, *J. Cell Biol.* **48**:455–472.

Lewis, W. H., 1947, Mechanics of invagination, *Anat. Rec.* **97**:139–156.

Lewis, W. H., and Lewis, M. R., 1924, Behavior of cells in tissue culture in: *General Cytology* (E. V. Cowdry, ed.), pp. 385–447, University of Chicago Press, Chicago.

Loewenstein, W. R., 1979, Junctional intercellular communication and the control of growth, *Biochim. Biophys. Acta* **560**:1–65.

Loewenstein, W. R., 1981, Junctional intercellular communication: The cell-to-cell membrane channel, *Physiol. Rev.* **61**:829–913.

Loewenstein, W. R., and Kanno, Y., 1964, Studies on an epithelial (gland) cell junction. I. Modification of surface membrane permeability, *J. Cell Biol.* **22**:565–586.

Loewenstein, W. R., and Kanno, Y., 1966, Intercellular communication and tissue growth: Lack of communication between cancer cells, *Nature (Lond.)* **209**:1248–1249.

Loewenstein, W. R., and Penn, R. D., 1967, Intercellular communication and tissue growth. II. Tissue regeneration, *J. Cell Biol.* **33**:235–242.

Loewenstein, W. R., Socolar, S. J., Higashino, S., Kanno, Y., and Davidson, N., 1965, Intercellular communication: Renal, urinary bladder, sensory, and salivary gland cells, *Science* **149:**295–298.

Louvard, D., 1980, Apical membrane aminopeptidase appears at site of cell–cell contact in cultured kidney epithelial cells, *Proc. Natl. Acad. Sci.* **77:**4132–4136.

Luciano, L., Thiele, J., and Reale, E., 1979, Development of follicles and occluding junctions between the follicular cells of thyroid gland. A thin-section and freeze-fracture study in the fetal rat, *J. Ultrastruct. Res.* **66:**164–181.

Macher, T. E., Erlij, D., and Wooding, F. B. P., 1972, Permeable junctional complexes. The movement of lanthanum across rabbit gall-bladder and intestine, *J. Cell Biol.* **54:**302–312.

Madara, J. L., Neutra, M. R., and Trier, J. S., 1981, Junctional complexes in fetal rat small intestine during morphogenesis, *Dev. Biol.* **86:**170–178.

Martinez-Palomo, A., Meza, I., Beaty, G., and Cereijido, M., 1980, Experimental modulation of occluding junctions in a cultured transport epithelium, *J. Cell Biol.* **87:**736–745.

Mathan, M., Moxey, P. C., and Trier, J. S., 1976, Morphogenesis of fetal rat duodenal villi, *Am. J. Anat.* **146:**73–92.

McNutt, N. S., Hershberg, R. A., and Weinstein, R. S., 1971, Further observation on the occurrence of nexuses in benign and malignant human cervical epithelium, *J. Cell Biol.* **51:**805–825.

Meeusen, R. L., and Cande, W. Z., 1979, N-ethylmeleimide-modified heavy meromyosin: A probe for actomyosin interactions, *J. Cell Biol.* **82:**57–65.

Meldolesi, J., Castiglioni, G., Parma, R., Nassivera, N., and DeCamilli, P., 1978, Ca^{++}-dependent disassembly and reassembly of occluding junctions in guinea pig pancreatic acinar cells. Effect of drugs, *J. Cell Biol.* **79:**156–172.

Messier, P. E., 1969, Effects of β-mercaptoethanol on the fine structure of the neural plate of the chick embryo, *J. Embryol. Exp. Morphol.* **21:**309–329.

Meyer, D. J., Yancey, S. B., and Revel, J.-P., 1981, Intercellular communication in normal and regenerating rat liver: A quantitative analysis, *J. Cell Biol.* **91:**505–523.

Meza, I., Ibarra, G., Sabanero, M., Martinez-Palomo, A., and Cereijido, M., 1980, Occluding junctions and cytoskeletal components in a cultured transporting epithelium, *J. Cell Biol.* **87:**746–754.

Millonig, G., and Giudice, G., 1967, Electron microscope study of the reaggregation of cells dissociated from sea urchin embryos, *Dev. Biol.* **15:**91–101.

Miyazaki, S.-I., Takahashi, K., Tsuda, K., and Yoshii, M., 1973, Analysis of non-linearity observed in the current voltage response of the tunicate embryo, *J. Physiol. (Lond.)* **238:**55–77.

Montesano, R., Friend, D. S., Perrelet, A., and Orci, L., 1975, *In vivo* assembly of tight junctions in fetal rat liver, *J. Cell Biol.* **67:**310–319.

Moore, A. R., 1930, On the invagination of the gastrula, *Protoplasma* **9:**25–33.

Moran, D., and Rice, R. W., 1976, Action of papaverine and ionophore A23187 on neurulation, *Nature (Lond.)* **261:**497–499.

Moskalewski, S., Thyberg, J., and Friberg, U., 1976, *In vitro* influence of colchicine on the Golgi complex in A- and B-cells of guinea pig pancreatic isles, *J. Ultrastruct. Res.* **54:**304–317.

Muir, A. R., 1967, The effects of divalent cations on the ultrastructure of the perfused rat heart, *J. Anat.* **101:**239–262.

Muir, A. R., and Peters, A., 1962, Quintuple-layered membrane junctions at terminal bars between endothelial cells, *J. Cell Biol.* **12:**443–448.

Nardi, J. B., 1981, Induction of invagination in insect epithelia: Paradigm for embryonic invagination, *Science* **214:**564–566.

New, D. A. T., 1959, The adhesive properties and expansion of the chick blastoderm, *J. Embryol. Exp. Morphol.* **7:**146–164.

Nuccitelli, R., and Erickson, C. A., 1983, Embryonic cell motility can be guided by physiological electric fields, *Exp. Cell Res.* **147:**195–201.

Odell, G. M., Oster, G., Alberch, P., and Burnside, B., 1981, The mechanical basis of morphogenesis. I. Epithelial folding and invagination, *Dev. Biol.* **85:**446–462.

Odland, G. F., 1958, The fine structure of the interrelationship of cells of the human epidermis, *J. Biophys. Biochem. Cytol.* **4:**529–538.

Odland, G., and Ross, R., 1968, Human wound repair. I. Epidermal regeneration, *J. Cell Biol.* **39:**135–151.

Orida, N., and Feldman, J. D., 1982, Directional protrusive pseudopodial activity and motility in macrophages induced by extracellular electrical fields, *Cell Motil.* **2:**243–255.

Osborn, M., Geisler, N., Shaw, G., Sharp, G., and Weber, K., 1982, Intermediate filaments, *Cold Spring Harbor Symp. Quant. Biol.* **46:**413–429.

Overton, J., 1962, Desmosome development in normal and reassociating cells, *Dev. Biol.* **4:**532–548.

Overton, J., 1974, Cell junctions and their development, *Prog. Surface Membrane Sci.* **8:**161–208.

Overton, J., 1977, Formation of junctions and cell sorting in aggregates of chick and mouse cells, *Dev. Biol.* **55:**103–116.

Overton, J., and DeSalle, R., 1980, Control of desmosome formation in aggregating embryonic chick cells, *Dev. Biol.* **75:**168–176.

Overton, J., and Shoup, J., 1964, Fine structure of cell surface specializations in the maturing duodenal mucosa of the chick, *J. Cell Biol.* **21:**75–85.

Plamer, J. F., and Slack, C., 1970, Some bio-electric parameters of early *Xenopus* embryos, *J. Embryol. Exp. Morphol.* **24:**535–553.

Pearce, T. L., and Zwaan, J., 1970, A light and electron microscopic study of cell behavior and microtubules in the embryonic chick lens using colcemid, *J. Embryol. Exp. Biol.* **23:**491–507.

Penn, R. D., 1966, Ionic communication between liver cells, *J. Cell Biol.* **29:**171–173.

Piatigorsky, J., Webster, H. D., and Wollberg, M., 1972, Cell elongation in the cultured embryonic chick lens epithelium with and without protein synthesis. Involvement of microtubules, *J. Cell Biol.* **55:**82–92.

Pictet, R. L., Clark, W. R., Williams, R. H., and Rutter, W. J., 1972, An ultrastructural analysis of the developing embryonic pancreas, *Dev. Biol.* **29:**436–467.

Pisam, M., and Ripoche, P., 1976, Redistribution of surface macromolecules in dissociated epithelial cells, *J. Cell Biol.* **71:**907–920.

Plagemann, P. G. W., Wohlhueter, R. M., Graff, J. C., and Marz, R., 1978, Inhibition of carrier-mediated and non-mediated permeation processes by cytochalasin B, in: *Cytochalasins—Biochemical and Cell Biological Aspects,* pp. 445–473, Elsevier/North-Holland, New York.

Politoff, A. L., 1977, Protein semiconduction: an alternative explanation of electrical coupling, in: *Intercellular Communication* (W. C. DeMello, ed.), pp. 127–143, Plenum Press, New York.

Poo, M.-M., and Robinson, K. R., 1977, Electrophoresis of concanavalin A receptors along embryonic muscle cell membrane, *Nature (Lond.)* **265:**602–605.

Poodry, C. A., and Schneiderman, H. A., 1971, Intercellular adhesivity and pupal morphogenesis in *Drosophila melanogaster,* *Arch. Entwicklungsmech. Org.* **168:**1–9.

Potter, D. D., Furshpan, E. J., and Lennox, E. S., 1966, Connections between cells of the developing squid as revealed by electrophysiological methods, *Proc. Natl. Acad. Sci.* **55:**328–336.

Radice, G. P., 1980, The spreading of epidermal cells during wound closure in *Xenopus* larvae, *Dev. Biol.* **76:**26–46.

Rambourg, A., and Leblond, C. P., 1967, Electron microscope observations on the carbohydrate-rich cell coat present at the surface of cell in the rat, *J. Cell Biol.* **32:**27–53.

von Recklinghausen, 1862, *Die Lymphgefässe und ihre Beziehung zum Bindegewebe,* August Hirschwald, Berlin.

Reese, T. S., and Karnovsky, M. J., 1967, Fine structural localization of a blood-brain barrier to exogenous peroxidase, *J. Cell Biol.* **34:**207–217.

Rhumbler, L., 1902, Zur Mechanik des Gastrulationsvorgänges, insbesonders der Invagination. Eine entwicklungsmechanische Studie, *Arch. Entwicklungsmech. Org.* **14:**401–476.

Richman, D. P., Stewart, R. M., Hutchinson, J. W., and Cavinoss, V. S., Jr., 1975, Mechanical model of brain convolution development. Pathologic and experimental data suggest a model based on differential growth within the cerebral cortex, *Science* **189:**18–21.

Rodriguez Boulan, E., and Sabatini, D. D., 1978, Asymmetric budding of viruses in epithelial monolayers: A model system for study of epithelial polarity, *Proc. Natl. Acad. Sci. USA* **75:**5071–5075.

Roth, S., 1983, The biochemistry of cell adhesion in vertebrates, in: *Cell Interactions and Development* (K. Yamada, ed.), pp. 77–98, Wiley, New York.

Roux, W., 1885, Beitrag zur Entwicklungsmechanik des Embryo, *Z. Biol.* **21**:411–526.

Rutishauser, U., Yahara, I., and Edelman, G. M., 1974, Morphology, motility and surface behavior of lymphocytes bound to nylon fibers, *Proc. Natl. Acad. Sci. USA* **71**:1149–1153.

Sanders, E. J., and Zalik, S. E., 1972, The blastomere periphery of *Xenopus laevis*, with special reference to intercellular relationships, *Arch. Entwicklungsmech. Org.* **171**:181–194.

Schaeffer, H. E., Schaeffer, B. E., and Brick, I., 1973, Effects of cytochalasin B on the adhesion and electrophoretic mobility of amphibian gastrula cells, *Dev. Biol.* **34**:163–168.

Schlafke, S., and Enders, A. C., 1967, Cytological changes during cleavage and blastocyst formation in the rat, *J. Anat.* **102**:13–32.

Schlesinger, A. B., 1958, The structural significance of the avian yolk in embryogenesis, *J. Exp. Zool.* **138**:223–258.

Sedar, A. W., and Forte, J. G., 1964, Effects of calcium depletion on the junctional complex between oxyntic cells of gastric glands, *J. Cell Biol.* **22**:173–188.

Shain, W. G., Hilfer, S. R., and Fonte, V. G., 1972, Early organogenesis of the embryonic chick thyroid. I. Morphology and biochemistry, *Dev. Biol.* **28**:202–218.

Sheridan, J. D., 1968, Electrophysiological evidence for low-resistance intercellular junctions in the early chick embryo, *J. Cell Biol.* **37**:650–659.

Smuts, M. S., Hilfer, S. R., Searls, R. L., 1978, Patterns of cellular proliferation during thyroid organogenesis, *J. Embryol. Exp. Morphol.* **48**:269–286.

Spooner, B. S., and Wessells, N. K., 1970, Effects of cytochalasin B upon microfilaments involved in morphogenesis of salivary epithelium, *Proc. Natl. Acad. Sci. USA* **66**:360–364.

Spratt, N. T., Jr., 1963, Role of the substratum, supracellular continuity, and differential growth in morphogenetic cell movements, *Dev. Biol.* **7**:51–63.

Staehelin, L. A., 1974, Structure and function of intercellular junctions, *Int. Rev. Cytol.* **39**:191–283.

Stern, C. D., 1982, Experimental reversal of polarity of chick embryo epiblast sheets *in vitro*, *Exp. Cell Res.* **140**:468–471.

Stump, R. F., and Robinson, K. R., 1982, Directed movement of *Xenopus* embryonic cells in an electric field, *J. Cell Biol.* **95**:331a.

Sugrue, S. P., and Hay, E. D., 1981, Response of basal epithelial cell surface and cytoskeleton to solubilized extracellular matrix molecules, *J. Cell Biol.* **91**:45–54.

Sun, T.-T., Shih, C., and Green, H., 1979, Keratin cytoskeletons in epithelial cells of internal organs, *Proc. Natl. Acad. Sci. USA* **76**:2813–2817.

Takeuchi, S., 1979, Wound healing in the cornea of the chick embryo. IV. Promotion of the migratory activity of isolated corneal epithelium in culture by the application of tension, *Dev. Biol.* **70**:232–240.

Thompson, D. W., 1945, *On Growth and Form*, Cambridge University Press, Cambridge.

Tilney, L. G., and Gibbins, J. R., 1969, Microtubules in the formation and development of the primary mesenchyme in *Arbacia punctalata*. II. An experimental analysis of their role in development and maintenance of cell shape, *J. Cell Biol.* **41**:227–250.

Trelstad, R. L., Hay, E. D., and Revel, J.-P., 1967, Cell contact during early morphogenesis in the chick embryo, *Dev. Biol.* **16**:78–106.

Trinkaus, J. P., 1951, A study of the mechanism of epiboly in the egg of *Fundulus heteroclitus*, *J. Exp. Zool.* **118**:269–319.

Trinkaus, J. P., 1976, On the mechanisms of metazoan cell movements, in: *The Cell Surface in Animal Embryogenesis and Development* (G. Poste and G. L. Nicolson, eds.), pp. 225–329, North-Holland, Amsterdam.

Trinkaus, J. P., 1982, Movement of clusters of pigmented epithelial cells *in vivo* during early morphogenesis of the teleost *Blennius pholis*, *J. Cell Biol.* **95**:325a.

Trinkaus, J. P., 1984a, Mechanism of *Fundulus* epiboly—A current view, *Am. Zool.* **24**:673–688.

Trinkaus, J. P., 1984b, *Cells into Organs: The Forces that Shape the Embryo*, 2nd ed., Prentice-Hall, Englewood Cliffs, New Jersey.

Trinkaus, J. P., and Lentz, T. L., 1967, Surface specializations of *Fundulus* and their relation to cell movements during gastrulation, *J. Cell Biol.* **32:**139–153.

Tsuchiya, W., Okada, Y., Yano, J., Inouye, A., Sasaki, S., and Doida, Y., 1981, Effects of cytochalasin B and local anesthetics on electrical and morphological properties of L cells, *Exp. Cell Res.* **133:**83–92.

Tupper, J. T., Saunders, J. W., Jr., and Edwards, C., 1970, The onset of electrical communication between cells in the developing starfish embryo, *J. Cell Biol.* **46:**187–191.

Van Blerkom, J., Manes, C., and Daniel, J. C., Jr., 1973, Development of preimplantation rabbit embryos *in vivo* and in vitro, *Dev. Biol.* **35:**262–282.

Waddington, C. H., and Perry, M. M., 1966, A note on the mechanism of cell deformation in the neural fold of the amphibia, *Exp. Cell Res.* **41:**691–693.

Warner, A. E., and Lawrence, P. A., 1973, Electrical coupling across developmental boundaries in insect epidermis, *Nature (Lond.)* **245:**47–48.

Warner, A. E., and Lawrence, P. A., 1982, Permeability of gap junctions at the segmental border in insect epidermis, *Cell* **28:**243–252.

Wartiovaara, J., 1966, Cell contacts in relation to cytodifferentiation in metanephrogenic mesenchyme *in vitro*, *Ann. Med. Exp. Biol. Fenn.* **44:**1–35.

Wessells, N. K., 1970, Mammalian lung development: Interactions in formation and morphogenesis of tracheal buds, *J. Exp. Zool.* **175:**455–466.

Wessells, N. K., and Evans, J., 1968, Ultrastructural studies of early morphogenesis and cytodifferentiation in the embryonic mammalian pancreas, *Dev. Biol.* **17:**413–436.

Wiley, L. M., 1982, Cavitation and nascent blastocoele formation mouse preimplantation embryos: Na/K-ATPase and cytoplasmic polarity, *J. Cell Biol.* **95:**148a.

Wilgram, G., Caulfield, J. B., and Madgic, E. B., 1964, A possible role of the desmosome in the process of keratinization. An electron-microscopic study of acantinolysis and dyskeratosis, in: *The Epidermis* (W. Montagna and W. C. Lobitz, Jr., eds.), pp. 275–301, Academic Press, New York.

Wiseman, L. L., and Strickler, J., 1981, Desmosome frequency: Experimental alteration may correlate with differential cell adhesion, *J. Cell Sci.* **49:**217–223.

Wolpert, L., and Mercer, E. H., 1963, An electron microscope study of the development of the blastula of the sea urchin embryo and its radial polarity, *Exp. Cell Res.* **30:**280–300.

Wood, R. L., 1959, Intercellular attachment in the epithelium of *Hydra* as revealed by electron microscopy, *J. Biophys. Biochem. Cytol.* **6:**343–352.

Wood, R. L., 1963, An electron microscope study of developing bile canaliculi in the rat, *Anat. Rec.* **151:**507–530.

Wrenn, J. T., and Wessells, N. K., 1969, An ultrastructural study of lens invagination in the mouse, *J. Exp. Zool.* **171:**359–367.

Wrenn, J. T., and Wessells, N. K., 1970, Cytochalasin B: Effects upon microfilaments involved in morphogenesis of estrogen-induced glands of oviduct, *Proc. Natl. Acad. Sci. USA* **66:**904–908.

Yee, A. G., and Revel, J.-P., 1978, Loss and reappearance of gap junctions in regenerating liver, *J. Cell Biol.* **78:**554–564.

Zackroff, R. V., Steinert, P., Aynardi Whitman, M., and Goldman, R. D., 1981, Intermediate filaments, *Cell Surface Rev.* **7:**57–97.

Zwaan, J., and Hendrix, R. W., 1973, Changes in cell and organ shape during early development of the ocular lens, *Am. Zool.* **13:**1039–1049.

Chapter 5

Endogenous Lectins and Cell Adhesion in Embryonic Cells

SARA E. ZALIK and NADINE C. MILOS

1. Introduction

In his classic studies on the amphibian blastula and gastrula, Holtfreter (1939) showed that the segregation of the embryonic germ layers taking place during the morphogenetic movements of gastrulation is accompanied by a change in the mutual affinities of the cells that constitute the presumptive germ layers. That these affinities were the result of discriminatory properties of the cells themselves was shown in later studies, by Townes and Holtfreter (1955). It was found, in recombination experiments, that dissociated cells from different regions of the amphibian gastrula would first adhere to each other in an indiscriminate fashion and then sort out into cell groups, each consisting of a similar cell type. These workers hypothesized that changes in the selectivity of cellular adhesion are due to alterations at the cell surface (see Fig. 27 in Townes and Holtfreter, 1955). Since that time, investigators have devoted considerable time and effort to studying the factors involved in cellular recognition.

The plasma membrane is currently regarded as an asymmetric fluid mosaic composed of lipids, phospholipids, and glycolipids arranged mostly in a bilamellar configuration. The lipids, being amphipatic molecules, are oriented with their polar groups facing the aqueous environment and their hydrophobic groups facing the interior of the bilayer (see Cook and Stoddart, 1973). Proteins and glycoproteins are also constituents of the plasma membrane, some of which span the lipid bilayers to varying degrees, depending on the extent of the interactions of their nonpolar groups with the internal hydrophobic environment of the bilayer. Other proteins and glycoproteins are associated only with the external polar regions of the bilayer *via* electrostatic interactions with the charged groups of the lipids; see Malhotra (1983) for a recent review of the subject. The carbohydrate groups of the glycolipids and glycoproteins are localized almost exclusively at the external face of the plasma membrane. There

SARA E. ZALIK • Department of Zoology, University of Alberta, Edmonton, Alberta T6G 2E9, Canada. NADINE C. MILOS • Department of Anatomy, University of Alberta, Edmonton, Alberta T6G 2H7, Canada.

is now evidence to support the idea that these cell surface saccharides play an important role in cellular recognition (see Cook and Stoddart, 1973; Hughes, 1976; Lackie, 1980; Sharon and Lis, 1981).

Oligosaccharides can exist in nature in many different structural forms, making them good candidates for carriers of biological information. The most common hexoses have three or four hydroxyl groups of almost equal chemical reactivity (potential sites for branch points), as well as two potential anomeric configurations, α and β. In oligosaccharides, overall structural information can reside in the number and position on the monomers, their anomeric configuration, their attached groups, and the occurrence of branch points, whereas in peptides information resides mainly in the number and position of the monomeric units (Sharon and Lis, 1981). It has been calculated that dipeptides and tripeptides of the same amino acid can exist in only one possible structural configuration, whereas the corresponding saccharide homopolymers can exist, respectively, in 11 and 176 different isomeric forms, respectively (Clamp, 1974). As a consequence of this, saccharide polymers carry more information per weight than do proteins (Sharon and Lis, 1981). Since carbohydrates are abundant at the cell surface, they have, therefore, emerged as possible candidates for playing a role in chemical sensing of the pericellular environment, as well as being responsible for a variety of cell interactions, among them cell-to-cell adhesion.

The idea that cell–cell interactions take place *via* the interlocking of complementary macromolecules present on apposing cells was first advanced by Tyler (1940, 1946) and Weiss (1947, 1950). Interactions between two proteins such as the antigen–antibody type were first suggested as mediators of the cellular adhesion between egg and sperm that occurs during fertilization (Tyler, 1946). Recent evidence, however, suggests that cell–cell and cell–substrate interactions take place between a protein and the carbohydrate portion of a glycoprotein or glycolipid (Hughes, 1976). These may involve the glycosyltransferase enzymes and the growing carbohydrate chain of the nascent glycoprotein (Roseman, 1970; Shur, 1982; Rauvala *et al.*, 1983). Alternatively, cell surface glycosidases, under conditions in which their lytic activity is inhibited but their carbohydrate-binding activity is retained, may also be molecular mediators in the formation of adhesions (Rauvala *et al.*, 1980; Hakomori, 1981).

This chapter deals with a third class of carbohydrate-binding molecules as candidates for intercellular adhesive bridges: the carbohydrate-binding proteins, or lectins, and their receptors, the carbohydrate portion of glycoproteins, proteoglycans, or glycolipids at the cell surface. The term lectin (from the Latin term *legere*: to select) was first introduced by Boyd and Shapleigh (1954a,b) to denote compounds present in some plant extracts that specifically agglutinate erythrocytes from different human blood groups and form precipitates with the saliva of secretor individuals of the same group. This term is now applied to a variety of carbohydrate-binding proteins of prokaryotic and eukaryotic origin that bear at least two carbohydrate-binding sites for a particular saccharide. As a consequence of this property, lectins are able to precipitate glycoconjugates and agglutinate cells, including erythrocytes of diverse vertebrate species (Goldstein and Hayes, 1978; Goldstein *et al.*, 1980). A lectin, by definition, should be of nonimmune origin and have no enzymatic activity under the operational conditions employed. This distinction may not be quite precise, since recent studies indicate that some plant and microbial lectins are endowed with glycosidase (Hankins and Shannon, 1978; Hankins *et al.*, 1980; Del Campillo *et al.*, 1981; Dey *et al.*, 1982, 1983) and protease activity (Finkelstein *et al.*, 1983), and some lectins from arthropods are glycosyltransferases (Vasta and Marchalonis, 1983). The remarkable ability of lectins to discriminate between specific monosaccharides is sometimes refined further, to the extent that some lectins distinguish between different isomeric configurations and specific linkages in a disaccharide. It is not surprising, therefore, that plant lectins have been used as probes for the elucidation of the chemical architecture of the cell surface (Cook and Stoddart, 1973; Nicolson, 1974).

Interest in studying a role of endogenous carbohydrate-binding proteins in cell interactions arose from the work of Rosen *et al.* (1973) with the differentiating slime mold *Dictyostelium discoideum*. These organisms exist in the vegetative state as single ameboid cells. After removal of the food source, the cells develop the ability to adhere to each other and form a multicellular aggregate with numerous intercellular contacts. Lectin activity was found to increase drastically in extracts obtained from aggregating cells, whereas extracts of vegetative cells were devoid of agglutinating activity. The lectin activity was inhibited by sugars bearing a D-galactose configuration, and affinity chromatography on a galactose-containing matrix was used to isolate the lectin. For further details on the slime mold lectin, the reader is referred to Frazier and Glaser (1979), Barondes (1981, 1984), and Bartles *et al.* (1982).

Research on slime molds stimulated a great deal of interest in the search for lectin–saccharide interactions in diverse biological processes involving specific recognition; these include bacterial attachment in *Rhizobium*–legume symbiosis (Dazzo, 1980; Dazzo and Truchet, 1983); mating interactions in cells of lower eukaryotes such as *Chlamydomonas* (Goodenough and Adair, 1980; Goodenough *et al.*, 1980; Adair *et al.*, 1982, 1983), *Fucus* (Bolwell *et al.*, 1979, 1980), and yeast (Crandall, 1978; Yanagishima and Yoshida, 1981); and animal fertilization (Vacquier, 1980; Glabe *et al.*, 1982). Other cell interactions involving carbohydrate-binding proteins include adhesion of pathogens to animal cell surfaces (Mirelman and Kobiler, 1981; Sharon *et al.*, 1981), the recognition and clearance of plasma glycoproteins from the circulation, and uptake of lysosomal hydrolases by fibroblasts (Neufeld and Ashwell, 1980; Ashwell and Harford, 1982; Sly, 1982; Monsigny *et al.*, 1983).

In this chapter, we limit our discussion to the evidence suggesting that lectins play a role in cellular adhesion of early embryonic cells. We shall deal with instances of cellular interactions affected by endogenous lectins in other systems only in cases where these studies could help us gain some insight into how lectins may be involved—at the molecular level—in the adhesive cell interactions that occur during early development.

2. Strategies for the Detection of Lectin Activity

2.1. Direct Evaluation

In this approach, extracts presumed to contain lectin activity or the purified lectin are allowed to interact directly with the *same cells* they are purported to affect *in vitro*. In most cases, the lectin extract is mixed with the cell suspension and is incubated for a certain time, usually at room temperature. The agglutination of the cells is then assessed either visually, microscopically, or by a variety of physical procedures such as changes in light scattering or detection of the decrease in single cells as measured with an electronic particle counter. This direct approach has been used to detect the sperm lectin involved in sea urchin fertilization (Vacquier and Moy, 1977), the sexual agglutinins of *Chlamydomonas* (Wiese and Wiese, 1978; Adair *et al.*, 1982) and yeast (Yanagishima and Yoshida, 1981), and the clover lectin that agglutinates *Rhizobium* (Dazzo and Brill, 1977; Dazzo *et al.*, 1978).

2.2. Indirect Evaluation

This approach uses intermediary indicator cells, usually erythrocytes from a variety of mammalian species. The erythrocyte is considered a relatively inert carrier of diverse classes of carbohydrate ligands, which are displayed at its surface, and act as receptors for the lectin(s) present in solution; the result of this interaction is an agglutination reaction that can be distinguished visually. The erythrocytes can be used fresh or may be sensitized by treatment with formaldehyde or with proteases such as trypsin (Nowak *et al.*, 1976). This enzyme probably enhances the agglutinability of the erythrocytes either by uncovering a lectin receptor or by allowing it a greater mobility within the cell membrane. Trypsinized erythrocytes can be used fresh or may be stabilized either by fixation with glutaraldehyde (Nowak *et al.*, 1976) or by suspension in azide (Zalik *et al.*, 1983).

In the simplest, most popular hemagglutination assay, the extract is diluted serially in the wells of a microtitration plate, mixed with a standard suspension of erythrocytes, shaken vigorously, and incubated for a suitable time interval. When erythrocytes agglutinate, they form a mat of diffuse outline covering the bottom of the well; those that do not agglutinate settle to form a distinctly delineated dot at the center of the well (Fig. 1). Agglutination is considered positive until a dilution of the extract is obtained at which the erythrocytes settle as a dot instead of a mat. The reciprocal of the highest dilution giving agglutination is the hemagglutination titer. When the protein concentration is determined, the specific activity of an extract in hemagglutination units (HU) per milligram protein can be determined. Titrations are usually performed in duplicate or triplicate, but because the lectin extract is serially diluted, only a semiquantitative measure of agglutination is obtained (Rosen *et al.*, 1973). More quantitative data can be collected if the number of nonaggluti-

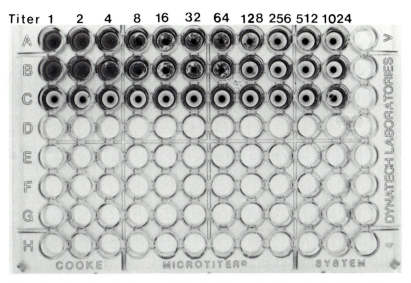

Figure 1. Titration plate showing hemagglutination titers of lectin extracts obtained from chick blastoderms. The rabbit erythrocytes used in these titrations have been trypsinized and stabilized in sodium azide. For titration, lectin extracts (25 μl) were serially diluted with 25 μl saline. This was followed by the addition of 0.5% bovine serum albumin in saline (25 μl). The final addition consisted of 25 μl of a 4% (v/v) suspension of erythrocytes followed by vigorous shaking. The titer in row A is 128, that of row B is 64; row C contains saline instead of lectin extract and serves as the control. Observe the clear dots formed by the erythrocyte suspension in row C and the diffuse mats formed in the presence of the lectin extracts in rows A and B.

nated erythrocytes is determined with a particle counter (Barondes *et al.*, 1978) or if the hemagglutination assay is assessed by changes in absorbance at 620 nm (Lis and Sharon, 1972). To determine the saccharide specificity of a particular lectin, different sugars or compounds bearing different terminal saccharide groups are incorporated into the medium used for the dilution of the lectin. The compound that inhibits hemagglutination at the lowest concentration is considered the one for which the lectin has the highest affinity.

An extension of this approach is the use of the rosette assay for determination of the carbohydrate-binding activity at the cell surface. If erythrocytes bind to the cells in such a manner as to resist all further manipulative procedures, a rosette (a cluster of erythrocytes surrounding the central cell) is said to have formed (Elliot *et al.*, 1974). If the binding of the erythrocytes by the test cell is inhibited or reversed by particular saccharides or glycoconjugates, it is assumed that rosette formation resulted from the interaction of a carbohydrate-binding protein at the surface of the test cell with a carbohydrate ligand displayed at the erythrocyte cell surface (Grabel *et al.*, 1979). A rosette is usually considered positive when a minimum of three or more erythrocytes are bound by the test cell. In this approach, care must be taken to conduct control experiments for the presence of carbohydrate-binding activity at the erythrocyte surface, since some surface proteins of red blood cells have been reported to display lectin activity (Bowles and Hanke, 1977).

More recently, inert carriers, coated with particular saccharides or glycoproteins, have also been used to detect saccharide-binding activity of cell surfaces. These include agarose, acrylamide, or latex beads. Particles of colloidal gold, coupled with particular sugars or glycoproteins, can also be used as tracers to detect carbohydrate-binding activity at the electron microscopic level (Kolb-Bachofen, 1981; Schlepper-Schäfer et al., 1981; Kolb-Bachofen et al., 1982).

When detecting lectin activity in tissue extracts by conventional hemagglutination assays, one should bear in mind that lectin binding by a particular type of erythrocyte may not necessarily indicate that this lectin mediates an event occurring at the cell surface in its tissue of origin. Unfortunately, many adult and embryonic animal lectins have been detected using the indirect approach, and there is little information regarding the biological significance of these molecules. Nevertheless, there are instances, such as that of the hepatic-binding protein of the mammalian liver, where the carbohydrate-binding activity involved in the clearance of proteins from the plasma is affected by some of the same factors that affect its hemagglutinating activity (Stockert et al., 1974). This suggests that, at least in some instances, the inference that a lectin detected by a hemagglutination assay has a role to play in an event involving cell surface carbohydrates may be justified.

Subsequent to its detection in tissue extracts, lectin activity can be purified by conventional chemical techniques such as differential precipitation, gel filtration, or affinity chromatography on sugars or glycoproteins immobilized on agarose or acrylamide beads, or on particular immunoabsorbants carrying specific sugar determinants. Formalinized erythrocytes have also been used as affinity adsorbants for lectin purification (Reitherman et al., 1973).

3. Animal Lectins: General Overview

From an operational point of view animal lectins can be divided into two main groups: (1) lectins that are either present normally in solution in the body fluids or are readily solubilized under physiological conditions of pH and ionic strength; and (2) lectins that require more drastic conditions for solubilization such as exposure of cells to media of high ionic strength, extremes of pH, presence of denaturing agents such as urea or guanidine hydrochloride, or treatment with detergents. Among the lectins that can be placed in the first group are those present in the body fluids of invertebrates such as gastropods, annelids, echinoderms, and arthropods (Yamada and Aketa, 1982; Komano et al., 1983; Vasta and Marchalonis, 1983); the galactoside-binding tissue lectins from sponges (Bretting et al., 1981a,b; Müller et al., 1983b), coelenterates (Koch et al., 1982), larvae, and pupae of some insects (Komano et al., 1983); the galactoside-binding lectins from the adult mammalian liver, spleen, heart, lung, and mammary gland (Barak Briles et al., 1979; Childs and Feizi, 1979; Powell, 1980; Powell and Whitney, 1980; Takeuchi et al., 1982); the galactoside-binding lectins of adult and embryonic avian intestine, liver, muscle, heart, brain, and retina (Kobiler and Barondes, 1977; Eisenbarth et al., 1978;

Mir-Lechaire and Barondes, 1978; Beyer *et al.*, 1980); the erythroid developmental agglutinin of the mammalian bone marrow (Harrison and Chesterton,
1980*a*); and the lectin from the epidermis of embryonic chick skin (Kitamura,
1981). In early embryos, readily solubilized lectin activity has been reported for
the mannan-binding lectin from cleaving embryos of *Rana pipiens* (Roberson
and Armstrong, 1980), the galactoside-binding lectin of the pregastrula and
gastrula chick blastoderm (Cook *et al.*, 1979; Zalik *et al.*, 1983), that of the eggs
and embryos of *Xenopus laevis* (Roberson and Barondes, 1982; Harris and
Zalik, 1982), and the galactose-sensitive lectin of the sea urchin egg (Sasaki and
Aketa, 1981). Further details on soluble animal lectins can be found in a recent
review (Barondes, 1984).

Examples of lectins belonging to group 2 are: bindin, the fucan-binding
lectin of the acrosome of the sea urchin sperm (Vacquier and Moy, 1977; Glabe
et al., 1982); the lectins of the mammalian and avian hepatocyte that bind
galactose and N-acetyl-glucosamine, respectively (Kawasaki and Ashwell,
1976; Lunney and Ashwell, 1976); the lectins from the membranes of platelets,
lymphocytes, and fibroblasts (Dysart and Edwards, 1977; Bowles and Rotman,
1978; Kieda *et al.*, 1978; Roff *et al.*, 1983); and the mannan-binding lectin of the
vitelline envelope of the early chick embryo (Rutherford and Cook, 1981, 1984).

It is interesting to note that some of the proteins of the extracellular matrix
implicated in cellular adhesion, such as fibronectin and laminin, also display
hemagglutinating activity. The fibronectin-induced agglutination of formalinized sheep erythrocytes is inhibited by amino sugars (Yamada *et al.*,
1975). Laminin agglutinates human and rabbit erythrocytes that have been
trypsinized and fixed with glutaraldehyde (Kennedy *et al.*, 1983; Ozawa *et al.*,
1983), as well as formalinized sheep erythrocytes (Kennedy *et al.*, 1983). There
is some disagreement as to the saccharide specificity of laminin. Using sheep
erythrocytes as the test system, its hemagglutinating activity is inhibited by
gangliosides and phosphatidic acid, whereas with rabbit erythrocytes, heparin,
and heparan sulfate are the best inhibitors (Ozawa *et al.*, 1983). It has been
suggested that laminin may recognize a specific molecular configuration of a
negative charge in a lipid or an oligosaccharide (Kennedy *et al.*, 1983).

3.1. Invertebrate Lectins

The presence of sperm agglutinating activity in sea urchin egg extracts was
recognized by Lillie early in this century (Lillie, 1913). It is now known that the
agglutination observed by Lillie was due to the release of bindin from the
acrosomal granule, induced by a component from the egg jelly (Vacquier,
1980). Since then, agglutinating activity has been found widely distributed in
tissue extracts and the body fluids of invertebrates.

3.1.1. Adult

The first isolation of an invertebrate lectin was achieved by Marchalonis
and Edelman (1968) with the purification of the sialic acid-binding lectin of the

horseshoe crab. At the same time, lectin activity was reported by Dodd *et al.* (1968) and subsequently by Khalap *et al.* (1970) in tissue homogenates of snails and marine sponges. This area of research has since expanded considerably, resulting in the discovery of lectins in the hemolymph and tissue extracts of a wide variety of invertebrates. The lectins found in the body fluids of invertebrates have been postulated to play a role similar to that of vertebrate immunoglobulins, i.e., self–non-self recognition (Vasta and Marchalonis, 1983). It is interesting to note that a large proportion of the lectins from tissue extracts have a specificity for sugars displaying a D-galactoside configuration. The β-D-galactoside-binding lectins of the marine sponges *Geodia cydonium* and *Axinella polypoides* and of the coelenterate *Eunicella cavolinii* have been hypothesized to play a role either in cellular adhesion (Müller and Müller, 1980; Müller, 1982; Müller *et al.*, 1983*a*) or in the organization of the spongin fibers of the extracellular matrix (Bretting *et al.*, 1983). A detailed discussion of the role of sponge lectins will be found in Section 5.1.

3.1.2. Embryonic

In developing systems of invertebrates, lectins have been isolated from the hemolymph of the larvae and pupae of insects. They have been postulated to be important during metamorphosis and to play a role in the stimulation of phagocytic activity of hemocytes for the clearance of debris from digested tissue. A β-D-galactoside-binding lectin has been isolated from the eggs of the sea urchin *Anthocidaris crassispina* (Table I). Immunofluorescence has been used to study the localization of this lectin; in the unfertilized egg, it appears to be distributed at random in the cytoplasm and to become localized in the egg cortex after fertilization. In other species of sea urchin, Oppenheimer and Meyer (1982) and Tonegawa (1982) found that the media in which embryos have been dissociated contain lectin activity specific, respectively, for galactosides and for endogenous glycoproteins bearing terminal mannose; however, the actual protein has not been characterized. Information on selected examples of invertebrate lectins is shown in Table I. The roles of lectins in early sea urchin development are discussed further in Chapter 6.

3.2. Vertebrate Lectins

As is also the case with invertebrate lectins, the majority of the experiments in which vertebrate lectins have been studied have made use of the indirect approach using, as indicators, normal or trypsinized erythrocytes from rabbit, sheep, or human origin. One of the first lectins to be isolated was that present in eel serum (Springer and Desai, 1971; Bezkoravainy *et al.*, 1971). Subsequently, Teichberg *et al.* (1975) reported the presence of a soluble lectin specific for β-D-galactosides in the electric organ and swimming muscle of the electric eel, in muscle of the embryonic and adult chicken, in a variety of rat tissues such as muscle, lung, kidney, lymphocytes, brain, and liver, and in several murine cell lines. Since then, the presence of a lectin with similar

Table I. Selected Examples of Invertebrate Lectins[a]

Source	Saccharide specificity	Nature (M_r)	Suggested function	Reference
Sponges				
1. *Geodia cydonium* tissue extracts; Lectin I	β-D-Galactose	Glycoprotein (15,000)	Self–non-self-recognition, sponge–symbiont interaction, synthesis of spongin fibers	Bretting et al. (1981)
		Trimer (13,800, 13,000, 12,200)	Cell adhesion	Müller et al. (1983b)
2. *Axinella polypoides* tissue extracts	β-D-Galactose	Protein Agglutinin I (21,000) Agglutinin II (15,500)	Synthesis of ground substance and spongin fibers	Bretting and Kabat (1976); Bretting et al. (1981a)
Coelenterates				
3. *Cerianthus membranaceus* (cerianthin) tissue extracts	D-Galactose	Protein (38,400)	Symbiosis, cell contact	Koch et al. (1982)
Arthropods				
4. *Limulus polyphemus* hemolymph	Sialic acid	Protein (22,500) (19,000)	Self–non-self-recognition	Marchalonis and Edelman (1968); Vasta and Marchalonis (1983) Roche and Monsigny (1974)

(continued)

Table I. (*Continued*)

Source	Saccharide specificity	Nature (M_r)	Suggested function	Reference
6. *Sarcophaga peregrina* hemolymph and fat body of larvae and pupae	D-Galactose	Protein subunits α (32,000) β (30,000)	Defence, humoral reaction, clearance of digested tissue during metamorphosis	Komano et al. (1983)
6′. *Bombyx mori* hemolymph of larvae and pupae	Glucuronic acid, heparin	Protein (260,000)	Metamorphosis, activation of phagocytes	Suzuki and Natori (1983)
Echinoderms				
7. *Strongylocentrotus purpuratus* sperm	Polymers of sulfated fucose, xylan β(-1→4)	Protein (30,500)	Adhesion of sperm to egg	Vacquier and Moy (1977); Glabe et al. (1982)
8. *Hemicentrotus pulcherimus* seminal plasma	L-Arabinose, D-xylose, D-galactose	Glycoprotein (140,000)	—	Yamada and Aketa (1982)
9. *Anthocidaris crassispina* eggs	Thiodigalactoside, D-galactose	Protein (11,500)	Fertilization, block of polyspermy	Sasaki and Aketa (1981)
10. *Strongylocentrotus purpuratus* blastula; suspension medium of dissociated cells	D-Galactose, N-acetyl-D-galactosamine	Protein not characterized	Adhesion	Oppenheimer and Meyer (1982)

[a]These lectins have been chosen to illustrate the wide distribution of these molecules among the invertebrates. The lectin of *Geodia cydonium* has been suggested to be involved in adhesive interactions in the adult sponge (see text), although the evidence for this is not conclusive. Lectins 5–10 may have some developmental significance in cell interactions (see text).

saccharide specificity has been reported in a number of tissues of adult and embryonic origin. Most of these lectins appear to be aggregates of protein subunits of low molecular weight that require sulfhydryl-reducing agents for their agglutinin activity and are independent of the presence of Ca^{2+} ions. However, there are some exceptions to this rule, such as the asialoglycoprotein receptor of the rabbit hepatocyte, which requires Ca^{2+} for its carbohydrate-binding as well as its hemagglutinin activities (Stockert *et al.*, 1974; Ashwell and Harford, 1982), the lectin from snake venom that requires Ca^{2+} and is inhibited by reducing agents (Gartner *et al.*, 1980), and the galactoside-binding lectin from the eggs and embryos of *Xenopus laevis*. This latter lectin requires Ca^{2+} and is independent of the presence of sulfhydryl-reducing agents for its hemagglutinating activity (Harris and Zalik, 1982; Roberson and Barondes, 1982).

3.2.1. Adult

Lectin activity has been reported in extracts of many mammalian and avian tissues—among them the chick liver, pancreas, intestine, and heart; the calf liver, heart, thymus, and spleen (Barak Briles *et al.*, 1979; Beyer *et al.*, 1979); the rat lung (Powell, 1980); and human fibroblasts, skeletal muscle, heart, and lung (Childs and Feizi, 1979; Powell, 1980; Roff *et al.*, 1983). Characteristics of these lectins are detailed in Table II.

Most of these proteins appear to be specific for galactosides bearing a β-D-configuration. In the lung of the newborn rat, galactoside-binding lectin activity is low at birth and increases 8–10-fold during differentiation of the alveoli and establishment of cholinergic innervation. It then decreases to minimal values in the adult lung (Powell, 1980). In the rat brain, hemagglutinin activity inhibited by galactosamine has also been detected in tissue extracts. This lectin undergoes changes in activity similar to that of the lung; peak lectin activity has been correlated with the peak period of synapse formation (Simpson *et al.*, 1977).

Of interest in the context of this chapter is the erythroid developmental agglutinin (EDA), a galactoside-binding lectin found in the bone marrow of the rabbit. In this system erythroid differentiation takes place in erythropoietic cells surrounding a macrophage nurse cell. The erythroid cells are released from the nurse cells during the enucleation stage, when the mature reticulocyte is formed. EDA is a protein with a relative molecular weight of 13,000; it can be obtained by washing suspensions of bone marrow cells with a saline solution containing a sulfhydryl-reducing reagent, EDTA, and lactose. When EDA is added to suspensions either of bone marrow cells or purified erythroblasts—both of them depleted of endogenous cell surface agglutinin by prior washing with 0.3 M lactose—the exogenously added EDA promotes aggregation almost exclusively of erythroblasts. Univalent antibodies against this lectin block the EDA-induced aggregation of erythroblasts. Immunofluorescence studies, using antibodies to this lectin, show that erythroblasts at all stages of differentiation are strongly labeled, whereas marrow reticulocytes display a weaker fluorescence. This suggests that EDA may mediate transitory intererythroblast adhe-

Table II. Selected Examples of Lectins from Adult Vertebrates and Chordates[a]

Source	Saccharide specificity	Nature (M_r)	Suggested function	Reference
Eel				
Anguilla rostrata	2- and 3-methyl-L-Fucose 2- and 3-methyl-D-Fucose 3-methyl-D-Galactose	Protein (10,000)	—	Springer and Desai (1971); Bezkorovaini et al. (1971)
Electrophorus electricus electric organ	β-D-Galactose	Protein (16,000)	Synaptogenesis	Teichberg et al. (1975); Teichberg (1978); Levi and Teichberg (1981)
Anguilla anguilla serum	L-Fucose	Protein (18,000)	—	Uhlenbruck et al. (1982)
Chick				
Intestine	β-D-Galactose	Protein (14,200 ± 300)	—	Beyer et al. (1980)
Pancreas	β-D-Galactose	Not characterized	—	Beyer et al. (1979)
Liver	(1) β-D-Galactose	Protein (15,900 ± 400)	—	Beyer et al. (1980)
	(2) N-acetyl-D-Galactosamine, heparin	Protein subunits (16,000; 13,000)	Binding of cells to, and organization of, intercellular matrix	Ceri et al. (1981)
Heart	β-D-Galactose	Protein (12,000)	—	Barak Briles et al. (1979)

Source / Tissue	Sugar	Protein (MW)	Function	Reference
Calf				
Spleen, heart, and muscle	β-D-Galactose	Protein (12,000)	—	Barak Briles et al. (1979)
Rat				
Mammary gland	β-D-Galactose	Glycoprotein (14,800)	Synthesis and transport of milk glycoproteins	Takeuchi et al. (1982)
Lung	β-D-Galactose	Protein (13,500)	Synaptogenesis, cholinergic innervation	Powell (1980); Powell and Whitney (1980)
Brain	Galactosamine	Not characterized	Synaptogenesis, cohesion	Simpson et al. (1977)
Rabbit				
Liver cell membrane	β-D-Galactose	Glycoprotein subunits A (48,000) B (40,000)	Receptor involved in the endocytosis and clearance of serum glycoproteins	Kawasaki and Ashwell (1976); Stockert et al. (1974)
Bone marrow erythroblasts (erythroid developmental agglutinin)	β-D-Galactose	Protein (13,000)	Erythroblast adhesion	Harrison and Chesterton (1980a,b)
Human				
Heart muscle	β-D-Galactose	Protein (13,500)	—	Childs and Feizi (1979)
Lung	β-D-Galactose	Protein (14,500)	—	Powell (1980)

[a]These lectins have been selected to illustrate the distribution of these molecules among different species of chordates and vertebrates. The molecular weights represent the subunit molecular weight. The galactose-binding lectins from chicken liver and intestine are mitogenic (Lipsick et al., 1980).

sion in the erythroblastic island during differentiation, up to the stage of the release of the mature reticulocyte (Harrison and Chesterton, 1980a,b).

The plasma membrane of avian and mammalian hepatocytes contains a series of receptors involved in the clearance and homeostasis of the serum glycoproteins *via* receptor-mediated endocytosis. The asialoglycoprotein receptor involved in the clearance of serum glycoproteins that display exposed terminal galactose groups is of interest. The isolated receptor from the rabbit hepatocyte is an aggregate and displays lectin activity toward human erythrocytes. This lectin is also mitogenic for human peripheral leukocytes (Novogrodsky and Ashwell, 1977) and as such is similar to many plant lectins (Lis and Sharon, 1981). It has been shown that, in culture, rat hepatocytes bind preferentially to galactose-derivatized surfaces. This cell attachment occurs via a large cluster of asialoglycoprotein receptors at the cell surface (Weigel, 1980), suggesting that some of the carbohydrate-specific receptors at the cell surface may perform several functions—among them, cell adhesion. It is important to point out, however, that not every carbohydrate-binding protein involved in the binding of serum glycoproteins exhibits lectin activity.

3.2.2. Embryonic

Changes in agglutinability with plant lectins occur during differentiation of various embryonic tissues (Moscona, 1971); cells of early embryos dissociated by mechanical means are also readily agglutinated by these proteins (Zalik and Cook, 1976; Fraser and Zalik, 1977; Oppenheimer, 1977; Phillips and Zalik, 1982). These findings stimulated the search for endogenous lectins that could, in principle, be involved in the control of the surface receptors of embryonic cells that are detectable with particular plant lectins. The characteristics and specificities of these embryonic lectins are shown in Table III.

As is the case with the adult lectins, many of the lectins from tissues of the chick embryo are specific for β-D-galactosides. These lectins show pronounced changes in activity during development. In the liver, activity rises between days 12 to 14 of development; in the brain, agglutinin activity reaches a peak in 12-day embryos, whereas in the heart, lectin activity reaches a maximum value at day 16 of incubation. The highest galactoside-binding lectin activity is present in the pectoral muscle of the 12 to 16-day–old chick embryo (Kobiler and Barondes, 1977). The galactoside-binding lectins have been purified and appear to be similar in all the tissues of the chick embryos examined. They are mitogenic (Lipsick et al., 1980), require sulfhydryl-reducing agents, are independent of Ca^{2+} for their agglutinin activity, and have a major component subunit of 15,000 M_r (Table III) (Kobiler et al., 1978). A similar lectin has been isolated from the thigh muscle; this agglutinin reaches maximal activity in the 13-day chick embryo (Den and Malinzak, 1977). The fact that highest lectin activities are found in the embryonic skeletal muscle has led some investigators to suggest that this protein may play a role in some of the processes of myogenesis, among them myoblast fusion. As will be discussed in Section 5.3, the evidence for this is still inconclusive.

The second lectin found in the pectoral muscle of the chick embryo is

Table III. Selected Examples of Lectins from Embryonic Tissue of Vertebrates[a]

Source	Saccharide specificity	Nature (M_r)	Suggested function	Reference
Mammals				
Mouse teratocarcinoma[a] Nulli SCC1 cell line	Polymannose; sulfated and nonsulfated polyfucose	Protein (56,000)	Adhesion	Grabel et al. (1981, 1983b)
Chick				
Skin (4–15 days) Cornea (6–12 days)	β-D-Galactose	Protein (16,500)	Condensation of dermal cells	Kitamura (1980, 1981)
Pectoral muscle (12–16 days)	Heparin N-acetyl-D-Galactosamine	Protein subunits (16,000) (13,000)	Binding of cells and organization of the intercellular matrix	Kobiler et al. (1979) Ceri et al. (1981) Min–Lechaire and Barondes (1978)
Thigh muscle (13 days)	β-D-Galactose	Protein (15,000)	—	Den et al. (1977)
Pectoral muscle (8–20 days)	β-D-Galactose	Protein (15,000)	—	Nowak et al. (1977)
Brain, liver and heart (8–20 days)	β-D-Galactose	Protein (15,000)	—	Kobiler et al. (1977, 1978)
Retina and spinal cord (8–22 days)	β-D-Galactose	Not characterized	Synaptogenesis, cell cohesion	Eisenbarth et al. (1978)
Kidney (9–19 days)	Amino sugars	Not characterized	—	Pitts et al. (1980)
Blastoderm Two-layered (stage 1)[b] Gastrula (stage 3)[b]	β-D-Galactosides	Proteins soluble (11,000) Particle-associated components (70,000) (72,000)	Cell adhesion	Cook et al. (1979), Zalik et al. (1983)
22-hr incubation	Polymannose	Not characterized	—	Rutherford and Cook (1981)

(continued)

Table III. (Continued)

Source	Saccharide specificity	Nature (M_r)	Suggested function	Reference
Vitelline membrane of the nonincubated egg	Polymannose	Protein multiaggregate (62,000)	—	Rutherford and Cook (1981, 1984)
Amphibians				
Xenopus laevis unfertilized ova, blastula, gastrula, and neurula	α-D-Galactose β-D-Galactose	Protein components (43,000)	Prevention of polyspermia	Wyrick et al. (1974); Greve and Hedrick (1978)
		(45,000)		Roberson and Barondes (1982)
		Protein (65,500)		Harris and Zalik (1982, 1985)
Rana pipiens blastula (32–64 cell)	Polymannose	Protein not characterized	Adhesion	Roberson et al. (1980)
Rana catesbiana ovarian oocytes	N-Acetyllactosamine D-Galactose	Protein (210,000)	—	Sakakibara et al. (1976)
Rana japonica ovarian oocytes	Gangliosides	Protein (13,500)	—	Sakakibara et al. (1979)
Fish				
Chinook salmon (*Oncorhynchus kisutch*) Ova, unfertilized	D-Galactose L-Rhamnose	Protein (122,000)[c]	Protection of eggs and embryos from bacterial infection	Voss et al. (1978)
Brown trout (*Salmo trutta*) (protectin) Ova, unfertilized	α-D-Galactose β-D-Galactose L-Rhamnose	Protein (200,000)[c]	Protection of eggs and embryos from bacterial infection	Anstee et al. (1973), Voak et al. (1974)

[a] The mouse teratocarcinoma is included because of the similarity between these cells and those of the early mouse embryo. Molelcular weights shown refer to the subunit molecular weight.

[b] The stages outlined by Hamburger and Hamilton (1951).

[c] Only the native molecular weight was available.

specific for N-acetyl-D-galactosamine. This agglutinin interacts strongly with the glycosaminoglycans heparin, dermatan sulfate, and heparan sulfate (Kobiler and Barondes, 1979; Ceri et al., 1981). To be sensitized to this lectin, trypsinized rabbit erythrocytes fixed with glutaraldehyde must be aged at low temperatures or treated with ethanol. Presumably, these treatments expose glycoproteins on the erythrocyte that were not previously available for interaction with this lectin. Material deposited on the substratum by myoblasts from 12-day pectoral muscle cultured in vitro interacts strongly with this lectin as evidenced by its strong inhibition of lectin-mediated agglutination. Enzymes that degrade dermatan sulfate remove the hemagglutination inhibitory effect of this substrate-attached material (Ceri et al., 1979); the lectin can also be isolated with the latter. This finding suggests that the N-acetyl-D-galactosamine-specific lectin could interact with glycosaminoglycans and be involved in the organization of the extracellular matrix.

Lectins are also present in fertilized and unfertilized eggs. Lectins specific for D-galactose and L-rhamnose have been isolated from eggs of the Chinook salmon and the brown trout. These lectins do not require sulfhydryl reducing agents and Ca^{2+} for their activity (Anstee et al., 1973; Voak et al., 1974; Voss et al., 1978). It has been suggested that they may provide a general defence mechanism against parasitic infection of the ovum. In the amphibians, a lectin specific for galactose and N-acetyllactosamine has been isolated from ovarian eggs of *Rana catesbiana* (Sakakibara et al., 1976). A lectin specific for cerebrosides and gangliosides has also been purified from *Rana japonica*. This lectin agglutinates transformed murine cells (Sakakibara et al., 1979).

The presence of a lectin has been reported in early embryos of *Rana pipiens* (Roberson and Armstrong, 1980). Homogenates from cleaving embryos display mannan-specific lectin activity. Dissociated cells from cleaving embryos also display mannan-specific binding activity at their surfaces. This is shown by their ability to form rosettes whose formation is inhibited by the same mannose-containing compounds that inhibit hemagglutination. In the superficial blastomeres of these embryos, the external pigmented surface is not in contact with other cells. On these cells, binding of the erythrocytes is localized almost entirely at the nonpigmented regions; these are the areas in which intercellular contact takes place. Although these results could merely be coincidental, the location of mannan-containing surface glycoproteins at the sites where cells contact each other, and the presence in homogenates of an endogenous lectin with a similar specificity, suggest a role for this lectin in some cell interactions during embryogenesis.

Eggs and embryos of *Xenopus laevis* contain a galactoside-binding lectin (Harris and Zalik, 1982; Roberson and Barondes, 1982). This lectin attains its highest activity during gastrulation and has a strong tendency to form aggregates (Harris and Zalik, 1985). According to Roberson and Barondes (1982), the lectin purified from ovarian eggs and embryos at the blastula stage has a relative molecular weight ranging from 43,000 to 45,000, while another laboratory reports a relative molecular weight of 65,500 (Harris and Zalik, 1985). The galactoside-specific lectin may bear some relationship to the agglutinin from the cortical granules of the unfertilized eggs from these embryos. During fertilization, a component released from the cortical granules interacts with a constituent of the jelly coat. The product of this interaction becomes incorporated into the external layer of the vitelline envelope, converting the latter into the fertilization envelope. Extracts of cortical granules can precipitate a sulfated component of the innermost jelly layer, and this precipitation is inhibited by D-galactose and α-methyl-galactoside. Antibodies produced against the innermost jelly component also interact with the fertilization envelope (Wyrick *et al.*, 1974; Greve and Hedrick, 1978). This suggests that the lectin located at the cortical granules may be released and translocated through the vitelline space to interact with a glycoprotein from the jelly coat; the product of this reaction subsequently becomes incorporated into the vitelline envelope to form the fertilization membrane.

Lectins specific for β-D-galactosides are also present in cells of the nonincubated and primitive streak-stage chick blastoderms (stages 1–5, Hamburger and Hamilton, 1951) (Cook *et al.*, 1979). One of these lectins is readily solubilized, has a subunit relative molecular weight of 11,000, and in aqueous solution appears to form aggregates of variable size. In the presence of lactose, this lectin appears to form dimers or trimers, since in Sepharose chromatography it elutes between ovalbumin (45,000 M_r) and chymotrypsinogen A (25,000 M_r). The soluble lectin is present in the area pellucida as well as in the ectoderm and endoderm of the area opaca (Cook *et al.*, 1979). A second particle-associated lectin is also galactoside-specific and has subunit relative molecular weights of 70,000 and 72,000 (Zalik *et al.*, 1983). The specific activity of this lectin is low to almost absent in extracts from the area pellucida. Experimental evidence suggests that the soluble lectin may affect adhesion of the endodermal cells of the area opaca (Milos and Zalik, 1982, 1983) (see Section 5.2.2). Lectin activity has also been extracted from the vitelline envelope of the fertilized nonincubated chick embryo. This lectin is specific for mannose polymers and requires detergents for its extraction. The hemagglutinin has a high tendency for self-association and appears to be composed of at least three glycoprotein components that differ in molecular weight (Rutherford and Cook, 1981, 1984) (see Table III).

The mouse teratocarcinoma is currently regarded as a model system for the study of early mammalian embryogenesis, since in this tumor cells of the embryonal carcinoma can differentiate into a variety of embryonic tissues (Pierce *et al.*, 1978; Martin, 1980). Cells from two cell lines can form rosettes with trypsinized glutaraldehyde-fixed rabbit erythrocytes. Rosette formation can be inhibited by pretreating the teratocarcinoma cell lines with polymannose- or polyfucose-containing molecules (Grabel *et al.*, 1979). Extracts from some cell lines contain lectin activity inhibited by mannose and fucose polymers, in particular those containing sulfated fucose (Grabel *et al.*, 1981). These same saccharide polymers also inhibit adhesion (Grabel *et al.*, 1983a) (see Chapter 9).

3.3. Cellular Localization of Animal Lectins

In order to play a role in cell adhesion, lectins must be present in some sort of association with the external cell membrane. Attempts to localize endogenous lectins have made use of labeled antibodies produced against these agglutinins. The presence of carbohydrate-binding activity at the cell surface has also been assessed either by the rosette assay or by the binding of glycosylated cytochemical markers. These latter molecules are labeled natural glycoproteins or synthetic neoglycoproteins consisting of carbohydrate derivatives attached covalently to selected proteins (Lee and Lee, 1982). By immunohistochemical techniques, galactoside-binding lectins have been localized in the cystosol and at the cell surface of differentiating muscle (Nowak et al., 1977; Barak Briles et al., 1979; Podleski and Greenberg, 1980; Barondes and Haywood-Reid, 1981) and in the cells of the optic tectum of the chick embryo (Gremo et al., 1978). In the adult rat lung, this lectin is present predominantly at the extracellular level in association with elastic fibers (Cerra et al., 1984). The erythroid developmental agglutinin is present extracellularly (Harrison and Chesterton, 1980a), and a similar situation is found with the pancreatic lectin of the adult chicken (Beyer et al., 1979). The galactose receptor of the mammalian hepatocyte represents an instance wherein the surface location of the lectin has been documented using the rosette assay (Kolb et al., 1979; Kolb-Bachofen, 1981; Schlepper-Schäfer et al., 1981), the binding of labeled glycoproteins and neoglycoproteins with terminal galactose groups (Weigel, 1980; Kolb-Bachofen, 1981), and by the direct isolation of plasma membranes (Ashwell and Harford, 1982). This receptor is also localized in intracellular membranes (Kolb-Bachofen, 1981; Harford and Ashwell, 1982). In some cases, lectins have been localized only in the cell interior within secretory granules such as in the intestinal goblet cells of the chick (Beyer et al., 1979) and in the spherulous cells of some species of sponges (Bretting and Koningsmann, 1979; Bretting et al., 1983).

In early embryonic cells, the galactoside-binding lectin of the sea urchin egg has been detected in the peripheral cytoplasm of the fertilized egg (Sasaki and Aketa, 1981). In fish oocytes, endogenous lectins are located in the cortical vesicles beneath the plasma membrane (Nosek et al., 1983). The mannan-binding lectin has been detected at the cell periphery of the blastomeres of the frog embryo (Roberson and Armstrong, 1980) and of some mouse teratocarcinoma cell lines (Grabel et al., 1979) by the rosette assay. In oocytes and embryos of X. laevis, studies using immunofluorescence indicate that the lectin is present intracellularly as well as extracellularly (Roberson and Barondes, 1983), and it has been suggested that the latter is due to secretion (Barondes, 1984). A similar localization of the chick blastoderm lectin has been found in cultures of endodermal cells from the area opaca of the primitive streak embryo (Zalik et al., 1982).

4. Cellular Adhesion

4.1. Importance of Cell Adhesion in Development

During gastrulation, as cells relocate within the embryo to reorganize into new groups that form the primary germ layers, cells form adhesive bonds that vary in their degrees of permanence; i.e., some may be transitory, whereas others may have a greater stability. Information on the nature of the molecular interactions that occur during the formation of different kinds of adhesive bonds is of crucial importance for our understanding of the biology of morphogenetic movements. The most obvious way to study this problem is to visualize directly the behavioral pattern of the different cell groups in gastrulating embryos after exposure to different experimental conditions. Direct visualization has only been used on transparent embryos of the sea urchin (see Chapter 6). Other embryos often have the added disadvantage of lack of transparency. This makes direct observation of cellular relocation more difficult; see Trinkaus (1976) for a detailed discussion of this topic.

Other means of investigating the molecular nature of cellular adhesion involve the study of this process *in vitro*. This somewhat arbitrary approach involves the prior separation of cells from their neighbors, followed by the determination of their adhesive properties under conditions that bear little resemblance to those present in the early embryo. It is generally assumed that some of the factors that affect the formation of adhesive bonds *in vitro* have relevance to those controlling adhesion within the embryo.

4.2. Cell Dissociation

With the exception of cells such as erythrocytes, lymphocytes, or macrophages, which carry on their specialized functions suspended in body fluids, most cells of metazoan animals associate with each other to form defined tissues. Unfortunately, the available methodology developed to study cell-to-cell or cell-to-substrate adhesion involves the study of single cells; consequently, most studies on cell adhesion necessitate the dissociation of cells from their parent tissue.

The least traumatic method for cell dissociation involves the removal of Ca^{2+} and Mg^{2+} from the medium followed by the application of a gentle shear force, usually accomplished by pipetting. This, however, can only be applied to very young embryos of chick (Milos et al., 1979), sea urchin (see Chapter 6), some teratocarcinoma cell lines (see Chapter 9), and some species of sponges (Humphreys, 1963; Burger et al., 1978). Most tissues and monolayers of established cell lines require more drastic methods for their dissociation. These involve the use of ion chelating agents such as EDTA, the calcium chelating agent ethylene-glycol-bis-(β-aminoethyl ether)-N,N-tetraacetic acid (EGTA), and the use of proteolytic enzymes such as trypsin and collagenase (Maslow, 1976; Grinnell, 1978; Garrod and Nicol, 1981). Chelating agents are presumed to act by binding to cations required for the maintenance of adhesive bonds, whereas proteolytic enzymes probably break down intercellular materials to different degrees. Cells are then dissociated by mechanical shearing forces applied by pipetting. Since these treatments alter cell surfaces to different extents, a recovery period is usually allowed after proteolytic dissociation. Although cells recover their adhesion after this period, the question remains as to how much of the regenerated cell surface resembles that present in the embryo and how much of it is synthesized as an adaptation to the culture conditions.

4.3. Methods for Assessing Cell Adhesion

4.3.1. Stationary Cultures

In this approach the adhesive behavior of cells and cell groups is determined under conditions in which the culture remains immobile during the duration of the assay. Cell behavior under these conditions reflects the chemical architecture of the cell surface as well as the cytoskeletal properties of the cells. In these studies, compounds to be tested can be adsorbed, chemically coupled to the substrate, or incorporated into the medium. Adhesion is assessed by determining the fraction of cells from a defined cell population that attaches to the substrate or by determining the ratio of attached cells per unit area. This approach has been followed in studies on the adhesive properties of fibronectin and laminin (Grinnell, 1978; Grinnell and Feld, 1979, 1982; Johansson et al., 1981; Kennedy et al., 1983), in studies on effects of plant lectins on cell-to-substrate adhesion and behavior (Hatten and François, 1981), and in experiments testing the effects of different carbohydrates immobilized on the substrate (Hatten, 1981). Stationary cultures have also been used to study the behavior and sorting of cell groups from different embryonic tissues (Trinkaus and Lentz, 1964; Garrod and Steinberg, 1973, 1975). These techniques permit direct observation of the adhesive behavior of cells in different conditions, and permanent records can be obtained using time-lapse cinematography (Maslow, 1976; Grinnell, 1978).

4.3.2. Motion-Mediated Adhesion

In this method, suspensions of single cells in their incubation medium are placed in a suitable vessel (e.g., conical flask, petri dish, multiculture well). The vial is then placed on a platform that is either rotated around its vertical axis or subjected to reciprocal motion at a constant temperature. Under these conditions, cells are allowed to collide with each other, and those collisions that result in the formation of intercellular adhesive bonds resistant to the shearing forces present in the moving culture medium will lead to the formation of cell aggregates. Cell suspensions are maintained at the desired temperature, and adhesion is measured by the decrease in single cells that occurs as cells form aggregates. The number of single cells in a suspension can be determined with a hemocytometer or with the electronic particle counter. Using this latter instrument, it is also possible to obtain a profile of aggregate size during the experiment. An alternative method for the detection of intercellular adhesion is to measure the decrease in light scattering of the suspending medium during aggregation (Beug and Gerisch, 1972). On the basis of this principle, an automatic instrument for continuous recording of cell aggregation has been designed (Thomas and Steinberg, 1980).

An alternative way to quantify cell adhesion is the collecting cell aggregate assay. In this technique, the capacity of cells in a suspension to adhere to a preformed group of cells is determined. The preformed cell group may be a confluent cell monolayer or a three-dimensional aggregate of a defined size (Roth *et al.*, 1971). In the latter case, the peripheral cells of the preformed aggregate represent the surfaces to which cells will adhere. In the collecting monolayer an aliquot of a cell suspension is allowed to attach to a substrate and form a monolayer. In the corresponding aliquot of the suspension, cells are labeled with fluorescein diacetate (Edelman, 1983) or with a radioactive precursor (Roseman *et al.*, 1974). Labeled cells are added to the vessel containing the cell monolayer, and the vessel is incubated under reciprocal or rotating motion. At the desired time interval, cells remaining in suspension are carefully removed, and the number of cells attached to the monolayer is determined by fluorescence microscopy (Edelman, 1983) or by measuring the radioactivity in lysates prepared from the monolayers with the attached labeled cells. The rationale and methodology associated with these assays is discussed in detail by Roseman *et al.* (1974).

Recently, surfaces carrying immobilized carbohydrate ligands have been used for the study of cell adhesion. These derivatized surfaces consist of glass coverslips, polystyrene tissue culture dishes, or polyacrylamide gels to which sugars, glycosides, or glycoproteins are chemically linked (Weigel *et al.*, 1978, 1979; Weigel, 1980). Cell suspensions are incubated on these surfaces under varying experimental conditions. Vessels are placed on a rotating platform or are kept stationary. In the latter case, at specified time intervals, cells are subjected to a mild shear force. The ratio of attached cells is measured indirectly by the amount of lactic dehydrogenase activity (a cytoplasmic enzyme) in lysates obtained from the attached cells (Weigel *et al.*, 1978; Weigel, 1980). The adhesion detected by this assay depends on the shear-resistant adhesive bonds formed by the carbohydrate-binding protein at the cell surface and the carbohydrate ligand on the derivatized substrate. The advantage of this assay is that it allows for concurrent studies of the kinetics of adhesion with the microscopical observation of cell behavior in long-term cultures. Such experiments indicate that cells can distinguish between substrata formed by different saccharide species (Weigel *et al.*, 1979; Weigel, 1980; Hatten, 1981).

5. What Do We Know about the Involvement of Lectins in Cell Interactions?

5.1. Sponges

From the point of view of their structure, sponges are organized into three cell layers. The innermost layer consists of flagellated cells, or **choanocytes;** these are involved in the circulation of water through the organism. The outer epithelial layer, or **pinacoderm,** is composed of **pinacocytes** and **porocytes;** the latter are cells containing a central pore through which water enters the interior of the sponge. Between the inner and outer cell layers, an intermediate layer, or

mesohyl, is located. It consists of a spongy extracellular matrix composed of collagen and spongin fibers, upon which a variety of cells—most of them highly mobile—are located (Müller, 1982). Among these, the wandering ameboid cells, or **archeocytes,** are the most prominent. Other cells present in this layer are the **collenocytes** and **lophocytes,** which secrete fibrillar collagen, and the **spongiocytes** and **sclerocytes,** which produce spongin and spicules, respectively. Cells with cytoplasmic inclusions are also localized in the intermediate layer; among these are the **globopherous cells, rhabdiferous cells,** and **spherulous cells.** While cells in the interior and exterior layers remain generally stationary with respect to each other, many cells in the mesohyl move through the extracellular matrix via the formation of transitory adhesive bonds (Junqua et al., 1974; Müller, 1982). Many species of sponge also live in symbiotic association with bacteria (Müller, 1982).

5.1.1. Lectins

Galstoff (1929) was the first investigator to report that sponges from different species contain substances that clump or agglutinate cells of other sponge species. Years later, agglutinin activity was found in the fluid employed for sponge cell dissociation, and the lectins were subsequently isolated from supernatants of dissociated sponge cells. These media were also shown to contain other factors promoting aggregation of homologous cells under the appropriate experimental conditions described in Section 5.1.2 (MacLennan and Dodd, 1967; Dodd et al., 1968). A large number of the lectins of sponges described so far appear to be galactose specific (Gold et al., 1974; Vaith et al., 1979; Bretting et al., 1981a,b; Müller et al., 1983b). In the sponge *Axinella polypoides,* at least three lectins have been isolated; two of these, lectin I and lectin II (15,000 and 21,000 M_r, respectively) have been characterized (Bretting and Kabat, 1976). In immunohistochemical studies using antibodies obtained against each of these two lectins, both have been localized in the spherulous cells; lectin I was present in spherulous cells with large vesicles and lectin II in spherulous cells with small vesicles. Both lectins were present in the vesicular content of these cells; small amounts of these lectins were also found in association with the extracellular spongin fibers (Bretting and Koningsmann, 1979; Bretting et al., 1983).

In the sponge *Geodia cydonium* at least two β-D-galactoside-binding lectins are present in tissue extracts (Vaith et al., 1979). The main lectin is a trimer (36,500 M_r) composed of different subunits (13,800, 13,000, and 12,200 M_r) (Müller et al., 1983b). The lectin can also be isolated from cell dissociation media endowed with cell aggregation-promoting activity (Vaith et al., 1979) and is mitogenic for human lymphocytes (Bretting et al., 1981b).

5.1.2. Cell Aggregation

In his classical studies, Wilson (1907) prepared cell suspensions from entire sponges or from fragments of these organisms. He observed that the cells could aggregate and would reorganize into miniature sponges. It is now known from the work of Humphreys (1963), Moscona (1968), and others (reviewed by Van de Vyer, 1975) that mixtures of cells from different sponge species are either able to recognize each other specifically and organize into separate aggregates, or adhere to each other, forming mixed aggregates, the cells of which subsequently sort out according to species. It is not surprising, therefore, that the species-specific aggregation and segregation displayed by sponge cells has been used as a model system to study specific cell adhesion, since these studies could have some relevance to the cell–cell recognition that occurs in embryogenesis.

As will be seen in the following discussion, results in this field vary significantly between laboratories. This is probably due, among other things, to particular ecological adaptations among individuals of the same species that could affect the display of different adhesive mechanisms (i.e., boring versus crust forms), age of the individual, as well as inherent variability in the adhesive mechanisms among species. In addition, differences in the particular dissociation protocols, as well as the methodology and parameters used for assessing adhesion, may contribute to this variability. These factors are discussed by Van de Vyer (1975), Burger *et al.* (1978), and Müller (1982).

Cells dissociated from the species *Microciona prolifera* in Ca^{2+},Mg^{2+}-free seawater were shown to release a high-molecular-weight factor that promotes species-specific adhesion (Humphreys, 1963). This factor has been purified and found to be a large proteoglycan complex (21×10^6 M_r) displaying a sunburst configuration consisting of a circular center 80 nm in diameter with 15 radiating arms, each 110 nm in length (Humphreys, 1963; Cauldwell *et al.*, 1973; Henkart *et al.*, 1973). This factor is irreversibly inactivated by 1 mM EDTA or EGTA, indicating that Ca^{2+} is required for its activity (Burger *et al.*, 1978). An **aggregation factor (AF)** has also been isolated from the Mediterranean sponge *Geodia cydonium*. In contrast to that of *Microciona*, the AF of *Geodia* is solubilized by placing cells or tissue fragments in Ca^{2+},Mg^{2+}-free seawater containing 20 mM EDTA and is quite stable in the continuous presence of EDTA. The AF of *Geodia* also has a higher protein content; electron microscopically, it appears as sunbursts with a center circle of 353 ± 8.5 nm in diameter and 25 ± 8 radiating arms, each of a length of 61 ± 12 nm (Müller and Zahn, 1973). During sponge cell adhesion, the AF is presumed to bind to a receptor present at the cell membrane. This receptor has been called **baseplate** in *Microciona* (Weinbaum and Burger, 1973; Burger *et al.*, 1978) or **aggregation receptor** (**AR**) in *Geodia* (Müller *et al.*, 1976). The binding of the AF to its receptor depends on the presence of terminal glucuronic acid. Whereas in *Microciona* the glucuronic acid must be present in the AF (Burger *et al.*, 1978; Jumblatt *et al.*, 1980), in *Geodia* the presence of glucuronic acid is required in the AR (Vaith *et al.*, 1979; Müller, 1982).

In *Microciona*, cell aggregation in the presence of AF appears to be a continuous phenomenon (Burger *et al.*, 1978; Jumblatt *et al.*, 1980). In *Geodia*, however, aggregation occurs in two stages; the formation of small aggregates (60–200 µm in diameter) called primary aggregates occurs in the absence of AF, whereas the formation of larger aggregates or secondary aggregates requires the presence of AF (Müller and Zahn, 1973). It therefore appears that two different adhesive mechanisms exist in *Geodia*. The first one results in the formation of small primary aggregates, requires Ca^{2+} and Mg^{2+}, and lacks species specificity; the mechanism involved in the formation of secondary aggregates is species specific and is mediated by AF (Müller and Zahn, 1973; Müller et al., 1976; Müller, 1982). In this species, Müller et al. (1979b) isolated another adhesion-related molecule that has been called **antiaggregation receptor (aAR)**. This molecule, extracted by solubilization of the membrane with a mixture of detergents, is a heat-stable glycoprotein (180,000 M_r) that contains a large number of galactose groups. The aAR binds to the aggregation factor, inhibiting its stimulating effect on aggregation. When added to an aggregating cell suspension of *G. cydonium*, aAR inhibits secondary adhesion and causes some dissociation of existing secondary aggregates. The inhibitory effect of aAR on adhesion can be removed by its pretreatment with β-galactosidase; other glycosidases, such as β-glucuronidase, β-glucosidase, or neuraminidase, do not affect the aggregation inhibitory activity of aAR. In aggregating cell suspensions, the inhibitory activity of aAR can be overcome by adding β-galactosidase to the assay (Fig. 2A). This finding suggests that the active site of aAR is a terminal galactose.

When cell suspensions of *Geodia* dissociated in Ca^{2+},Mg^{2+}-free medium are permitted to aggregate in the presence of AF, approximately 75% of the cells form aggregates; these cells have been called aggregation-susceptible cells. The remaining 25% of the cells do not form secondary aggregates and have been termed aggregation-deficient cells. Aggregation-susceptible cells can be converted to aggregation-deficient cells by incubation with aAR. Treatment of aggregation-deficient cells with β-galactosidase converts them to aggregation-competent cells, allowing them to form large aggregates in the presence of AF (Müller et al., 1979b). Müller and Müller (1980) suggest that the aAR, carrying terminal galactose, weakens cell–cell adhesion by interfering with the formation of the AF–AR complex. The aAR is present in an active state in the aggregation-deficient cells; removal of the galactose from this factor inactivates its inhibitory activity (Müller et al., 1979b; Müller and Müller, 1980; Müller, 1982).

As mentioned previously, the main β-D-galactoside-binding lectin has been purified from supernatants of disaggregated cells of *G. cydonium* (Müller et al., 1979a; Vaith et al., 1979). This lectin forms precipitates with the aAR, converts aggregation-deficient to aggregation-susceptible cells, and overcomes the inhibition of aggregation elicited by aAR (Fig. 2B). Neither the lectin nor galactose by itself affects the response of aggregation-susceptible cells to AF. The galactose-specific lectin restores adhesion of aggregation-deficient cells by virtue of its inactivation of the aAR; this occurs *via* masking of the galactose in aAR. This masking results in the loss of the recognition site in the aAR for the AF. As a consequence, the AF is now free to bind with its AR. Since the aAR interferes with specific adhesion by binding to the AF and making it inaccessible to the AR, the endogenous lectin would provide a mechanism for controlling deadhesion and modulating the formation of transitory adhesive bonds (Fig. 3).

Immunofluorescence studies as well as direct determination of lectin ac-

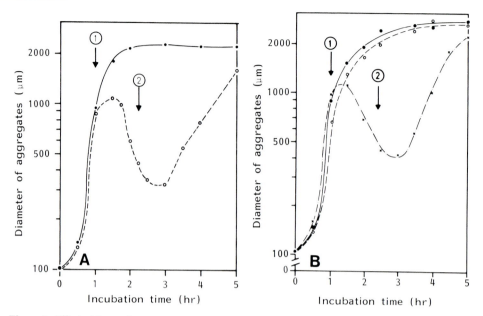

Figure 2. Effect of the antiaggregation receptor and the endogenous galactose-binding lectin on cell aggregation in *Geodia cydonium*. (A) Effect of the antiaggregation receptor. Cells were allowed to aggregate in the presence of aggregation factor alone or with the addition of either antiaggregation receptor or β-galactosidase. Solid line with closed circles: control assay; dotted line with open circles: experimental assay. In the experimental assay, aggregation was initiated by aggregation factor and at the time shown by arrow ① , antiaggregation receptor (100 μl) was added to the suspension. Observe that the antiaggregation receptor inhibits adhesion and causes aggregate dissociation. At the time shown by arrow ② , β-galactosidase was added to the assay; observe that the inhibitory effect of the antiaggregation receptor on cell adhesion is removed by this enzyme. (From Müller *et al.*, 1979b.) (B) Effect of the antiaggregation receptor and the endogenous lectin on cell aggregation. Cell suspensions were allowed to aggregate in the presence of aggregation factor. Solid line with closed circles: control assay; dashed line with open circles: cell suspensions in the presence of lectin (10 μg protein in 5 μl); dashed and dotted line with X's: cell suspension subjected to the action of the antiaggregation receptor. In this latter case, at time ① , antiaggregation receptor (100 μl) was added to the assay; at time ② , the endogenous lectin (5 μl) was added. Observe that the addition of lectin overcomes the inhibitory effect of the antiaggregation receptor. This effect is similar to the one observed after β-galactosidase treatment. (From Müller *et al.*, 1979a.)

tivity in isolated cell populations indicate that the lectin is produced primarily in mucous or spherulous cells (Müller *et al.*, 1981b; Müller, 1982). Müller *et al.* (1981b) suggest that mucoid cells in primary aggregates release lectin into the extracellular space, which can be detected by immunofluorescence in primary aggregates. As a consequence of the charging with lectin of other cells in the aggregate (i.e., archeocytes and choanocytes), aggregation is stimulated. This is substantiated by the fact that the lectin is present in its highest concentration in mucous or spherulous cells and that incubation of aggregation-deficient cells with mucous cells converts the latter into aggregation-competent cells. Con-

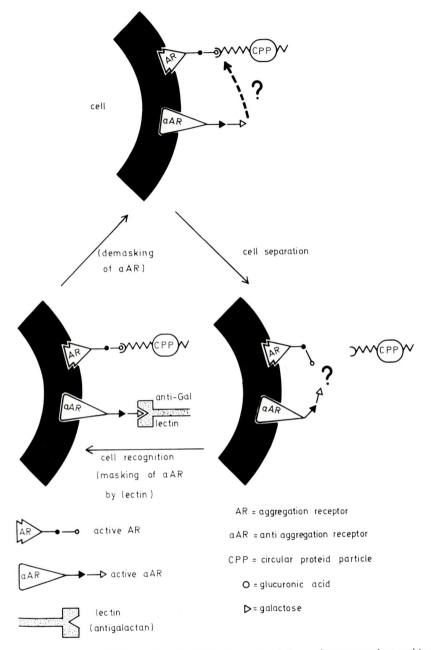

Figure 3. Tentative model illustrating the biological role of the endogenous galactose-binding lectin of *Geodia cydonium* on cell aggregation. The antiaggregation receptor (aAR) carrying a terminal galactose "weakens" cell–cell adhesion by interfering with the formation of the aggregation factor–aggregation receptor complex. The aggregation factor is represented in this model as CPP (circular protein particle, the "sunburst" structure carrying the aggregation factor). The lectin binds to the terminal galactose of the antiaggregation receptor, thereby inactivating this molecule and allowing for cell adhesion. (From Müller and Müller, 1980.)

comitantly, the percentage of *Geodia* cells capable of forming rosettes increases as cells become aggregation competent (Müller *et al.*, 1981*b*).

Although the results reported by Müller's group pose a plausible working model to probe further the molecular basis for adhesion in sponges, the actual localization of the lectin in cells and cell aggregates remains controversial. Using immunohistochemical techniques with glutaraldehyde- and formalde-hyde-fixed cells and tissues of *Axinella polypoides*, Bretting and Konigsmann (1979) and Bretting *et al.* (1983) localized the endogenous lectins mainly with-in the granules of the spherulous cells and, to a certain extent, in association with the spongin fibers. These workers suggest that in the Ca^{2+},Mg^{2+}-free medium used for cell dissociation, lectin leaks out of the spherulous cells and the association of lectin with the cell surface may be artifactual. Little is known of the mechanisms that control the release of lectin by spherulous cells in different sponge species. Before a role of the β-galactoside binding lectin in adhesion can be established firmly, it remains to be determined whether the release of lectin by the cells of *Geodia* is due to physiological secretion or to cellular leakage due to stress. It is also possible that sponge lectins could be involved in other phenomena such as self–non-self recognition and pathogen rejection (Van de Vyer, 1975; Vaith *et al.*, 1979), as well as the establishment of symbiotic relationships with bacteria (Müller *et al.*, 1981*a*).

5.2. The Early Chick Embryo

In early embryos, one of the cell groups in which adhesion has been stud-ied in some detail is the extraembryonic endoderm of the gastrulating chick embryo. These cells are present as a multilayered cellular array at the ventral surface of the area opaca, adjacent to the yolk and below the ectoderm (Fig. 4). While those cells present in the area pellucida undergo the morphogenetic movements of gastrulation with the formation of the primitive streak (see Chap-ter 12), they are accompanied by the epibolic spreading of the cells of the extraembryonic endoderm of the area opaca (Bellairs, 1971; Nicolet, 1971). These cells will eventually surround the yolk and differentiate into the yolk sac endoderm (Romanoff, 1960). The latter cell group is organized as an absorptive and secretory epithelium in direct contact with the yolk and associated with the extraembryonic mesoderm that supports the extraembryonic blood vessels of the yolk sac.

Cells of the extraembryonic endoderm acquire surface characteristics that determine them as extraembryonic early in development. This is shown by experiments in which presumptive extraembryonic endoderm cells sort out in cell aggregates produced by cell suspensions prepared from whole blastoderms at stages before incubation (Zalik and Sanders, 1974; Eyal-Giladi *et al.*, 1975) (see Fig. 5 A and B). Besides sorting out from other cell types, the cell groups formed by the extraembryonic endoderm tend to cavitate and differentiate into fluid-filled vesicles surrounded by cells that resemble histologically the yolk sac endoderm (Fig. 5C–E). These same characteristics are shown by aggregates

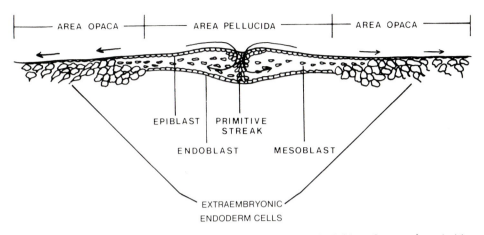

Figure 4. Diagrammatic representation of a cross section of a chick blastoderm at the primitive streak stage. The direction of the morphogenetic movements of the presumptive germ layers is shown by arrows. The cells of the extraembryonic endoderm of the area opaca will spread peripherally around the yolk sac and differentiate into the absorptive and secretory epithelium of the yolk sac. (From Milos et al., 1984.)

prepared from purified cell suspensions of the endodermal cells of the area opaca of gastrulating chick embryos, indicating that the ability to form aggregates that cavitate is intrinsic to the cells themselves and does not depend on interactions with other cell layers (Figs. 6 and 7).

5.2.1. Lectins

Cell suspensions from early chick blastoderms, obtained by mechanical dissociation in Ca^{2+},Mg^{2+}-free media, are highly agglutinable with a variety of lectins such as concanavalin A (specific for D-mannose or D-glucose), wheat germ agglutinin (N-acetyl-D-glucosamine and N-acetylneuraminic acid), and *Ricinus communis* agglutinin (D-galactose) (Zalik and Cook, 1976). This agglutinability is also detected in cells obtained from different regions of the blastoderm (Phillips and Zalik, 1982). The fact that lectin-receptor glycoproteins are present in an agglutinable condition at the cell surface led to the search for endogenous lectins that could, in principle, interact with the surface glycoproteins of early embryonic cells. Two galactoside-binding lectins isolated from primitive streak blastoderms differ in extractability and molecular weight (see Table III). Lectin activity specific for D-galactose is also released by isolated extraembryonic endodermal cells (Milos and Zalik, 1982), whereas mannan-specific lectin activity can be detected in the saline solutions in which freshly isolated chick blastoderms have been stored (Rutherford and Cook, 1981) and in membrane preparations from chick blastoderms (Rutherford and Cook, 1984). A mannan-binding lectin has been isolated from the vitelline membranes of nonincubated fertile eggs (Table III).

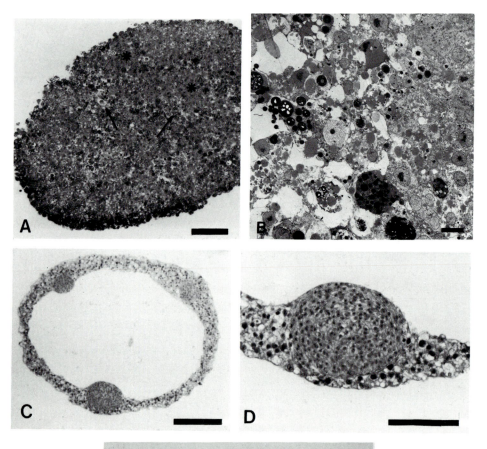

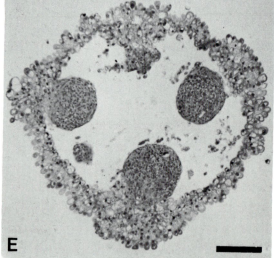

5.2.2. Cellular Adhesion

Pure cell suspensions from the extraembryonic endoderm cells of the area opaca can be obtained by differential mechanical dissociation of the area opaca of gastrulating chick blastoderms in Ca^{2+}, Mg^{2+}-free saline (Milos et al., 1979). When these cells are cultured in rotating flasks, they adhere to each other to form aggregates that cavitate and eventually differentiate to form vesicles surrounded by cells resembling the yolk sac epithelium (see Figs. 6 and 7). These cells also exhibit calcium-dependent as well as calcium-independent adhesive mechanisms, similar to the situation found in other cell types (see Chapter 8). Calcium-independent adhesion is demonstrated in the endodermal cells by the retention of a sizeable proportion of the original control adhesion (60%) in media devoid of calcium and containing EGTA (Milos and Zalik, 1981a). When cells are treated with β-galactosidase, their adhesion is decreased compared with that of control nontreated cells. A sizeable proportion of the adhesion of galactosidase-treated cells is calcium independent. Although it cannot be ruled out that β-galactosidase treatment does not affect the calcium-dependent adhesive mechanisms, the results suggest that surface galactose groups are involved in intercellular adhesive bonds that do not require calcium for their formation (Milos and Zalik, 1983). This premise is also supported by the fact that the incorporation of lectin into the medium decreases calcium-independent adhesion (Milos and Zalik, 1983).

The effects of the lectin can also be observed with cells in stationary culture. Under these conditions, addition of the soluble lectin to the medium decreases cell adhesion to the substratum and promotes the adoption of a fibroblastlike shape in contrast to the epithelial morphology exhibited by control cells (Milos and Zalik, 1981b). A most interesting observation is that cells release lectin activity into the extracellular medium, which is associated with a decrease in the basal levels of adhesion (Fig. 8). This inhibition of aggregation can be overcome by hapten inhibitors of the galactoside-binding lectin(s), indicating that the released lectin may be causing the decreased aggregation (Milos and Zalik, 1982). This lectin is also present in the vesicular contents of the cavitated aggregates (Milos et al., 1983; Milos and Zalik, 1985). Using an antiserum prepared against the two galactose-binding lectins, immunofluorescence

---

Figure 5. Cell sorting and cavitation shown by aggregates formed by dissociated cells from whole blastoderms at stages before incubation. (A) Aggregate obtained after 24 hr of culture; observe the initial sorting out into closely packed cells (star) and the loosely packed cells, the latter representing the presumptive extraembryonic endoderm (arrow). (B) Higher magnification showing the arrangement of the loosely packed cells (left) and the closely packed cells (right). (C, D) With subsequent culture, the loosely packed cells cavitate and surround a central cavity. The closely packed cells have sorted out together to form nodules within the aggregate. D shows the sorted cell groups at higher magnification. (E) Aggregate cultured for 7 days shows the continuous layer of cells surrounding a central cavity; this layer has differentiated from the loosely packed cells of the extraembryonic endoderm. The cells resemble histologically the endodermal lining of the yolk sac. The closely packed cells represent the rest of the cells of the blastoderm; they have sorted into cellular aggregates that protrude into the central cavity. (From Zalik and Sanders, 1974.) Scale bars: 100 μm (A), 10 μm (B), 200 μm (C), 100 μm (D), and 50 μm (E).

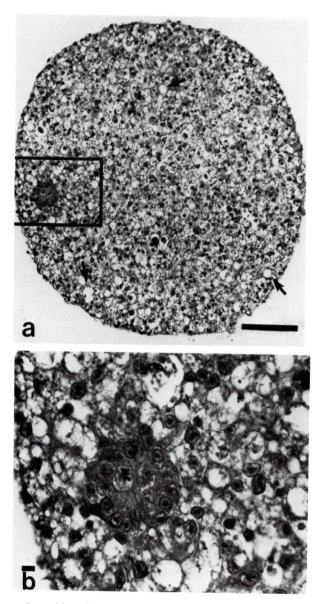

Figure 6. Aggregates formed by cell suspensions of extraembryonic endoderm from the area opaca of primitive streak stage chick blastoderms. (a), An early aggregate with a spongy texture. Note the smooth outer surface and the small spaces interspersed through the aggregate. (b) The boxed area is shown below at higher magnification; it contains a small clump of contaminating ectodermal cells. (c) Electron micrograph of an early aggregate. Here, cells are generally round but extend long cytoplasmic projections that interdigitate. Cell-to-cell contact is sporadic and limited to localized areas of apposing cell surfaces (arrows). Large intercellular spaces are abundant, giving a spongy appearance to the aggregate.   A, type A yolk (lipid and lipoprotein); B, type B yolk (lipid); C, complex

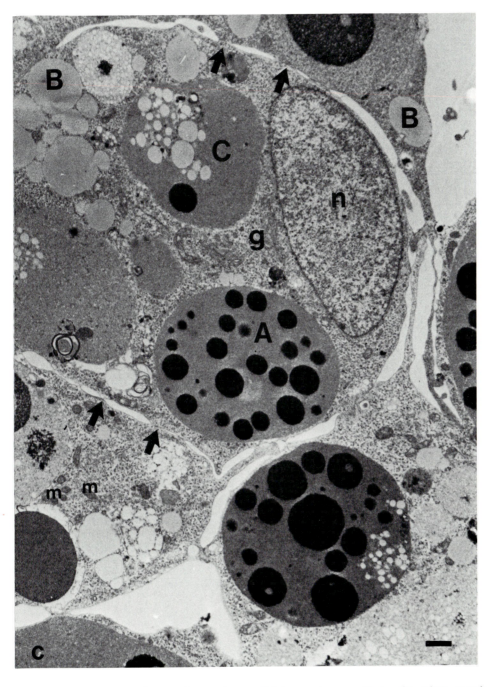

yolk (lipid, lipoprotein and protein); n, nucleus; g, Golgi apparatus; m, mitochondria. (d) Cavitated aggregate; a large, smooth central cavity has formed, the cell mass is vacuolated, and several clumps of contaminating cells are present (arrow). The large spheres are yolk platelets. Scale bars: 1 μm (a), 25 μm (b), 100 μm (c), and 200 μm (d). (From Milos et al., 1979, 1984.)

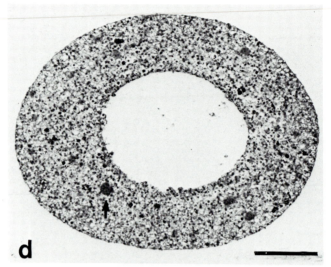

Figure 6. (*continued*)

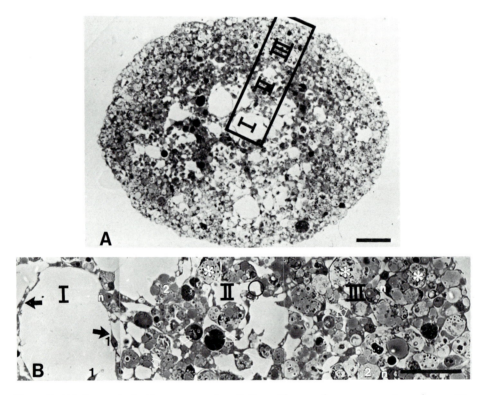

Figure 7. (A) Aggregate in the process of cavitation. Observe the numerous irregular cavities forming in the interior of the aggregate. (B) Enlargement of the area shown in A. I: central area— cells bordering the internal cavities have an elongated shape and contain little yolk (black arrow); II: intermediate area showing a lacy type of intercellular arrangement; III: peripheral region showing a more compact cellular association. Yolk platelets have coalesced into larger structures (*), and type B yolk is being extruded by some cells (white arrows). n, nucleus. Scale bars: 100 μm. (From Milos *et al.*, 1984.)

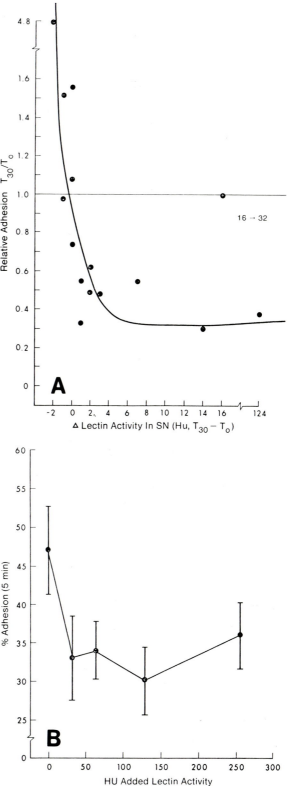

Figure 8. Effect of the endogenous β-D-galactoside-binding lectin of the early chick blastoderm on the adhesion of cells from the extraembryonic endoderm. (A) Relationship of changes in lectin activity to changes in adhesion in the cell suspensions. Freshly prepared cell suspensions were divided into aliquots of 1.5×10^5 cells in 250 μl and incubated at 22°C for 30 min. Two aliquots were prepared for each time point. In one aliquot, adhesion was assessed with the Coulter counter after rotation-mediated aggregation. The second aliquot supplied the supernatant tested for hemagglutinating activity. Data points above 1 on the ordinate are those at which more adhesion was observed after incubation; data points below 1 are those at which a lower adhesion was observed after 30 min in both cases as compared with the control cells, in which adhesion was assessed immediately after preparation of the cell suspension. In the experiment not included in the curve, adhesion was minimal (4%) at the outset and did not change after 30 min. The lectin titer in this experiment, initially 16, increased to 32. Observe that aggregation decreases as the lectin activity in the medium increases. (B) Effect of the purified blastoderm lectin on cell aggregation. Cells were incubated in the presence of increasing amounts of lectin. Incorporation of lectin into the medium decreases cellular adhesion.

studies indicate that these two lectins are present inside the cell as well as extracellularly in association with the cell surface (Zalik *et al.*, 1982).

At the present time the working hypothesis for a role of the galactose-binding lectins in adhesion in this cell system is as follows: Cells can release a form of β-ᴅ-galactoside-binding lectin that may bind competitively to cell surface galactose groups involved in forming adhesive bonds. This masking of surface galactose groups by the lectin may cause the dissociation of the galactose–receptor interactions between surfaces of adjacent cells and permit adhesive bonds to loosen in localized areas of the cell surfaces. These localized deadhesions could produce cellular rearrangements and movements such as that of epiboly as well as aggregate cavitation, implicating that these lectins may be involved in the formation of transitory adhesive bonds during cell relocation. A dual antagonistic role in adhesion has been also recently suggested for fibronectin (Yamada and Kennedy, 1984). It is tempting to speculate that the endogenous receptor for the cell-surface galactose groups that can be detected by *Ricinus communis* lectin is itself a galactose-binding protein. The evidence for this assumption is still missing and will have to await the isolation of the galactose-binding protein(s) from purified plasma membrane preparations.

5.3. Myogenesis

During the development of skeletal muscle, the precursor myogenic mesenchymal cells proliferate and give rise to myoblasts. The latter cells withdraw from the cell cycle and fuse with each other to give rise to multinucleate myotubes. Fusion of myoblasts with multinucleate myotubes also takes place, and young myotubes may also fuse with each other (Holtzer, 1970; Bischoff, 1978). The cell interactions involved in myotube formation have been studied in primary cultures of myoblasts from pectoral muscle of the 11-day chick embryo (Okazaki and Holtzer, 1965), the leg muscle of the 11–12-day chick embryo (Konigsberg, 1961; Konigsberg and Hauschka, 1965), and the thigh muscle of the newborn rat (Yaffe and Feldman, 1965). The pectoral muscle arises from the myotomal regions of the trunk somites (Chevallier, 1979), and the hind limb muscles appear to be of somitic as well as somatopleural origin (Chevallier *et al.*, 1977; Chevallier, 1979). Using stationary cultures, it has been found that myoblast fusion exhibits tissue specificity; i.e., myoblasts will fuse with other myoblasts independently of the species of origin but will not fuse with fibroblasts, chondrocytes, hepatocytes, or kidney cells (Okazaki and Holtzer, 1965; Yaffe and Feldman, 1965; Holtzer, 1970).

The relationship between cell adhesion and cell fusion has been studied using myoblast suspensions from 11-day chick embryo pectoral muscle cultured under conditions of stationary culture (Bischoff and Lowe, 1974; Bischoff, 1978) or subjected to rotation-mediated aggregation (Knudsen and Horwitz, 1977, 1978). In both situations, it has been found that at least two sequential steps in cell interaction precede myotube formation: **cell adhesion** and **mem-**

brane union. The initial event in cell adhesion requires calcium and involves cell recognition. This has been shown by the finding that there is a calcium-dependent preferential adhesion of myoblasts to each other or to young myotubes with exclusion of nonmyogenic cell types (Bischoff and Lowe, 1974; Knudsen and Horwitz, 1977, 1978; Bischoff, 1978). The adhesive bonds formed during the initial stage of myoblast adhesion are relatively weak, since the early aggregates obtained using rotation-mediated aggregation can be dissociated easily either by shearing or by treatment with EDTA. With subsequent time in culture, aggregates become resistant—first to dissociation by EDTA and somewhat later to dissociation by trypsin.

The acquisition by aggregates of resistance to EDTA dissociation is thought to indicate a qualitative change in the nature of the adhesive bonds. Since EDTA-resistant aggregates can be dissociated by trypsin, it is assumed that they are still composed of individual cells and that fusion has not yet occurred. The onset of membrane union is thought to occur when aggregates become refractory to dissociation by trypsin. Correlated with this is the presence of multinucleated cells and the transformation of aggregates into smooth-surfaced multinucleated myoballs (Knudsen and Horwitz, 1977, 1978). Membrane union also requires calcium, since it can be reversibly inhibited by the removal of calcium or the addition of EGTA (Shainberg *et al.*, 1969; Bischoff, 1978; Knudsen and Horwitz, 1978). Although some conditions, such as the absence of calcium and the presence of metabolic inhibitors, affect adhesion as well as fusion, others appear to have differential effects; i.e., cytochalasin B, colchicine, and magnesium have little effect on aggregation but inhibit fusion (Knudsen and Horwitz, 1977, 1978).

5.3.1. Lectins

The main lectin of skeletal muscle is a β-D-galactoside-specific agglutinin. This lectin has been isolated from the pectoral and hind limb muscles of the chick embryo (see Table III). In the pectoral muscle, the highest lectin activities are found between days 10 and 16 of development, whereas for the hindlimb the highest lectin activities are found in the 13-day embryo; both lectins decrease with further development (Den *et al.*, 1976; Nowak *et al.*, 1976; Kobiler and Barondes, 1977). The embryonic stages with the highest lectin activities correspond to the stages of myogenesis when mitotic division decreases and the onset of myoblast fusion occurs (Herrmann *et al.*, 1970). A similar galactose-specific lectin has been found in two myogenic cell lines derived from hind limb myoblasts from the newborn rat: L_6 and L_8 (Podleski *et al.*, 1979; Kaufman and Lawless, 1980). The L_6 line was derived by methylcholanthrene treatment (Yaffe, 1968, 1969), whereas the L_8 line was selected without prior exposure to carcinogen (Kaufman and Parks, 1977). In the L_6 line, lectin activity reaches maximum values at the stage when myoblast fusion is evident (Gartner and Podleski, 1976; Nowak *et al.*, 1976).

A second lectin has been found in extracts of the 15–16-day pectoral muscle of the chick embryo. This lectin is detectable using a preparation of trypsinized rabbit erythrocytes fixed with glutaraldehyde that have been further sensitized by aging or by treatment with ethanol. The agglutinin is specific for N-acetyl-D-galactosamine; the most effective inhibitors of the agglutination induced by this lectin are heparin, heparan sulfate, and dermatan sulfate (Mir-Lechaire and Barondes, 1978; Ceri et al., 1979, 1981; Kobiler and Barondes, 1979) (see Table III). In myoblast cultures at the time of myoblast fusion, this lectin is released into the medium; by contrast, the galactose-specific lectin is only detectable in cell extracts and is absent from the medium at this time. Interestingly, the agglutination induced by the N-acetyl-D-galactosamine-specific lectin is also inhibited by the glycosaminoglycans extracted from chick embryonic pectoral muscle and by the substrate-attached material produced by cultured myoblasts from this tissue. It has been suggested that the heparin-binding lectin may play a role in cell surface interactions with glycosaminoglycans (Ceri et al., 1979).

5.3.2. Myoblast Adhesion and Fusion

The involvement of cell surface glycoconjugates in myotube formation was suggested by experiments indicating that Con A inhibited fusion of cultured myoblasts (Den et al., 1975; Sandra et al., 1977). The finding of an endogenous β-D-galactoside-binding lectin in developing muscle whose activity increased significantly at the stages of myoblast fusion stimulated further interest in the role that this lectin might play in cell interactions in this system. In most studies, the approach to this problem has been to investigate the effect of this lectin and its saccharide inhibitors on fusion of myoblasts maintained in stationary cultures. In some experiments, myoblasts are maintained in culture under conditions that inhibit fusion, such as low Ca^{2+} concentrations (0.16–0.29 mM), or presence of low concentrations of EGTA. In these fusion-blocked cultures, myoblasts proliferate but are prevented from fusing. When the normal concentration of Ca^{2+} is restored in the medium (1.4–2.0 mM), fusion occurs within 3–4 hr to reach the level of control cultures maintained in the presence of this cation (Shainberg et al., 1969; Bischoff, 1978).

At the present time, results obtained by different laboratories on the effect of the galactose-specific lectin and its saccharide inhibitors on myoblast fusion are inconsistent. In myoblast cultures from 11-day chick pectoral muscle maintained in the continuous presence and replenishment of picomolar concentrations of lectin in the medium, myoblast fusion is inhibited, and the inhibitory effects of the lectin are overcome by the addition of lactose to the medium (MacBride and Przybylski, 1980). A considerable retardation of fusion was also observed in cells exposed to lactose (0.15 M) for 48 hr, but this inhibition was overcome with time; other sugars, such as α-methyl-D-mannoside, α-methyl-D-glucoside, and α-L-fucose, at equivalent concentrations were not inhibitory. In L_6 and L_8 myoblasts, thiodigalactoside inhibits myoblast fusion only when cells are subject to long-term exposure (3–7 days) to this disaccharide (Gartner and Podleski, 1976; Kaufman and Lawless, 1980). Short-term exposure of cultures either to the lectin or to thiodigalactoside did not significantly affect myoblast fusion (Den *et al.*, 1976; Knudsen and Horwitz, 1978; Kaufman and Lawless, 1980; Den and Chin, 1981). Similarly, short-term exposure of cultures of embryonic chick thigh myoblasts to antilectin antibodies did not affect fusion (Den and Chin, 1981).

The cellular localization of the galactoside-binding lectins has been investigated primarily using immunohistochemical approaches. In the embryonic chick pectoral muscle *in situ*, the localization of this lectin is mainly intracellular, although some lectin is consistently found at the cell periphery. In contrast, after birth, the lectin is found predominantly at the cell periphery (Barondes and Haywood-Reid, 1981). In cell cultures of myoblasts and myoblast cell lines, the lectin is localized predominantly in the cytoplasm. In most of these experiments, however, some lectin is also present in clusters or spots at the cell surface (Nowak *et al.*, 1977; Kaufman and Lawless, 1980; Podleski and Greenberg, 1980), and specific carbohydrate-binding activity at the cell surface has been detected by a modification of the rosette assay (Gartner and Podleski, 1975). However, since so much of the lectin is intracellular, the possibility remains of leakage of lectin into the medium with nonspecific binding to the cell periphery. As a result, some investigators have been reluctant to regard the immunohistochemical staining at the cell periphery as an indicator of an actual localization of the lectin at the cell surface (Kaufman and Lawless, 1980; Den and Chin, 1981).

As mentioned previously, myoblast fusion is the final result of a series of cell interactions beginning with cell recognition and adhesion. Unfortunately, research on the effects of the galactoside-binding lectin and its saccharide inhibitors on cell adhesion is scanty. When aggregation is measured by the rotating aggregate assay, myoblasts appear to have some Ca^{2+}-independent adhesion (about 20% of the control value, see Fig. 1 in Knudsen and Horwitz, 1977). In this system, incorporation of thiodigalactoside into the medium either inhibits or stimulates, to a small degree, myoblast adhesion (Knudsen and Horwitz, 1978, Table I). This situation is similar to that encountered with cells of the extraembryonic endoderm of the gastrulating chick blastoderm (Milos and Zalik, 1982). It would be of interest to determine if the galactose-binding lectin has an effect on the Ca^{2+}-independent adhesion in myoblasts. That this Ca^{2+}-independent adhesion may be of importance is suggested by the fact that myoblasts cultured in the presence of EGTA for several hours form aggregates whose cells can fuse when calcium is added to the medium (Knudsen and Horwitz, 1978), suggesting that under some conditions Ca^{2+}-independent adhesive bonds could replace Ca^{2+}-dependent adhesive bonds.

It is also known that at a certain stage in embryonic development the surface of the myotube becomes refractory to fusion (Bischoff and Holtzer, 1969). It would be of interest to determine whether the endogenous galactose-specific lectin plays a role in this process by masking some adhesive sites. This type of role has been suggested for the galactose-binding lectin of the chick blastoderm (see Section 5.2.2). The localization of lectin at discrete sites in the developing myotube could regulate its length and inhibit branching, as suggested by recent observations of MacBride and Przybylski (1981). Branched myotubes are not found *in vivo*, but it is known that extensive branching occurs in myotubes formed *in vitro* (Okazaki and Holtzer, 1965; Herrmann *et al.*, 1970). The precipitous synchronized fusion that occurs in myoblast cultures during a time span of several hours *in vitro* may not allow for the analysis of the highly regulated sequential events of myoblast and myotube fusion that occur in the embryonic muscle *in vivo* during a time span of several weeks (Herrmann *et al.*, 1970; Bischoff, 1978).

6. Concluding Remarks

It is evident that lectins are widely distributed in the plant and animal kingdoms. The presence of endogenous lectins in developing systems has been documented, and in some tissues marked changes in lectin activity occur during particular developmental events. This chapter discusses the existing evidence implicating a role for endogenous lectins in cell adhesion in animal systems. One can only say that in the systems in which this problem has been examined, the results point to an indirect role for endogenous lectins in cell interactions—perhaps in the modulation of cell adhesion. It is clear that research on this subject is still in its infancy, and gaps in our knowledge remain. It will be particularly important to establish unequivocally the localization of the lectins at the cell surface as well as the mechanisms that regulate the externalization of these molecules from the interior of the cell.

ACKNOWLEDGMENTS. The original research reported in this chapter was supported by grants from the Medical Research Council of Canada and from the Natural Sciences and Engineering Research Council of Canada to S. E. Z. This work was also supported in part by grants from the Alberta Heritage Foundation for Medical Research to S. E. Z. and by a postdoctoral fellowship to N. C. M. We thank Dr. R. G. MacBride and Dr. R. J. Przybylski for allowing us to see their manuscript before publication. We are grateful to Dr. W. E. G. Müller for giving us permission to reproduce some of his published figures. We extend our appreciation to Irene Ledsham and Eva Dimitrov for their assistance. S. E. Z. thanks Mrs. Kay Baert for her valuable suggestions in the typing of the manuscript. N. C. M. is a Research Scholar of the Alberta Heritage Foundation for Medical Research.

References

Adair, W. S., Monk, B. C., Cohen, R., Hwang, C., and Goodenough, U. W., 1982, Sexual agglutinins from *Chlamydomonas* flagellar membrane. Partial purification and characterization, *J. Biol. Chem.* **257**:4593–4602.

Adair, W. S., Hwang, C., and Goodenough, U. W., 1983, Identification and visualization of the sexual agglutinin of the mating-type plus flagellar membrane of *Chlamydomonas*, *Cell* **33**:183–193.

Anstee, D. M., Holt, P. D. J., and Pardoe, G. I., 1973, Agglutinins from fish ova defining blood group B and P, *Vox Sang* **25**:347–360.

Ashwell, G., and Harford, J., 1982, Carbohydrate-specific receptors of the liver, *Annu. Rev. Biochem.* **51**:531–554.

Barak Briles, E., Gregory, W., Gletcher, P., and Kornfeld, S., 1979, Comparison and properties of β-galactoside-binding lectins of tissues of calf and chicken, *J. Cell Biol.* **81**:528–537.

Barondes, S. H., 1981, Lectins: Their multiple endogenous cellular functions, *Annu. Rev. Biochem.* **50**:207–231.

Barondes, S. H., 1984, Soluble lectins: A new class of extracellular proteins, *Science* **223**:1259–1264.

Barondes, S. H., and Haywood-Reid, P. L., 1981, Externalization of an endogenous chicken muscle lectin with *in vivo* development, *J. Cell Biol.* **91**:568–572.

Barondes, S. H., Rosen, S., Frazier, W. A., Simpson, D., and Haywood, P., 1978, Dictyostelium discoideum agglutinins, *Methods Enzymol.* **50**(C):306–312.

Bartles, J. R., Frazier, W. A., and Rosen, S., 1982, Slime mold lectins, *Int. Rev. Cytol.* **75**:61–99.

Bellairs, R., 1971, *Developmental Processes in Higher Vertebrates*, Logos Press, London.

Beug, H., and Gerisch, G., 1972, A micromethod for routine measurement of cell agglutination and dissociation, *J. Immunol. Methods* **2**:49–57.

Beyer, E. C., Tokuyasu, K. T., and Barondes, S. H., 1979, Localization of an endogenous lectin in chicken liver, intestine and pancreas, *J. Cell Biol.* **82**:565–571.

Beyer, E. C., Zweig, S. E., and Barondes, S. H., 1980, Two lactose-binding lectins from chick tissues, *J. Biol. Chem.* **255**:4236–4239.

Bezkorovainy, A., Springer, G. F., and Desai, P. S., 1971, Physicochemical properties of the eel anti-human blood-group H(O) antibody, *Biochemistry* **10**:3761–3764.

Bischoff, R., 1978, Myoblast fusion, in: *The Cell Surface in Animal Embryogenesis and Development* (G. Poste and G. L. Nicolson, eds.), pp. 128–181, Elsevier/North-Holland, Amsterdam.

Bischoff, R., and Holtzer, H., 1969, Mitosis and the processes of differentiation of myogenic cells *in vitro*, *J. Cell Biol.* **41**:188–200.

Bischoff, R., and Lowe, M., 1974, Cell surface components and the interaction of myogenic cells, in: *Exploratory Concepts in Muscular Dystrophy* (A. T. Milhorat, ed.), pp. 17–31, Excerpta Medica, New York.

Bolwell, G. P., Callow, J. A., Callow, M. E., and Evans, L. V., 1979, Fertilization in brown algae. II. Evidence for lectin-sensitive complementary receptors involved in gamete recognition in *Fucus serratus*, *J. Cell Sci.* **36**:19–30.

Bolwell, G. P., Callow, J. A., and Evans, L. V., 1980. Fertilization in brown algae. III. Preliminary characterization of putative gamete receptors from eggs and sperm of *Fucus serratus*, *J. Cell Sci.* **43**:209–224.

Bowles, D. H., and Hanke, D. E., 1977, Evidence for lectin activity associated with glycophorin, the major glycoprotein in human erythrocyte membranes, *FEBS Lett.* **82**:34–39.

Bowles, D., and Rotman, A., 1978, Agglutination activity associated with a glycoprotein extract of human platelet plasma membranes, *FEBS Lett.* **90**:283–285.

Boyd, W. C., and Shapleigh, E., 1954a, Specific precipitating activity of plant agglutinins (lectins), *Science* **119**:419.

Boyd, W. C., and Shapleigh, E., 1954b, Antigenic relation of blood group antigens as suggested by tests with lectins, *J. Immunol.* **73**:226–231.

Bretting, H., and Kabat, E. A., 1976, Purification and characterization of agglutinins from the sponge *Axinella polypoides* and a study of their combining sites, *Biochemistry* **15**:3228–3236.

Bretting, H., and Koningsmann, R., 1979, Investigation of lectin-producing cells in the sponge *Axinella polypoides* (Schmidt), *Cell Tissue Res.* **201**:487–497.

Bretting, H., Donadey, C., Vacelet, J., and Jacobs, G., 1981a, Investigations on the occurrence of lectins in marine sponges with special regard to some species of the family Axinellidae, *Comp. Biochem. Physiol.* **70B**:69–76.

Bretting, H., Phillips, S. G., Klumpart, H. J., and Kabat, E. A., 1981b, A mitogenic lactose-binding lectin from the sponge *Geodia cydonium*, *J. Immunol.* **127**:1652–1658.

Bretting, H., Jacobs, G., Donadey, C., and Vacelet, J., 1983, Immunohistochemical studies on the distribution and function of the β-galactose-specific lectins in the sponge *Axinella polypoides* (Schmidt), *Cell Tissue Res.* **229**:551–571.

Burger, M. M., Burkart, W., Weinbaum, G., and Jumblatt, J., 1978, Cell–cell recognition: Molecular aspects. Recognition and its relation to morphogenetic processes in general, in: *Cell–Cell Recognition*, Vol. XXXII (A. Curtis, ed.), *Symposia of the Society for Experimental Biology*, pp. 1–24, Cambridge Press, Cambridge.

Cauldwell, G. B., Henkart, P., and Humphreys, T., 1973, Physical properties of sponge aggregation factor. A unique proteoglycan complex, *Biochemistry* **12**:3051–3055.

Ceri, H., Shadle, P. J., Kobiler, D., and Barondes, S. H., 1979, Extracellular lectin and its glycosaminoglycan inhibitor in chick muscle cultures, *J. Supramol. Struct.* **11**:61–67.

Ceri, H., Kobiler, D., and Barondes, S. H., 1981, Heparin-inhibitable lectin. Purification from chicken liver and embryonic chicken muscle, *J. Biol. Chem.* **256**:390–394.

Cerra, R. F., Haywood-Reid, P. L., and Barondes, S. H., 1984, Exogenous mammalian lectin localized extracellularly in lung elastic fibers, *J. Cell Biol.* **98**:1580–1589.

Chevallier, A., 1979, Role of the somitic mesoderm in the development of the thorax in bird embryos, *J. Embryol. Exp. Morphol.* **49**:73–88.

Chevallier, A., Kieny, M., and Mauger, A., 1977, Limb–somite relationship: Origin of the limb musculature, *J. Embryol. Exp. Morphol.* **41**:245–258.

Childs, R. A., and Feizi, T., 1979, β-Galactoside-binding muscle lectins of man and monkey show antigenic cross-reactions with those of bovine origin, *Biochem. J.* **183**:755–758.

Clamp , J. R., 1974, Analysis of glycoproteins, in: *The Metabolism and Function of Glycoproteins*, Biochemical Society Symposium, Vol. 40 (R. M. S. Smellie and J. G. Beeley, eds.), pp. 3–16, William Clowes and Sons Ltd., London.

Cook, G. M. W., and Stoddart, R. W., 1973, *Surface Carbohydrates of the Eukaryotic Cell*, Academic Press, New York.

Cook, G. M. W., Zalik, S. E., Milos, N., and Scott, V., 1979, A lectin which binds specifically to β-D-galactoside groups is present at the earliest stages of chick embryo development, *J. Cell Sci.* **38**:293–304.

Crandall, M., 1978, Mating-type interactions in yeast, in: *Cell–Cell Recognition, Vol. XXXII, Symposia of the Society for Experimental Biology*, pp. 105–119, Cambridge Press, Cambridge.

Dazzo, F. B., 1980, Lectins and their saccharide receptors as determinants of specificity in Rhizobium-legume symbiosis, in: *The Cell Surface: Mediator of Developmental Processes* (S. Subtelny and N. Wessels, eds.), pp. 277–304, Academic Press, New York.

Dazzo, F. B., and Brill, W. J., 1977, Receptor site on clover and alfalfa roots for Rhizobium, *Appl. Environ. Microbiol.* **33**:132–136.

Dazzo, F. B., and Truchet, G. L., 1983, Interactions of lectins and their saccharide receptors in the Rhizobium-legume symbiosis, *J. Membrane Biol.* **73**:1–16.

Dazzo, F. B., Yanke, W. E., and Brill, W. J., 1978, Trifolin: A Rhizobium recognition protein from white clover, *Biochim. Biophys. Acta* **539**:276–286.

Del Campillo, E., Shannon, L. M., and Hankins, C. N., 1981, Molecular properties of the enzymic phytohemagglutinin of mung bean, *J. Biol. Chem.* **256**:7177–7180.

Den, H., and Chin, J. H., 1981, Endogenous lectin from chick embryo skeletal muscle is not involved in myotube formation *in vitro*, *J. Biol. Chem.* **264**:8069–8073.

Den, H., and Malinzak, D. A., 1977, Isolation and properties of the β-D-galactoside-specific lectin from chick embryo thigh muscle, *J. Biol. Chem.* **252**:5444–5448.

Den, H., Malinzak, D. A., Keating, H. J., and Rosenberg, A., 1975, Influence of Concanavalin A, wheat germ agglutinin, and soybean agglutinin on the fusion of myoblasts *in vitro*, *J. Cell Biol.* **67**:826–834.

Den, H., Malinzak, D. A., and Rosenberg, A., 1976, Lack of evidence for the involvement of a β-D-galactosyl-specific lectin in the fusion of chick myoblasts, *Biochem. Biophys. Res. Commun.* **69**:621–627.

Dey, P. M., Naik, S., and Pridham, J. P., 1982, The lectin nature of α-galactosidases from *Vicia faba* seeds, *FEBS Lett.* **150**:233–237.

Dey, P. M., Del Campillo, E. D., and Pont Lezica, P., 1983, Characterization of a glycoprotein α-galactosidase from lentil seeds (*Lens culinaris*), *J. Biol. Chem.* **258**:923–929.

Dodd, R. Y., MacLennan, A. P., and Hawkins, D. C., 1968, Haemagglutinins from marine sponges, *Vox Sang* **15**:386–391.

Dysart, J., and Edwards, J. G., 1977, A membrane-bound hemagglutinin from cultured hamster fibroblasts, *FEBS Lett.* **75**:96–100.

Edelman, G. M., 1983, Cell adhesion molecules, *Science* **219**:450–457.

Eisenbarth, G. S., Ruffolo, R. S., Walsch, F. S., and Nirenberg, M., 1978, Lactose sensitive lectin of chick retina and spinal cord, *Biochem. Biophys. Res. Commun.* **83**:1246–1252.

Elliott, E. V., Kerbel, R. S., and Phillips, B. J., 1974, Rosette formation by a mouse fibroblast cell line, *Nature (Lond.)* **248**:514–515.

Eyal-Giladi, H., Kochav, S., and Yerushalmi, S., 1975, The sorting-out of thymidine-labelled chick hypoblast cells in mixed epiblast-hypoblast aggregates, *Differentiation* **4**:57–60.

Finkelstein, R. A., Boesman-Finkelstein, M., and Holt, P., 1983, *Vibrio cholerae* hemagglutinin/lectin/protease hydrolyzes fibronectin and ovomucin: F. M. Burnet revisited, *Proc. Natl. Acad. Sci. USA* **80**:1992–1995.

Fraser, B. R., and Zalik, S. E., 1977, Lectin-mediated agglutination of amphibian embryonic cells, *J. Cell Sci.* **27**:227–243.

Frazier, W., and Glaser, L., 1979, Surface components and cell recognition, *Annu. Rev. Biochem.* **48**:491–523.

Galstoff, P. S., 1929, Heteroagglutination of dissociated sponge cells, *Biol. Bull.* **57**:250–260.

Garrod, D. R., and Nicol, A., 1981, Cell behaviour and molecular mechanisms of cell–cell adhesion, *Biol. Rev.* **56**:199–242.

Garrod, D. R., and Steinberg, M. S., 1973, Tissue-specific sorting-out in two dimensions in relation to contact inhibition of cell movement, *Nature (Lond.)* **244**:568–569.

Garrod, D. R., and Steinberg, M. S., 1975, Cell locomotion within a contact-inhibited monolayer of chick embryonic liver parenchyma cells, *J. Cell Sci.* **18**:405–425.

Gartner, T. K., and Podleski, T. R., 1975, Evidence that a membrane-bound lectin mediates fusion of L-6 myoblasts, *Biochem. Biophys. Res. Commun.* **67**:972–978.

Gartner, T. K., and Podleski, T. R., 1976, Evidence that the types and specific activity of lectins control fusion of L-6 myoblasts, *Biochem. Biophys. Res. Commun.* **70**:1142–1149.

Gartner, T. K., Stocker, K., and Williams, D. C., 1980, Thrombolectin: A lectin isolated from *Bothrops atrox* venom, *FEBS Lett.* **117**:13–16.

Glabe, C. G., Grabel, L. B., Vacquier, V. D., and Rosen, A., 1982, Carbohydrate specificity of the sea urchin sperm bindin: A cell surface lectin mediating sperm-egg adhesion, *J. Cell Biol.* **94**:123–128.

Gold, E. R., Phelps, C. F., Khalap, S., and Balding, P., 1974, Observations on *Axinella* sp. hemagglutinin, *Ann. NY Acad. Sci.* **234**:122–128.

Goldstein, I. J., and Hayes, C. E., 1978, The lectins, carbohydrate binding proteins of plants and animals, *Adv. Carbohydrate Chem. Biochem.* **35**:127–340.

Goldstein, I. J., Hughes, C. R., Monsigny, M., and Sharon, N., 1980, What should be called a lectin?, *Nature (Lond.)* **285**:66.

Goodenough, U. W., and Adair, W. S., 1980, Membrane adhesions between *Chlamydomonas* gametes and their role in cell interactions, in: *The Cell Surface Mediator of Developmental Processes* (S. Subtelny and N. K. Wessels, eds.), pp. 101–112, Academic Press, New York.

Goodenough, U. W., Adair, W. S., Caligor, E., Forest, C. L., Hoffman, J. L., Mesland, D. A. M., and Spath, A., 1980, Membrane-membrane and membrane-ligand interactions in *Chlamydomonas* mating, in: *Membrane–membrane Interactions* (N. B. Gilula, ed.), pp. 121–152, Raven Press, New York.

Grabel, L. B., Rosen, S. D., and Martin, G., 1979, Teratocarcinoma stem cells have a cell surface carbohydrate-binding component implicated in cell–cell adhesion, *Cell* **17**:477–484.

Grabel, L. B., Glabe, C. G., Singer, M. S., Martin, G. R., and Rosen, S. D., 1981, A fucan specific lectin on teratocarcinoma stem cells, *Biochem. Biophys. Res. Commun.* **102**:1165–1171.

Grabel, L. B., Singer, M. S., Martin, G. R., and Rosen, S. D., 1983a, Teratocarcinoma stem cell adhesion: The role of divalent cations and a cell surface lectin, *J. Cell Biol.* **96**:1532–1537.

Grabel, L. B., Singer, M. S., Martin, G. R., and Rosen, S. D., 1983b, Purification of a putative cell adhesion mediating lectin from teratocarcinoma stem cells and its possible role in differentiation, *J. Cell Biol.* **97**:251A.

Gremo, F., Kobiler, D., and Barondes, S. H., 1978, Distribution of an endogenous lectin in the developing chick optic tectum, *J. Cell Biol.* **79:**491–499.

Greve, L. C., and Hedrick, J. L., 1978, An immunochemical localization of the cortical granule lectin in fertilized and unfertilized eggs of *Xenopus laevis, Gamete Res.* **1:**13–18.

Grinnell, F., 1978, Cellular adhesiveness and extracellular substrata, *Int. Rev. Cytol.* **53:**56–144.

Grinnell, F., and Feld, M. K., 1979, Initial adhesion of human fibroblasts in serum-free medium: Possible role of secreted fibronectin, *Cell* **17:**117–129.

Grinnell, F., and Feld, M. K., 1982, Fibronectin adsorption on hydrophilic and hydrophobic substances detected by antibody binding and analyzed during cell adhesion in serum-containing medium, *J. Biol. Chem.* **257:**4888–4893.

Hakomori, S., 1981, Glycosphingolipids in cellular interaction, differentiation and oncogenesis, *Annu. Rev. Biochem.* **50:**733–764.

Hamburger, V., and Hamilton, H. J., 1951, A series of normal stages in the development of the chick embryo, *J. Morphol.* **88:**49–92.

Hankins, C. N., and Shannon, L. M., 1978, The physical and enzymatic properties of a phytohemagglutinin from mung beans, *J. Biol. Chem.* **253:**7791–7797.

Hankins, C. N., Kindinger, J. I., and Shannon, L. M., 1980, Legume α-galactosidases which have hemagglutinin properties, *Plant Physiol.* **65:**618–622.

Harford, J., and Ashwell, G., 1982, The hepatic receptor for asialoglycoproteins, in: *The Glycoconjugates,* Vol. IV-B (M. I. Horowitz, ed.), pp. 27–56, Academic Press, New York.

Harris, H. L., and Zalik, S. E., 1982, The presence of an endogenous lectin in early embryos of *Xenopus laevis, Arch. Dev. Biol.* **191:**208–210.

Harris, H., and Zalik, S. E., 1985, Studies on the endogenous galactose-binding lectin during early development of the embryo of *Xenopus laevis, J. Cell Sci.*(in press).

Harrison, F. L., and Chesterton, C. J., 1980a, Erythroid developmental agglutinin is a protein mediating specific cell–cell adhesion between differentiating rabbit erythroblasts, *Nature (Lond.)* **286:**502–504.

Harrison, F. L., and Chesterton, C. J., 1980b, Factors mediating cell recognition and adhesion. Galaptins, a recently discovered class of bridging molecules, *FEBS Lett.* **122:**157–165.

Hatten, M. E., 1981, Cell assembly patterns of embryonic mouse cerebellar cells on carbohydrate-derivatized polylysine culture substrata, *J. Cell Biol.* **89:**54–61.

Hatten, M. E., and François, A. M., 1981, Adhesive specificity of developing cerebellar cells on lectin substrata, *Dev. Biol.* **87:**102–113.

Henkart, P., Humphreys, S., and Humphreys, T., 1973, Characterization of sponge aggregation factor. A unique proteoglycan complex, *Biochemistry* **12:**3045–3051.

Herrmann, H., Heywood, S. M., and Marchok, A. C., 1970, Reconstruction of muscle development as a sequence of macromolecular syntheses, in: *Current Topics in Developmental Biology,* Vol. 5 (A. A. Moscona and A. Monroy, eds.), pp. 181–280, Academic Press, New York.

Holtfreter, J., 1939, Gewebeaffinität, ein Mittel der embryonalen Formbilding. Tissue affinity, a means for embryonic morphogenesis, in: *Foundations of Experimental Embryology* (B. Willier and J. M. Oppenheimer, eds.), pp. 186–225, Prentice-Hall, Englewood, Cliffs, New Jersey; *Arch. Exp. Zellforschung.* **23:**169–209.

Holtzer, H., 1970, Myogenesis, in: *Cell Differentiation* (O. A. Schjeide and J. De Vellis, eds.), pp. 476–503, Van Nostrand, New York.

Hughes, R. C., 1976, *Membrane Glycoproteins,* Butterworth, London.

Humphreys, T., 1963, Chemical dissolution and *in vitro* reconstruction of sponge cell adhesion. I. Isolation and functional characterization of the components involved, *Dev. Biol.* **8:**27–47.

Johansson, A., Kjellen, L., Höok, M., and Timpl, R., 1981, Substrate adhesion of rat hepatocytes: A comparison of laminin and fibronectin as attachment proteins, *J. Cell Biol.* **90:**260–264.

Jumblatt, J. E., Schlup, V., and Burger, M. M., 1980, Cell recognition: Specific binding of *Microciona* sponge aggregation factor to homotypic cells and the role of calcium ions, *Biochemistry* **19:**1038–1042.

Junqua, S., Robert, L., Garrone, R., Pavans de Ceccatty, M., and Vaselet, J., 1974, Biochemical and morphological studies on collagens in horny sponges. *Ircinia* filaments compared to spongin, *Connect. Tissue Res.* **2:**193–203.

Kaufman, S. J., and Lawless, M. L., 1980, Thiodigalactoside binding lectin and skeletal myogenesis, *Differentiation* **16**:41–48.

Kaufman, S. J., and Parks, C. M., 1977, Loss of control and differentiation in the fu-l variant of the L_8 line of rat myoblasts, *Proc. Natl. Acad. Sci. USA* **74**:3888–3892.

Kawasaki, T., and Ashwell, G., 1976, Chemical and physical properties of an hepatic membrane protein that binds asialoglycoproteins, *J. Biol. Chem.* **251**:1296–1302.

Kennedy, D. W., Rohrbach, D. H., Martin, G. R., Momoi, T., and Yamada, K. W., 1983, The adhesive glycoprotein laminin is an agglutinin, *J. Cell Physiol.* **114**:257–262.

Khalap, S., Thompson, T. E., and Gold, E. R., 1970, Haemagglutination and haemagglutination inhibition reactions of extracts from snails and sponges. Agglutination of human and various animal red cells. Its inhibition by sugars and amino sugars, *Vox Sang* **18**:501–526.

Kieda, M. T., Bowles, D., Amiram, D., and Sharon, N., 1978, Lectins in lymphocyte membranes, *FEBS Lett.* **94**:391–396.

Kitamura, K., 1980, Changes in lectin activity during the development of the embryonic chick skin, *J. Embryol. Exp. Morphol.* **59**:59–69.

Kitamura, K., 1981, Distribution of the endogenous β-D-galactoside-specific lectin, fibronectin and Type I and III collagens during dermal condensation in chick embryos, *J. Embryol. Exp. Morphol.* **65**:41–56.

Knudsen, K. A., and Horwitz, A. F., 1977, Tandem events in myoblast fusion, *Dev. Biol.* **58**:328–338.

Knudsen, K. A., and Horwitz, A. F., 1978, Differential inhibition of myoblast fusion, *Dev. Biol.* **66**:294–308.

Kobiler, D., and Barondes, S. H., 1977, Lectin activity from embryonic chick brain, heart and liver. Changes with development, *Dev. Biol.* **60**:326–330.

Kobiler, D., and Barondes, S. H., 1979, Lectin from embryonic chick muscle that interacts with glycosaminoglycans, *FEBS Lett.* **101**:257–261.

Kobiler, D., Beyer, R. C., and Barondes, S. H., 1978, Developmentally regulated lectins from chick muscle, brain and liver have similar chemical and immunological properties, *Dev. Biol.* **64**:265–272.

Koch, O. M., Lee, C. K., and Uhlenbrück, G., 1982, Cerianthin lectins: A new group of agglutinins from *Cerianthus membranaceus* (Singapore), *Immunobiology* **163**:53–62.

Kolb, H., Kolb-Bachoffen, V., and Schlepper-Schäffer, J., 1979, Cell contacts mediated by D-galactose-specific lectins in liver cells, *Biol. Cell.* **36**:301–308.

Kolb-Bachoffen, V., 1981, Hepatic receptor for asialoglycoproteins ultrastructural demonstration of ligand-induced microaggregation of receptors, *Biochim. Biophys. Acta* **645**:293–299.

Kolb-Bachoffen, V., Schlepper-Schäffer, J., Vogell, W., and Kolb, H., 1982, Electron microscopic evidence for an asialoglycoprotein receptor in Kupffer cells. Localization of lectin-mediated endocytosis, *Cell* **29**:859–866.

Komano, H., Nozawa, R., Mizuno, D., and Natori, S., 1983, Measurement of *Sarcophaga peregrina* lectin under various physiological conditions by radioimmunoassay, *J. Biol. Chem.* **258**:2143–2147.

Konigsberg, I. R., 1961, Some aspects of myogenesis in vitro, *Circulation* **24**:447–457.

Konigsberg, I. R., and Hauschka, S. D., 1965, Cell and tissue interactions in the reproduction of cell type, in: *Reproduction: Molecular, Subcellular, and Cellular* (M. Locke, ed.), pp. 243–290, 24th Symposia of the Society for Developmental Biology, Academic Press, New York.

Lackie, J. M., 1980, The structure and organization of the cell surface, in: *Membrane Structure and Function*, Vol. I (E. E. Bittar, ed.), pp. 73–102, John Wiley and Sons, New York.

Lee, Y., and Lee, R. T., 1982, Neoglycoproteins as probes for binding and cellular uptake of glycoconjugates, in: *The Glycoconjugates*, Vol. IV, Part B (M. Horowitz, ed.), pp. 57–84, Academic Press, New York.

Levi, G., and Teichberg, V. I., 1981, Isolation and physiochemical characterization of electrolectin, a β-D-galactoside-binding lectin from the electric organ of *Electrophorus electricus*, *J. Biol. Chem.* **256**:5735–5740.

Lillie, F. R., 1913, The mechanism of fertilization, *Science* **38**:524–528.

Lipsick, J. S., Beyer, E. C., Barondes, S. H., and Kaplan, N. D., 1980, Lectins from chicken tissues are mitogenic for THY-1 negative murine spleen cells, *Biochem. Biophys. Res. Commun.* **97:**56–61.

Lis, H., and Sharon, N., 1972, Soybean (*Glycine max*) agglutinin, *Methods Enzymol.* **38:**360–368.

Lis, H., and Sharon, N., 1981, Lectins in higher plants, in: *The Biochemistry of Plants*, Vol. 6 (A. Marcus, P. K. Stumpf, and E. E. Conn, eds.), pp. 371–447, Academic Press, New York.

Lunney, J., and Ashwell, G., 1976, A hepatic receptor of avian origin capable of binding specifically modified glycoproteins, *Proc. Natl. Acad. Sci. USA* **73:**341–343.

MacBride, R. G., and Przybylski, R. J., 1980, Purified lectin from skeletal muscle inhibits myotube formation *in vitro, J. Cell Biol.* **85:**617–625.

MacBride, R. G., and Przybylski, R. J., 1981, The role of endogenous lactose-binding lectins in chick skeletal myogenesis, *J. Cell Biol.* **91:**355a.

MacLennan, A. P., and Dodd, R. Y., 1967, Promoting activity of extracellular materials on sponge cell aggregation, *J. Embryol. Exp. Morphol.* **17:**473–480.

Malhotra, S. K., 1983, *The Plasma Membrane*, John Wiley and Sons, New York.

Marchalonis, J. J., and Edelman, G. M., 1968, Isolation and characterization of a hemagglutinin from *Limulus polyphemus, J. Mol. Biol.* **32:**453–465.

Martin, G. R., 1980, Teratocarcinomas and mammalian embryogenesis, *Science* **209:**768–775.

Maslow, D. E., 1976, *In vitro* analysis of surface specificity in embryonic cells, in: *The Cell Surface in Animal Embryogenesis and Development* (G. Poste and G. L. Nicolson, eds.), pp. 697–745, Elsevier/North-Holland, Amsterdam.

Milos, N., and Zalik, S. E., 1981a, Mechanisms of adhesion among cells of the early chick blastoderm. Role of calcium ions in the adhesion of extraembryonic endoderm cells, *Arch. Dev. Biol.* **190:**139–142.

Milos, N., and Zalik, S. E., 1981b, Effect of the β-D-galactoside-binding lectin on cell to substratum and cell to cell adhesion of cells from the extraembryonic endoderm of the early chick blastoderm, *Arch. Dev. Biol.* **190:**259–266.

Milos, N., and Zalik, S. E., 1982, Mechanisms of adhesion among cells of the early chick blastoderm. Role of the β-D-galactoside-binding lectin in the adhesion of extraembryonic endoderm cells, *Differentiation* **21:**175–182.

Milos, N., and Zalik, S. E., 1983, Calcium-independent adhesion of extra-embryonic endoderm cells from the early chick blastoderm is inhibited by the β-D-galactoside-binding lectin and by β-galactosidase, *Cell Diff.* **12:**341–347.

Milos, N., and Zalik, S. E., 1985, Release of β-D-galactoside-binding lectins into the cavities of chick extraembryonic endoderm cells, *Cell Diff.* (in press).

Milos, N., Zalik, S. E., and Phillips, R., 1979, An analysis of the aggregation and morphogenesis of area opaca endoderm cells from the primitive streak chick embryo, *J. Embryol. Exp. Morphol.* **51:**121–135.

Milos, N., Ledsham, I. M., Boerner, S. L., and Zalik, S. E., 1983, Presence of the soluble galactoside-binding lectin in the fluid contents of cavitating aggregates of extraembryonic endoderm cells from gastrulating chick embroys, *J. Cell Biol.* **97:**320a.

Milos, N., Zalik, S. E., Sanders, E. J., and Ledsham, I., 1984, Intercellular relationships during cavitation of aggregates of extraembryonic endoderm cells from gastrulating chick embryos, *J. Embryol. Exp. Morphol.* **83:**43–61.

Mir-Lechaire, F. J., and Barondes, S. H., 1978, Two distinct developmentally regulated lectins in chick embryo muscle, *Nature (Lond.)* **272:**256–258.

Mirelman, D., and Kobiler, D., 1981, Adhesion properties of *Entamoeba histolytica* in: *Adhesion and Microorganism Pathogenicity*, Ciba Foundation Symposium No. 80, pp. 17–30. Pitman Books, London.

Monsigny, M., Kieda, C., and Roche, A. C., 1983, Membrane glycoproteins, glycolipids and membrane lectins in normal and malignant cells, *Biol. Cell.* **47:**95–110.

Moscona, A. A., 1968, Cell aggregation: Properties of specific cell-ligands and their role in the formation of multicellular systems, *Dev. Biol.* **18:**250–277.

Moscona, A. A., 1971, Embryonic and neoplastic cell surfaces: Availability of receptors for Concanavalin A and wheat germ agglutinin, *Science* **171:**905–906.

Müller, W. E. G., 1982, Cell membranes in sponges, *Int. Rev. Cytol.* **77:**129–181.

Müller, W. E. G., and Müller, I., 1980, Sponge cell aggregation, *Mol. Cell. Biochem.* **29**:131–143.

Müller, W. E. G., and Zahn, R. K., 1973, Purification and characterization of a species-specific aggregation factor in sponges, *Exp. Cell Res.* **80**:95–104.

Müller, W. E. G., Müller, I., Zahn, R. K., and Kurelec, B., 1976, Species specific aggregation factor in sponges. VI. Aggregation receptor from the cell surface, *J. Cell Sci.* **21**:227–241.

Müller, W. E. G., Kurelec, B., Zahn, R. K., Müller, I., Vaith, P., and Uhlenbruck, G., 1979a, Aggregation of sponge cells. Function of a lectin in its homologous biological system, *J. Biol. Chem.* **254**:7479–7481.

Müller, W. E. G., Zahn, R. K., Kurelec, B., Müller, I., Vaith, P., and Uhlenbruck, G., 1979b, Aggregation of sponge cells. Isolation and characterization of an inhibitor of aggregation receptor from the cell surface, *Eur. J. Biochem.* **97**:585–591.

Müller, W. E. G., Zahn, R. K., Kurelec, B., Lucu, C., Müller, I., and Uhlenbruck, G., 1981a, Lectin, a possible basis for symbiosis between bacteria and sponges, *J. Bacteriol.* **145**:548–558.

Müller, W. E. G., Zahn, R. K., Müller, I., Kurelec, B., Uhlenbruck, G., and Vaith, P., 1981b, Cell aggregation in the marine sponge *Geodia cydonium*. Identification of lectin-producing cells, *Eur. J. Cell Biol.* **24**:28–35.

Müller, W. E. G., Conrad, J., Schröder, C., Zahn, R. K., Kljajić, Z., Müller, I., and Uhlenbruck, G., 1983a, Cell–cell recognition system in gorgonians: Description of the basic mechanism, *Mar. Biol.* **76**:1–6.

Müller, W. E. G., Conrad, J., Schröder, C., Zahn, R. K., Kurelec, B., Dreesback, K., and Uhlenbruck, G., 1983b, Characterization of the trimeric, self-recognizing *Geodia cydonium* lectin I, *Eur. J. Biochem.* **133**:263–267.

Neufeld, E. F., and Ashwell, G., 1980, Carbohydrate recognition system for receptor mediated pinocytosis, in: *The Biochemistry of Glycoproteins and Proteoglycans* (W. J. Lennarz, ed.), pp. 241–266, Plenum Press, New York.

Nicolet, G., 1971, Avian gastrulation, in: *Advances in Morphogenesis*, Vol. 9 (M. Abercrombie, J. Brachet, and T. King, eds.), pp. 231–262, Academic Press, New York.

Nicolson, G. L., 1974, The interactions of lectins with animal cell surfaces, *Int. Rev. Cytol.* **39**:89–190.

Nosek, J., Krajhanzl, A., and Kacourek, J., 1983, Immunofluroescence localization of lectins present in fish ovaries, *Histochemistry* **79**:131–139.

Novogrodsky, A., and Ashwell, G., 1977, Lymphocyte mitogenesis induced by a mammalian liver protein that specifically binds desialylated glycoproteins, *Proc. Natl. Acad. Sci. USA* **74**:676–678.

Nowak, T. P., Haywood, P. L., and Barondes, S. H., 1976, Developmentally regulated lectin in embryonic chick muscle and a myogenic cell line, *Biochem. Biophys. Res. Commun.* **68**:650–657.

Nowak, T. P., Kobiler, D., Roel, L., and Barondes, S. H., 1977, Developmentally regulated lectin from embryonic chick pectoral muscle, *J. Biol. Chem.* **252**:6026–6030.

Okazaki, K., and Holtzer, H., 1965, An analysis of myogenesis *in vitro* using fluorescein-labelled antimyosin, *J. Histochem. Cytochem.* **13**:726–739.

Oppenheimer, S., 1977, Interactions of lectins with embryonic cell surfaces, in: *Current Topics in Developmental Biology*, Vol. 11 (A. A. Moscona and A. Monroy, eds.), pp. 1–16, Academic Press, New York.

Oppenheimer, S., and Meyer, J. T., 1982, Carbohydrate specificity of sea urchin blastula adhesion component, *Exp. Cell Res.* **139**:451–454.

Ozawa, M., Sato, M., and Maromatsu, T., 1983, Basement membrane glycoprotein laminin is an agglutinin, *J. Biochem.* **94**:479–485.

Phillips, J. R., and Zalik, S. E., 1982, Differential lectin-mediated agglutinabilities of the embryonic and first extraembryonic cell line of the early chick embryo, *Arch. Dev. Biol.* **191**:234–240.

Pierce, G. B., Shikes, R., and Fink, L. M., 1978, *Colon Cancer: A Problem of Developmental Biology*, Prentice-Hall, Englewood Cliffs, New Jersey.

Pitts, M. J., and Yang, C. H. D., 1980, Isolation of a developmentally regulated lectin from chick embryo, *Biochem. Biophys. Res. Commun.* **95**:750–757.

Podleski, T. R., and Greenberg, I., 1980, Distribution and activity of endogenous lectin during myogenesis as measured with antilectin antibody, *Proc. Natl. Acad. Sci. USA* **77**:1054–1058.

Podleski, T. R., Greenberg, I., and Nichols, S. C., 1979, Studies on lectin activity during myogenesis, *Exp. Cell Res.* **122**:305–316.

Powell, J. T., 1980, Purification and properties of lung lectin. Rat lung and human lung β-galactoside-binding proteins, *Biochem. J.* **187**:123–129.

Powell, J. T., and Whitney, P. L., 1980, Post natal development of rat lung. Changes in lung lectin, elastin, acetylcholinesterase and other enzymes, *Biochem. J.* **188**:1–8.

Rauvala, H., Carter, W. G., and Hakomori, S., 1980, Studies on cell adhesion and recognition . I. Extent and specificity of cell adhesion triggered by carbohydrate-reactive proteins (glycosidases and lectins) and by fibronectin, *J. Cell Biol.* **88**:127–137.

Rauvala, H., Prieels, J. P., and Finne, J., 1983, Cell adhesion mediated by a purified fucosyltransferase, *Proc. Natl. Acad. Sci. USA* **80**:3991–3995.

Reitherman, R. W., Rosen, S. D., and Barondes, S. H., 1973, Lectin purification using formalinised erythrocytes as a general affinity adsorbant, *Nature (Lond.)* **248**:509–510.

Roberson, M. M., and Armstrong, P. B., 1980, Carbohydrate-binding component of amphibian embryo cell surfaces: Restriction to surface regions capable of cell adhesion, *Proc. Natl. Acad. Sci. USA* **77**:3460–3463.

Roberson, M. M., and Barondes, S. H., 1982, Lectin from embryos and oocytes of *Xenopus laevis*, *J. Biol. Chem.* **257**:7520–7524.

Roberson, M. M., and Barondes, S. H., 1983, *Xenopus laevis* lectin is localized at several sites in *Xenopus* oocytes, eggs and embryos, *J. Cell Biol.* **97**:1875–1881.

Roche, A. C., and Monsigny, M., 1974, Purification and properties of limulin: a lectin (agglutinin) from hemolymph of *Limulus polyphemus*, *Biochim. Biophys. Acta* **371**:242–254.

Roff, C. G., Rosevear, P. R., Wang, J. L., and Barker, R., 1983, Identification of carbohydrate-binding proteins from mouse and human fibroblasts, *Biochem. J.* **211**:625–629.

Romanoff, A. L., 1960, *The Avian Embryo*, Macmillan, New York.

Roseman, S., 1970, The synthesis of complex carbohydrates by multiglycosyltransferase systems and their potential function in intercellular adhesion, *Chem. Phys. Lipids* **5**:270–297.

Roseman, S., Rottman, W., Walther, B., Ohman, R., and Umbreit, J., 1974, Measurement of cell–cell interactions, *Methods Enzymol.* **32**(B):597–611.

Rosen, S., Kafka, J. A., Simpson, D. L., and Barondes, S. H., 1973, Developmentally regulated, carbohydrate-binding protein in *Dictyostelium discoideum*, *Proc. Natl. Acad. Sci. USA* **70**:2554–2557.

Roth, S., McGuire, E. J., and Roseman, S., 1971, An assay for intercellular adhesive specificity, *J. Cell Biol.* **51**:525–535.

Rutherford, N. G., and Cook, G. M. W., 1981, Isolation and characterization of a mannan-binding protein associated with the early chick embryo, *FEBS Lett.* **136**:105–110.

Rutherford, N. G., and Cook, G. M. W., 1984, Endogenous lectins in the developing chick embryo, in: *Matrices and Differentiation* (J. R. Hinchliffe and R. B. Kemp, eds.), pp. 129–146, Alan R. Liss, New York.

Sakakibara, F., Takayanagi, G., Kawauchi, H., Watanabe, K., and Hakomori, S., 1976, An anti-A-like lectin of *Rana catesbiana* eggs showing unusual reactivity, *Biochim. Biophys. Acta* **444**:386–395.

Sakakibara, F., Kawauchi, H., Takayanagi, G., and Ise, H., 1979, Egg lectin of *Rana japonica* and its receptor glycoprotein of Ehrlich tumour cells, *Cancer Res.* **39**:1347–1352.

Sandra, A., Leon, M. A., and Przbylski, R. J., 1977, Suppression of myoblast fusion by Concanavalin A; a possible involvement of membrane fluidity, *J. Cell Sci.* **28**:251–272.

Sasaki, H., and Aketa, K., 1981, Purification and distribution of a lectin in sea urchin (*Anthocidaris crassispina*) egg before and after fertilization, *Exp. Cell Res.* **135**:15–19.

Schlepper-Schäffer, J., Friedrich, E., and Kolb, H., 1981, Galactosyl-specific receptor on liver cells: Binding site for tumor cells, *Eur. J. Cell Biol.* **25**:95–102.

Shainberg, A., Yagil, G., and Yaffe, D., 1969, Control of myogenesis *in vitro* by Ca^{2+} concentration in nutritional medium, *Exp. Cell Res.* **58**:163–167.

Sharon, N., and Lis, H., 1981, Glycoproteins: Research booming on long ignored, ubiquitous compounds, *Chem. Eng. News* **59:**21–44.

Sharon, N., Eshdat, Y., Silverblatt, F. J., and Ofek, I., 1981, Bacterial adherence to cell surface sugars, in: *Adhesion and Microorganism Pathogenicity*, Ciba Foundation Symposium, No. 80, pp. 119–135, Pitman Books, London.

Shur, B. D., 1982, Cell surface glycosyltransferase activities during fertilization and early embryogenesis, in: *The Glycoconjugates*, Vol. III (M. Horowitz, ed.), pp. 145–185, Academic Press, New York.

Simpson, D., Thorne, D., and Loh, H., 1977, Developmentally regulated lectin in the neonatal rat brain, *Nature (Lond.)* **266:**367–369.

Sly, W. S., 1982, The uptake and transport of lysosomal enzymes, in: *The Glycoconjugates*, Vol. IV (M. I. Horowitz, ed.), pp. 3–25, Academic Press, New York.

Springer, G. F., and Desai, P. R., 1971, Monosaccharides as specific precipitinogens in eel anti-human blood-group H(O) antibody, *Biochemistry* **10:**3749–3760.

Stockert, R. J., Morell, A. G., and Scheinberg, I. H., 1974, Mammalian hepatic lectin, *Science* **186:**365–366.

Suzuki, T., and Natori, S., 1983, Identification of a protein having hemagglutinating activity in the hemolymph of the silkworm *Bombyx mori, J. Biochem.* **93:**583–590.

Takeuchi, M., Yoshikawa, M., Sasaki, R., and Chiba, H., 1982, Partial purification and characterization of the rat mammary gland lectin, *Agric. Biol. Chem.* **46:**2741–2747.

Teichberg, V. I., 1978, Electrolectins: β-D-galactoside-binding proteins, *Methods Enzymol.* **50:**291–302.

Teichberg, V. I., Silman, I., Beitsch, D. D., and Resheff, G., 1975, A β-D-galactoside-binding protein from electric organ tissue of *Electrophorus electricus, Proc. Natl. Acad. Sci. USA* **72:**1383–1387.

Thomas, W. A., and Steinberg, M., 1980, A twelve-channel automatic device for continuous recording of cell aggregation by measurement of small angle light-scattering, *J. Cell Sci.* **41:**1–18.

Tonegawa, Y., 1982, Cell aggregation factor and endogenous lectin in sea urchin embryos, *Cell Diff.* **11:**335–337.

Townes, P. L., and Holtfreter, J., 1955, Directed movements and selective adhesion of embryonic amphibian cells, *J. Exp. Zool.* **128:**53–120.

Trinkaus, J. P., 1976, On the mechanism of metazoan cell movements, in: *The Cell Surface in Animal Embryogenesis and Development* (G. Poste and G. L. Nicolson, eds.), pp. 225–329, Elsevier/North-Holland, Amsterdam.

Trinkaus, J. P., and Lentz, T. L., 1964, Direct observation of type specific segregation in mixed cell aggregates, *Dev. Biol.* **9:**115–136.

Tyler, A., 1940, Agglutination of the sea urchin eggs by means of a structure extracted from eggs, *Proc. Natl. Acad. Sci. USA* **26:**249–256.

Tyler, A., 1946, An auto-antibody concept of cell structure, growth and differentiation, *Growth* **10:**7–19, (Suppl.)

Uhlenbruck, G., Janssen, E., and Javeri, S., 1982, Two different anti-galactan lectins in eel serum, *Immunobiology* **163:**36–47.

Vacquier, V. D., 1980, The adhesion of sperm to sea urchin eggs, in: *The Cell Surface: Mediator of Developmental Processes* (S. Subtelny and N. K. Wessells, eds.), pp. 151–185, Academic Press, New York.

Vacquier, V. D., and Moy, W. G., 1977, Isolation of bindin: The protein responsible for adhesion of sperm to sea urchin eggs, *Proc. Natl. Acad. Sci. USA* **74:**2456–2460.

Vaith, P., Uhlenbruck, G., Müller, W. E. G., and Holz, G., 1979, Sponge aggregation factor and sponge hemagglutinin: Possible relationship between two different molecules, *Dev. Comp. Immunol.* **3:**399–416.

Van de Vyer, G., 1975, Phenomena of cellular recognition in sponges, in: *Current Topics in Developmental Biology*, Vol. 10 (A. A. Moscona and A. Monroy, eds.), pp. 123–140, Academic Press, New York.

Vasta, G. R., and Marchalonis, J. J., 1983, Humoral recognition factors in the Arthropoda. The specificity of Chelicerata serum lectins, *Am. Zool.* **23:**157–171.

Voak, D., Todd, G. M., and Pardoe, G. I., 1974, A study of the serological behaviour and nature of anti-B/P/P^k activity of Salmonidae roe protectins, *Vox Sang* **26:**176–188.

Voss, E. W., Fryer, J. L., and Banowetz, G. M., 1978, Isolation, purification and partial characterization of a lectin from chinook salmon ova, *Arch. Biochem. Biophys.* **186:**25–34.

Weigel, P. H., 1980, Rat hepatocytes bind to synthetic galactoside surfaces via a patch of asialoglycoprotein receptors, *J. Cell Biol.* **87:**855–861.

Weigel, P. H., Schmell, E., Lee, Y. C., and Roseman, S., 1978, Specific adhesion of rat hepatocytes to β-galactosides linked to acrylamide gels, *J. Biol. Chem.* **253:**330–333.

Weigel, P. H., Schnaar, R. L., Kuhlenschmidt, M. S., Lee, R. T., and Roseman, S., 1979, Adhesion of hepatocytes to immobilized sugars, *J. Biol. Chem.* **254:**10830–10838.

Weinbaum, G., and Burger, M. M., 1973, A two component system for surface guided reassociation in animal cells, *Nature (Lond.)* **244:**510–512.

Weiss, P., 1947, The problem of specificity in growth and development, *Yale J. Biol. Med.* **19:**235–278.

Weiss, P., 1950, Perspectives in the field of morphogenesis, *Q. Rev. Biol.* **25:**177–198.

Wiese, L., and Wiese, W., 1978, Sex cell contact in *Chlamydomonas*, a model for cell recognition, in: *Cell–Cell Recognition*, Symposium for the Society of Experimental Biology, Vol. 32 (C. A. Curtis, ed.), pp. 83–104, Cambridge University Press, Cambridge.

Wilson, H. V., 1907, On some phenomena of coalescence and regeneration in morphogenesis, *J. Exp. Zool.* **5:**245–258.

Wyrick, R. E., Nishihara, T., and Hedrick, J., 1974, Agglutination of jelly coat and cortical granule components and the block of polyspermy in the amphibian *Xenopus laevis*, *Proc. Natl. Acad. Sci. USA* **71:**2067–2071.

Yaffe, D., 1968, Retention of differentiated potentialities during prolonged cultivation of myogenic cells, *Proc. Natl. Acad. Sci. USA* **61:**477–483.

Yaffe, D., 1969, Cellular aspects of muscle differentiation *in vitro*, in: *Current Topics in Developmental Biology*, Vol. 4 (A. A. Moscona and A. Monroy, eds.), pp. 37–77, Academic Press, New York.

Yaffe, D., and Feldman, M., 1965, The formation of hybrid multinucleated muscle fibers from myoblasts of different genetic origin, *Dev. Biol.* **11:**300–317.

Yamada, K. W., and Kennedy, D. W., 1984, Dualistic nature of adhesive protein function: Fibronectin and its biologically active peptide fragments can autoinhibit fibronectin function, *J. Cell Biol.* **99:**29–36.

Yamada, K. W., Yamada, S. S., and Pastan, I., 1975, The major cell surface glycoprotein of chick embryo fibroblasts is an agglutinin, *Proc. Natl. Acad. Sci. USA* **72:**3158–3162.

Yamada, Y., and Aketa, K., 1982, Purification and partial characterization of hemagglutinins in seminal plasma of the sea urchin *Hemicentrotus pulcherrimus*, *Biochim. Biophys. Acta* **709:**220–226.

Yanagishima, N., and Yoshida, K., 1981, Sexual interactions in *Saccharomyces cerevisiae* with special reference to the regulation of sexual agglutinability, in: *Sexual Interactions in Eukaryotic Microbes* (D. H. O'Day and P. A. Horgen, eds.), pp. 261–294, Academic Press, New York.

Zalik, S. E., and Cook, G. M. W., 1976, Comparison of embryonic and differentiating cell surfaces. Interactions of lectins with plasma membrane components, *Biochim. Biophys. Acta* **419:**119–136.

Zalik, S. E., and Sanders, E. J., 1974, Selective cellular affinities in the unincubated chick blastoderm, *Differentiation* **2:**25–28.

Zalik, S. E., Milos, N., and Ledsham, I., 1982, Distribution of a galactose-specific lectin in endoderm cells from early chick embryos, *Cell Tissue Res.* **225:**223–228.

Zalik, S. E., Milos, N., and Ledsham, I., 1983, Distribution of two β-D-galactoside-binding lectins in the gastrulating chick embryo, *Cell Diff.* **12:**121–127.

Chapter 6

Cell–Cell Interactions during Sea Urchin Morphogenesis

EVELYN SPIEGEL and MELVIN SPIEGEL

1. Introduction

The dissociation of sea urchin embryos into a cell suspension and their subsequent reaggregation provide a model system for studying cell–cell interactions during morphogenesis. Herbst (1900) first showed that sea urchin blastomeres could be dissociated with calcium-free seawater and—when returned to normal seawater—could reaggregate and develop into abnormal larvae. These experiments demonstrated that the hyaline layer surrounding the embryo holds the cells together and that calcium plays a role in cell adhesion. Giudice and Mutolo (1970) used this system to study the role of cell–cell interactions in the appearance of new metabolic patterns in development. This chapter reports on more recent work from our own and other laboratories using reaggregating cells to study cell adhesion and other cell–cell interactions during development. It also reports on our use of intact embryos to study the extracellular matrices and cell junctions on the external and internal surfaces during sea urchin morphogenesis.

2. Morphological Studies of Cell Reaggregation

2.1. Development of Aggregates

The principal method used for the dissociation of embryos into a suspension of blastomeres is to remove the hyaline layer by washing embryos in Ca^{2+},Mg^{2+}-free seawater, followed by gentle aspiration using a Pasteur pipette, which breaks the cell junctions. After return to normal seawater, the cells can reaggregate and develop (Spiegel and Spiegel, 1975). There are at least three different pathways of development by which reaggregating cells from 16-

EVELYN SPIEGEL and MELVIN SPIEGEL • Department of Biological Sciences, Dartmouth College, Hanover, New Hampshire 03755.

cell-stage embryos can form pluteus larvae. Several species used, including *Arbacia punctulata, Lytechinus pictus, Strongylocentrotus drobachiensis,* and *Strongylocentrotus purpuratus,* are all capable of reaggregating to form almost normal plutei.

One pathway of aggregate development is from clusters of cells and is the one that most resembles normal development. The dissociated cells settle, adhere to the substratum, and immediately begin to move about. Figure 1,

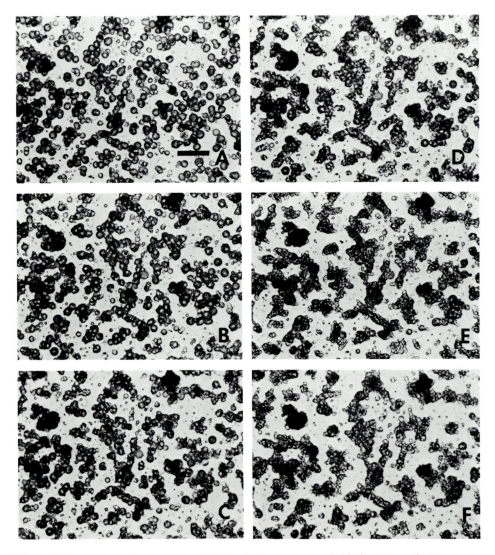

Figure 1. Time course of reaggregation within a single microscope field of mixtures of micromeres, mesomeres, and macromeres isolated from *Arbacia punctulata* 16-cell–stage embryos and recombined in the cell number ratio of 4 : 8 : 4 respectively. Photographs were taken at 0.5-hr intervals, with (A) taken 4 min after cells were returned to seawater and reaggregation was initiated. Scale bar: 100 μm. (From Spiegel and Spiegel, 1975.)

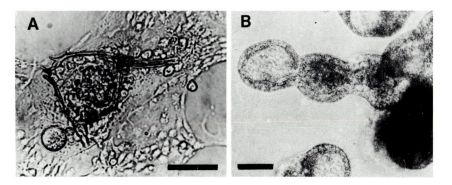

Figure 2. (A) Pluteus formation from *Arbacia* tissue culture phase; "A" frame of skeleton is surrounded by flattened fibroblastlike cells. (B) Chain of beads formation showing swellings that resemble blastulae, i.e., a single layer of cells enclosing a central cavity. Scale bars: 50 μm. (From Spiegel and Spiegel, 1975.)

photographed at half-hour intervals, illustrates the time course of reaggregation. Small clumps of cells are formed, and, as new cells are added, the aggregates gradually become larger. During reaggregation, the cells divide at a rate comparable to that in the normal embryo. Not only do individual cells move, but large aggregates of as many as 50–100 cells are capable of movement. Aggregates increase in size by the adherence of cell to cell, the addition of free cells to aggregates, and the fusion of aggregate to aggregate. With time, the aggregates round up, become ciliated, lose their adhesion to the dish, and swim off, looking very much like normal blastulae.

The second type of aggregate development occurs when the concentration of cells is very high. The cells settle out, forming a dense mat. A branching network of cells forms, and swellings of beads appear (Fig. 2B). The beads resemble blastulae (i.e., a single layer of cells enclosing a central cavity). The chains of blastulae become ciliated, and individual blastulae frequently break away from the chain, gastrulate, and form almost normal plutei.

The third pathway of aggregate development is even more unusual. A large aggregate of 1–2-mm diameter reattaches to the substratum and spreads out over the surface as a monolayer of cells. Many of the cells resemble fibroblasts in appearance, and the sheet of cells resembles a tissue culture preparation. Skeletal components are laid down, and typical A frames of the pluteus become visible in many parts of the monolayer. Cells appear to be piling up around the skeleton, and a pluteus begins to take shape (Fig. 2A). The pluteus-like larva eventually detaches and swims off, leaving a gap in the monolayer.

2.2. Morphological Specificity of Cell Adhesion

In order to study the specificity of cell adhesion, mixtures of cells of two different species were combined. *Arbacia punctulata* cells are a densely pigmented dark red and are easily distinguishable from *Lytechinus pictus* cells,

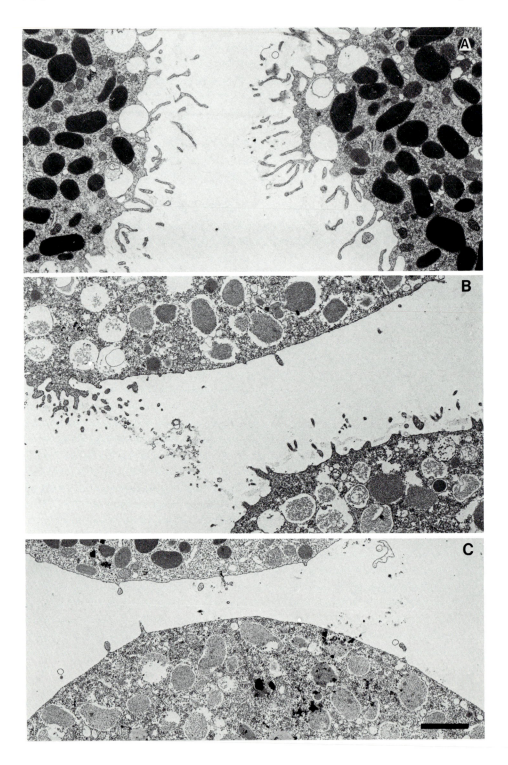

which are a pale transparent yellow in color. Initially, the cells of both species can adhere to each other. With time, however, the cells of the two species sort out from each other, eventually forming aggregates composed of either one or the other species, but never of both. Thin sections of aggregates were examined with transmission electron microscopy (TEM). Cells of the same species were found to make initial contact with each other by forming microvilli and secreting hyaline layer material (Fig. 3A,B). When cells of different species approach each other, microvilli are not formed, and hyaline material is not secreted (Fig. 3C). These results indicate that there may be a mutually inhibitory effect between the unlike cells or that like cells mutually stimulate the formation of microvilli and hyaline layer.

During normal cleavage of *Arbacia* embryos, microvilli are formed between apposing cell surfaces (Fig. 4A). In embryos of the sand dollar, *Echinarachnius parma*, cleaving cells form scalloped borders instead of microvilli (Fig. 4B). These two species were used to study further the species-specific morphology of cell adhesion (Spiegel and Spiegel, 1978a). Reaggregating cells of each species cultured separately displayed the same cell surface morphology as in normal cleavage: *Arbacia* reaggregating cells formed microvilli and *Echinarachnius* reaggreating cells formed scalloped borders. In a mixed culture containing both species of blastomeres, neither microvilli nor a scalloped border formed between blastomeres of the two different species. This may be another example of the mutual inhibition of cell adhesion by unlike cells or—alternatively—the stimulation of like cells by each other. Although sea urchin and sand dollar cells apparently adhere to one another early in reaggregation, the formation of hyaline material or cell junctions was not observed. After 8–10 hr, clusters were composed of either one species or the other, but never of both. It appears that the cells of the two different species recognize each other as being different and sort out. Another way of expressing the sorting out behavior may be that cells unable to form intercellular adhesive substances or cell junctions are not able to adhere to each other permanently.

Cell adhesion also appears to be cell type-specific, as shown in the following experiment (Spiegel and Spiegel, 1978b). *Arbacia* embryos at the 8-cell stage were stained with the vital dye Nile blue sulfate. At the 16-cell stage, the embryos were dissociated, and the different cell types were separated. Stained micromeres were then combined with unstained mesomeres and macromeres in the normal ratio of 4 : 8 : 4 and allowed to reaggregate. The stained micromeres gradually sorted out so that in 8-hr reaggregates there was usually only a single mass of stained cells. These results support similar work described for vertebrate cells, in which sorting out according to cell type also occurs (Townes and Holtfreter, 1955).

Figure 3. (A) Reaggregating *Arbacia* cells 5 min after return to normal seawater and initiation of reaggregation. (B) Reaggregating *Lytechinus* cells 5 min after return to normal seawater and initiation of reaggregation. (C) Mixed culture of reaggregating cells showing *Arbacia* cells (top) and *Lytechinus* cells (bottom); microvilli and hyaline material have not formed at 30 min after initiation of reaggregation. Scale bar: 2 μm. (From Spiegel and Spiegel, 1975.)

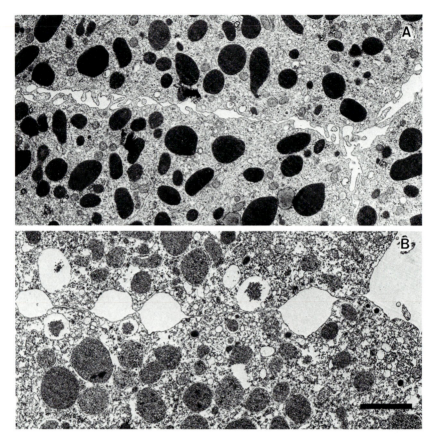

Figure 4. (A) *Arbacia* 4-cell stage showing microvilli formed between two apposing cell surfaces. (B) *Echinarachnius* 4-cell stage showing scalloped border between two apposing cell surfaces. Scale bar: 2 μm. (From Spiegel and Spiegel, 1978.)

Figure 5. (A) *Lytechinus* 2-cell stage covered with dense meshwork of hyaline layer in which microvilli are embedded. (B) Four blastomeres of an 8-cell *Lytechinus* embryo, which became detached during processing. The meshwork layer surrounds the outer cell surfaces of the embryo. The inner cell surfaces have fewer microvilli. (C) *Lytechinus* 16-cell stage, beginning the fifth cleavage. The four large macromeres are behind the four small micromeres. Behind the macromeres are the eight mesomeres. Remnants of the fertilization envelope are attached to some of the cells. (D) A suspension of *Lytechinus* cells dissociated at the 16-cell stage, showing a micromere, a mesomere, and a macromere. The dense meshwork of hyaline layer and microvilli has been partially stripped off and is seen as light, fluffy clumps or threads, pieces of which are attached to cells or to the substrate. (E) A cleaving blastomere of a late 4-cell stage of *Lytechinus* that was processed for SEM and separated from the other blastomeres by touching it with a needle. Part of the outer meshwork layer was broken off where the needle touched the embryo (arrow). (F) A higher magnification of part of E showing that the outer meshwork layer consists of a dense, structured hyaline layer with microvilli embedded within it. Scale bars: 10 μm (A–E); 5 μm (F). (From Spiegel and Spiegel, 1978.)

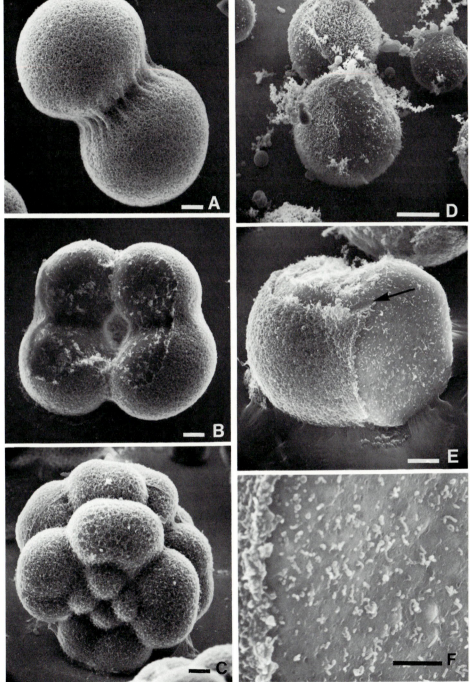

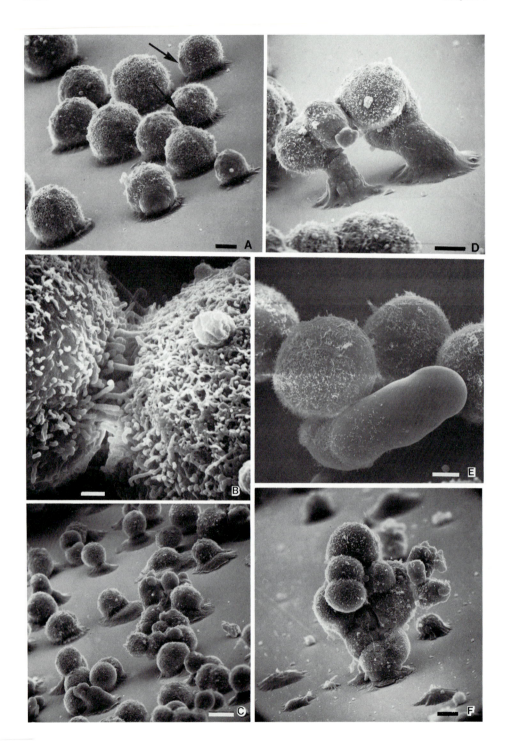

2.3. Scanning Electron Microscopy of Reaggregating Cells

In order to examine the changes occurring at the cell surface in more detail, scanning electron microscopy (SEM) was used (Spiegel and Spiegel, 1977). The first step was to observe the surface of normal embryos. Figure 5A shows a *Lytechinus* embryo covered by a dense meshwork of hyaline layer, in which microvilli are embedded. Figure 5B shows two mesomeres above two macromeres from an 8-cell-stage; the other four cells became detached during processing, exposing the inner surface of the embryo. The hyaline layer surrounds the embryo but does not penetrate between the cells. The plasma membrane surfaces are not completely smooth, however; they have fewer and shorter microvilli. Figure 5E is a cleaving blastomere that was separated from a 4-cell-stage embryo, which had been processed for SEM by gently touching it with a fine glass needle. The needle also chipped off part of the hyaline layer and exposed the cell surface. Figure 5F is a higher magnification of an area in Figure 5E. Figure 5 clearly shows that the hyaline layer, which appears to be an empty, clear layer in the light microscope and in some TEMs of thin sections, is actually a dense, structured layer.

Figure 5C is a 16-cell-stage *Lytechinus* embryo that has started the fifth cleavage. It shows the hyaline layer closely adhering to all three cell types. Figure 5D shows representative cells of a cell suspension of *Lytechinus* that was dissociated at the 16-cell stage. The dense meshwork of hyaline layer and microvilli has been partially stripped off and is seen as light, fluffy clumps or threads, pieces of which are attached to cells or the substrate. The three cell types—a micromere, a mesomere, and a macromere—are shown; these cells when returned to normal seawater immediately begin to move about and reaggregate.

Within 5 min after reaggregation has begun, ruffles and filopodia have formed in all three cell types (Fig. 6A). Within 15 min, microvilli can be seen extending between closely adhering cells (Fig. 6B). After 30 min, the ruffles have grown larger, enabling the cells to move rather quickly and extensively over the substratum (Fig. 6C). Blunt-shaped pseudopodia, which also enable

---

Figure 6. (A) Dissociated *Arbacia* cells at 5 min after reaggregation has begun. Long, thin filopodial strands attach the cells to the substrate, and ruffles have formed. Arrows indicate areas where cells were attached in the intact embryo. The different sizes of blastomeres represent the three cell types present at the 16-cell stage. (B) *Arbacia* cells at 15 min after reaggregation has begun. Microvilli extend between adhering cells. (C) *Arbacia* cells at 30 min after reaggregation has begun. The ruffles (lamellipodia) have grown larger and form a "skirt" surrounding the cell or are oriented in one direction, which appears to be the direction of movement. (D) *Arbacia* cells at 30 min after reaggregation has begun. A stalk has formed on two adhering cells, which raises the cells above the substrate. The stalks are free of microvilli except in the area nearest to the cell proper. (E) *Lytechinus* cells at 30 min after reaggregation has begun. One of the cells has formed a large, blunt pseudopodlike structure, which is mostly free of microvilli except in the area nearest to the cell proper. (F) A cluster of *Arbacia* cells formed after 2 hr of reaggregation. The cluster appears to have developed by adding cells upward. The ruffles and stalks left on the substrate may have been left behind as cells were added to the cluster or because the cells attached to them were washed away during processing. Scale bars: 10 μm (A,D–F); 2 μm (B); 25 μm (C). (From Spiegel and Spiegel, 1977.)

the cells to move, are seen to be enormous structures compared with the size of the cell itself (Fig. 6E). Another unusual structure that is formed is a stalk that raises cells above the surface of the dish (Fig. 6D). After several hours the aggregates, which were originally composed of a flat layer of cells, raise up off the surface to form a compact mass of cells. One such cluster is shown in Fig. 6F, which was taken after 2 hr of reaggregation. It is not known whether the ruffles and stalk remaining on the substrate were left behind as cells were added upward to the cluster or because the cells attached to them were washed away during processing for SEM.

3. The Internal Clock of Embryonic Development in Sea Urchins

3.1. Reaggregation Studies

A number of studies on sea urchin embryos have suggested that certain developmental events may be governed by an internal clock. The existence of a "micromere clock" has been demonstrated by several investigators. Micromeres appear to be the pacemakers of mitotic activity during early cleavage (Rustad, 1960; Dan and Ikeda, 1971; Parisi *et al.*, 1978). Giudice and co-workers (1962, 1970; Pfohl and Giudice, 1967) suggested that an internal clock might also be operative during later stages of development. They investigated the ability of sea urchin reaggregating cells to carry out developmental commitments made before and after dissociation of the embryos. The time required for reaggregated larvae to reach a given stage of differentiation, or to express alkaline phosphatase activity, depends on the embryonic stage at which the cells are dissociated. The later the embryonic stage at which the cells are dissociated the shorter the time it takes reaggregates to reach a given stage of differentiation or to develop alkaline phosphatase activity.

To investigate whether an internal clock does regulate the timing of later development in the sea urchin embryo, we followed the development of the second pathway of reaggregation (Spiegel and Spiegel, 1980). This type of aggregate development is more complex than the first pathway and would therefore provide more specific morphological criteria for the successful development of a second reaggregation. Figure 7 presents the development of chains of beads in *Arbacia* at 7 and 12 hr of reaggregation. Figure 7A illustrates the blastulalike aggregates, and Figure 7B shows the cilia and meshwork sheath surrounding the aggregates. Figure 7C demonstrates the early gastrulalike beads formed at 12 hr, and Figure 7D shows an embryo with several foci of invagination, one of the abnormalities encountered in reaggregating cells. At 33−36 hr after reaggregation was initiated, pluteus development ranged from an almost normal appearance (Fig. 8A) to the abnormal (Fig. 8B,C). Occasionally, multiple plutei formed due to failure of the "beads" in the chain to detach from each other.

When cells of the first reaggregation had developed for 7 hr and reached

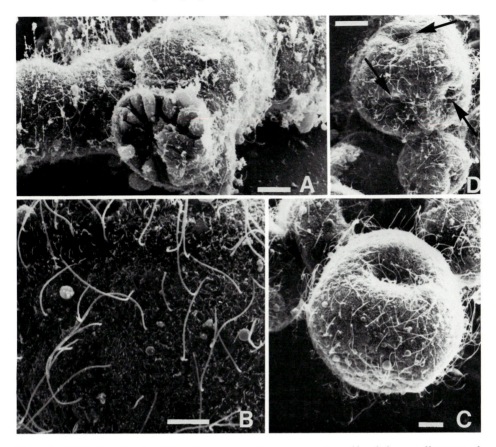

Figure 7. (A) Chain of beads at 7 hr after reaggregation; cross section of bead shows cells arranged around a central cavity as in a blastula. The outer surface of the chain of beads is covered by a meshwork sheath of hyaline layer with microvilli embedded within it. Cilia protrude through the meshwork sheath. (B) Higher magnification of surface of 7 hr reaggregate showing meshwork sheath and cilia. (C) Early gastrulae at 12 hr after reaggregation illustrating the size differences seen in reaggregated embryos. The large one is approximately three times the volume of the smaller one (in the upper right corner). (D) Early gastrulae at 12 hr after reaggregation. The smaller embryo appears to be normal. The larger embryo has three foci of invagination (arrows). Scale bars: 10 μm (A,C); 5 μm (B); 20 μm (D). (From Spiegel and Spiegel, 1980.)

the stage shown in Fig. 7A, the chains of beads were dissociated into a cell suspension that was allowed to reaggregate a second time. Figure 9A–D shows the appearance of the reaggregates at 15 min, 2 hr, 7 hr, and 13 hr, respectively, after the second reaggregation was initiated. Development of the second reaggregation is the same as that of the first, with similar types of abnormalities and with a majority of embryos developing into almost normal plutei.

The surprising result of these observations was to find that the developmental rates of intact control embryos and embryos resulting from both the first and second reaggregations were identical. At first glance, it might have been expected that control and first reaggregation embryos should develop at the

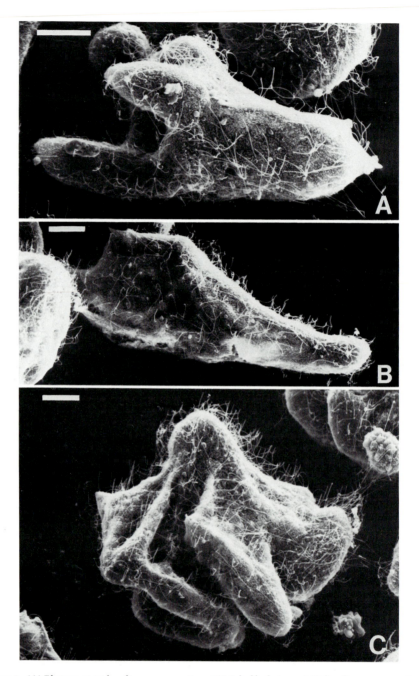

Figure 8. (A) Pluteus at 36 hr after reaggregation. (B) A half-pluteus at 36 hr after reaggregation. (C) Compound pluteus at 36 hr after reaggregation, resulting from gastrula with multiple foci of invagination. Scale bars: 20 μm (A–C). (From Spiegel and Spiegel, 1980.)

same rate, since the cells of the intact 16-cell-stage embryos are presumably identical to the cell suspension used for the first reaggregation. There are, however, morphological differences between them. Dissociation disrupts the spatial organization of the blastomeres and removes the meshwork layer consisting of the hyaline layer and microvilli embedded within it, and the reaggregating cells form ruffles and lamellipodia, which enable them to move. The time required for the cells to reassemble into groups is another difference. The results suggest that the reaggregating cells may actually be developing at a faster rate than the controls, since cells of the latter are already assembled into groups. During the first few hours, they have regenerated hyaline and microvilli and formed multicellular groups, but they reach the beginning of gastrulation at the same time as do control embryos.

In addition to the differences between cells of the first reaggregation and intact embryos, the cells of the second dissociation have two additional differences: They are much smaller, and they are derived from aggregates that have developed for a relatively long period. One might expect that the smaller cells, with their smaller lamellipodia, would require a longer time to move across the substratum and form cell groups. However, the cells of the second reaggregation take no longer to form gastrulae and plutei than do cells of the first reaggregation and the controls.

It is difficult to account for the ability of reaggregating cells to make up for time lost in the initial stages of reaggregation and to achieve the level of normal development that they frequently do. There is, however, some evidence indicating that cell behavior during reaggregation is not completely random. Cells sort out according to cell type so that each blastomere type tends to recognize and adhere to others of the same kind (Spiegel and Spiegel, 1978b). Cells in reaggregates, therefore, would tend to be arranged in smaller groups of like cells—macromeres, mesomeres or micromeres. Degrees of abnormality could be caused by the lack of a minimal number of cells within each group (Fig. 8B), the positional arrangements of these cell groups in relation to each other (Fig. 7D), and the numbers of such cell groups involved in the formation of one embryo (Fig. 8B,C). Abnormal development could also be caused by an unbalanced ratio of the different cell types within a cell group.

Giudice and colleagues have shown that, in many cases when development is interrupted by dissociation, it can be resumed immediately when cells reaggregate and cell interaction is reestablished. They suggest that dedifferentiation does not occur during the dissociation–reaggregation processes. The lack of dedifferentiation and the ability of the cells to sort out according to cell type could account, at least in part, for the ability of the cells to make up for time lost during reaggregation.

The observations described in this section indicate that specific developmental events are programmed to take place at specific times. This programming is followed even after the embryo has been completely disoriented by dissociating the cells from one another. As soon as the cells reaggregate and cell interaction is reestablished, they resume the program of differentiation.

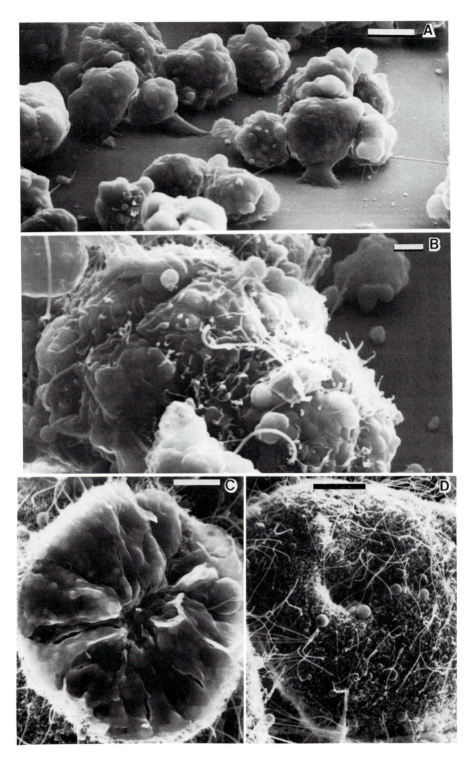

4. The Extracellular Matrix in Development

4.1. The Hyaline Layer

The importance of the hyaline layer in early sea urchin development has long been recognized. The firm attachment of the cells to the hyaline layer maintains the shape of the embryo and the orientation of the cells during cleavage and throughout gastrulation and larval development (see Chapter 3). Early biochemical analyses of the hyaline layer (Immers, 1961; Citkowitz, 1971, 1972) showed that it contains mucopolysaccharides and proteins. The major protein in the hyaline layer is hyalin, a calcium-insoluble protein (Kane and Stephens, 1969; Stephens and Kane, 1970). In the older literature, glycosaminoglycans (GAG) and proteoglycans (PG) were called acid mucopolysaccharides and mucoproteins, respectively (Hay, 1978a). The abundant literature reporting the presence of mucopolysaccharides and proteins in the hyaline layer suggested to us that these might be analogous to the GAG and PG of extracellular matrices found in vertebrate cells (Slavkin and Greulich, 1975; Vaheri et al., 1978; Subtelny and Wessells, 1980; Hay, 1981). In addition, these components could, perhaps, be visualized in the sea urchin embryo at the ultrastructural level using the same specimen preparation methods as those used with vertebrate cells (Hay, 1978a; Markwald et al., 1978, 1979).

Among the best methods developed for the preservation and staining of extracellular matrix (ECM) are those using the cationic dyes ruthenium red (Luft, 1971a,b) or Alcian blue (Scott and Dorling, 1965; Weston et al., 1978). The methods are based on the ability of these dyes to bind to large, charged polyanions such as GAG. They also help retain matrix components that are lost during conventional fixation. When these methods of specimen preparation are used, plus specific enzyme degradation of matrix components, it is possible to detect certain types of ECM macromolecules at the ultrastructural level. For example, GAG are seen as microfibrils of various sizes, which form a meshwork of interconnected filamentous material that is sensitive to testicular hyaluronidase. PG are seen as dense granules, from 10–50 nm in diameter, that are digested by testicular hyaluronidase and by trypsin. Collagen is seen as striated glycoprotein fibrils that can be digested by collagenase. GAG, PG, and collagen are the major classes of macromolecules that define a cell surface layer as an ECM (Hay, 1978a, 1981; Vaheri et al., 1978; Subtelny and Wessells, 1980; Slavkin and Greulich, 1975). The staining methods do not identify the specific molecular species of GAG, PG, or collagen; only the appropriate biochemical analyses can determine this. Studies of sea urchin embryos have shown that

---

Figure 9. (A) Fifteen min after second dissociation, cells are stripped of meshwork sheath and cilia. Lamellipodia and filopodia have formed, and cells are already in small groups. (B) Microvilli and cilia begin to reappear, 2 hr after the second reaggregation began. (C) Chains of beads have formed, 7 hr after the second reaggregation. The blastulae show a circular arrangement of cells around a central cavity. The meshwork sheath of hyaline layer with microvilli embedded within it has regenerated and covers the surface. Cilia have also regenerated. (D) Early gastrula at 13 hr after second reaggregation. Scale bars: 5 μm (A,C); 2 μm (B); 10 μm (D). (From Spiegel and Spiegel, 1980.)

they synthesize PG (Oguri and Yamagata, 1978; Kinoshita and Saiga, 1979; Kinoshita and Yoshii, 1979), GAG (Akasaka and Terayama, 1980, 1982, 1983), and collagen (Golob et al., 1974; Pucci-Minafra et al., 1972; Gould and Benson, 1978).

We have used the ruthenium red method, with specific enzyme degradation, for Arbacia embryos and reaggregating cells in order to obtain ultrastructural information concerning the hyaline layer (Spiegel and Spiegel, 1979). Figure 10A is from a gastrula stage fixed in glutaraldehyde/paraformaldehyde and postfixed in osmium tetroxide in the conventional way. Some wispy material appears irregularly, but it shows the hyaline layer area to be relatively clear. Figure 10B is from an embryo fixed with ruthenium red, which shows the hyaline layer to be a complex, structured entity. Higher magnifications of the hyaline layer in Fig. 10C and D, which are of the 16-cell and gastrula stages, respectively, show that the structures have a high degree of spatial order. The structures shown in Figure 10C were seen in normal embryos shortly after fertilization through the pluteus stage and, in reaggregates, from approximately a few minutes after the beginning of reaggregation through the development of plutei.

To identify these structures, isolated hyaline layers were fixed, treated with testicular hyaluronidase or collagenase, and postfixed in osmium tetroxide containing ruthenium red (Spiegel and Spiegel, 1979). The major structures in the hyaline layer, as visualized by these methods, are a large filamentous component and 30-nm granules, which were sensitive to hyaluronidase. In addition, an unusual fibrillar structure was seen, which is striated and was digested by collagenase but not by hyaluronidase. Since it is a striated glycoprotein and is digested by collagenase, it may be a collagen-containing fibril. It does not resemble the usual vertebrate collagen fibril, since it has a 4.5-nm periodicity. It has been shown, however, that marine invertebrate collagen is often quite different in appearance from vertebrate collagen (Murray et al., 1981). Using the same criteria and terminology as that for vertebrate cells, it appears that the filamentous component consists of GAG, the 30-nm granules are PG, and the striated fibrils are collagen-containing fibrils. The hyaline layer was therefore shown to be an extracellular matrix (Spiegel and Spiegel, 1979) similar to those found in vertebrate cells (Hay, 1981; Vaheri et al., 1978; Subtelny and Wessells, 1980; Slavkin and Greulich, 1975). The roles ascribed to the hyaline layer (Dan, 1960; Gustafson and Wolpert, 1967; Citkowitz, 1971; Herbst, 1900) are also similar to those of ECMs in vertebrate cells.

We have examined five species of sea urchin embryos, all of which contain similar morphological components in the hyaline layer. Striated collagenlike fibrils were seen in only A. punctulata and, to a lesser extent, in S. purpuratus, but they have also been seen in TEMs of two other sea urchin embryos, Psammechinus miliaris (Afzelius, 1956) and Clypeaster japonicus (Endo, 1961). Several collagen types have recently been demonstrated in sea urchin embryos using immunofluorescence procedures. Collagen types I, III, and IV have been reported in eggs and embryos of Arbacia punctulata and Lytechinus variegatus (Wessel et al., 1984). We have also found type IV collagen in the basal lamina of

Arbacia and *Sphaerechinus granularis* with this method (unpublished observations). Collagen fibrils with a 61-nm periodicity have been identified within the blastocoel matrix of sea urchin embryos (Crise-Benson and Benson, 1979). The immunofluorescence method has also been used to show the presence of heparan sulfate proteoglycan (Wessel *et al.*, 1984).

4.2. Fibronectin and Laminin

The sea urchin ECM can be compared with the ECM of vertebrate cells. In the sea urchin, the ECM maintains the cohesion and orientation of cells during cleavage stages and through gastrulation (Dan, 1960; Gustafson and Wolpert, 1967; Citkowitz, 1971; Spiegel and Spiegel, 1979). Gastrulation involves (1) cell motility, as cells migrate and change their positions; (2) changes in cell adhesion, as micromeres detach and move into the blastocoel; (3) changes in the shapes of cells and of the embryo itself; and (4) changes in the appearance and organization of the ECM. Similar changes have also been observed in cultured cells, and it has been suggested that the ECM contributes to their orderly growth and positioning (Hay, 1981). It may also be involved in other cell–cell interactions such as cell recognition and cell communication (Vaheri *et al.*, 1978).

A major component found on cell surfaces of many cultured vertebrate cells and in the basal laminae and connective tissue matrices of vertebrate cells is fibronectin, a high-molecular-weight non-collagenous glycoprotein (Hynes, 1973; Ruoslahti *et al.*, 1973; Yamada and Weston, 1974; Vaheri and Mosher, 1978). The loss of fibronectin from the surfaces of transformed cell lines has been associated with tumorigenicity leading to a decrease in cell adhesion, cell spreading, and cell motility, accompanied by an altered cellular morphology (Hynes, 1973; Yamada and Olden, 1978; Vaheri and Mosher, 1978). The addition of fibronectin restores these cells to a more normal morphology and behavior (Yamada *et al.*, 1976; Ali *et al.*, 1977). Fibronectin has also been found in the ectoderm of early chick embryos and has been proposed to facilitate morphogenetic movements during early development (see Chapter 12). In addition, it has been found in Reichert's membrane of mouse embryos, a basal laminalike structure that surrounds the embryo (Wartiovaara *et al.*, 1980). These analogies in cell behavior between cultured cells and developing embryos, the presence of an ECM in both sea urchin embryos and cultured cells, and the finding of fibronectin in vertebrate embryos led us to investigate the possibility that fibronectin might also be present in an invertebrate embryo such as the sea urchin.

Using the immunofluorescence method, we found that fibronectin is present on cell surfaces and between the cells of both living and fixed sea urchin embryos, as shown in Fig. 11 (Spiegel *et al.*, 1980). During gastrulation there appears to be a decrease in fibronectin on the outer cell surfaces, as indicated by the narrower rim of fluorescence in Fig. 11C as compared with Fig. 11D, which is an earlier stage. Because the cells in the ectodermal layer are elongating as they migrate inward at the blastopore, this observation could mean that

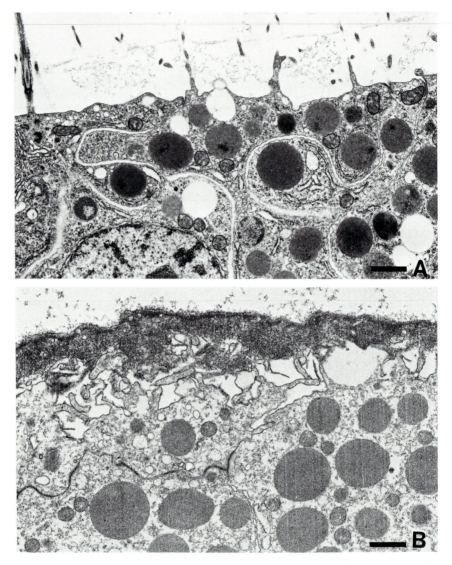

Figure 10. (A) *Arbacia* gastrula fixed in glutaraldehyde–paraformaldehyde and postfixed in osmium tetroxide without the presence of ruthenium red. The hyaline layer surrounding the embryo
appears to be relatively clear except for some wispy material and a few microvilli. (B) *Arbacia*
gastrula fixed with ruthenium red present during fixation and postfixation. The hyaline layer is
seen as a dense, structured layer as compared with (A). (C) High magnification of *Arbacia* hyaline
layer at the 16-cell stage, showing outer row of vesicles (V), microvilli (MV), filamentous meshwork
layer (FL), 30-nm granules (G), crystalline lattice configurations (CL), striated fibrils (SF), and
plasma membrane (PM). (D) High magnification of *Arbacia* hyaline layer at the gastrula stage
showing, in addition to components seen in (C), an additional outer, thin, filamentous layer
enclosing a row of widely spaced dense bodies (arrows). Scale bars: 1 μm (A,B); 200 nm (C,D). (C
and D from Spiegel and Spiegel, 1979.)

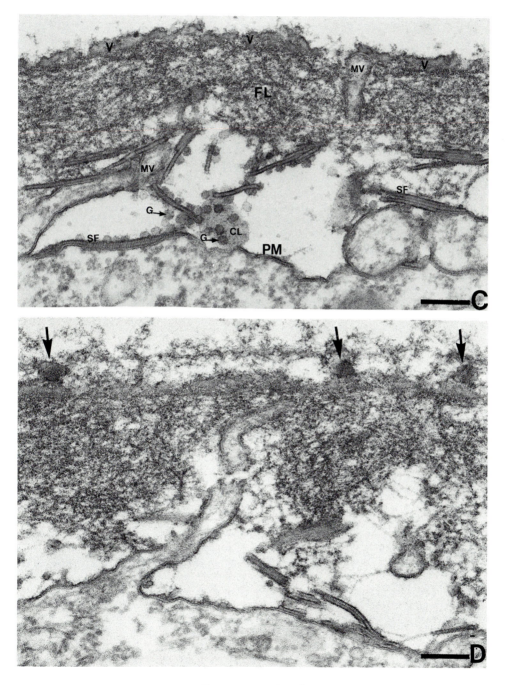

Figure 10. (continued)

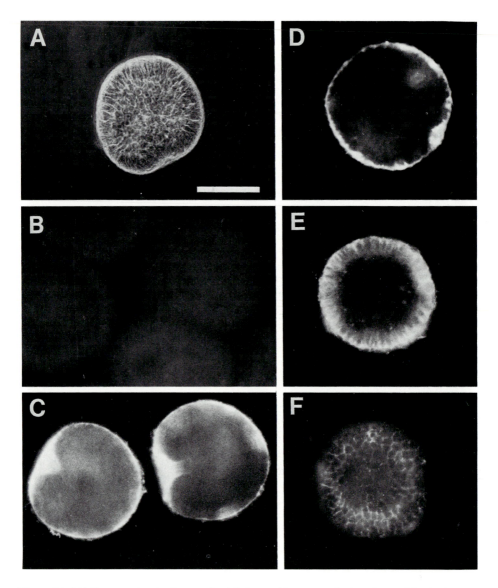

Figure 11. (A) Phase-contrast micrograph of *Sphaerechinus granularis* embryo at beginning gastrula stage. Scale bar: 50 μm. (B) Control embryo incubated with preimmune serum and showing no fluorescent staining. (C) Early gastrula showing fluorescent pattern of fibronectin staining migrating further inward as the cells migrate inward at the blastopore. (D) Beginning gastrula, focused at equator of embryo, showing fluorescent staining of fibronectin on the exterior cell surfaces. Fluorescence is more pronounced at the blastoporal lip where the cells are beginning to migrate inward (at 4 o'clock). (E) Same embryo as in D, focused slightly away from equator, showing fluorescent staining between cells in addition to exterior cell surface staining. (F) Same embryo as in D and E, focused farther away from equator. This shows a honeycomb pattern of fluorescent staining indicating the presence of fibronectin between the cells. (From Spiegel *et al.*, 1980. Reproduced from *The Journal of Cell Biology*, 1980, Vol. 87, pp. 309–313, by copyright permission of the Rockefeller University Press.)

the fibronectin layer is also elongating or stretching to cover the increased surface area. The fluorescence has extended further in the invaginating area in Fig. 11C as compared with Fig. 11D, which indicates an increase in fibronectin as the cells migrate inward. The presence of fibronectin on invaginating cells suggests its involvement in morphogenesis during early development.

Since fibronectin is closely apposed to the cell surfaces, it is necessary to remove the surrounding ECM (hyaline layer) with Ca^{2+}, Mg^{2+}-free seawater (CMFSW) to expose the cell surfaces to antiserum. If embryos are exposed to CMFSW for too long, cells become loose from the embryo. These loose cells do not exhibit fluorescent staining, whereas their still-adhering neighbor cells do. These results indicate that fibronectin can be removed from the cell surface with CMFSW and that it is involved in cell adhesion in the sea urchin embryo (Spiegel et al., 1980).

Using immunofluorescence methods, Katow et al. (1982) found fibronectin on the surfaces of primary mesenchyme cells in the mid-mesenchyme blastula stage only when these cells are migratory. It was not detected in the blastocoel or on the basal surface of the blastula wall. More recently, using the immunofluorescence procedure on paraffin sections, Wessel et al. (1984) detected fibronectin in the basal lamina and also associated with primary mesenchyme cells. On the other hand, we were unable to detect fibronectin when the antibody was microinjected into the living embryo (Spiegel and Burger, 1982). Although the antibodies used by all three groups were raised against human plasma fibronectin, the antisera were different, the species were different, and the techniques for specimen preparation varied from the use of living embryos to fixed material and paraffin sections. These experiments demonstrate that much remains to be studied regarding the interactions of ECM components with each other and with cell surfaces. The use of antibodies raised against sea urchin antigen should help resolve this dilemma.

Laminin is another noncollagenous, high-molecular-weight glycoprotein found in basal laminae of a number of organisms (Timpl et al., 1979), including those of mouse embryos (Leivo et al., 1980; Wartiovaara et al., 1980). It is thought to be involved in adhesion of cells to the basal lamina and may establish cell polarity by linking cells to type IV collagen, which is also found in basal laminae (Kleinman et al., 1981; Terranova et al., 1980; Hogan, 1981). The role of laminin in development is not as yet clearly understood, but it is thought to be involved in increased cell adhesion during early development (Ekblom et al., 1980a; Timpl, 1980) and in the induction of epithelial tubules in mouse kidney development (Ekblom et al., 1980b). Using the immunofluorescence method, we found that laminin is present on cell surfaces and between the cells of sea urchin embryos (Spiegel et al., 1983). The staining pattern is the same as that found for fibronectin. Early studies showed that the basal lamina of the sea urchin embryo, which lines the blastocoel, contains mucopolysaccharide (Okazaki and Niijima, 1964). This finding is supported by immunofluorescence studies (Spiegel et al., 1983; Wessel et al., 1984) and ruthenium red staining of isolated basal laminae (Spiegel et al., 1983). In addition to fibronectin and laminin, types I, III, and IV collagen and heparan sulfate proteoglycan have also been localized in the basal lamina (Wessel et al., 1984).

At the ultrastructural level, the basal lamina is composed of fibrils of various sizes decorated with 30-nm granules (Solursh and Katow, 1982; Spiegel et al., 1983), which are similar to the hyaluronidase-sensitive PG granules found in the hyaline layer ECM (Spiegel and Spiegel, 1979). It appears as though the same classes of ECM components, therefore, are present in both the hyaline layer and basal lamina ECMs. The similarity of sea urchin ECM components to their vertebrate ECM counterparts has also been pointed out by other investigators (Wessel et al., 1984).

Two major functions have been ascribed to basal laminae in adult tissues: (1) They act as semipermeable filters, and (2) they form a supporting and/or boundary structure (Kefalides et al., 1979). Basal laminae in developing systems appear to play similar roles. They are thought to act as membranes that monitor or regulate the microenvironment of differentiating cells (Hay, 1978b; Bernfield and Banerjee, 1982) and as substrata that help guide the morphogenetic movements of cells and tissues (Hay, 1978b; Bernfield and Banerjee, 1982; Lash and Vasan, 1977). The sea urchin basal lamina also appears to fulfill these roles; it serves as a substratum for primary mesenchyme cell ingression (Katow and Solursh, 1980) and may guide these cells along the surface of the blastular wall to form the characteristic arrangement of primary mesenchyme cells (Oguri and Yamagata, 1978; Katow and Solursh, 1980). In addition, some early experiments indicate that the sea urchin basal lamina acts as a semipermeable filter (Dan, 1960).

The work described above indicates that there is a similarity of structural components and functional roles of the hyaline layer ECM and the basal lamina ECM. On the basis of our observations presented here and in earlier studies (Dan, 1960; Gustafson and Wolpert, 1967), we suggest that (1) the outer ECM (hyaline layer) forms a shell-like structure surrounding the sea urchin embryo, (2) fibronectin and laminin are closely apposed to the cell surface and form a continuous matrix surrounding the individual cells of the embryo, and (3) the outer ECM (hyaline layer) is linked to the inner ECM (basal lamina) through the fibronectin–laminin network. Such a network could coordinate the various activities of the embryo. The functions of structural support, positional relationships, guidance of cell movements, cell adhesion, and changes in cell shape could be reinforced and coordinated if both ECMs acted together.

5. The Biochemistry of Cell Adhesion

5.1. Experiments on the Living Embryo

Many biochemical studies on cell adhesion have used reaggregating cells. Although such reaggregation studies contribute to our understanding of the mechanisms of cell adhesion, it is important to determine whether the information obtained is relevant to normal embryogenesis. Indeed, Edwards (1976) pointed out that reaggregation is very remote from anything that occurs in vivo, and Garrod and Nicol (1981) questioned whether the adhesive interactions studied in aggregation experiments bear any resemblance to "normal" adhesive processes. However, experimental approaches involving the use of the intact living embryo to study cell adhesion are relatively rare. The living echinoderm embryo offers unique advantages for such studies.

In a notable series of investigations on sea urchin gastrulation, Gustafson and Wolpert (1967) and Dan and Okazaki (1956) demonstrated that secondary mesenchyme cells from the advancing archenteron extend long, narrow filopodia, which probe the surface of the blastocoel, rejecting some surface contacts before adhering to other cells. After specific cell adhesions are made, contractions of the filopodia pull the leading tip of the elongating archenteron to the opposite wall of the blastocoel. In most species, invagination of the archenteron is preceded by a migration of the primary mesenchyme cells (which develop from the four micromeres of the 16-cell stage) into the blastocoel. They form a primary mesenchyme ring, which eventually produces a larval skeleton (see Chapter 10). All these morphological events can easily be seen and studied in the living, intact, transparent embryo. Furthermore, isolated micromeres or primary mesenchyme cells can differentiate to form a larval skeleton when cultured in seawater containing horse serum (Okazaki, 1975; Harkey and Whiteley, 1980).

We have used these observations as the basis for a study of the biochemistry and morphology of cell adhesion that focuses on the events occurring during gastrulation (Spiegel and Burger, 1982). We have microinjected a variety of compounds into the blastocoel of living *Sphaerechinus granularis* and *Lytechinus pictus* gastrulae and studied their effects on filopodial extensions, cell movements, and cell adhesion. Initially, two proteolytic enzymes, trypsin and pronase, were injected as well as the enzyme collagenase. All three enzymes caused the release of attached secondary mesenchyme filopodia and prevented new attachments or reattachments (Fig. 12A–E). It is not known whether the filopidia were released from attachment to the underlying basal lamina or from the plasma membrane of the ectodermal cells. Within 4 min, all the filopodia lose their attachments, and the archenteron (which has now lost its anchoring points on the animal hemisphere side of the blastocoel) retracts and ultimately forms a small knob of collapsed cells at the vegetal pole (Fig. 12E). The enzymes have no obvious effect on other cell–cell adhesions. The effect of the two proteases is probably due to partial digestion of the basal lamina, which is also sensitive to pancreatin (Okazaki and Niijima, 1964), a mixture of proteases. Type IV collagen has been found in the sea urchin basal lamina (Wessel *et al.*, 1984) and may account for the collagenase effect. A protease-free collagenase gave similar results (unpublished observations). None of the enzymes had any effect on cell motility. The embryos are apparently normal except for this effect and, occasionally, filopodial reattachment and archenteron elongation may begin again.

These results support the proposal (Wolpert and Gustafson, 1961; Dan and Okazaki, 1956) that archenteron extension is due to contraction of the filopodia of the secondary mesenchyme cells and not to a further invagination of the archenteron. Further confirmation comes from experiments in which we sev-

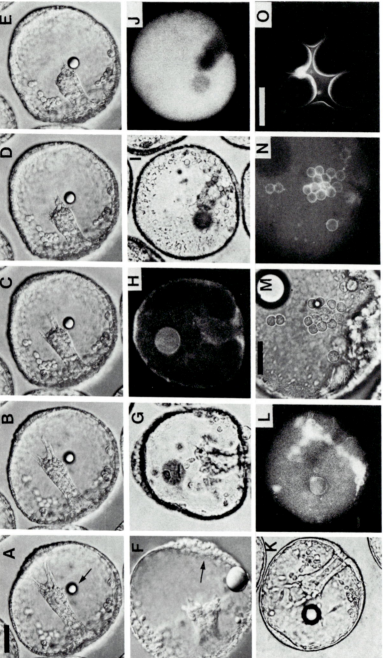

Figure 12. Effects of various compounds after microinjection (100–150 pl). (A–E) Pronase (1 mg/ml artificial seawater). Same embryo at 0 (A), 3 (B), 4 (C), 6 (D), and 11 (E) min after injection into blastocoel. Arrow points to oil droplet also injected to identify embryo. Control embryos injected with artificial seawater and oil droplet develop normally to pluteus stage. (F) Con A (1 mg/ml), 43 min after injection. Arrow points to dissociating cells. (G, H) Fluorescein isothiocyanate (FITC)- conjugated Con A (0.8 mg/ml), both 4 min after injection: interference optics (G); same embryo by ultraviolet (UV) illumination (H). (I,J) FITC-conjugated *Dolichos biflorus* (2.5 mg/ml); both 5 min after injection: interference optics (I); same embryo by UV illumination (J). (K,L) FITC-conjugated Con A (1.5 mg/ml) plus α-methylmannose (5×10^{-2} M), both 6 min after injection: interference optics (K); same embryo by UV illumination (L). (M,N) FITC-conjugated WGA (4.1 mg/ml), both 4 min after injection: interference optics (M); same embryo by UV illumination (N). (O) FITC-conjugated *Ulex europaeus* (1.8 mg/ml) 3 min after injection into seawater surrounding mid-gastrula stage embryos, UV illumination to show binding to external surface. (A–E) *Sphaerechinus granularis* embryos; (F–O) *Lytechinus pictus* embryos. Scale bars: 25 μm (A–L), 10 μm (M,N), 100 μm (O). (From Spiegel and Burger, 1982.)

ered the filopodia mechanically, which was followed by immediate retraction of the filopodia and of the archenteron tip toward the vegetal pole.

It was recently suggested that in *Anthocidaris*, contraction of secondary mesenchyme filopodia may not play a role in archenteron elongation, since in this species the filopodia are poorly developed and the blastocoel is very small (Amemiya et al., 1982a,b). Interestingly, it has been suggested that in the starfish, *Pisaster ochraceus*, material in the blastocoel that stains for polysaccharide and that forms webs may guide the movement of mesenchyme cells (Crawford and Chia, 1982). In the sea urchin, similar material in the blastocoel consists of 30-nm granules attached to fine fibers forming a thick, rough-surfaced, fibrous structure (see Chapter 10).

Lectins (see Chapter 5) can be used to determine whether cell–cell associations are dependent on a carbohydrate-binding mechanism. Inhibition of a function by a lectin implies that the lectin binds to a carbohydrate (e.g., the carbohydrate moiety of a glycoprotein), thereby preventing the carbohydrate from functioning in adhesion. Since the lectin Concanavalin A (Con A) is known to have an inhibitory effect on embryonic echinoderm cell adhesion (Lallier, 1972; Kyoizumi and Kominami, 1980), we microinjected Con A into mid-gastrula-stage embryos. About ½ hr after injection, detachment of filopodia and retraction of the archenteron occur. Although filopodia continue to re-form, probe, and change directions, adhesion to the blastocoel wall does not occur. Forty-five min after injection, in the region in which the filopodia were originally attached, the ectodermal cells round up and lose their adhesiveness to one another (Fig. 12F, arrow); their cilia continue to beat. Adhesions between ectodermal cells in the remainder of the embryo are unaffected. Still later, cells at the archenteron tip (but not at the base) round up and lose their attachment to one another. Other lectins (Fig. 12I,J) were also injected without these effects. When α-methyl mannose (the sugar that specifically inhibits Con A binding) is co-injected with Con A, the effect of Con A on filopodial attachment is prevented.

The sites of Con A binding were explored by injection of fluorescein-conjugated Con A (FITC–Con A) into mid-gastrulae. After 2–5 min, a pronounced fluorescence was noted at the inner surface of the animal hemisphere wall at the place where the secondary mesenchyme filopodia normally adhere and attach, but not elsewhere (Fig. 12G,H). During this period, secondary mesenchyme filopodia begin to detach as described above. The specific binding of FITC-Con A is prevented by co-injection of α-methyl mannose (Fig. 12K,L). FITC-succinyl Con A and FITC-*Lens culinaris* lectins (which have a sugar specificity similar to Con A) bind weakly to the inner surface of the animal hemisphere wall. *Ricin* 60 binds weakly to the sides of the archenteron. When FITC-wheat germ agglutinin (FITC-WGA) was injected into the blastocoel (Fig. 12M,N), binding was uniquely observed on the surface of the cells of the primary mesenchyme ring.

We have also demonstrated specific lectin binding to the external surface of the embryo. For example, FITC-conjugated *Ulex* (Fig. 12O), *Lens* and *Ricin* 60 lectins showed strong binding to the external surface; wheat germ and soybean agglutinin exhibited weak external binding; no external binding was detected with *Arachis, Ricin* 120, and Con A. We have no information as to whether binding of lectins to the external surface is to the hyaline layer or directly to the external cell surfaces.

There are a number of related lectin-binding studies of early sea urchin development. Lallier (1972) observed that Con A dissociates embryos and inhibits the reaggregation of dissociated cells. Oppenheimer (1977) reported that carbohydrate lectin-binding sites become less agglutinable during sea urchin development. Furthermore, significantly more capping of receptor sites was found on micromeres than on mesomeres or macromeres, which may be related to the fact that micromere descendants are migratory during development. Katow and Solursh (1982) described temporal changes in the distribution of Con A-binding sites on the basal surface of the ectoderm.

A number of sea urchin studies not relying on lectins suggest that glycoproteins play a role in cell adhesion, cell movement, and the events of gastrulation. A correlation has been made between sulfated polysaccharide synthesis, cell surface morphology, and cell movement (Karp and Solursh, 1974). Glycosyltransferases that are located on the surfaces of eggs and embryos have been implicated in cell adhesion (Schneider and Lennarz, 1976). A number of specific saccharides (e.g., D-galactose and N-acetyl-D-galactosamine) inhibit reaggregation (Schneider *et al.*, 1978). Tunicamycin, which inhibits glycosylation of proteins, blocks primary mesenchyme cell migration (Schneider *et al.*, 1978). In the sea urchin, it appears to inhibit the synthesis of a class of sulfated glycoproteins but does not inhibit sulfated proteoglycans (Heifitz and Lennarz, 1979). On the other hand, development ceases at the blastula stage if proteoglycan synthesis is inhibited by either aryl-β-D-xyloside, sodium selenate or 2-deoxy-D-glucose (Kinoshita and Saiga, 1979; Kinoshita and Yoshii, 1979). The inhibition by xyloside (but not by selenate or 2-deoxy-D-glucose) could be prevented by addition of proteoglycan.

Sea urchin embryos cultured in sulfate-free seawater fail to gastrulate. Ascorbate and α-ketoglutarate, activators of protocollagen proline hydroxylase, can cause the formation of an archenteron in sulfate-free seawater-cultured embryos (Mizoguchi and Yasumasu, 1982, 1983). The effect is inhibited by α,α'-dipyridyl, an inhibitor of protocollagen proline hydroxylase. Normal blastulae treated with ascorbate and α-ketoglutarate develop into abnormal embryos with a large exogut. These observations suggest that archenteron formation is supported, in part, by collagen synthesis.

Historically, antibodies have been used frequently to identify antigenic changes during early development, although the precise biochemical identification of the antigens in question has usually not been carried out. For example, McClay and co-workers, using hybrid sea urchin embryos, demonstrated the appearance of new cell surface antigens during development, which is correlated with a change in adhesive specificity of the hybrid cells (McClay, 1979; McClay *et al.*, 1977). More recently, McClay *et al.* (1983) and Wessel *et al.* (1984) used monoclonal antibodies made against cell membrane and extracellular matrix to localize antigens during sea urchin development. The initial results suggest that these monoclonal antibodies will be extremely useful in analyzing the molecular basis of cell adhesion in the embryo.

5.2. Protein Composition of the Extracellular Matrix and Cell Surface

We have used a number of approaches to study further the biochemical composition of the cell surface and/or the extracellular matrix. A more direct approach has been to isolate the cell surface with a high degree of purity, to analyze it biochemically, and to describe changes taking place during reaggregation, cell movement, and other events. Although the attempt has been made, it has not been possible to isolate the embryonic cell surface with sufficiently high purity. Another approach is to attempt to label specifically only the cell surface or ECM proteins with an isotope such as ^{125}I and not label internal proteins. Thus, by definition, proteins containing the label are cell surface or ECM proteins. We have been successful in labeling the surface of ECM proteins with [^{125}I]chloroglycoluril, the IODOGEN procedure of Fraker and Speck (1978). Autoradiography indicates that the label is associated only with the surface and cannot be detected within the cell.

The basic experimental procedure is to label the external surface proteins with ^{125}I during normal development or reaggregation. Proteins that are soluble in sodium dodecyl sulfate (SDS) buffer are subjected to one-dimensional polyacrylamide gel electrophoresis (PAGE), which separates proteins according to size. It is important to note that all soluble proteins of the cell—internal and external proteins—are subjected to the gel electrophoresis. When we stain the gel with Coomassie blue to locate the resolved proteins, it is usually impossible to distinguish any protein changes taking place in normal development or during reaggregation. There are too many proteins present to permit adequate resolution and identification, and most proteins stained are probably internal. Furthermore, the staining procedure obviously cannot distinguish the external surface and/or ECM proteins from internal proteins. Thus, in order to identify the external proteins (labeled by the [^{125}I]chloroglycoluril procedure) the gel is subjected to autoradiography. The internal proteins are not labeled and do not obscure the labeled external proteins, which can be identified readily on the developed and fixed autoradiogram.

Using these procedures, McCarthy and Spiegel (1983a) reported the following results. If intact *Arbacia* blastulae are iodinated, more than 30 proteins can be identified on the external surface with the major proteins having relative molecular weights of 290,000, 175,000, 145,000, 110,000, 70,000, and 50,000 (Fig. 13c). The 290,000-M_r protein is hyalin, the major component of the hyaline layer (Stephens and Kane, 1970). If blastulae are first dissociated into single cells by washing in Ca^{2+},Mg^{2+}-free seawater containing EDTA (CMF + EDTA) and then iodinated, additional surface proteins are labeled (Fig. 13d).

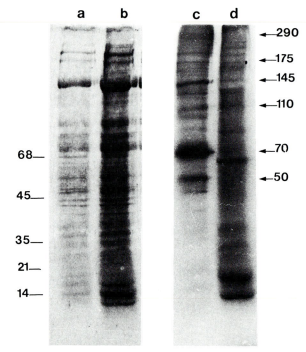

Figure 13. Coomassie blue staining pattern (a,b) and corresponding autoradiograph (c,d) of an SDS-urea gradient gel. Proteins stained by Coomassie blue in (a) intact *Arbacia punctulata* blastula embryos and (b) blastula cells dissociated in CMF + EDTA. Proteins iodinated on (c) intact blastula and (d) dissociated cells labeled after dissociation. Markers indicate approximate molecular weights ($\times$ 10^3). (From McCarthy and Spiegel, 1983a.)

Although more surface proteins could be labeled after dissociation than were iodinated in the intact embryo, most of the proteins labeled in the intact embryo could no longer be detected. This result indicates that most proteins labeled in the intact embryo are components of the hyaline layer, which is removed by the dissociation procedure. The additional proteins labeled in dissociated cells, as compared with the whole embryo, may be proteins associated with the apical, lateral, and basal cell surfaces of the blastula cells, which are now exposed to iodination.

To identify more clearly the iodinated proteins of the hyaline layer, embryos were labeled and then dissociated by three successive washes of CMF + EDTA. The three CMF + EDTA washes and the dissociated cells were then run on SDS-urea gradient gels and autoradiographed. Each of the washes contained a different subset of the iodinated proteins present in the intact embryo sample. Hyalin and the 70,000-M_r proteins are removed in the first two washes; the third wash removes the 175,000-, 145,000-, 100,000-, and 50,000-M_r proteins. Not all the labeled proteins are removed by the CMF + EDTA. It is attractive to suggest that these results indicate that the proteins of the hyaline layer are not uniformly distributed in the hyaline layer; i.e., some are located closer to the embryo cell surface, and others are located closer to the external surface. In-

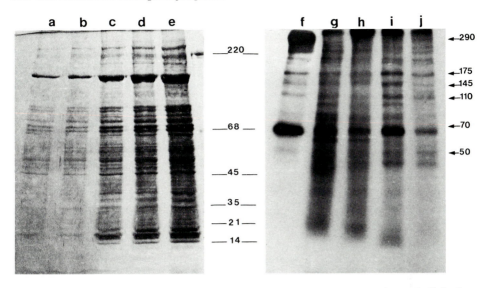

Figure 14. Coomassie blue staining pattern (a–e) and corresponding autoradiograph (f–j) of an SDS–urea gradient polyacrylamide gel. Iodinated (a,f) intact *Arbacia punctulata* blastula embryos and (b,g) iodinated blastula cells, which had been dissociated in CMF + EDTA and labeled after dissociation but before reaggregation. Cells iodinated (c,h) after 1 hr of reaggregation, (d,i) after 3 hr of reaggregation, and (e,j) after 7 hr of reaggregation. Markers indicate approximate molecular weights ($\times 10^3$). (From McCarthy and Spiegel, 1983a.)

deed, it has been suggested that, in *S. purpuratus*, hyalin is located on the external surface of the hyaline layer (Hall and Vacquier, 1982). It is also possible, however, that the results indicate that the hyaline layer proteins have different solubilities in CMF + EDTA.

In reaggregating *Arbacia* cells iodinated at various time intervals, the same electrophoretic pattern of labeled proteins formed in the intact embryo is detected on the surface of aggregates by 7 hr, although initially they are quite different (Fig. 14). This result suggests that during reaggregation, the hyaline layer is reconstructed as in the intact embryo. Many of the proteins that could be labeled at the start of reaggregation can no longer be labeled after 7 hr of reaggregation. This result may be due to the accumulation of hyaline layer material over the aggregate surface so that proteins close to the cell surface are no longer available for labeling by the [^{125}I]chloroglycoluril technique.

McCarthy and Spiegel (1983a) also compared the composition of the iodinated surface proteins of intact blastula-stage embryos of *Arbacia*, *S. purpuratus*, and *S. drobachiensis*. Each species has a distinct pattern of iodinated surface proteins, which suggests that there are species-specific protein differences in the hyaline layer and embryonic surface. This work should be extended to an analysis of species differences of cell surface proteins. Once resolved and identified, it should be possible to determine which protein(s), if any, is/are involved in the initial adhesion, followed by the sorting out of cells according to species, which occurs during reaggregation of a cell suspension of two different species.

Changes in the organization of the hyaline layer during early development have been observed in electron micrographs (Spiegel and Spiegel, 1979; Wolpert and Mercer, 1963). In an effort to determine whether the observed morphological changes reflect underlying changes in the protein composition of the hyaline layer, McCarthy and Spiegel (1983a) iodinated developmental stages of *Arbacia* and *S. purpuratus* embryos. In *Arbacia*, the major protein components are very similar in the 16-cell, mesenchyme blastula, and late gastrula stages. However, additional proteins of 58,000 and 48,000 M_r are labeled on the 16-cell stage compared with the blastula and gastrula stages. A 50,000-M_r protein present in the later stages is not present in the 16-cell stage embryo. In *S. purpuratus* embryos, the major proteins of the hyaline layer are also conserved in early development. Proteins can be detected in the 2-cell stage that cannot be detected in later stages. These developmental changes may reflect changes in the protein composition of the hyaline layer; alternatively, proteins may become inaccessible to surface iodination due to changes in the hyaline layer during development.

More recently, DeSimone and Spiegel (1984a) used these procedures to study the cell-type specificity of surface proteins of the 16-cell-stage embryo. In *Arbacia*, four high-molecular-weight proteins are present on the surface of isolated micromeres but not on mesomere–macromere fractions. In *S. drobachiensis*, a micromere-specific protein of 133,000 M_r is present, which binds to WGA but not to Con A. If FITC-WGA is injected into *S. drobachiensis* hatched blastulae, the basal lamina and the intercellular material between the epithelial cells bind WGA (D. W. DeSimone and M. Spiegel, unpublished results). In mesenchyme blastulae, WGA binding is limited to primary mesenchyme cells. In *S. purpuratus* (DeSimone and Spiegel, 1984b) two high-molecular-weight micromere-specific proteins are found on fertilized eggs, 2-, 4-, and 8-cell-stage blastomeres. By the 16-cell stage they are found on micromeres but not on mesomeres or macromeres. By contrast, two low-molecular-weight micromere-specific cell surface proteins are first observed at the 16-cell stage.

5.3. Aggregation Factors and the Cell Surface

A number of investigators in the 1970s isolated "factors" that accelerate or enhance reaggregation of dissociated sea urchin embryonic cells (Kondo and Sakai, 1971; Tonegawa, 1973). Whether or not the factor is hyalin is open to question. More recently, interest has increased in the search for aggregation factor(s). For example, Noll *et al.* (1979) reported that dissociated sea urchin blastula cells, after extraction with seawater containing butanol, fail to reaggregate when returned to seawater. However, these cells can reaggregate and form larvae if soluble proteins from the butanol extract or proteins extracted from purified blastula membranes are added to cultures of these cells. Furthermore, Fab antibodies made against membrane preparations block the reaggregation of dissociated cells, but after mixture with butanol extracts they no longer block reaggregation. The implication is that the butanol extract contains components necessary for cell adhesion that may be in or at the cell surface. Although the reaggregation-promoting ability of the butanol extract is neither stage specific nor species specific, the inhibition of reaggregation by Fab fragments is species-specific (Noll et al., 1981; Vittorelli et al., 1982). Oppenheimer and Meyer (1982a,b) have reported the isolation of a species-specific and stage-specific adhesion-promoting component by disaggregation of intact sea urchin embryo cells with CMF. The aggregation-promoting factor is a hemagglutinin and binds cells by D-galactose-like and N-acetyl-D-galactosamine-like residues. Tonegawa (1973, 1982) also extracted an aggregation factor from sea urchin embryos. It is a large acid–sugar–protein complex containing sialic acid, uronic acid, and sulfate, that reacts with both calcium and Con A. Finally, Akasaka and Terayama (1982, 1983) obtained, by EDTA extraction of mid-gastrulae, a sulfated fucogalactan–protein conjugate, which promotes the reaggregation of dissociated cells.

Thus, evidence is accumulating for a role of an aggregation factor(s), which enhances reaggregation of dissociated sea urchin embryonic cells and has either stage or species specificity, or both. The precise role, if any, for the specificity of such factors is difficult to understand, since it is well known that cells from different developmental stages of the same species of sea urchin initially adhere to one another as do cells from different species. It is only much later in time that species specificity of reaggregation is exhibited by sorting out.

McCarthy and Spiegel (1983b) studied the proteins extracted from Arbacia blastula cells using seawater containing low concentrations of butanol. We confirmed Noll's report that butanol-extracted cells are not able to reaggregate unless the butanol-extracted proteins are added to the culture. The extent of reaggregation is related to the protein concentration of the butanol-extracted material. Blastula cells were also extracted with CMF + EDTA. Addition of the CMF + EDTA extract to butanol-extracted cells also enhanced reaggregation to an extent similar to the addition of butanol extract. This effect is also concentration dependent.

We found that human plasma fibronectin and bovine serum albumin also enhance reaggregation in a concentration-dependent manner. With ovalbumin there was less enhancement. Since the aggregates can develop into ciliated swimming blastulalike and gastrulalike embryos, reaggregation by these proteins was not a nonspecific agglutination but rather a reformation of normal cell adhesions. These results indicate that the enhancement of reaggregation by butanol-extracted proteins may be nonspecific, since all three of the proteins tested enhanced reaggregation.

McCarthy and Spiegel (1983b) also investigated the composition of the butanol extract by SDS-urea PAGE. The Coomassie blue staining pattern of the butanol extract displays a complex mixture of proteins. In order to determine which proteins were extracted by butanol from the surface/ECM of blastula cells, intact blastulae and blastulae that had been dissociated into single cells by CMF + EDTA were iodinated. A portion of the labeled dissociated cells was then extracted with butanol. A comparison of the labeled proteins of intact

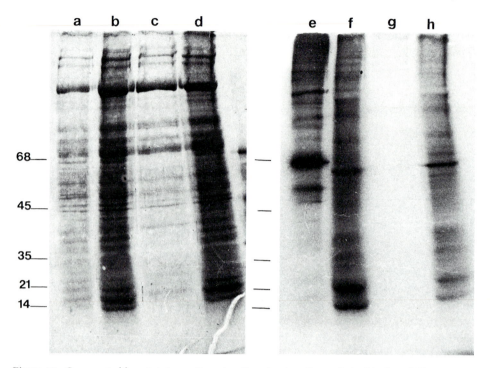

Figure 15. Coomassie blue staining pattern (a–d) and autoradiograph (e–h) of an SDS–urea gradient polyacrylamide gel. Coomassie blue stained proteins of intact *Arbacia punctulata* embryos (a). CMF + EDTA-dissociated cells (b). Three percent butanol extract of dissociated cells (c) and butanol-extracted cells (d). Proteins iodinated in intact *Arbacia punctulata* embryos (e). CMF + EDTA-dissociated cells (f). Three percent butanol extract of dissociated cells (g) and butanol-extracted cells (h). (From McCarthy and Spiegel, 1983*b*.)

embryos with those of the CMF + EDTA-dissociated cells indicates that they are quite different due to the removal of the hyaline layer proteins and an exposure of the cell surfaces, after dissociation, to iodination. Butanol extraction removes only a few iodinated proteins from the surface of the labeled blastula cells (Fig. 15). Furthermore, most of the labeled proteins remain on the cell surface after the cells have been extracted in butanol.

Noll and colleagues (1979, 1981) suggested that a cell surface protein(s) is removed by butanol extraction of dissociated sea urchin cells. Replacement of these proteins promotes reaggregation of cells and the formation of plutei. The results reported by McCarthy and Spiegel (1983*b*) demonstrate that butanol-extracted proteins are not specific for the enhancement of reaggregation. Furthermore, surface iodination experiments demonstrated that few, if any, surface-labeled proteins are detected in a butanol extract. However, it is possible that (1) the major surface proteins removed by butanol treatment are not labeled by the chloroglycoluril technique, or (2) most of the proteins removed from the cells by butanol extraction are located in the cell interior. The acrylamide gel electrophoretic pattern of the butanol extract contains many uniodinated bands, but iodinated surface proteins remain on the surface after butanol extraction. Although the results may be attributable to species differences, they are consistent with the view that most proteins removed from sea urchin embryonic cells by butanol extraction are internal cell proteins.

6. Junction Formation during Development

In addition to the macromolecular components described above, certain structural components should be included in a discussion of cell–cell interactions and especially cell adhesion—namely, the cell junctions. The investigation of cell junctions has been an active area of research in recent years, resulting in detailed information regarding their structure and function (Staehelin, 1974). Several earlier workers observed septate junctions in sea urchin embryos (Dan, 1960; Wolpert and Mercer, 1963; Balinsky, 1959; Chang and Afzelius, 1973; Gilula, 1973). Recently developed techniques, however, have permitted study of the formation of several types of junctions during morphogenesis of the sea urchin embryo (Spiegel and Howard, 1983). These methods include the use of freeze fracture and lanthanum impregnation as well as conventional thin-section techniques (Lane, 1982). In examining embryos of *Arbacia punctulata* and S*trongylocentrotus drobachiensis,* we found them to be similar in all respects.

6.1. Desmosomes

Figure 16A is of an early gastrula stage showing the outer cell surfaces of two adjoining cells surrounded by the ECM layer. The apical edges of the cells show two types of junctions. The most apical is the **belt desmosome,** or **zonula adhaerens,** which is composed of dense intracellular plaques on opposite edges of adjacent cells. Closely following the desmosome is the **septate junction,** which has a ladderlike appearance. Figure 16B shows **spot desmosomes** composed of small, dense intracellular plaques found at various points along the cell membranes of adjoining cells. Figure 16C shows **hemidesmosomes** attaching cells to the basal lamina and appearing as a single, dense, intracellular plaque with no opposite plaque on the basal lamina. Both spot and hemidesmosomes are localized, or focal, junctions as compared with belt desmosomes, which encircle the cells. The function of the desmosomes is the attachment of cells, either to each other or to the basal lamina (Staehelin, 1974).

Figure 16D is a higher magnification of the belt desmosome, which shows that, in addition to the dense intracellular plaques on adjoining cells, there is some extracellular, less dense material between the plaques. The embryonic belt desmosome does not show microfilaments emanating from the plaques as seen in the adult desmosome (Spiegel and Howard, 1983). An increased intercellular density can be seen where two spot desmosomes are forming (Fig. 16E).

The development of belt desmosomes begins as early as the 4-cell stage. By the 16-cell stage, the plaques are thicker and denser, but the desmosome does

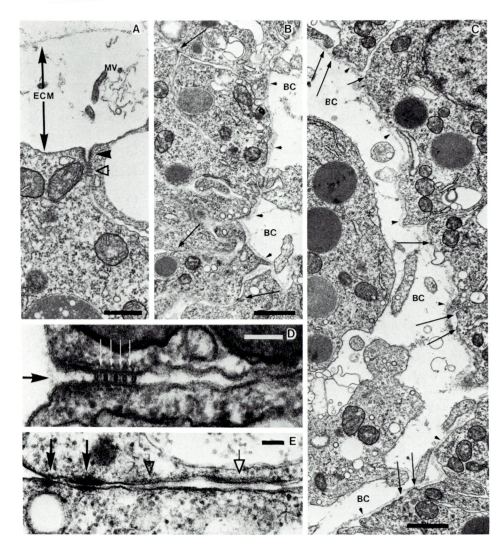

Figure 16. (A) Outer cell surfaces of two adjoining cells showing belt desmosome (black arrowhead) and septate junction (open arrowhead). Extracellular matrix (ECM) surrounds cells, and several microvilli (MV) are seen within the ECM. (B) Blastocoel cavity (BC) in interior of embryo is lined with basal lamina (arrowheads); spot desmosomes are indicated by arrows. (C) Blastocoel cavity (BC) in interior of embryo is lined with basal lamina (arrowheads); hemi-desmosomes are indicated by arrows. (D) High magnification of junctional area showing desmosome and septate junction with extracellular dense material within desmosome (black arrow) and between the septa (white arrows). (E) Four-cell stage with spot desmosomes (black arrows) and intercellular dense areas (open arrows) where additional spot desmosomes appear to be forming. Scale bars: 500 nm (A); 1 μm (B,C); 100 nm (D,E). (From Spiegel and Howard, 1983.)

not become a true zonula adhaerens, surrounding each cell, until the early blastula stage. At this time also, the hemidesmosomes appear. Spot desmosomes are already numerous at the 4-cell stage and seem to decrease in number during development.

When sea urchin eggs are fertilized, they become surrounded almost immediately by a thick layer of ECM. The ECM is enclosed by a fertilization envelope, which serves as an additional protective layer. The embryos are sessile during early development up to the early blastula stage. At this point, they have developed cilia and subsequently hatch out of the fertilization envelope to become swimming embryos throughout the rest of embryogenesis. The timing of development of the desmosomes is well adapted to this sequence of events. It is at the early blastula stage that belt desmosomes are fully developed and attach cells to each other at their apical ends. At this stage also, the inner blastocoel cavity has formed, with the concomitant formation of the basal lamina. The hemidesmosomes are found attaching the basal lamina to the basal cell surfaces at this time. It appears, therefore, that as soon as the belt and hemidesmosomes are fully developed and functional, hatching of the embryo from the fertilization envelope takes place. Spot desmosomes may be temporary adhesive devices that are most prevalent during very early development and tend to decrease in number as the belt and hemidesmosomes increase and mature.

6.2. Septate Junctions

Two kinds of septate junctions are found in sea urchin embryos. The straight, unbranched septate junction is present throughout morphogenesis, from the 4-cell stage to the pluteus. In Figure 16D, the extracellular material seen between the desmosomal plaques extends into the interseptal spaces of the septate junction. The number of septa increases during development, starting with two to four septa in the 4-cell stage to as many as 20 in the pluteus. Figure 17A shows that the septa are unbranched; i.e., each septum crosses the entire intercellular space without crossing or linking to any other septum. Using the lanthanum impregnation method (Fig. 17B), two additional properties of this type of septate junction can be seen: (1) The septa appear as straight flat lines, in contrast to the pleated septa described below, and (2) what appears to be a single septum, as seen in Figure 17A, is actually composed of two thin septal sheets that are very close together, forming a double septum (Fig. 17B) (see also Fig. 20A,B). In freeze-fracture replicas, the straight septate junction is seen as rows of 8–11-nm particles on the E face (Fig. 17C). Corresponding grooves are seen on the P face, which are in register with the E face particles. The replica also indicates that the straight septate does not extend very far in a basal direction but remains rather localized at the apical cell border.

The second type of septate junction found in the sea urchin embryo is the pleated, anastomosing septate junction, which was first described in the adult sea urchin stomach (Green, 1978, 1981; Green et al., 1979). The presence of this

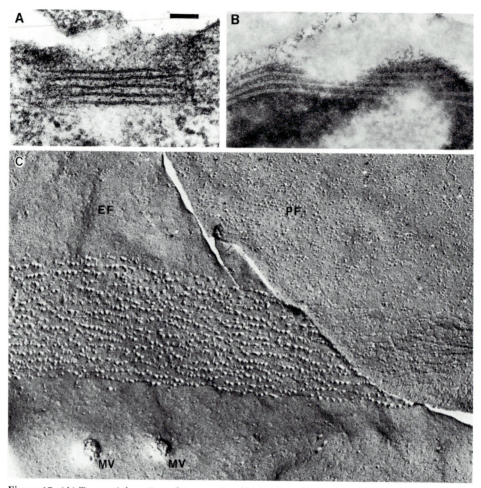

Figure 17. (A) Tangential section of conventionally fixed mid-gastrula stage showing septa to be unbranched, i.e., crossing entire intercellular space without crossing or linking to another septum. Scale bar: 100 nm. (B) Tangential section of lanthanum-impregnated mid-gastrula stage showing septa to be double-layered and smooth. (C) Freeze-fracture replica of pluteus stage showing 8–11 nm particles in rows on E face (EF) and corresponding grooves on P face (PF). Apical end of cell is at bottom of micrograph. MV, microvilli. (From Spiegel and Howard, 1983.)

type of septate junction in the embryonic digestive system is shown in the pluteus stage in Figure 18A and is similar to that found in the adult intestine. Lanthanum impregnation shows that the septa are single layered and pleated and form an anastomosing pattern. Freeze-fracture images show that the intramembrane particles are 6–9 nm in diameter and fracture onto the P face (Fig. 18B). The junctions in the pluteus embryo, as seen in both the lanthanum and freeze-fracture images, are not as well formed as those seen in the adult.

The transition from straight, unbranched, double septum to pleated, anastomosing, single-septum septate junction takes place only in cells that have invaginated to the interior of the embryo to form the embryonic digestive sys-

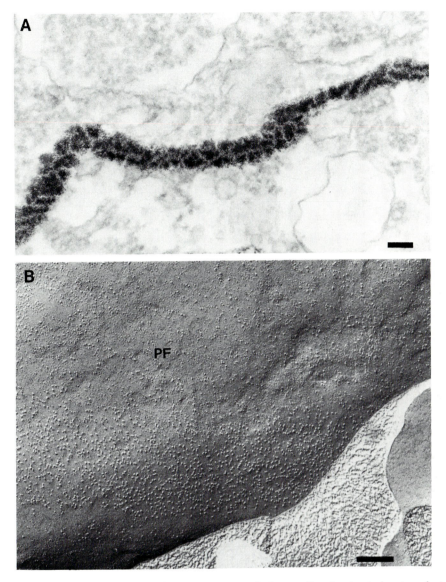

Figure 18. (A) Thin section of lanthanum-impregnated pluteus digestive tract showing pleated, anastomosing single septum septate junction. (B) Freeze-fracture replica of anastomosing septate junction of early pluteus digestive tract showing 6–9 nm particles on P face (PF). The band of P face particles is beginning to form the network pattern characteristic of this junction. Apical end of cell is at bottom of micrograph. Scale bars: 100 nm (A); 250 nm (B). [(B) from Spiegel and Howard, 1983.]

tem. This process appears to involve the degeneration of the straight septate junction and its replacement by the newly formed pleated septate junction. The transition is a gradual one that begins at gastrulation, but it is not completed until the end of larval development, when metamorphosis and feeding begin.

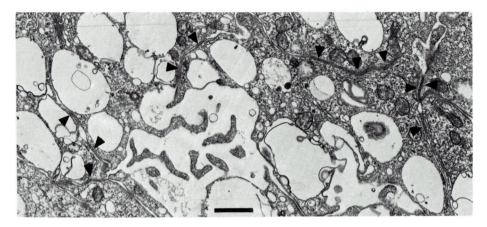

Figure 19. Invaginated cells of mid gastrula showing increased intercellular density (arrowheads) associated with transition from straight to pleated septate junctions. Scale bar: 1 μm. (From Spiegel and Howard, 1983.)

The first indication that anastomosing septates are forming appears when the intercellular space begins to show an increase in dense material and fewer "ladders" of the straight septates. This is shown in the invaginated cells of a mid-gastrula in Figure 19. The intercellular density could be due to the breakdown of the straight septate junctions or to the observation stated earlier that the formation of junctions appears to be accompanied by a localized increase in intercellular density. Before gastrulation, therefore, only one type of septate junction is present; after gastrulation, two types of septate junctions are present. The straight, unbranched double septum septate is found in the outer epithelial cells throughout morphogenesis. After gastrulation, the pleated, anastomosing, single septum septate is found only in the endothelial cells of the digestive tract.

6.3. Tricellular Junctions

A **tricellular junction** is one in which the bicellular junctions of three adjoining cells meet. Lanthanum impregnation shows that the tricellular junction is composed of a series of diaphragms made up of a central vesicular part with thin arms that connect it to the three bicellular junctions (Graf et al., 1982; Noirot-Timothee et al., 1982). This is clearly demonstrated in Figure 20A,B,

---

Figure 20. (A) Lanthanum-stained thin section of mid-gastrula stage showing two tricellular junctions. The one on the left is a tangential section showing the orientation of the double septa. The one on the right is cut through the center and shows several diaphragms (arrowheads), which are made up of a central vesicular part with thin arms that are connected to three bicellular junctions. (B) Lanthanum-stained thin section showing septa oriented parallel to cell apices and converging toward juncture line in center. Septa tend to become oriented parallel to juncture line as they proceed basally. (C) Freeze-fracture replica of mid-gastrula stage showing E face (EF) particle rows converging from both sides toward deep vertical furrow (black arrow) and complementary P face

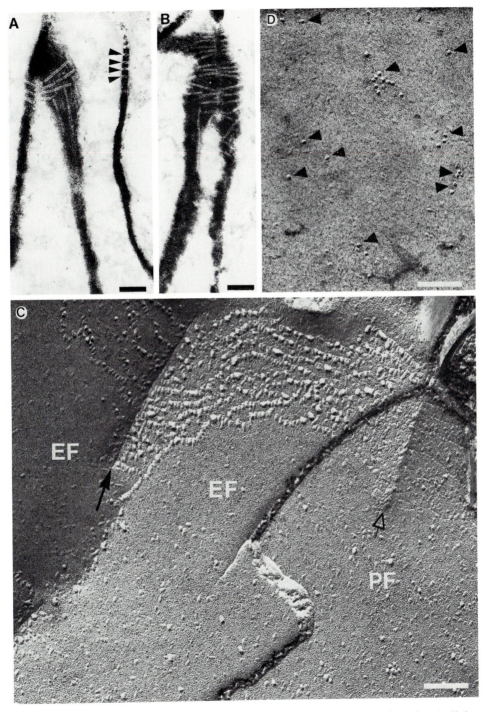

(PF) grooves converging toward ridge (open arrow); vertical furrow and ridge form the tricellular junction. (D) Freeze-fracture replica of blastula stage showing irregular clusters of particles and individual particles measuring 12–14 nm on E face (arrowheads), with possible central pore or channel, suggesting gap junction particles. Scale bars: 100 nm (A–D). (From Spiegel and Howard, 1983.)

lanthanum tracer thin sections of a sea urchin embryonic tricellular junction; the septa are oriented parallel to the cell apices at the apical end of the junction but tend to become oriented parallel to the juncture line as they proceed basally. In addition, the septa in Figure 20B appear to converge from both sides toward the juncture line in the center in this micrograph. This lanthanum image can be compared to the freeze-fracture image of an embryonic sea urchin tricellular junction (Fig. 20C), in which the E face particle rows appear to converge from both sides toward the deep vertical furrow. The P face shows the complementary ridge with shallow grooves converging toward it. The vertical furrow and ridge indicate a juncture line, as seen in the lanthanum tracer image, and identify this structure as a tricellular junction.

6.4. Gap Junctions

Our evidence for the presence of gap junctions in sea urchin embryos is incomplete because of the difficulty of obtaining clear data from either lanthanum tracer thin sections or freeze-fracture replicas. There are indications of their presence, however, as shown in Figure 20D. Freeze-fracture replicas show very small clusters of irregularly-spaced particles and individual particles on the E face, which are 12–14 nm in diameter. These particles are larger than those forming the septate junctions and are in the size range of gap junction particles described for other tissues (Lane and Swales, 1980). Complementary P face pits have not been observed, probably due to the small size of the particle clusters, their irregular arrangement, or the angle of shadowing. Some of the particles appear to have a small pore in the center, but this effect could be due to carbon scattering. The use of lanthanum tracer does not show the hexagonal array of junctional areas usually seen in tangential sections of gap junctions. It is possible that an individual connexon constitutes a gap junction, but further studies are needed to establish unequivocally the presence of gap junctions in the sea urchin embryo.

The functions of most junctional types are generally agreed upon (Staehelin, 1974; Lane, 1982). Desmosomes are involved in cell adhesion, gap junctions in intercellular communication, and tight junctions as permeability barriers. The function of septate junctions, however, is still not clear. They are thought to be adhesive structures, as indicated by their location and structure. It has also been suggested that septate junctions may function as selective, semipermeable barriers due to the presence of ECM within the septal scaffolding (Wood and Kuda, 1980; Spiegel and Howard, 1983). Additional studies are needed to ascertain the role(s) of septate junctions.

7. Conclusions

To understand cell adhesion in the sea urchin embryo, we have attempted to define the various components of the cell surfaces, including the outer ECM (hyaline layer), the inner ECM (basal lamina), and the mesenchyme cells. In addition, we have identified the various types of intercellular junctions, which are also components of the cell surfaces. This work has provided us with an overview of the microecology of the cell surfaces in embryonic sea urchin cells and points out similarities to cell surfaces of vertebrate embryos and cultured cells. The special advantage of the sea urchin embryo, however, is that it is a relatively simple, independently developing entity that can be studied as a model of cell–cell interactions during morphogenesis.

ACKNOWLEDGMENTS. The original research reported in this chapter was supported by grants from the National Science Foundation. The authors are indebted to Marjorie Audette, Louisa Howard, and Kendra Ballou for help in the preparation of the manuscript.

References

Afzelius, B. A., 1956, The ultrastructure of the cortical granules and their products in the sea urchin egg as studied with the electron microscope, *Exp. Cell Res.* **10**:257–285.

Akasaka, K., and Terayama, H., 1980, General pattern, $^{35}SO_4$ incorporation and intracellular localization of glycans in developing sea urchin (*Anthocidaris*) embryos, *Dev. Growth Diff.* **22**:749–762.

Akasaka, K., and Terayama, H., 1982, Sulfated fucogalactan–protein conjugate present in EDTA extract from *Anthocidaris* embryos, *J. Fac. Sci. Univ. Tokyo Ser. IV* **15**:181–189.

Akasaka, K., and Terayama, H., 1983, Sulfated glycan present in the EDTA extract of *Hemicentrotus* embryos, *Exp. Cell Res.* **146**:177–185.

Ali, I. U., Mautner, V., Lanza, R., and Hynes, R. O., 1977, Restoration of normal morphology, adhesion and cytoskeleton in transformed cells by addition of a transformation-sensitive protein, *Cell* **11**:115–126.

Amemiya, S., Akasaka, K., and Terayama, H., 1982a, Scanning electron microscopical observations on the early morphogenetic processes in developing sea urchin embryos, *Cell Diff.* **11**:291–293.

Amemiya, S., Akasaka, K., and Terayama, H., 1982b, Scanning electron microscopy of gastrulation in a sea urchin (*Anthocidaris crassispina*), *J. Embryol. Exp. Morphol.* **67**:27–35.

Balinsky, B. I., 1959, An electron microscope investigation of the mechanism of adhesion of the cells in a sea urchin blastula and gastrula, *Exp. Cell Res.* **16**:429–433.

Bernfield, M., and Banerjee, S. B., 1982, The turnover of basal lamina glycosaminoglycan correlates with epithelial morphogenesis, *Dev. Biol.* **90**:291–305.

Chang, D. C., and Afzelius, B. A., 1973, Electron microscope study on membrane junctions of *Arbacia punctulata* blastomeres, *Biol. Bull.* **145**:428.

Citkowitz, E., 1971, The hyaline layer: Its isolation and role in echinoderm development, *Dev. Biol.* **24**:348–362.

Citkowitz, E., 1972, Analysis of the isolated hyaline layer of sea urchin embryos, *Dev. Biol.* **27**:494–503.

Crawford, B. J., and Chia, F. S., 1982, Genesis and movement of mesenchyme cells in embryos of the starfish, *Pisaster ochraceus*, in: *International Echinoderm Conference* (J. M. Lawrence, ed.), pp. 505–511, A. A. Balkema, Rotterdam.

Dan, K., 1960, Cyto-embryology of echinoderms and amphibia, *Int. Rev. Cytol.* **9**:321–367.

Dan, K., and Ikeda, M., 1971, On the system controlling the time of micromere formation in sea urchin embryos, *Dev. Growth Diff.* **13**:285–301.

Dan, K., and Okazaki, K., 1956, Cyto-embryological studies of sea urchins. III. Role of the secondary mesenchyme cells in the formation of the primitive gut in sea urchin larvae, *Biol. Bull.* **110**:29–42.

DeSimone, D. W., and Spiegel, M., 1984a, Micromere-specific cell surface proteins of sixteen-cell stage sea urchin embryos, *Exp. Cell Res.* **156:**7–14.

DeSimone, D. W., and Spiegel, M., 1984b, The detection of micromere-specific cell surface proteins on the fertilized sea urchin egg surface, *J. Embryol. Exp. Morphol. (Suppl.)* **82:**82.

Edwards, J., 1976, Intercellular Adhesion, in: *New Techniques in Biophysics and Cell Biology* (R. Pain and B. J. Smith, eds.), pp. 1–27, Wiley, New York.

Ekblom, P., Alitalo, K., Vaheri, A., Timpl, R., and Saxen, L., 1980a, Induction of a basement membrane glycoprotein in embryonic kidney: Possible role of laminin in morphogenesis, *Proc. Natl. Acad. Sci. USA* **77:**485–489.

Ekblom, P., Lehtonen, E., Timpl, R., and Saxen, L., 1980b, Matrix proteins in kidney development, *Eur. J. Cell Biol.* **22:**437.

Endo, Y., 1961, Changes in the cortical layer of sea urchin eggs at fertilization as studied with the electron microscope, *Exp. Cell Res.* **25:**383–397.

Fraker, P. J., and Speck, J. C., Jr., 1978, Protein and cell membrane iodinations with a sparingly soluble chloroamide, 1,3,4,6-tetrachloro-3a,6a-diphenylglycoluril, *Biochem. Biophys. Res. Commun.* **80:**849–857.

Garrod, D. R., and Nicol, N., 1981, Cell behaviour and molecular mechanisms of cell–cell adhesion, *Biol. Rev.* **56:**199–242.

Gilula, N. B., 1973, Development of cell junctions, *Am. Zool.* **13:**1109–1117.

Giudice, G., 1962, Restitution of whole larvae from disaggregated cells of sea urchin embryos, *Dev. Biol.* **5:**402–411.

Giudice, G., and Mutolo, V., 1970, Reaggregation of dissociated cells of sea urchin embryos, *Adv. Morphol.* **8:**115–158.

Golob, R., Chetsanga, C. J., and Doty, P., 1974. The onset of collagen synthesis in sea urchin embryos, *Biochem. Biophys. Acta* **349:**135–141.

Gould, D., and Benson, S. C., 1978, Selective inhibition of collagen synthesis in sea urchin embryos by a low concentration of actinomycin D, *Exp. Cell Res.* **112:**73–78.

Graf, F., Noirot-Timothee, C., and Noirot, C., 1982, The specialization of septate junctions in regions of tricellular junctions. I. Smooth septate junctions, *J. Ultrastruct. Res.* **78:**136–151.

Green, C., 1978, Variations of septate junction structure in the invertebrates, in: *Ninth International Conference on Electron Microscopy*, Vol. 7 (J. M. Sturgess, ed.), pp. 338–339, Imperial Press, Toronto.

Green, C., 1981, Fixation-induced intramembrane particle movement demonstrated in freeze-fracture replicas of a new type of septate junction in echinoderm epithelia, *J. Ultrastruct. Res.* **75:**11–22.

Green, C., Bergquist, P. R., and Bullivant, S., 1979, An anastomosing septate junction in endothelial cells of the phylum Echinodermata, *J. Ultrastruct. Res.* **67:**72–80.

Gustafson, T., and Wolpert, L., 1967, Cellular movement and contact in sea urchin morphogenesis, *Biol. Rev.* **42:**442–498.

Hall, H. G., and Vacquier, V. D., 1982, The apical lamina of the sea urchin embryo: Major glycoproteins associated with the hyaline layer, *Dev. Biol.* **89:**168–178.

Harkey, M. A., and Whiteley, A. H., 1980, Isolation, culture and differentation of echinoid primary mesenchyme cells, *Arch. Dev. Biol.* **189:**111–122.

Hay, E. D., 1978a, Fine structure of embryonic matrices and their relation to the cell surface in ruthenium red fixed tissues, *Growth* **42:**399–423.

Hay, E. D., 1978b, Role of basement membranes in development and differentiation, in: *Biology and Chemistry of Basement Membranes* (N. A. Kefalides, ed.), pp. 119–136, Academic Press, New York.

Hay, E. D. (ed.), 1981, *Cell Biology of Extracellular Matrix*, Plenum Press, New York.

Heifitz, A., and Lennarz, W. J., 1979, Biosynthesis of N-glycosidically linked glycoproteins during gastrulation of sea urchin embryos, *J. Biol. Chem.* **254:**6199–6127.

Herbst, C., 1900, Über das Auseinandergehen von Furchungs und Gewebezellen in Kalk-freiem Medium. *Arch. Entwicklungsmech. Org.* **9:**424–463.

Hogan, B., 1981, Laminin and epithelial cell attachment, *Nature (Lond.)* **290:**737–738.

Hynes, R. O., 1973, Alteration of cell surface proteins by viral transformation and by proteolysis, *Proc. Natl. Acad. Sci. USA* **70:**3170–3174.

Immers, J., 1961, Comparative study of the localization of incorporated ^{14}C-labeled amino acids and $^{35}SO_4$ in the sea urchin ovary, egg and embryo, *Exp. Cell Res.* **24:**356–378.

Kane, R. E., and Stephens, R. E., 1969, A comparative study of the isolation of the cortex and the role of the calcium insoluble protein in several species of sea urchin eggs, *J. Cell Biol.* **41:**133–144.

Karp, G. C., and Solursh, M., 1974, Acid mucopolysaccharide metabolism, the cell surface, and primary mesenchyme cell activity in the sea urchin embryo, *Dev. Biol.* **41:**110–123.

Katow, H., and Solursh, M., 1980, Ultrastructure of primary mesenchyme cell ingression in the sea urchin *Lytechninus pictus, J. Exp. Zool.* **213:**231–246.

Katow, H., and Solursh, M., 1982, In situ distribution of concanavalin A-binding sites in mesenchyme blastulae and early gastrulae of the sea urchin *Lytechninus pictus, Exp. Cell Res.* **139:**171–180.

Katow, H., Yamada, K. M., and Solursh, M., 1982, Occurrence of fibronectin on the primary mesenchyme cell surface during migration in the sea urchin embryo, *Differentiation* **22:**120–124.

Kefalides, N. A., Alper, R., and Clark, C. C., 1979, Biochemistry and metabolism of basement membranes, *Int. Rev. Cytol.* **61:**167–228.

Kinoshita, S., and Saiga, H., 1979, The role of proteoglycan in the development of sea urchins. I. Abnormal development of sea urchin embryos caused by the disturbance of proteoglycan synthesis, *Exp. Cell Res.* **123:**229–236.

Kinoshita, S., and Yoshii, K., 1979, The role of proteoglycan synthesis in the development of sea urchins. II. The effect of administration of exogenous proteoglycan, *Exp. Cell Res.* **124:**261–269.

Kleinman, H. K., Klebe, R. J., and Martin, G. R., 1981, Role of collagenous matrices in the adhesion and growth of cells, *J. Cell Biol.* **88:**473–485.

Kondo, K., and Sakai, H., 1971, Demonstration and preliminary characterization of reaggregation-promoting substances from embryonic sea urchin cells, *Dev. Growth Diff.* **13:**1–14.

Kyoizumi, S., and Kominami, T., 1980, Inhibitory effect of concanavalin A on the cell-to-cell adhesion during early development of the starfish, *Asterina pectinifera, Exp. Cell Res.* **128:**323–331.

Lallier, R., 1972, Effects of concanavalin A on the development of sea urchin eggs, *Exp. Cell Res.* **72:**157–163.

Lane, N., 1982, Studies on invertebrate cell interactions—Advantages and limitations of the morphological approach, in: *The Functional Integration of Cells in Animal Tissues* (J. D. Pitts and M. E. Finbow, eds.), pp. 81–103, Cambridge University Press, Cambridge.

Lane, N. J., and Swales, L. S., 1980, Dispersal of junctional particles, not internalization, during the in vivo disappearance of gap junctions, *Cell* **19:**579–686.

Lash, J. W., and Vasan, N. S., 1977, Tissue interactions and extracellular matrix components, in: *Cell and Tissue Interactions* (J. W. Lash and M. M. Burger, eds.), pp. 101–114, Raven Press, New York.

Leivo, I., Vaheri, A., Timpl, R., and Wartiovaara, J., 1980, Appearance and distribution of collagens and laminin in the early mouse embryo, *Dev. Biol.* **76:**100–114.

Luft, J. H., 1971a, Ruthenium red and violet. I. Chemistry, purification, methods of use for electron microscopy and mechanism of action, *Anat. Rec.* **171:**347–368.

Luft, J. H., 1971b, Ruthenium red and violet. II. Fine structural localization in animal tissues, *Anat. Rec.* **171:**369–416.

Markwald, R. R., Fitzharris, T. P., Bank, H., and Bernanke, D. H., 1978, Structural analyses on the matrical organization of glycosaminoglycans in developing endocardial cushions, *Dev. Biol.* **62:**292–316.

Markwald, R. R., Fitzharris, T. P., and Bernanke, D. H., 1979, Morphologic recognition of complex carbohydrates in embryonic cardiac extracellular matrix. *J. Histochem. Cytochem.* **27:**1171–1173.

McCarthy, R. A., and Spiegel, M., 1983a, Protein composition of the hyaline layer of sea urchin embryos and reaggregating cells, *Cell Diff.* **13:**93–102.

McCarthy, R. A., and Spiegel, M., 1983b, The enhancement of reaggregation of sea urchin blastula cells by exogenous proteins, *Cell Diff.* **13:**103–114.

McClay, D. R., 1979, Surface antigens involved in interactions of embryonic sea urchin cells, *Curr. Top. Dev. Biol.* **13**:199–236.

McClay, D. R., Chambers, A. R., and Warren, R. H., 1977, Specificity of cell–cell interactions in sea urchin embryos. Appearance of new cell surface determinants at gastrulation, *Dev. Biol.* **56**:343–355.

McClay, D. R., Cannon, G. W., Wessel, G. M., Fink, R. D., and Marchase, R. M., 1983, Patterns of antigenic expression in early sea urchin development, in: *Time, Space, and Pattern in Embryonic Development* (W. R. Jeffery and R. A. Raff, eds.), pp. 157–169, Alan R. Liss, New York.

Mizoguchi, H., and Yasumasu, I., 1982, Archenteron formation induced by ascorbate and α-ketoglutarate in sea urchin embryos kept in SO_4-free artificial sea water, *Dev. Biol.* **93**:119–125.

Mizoguchi, H., and Yasumasu, I., 1983, Effect of α,α' dipyridyl on exogut formation in vegetalized embryos of the sea urchin, *Dev. Growth Diff.* **25**:57–64.

Murray, L. W., Tanzer, M. L., and Cooke, P., 1981, *Nereis* cuticle collagen: Relationship of fiber ultrastructure to biochemical and biophysical properties, *J. Ultrastruct. Res.* **76**:27–45.

Noirot-Timothee, C., Graf, F., and Noirot, C., 1982, The specialization of septate junctions in regions of tricellular junctions. II. Pleated septate junctions, *J. Ultrastr. Res.* **78**:152–165.

Noll, H., Matranga, V., Cascino, D., and Vittorelli, L., 1979, Reconstitution of membranes of embryonic development in dissociated blastula cells of the sea urchin by reinsertion of aggregation-promoting membrane proteins extracted with butanol, *Proc. Natl. Acad. Sci. USA* **76**:288–292.

Noll, H., Matranga, V., Palma, P., Cutrono, F., and Vittorelli, L., 1981, Species-specific dissociation into single cells of live sea urchin embryos by Fab against membrane components of *Paracentrotus lividus* and *Arbacia lixula*, *Dev. Biol.* **87**:229–241.

Oguri, K., and Yamagata, T., 1978, Appearance of proteoglycans in developing sea urchin embryos, *Biochem. Biophys. Acta* **541**:385–393.

Okazaki, L., 1975, Spicule formation by isolated micromeres of the sea urchin embryo, *Am. Zool.* **15**:567–581.

Okazaki, K., and Niijima, L., 1964, Basement membrane in sea urchin larvae, *Embryology* **8**:89–100.

Oppenheimer, S. B., and Meyer, J. T., 1982a, Isolation of species-specific and stage-specific adhesion promoting component by disaggregation of intact sea urchin embryo cells, *Exp. Cell Res.* **137**:472–476.

Oppenheimer, S. B., and Meyer, J. T., 1982b, Carbohydrate specificity of sea urchin blastula adhesion component, *Exp. Cell Res.* **139**:451–455.

Parisi, E., Filosa, S., DePetrocellis, B., and Monroy, A., 1978, The pattern of cell division in the development of the sea urchin, *Dev. Biol.* **65**:38–49.

Pfohl, R. J., and Giudice, G., 1967, The role of cell interaction in the control of enzyme activity during embryogenesis, *Biochem. Biophys. Acta* **142**:263–266.

Pucci-Minafra, I., Casano, C., and LaRosa, C., 1972, Collagen synthesis and spicule formation in sea urchin embryos, *Cell Diff.* **1**:157–165.

Ruoslahti, E., Vaheri, A., Kuusela, P., and Linder, E., 1973, Fibroblast surface antigen: A new serum protein, *Biochem. Biophys. Acta* **322**:352–358.

Rustad, R., 1960, Dissociation of the mitotic time schedule from the micromere "clock" with X-rays, *Acta Embryol. Morphol. Exp.* **3**:155–158.

Schneider, E. G., and Lennarz, W. J., 1976, Glycosyl transferases of eggs and embryos of *Arbacia punctulata*, *Dev. Biol.* **53**:10–20.

Schneider, E. G., Nguyen, H. T., and Lennarz, W. J., 1978, The effect of tunicamycin, an inhibitor of protein glycosyation, on embryonic development in the sea urchin, *J. Biol. Chem.* **253**:2348–2355.

Scott, J. E., and Dorling, J., 1965, Differential staining of acid glycosaminoglycans (mucopolysaccharides) by alcian blue in salt solutions, *Histochemistry* **5**:221–233.

Slavkin, H. C., and Greulich, R. C. (eds.), 1975, *Extracellular Matrix Influences on Gene Expression*, Academic Press, New York.

Solursh, M., and Katow, H., 1982, Initial characterization of sulfated macromolecules in the blastocoels of mesenchyme blastulae of *Stronglyocentrotus purpuratus* and *Lytechinus pictus*, *Dev. Biol.* **94**:326–336.

Spiegel, E., and Howard, L., 1983, Development of cell junctions in sea urchin embryos, *J. Cell Sci.* **62:**27–48.

Spiegel, E., and Spiegel, M., 1977, A scanning electron microscope study of early sea urchin reaggregation, *Exp. Cell Res.* **108:**413–420.

Spiegel, E., and Spiegel, M., 1979, The hyaline layer is a collagen-containing extracellular matrix in sea urchin embryos and reaggregating cells, *Exp. Cell Res.* **123:**434–441.

Spiegel, E., and Spiegel, M., 1980, The internal clock of reaggregating embryonic sea urchin cells, *J. Exp. Zool.* **213:**271–281.

Spiegel, E., Burger, M. M., and Spiegel, M., 1980, Fibronectin in the developing sea urchin embryo, *J. Cell Biol.* **80:**309–313.

Spiegel, E., Burger, M. M., and Spiegel, M., 1983, Fibronectin and laminin in the extracellular matrix and basement membrane of sea urchin embryos, *Exp. Cell Res.* **144:**47–55.

Spiegel, M., and Burger, M. M., 1982, Cell adhesion during gastrulation. A new approach, *Exp. Cell Res.* **139:**377–382.

Spiegel, M., and Spiegel, E., 1975, The reaggregation of dissociated embryonic sea urchin cells, *Am. Zool.* **15:**583–606.

Spiegel, M., and Spiegel, E., 1978a, The morphology and specificity of cell adhesion of echinoderm embryonic cells, *Exp. Cell Res.* **117:**261–268.

Spiegel, M., and Spiegel, E., 1978b, Sorting out of sea urchin embryonic cells according to cell type, *Exp. Cell Res.* **117:**269–271.

Staehelin, L. A., 1974, Structure and function of intercellular junctions, *Int. Rev. Cytol.* **39:**191–283.

Stephens, R. E., and Kane, R. E., 1970, Some properties of hyalin. The calcium-insoluble protein of the hyalin layer of the sea urchin egg, *J. Cell Biol.* **44:**611–617.

Subtelny, S., and Wessells, N. K. (eds.), 1980, *The Cell Surface: Mediator of Developmental Processes,* Academic Press, New York.

Terranova, V. P., Rohrbach, D. H., and Martin, G. R., 1980, Role of laminin in the attachment of PAM 212 (epithelial) cells to basement membrane collagen, *Cell* **22:**719–726.

Timpl, R., 1980, Structural components of basement membranes, *Eur. J. Cell Biol.* **22:**425.

Timpl, R., Rohde, H., Robey, P. G., Rennard, S. I., Foidart, J. M., and Martin, G. R., 1979, Laminin—A glycoprotein from basement membranes, *J. Biol. Chem.* **254:**9933–9937.

Tonegawa, Y., 1973, Isolation and characterization of a particulate cell-aggregation factor from sea urchin embryos, *Dev. Growth Diff.* **14:**337–352.

Tonegawa, Y., 1982, Cell aggregation factor and endogenous lectin in sea urchin embryos, *Cell Diff.* **11:**335–337.

Townes, P. L., and Holtfreter, J., 1955, Directed movements and selective adhesion of embryonic amphibian cells, *J. Exp. Zool.* **128:**53–120.

Vaheri, A., and Mosher, D. F., 1978, High molecular weight, cell surface-associated glycoprotein (fibronectin) lost in malignant transformation, *Biochem. Biophys. Acta* **516:**1–25.

Vaheri, A., Alitala, K., Hedman, K., Keski-Oja, J., Kurkinen, M., and Wartiovaara, J., 1978, Fibronectin and the pericellular matrix of normal and transformed adherent cells, *Ann. N.Y. Acad. Sci.* **312:**343–353.

Vittorelli, L., Matranga, V., Cervello, M., and Noll, H., 1982, Aggregation factors of sea urchin embryonic cells, in: *Embryonic Development, Part B:Cellular Aspects* (M. M. Burger, ed.), pp. 211–221, Alan R. Liss, New York.

Wartiovaara, J., Leivo, I., and Vaheri, A., 1980, Matrix glycoproteins in early mouse development and in differentiation of teratocarcinoma cells, in: *The Cell Surface: Mediator of Developmental Processes* (S. Subtelny and N. K. Wessells, eds.), pp. 305–324, Academic Press, New York.

Wessel, G. M., Marchase, R. B., and McClay, D. R., 1984, Ontogeny of the basal lamina in the sea urchin embryo, *Dev. Biol.* **103:**235–245.

Weston, J. A., Derby, M. A., and Pintar, J. E., 1978, Changes in the extracellular environment of neural crest cells during their early migration, *Zoon* **6:**103–113.

Wolpert, L., and Gustafson, T., 1961, Studies of the cellular basis of morphogenesis of the sea urchin embryo. The formation of the blastula, *Exp. Cell Res.* **25:**374–382.

Wolpert, L., and Mercer, E. H., 1963, An electron microscope study of the development of the blastula of the sea urchin and its radial polarity, *Exp. Cell Res.* **30:**280–300.

Wood, R. L., and Kuda, A. M., 1980, Formation of junctions in regenerating *Hydra:* Septate junctions, *J. Ultrastruct. Res.* **70:**104–117.

Yamada, K. M., and Olden, K., 1978, Fibronectins—adhesive glycoproteins of cell surface and blood, *Nature (Lond.)* **275:**179–184.

Yamada, K. M., and Weston, J. A., 1974, Isolation of a major cell surface glycoprotein from fibroblasts, *Proc. Natl. Acad. Sci. USA* **71:**3492–3496.

Yamada, K. M., Yamada, S. S., and Pastan, I., 1976, Cell surface protein partially restores morphology, adhesiveness and contact inhibition of movement to transformed fibroblasts, *Proc. Natl. Acad. Sci. USA* **73:**1217–1221.

Chapter 7

The Cellular Basis of Amphibian Gastrulation

RAY E. KELLER

1. Introduction to Amphibian Gastrulation

1.1. Amphibian Gastrulation as a Morphogenetic System

Amphibian gastrulation is a complex integration of local cellular behavior to produce a supracellular system that, in turn, constrains and organizes the behavior of individual cells. Such behavior has fascinated and challenged embryologists for over a hundred years and has also perplexed some of them to the point of thinking it not reducible to part-processes. These thoughts were expressed by Walter Vogt (translated in Spemann, 1938), who did more than anyone to characterize the early morphogenesis of amphibians:

> It does not appear at all as if cells were walking in the sense, that single part movements were combining to form the movements of the masses; for even the most natural and plausible explanation by means of amoeboid moving of single cells fails utterly. We evidently have not the wandering of cells before us, but rather a passive obedience to a superior force.

But this characteristic of amphibian gastrulation may be more of an advantage than a disadvantage. The principal goal in establishing a cellular basis of morphogenesis is understanding how the structural properties and motile behavior of individual cells produce the macroscopic distortions of the whole embryo in the proper proportion and geometry. Such an integration of cellular and supracellular levels of organization is probably best studied where it is highly developed and easily approached experimentally. Because of their numbers, availability, and size, amphibian embryos are ideal for microsurgical manipulation, cinemicrographic and morphometric analysis of cell behavior, electrophysiology, cell lineage tracings, pattern formation, and biochemical and immunological analyses of molecular constituents functioning in early devel-

RAY E. KELLER • Department of Zoology, University of California—Berkeley, Berkeley, California 94720.

opment. The rationale and promise of applying cell biology to amphibian gastrulation were set forth by Johannes Holtfreter more than 40 years ago (1943b):

> Yet the most astonishingly integrated character of the gastrulation process as a whole has kept the investigators wondering if there might not exist a "superior force" supervising and directing the single part-movements. . . . However, the results of a more intimate analysis of the problem here at hand refutes an over-hasty skepticism. The following sections give evidence that the directed movements of embryonic regions can actually be traced back to basic facilities of the single cells and to their specific response to changes of environment. The unitarian character of their combined effort is mainly the result of the predisposed arrangement of cells with a locally different kinetic behavior.

1.2. Objectives of a Cellular Analysis of Gastrulation

My objective here is to examine our progress in defining local, specific types of cell behavior, such as polarity, shape change, protrusive activity, adhesion, and traction, that can be shown, in the context of the gastrula, to produce gastrulation. The relevant processes also involve subcellular and supracellular levels of biological organization and developmental phenomena usually considered apart from morphogenesis. On the subcellular side, the aspects of cell phenotype relevant to morphogenesis include specialized intercellular junctions, ion channels, structural and regulatory components of the cytoskeleton, and cell surface macromolecules. The synthesis and function of these constituents may involve regulatory events at several levels (e.g., transcription, translation, and assembly) and control by pattern specification events involving cytoplasmic localization, cell lineage, and cellular interactions. On the supracellular side, local, motile behavior can be interpreted and exercised effectively only within the context of a mechanical system of other cells and extracellular matrix. It is, in part, dependent on and controlled by a supracellular system of extracellular ionic and macromolecular conditions and social interactions that emerge only at the cell population level. Thus, the solution to the problem of morphogenesis lies in many aspects of developmental, cellular, and molecular biology. Here, I will discuss what is known about how cells function in amphibian gastrulation and also attempt to identify what is not understood about this problem.

Amphibian gastrulation is sufficiently complex in its anatomy, in its geometry, and in its hierarchical organization of cell behavior that it is useful at the outset to discuss what information might constitute its cellular basis.

First, it is important to describe the paths followed by cells during the course of gastrulation. Movements of amphibian cells have been traced by recognition of anatomical relationships and histological characteristics at successive stages (Vogt, 1929a,b; Nieuwkoop and Florshutz, 1950), by marking cells with vital dyes (Goodale, 1911a,b; Vogt, 1923a,b, 1925, 1929b; Pasteels, 1942; Keller, 1975, 1976), with radioisotope labeling (Sirlin, 1956), and with injection of cell lineage-restricted tracers such as horseradish peroxidase (HRP) (Hirose and Jacobson, 1979) and fluoresceinated lysine–dextran (Gimlich and

Cooke, 1983). Natural pigment markers between species have been used (Spemann and Mangold, 1924), and cells can be followed directly with time-lapse cinemicrography (Keller, 1978). Movements were also plotted by following exovates from punctures, a technique that led to both inaccurate (Ikedo, 1902) and accurate (King, 1902a) interpretations. Despite these efforts, our knowledge of the paths of cell movement is barely adequate, and then only for a few species of amphibians. A precise description of where cells move is important. Without it, the problem is not clearly defined, and thus its analysis is likely to be naive.

Second, it is important to know what mechanical forces bring about displacement of cells and what relationship cell behavior bears on generation of these forces. Gastrula cells both contribute to, and in turn are moved by, a macroscopic system of forces. Unfortunately, the magnitude, direction, and source of these forces are usually either assumed or ignored. To what degree do cells in a given region actively bring about their own displacement? To what degree are they moved by virtue of their membership in an aggregate that transmits mechanical stresses generated elsewhere? The relationship between the behavior of a cell and its displacement is not always direct and may, in fact, be indirect in most cases. A cell may use some form of protrusive activity to exert tractional forces on a substratum and thus translocate or migrate with respect to it (Fig. 1a). A cell may be displaced as a function of its membership in a contiguous epithelium that distorts as a result of active change in the shape of

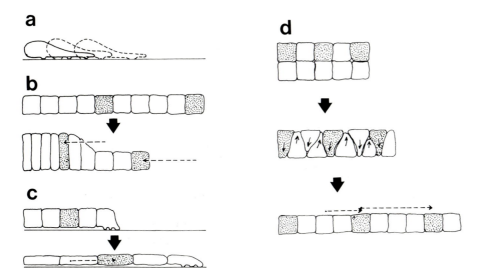

Figure 1. Relationships between a cell's behavior and its displacement. A cell may form protrusions that adhere to a substratum and exert traction sufficient to displace the cell (a). Individual cells in an epithelium may be displaced as a result of change in the shape of cells in the epithelium, whether the displaced cell participates in the shape change or not (b). Traction of cells at the margin of an epithelial sheet may result in stretching of the sheet and passive displacement of its cells (c). Individual cells in a cell population are displaced as individuals intercalate between one another to form a longer, narrower array (d).

all or some of its constituent cells. In this situation, the displacement of a cell is a function of its initial position and the geometry and mechanical properties of the aggregate, whether it actively contributes or not (Fig. 1b). A cell may be passively displaced as a result of forces generated elsewhere, such as expansion of an epithelium brought about by traction at its margin (Fig. 1c). Finally, a cell may change position as a function of a distortion of the cell population, driven by an active rearrangement of cells to produce a narrower, longer array (Fig. 1d). Here again, an individual cell exercises a local behavior, but its displacement in the embryo is a function of its context in the cell population. There are other possibilities as well, e.g., cells may generate tractional forces that organize the extracellular matrix such that it patterns the movements of other cells (see Stopak and Harris, 1982). It is probable that all these relationships occur in gastrulation; therefore it is useful to take a broad view of the possible relationships between behavior of a cell and its displacement at the outset.

Third, it is important to know what cellular properties bring about specific patterns of morphogenetic behavior and how they do so. Several general categories of cellular characteristics are thought to be important in morphogenesis. It has been argued that properties of the cell surface, such as selective (Townes and Holtfreter, 1955) or differential (Steinberg, 1964, 1970) adhesion between cell types could drive the cell sorting and tissue rearrangements seen in culture that mimic morphogenetic events. Gustafson and Wolpert (1967) postulated that differences in adhesion at various sides of cells could cause changes in their shape and thus result in tissue deformation. Also, changes in cell position or shape may be direct or indirect results of mechanical forces generated by an internal cytoskeleton, expressed as change in cell shape or as the exertion of tractional forces on a substratum. But it is probably more useful to view the cell behavior that is important in gastrulation as arising from interdependent activities of the cell surface and the cytoskeleton, perhaps regulated by specific cell surface receptors or ion channels that serve as transducers of extracellular signals. Therefore, our analysis is best served by Holtfreter's broader concept of tissue affinities as a morphogenetic mechanism (Holtfreter, 1939) and his notions about the role of changing patterns of motile behavior and structural organization (Holtfreter, 1943a,b,c, 1944, 1946a,b,c,d, 1947a,b, 1948) in morphogenesis.

Fourth, it is important to demonstrate experimentally a cellular mechanism of morphogenesis. Morphogenesis attracted attention early and thus has a long history. Many mechanisms have been set forth. A few were supported to some degree by fact, but most were set forth as possible or plausible explanations and by no means constituted proven or satisfactory explanations, even to their proponents. A selected few have been cited and recited with increasing confidence and have become standard fixtures, first in reviews and finally in textbooks. As a result, it is commonly believed that cell function in morphogenesis is understood and all that remains is to work out molecular details. However, a cellular mechanism of morphogenesis requires a demonstration that a specific and defined cellular behavior does, in fact, generate a pattern of mechanical stresses that produces the observed distortion of the embryo. By this criterion, no morphogenetic process—and especially amphibian gastrulation—is satisfactorily understood at the cell or cell population level.

2. Blastula Structure and Organization

Gastrulation depends on the "predisposed arrangement of cells of locally different kinetic behavior" (Holtfreter, 1943b; see also Løvtrup, 1966). Such organization arises from pattern-generating processes and morphogenetic movements in the pregastrula stages.

2.1. Establishment and Expression of Axes in *Xenopus laevis*

This subject is best understood in *Xenopus laevis* (see Gerhart *et al.*, 1983a,b). Other amphibians differ in details, but many show some expression of axes in blastula stages that are predictive of gastrulation events and the organization of the body plan (see Gerhart, 1980).

The unfertilized egg of *Xenopus* has an animal–vegetal axis and polarity (Fig. 2a). The dark **animal hemisphere** contains melanin pigment in its cortical region, the meiotic spindle, cytoplasm rich in small yolk platelets, fat droplets, glycogen granules, ribosomes, mitochondria, and nuclear sap from germinal vesicle breakdown. By contrast, the light **vegetal hemisphere** contains little pigment and cytoplasmic constituents but has yolk platelets in large number and size (see Klag and Ubbels, 1975). The "animal" hemisphere (or animal pole) was so named because it gave the impression of actively participating in morphogenesis and formation of the animal body, whereas the "vegetal" hemisphere appeared to be passively pushed inside and formed the short-lived nutritive cells of the gut. Actually this impression is erroneous, and the animal–vegetal axis bears no simple relationship to the axes of the body plan that will emerge from gastrulation and neurulation. But this terminology is embedded in the literature, and it happens to be useful for describing egg structure and the morphogenetic movements that follow.

At fertilization, sperm entry results in events that introduce a second set of asymmetries in the egg normally predictive of where gastrulation will begin and where the dorsal side of the animal will form (see Gerhart, 1980; Elinson, 1980; Stewart-Savage and Grey, 1982; Gerhart *et al.*, 1983b). Sperm entry, which occurs at a random position around the animal pole, results in a complex rearrangement of cortical and cytoplasmic constituents and pigment with respect to the **sperm entry point** (**SEP**), such that the side opposite the SEP is generally lighter in pigmentation (Palacek *et al.*, 1978) (Fig. 2b). Changes also occur in the plasma membrane; animal–vegetal differences in its lipid mobility increase on fertilization (Dictus *et al.*, 1984). Pattern-determining events occur during this period that, in normal development, bias the formation of the dorsal side of the animal such that it usually lies on the lightly pigmented half of the

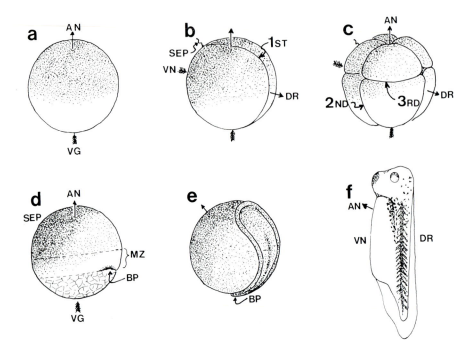

Figure 2. Emergence of the major axes and embryonic regions relevant to gastrulation and formation of the body plan. The oocyte (a) has an animal–vegetal axis (AN–VG). After fertilization, the
dorsoventral (DR–VN) axis is formed relative to the sperm entry point (SEP) and is expressed as
lighter pigmentation on the dorsal side (b). These features are preserved through cleavage (c) to the
early gastrula (d), when the blastopore (BP) forms on the dorsal side just below the marginal zone
(MZ). During gastrulation and neurulation (e), the dorsal side elongates at the expense of the
ventral side and forms the axial structures characteristic of the dorsal sector of the vertebrate body
plan (f).

egg opposite the SEP (Fig. 2b–d). Changes in yolk distribution are such that the
dark animal hemisphere of the fertilized egg becomes less dense than the vegetal and thus is oriented uppermost by gravity.

The first cleavage is meridional and tends to pass through the SEP, the
animal–vegetal axis, and the approximate midline of the future animal (Fig.
2b). The second is also meridional but perpendicular to the first, and it usually
divides the egg into two light-colored dorsal blastomeres and two darker ventral blastomeres. The third is latitudinal and nearer the animal pole (Fig. 2c).
During cleavage the dorsoventral difference in pigmentation of the animal
hemisphere persists. Cleavage is slower in the vegetal half, and the resulting
blastomeres show an animal-to-vegetal gradient in size. The animal–vegetal
pigment boundary loses its sharpness and moves vegetally as the pregastrular
movements and swelling of the blastocoel occur (see Section 2.2). This equatorial transition zone in cell size and pigmentation marks the approximate

position of a third region, the **marginal zone.** It is torus shaped and lies just to the animal side of the large, lightly pigmented cells of the vegetal region (Fig. 2d). The blastopore first forms at the vegetal edge of the marginal zone, usually opposite the site of sperm entry on what will become the dorsal side of the animal (Fig. 2d). The marginal zone was so named because it lies at the upper margin of the blastopore.

The first overt expression of the underlying dorsoventral and anteroposterior asymmetries that follow from fertilization is the behavior of the marginal zone during gastrulation. The dorsal marginal zone initiates and dominates the morphogenetic movements of gastrulation. It performs the morphogenetic movement of **convergent extension,** which moves cells to the dorsal side of the embryo and lengthens it at the expense of the ventral region (Vogt, 1929b) (Fig. 2d–f). As the "organizer," the dorsal marginal zone displays a pattern of self-differentiation and tissue interactions that organizes the dorsoventral and anteroposterior array of structures characteristic of the vertebrate body plan (see Spemann, 1938). The dorsal marginal zone expresses an anteroposterior polarity in its morphogenetic behavior (Holtfreter, 1938a, b) and in the formation of an axial array of head and tail structures (see Spemann, 1938; Hall, 1937; Hama, 1978; Kaneda and Hama, 1979) (Fig. 2e,f). It controls the differentiation of the dorso–ventral array of mesodermal structures (see Cooke, 1972–1979), induces the formation of the neural structures from ectoderm, and can organize a second set of axial structures when transplanted to the ventral side of another embryo (Spemann and Mangold, 1924; Spemann, 1938; Gimlich and Cooke, 1983).

The development of dorsoventral differences in the gastrulation movements and in the axial organization that follows depend on the cytoplasmic and cortical rearrangements mentioned above or on events associated with them. There is a critical period at 0.4–0.8 of the time between fertilization and first cleavage in which the site of dorsal morphogenesis and differentiation can be reversed or doubled (twinning) by misorientation with respect to gravity (see Gerhart et al., 1983b) and by directional centrifugation (see Black and Gerhart, 1985, 1986). Morphogenetic behavior characteristic of the dorsal sector and differentiation of dorsal and anterior tissues can be suppressed progressively with increasing doses of ultraviolet radiation (see Scharf and Gerhart, 1983; Grant and Wacaster, 1972; Manes and Elinson, 1980; Malacinski et al., 1977) and, with cold and pressure, probably by disrupting the cytoskeletal filament systems (Scharf and Gerhart, 1983). It is not known how the pattern-specification machinery establishes and maintains polarity from this critical period, just after fertilization, to the onset of gastrulation, although cell lineage and interactions between cells may both be involved (see Nieuwkoop, 1977; Kageura and Yamana, 1983; Gimlich and Gerhart, 1984). Whatever their mechanism, perturbation of these pattern-forming events is expressed at gastrulation primarily as a change in the degree and location of the two major morphogenetic patterns of the marginal zone—narrowing and lengthening of its dorsal posterior sector and migration and spreading of its anterior and ventral sectors (see Section 9.2).

2.2. The Role of Cleavage in Organizing the Gastrula

The presence of a blastocoel and two populations of blastomeres—a super-ficial layer of epithelial cells and a deep population of nonepithelial cells—figure strongly in the organization and execution of gastrulation. Formation of these structures is accomplished in one stroke by insertion of new plasma membrane and formation of specialized intercellular junctions during cleavage furrow formation.

The first cleavage furrow in *Xenopus* moves inward about one-third the diameter of the egg in the animal region and one-sixth in the vegetal (Fig. 3a–a''). From this point on, the furrow is deepened by the addition of new mem-brane as vesicles in the cytoplasm fuse with the walls of the furrow (Fig. 3a) (Kalt, 1971a,b; Bluemink and deLaat, 1973, 1977; Sanders and Singal, 1975). The new furrow membrane, the vesicle membrane, and the Golgi membrane all have a Na–K-ATPase, based on the presence of thiamine pyrophosphatase activity; this fact suggests an origin of this membrane in the Golgi and insertion into the furrow by exocytosis of the vesicles (Sanders and Singal, 1975). The separating blastomeres are initially unattached, but adhering close junctions impermeable to ruthenium red soon form between apposing walls of the furrow near the junction of the original surface membrane and the new membrane (see Kalt, 1971b; Sanders and Zalik, 1972; Singal and Sanders, 1974a,b; Sanders and Singal, 1975) (Fig. 3a''). The resulting blastomeres have an outer surface derived from the **primary membrane** of the egg (also called **original** or **preexist-ing membrane**) and a **secondary membrane** (also referred to as **nascent** or **subjunctional membrane**), which was added below the circumapical system of sealing tight junctions that develop in the region of the original close contacts (Fig. 3b,c). The outer primary membrane is relatively impermeable to most ions, as are the tight junctions (see Slack *et al.*, 1973; Morrill *et al.*, 1975), and thus the blastocoel becomes a *milieu interieur*, largely independent from the exterior. Tangential (periclinal) divisions at the early midblastula (stage 7) of *Xenopus* produce blastomeres lacking the primary surface at the inner ends of the heretofore monolayered epithelium (Nieuwkoop and Faber, 1967) (Fig. 3c). Thus, a deep nonepithelial cell population is produced at the expense of the superficial epithelium. The newly formed deep cells and the inner basal ends of the superficial cells share the property of having secondary plasma mem-brane (Fig. 3c–d).

The insertion of vesicles into the cleavage furrow may have a dual function in blastocoel formation. The Na–K-ATPase of the secondary vesicle-derived membrane pumps sodium out of the cells into the extracellular (blastocoelic) space (Slack and Warner, 1973; Morrill *et al.*, 1975; deLaat *et al.*, 1976; Blue-mink and deLaat, 1977). In addition, acidic mucosubstances in the vesicles are deposited in the furrow (prospective blastocoel) during fusion with the furrow wall (Kalt, 1971a,b) (Fig. 3a'') and may function in its enlargement. As these processes increase solute concentration in the blastocoel, it swells by osmotic uptake of water across the tight junctions. Blastocoel enlargement can be slowed or prevented by ouabain inhibition of the Na–K pump (see Slack and

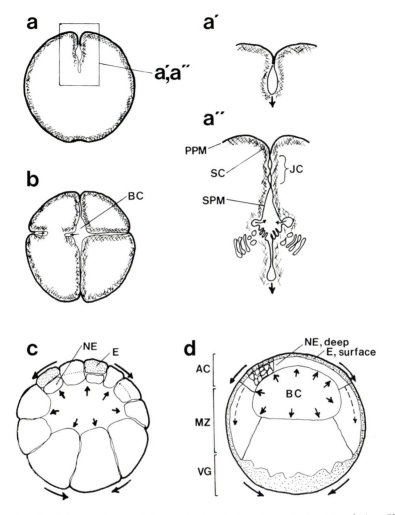

Figure 3. The role of cleavage in organizing anatomical features important to gastrulation. Cleavage occurs partly by furrow formation (a, a′) and partly by addition of new membrane to the furrow walls by vesicle fusion with the walls (a″). A junctional complex (JC) forms near the apical edges of the blastomeres and separates the forming blastocoel (BC) from the exterior (a″, b). Tangential cleavages form a superficial epithelial (E) and a deep nonepithelial (NE) cell population (c,d). The short arrows indicate blastocoel expansion, the long arrows show expansion of the animal cap (AC) and shrinkage of the vegetal region (VG), and the dashed arrow shows vegetal movement of the marginal zone (MZ) in the blastula stage (c,d). The apical cell surfaces are derived from the original or primary plasma membrane (PPM, heavy lines), whereas the deep surfaces of both superficial and deep cells are nascent or secondary plasma membrane (SPM, light lines). SC, subcortical cytoplasm. (Figures a′ and a″ redrawn from Singal and Sanders, 1974a, with permission.)

Warner, 1973; Morrill *et al.*, 1975), with 0.1 M NaCl (see Holtfreter, 1943b), and with hypertonic sucrose (see Tuft, 1957, 1962, 1965; Zotin, 1965; Morgan, 1906a). Tuft (1965) argued that active pumping of water might also be involved, but no mechanism for this has been demonstrated.

The development of a superficial epithelium and of a deep nonepithelial cell population figures prominently in the organization of the gastrula. First, its nonadhesive outer surface prevents adhesion of apposed surfaces of the epithelium (e.g., the roof and the floor of the archenteron) during gastrulation. Only the lateral and basal surfaces of the superficial cells and all surfaces of deep cells have the ability to make protrusions, participate in motile activity, and make adhesions (Holtfreter, 1943a,b). The outer (apical) surfaces of superficial cells differ from their lateral and basal surfaces in lectin receptors (see Chapter 5; Roberson and Armstrong, 1980; Roberson et al., 1980) and particle density in freeze-fracture studies (Bluemink et al., 1976, Sanders and Decaprio, 1976). Second, the deep mesenchymal cells and superficial epithelial cells have different behavior, mechanical properties, morphogenetic function, and capacities for differentiation (Holtfreter, 1943a,b; Detlaff, 1983; Keller, 1981; Keller et al., 1985) (see Sections 4 and 5). Third, epithelial–mesenchymal transitions are used as a morphogenetic mechanism in gastrulation of some amphibian species, such as the axolotl, but not in others, such as Xenopus (see Sections 3.2, 3.4, and 11.1). Fourth, the prospective tissues are not distributed between deep and superficial layers in the same way in all amphibians; i.e., mesoderm is found in the superficial marginal zone of at least some urodeles and perhaps some anurans (Vogt, 1929b) but not the anuran Xenopus laevis (Nieuwkoop and Florshutz, 1950; Keller, 1975, 1976) (see Section 3).

Significant growth does not occur during either the blastula stages or the gastrula stages that follow. The increase in cellular volume through gastrulation is about 6%, although the volume of the embryo increases by more than this, due to inflation of the blastocoel (Briggs, 1939; Tuft, 1962, 1965). It appears that cleavage must provide a certain number of cells or reduce cells to a certain size in order for the gastrulation events to occur normally. Blockage of division with mitomycin C at the mid-blastula stage of Xenopus results in a halt of gastrulation at the mid-gastrula stage, but blockage at the late blastula stage yields a neurula with a shortened dorsal aspect (Cooke, 1973a). Such failures may be attributable to a cell size too large or a cell number too small to accomplish the cell rearrangements necessary for gastrulation (Cooke, 1973a) (see Sections 4.2 and 5.5).

Beyond providing the appropriate cell size and number, continued cell division does not appear to be a major morphogenetic force in gastrulation because of the lack of significant cell growth (see Tuft, 1962), the slowing of division in the gastrula stages, and lack of obvious correlation between orientation of division planes and morphogenetic movements. There is one unconfirmed report of a high local rate of division (Bragg, 1938). The division rate of gastrula superficial cells appears to be proportional to their rate of spreading (epiboly) such that the mean area they occupy in the epithelium is nearly constant (Keller, 1978). Such division may be a result, rather than a cause, of spreading and may reflect the same sort of relationship between spreading and division seen in cultured tissue cells (Moscona and Folkman, 1981). See Holtfreter (1943b) for a critique of earlier ideas about the role of division in early morphogenesis.

2.3. The Cytoskeleton of the Early Embryo

A force-generating mechanochemical system is in place and functions in the early embryo (see Franke *et al.*, 1976; Clark *et al.*, 1978; Campanella and Gabbiani, 1980; Columbo *et al.*, 1980; Franz *et al.*, 1984; Ezzell *et al.*, 1984). The cortical region of the egg shows several global patterns of contraction that may function in establishing early biases in axis determination (Hara *et al.*, 1977; Elinson, 1980, 1984; Kirschner *et al.*, 1980; Stewart-Savage and Grey, 1982; Scharf and Gerhart, 1983), and a similar cytoskeleton may organize the metachronal cleavage pattern (Hara, 1971, Satoh, 1977) and pseudogastrulation (see Section 9.3). The relationship of the cytoskeletal constituents involved in these early events to those functioning in the motile behavior of cells during gastrulation is unknown, nor is it known whether the activities of the two are regulated in the same manner. The cytoskeleton is discussed in detail in Chapter 1.

2.4. Pregastrular Morphogenetic Movements

Concurrent with the process of cleavage, particularly during the mid- and late-blastula stages, the animal region shows an increase in area, known as **epiboly**, and a concurrent thinning of the blastocoel roof (Morgan, 1960b; Vogt, 1929b; Keller, 1978, 1980) (Fig. 3a–d). These changes occur during the increase in volume of the blastocoel (Briggs, 1939; Tuft, 1965) and result in an increased diameter of the embryo (Fig. 3c–e). Concurrently, the vegetal region decreases in area (Ballard, 1955; Keller, 1978). Schechtman (1934), working with the California newt, *Taricha torosus* (which was named *Triturus torosus* at the time), argued that individual superficial cells in this region move inside at the vegetal pole by unipolar ingression. This conclusion was based on the movement of vital dye applied to the surface into the interior, a phenomenon also observed by Nicholas (1945) and Nieuwkoop (1947). Direct observation of *Rana pipiens* blastulae (Ballard, 1955) and time-lapse cinemicrography of *Xenopus* blastulae (Keller, 1978) showed that superficial cells do not leave the surface and that the vegetal shrinkage is accounted for by decreases in area of the apices of cells in this region. The inward movement of vital stain in several species of amphibian is due to the movement of cortical and subcortical cytoplasm inward with the cleavage furrows and occurs in all sectors, not just the vegetal region (Ballard, 1955; also see Harris, 1964).

If the decrease in the thickness of the deep region of the animal pole in the blastula stages of *Xenopus* is accounted for by epiboly, the deep region undergoes spreading at a greater rate than the superficial layer (Keller, 1980). Thus, the marginal zone may gain cells at the expense of the animal region during blastulation (dashed arrows, Fig. 3d). This occurs in many amphibians, as was found by T. H. Morgan (1906b) in *Rana palustris*, *Rana sylvatica*, and *Bufo lentiginosus*, and in the sturgeon, which gastrulates much like *Xenopus* (see Ballard and Ginsberg, 1980). These pregastrular movements are reflected in the fact that fate maps of blastula stages (Vogt, 1929b; Nakamura and Kishiyama,

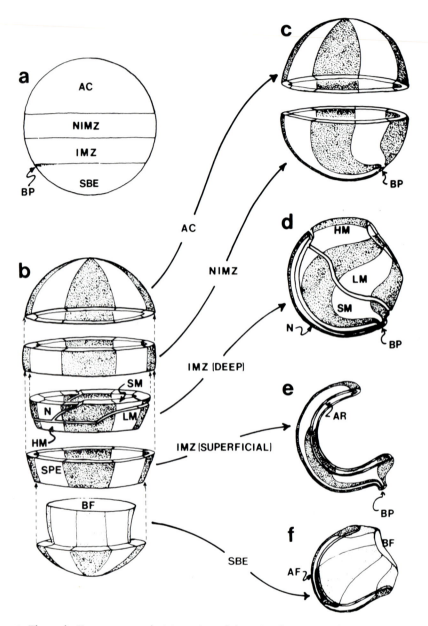

Figure 4. The early *Xenopus* gastrula (a) consists of the animal cap (AC), the noninvoluting marginal zone (NIMZ), the involuting marginal zone (IMZ), and the sub-blastoporal endoderm (SBE). These are shown in an exploded view of the early gastrula (b), and each is shown separately after gastrulation is complete (c–f). The dorsal sides of all figures are to the left, and c through f show successive layers of the neurula, from outside (c) to inside (f). During gastrulation (b–c), the AC expands uniformly and the NIMZ mostly on the dorsal side to form the entire outer shell of the neurula, consisting of prospective neural and epidermal ectoderm (c). In the early gastrula (b), the deep sector of the IMZ is a torus consisting of prospective notochord (N), somitic mesoderm (SM), head mesoderm (HM), and lateral–ventral mesoderm (LM). During gastrulation, the torus turns

1971; Hirose and Jacobson, 1979) are different from those of the early gastrula (Vogt, 1929*b*; Keller, 1975, 1976; see Vogt, 1929*b* for direct comparison). The precise disposition of prospective areas in the early blastula is uncertain because the limit to the resolution of most marking techniques is the size of blastomeres. Moreover, the first deep blastomeres of the animal cap are often loosely attached to one another, and their dislocation and repositioning may make a reliable fate map unlikely (R. E. Keller, unpublished observations; see Morgan, 1906*b*).

Little is known of the mechanisms underlying these pregastrular movements. Hydrostatic pressure from blastocoel fluid accumulation (Tuft, 1962, 1965) may expand the roof of the blastocoel, but the motile properties of cells may also function in blastular epiboly (Keller, 1980). The evidence for autonomous epiboly in the gastrula stages and possible cellular mechanisms underlying it are discussed in Section 4. Gastrular and blastular epiboly occur at roughly the same rate (Keller, 1978), but it is not known to what degree, if any, they resemble one another in mechanism.

3. Prospective Germ Layers and Major Morphogenetic Movements

Germ layer organization and morphogenetic movements differ among amphibians (see Section 10). I shall describe the situation in the anuran *Xenopus laevis* and the urodele *Ambystoma mexicanum* in detail, as they represent two basic patterns of amphibian gastrulation and are the best understood examples.

3.1. Prospective Areas of *Xenopus laevis*

The prospective regions at the onset of gastrulation of the *Xenopus*, based on vital dye mapping (Keller, 1975, 1976), are shown in Figure 4. This fate map is an exploded view of the embryo, showing the deep regions as well as the superficial epithelium. I have made a distinction between the **involuting marginal zone,** which will involute in the course of gastrulation, and the **noninvoluting marginal zone,** which does not involute. The **limit of involution** demarcates the two and lies at the margin of the blastopore at the end of gastrulation (Fig. 4). Note that the superficial layer of the involuting marginal zone

---

inside out to form the mesodermal mantle (d). The HM and LM spread toward animal and ventral regions, whereas the N and, to a lesser extent, the SM, converge dorsally and elongate to form the dorsal, axial array of notochord and somites (d). The superficial layer of the IMZ consists of prospective endoderm (suprablastoporal endoderm, SPE) and lies outside the deep mesodermal torus (b). During gastrulation, it turns inside out, converges dorsally, and elongates with the underlying mesoderm, to form the archenteron roof (AR) at the end of gastrulation (e). The large mass of subblastoporal endoderm (SBE) is pushed inside and covered over by the IMZ during gastrulation to form the archenteron floor (AF). BP, blastopore; BF, blastocoel floor.

consists of prospective endoderm of the archenteron roof, whereas the deep region consists of prospective mesoderm. The mesoderm consists of prospective notochord, somites, and anterior and lateral mesodermal derivatives such as head, heart, and lateral plate mesoderm (Fig. 4b). Also note that the endodermal cells of the prospective archenteron roof lie above the prospective blastopore and are called **suprablastoporal endodermal cells,** whereas the endodermal cells of the prospective archenteron floor lie below it and are called **sub-blastoporal endodermal cells.** The **animal cap** and noninvoluting marginal zone consist of prospective ectoderm in both the deep and superficial layers.

3.2. Morphogenetic Movements of *Xenopus laevis*

Gastrulation begins with the formation of the "bottle" or "flask" cells on the dorsal side about 55° from the vegetal pole (Fig. 5a). These form by apical constriction of the vegetal-most suprablastoporal endodermal cells, although a few of the adjacent sub-blastoporal endodermal cells may participate as well (Keller, 1981). As bottle cells decrease their apparent apical area, the involuting marginal zone is pulled vegetally toward them, and a **blastoporal pigment line** is formed, due to the concentration of pigment in the constricted apices. Soon afterward, a depression, the **blastoporal groove,** is formed (Fig. 5a). Internally, the dorsal, vegetal edge of the torus of prospective mesoderm turns inward, back on itself, in a movement called **involution** (Fig. 5a). This movement was called *Umschlag* or *Einrollung* (inrolling) by Vogt (1929*b*). In subsequent stages, the turned-in lip of the torus moves animalward (*Vordringen* of Vogt, 1929*b*) as it is added to from behind by material that rolls (involutes) over an **inner blastoporal lip** (Fig. 5a–d). This process of involution begins dorsally coincident with the formation of the bottle cells and continues laterally and ventrally (Fig. 5a–d). Involution involves a turning of the torus of prospective mesoderm outside-in to form the definitive **mesodermal mantle** (Fig. 4b–e). The suprablastoporal endoderm is attached to the outside of the mesodermal torus and involutes with it to form the archenteron roof (Fig. 4b–f). As the involuting marginal zone undergoes involution, the noninvoluting marginal zone and animal cap undergo spreading, or epiboly, and occupy the positions vacated by the involuting marginal zone (Fig. 4b–c,d; Fig. 5a–d). Epiboly of the animal cap occurs at the same rate as in the blastula stages and is uniform in all directions (Keller, 1978) (Fig. 4b–c).

In the dorsal noninvoluting and involuting marginal zones, epiboly occurs at a rate twice that of the animal cap, and it is anisotropic (Keller, 1978) (Fig. 4b–d). During epiboly, both regions show narrowing (**convergence**) toward the dorsal midline and meridional lengthening (**extension**) at the dorsal mid-line in the course of gastrulation (Fig. 4b–d,e) (Keller, 1975, 1976, 1978). Because the narrowing is always accompanied (during gastrulation) by an increase in length, convergence and extension will be referred to as **convergent extension.** Extension was called *Staffelung* and *Streckung* and convergence *Raffung* by Vogt (1929*b*). Material of the involuting marginal zone moves vegetally and

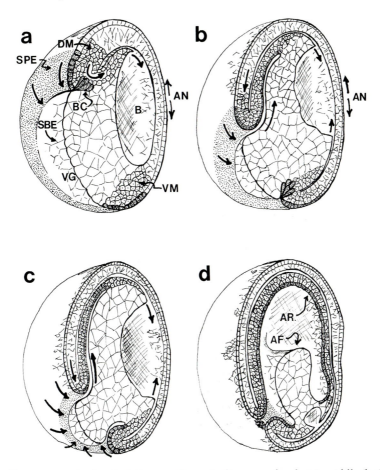

Figure 5. The movements of gastrulation are shown in diagrams of early (a), middle (b,c), and late (d) gastrulae, sectioned mid-sagitally and positioned with the vegetal pole (VG) toward the observer and one-quarter turn to the left. The major movements are indicated by arrows and are described in the text. AN, animal pole; AR, archenteron roof; AF, archenteron floor; B, blastocoel; BC, bottle cells; DM, dorsal mesoderm; SPE, suprablastoporal endoderm; SBE, sub-blastoporal endoderm; VG, vegetal pole.

approaches the lip as part of the process of convergent extension, and there it passes over the lip and out of sight during involution. The lip itself moves vegetally and decreases circumference (Fig. 5a–d) in a movement called **constriction** (*Konzentrisches Urmundschluss* of Vogt, 1929b). Involution follows from the fact that the involuting marginal zone approaches the blastoporal lip faster than the lip advances. Constriction of the blastoporal lips around the yolk plug and closure of the blastopore is coincident with, and related to, convergent extension. Each sector of the lip is narrowing, because of its convergence and moving vegetally because of its extension. These facts conspire to decrease the circumference and diameter of the blastopore (Fig. 5a–d). In order to describe them initially, extension, convergence, constriction of the blastopore, and involution are treated as separate processes, but evidence presented in Section 5 shows that they are, in fact, inseparable aspects of one basic cellular process.

The degree of convergent extension and the proportion of the marginal zone doing it decreases progressively in lateral and ventral sectors of the gastrula, and it is the dorsal and most posterior region of the marginal zone that undergoes the greatest convergent extension (Keller, 1975, 1976, 1978) (Fig. 4b–e). The anterior dorsal mesoderm does not show convergent extension, and the ventral mesoderm shows a spreading or **divergence** from the ventral midline that is complimentary to the convergence of the dorsal, posterior mesoderm (Fig. 4b–e). Because more convergent extension occurs dorsally than ventrally, more material moves over the dorsal lip than over the lateral and ventral lips, and the dorsal lip also moves farther across the sub-blastoporal endoderm (yolk plug) than do the lateral and ventral lips. Thus, the blastopore closes eccentrically, on the ventral side of the vegetal pole. Consequently, the dorsal aspect of the animal is lengthened, and the ventral aspect is shortened at the close of gastrulation (Keller, 1975) (Figs. 4b–d,e,f and 5). This dorsoventral asymmetry is continued in the neurula stages with the formation of the dorsal axial structures (the notochord, somites, and neural tube) and results directly from the dorsoventral specialization of morphogenetic behavior—a convergent extension dorsally and divergent spreading ventrally—within the torus of prospective mesoderm during gastrulation.

The sub-blastoporal endoderm forms the floor of the archenteron. It appears to sink into the gastrula in a movement called *Einstulpung* (insinking) by Vogt (1929b), but, in fact, it moves very little and is covered over as the constricting blastoporal lips roll vegetally across it (Figs. 4b–f and 5a–d).

3.3. Prospective Areas of *Ambystoma mexicanum*

The following is based principally on the work of Vogt (1929b) on several species of the European newts *Triturus* (*Triton*) *alpestris* and *cristatus*, and *Pleurodeles waltlii* and on the work of Lundmark (1986) on the Mexican axolotl, *Ambystoma mexicanum*. Other useful references are the works of Brachet (1903), Nakamura (1942), Pasteels (1942), and especially Eycleshymer (1895). The European newts and Mexican axolotl gastrulae are similar except that the former has less yolky endoderm than does the axolotl; thus the blastopore and the marginal zone form about 15–20° farther from the vegetal pole in the latter (see Holtfreter, 1938a).

In general character, the movements of epiboly, convergence, extension, and involution in *Ambystoma* are similar to those seen in *Xenopus*, but the locations of the prospective areas are not—especially with respect to the deep and superficial layers (Fig. 6). The superficial and deep regions of the involuting marginal zone consist of prospective mesoderm with the notochord, somitic mesoderm, and lateral mesoderm arrayed, in that order, from dorsal to ventral (Fig. 6b). In addition, the superficial layer contains a small area of suprablastoporal endoderm in its dorsal sector (Fig. 6b); this endoderm covers part of the anterior, dorsal edge of the mesodermal mantle (Vogt, 1929*b*). Farther laterally, there is no suprablastoporal endoderm (Fig. 6b). The sub-blastoporal region consists of prospective endoderm that will ultimately form most of the lining of the archenteron (Fig. 6b). Archenteron formation involves removal of ventral somitic and notochordal mesoderm from the superficial layer in two steps described in Section 3.4. In the axolotl, the gastrula wall is relatively thin, the prospective areas cover greater area, and the yolky vegetal core of endoderm is relatively small when compared with these features in *Xenopus*.

The presence of mesoderm in the superficial layer of urodeles was challenged by Løvtrup (1975), who suggested that Vogt had unknowingly stained the deep layer, through the superficial layer, with vital dyes. Therefore, the mesodermal anlage shown on the surface were actually deep, as in *Xenopus*. Lundmark (1986), using fluorescinated peptide and dextran cell lineage tracers, and Smith and Malacinski (1983), using Bolton-Hunter reagent, found that this mesoderm is, in fact, in the superficial layer.

3.4. Morphogenetic Movements of *Ambystoma mexicanum*

In the axolotl, the gastrulation movements on the animal side of the limit of involution are similar to those of *Xenopus*. The animal cap undergoes epiboly, and the dorsal sector of the noninvoluting marginal zone undergoes convergent extension (Fig. 6b,c,d). The degree of epiboly in the axolotl is greater than in *Xenopus* because initially the region above the limit of involution occupies a smaller percentage of the total embryonic surface area (compare Fig. 6a with Fig. 4a). By contrast, below the limit of involution there are major differences. The deep component of the mesoderm undergoes involution and then moves anteriorly and ventrally, where it forms the leading edge of the mesodermal mantle (Fig. 6b,d). The superficial component of the somitic mesoderm and ventrolateral mesoderm is removed from the superficial layer by ingression of cells at the lateral and ventral regions of the blastopore during gastrulation. These cells move deep to join the mesodermal mantle, and there they contribute to the somitic mesoderm and the posterior, ventral mesoderm (Fig. 6b,d). Only two regions of the superficial epithelium of the involuting marginal zone remain superficial during involution: the notochord and the small patch of suprablastoporal endoderm in the dorsal sector, which involute to form the entire roof of the archenteron of the early neurula (Fig. 6b,e). Ingression of the somitic and ventral mesoderm brings the lateral boundaries of the notochord into apposition with the sub-blastoporal endoderm (Fig. 6e). The sub-blastoporal endoderm sinks inside, much as it does in *Xenopus*, but its lateral edges move dorsally and meet the lateral edges of the notochord as the intervening somitic and lateral–ventral mesoderm are removed. The notochord is not removed from the surface until neurulation (King, 1903; Ruffini, 1925; Lofberg, 1974; Brun and Garson, 1984; Lundmark, 1986) (see Section 11.1). The complicated removal of mesoderm from the surface is best understood

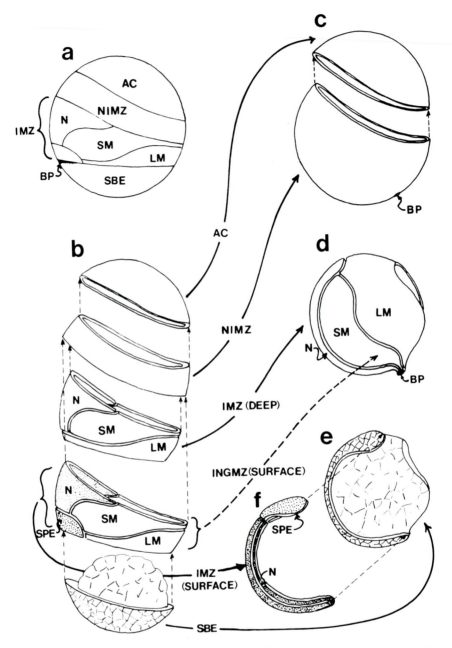

Figure 6. The typical urodele early gastrula (a) consists of the animal cap (AC), the noninvoluting marginal zone (NIMZ), the involuting marginal zone (IMZ), and the sub-blastoporal endoderm (SBE). In the urodele the prospective notochord (N), somitic mesoderm (SM) and lateral–ventral mesoderm (LM) are exposed, on the surface. These regions are shown in an exploded view of the early gastrula (b), and their positions in the neurula are shown separately, as layers, from the outside (c) to the inside (e,f). The dorsal sides are to the left in all diagrams. The AC expands uniformly, and the NIMZ expands mostly on the dorsal side to form the outer shell of the neurula, consisting of prospective neural and epidermal ectoderm (b,c). The deep sector of the IMZ invo-

by mapping hoops of tissue at several anteroposterior levels back on to the surface of the early gastrula (Fig. 7). Each hoop consists of a suprablastoporal notochordal component (SP–N), a suprablastoporal somitic or lateral–ventral mesodermal component (SP–S), and a sub-blastoporal endodermal component (SBE). During involution, at the corners of the blastopore, the lateral edge of the notochord is brought into apposition with the corresponding part of the sub-blastoporal endoderm as the intervening suprablastoporal somitic mesoderm is removed by ingression of cells into the deep layer and their subsequent migration laterally and anteriorly (Fig. 7a,b). The forming gastrocoel thus consists initially of notochord dorsally and sub-blastoporal endoderm ventrally (Fig. 7c,d). In the neurula stage, notochord is then removed from the lining of the gastrocoel by a complex and poorly understood process that involves narrowing, elongation, and ingression of cells (see Section 5.6) such that the corresponding parts of the sub-blastoporal endoderm come into apposition and form the definitive archenteron (Fig. 7e,f,g). The lateral and ventral mesoderm also moves into the deep region by ingression of cells at the lateral and ventral sectors of the blastopore during gastrulation (Fig. 7a,c,e).

Corresponding hoops, mapped onto the early gastrula of *Xenopus*, show the similarity of distortion of the circumblastoporal region to that seen in the axolotl (Fig. 7h–k); both decrease the length of the suprablastoporal component as the length of each hoop decreases. In *Xenopus* this is solely a matter of narrowing (convergence) and a corresponding increase in length (extension) of the suprablastoporal endoderm. In the axolotl, the decreasing length of the hoops involves (1) convergence (and extension) of the notochord, (2) removal of somitic mesoderm by ingression during gastrulation, and (3) ingression of the notochord in the neurula. The geometric similarity of these systems reflects a similar mechanical function in gastrulation (see Section 5.6).

3.5. Tail Formation in Amphibians

The gastrulation movements up to the closure of the blastopore involve formation of only the head and anterior trunk of the larva (see Vogt, 1929b; Muchmore, 1951). The posterior trunk and tail are formed by postgastrular events that are poorly understood and controversial (see Bijtel, 1958; Vogt, 1939; Holmdahl, 1947). The consensus of opinion is that the tail somites are located in the posterior neural plate. The eventual tip of the neural tube is near

---

lutes and contributes to the formation of the mesodermal mantle (d). The prospective somitic (SM) and lateral–ventral (LM) mesoderm in the superficial layer of the IMZ leave the superficial layer by ingression of individual cells at the lateral aspects of the blastopore and join the mesodermal mantle during gastrulation (b,d). Thus, this region is designated the ingressing marginal zone (INGMZ). By contrast, the prospective notochord (N) and suprablastoporal endoderm (SPE) in the superficial layer of the IMZ involute to form the roof of the gastrocoel of the neurula (b,f). The SBE sinks inside and temporarily forms the floor of the gastrocoel (b,e). The notochord is removed from the superficial layer during the neurula stages (see Fig. 7); then the edges of the SBE meet to form, with the small patch of SPE, a continuous endodermal lining of the archenteron.

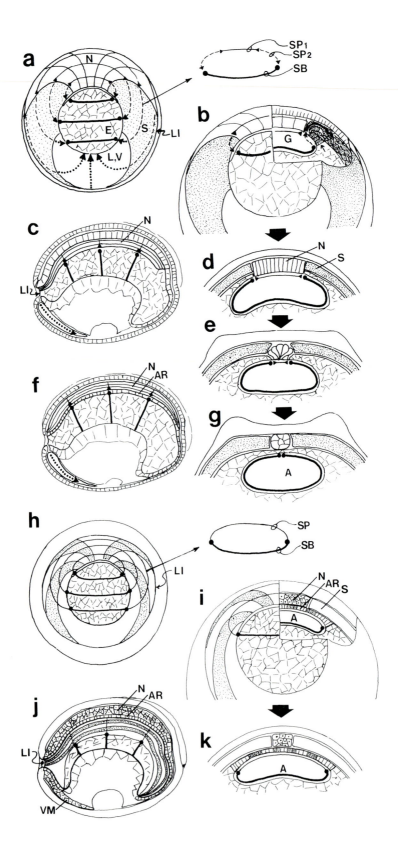

the anterior border of the prospective tail somites. The notochord elongates posteriorly, and the posterior mesodermal part of the "neural" plate folds along the side of the notochord and neural tube to form the tail somites (Bijtel, 1930, 1931; Nakamura, 1938, 1947; Pasteels, 1942; Spofford, 1945, 1948; Chuang, 1947; Smithberg, 1954; Bijtel, 1958). Bijtel (1958) interpreted the results of notochord extirpation experiments to mean that extension of the tail is due to the neural tube and somites, with the notochord serving as a support once the tail is extended. The complex phenomenon of tail formation cannot be covered here, and neither tail mesoderm nor its movements are represented on the fate maps. Suffice it to say that many issues in posterior trunk and tail formation are unresolved and warrant further investigation.

3.6. The Gastrula as a Mosaic of Regional Processes

In an effort to reduce gastrulation to simpler part-processes, microsurgical isolation and rearrangment of parts of the gastrula have been used to identify regional patterns of autonomous morphogenetic behavior, called **formative tendencies** (see Spemann, 1938; Vogt, 1922a; Holtfreter, 1929, 1934; Schechtman, 1942). The early gastrula has been viewed as a mosaic of these formative tendencies, acting in concert to bring about gastrulation. What are these formative tendencies, and what is the evidence that they are autonomous active processes?

The most powerful of these processes is the "stretching" and "narrowing" (extension and convergence) of the marginal zone, particularly in the dorsal sector (Vogt, 1922a,b). The evidence for this was as follows: (1) the fact that a dorsal marginal zone grafted to another region of the embryo will elongate and narrow to form a long excrescence (Spemann, 1902; O. Mangold, 1920; Lehman, 1932); (2) the marginal zone normally constricts below the yolk mass but, when it is prevented from doing so by frontal constriction of the embryo with a

---

Figure 7. The removal of the mesoderm from the superficial layer and the formation of the archenteron in urodeles can be visualized by representing successive anterioposterior levels of the embryo by hoops mapped onto the vegetal and marginal zone region of the early gastrula (a). Each hoop consists of a suprablastoporal, notochordal component (SP–1), a suprablastoporal somitic or lateral mesodermal component (SP–2), and a sub-blastoporal, endodermal component (SB). The SP–2 sector is removed at the lateral region of the blastopore during gastrulation by ingression of individual cells and their migration laterally into the deep region (b). Such ingression brings the lateral boundaries of the notochord in apposition with the edges of the sub-blastoporal endoderm to form a gastrocoel with an endodermal floor and notochordal roof, as seen in midsaggital view (c) and in transverse view (d). This is the state of affairs at the end of gastrulation (compare Fig. 6e,f). During the neurula stages, the notochordal roof is removed by ingression of the notochordal cells (e) to form a definitive archenteron lined with sub-blastoporal endoderm (f,g). By contrast, the same hoops mapped on the vegetal aspect of *Xenopus* fall only on suprablastoporal (SP) and sub-blastoporal (SB) endoderm of the early gastrula (h). As the SP component of each hoop narrows during convergent extension, the hoops are involuted (i) to form the archenteron roof (AR) in the neurula stage (j, k). Hoops falling in more posterior regions undergo the greatest narrowing, with the limit of involution (LI) being the extreme. A, archenteron; AR, archenteron roof; G, gastrocoel; L, lateral mesoderm; N, notochord; S, somitic mesoderm; V, ventral mesoderm.

ligature (Spemann, 1902) or by removal of the blastocoel roof (Vogt, 1922b), its tendency to narrow was found to be powerful enough to squeeze a groove in the yolk mass and nearly pinch it into two parts; and (3) when isolated by explantation into culture (Schechtman, 1942; Holtfreter, 1938a,b, 1944) or by exogastrulation (Holtfreter, 1933b), this region elongates and narrows autonomously.

The second formative movement is the epiboly of the animal region. When prospective ventral epidermis (animal cap) is explanted in culture, it appears to expand and throw itself into folds (O. Mangold, 1923; Spemann, 1931). When separated from the rest of the gastrula in exogastrulae, it also appears to spread and generate folds (Holtfreter, 1933b).

Third, the superficial epithelial cells at the vegetal end of the marginal zone have the property of forming bottle-shaped cells, invaginating, and forming the blastoporal groove. The blastoporal groove forms independently and some distance below the groove formed by constriction of marginal zone in an embryo without a blastocoel roof and hence, is independent of convergence and extension (Vogt, 1922b). The way the bottle cells bring about groove formation was thought to involve a tendency to invade the gastrula interior (Rhumbler, 1902; Holtfreter, 1943a,b) or a deformation of the superficial epithelium as a result of their change in shape (Rhumbler, 1902; Lewis, 1947).

Fourth, the cells fed over the lip of the blastopore have a tendency to migrate inward and toward the animal pole, the *Vordringen* described by Vogt (1929b). There are several examples of this behavior in gastrulation. Holtfreter (1943a,b) viewed bottle cell function primarily in terms of migration inward—an invasion. In urodeles, the ventral mesodermal cells become bottle-shaped, lose their epithelial character, leave the superficial layer, and migrate inward as individual cells to form the ventral mesodermal mantle (Holtfreter, 1943b). Also, according to Holtfreter (1943b), the dorsal mesoderm invaginates, loses its epithelial character, and forms a scalelike array of cells on the roof of the blastocoel.

The gastrula was perceived as a mosaic of these formative tendencies acting in concert to produce gastrulation (Spemann, 1938). To summarize, stretching and narrowing of the marginal zone pushes material vegetally to the region of "invagination," where it rolls inside (involution), and then migrates toward the animal pole. Simultaneously, the animal region spreads, allowing the marginal zone to move vegetally. But few major elements in this scenario were ever tested rigorously, and the cellular behaviors underlying them were never resolved in any satisfactory way. Recent investigations bear on the cellular basis of these formative tendencies and their function in gastrulation, revealing a sophisticated mechanical design and spatial ordering of cellular behavior.

4. Mechanism and Function of Epiboly in Gastrulation

How do cells bring about expansion of the animal cap, and how does epiboly function in gastrulation? The facts pertaining to these questions differ between the amphibians that have been studied in detail—the anuran *Xenopus laevis* on one hand and the urodeles *Ambystoma mexicanum* and *Taricha torosus* on the other.

4.1. Cell Behavior during Epiboly

The cellular changes during epiboly in *Xenopus* are shown diagrammatically in Figure 8a. Time-lapse cinemicrography (Keller, 1978) shows that superficial cells of the animal region expand the area they occupy in the superficial layer and divide anticlinally (spindle parallel to the surface). Their rate of expansion accounts for the macroscopic distortion seen by vital dye mapping (Keller, 1975). Outward (radial) intercalation of deep cells into the superficial layer does not occur, nor do superficial cells move inward to join the deep region (Keller, 1978). The superficial layer thins in proportion to its increased area during gastrular epiboly (Keller, 1980).

The deep cells are not visible from the outside, and thus less is known about their behavior, although some aspects of their change in shape, packing pattern, and number of layers during the course of gastrulation are revealed by a morphometric analysis of scanning electron micrographs (SEMs) of gastrulae fractured at successive stages of development (see Keller, 1980). The deep cells are pleiomorphic and are connected to one another by small filiform and lamelliform protrusions. These appear at stage 8 and become numerous by the onset of gastrulation (Fig. 9a–f). The number of layers of deep cells first increases as division occurs during the blastula stage; it then decreases from about two at the onset of gastrulation to one in the late gastrula (Fig. 9f). Morphometric analysis shows that several layers of deep cells interdigitate (intercalate) radially (along radii of the embryo) during epiboly to form fewer layers of cells of greater area (Fig. 8a).

The inner ends of the innermost layer of deep cells are flattened and have long filiform protrusions lapping across the surfaces of their neighbors (Fig. 9g). It is this surface that postinvolution deep cells are thought to use as a substratum for their migration animalward (see Section 6). These inner ends form a

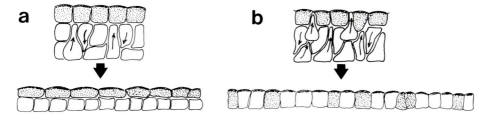

Figure 8. The contrasting cell behavior during epiboly is shown diagrammatically in sectional view. In *Xenopus* (a), epiboly occurs by spreading and division of superficial epithelial cells (shaded) and radial intercalation of deep cells to form fewer layers (usually one) of greater area. In the axolotl and similar urodeles (b), epiboly occurs by intercalation of deep cells between superficial cells, as well as between themselves, to form a single layer of greater area.

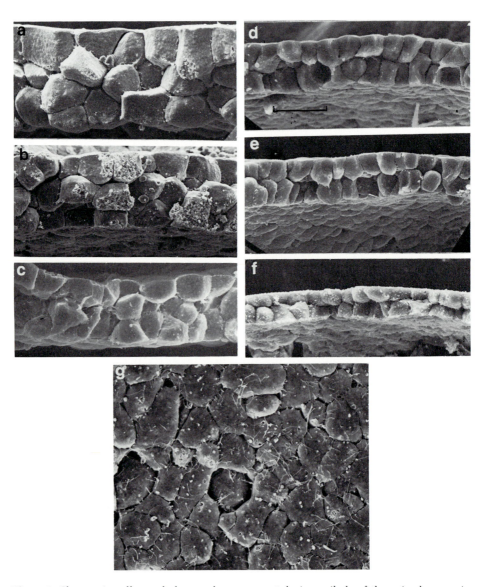

Figure 9. Changes in cell morphology and arrangement during epiboly of the animal cap region (blastocoel roof) of *Xenopus* are shown by scanning electron micrographs at the late mid-blastula (a), late blastula (b), and early (c) through the late gastrula stages (d–f). The inner surface of the blastocoel roof of the gastrula (g) shows the tightly packed array and overlapping filiform protrusions characteristic of this surface. Scale bar: 50 μm (d); all micrographs are at the same magnification.

stable surface bounding the blastocoel, and cells moving upward, on the inside, do not violate the integrity of this boundary. This is also the surface on which fibronectin is concentrated (see Section 8.1).

Epiboly in urodeles (*Ambystoma* and *Taricha* species) has not been studied in detail, but the available evidence suggests that it involves similar events with one addition: intercalation of deep cells into the superficial layer (Fig. 8b). Histological sections of *Taricha* (Daniel and Yarwood, 1939) show a multi-layered blastocoel roof at the late blastula stage and a single-layered roof at the end of gastrulation. Holtfreter, Ross Harrison, and others, referring principally to species of *Ambystoma*, suggested that the basis of the decreased pigmentation in the animal cap during epiboly is the intercalation of unpigmented deep cells into the superficial layer (see Holtfreter, 1943b). Smith and Malacinski (1983) showed this to be true for the axolotl by labeling original surface cells with Bolton-Hunter reagent and finding unlabeled superficial cells at later stages. Goodale's (1911b) histological sections clearly show that the blastocoel roof of another urodele, *Sclerpes bilineatus*, is decreased in the number of layers of cells and in its thickness, probably by radial intercalation. Jordan's (1893) illustrations show the same to be true of the spotted newt, *Dimictylus viridescens*. Urodeles, in general, show greater epiboly than anurans (see Vogt, 1929b), which may account for the process of intercalation occurring in the superficial layer, as well as in the deep layer. The first good evidence for epiboly by intercalation of several layers of cells was in an excellent paper on gastrulation of the lamprey *Petromyzon* by Hatta (1907).

How the deep cells enter the epithelium and become epithelial is not understood, particularly since the resident epithelial cells are bound together circumapically by the relatively high resistance junctional complex. This paradox warrants additional investigation, as do other aspects of epithelial–nonepithelial transitions in the gastrula that have been ignored since Holtfreter discovered them (see Section 9.1).

4.2. What Forces Drive Epiboly?

Radial interdigitation of cells could occur as a force-producing process whereby cells actively wedge between one another, along radii of the gastrula, thereby generating a pushing or spreading force in the tangential direction intrinsic to the tissue itself (Fig. 10a). Alternatively, the superficial layer could be stretched by tension generated elsewhere, and the deep cells could rearrange to take up the newly available space on its inner surface (see Keller, 1980) (Fig. 10b) or, in the case of the urodeles, intercalate into the superficial layer itself.

Thus, the first important question about mechanisms is whether expansion of the animal cap is independent of external forces or is stretched by tension generated elsewhere. Evidence for the former is not strong. Pieces of animal region, isolated by microsurgery or by exogastrulation, do throw themselves into folds, as recounted earlier, but time-lapse films of exogastrulation of *Xenopus* show this to be accompanied by apical constriction of superficial

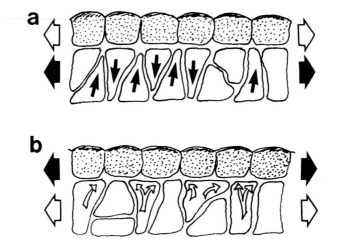

Figure 10. Epiboly in *Xenopus* may be driven by local cellular activity (a) if active intercalation of deep cells (solid arrows) brings about passive stretching (open arrows) of the attached, superficial epithelium. Alternatively, it may be a passive process (b) in which external tensions stretch the superficial layer (solid arrows) and the deep cells rearrange on its inner surface (open arrows), as space becomes available, without contributing to the forces that generate the spreading.

cells, an actual decrease in surface area, and an abnormal ingression of super-ficial cells (R. E. Keller, unpublished data) (see Section 9.3). The explanted animal cap of the early gastrula shows expansion, but it is accompanied by fluid accumulation in blastocoel-like vesicles (Keller *et al.*, 1985) (see Section 5.1). This finding suggests that the hydrostatic pressure that functions in blastular epiboly may do the same in gastrular epiboly. Puncture of the blastocoel roof results in its immediate and rapid collapse throughout gastrulation, also sug-gesting that hydrostatic pressure supports, and perhaps expands, the roof. The animal cap may be stretched, to some degree, by the circumblastoporal region of the gastrula. When the blastocoel is filled with gelatin to prevent its usual occupation by involuted material, the circumblastoporal region performs the powerful convergent extension movements normally occurring in that region and stretches the animal cap down over the large mass of gelatin to the point of tearing it (Eakin, 1933, 1939) (see Section 5.5).

Thus, the degree to which epiboly of the animal cap depends on stretching by blastocoel hydrostatic pressure, on tension generated by the circumblast-oporal region, and on local motile behavior of its cells—perhaps by forceful intercalation of cells—is not known. All three may make a contribution in normal development, although any of the three may prove sufficient.

5. Mechanism and Function of Convergent Extension

Recent work on *Xenopus* suggests that convergent extension does not func-tion as previously believed. This section discusses this evidence first and reviews the evidence from the classic works, followed by a description of recent work on the cellular behavior underlying convergent extension.

5.1. Morphogenetic Behavior of the Dorsal Sector of *Xenopus*

The dominant role of convergent extension in gastrulation and its relationship to other, ancillary processes can be best understood by beginning with the morphogenetic properties shown by various regions of the dorsal sector of the gastrula as cultured explants or when reoriented in the embryo. In recent experiments (Keller *et al.*, 1985), the dorsal sectors of two early gastrulae were cut out and sandwiched together, with their deep surfaces in contact (Fig. 11a–b). Four regions can be recognized in the explants, based on their subsequent morphogenetic behavior and differentiation: the leading (head) mesoderm, the involuting marginal zone, the noninvoluting marginal zone, and the animal cap (Fig. 11b). The epithelial layer of the explants heals across their edges; by the mid-gastrula stage, the involuting marginal zone begins to narrow and extend (Figs. 11c–d and 12a). The noninvoluting marginal zone also begins to narrow and extend (Fig. 11c–d), and elongation and narrowing of both these regions continues through the late gastrula stage (Figs. 11d and 12a) and into the late neurula stage (Figs. 11e and 12b). The animal cap and leading mesoderm remain knoblike, with the former expanding in size (Fig. 11c–d). Histological sections and SEM examination of explants of this stage show that the involuting marginal zone has differentiated into a central notochord, flanked on either side by somites (Figs. 11e and 12c). The noninvoluting marginal zone forms a compact mass of small pleiomorphic, undifferentiated cells that are probably equivalent to the notoplate (see Jacobson, 1982) of normal embryos. The animal cap invariably contains fluid-filled cavities and is otherwise similar to the noninvoluting marginal zone in histology (Fig. 11e). The leading head mesoderm forms a ball of mesenchymal cells. If these regions are cut apart shortly after the explant is made and cultured separately (Fig. 11f), each undergoes the same morphogenesis and tissue differentiation seen in intact whole explants (Fig. 11g).

The development of these explants can be related to the properties of the intact gastrula. In explants, the involuting and noninvoluting marginal zones form a pair of convergent extension machines placed end to end and acting coordinately. In the intact gastrula, these two regions elongate in parallel: The involuting marginal zone extends after it involutes, in the postinvolution position, and the noninvoluting marginal zone extends on the outside and pushes toward the blastoporal lip (see Section 5.2). In explants, they act in series, which is equivalent to exogastrulation. The animal cap inflates blastocoel-like vesicles; it corresponds to the ventral epidermis and perhaps some of the prospective anterior neural plate in the gastrula. The leading mesoderm forms prechordal mesoderm and also often contains pharyngeal endoderm (see Keller, 1975, 1976). The character of each region and the boundaries between them must be firmly fixed by the onset of gastrulation; the behavior and differentiation of each sector are reliable and show sharp boundaries.

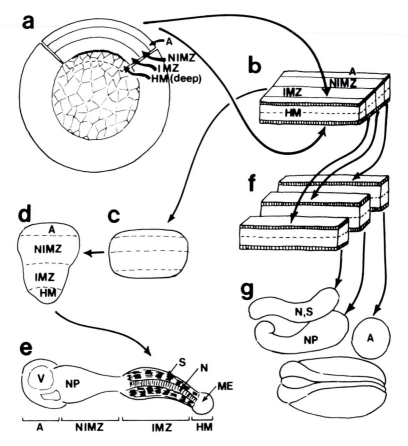

Figure 11. The animal cap (A), noninvoluting marginal zone (NIMZ), involuting marginal zone (IMZ), and head mesoderm (HM) of the dorsal sector of the early gastrula (a) each show highly stereotyped behavior and differentiation when two such regions are excised (a), sandwiched with their inner surfaces together (b), and allowed to develop through the early gastrula (c) and late gastrula (d) stages, to the late neurula stage (e). If these regions are separated after sandwiching (f) and allowed to develop independently to the late neurula stage (g), they show the same patterns of development. C, control neurula; ME, mesenchyme; N, notochord; NP, notoplate; S, somitic mesoderm; V, fluid-filled vesicle.

5.2. The Function of Convergent Extension in Gastrulation

It has long been known that the capacity of the marginal zone to constrict is the most powerful force-producing system in gastrulation, but it was thought that convergent extension of the marginal zone pushes material toward the blastoporal lip (see Section 5.4). The above experiments suggest that the situation may be more complex, with two regions of convergent extension. Additional experiments on *Xenopus* suggest a major revision in our understanding of how convergent extension functions in gastrulation. When the dorsal involuting marginal zone is grafted into some region that normally does not show

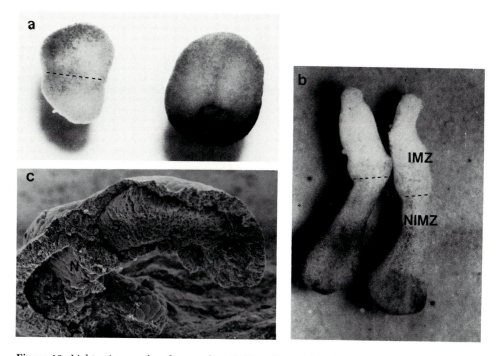

Figure 12. Light micrographs of an explant (left) and control embryo (right) at the late gastrula stage (a) and of explants at the late neurula stage (b). Scanning electron micrograph (c) shows the internal structure of the IMZ region of an explant at the late neurula stage. IMZ, involuting marginal zone; NIMZ, noninvoluting marginal zone; N, notochord; S, somitic mesoderm.

convergent extension, such as the animal pole, reoriented *in situ*, or isolated *in situ*, it undergoes convergent extension and forcibly distorts the surrounding uncooperative tissue (Keller, 1984) (Fig. 13). But it begins to do so only at the mid-gastrula stage (Fig. 13) when that part of it in convergent extension normally would have involuted. This behavior mimics the timing of the onset of convergent extension in the explants. Thus, convergent extension is largely an event of the second half of gastrulation, and it is exercised by the involuting marginal zone only at or beyond the point of its involution. In contrast to the involuting marginal zone, the noninvoluting marginal zone will not show the same degree of convergent extension when reoriented or moved about in the gastrula (Fig. 13) as it does in explants (Figs. 11 and 12). It may be that this region is mechanically weaker than the involuting marginal zone and cannot extend if resisted by surrounding tissues. Alternatively, it may be sensitive to its position relative to neighboring tissues. Failure of the noninvoluting region to extend when manipulated in the gastrula suggested that active convergent extension is solely a postinvolution process (Keller, 1984), a contention that the behavior of the explants shows is not true (Keller *et al.*, 1985).

Reorientation of the involuting marginal zone of the early gastrula 90°, *in situ*, prevents its involution (Fig. 13e–g). It remains essentially the same shape until the midgastrula stage, and then it extends in its normal direction, 90° to its

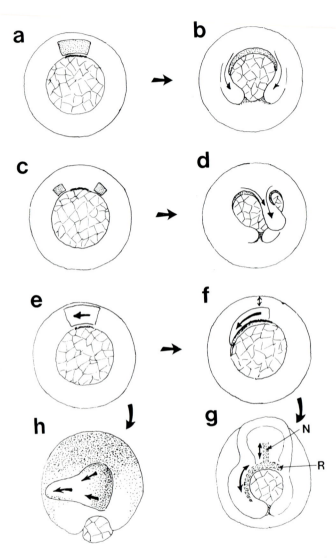

Figure 13. The behavior of the marginal zone of *Xenopus*, after experimental manipulation, is shown diagrammatically. If a patch of a region above the limit of involution (shaded area) is grafted to the dorsal, involuting marginal zone (a), the patch fails to involute and the involuting marginal zone on either side extends vegetally during gastrulation (b). If the dorsal, involuting marginal zone is isolated, *in situ*, in the early gastrula (c) by patches of such material that will not involute (shaded areas), the isolated sector extends vegetally and narrows without involuting (d). If a patch of the dorsal, involuting marginal zone is rotated 90° in the early gastrula (e), the rotated patch extends laterally around the blastopore without involuting (f,g). The larger patch, including noninvoluting marginal zone, rotated the same way, will dominate the behavior of the dorsal sector of the embryo and extend above the surface of the embryo (h). N, normal notochord from the marginal zone already involuted when the rotation was done; R, notochord from rotated patch of involuting marginal zone.

usual orientation, and moves around the lateral lip of the blastopore (Fig. 13f) (Keller, 1984; also see Bautzman, 1933). Sections of these embryos show a short, normal notochord (that had involuted before the reorientation) under a broad, anterior neural plate. Posteriorly, near the open blastopore, it meets the redirected notochord that extends around the left side of the blastopore (Fig. 13h). The fact that rotation will block involution of material that normally involutes suggests that there is something important about the orientation of material presented for involution. By contrast, 180° rotations of tissue patches lying below the limit of involution will allow involution of the patch (Keller, 1984; see Bautzmann, 1933). If patches of dorsal marginal zone are exchanged across the dorsal midline, with their orientation preserved, extension, convergence, and involution are not disrupted. Thus, it is not some mediolateral directional property of the involuting marginal zone that is important for execution of these movements, but rather the fact that its circumferential organization is different from its meridional organization.

Convergent extension and involution are both dependent on the activities of the deep region rather than on the superficial layer. If animal pole superficial layer is substituted for the superficial layer of the dorsal marginal zone of the early gastrula, extension, convergence, and involution still occur (Keller, 1981, 1984). Likewise, if the superficial endoderm of explants of dorsal marginal zone is replaced with animal region superficial ectoderm, convergence and extension of the explant occur (Keller et al., 1985). Rotation of the superficial epithelium of the marginal zone also has no effect. However, if a patch of deep cells from anywhere above the limit of involution is grafted to the deep region of the marginal zone of the early gastrula, it will fail to extend, converge, or involute, and it stops just at the point of joining the postinvolution deep mesodermal mantle (Keller, 1981, 1984) (Fig. 13).

Schechtman (1942) grafted pieces of animal cap material on either side of the dorsal marginal zone and found that the latter extended outward to form an excrescence. He interpreted this to mean that it required continuity with the lateral marginal zones; in fact, he was making local blockages of involution. There are some cases where tissue above the prospective limit of involution appears to have involuted and, in the process, acquired a new fate (see Tondury, 1936). Once involuted, the postinvolution mesodermal mantle–archenteron roof can not involute again. If it is cut out and grafted into another embryo, just above the dorsal lip, the graft, which is very stiff, rides up on the lip much like a canoe on the lip of a waterfall and stops, seemingly unable to make the corner (Keller et al., 1985). On extirpation, the immediate response of the pre- and postinvolution regions is to fold apart, indicating that there is stress that tends to pull the preinvolution material over the lip (see Panel 2 of Fig. 2 in Belousov et al., 1975).

These experiments suggest that (1) Only cells below the limit of involution (in the involuting marginal zone) can accomplish involution; (2) the nature of the deep region is critical to the performance of extension, convergence, and involution, whereas that of the superficial layer is not; (3) the critical step in involution occurs as the deep cells are passing over the lip and attempting to join the mesodermal mantle; (4) convergent extension and involution involve a process that differs in its circumferential and meridional aspects; (5) convergent extension occurs in the noninvoluting and involuting marginal zones but after involution of the latter region; and (6) convergent extension begins in the midgastrula stage and continues through the neurula stages and thus is a process that the gastrula and the neurula hold in common.

Convergent extension of the involuting marginal zone is normally associated with notochord and somite differentiation in the late gastrula and neurula stages. But *Xenopus* embryos that were treated with ultraviolet light between fertilization and first cleavage may have a reduced or absent notochord and the somitic mesoderm fused beneath the neural tube, and yet these embryos may show what appears to be normal early neural development and axial elongation (Malacinski and Youn, 1981). Lithium treatment also results in defective notochord development, including its absence and medially fused somites in its place, but in this case notochordal defects are usually accompanied by small short neural plates and some degree of failure in anterio–posterior extension (Lehman, 1937, 1938; Lehman and Ris, 1938). Several studies show lack of extension in absence of notochord (Kitchen, 1938, 1949; Horstadius, 1944; Nieuwkoop, 1947). But other dorsal axial structures have also been shown to elongate autonomously: the notoplate during neurulation (Jacobson, 1982), the neural and mesodermal structures (other than the notochord) during tail formation (Bijtel, 1958), and the noninvoluting marginal zone described above. Later stages of notochord development involve formation of a fibrous sheath and vacuolation of cells inside it to produce a stiff, elongating rod (Mookerjee, 1953; Mookerjee *et al.*, 1953).

The autonomous convergent extension of the noninvoluting marginal zone challenges old notions about the developmental potential of this region. It maps to the central part of the prospective neural plate and would normally form the floor of the neural tube (see Keller, 1975, 1976). It does not form neural structures in sandwich explants, even when maintained to the equivalent of the swimming-tadpole stage. Differentiation of neural structures would not be expected because the tissue was excised at the early gastrula, and thus presumably before neural induction. Why then does it show convergent extension like that seen in the central region of the neural plate?

The resolution of this paradoxical behavior may be that the dorsal, noninvoluting marginal zone consists of prospective notoplate. Jacobson (1982), working with *Taricha torosus* has found that the central posterior region of the "neural plate," lying immediately above the notochord, normally shows coordinate narrowing and lengthening (convergent extension) and can display this distortion independant of the notochord after the early neurula stage; he designates this region as the **notoplate** to distinguish it from the surrounding neural plate. The notoplate does not differentiate into neural tissue, but forms the nonneural floor plate of the neural tube (Jacobson, personal communication). Perhaps in *Xenopus* the notoplate can self-differentiate autonomously from the noninvoluting marginal zone even at the early gastrula stage. This would account for the convergent extension of this region and also for its failure to form neural structures.

Convergence is not accompanied by extension in all situations. Dorsal convergence of the somitic mesoderm in the neurula stages is accompanied by thickening rather than lengthening (see Keller, 1976), and a part of the convergence in gastrulation of urodelean amphibians is accompanied by removal of cells from a layer rather than by extension (see Section 5.6).

5.3. Convergent Extension as the "Main Engine" of Gastrulation

The above experiments show that the complex of movements around the amphibian blastopore (the movement of cells toward the blastopore, involution, postinvolution movement of the mesodermal mantle, and the constriction of the blastopore to closure) are all brought about by a single process (convergent extension) acting in the unique geometry of the circumblastoporal region (Keller, 1984; Keller *et al.*, 1985). This constitutes the **main engine of gastrulation.** It is set up for its proper function in the first half of gastrulation by the activities of the bottle cells (see Section 7) and by the invasive, migratory behavior of the leading tongue of mesodermal cells (Section 6). These processes result in the turning in of the dorsal, vegetal edge of the mesodermal torus by the midgastrula stage (Fig. 14a). At the mid-gastrula stage, the powerful convergent extension seen in the explants of the involuting marginal zone begins in this involuted lip of the torus, resulting in a decrease in its circumferential extent (width) and an increase in its meridional extent (length) (Fig. 14a). This

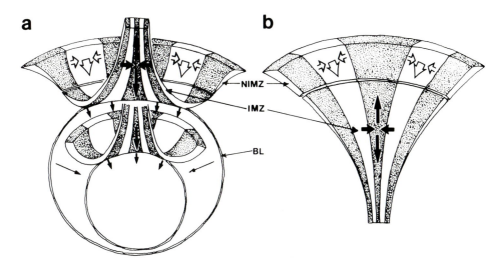

Figure 14. The function of convergent extension of the involuting marginal zone (IMZ) and noninvoluting marginal zone (NIMZ) in normal gastrulation (a) and exogastrulation (b) is shown diagrammatically as seen from the inside of the circumblastoporal region of the gastrula. The IMZ undergoes convergent extension (solid arrows) in the postinvolution position and thus simultaneously pushes vegetally and constricts the blastopore (BL). On the outside, the NIMZ undergoes convergent extension (open arrows) and thus pushes vegetally, toward the blastopore. If the IMZ is not turned over the lip and directed inward by the early mid-gastrula stage, its convergent extension movement is directed outward, in series with that of the NIMZ (b) rather than parallel to it (a).

behavior, in one stroke, forms a constriction ring just inside the blastoporal lip and pushes vegetally on the inside of the blastoporal lip (Fig. 14a). Preinvolution cells of the involuting marginal zone are recruited to join in the convergent extension process as they enter the postinvolution region. The noninvoluting marginal zone begins its convergent extension at a faster rate and feeds cells vegetally to take up positions vacated by the involuting marginal zone (Fig. 14a). Gastrulation stops when cells of the noninvoluting marginal zone reach the point of involution.

The convergent extension of the noninvoluting and involuting marginal zones in explants is sufficient to form twice the length of the dorsal aspect of either the gastrula or the neurula, even when separated (Fig. 11). In the explants, these two regions act in series, end to end. In the intact gastrula and neurula, they act in parallel, one member of the pair pushing posteriorly from the inside and narrowing the blastopore and the other feeding material from the outside, toward the blastopore. If the ancillary processes of bottle cell formation and leading mesodermal cell migration do not occur properly, the lower edge of the mesodermal torus will not be tucked under, and the powerful convergent extension will occur in series, as in the explants, forming an exogastrula or an excrescence (Fig. 14b).

The power and efficiency of the convergent extension machinery in producing gastrulation can be visualized by imagining the convergent extension process shortening the circumblastoporal segments of the hoops mapped onto the early gastrula (Fig. 7h). In doing so, the convergent extension machinery forms sucessive hoops of tension that act, at the inside of the blastoporal lip, to roll the involuting marginal zone down and over the lip. Note that the degree of convergent extension of the involuting marginal zone increases progressively from a minimum at its vegetal (prospective anterior) end, to a maximum at its animal (prospective posterior) end, which is the limit of involution. This same pattern of convergent extension is reiterated, by reflection across the limit of involution, in the noninvoluting marginal zone (animalward from the involuting marginal zone). Note also that in passing from dorsal to ventral sectors of both the involuting and noninvoluting marginal zones, convergent extension becomes progressively restricted to regions lying near the limit of involution, which will form posterior regions of the embryo (Fig. 4d,e).

5.4. Classic Evidence for the Role of Convergent Extension

Convergent extension was described and recognized as the major force-generator in gastrulation long ago, but it has been ignored in the recent literature. Convergent extension emerged near the turn of the century in competition with the related idea of **concrescence** to explain the closure of the blastopore and formation of the embryo. According to the theory of concrescence, the embryo is formed from right and left sets of primordia distributed on each side of the blastopore, by progressive apposition and fusion of the lateral lips of the blastopore. The neural groove was thought to represent the line of fusion.

Support for this idea came from the fact that mesodermal and neural structures are distributed on either side of a large open blastopore in "ring embryos" produced by various mainipulations of the marginal zone (Morgan, 1897; Todd, 1904; Bautzmann, 1933; Hertwig, 1892; Goerttler, 1926; see also Schechtman, 1942; Schenk, 1952; Keller, 1984). But the movements of exovates formed by pricking the embryo (King, 1902a), and vital dye marks (Goodale, 1911a,b; Vogt, 1929b; Keller, 1975, 1976) show that the anterio–posterior dimensions of the embryo map along the meridians of the marginal zone, as predicted by convergent extension, rather than circumferentially, around the blastopore, as predicted by concrescence.

The power of convergent extension in gastrulation was attributed primarily to the capacity of the marginal zone to constrict the yolk mass in experimental situations (Vogt, 1922b), although in normal development it was thought to occur below the yolk mass and help push it inside (Spemann, 1902; Vogt, 1922a). In one of the best (and most ignored) papers on gastrulation, Schechtman (1942) showed that the dorsal marginal zone has the autonomous capacity to extend and converge. He also found that it must be tucked under by the leading mesoderm in order to function effectively in gastrulation; if not, exogastrulation or excrescences would result (see Fig. 25, Section 6.3). Explants of the dorsal marginal zone (Holtfreter, 1938a,b; Schechtman, 1942; Ikushima, 1958, 1959, 1961; Ikushima and Maruyama, 1971) show some tendency to elongate and narrow. Many found that if the dorsal sector of the gastrula or adjacent tissue was perturbed in a variety of ways by microsurgery, it would form elongated, narrowed structures often refered to as "excrescences" (Mangold, 1920; Spemann, 1931; Lehman, 1932; Holtfreter, 1933b, 1938a,b; Schechtman, 1942). These excrescences were taken as evidence of the intrinsic capacity of the marginal zone to show convergent extension, although much of it appeared to occur in the neurula stage (Holtfreter, 1944; see Goerttler, 1925a,b; Holtfreter, 1933a). Holtfreter also realized that not all the excrescences in the literature were equivalent; indeed, the appearance of the excrescences in the above papers suggests that the behavioral characteristic of both involuting and noninvoluting marginal zones was seen and lumped together as a common phenomenon. Convergent extension was thought to push material toward the blastoporal lip, but Holtfreter (1944) suggested that it might also be a postinvolution process, which is, in fact, the case.

Convergent extension has been ignored, probably because other phenomena, such as tissue affinity and cell recognition (Holtfreter, 1939), and competing cell behaviors, such as bottle cell formation (Holtfreter, 1943a,b; Baker, 1965; Perry and Waddington, 1966), and cell migration (Holtfreter, 1943b, 1944; Nakatsuji, 1974; Keller and Schoenwolf, 1977), all involve concepts of cell function in morphogenesis that were more palatable and geometries that were easier to understand. Complex or not, convergent extension accounts for the bulk of the distortion of the gastrula and must be understood. Recent evidence shows that it is based on a relatively unappreciated notion of how cells function in morphogenesis—intercalation of cells to form a longer, narrower array.

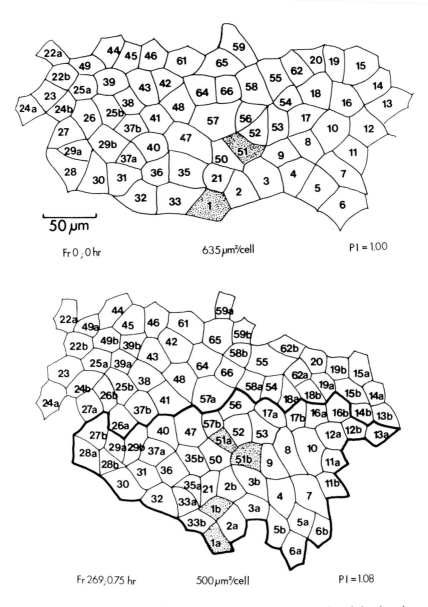

Figure 15. Tracings of superficial cells from time-lapse cinemicrography of the dorsal marginal zone of *Xenopus* show rearrangement of cells to form a longer, narrower array. Individual cells are identified by number. The frame number, time elapsed, mean area occupied in the epithelium by the cell apices, and an index of expansion are shown from left to right at the bottom of each figure. (From Keller, 1978, with permission.)

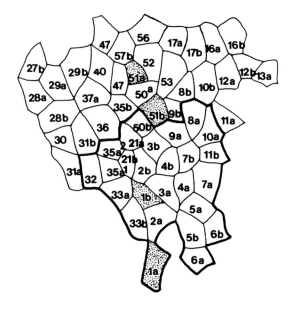

Fr 527; 1.46 hr   530 μm²/cell   P I = 1.39

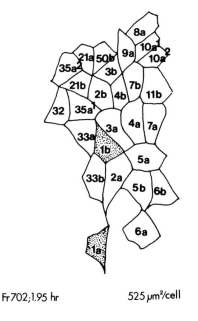

Fr 702; 1.95 hr   525 μm²/cell   P I = 1.51

Fig. 15 (continued)

5.5. Cellular Behavior during Convergent Extension

Waddington (1940, page 109) pointed out that the extraordinary elongation of the dorsal sector of the gastrula is not reflected in a permanent and proportionate change in cell shape and therefore must involve cell rearrangement. Recent work shows that, in fact, cell rearrangement is the principal cellular behavior occuring during convergent extension.

The superficial epithelial cells do not appear to undergo convergent extension actively; rather, they passively accomodate the distortion of the deep region beneath (Keller, 1978, 1981). Time-lapse cinemicrography of superficial cells in the dorsal marginal zone of *Xenopus* (Keller, 1978) shows that these cells exchange neighbors and rearrange to form fewer columns of less width and greater length (Fig. 15a–c). During rearrangement, their mean length–width ratio increases from about 1.0 to 1.6 and then returns to 1.0 at the end of gastrulation. This behavior suggests that the superficial cells initially accomodate external forces by stretching, and when the intercellular, circumapical adhesions can no longer resist the shear forces between the stretched cells, they slide past one another and return to their original shape. The cells maintain the apical intercellular contact typical of epithelial cells, and lesions between cells are rare even under electron microscopic resolution (Keller and Schoenwolf, 1977). Marginal zone cells are apparently specialized to permit rearrangement; animal pole superficial cells, grafted to the marginal zone, are reluctant to rearrange and thus are stretched to a large mean length–width ratio (R. E. Keller, unpublished data). Immediate deformation of the superficial marginal zone on extirpation suggests that it is under tension (Beloussov *et al.*, 1975).

Rearrangement of epithelial cells occurs among the marginal and submarginal cells of the teleost enveloping layer, as this layer is distorted during gastrular epiboly (Kageyama, 1982; Keller and Trinkaus, 1982), among corneal epithelial cells during wound healing (Honda *et al.*, 1982), and among *Drosophila* imaginal disc cells during disc evagination (see Chapter 4 and Fristrom, 1976). It also occurs during regeneration in *Hydra* (Graf and Gierer, 1980), during newt neurulation (Jacobson and Gordon, 1976), and in a mesenchymal cell population during pronephric duct migration (elongation) in amphibians (Poole and Steinberg, 1981). In teleost gastrulation, exchange of neighbors occurs while a physiological barrier consisting of tight junctions is maintained (Keller and Trinkaus, 1982). In *Drosophila* it occurs in the presence of a septate junctional complex (Fristrom, 1982). In *Rana pipiens*, tight junctions are progressively disassembled in the course of neurulation, beginning with the formation of the neural plate (Decker, 1981), suggesting that (at least in neurulation) the rearrangement may occur after breakdown of zonula occludens. It is important to look for such specializations or breakdown of tight junction structure in the amphibian gastrula, particularly in light of the fact that some cells, such as those of the marginal zone, can rearrange whereas neighboring cells cannot. The general problem of how epithelial cells passively accommodate (Keller, 1978) or drive (Fristrom and Chihara, 1978) morphogenetic distortions is poorly understood and needs more study (see Chapter 4).

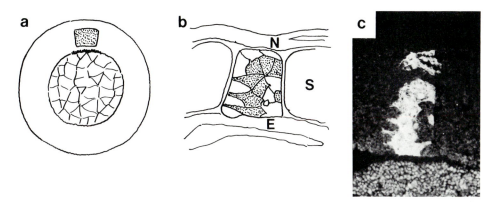

Figure 16. Circumferential intercalation of deep cells of the involuting marginal zone was established by grafting a patch of dorsal marginal zone cells, labeled with the cell lineage tracer fluorescein–lysine–dextran, to the corresponding region of an unlabeled embryo (a) and monitoring intercalation of labeled and unlabeled cells in the notochord of the neurula. The results are shown diagramatically (b) and in a fluorescence light micrograph (c). E, endoderm of archenteron roof; N, neural plate; S, somite.

There is now direct evidence that circumferential (medio–lateral) intercalation of deep cells also occurs in the course of convergent extension. Medio–lateral cell mixing was monitored (Keller *et al.*, 1985) by marking populations of cells with a new cell lineage marker suitable for use in amphibians; it is a fluorescein–lysine–dextran (FLDx) made by Gimlich and Braun (see Gimlich and Cooke, 1984). Patches of deep and superficial cells from labeled embryos were grafted to the corresponding region of unlabeled host embryos. The boundary of labeled and unlabeled cells becomes progressively feathered due to intercalation of deep cells in the circumferential direction in the course of gastrulation and neurulation (Fig. 16).

Radial intercalation of cells also occurs during convergent extension, at least in the noninvoluting marginal zone. Morphometric data (Keller, 1980) show that the noninvoluting marginal zone decreases from nearly six layers at the outset of gastrulation to two at the end (Fig. 17a–d), probably by radial intercalation to form fewer layers of greater area (Keller, 1980). No corresponding analysis of the involuting marginal zone has been done, though it appears to be reduced in thickness from six to about three to four layers. The combination of radial and circumferential intercalation transforms the relatively thick and laterally spread tissues comprising the convergent extension machinery of *Xenopus* into a thinner, elongate mass on the dorsal side of the neurula.

The details of how deep cells intercalate circumferentially (or radially) are now being studied by direct observation of explants of the marginal zone, in culture, using a culture medium that approximates the ionic composition of the blastocoel fluid (Keller *et al.*, 1985; see Section 8.4). Deep cells actively intercalate between one another to form a longer, narrower array, using a complex pattern of protrusive activity that involves cytoplasmic flow. At the gastrula stage, intercalation was observed to occur in both the circumferential (Fig. 18a)

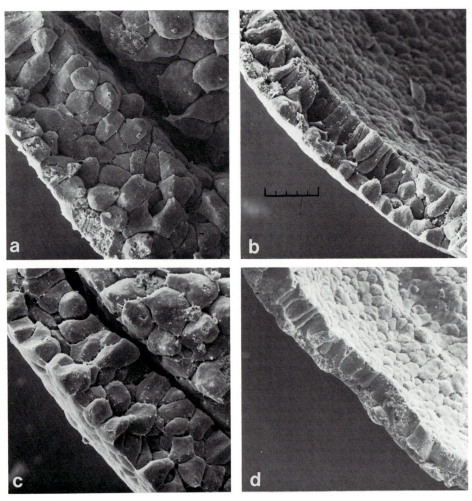

Figure 17. Scanning electron micrographs of mid-saggital fractures of the dorsal marginal zone show changes in cell morphology and arrangement from the early gastrula stage (a), through the early middle (b), late middle (c), to the late gastrula (d) stages. Scale bar: 50 μm. (From Keller, 1980, with permission.)

and radial (Fig. 18b) directions. In addition, circumferential intercalation of notochordal cells (Fig. 18c) was observed as part of the process of convergent extension in the neurula stage.

Exactly how intercalation is related to the generation of the mechanical forces that drive convergent extension is not yet clear, nor is it clear what principles organize and regulate cell behavior during this process. Holtfreter (1944) suggested that the original tissue organization is necessary for convergent extension. He found that an explant of dorsal marginal zone normally would extend and form notochord tissue, but if the component cells were disaggregated in alkaline or calcium-free media and reaggregated, it would differentiate into notochord tissue but would not elongate. Beloussov (1978,

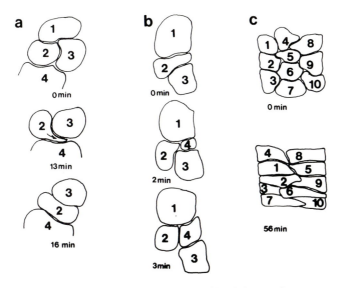

Figure 18. Time-lapse videomicrography of the deep cells of the involuting marginal zone, explanted in culture, shows circumferential intercalation (a) and radial intercalation (b) during convergent extension and intercalation of notochord cells in the neurula (c). The axis of extension is vertical and that of convergence is horizontal in all figures.

1980) suggested that the polarized cell behavior found in the elongating axial rudiments from the late gastrula stage onward may be transmitted from cell to cell by tensile fields. Luchinskaya and Belousov (1974) argued that such "contact polarization" involves a repatterning of the cytoskeleton. LeBlanc and Brick (1981a) observed that cells above the dorsal lip are polarized in their spreading behavior in culture. But the local autonomy exhibited by the marginal zone seems to rule out long-range mechanical tension as an initiator and organizer of cell behavior (Cooke, 1975), and the same follows from the patch rotation experiments described above (Keller, 1984).

5.6. Convergent Extension in the Urodele

The urodele marginal zone also shows the capacity of autonomous convergent extension (see Vogt, 1922a, 1929b; Holtfreter, 1938a, 1944; Lewis, 1952; Ikushima, 1961; Lundmark, 1986), but whether it shows the same spatial distribution and involves the same cell behavior seen in *Xenopus* is not known. The geometric and mechanical aspects of the function of convergent extension in *Ambystoma* resemble those found in *Xenopus*, in that convergent extension of the notochord decreases the length of the suprablastoporal hoops (Fig. 7a), thereby contributing to the constriction forces that tend to roll the marginal zone over the lip (Lewis, 1948, 1952; Lundmark, 1986). But, in addition, ingression of a large area of prospective somitic mesoderm at the lateral aspects of the blastoporal groove, during involution (Fig. 7b), greatly decreases the length of these suprablastoporal hoops, and thus also contributes to the constriction forces (see Lundmark, 1986). Thus, the same forces are generated in both *Xenopus* and the axolotl, but in the former they are generated entirely by active convergent extension of the deep mesoderm, whereas in the latter they are generated by ingression of the somitic mesoderm as well as by convergent extension of the notochordal region.

How does this ingression of somitic mesoderm occur? There is a common and mistaken impression, based on pre-Vogtian work and purveyed by textbooks too numerous to mention, that removal of mesoderm from the superficial layer involves migration of lateral endodermal crests dorsally across the mesodermal roof of the gastrocoel and that there is some sort of shear zone at the dorsal, free edge of these endodermal crests. Vogt (1929*b*) showed that during involution, at the lateral aspects of the blastopore, the somitic mesoderm diverges laterally, whereas the endoderm converges dorsally to bound the notochord; the notochord is then removed from the superficial layer in the neurula stages (Vogt, 1929*b*) (see Fig. 7). Hamburger (1960) gives a clear description of Vogt's results. But this elusive line of divergence of mesodermal and endodermal movements was never clearly demonstrated and characterized.

Lundmark (1986) has clarified this situation. She found that somitic mesoderm forms bottle-shaped cells, which leave the superficial layer and then migrate laterally in the deep region as individual, mesenchymal cells or as members of a stream (Fig. 19). There is no zone of shear as such. Thus, the bottle cells in the lateral sectors of the urodele are an ingressing cell popula-

Figure 19. Scanning electron micrography shows the relationship between the prospective notochord (N), sub-blastoporal endoderm (SBE), and somitic mesodermal cells (SM) that have left the superficial layer and have joined the deep mesodermal mantle, as viewed from inside the gastrocoel, looking posteriorly (a). In dissected gastrulae, the point where the bottle-shaped, ingressing mesodermal cells leave the superficial layer in the lateral region of the blastoporal groove (open arrow), and the resulting stream of migrating mesodermal cells (solid arrow) are visible. (Micrographs courtesy of Cathy Lundmark.)

tion, similar to that seen in the primitive streak of the chick (see Chapter 12), and therefore represent a cellular morphogenetic process entirely different from that represented by the bottle cells in *Xenopus*. After somitic mesoderm is removed, the lateral margins of the notochord bound the sub-blastoporal endoderm, and the notochord is subsequently removed from the superficial layer during neurulation by a process that involves simultaneous convergent extension and constriction of the apices of the cells to the point of ingression (Figs. 6e,f and 7). As the notochord narrows, the endodermal "wings", bounding it laterally, meet in the midline (Ruffini, 1925; Vogt, 1929*b*; Lofberg, 1974; Brun and Garson, 1984; Lundmark, 1986). As in *Xenopus*, the notochord of the axolotl undergoes radial intercalation of cells. Before and during its involution the notochordal region consists of several deep layers and a superficial layer. But in the late gastrula, the notochord is one cell thick, probably due to radial intercalation of deep cells into the superficial layer to form a single-layered epithelium (Lundmark, 1986).

Convergent extension in urodeles is no less powerful than that in *Xenopus*. In the urodele *Taricha torosus*, migration of the cells into the blastocoel can be prevented by filling the blastocoel with injection of gelatin. In these embryos, the powerful convergent extension machinery in the circumblastoporal region appears to pull cells vegetally, to the point of tearing the animal cap, and there organizes the axial structures of an embryo (Eakin, 1933, 1939). Likewise, this powerful circumblastoporal behavior persists and brings about gastrulation after removal of the blastocoel roof (Lewis, 1952).

6. Migration of Involuted Mesodermal Cells

The cells of the mesodermal torus have a dual character, and the convergent extension of the dorsal posterior mesoderm of *Xenopus* and the notochord of the axolotl is not characteristic of mesodermal cells in the anterior leading edge and ventral parts of the mesodermal mantle. In both amphibians, but particularly in the axolotl, this mesoderm appears to migrate as a population of individual cells or as members of a cells stream, which are dependent on an exogenous substratum.

6.1. Evidence for Active Mesodermal Cell Migration

Involuted mesodermal cells have a morphology that suggests that they actively migrate on the inner surface of the blastocoel roof. Holtfreter (1944) found them to be in a leaf-like array with pointed ends animalward. Interest in their active migration was rekindled by recent work on urodeles (Nakatsuji, 1975*a*) and anurans (Nakatsuji, 1974, 1975*b*, 1976; Keller and Schoenwolf, 1977), which showed with SEM that these cells have morphologies and contacts consistent with migration. They are rotund, separated by intercellular spaces, and are connected to one another and to the overlying blastocoel roof by

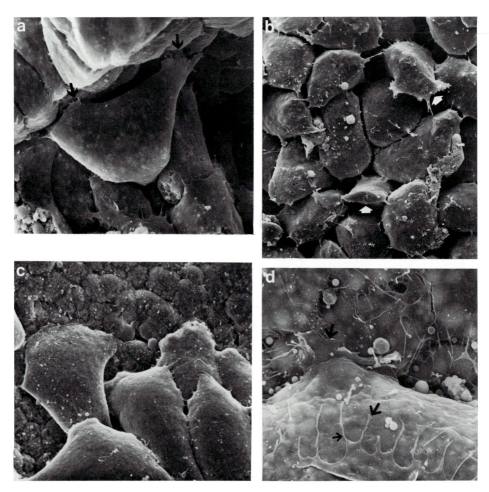

Figure 20. Scanning electron microscopy of a *Xenopus* gastrula shows an involuted mesodermal cell (M) attached to the inner surfaces of the deep cells of the gastrular wall by protrusions (pointers) of its margin (a). The outer surface of the involuted mesodermal mantle, exposed by removal of the overlying gastrular wall (b) shows the pleiomorphic cell shape, large intercellular spaces, and lamelliform and filiform protrusions (pointers) that characterize this cell population. In the axolotl, the migrating mesodermal cells (M) have lamelliform protrusions (pointers) spread on the inner surface of the gastrular wall (c,d) and on one another (d). Figure a is from Keller and Schoenwolf, and c and d are from Lundmark *et al.*, 1984, with permission from *Scanning Electron Microscopy*, 1984, Part III, pp. 1291–1296.

large lamelliform or filiform protrusions (Fig. 20). There are differences between the axolotl and *Xenopus* mesodermal cells that may reflect differences in organization of these migrating cell population. When the gastrular wall is peeled off to expose the mesodermal mantle, the mantle usually remains intact in *Xenopus* (Fig. 20b) (Keller, 1984). By contrast, the mesodermal cells in *Ambystoma* usually, but not always, maintain their association with the blastocoel roof and separate from one another (Fig. 20c). These facts suggest that in *Xenopus* the migrating mesoderm forms a cohesive population of cells on the periphery of the endodermal core, which exerts traction on the overlying blastocoel roof and thus pulls it vegetally. By contrast, the corresponding mesoderm in the axolotl appears loosely organized as individual cells whose movement is best thought of as a migration animalward on the roof of the blastocoel (see Nakatsuji, 1984). The best evidence for active migration in the gastrula comes from time-lapse cinemicrography of opened gastrulae in culture; although the details of migration could not be observed at the low magnifications used, the films show what appear to be successive waves of mesodermal cells advancing across the blastocoel substratum (Kubota and Durston, 1978).

6.2. The Substratum for Migration *in Vivo* and in Culture

The inner surface of the overlying preinvolution material forms a special surface that, at a minimum, serves to maintain a discontinuity along which shear can occur (Keller and Schoenwolf, 1977). Any postinvolution cell placed on this surface does not penetrate it but spreads laterally on it; by contrast, preinvolution deep cells, when placed on this surface, actively and quickly intercalate themselves between the cells of the substratum (R. E. Keller, unpublished data). This property of the inner surface of the blastocoel roof insures that no postinvolution cell can invade its integrity but will spread on it.

It is likely that fibronectin contributes to maintaining this discontinuity and perhaps serves as a substratum for migration. The inner surface of the blastocoel roof shows a stronger immunofluorescence with fluorescein-labeled antibodies to fibronectin than any other region (Lee *et al.*, 1982, 1984; Boucaut *et al.*, 1984a) (Fig. 21a,b). Moreover, monovalent antibodies to fibronectin, injected into *Pleurodeles* embryos, result in collapse and wrinkling of the blastocoel and other abnormalities of gastrulation (Boucaut *et al.*, 1984a), which may be due to the inhibition of migration of cells on the roof of the blastocoel (Fig. 21d,e). The same occurs when a synthetic peptide, perhaps containing the cell recognition sequence of fibronectin, is injected (Boucaut *et al.*, 1984b). When the nonadhesive outer (apical) surface of the blastocoel roof is turned inside, mesodermal cells cannot use it as a substratum (Fig. 21c) and it does not stain for fibronectin (Fig. 21b) (Boucaut *et al.*, 1984a).

The inner surface of the blastocoel roof bears a fibrillar extracellular matrix that is well developed in the urodeles and less so in the anurans (Nakatsuji *et al.*, 1982; Nakatsuji and Johnson, 1983a,b; Nakatsuji, 1984) (Fig. 22). These fibrils are fibronectinlike in size and appearance, but at present there is no direct evidence that they are fibronectin or contain fibronectin. In the axolotl, these fibrils appear to be aligned along the animal–vegetal meridians of the embryo (Nakatsuji *et al.*, 1982). Explanted pieces of the gastrular ectodermal layer of the axolotl deposit material on plastic coverslips that enables mesodermal cells subsequently cultured on the coverslip to adhere and migrate, whereas without this conditioning, they could do neither (Nakatsuji and Johnson, 1983a). Moreover, the migration of these cells is oriented parallel to the direction of the alignment of the fibrils deposited by the ectodermal layer. Mechan-

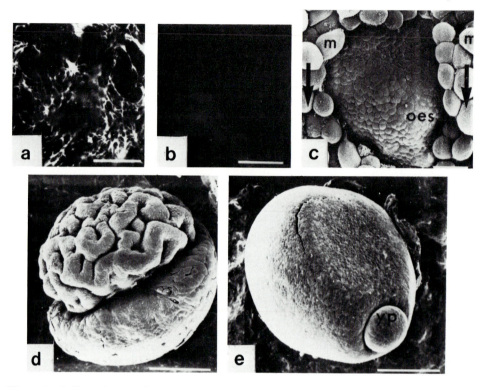

Figure 21. Indirect immunofluorescence staining of the inner surface of the blastocoel roof of *Pleurodeles waltlii* with antibody to fibronectin shows a fibronectin-rich matrix (a). By contrast, no staining for fibronectin is visible on the outer surface of the blastocoel roof (b). When a patch of the blastocoel roof of the *Pleurodeles* gastrula is turned inside out, the mesodermal cells (M) cannot adhere to the nonadhesive outer ectodermal surface (OES) of the reversed patch (c). Injection of monovalent antifibronectin antibodies into the blastocoel results in abnormal gastrulation (d). Control embryos injected with monovalent preimmune antibodies develop normally (e). Scale bars: 20 μm (a,b); 320 μm (d,e). (From Boucaut *et al.*, 1984a. Reprinted from *Nature*, 1984, Vol. 307, pp. 364–367 by copyright permission of Macmillan Journals Limited.)

ical tension on explanted ectodermal layer appears to align the fibrils and the cell movement that they support (Nakatsuji and Johnson, 1984). These experiments show that the mesodermal cells can use the fibrillar matrix as a substratum and that aligned matrix can orient their migration in culture. It is not known whether the alignment of fibrils and contact guidance of mesodermal cells *in vivo* is necessary for gastrulation.

Kubota and Durston (1978) argued, on several grounds, that contact inhibition of movement directs the movements of mesodermal cells into the gastrula. First, the leading mesodermal cells advance first, followed by waves of those behind, suggesting that the leaders must move before those following can do so. Second, the cells in opened gastrulae will migrate beyond their usual stopping point and move into the ventral marginal zone, counter to their usual direction of migration in this region. Third, isolated cells of the margin of the leading edge migrate aimlessly until contacted again by the group. Fourth, groups of cells placed on the blastocoel roof spread radially regardless of orientation of the roof. The last three observations argue against the substratum being important in guidance. On the other hand, a contact guidance system may not survive in opened gastrulae; the alignment may depend on stretching of the blastocoel roof by epiboly (Nakatsuji, 1984) or, more likely, by the convergent extension machinery described above. Amphibian mesodermal cells do show contact inhibition of movement in culture (Johnson, 1976a; Nakatsuji and Johnson, 1982). If they show contact inhibition *in vivo*, the blastocoel is the only cell-free space available. In summary, it is not clear whether contact guidance, contact inhibition, or perhaps additional mechanisms, function to direct mesoderm migration in the gastrula.

6.3. Function of Mesodermal Migration in Gastrulation

What is the role of mesodermal cell migration in gastrulation? It is perhaps easier to establish what it does not do in gastrulation. It does not function in the performance of the convergent extension movements of the dorsal, posterior sector; the full length of the dorsal side of the gastrula and the axial structures contained therein can be generated by the sandwich explants, which do not use any substratum, including the blastocoel roof (Fig. 11f,g). It is also unlikely to contribute to constriction of the blastopore (Schechtman, 1942; Keller, 1984). In some anurans, it appears to have little importance for the closure of the blastopore, involution, or organization of the mesodermal mantle. For example, Holtfreter (1933b) removed the roof of the blastocoel of the early gastrula of *Hyla*, and the major distortions of gastrulation proceeded in a remarkably normal manner (Fig. 23). Even in the urodele *Taricha*, convergent extension of the circumblastoporal region seems to be able to compensate for failure of animalward migration by pulling the animal cap vegetally when migration into the blastocoel is blocked with gelatin (see Eakin, 1939) (Fig. 24). The same is true of *Ambystoma* gastrulae from which the blastocoel roofs have been removed (Lewis, 1952).

In anurans, migration of the leading, (prospective head) mesoderm in the dorsal sector of the mesodermal torus reorients the convergent extension machinery in the first half of gastrulation. These cells are the first to involute, and they do so during the formation of the bottle cells (see Section 7). Schechtman (1942) found that in the anuran *Hyla regilla* these cells have a capacity to involute and migrate on the blastocoel roof, whereas more posterior, prospective chordamesoderm has no tendency to do so; it must be turned inside by the leading mesoderm or lateral mesoderm or else it extends into space, forming an excrescence (Fig. 25). If the head mesoderm is given an opportunity to turn inside the sandwich explants of *Xenopus*, it migrates along between the two halves and reorients the involuting marginal zone such that it and the noninvoluting marginal zone undergo coordinate convergent extension in parallel, one inside the other, as *in vivo* (Keller *et al.*, 1985).

The remaining cells of the leading mesoderm in lateral and ventral sectors

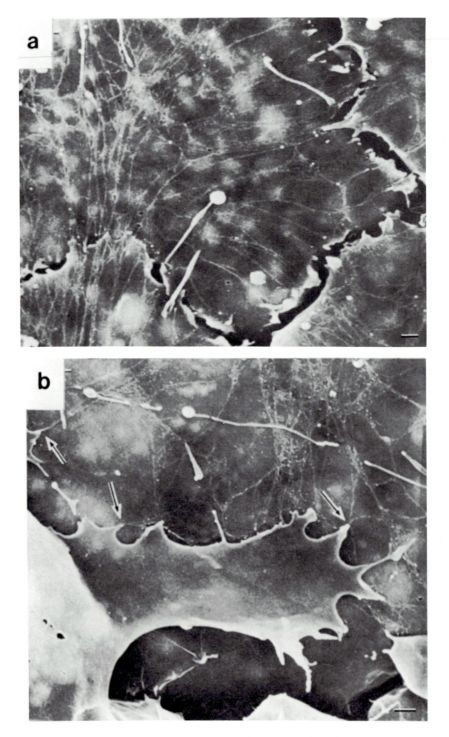

Figure 22. The inner surface of the blastocoel roof bears many fine fibrils (a). The lamelliform protrusions of the migrating mesodermal cells are often found in close association with these fibrils (b). Scale bars: 1 μm. (From Nakatsuji et al., 1982, with permission.)

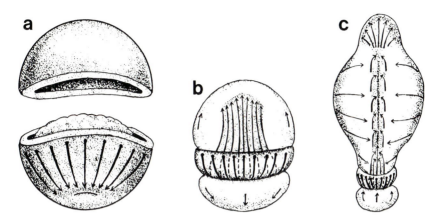

Figure 23. Removal of the entire ectoderm (animal cap) of the early gastrula of *Hyla* (a) does not prevent invagination (or involution) of the endoderm and mesoderm (b,c). The arrows show the morphogenetic movements. Note the dorsal convergence and extension in the gastrula (b) and neurula (c) stages. (From Holtfreter, 1933*b*, with permission.)

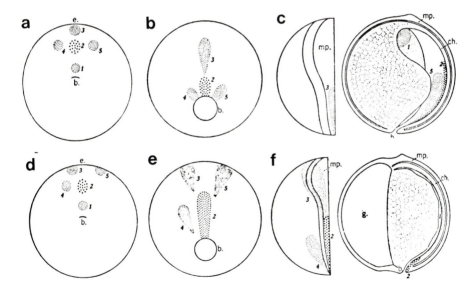

Figure 24. Vital dye marks show the morphogenetic movements in a normal embryo of *Taricha torosus* (a–c) and an embryo in which the blastocoel was filled with gelatin (g) at the early gastrula stage (d–f). Note that even when migration into the blastocoel is blocked, involution occurs, and an embryonic axis, including a notochord (chordamesoderm, ch) and neural plate (medullary plate, mp), are formed, although farther vegetally than in the normal embryo. c and f show dorsal and mid-saggital views; all others are vegetal views. (From Eakin, 1939, with permission.)

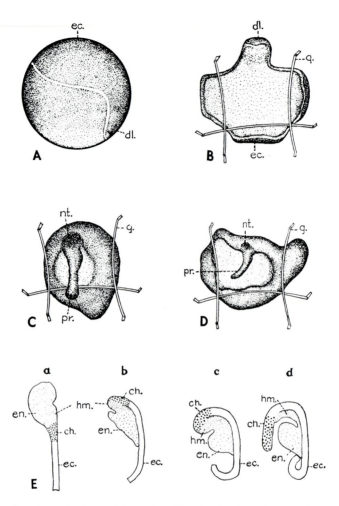

Figure 25. The dorsal marginal zone (dorsal lip, dl) and animal cap (prospective ectoderm, ec) of the early gastrula of *Hyla* were excised (A) and pinned down with foil girders (g) to a wax substratum (B). The dorsal marginal zone elongated and narrowed to form a chordal projection (pr) in subsequent development (C, D). Diagrams of the explant (E) at successive stages (a–d) show that the leading endoderm (en) and head mesoderm (hm) spread on the inner surface of the ectoderm (ec) but that the prospective chordamesoderm (ch) extended into space to form the proboscislike structure. (From Schechtman, 1942, with permission.)

of the gastrula may contribute to their own relative movement animalward by exerting traction on the overlying epidermal ectoderm. In anurans, most of this mesoderm originates in the deep region of the gastrula and forms a collar around the central core of subblastoporal endoderm. These cells appear to be more strongly associated with one another than with the overlying epidermis, and thus it may be more accurate to view them as maintaining stable relationships with one another, simultaneously exerting traction on the epidermis and displacing it vegetally, than to view them as a migratory population in the usual sense.

In urodeles, the migrating mesodermal cells are less cohesive as a population and seem to be organized more as individual cells on a substratum—the overlying ectoderm—than in anurans (Nakatsuji, 1984). In normal development, it is tempting to describe the movement of these cells as a migration of individuals on a substratum. However, they do show enough cohesion among themselves to carry their neighbors across a nonadhesive patch of reversed gastrular wall placed in their path (Boucaut et al., 1984a) (Fig. 21c). In the newt, Waddington (1939), using a magnet and steel ball, roughly estimated the pressure exerted by the mesoderm moving upward inside the gastrula at 0.34 mg/mm². It is not known how much of this is due to active migration on the roof of the blastocoel and how much is due to convergent extension.

7. Bottle Cell Function in Gastrulation

Two facts assure that the bottle cells will attract and hold the attention of embryologists interested in gastrulation: They are found at the site of blastopore formation and they show dramatic behavior. But recent experiments show that their function is probably more limited and yet more complex than once thought. Moreover, it is likely that not all bottle cells have the same function.

7.1. Classic Views of Bottle Cell Function

The common view of bottle cell function is either that the changes in shape involved in their formation cause a bending of a cell sheet (an invagination), or that they have an invasive behavior and actively migrate into the interior, pulling all else behind them. The first idea is part of the more general concept of how epithelial cell shape changes can result in morphogenesis of the sheet; this was formally set forth by Lewis (1947) and has been invoked to explain part of sea urchin gastrulation (Gustafson and Wolpert, 1967), vertebrate neurulation (see Jacobson, 1982), and *Volvox* inversion (Viamontes et al., 1979), to name only several of many examples. The second idea comes from Holtfreter's experiments. He observed that bottle cells dissociated from their neighbors by alkali (pH 9) would actively spread and migrate with their basal ends leading when put into culture on glass. He also observed that these cells were normally connected to all other superficial cells by a common syncytial "surface coat" (see Section 8.1). Thus, when the bottle cells invade the gastrula, they pull the superficial cells around them inside as well. In fact, bottle cells placed on endodermal substrata will form a pit resembling a blastopore (Holtfreter, 1943b). Variants of this idea were held by Rhumbler, Ruffini, and others before Holtfreter (see Holtfreter, 1943a,b).

Several facts argue against these notions. First, the bottle cells are part of an epithelium with a nonadhesive apical surface; if they are able to stick and migrate on any substratum, they must use their basal ends. The polarity demonstrated in culture is consistent with their epithelial nature and in itself does not argue strongly that they function this way *in vivo*. Second, the geometry of the gastrula is not consistent with them playing so large a role in the displacements of gastrulation. In relationship to cells around them, they do not move very far in *Xenopus* (see Keller, 1981). Their shape change could account for simple bending of the epithelium to form the initial blastoporal groove, but that is only a small part of the total depth of the definitive archenteron, most of which is formed as a result of involution (see Keller, 1981). Indeed, the groove that they form when cultured on endoderm is on the order of one cell dimension in depth, about like the blastoporal groove in the early gastrula. Third, they disappear midway through gastrulation on the dorsal sector of the gastrula, long before movements in this region have stopped (Keller, 1981; see Daniel and Yarwood, 1939). Fourth, their removal truncates the archenteron but leads to surprisingly little disruption of gastrulation (Cooke, 1975; Keller, 1981).

7.2. Bottle Cell Behavior in *Xenopus*

Time-lapse cinemicrography (Keller, 1981) shows that bottle cells form by apparent decrease in area of their apices, beginning about a half hour before what is usually called the onset of gastrulation (Fig. 26a–b). The dark cells identified in these figures lost an average of 87% of the area that they occupied in the superficial layer in the preceding 19 minutes. Because the apical plasma membrane is thrown into microfolds (Fig. 26d), the actual area may not be decreased (Keller, 1981). Numerous vesicles are found in the necks of the bottle cells (Baker, 1965; Perry and Waddington, 1966) and may represent surface membrane taken up as the apices shrink, although there is no direct evidence for this. As the apices decrease in apparent area, the cells elongate radially and form a shape that resembles bottles or flasks made by glassblowers in the days when these cells were named (Fig. 26b). As bottle cell formation proceeds laterally, the torus-shaped region of prospective bottle cells is compacted into a dark line (the **blastoporal pigment line**), which represents the bottle cells apices and acquires its darkness from condensation of pigment (Fig. 26c). The apparent high concentration of extracellular matrix in regions of bottle cell formation (Moran and Mouradian, 1975; Monroy *et al.*, 1976) is likewise due to condensation and is probably a result rather than a cause of morphogenesis. Shrinkage of the apices is biased in the meridional (animal–vegetal) direction (cells 18, 20, 21, Fig. 26a). As a result, the apices of fully formed bottle cells are elongated circumferentially, and the primary distortion of the embryo that follows is a decrease in meridional extent of the torus of suprablastoporal endodermal bottle cells rather than a circumferential constriction (Keller, 1981; Hardin and Keller, 1986). The population of cells forming bottle cells is contiguous, but not all members begin bottle cell formation at the same time; thus, some light-colored cells with large apices are isolated among darker cells with small apices that have already begun apical constriction. Methods available 20 years ago (Baker, 1965; Perry and Waddington, 1966) suggested that bottle cell apices contain microfilaments, and the birefringence of their elongated necks suggested alignment of filament systems (Perry and Waddington, 1966), but nothing has been added to our knowledge of the cytoskeletal organization of these cells since these works appeared.

The bottle cells seem to disappear beginning at the mid-gastrula stage in the dorsal sector where they had first formed and thereafter in lateral and ventral sectors of the original blastoporal pigment line. This disappearance is observed in Xenopus (Keller, 1981; Hardin and Keller, 1986) as well as other species of amphibians (see Daniel and Yarwood, 1939) and is due to the loss of their characteristic shape. They respread to form a large part of the archenteron; as they respread, the blastoporal pigment line disappears (Fig. 26e,f). Vital dye mapping (Keller, 1975) shows that the bottle cell region of the early gastrula (the blastoporal pigment line) forms a large area of the lining of the peripheral archenteron in the neurula. When bottle cells are removed, the peripheral archenteron is truncated, with dorsal bottle cells corresponding to the anterior archenteron, lateral bottle cells to the middle archenteron, and ventral bottle cells to the posterior archenteron (Keller, 1981). These facts suggest that the small area represented by the blastoporal pigment line (the bottle cells) re-spreads to form a large region of epithelial, endodermal lining of the archen-teron, beginning in the dorsal sector at stage 10.5 (midgastrula) and spreading laterally and finally ventrally in the neurula stage (Fig. 27a,b) (see Keller, 1981; Hardin and Keller, 1986). The formation of bottle cells and their subsequent respreading occurs autonomously, as shown by the behavior of cultured bottle cells isolated at the early gastrula stage in Xenopus (Fig. 27c,d). This pattern of behavior may be fixed in the blastula stage (Doucet–de Brune, 1973).

7.3. Bottle Cell Function in Gastrulation of *Xenopus*

A comparison of the behavior of explanted bottle cells with that *in vivo* yields clues as to their role in gastrulation. Both the radial (apical–basal) elongation and the bias of contraction of the apices in the meridional direction reflect mechanical constraints of adjacent regions of the embryo rather than intrinsic properties of bottle cells. Prospective bottle cells excised from late blastula-stage embryos undergo uniform contraction of their apices in all direc-tions and form a pit (Fig. 27d) rather than the groove formed *in vivo* (Fig. 27c). They also become rotund and wedge one another apart at their basal ends (Fig. 29d) rather than elongating radially as they do *in vivo* (Fig. 27c).

Why do these cells behave differently when removed from the gastrula? The failure of bottle cells to decrease the circumferential aspect of their apices is due to mechanical resistance offered by the large core of central endodermal cells. When the endodermal core is removed before bottle cell formation, the bottle cell apices contract uniformly, as in culture (Hardin and Keller, 1986); by contrast, removal of the other adjacent region—the marginal zone—does not change the usual bias towards meridional contraction. This suggests that bottle cell apices normally contract with equal tension in all directions, but that great resistance is offered by the sub-blastoporal endodermal core. Thus, most of the apical contraction is channelled into the meridional direction, and the less

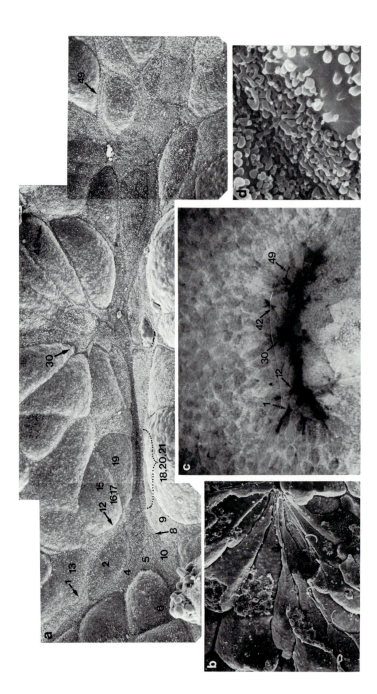

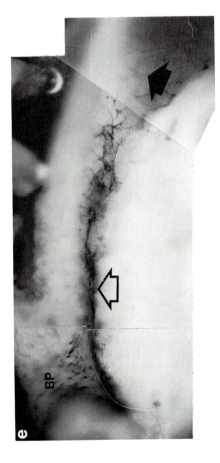

Figure 26. A scanning electron micrograph (a) and a corresponding light micrograph (b) show the apices of the bottle cells in various degrees of contraction to form the darkly pigmented blastoporal pigment line (b). The bottle cells have elongated narrow necks (arrow) and dilated basal (deep) ends (c). Their apices are covered with microfolds (d). Later, in the late gastrula and early neurula stages (e) the bottle cell apices, represented by the darkened blastoporal pigment line (open arrow) respread to form a large area of the peripheral archenteron lining (solid arrow). BP, blastopore. (Figures a, b, d, and e from Keller, 1981, and c from Lundmark et al., 1984, with permission.)

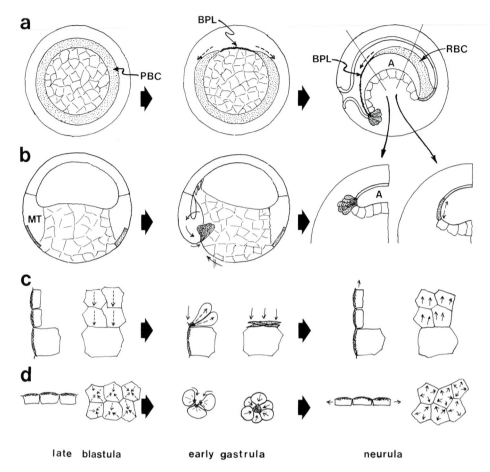

Figure 27. The behavior and function of bottle cells in the context of the embryo are shown from the late blastula stage (left) to the neurula stage (right), as seen from their apical surfaces (a) and from sectional views (b). Detailed diagrams show differences between bottle cell behavior *in vivo* (c) and in culture (d) at corresponding stages of development; sectional views are shown first, followed by surface (apical) views at each stage. In the late blastula stage, the prospective bottle cells (PBC) constrict apically and become bottle shaped, resulting in the formation of the blastoporal pigment line (BPL) and rotation of the mesodermal torus (MT) by the early gastrula stage (a,b). From the late gastrula stage onward, the bottle cells respread to form a large area of the peripheral archenteron (a, b). The apical constriction and respreading is largely in the original meridional (animal–vegetal) direction *in vivo* (c), whereas it is uniform in culture (d). A, archenteron; BPL, blastoporal pigment line; MT, mesodermal torus; RBC, respread bottle cells; PBC, prospective bottle cells.

resistant marginal zone above the forming bottle cells is pulled vegetally (Fig. 27a). The apical constriction forces cytoplasm deep, toward the basal ends of the cell, and the bottle cells form wedges, lying at the periphery of the central core of endoderm. The wedging of cytoplasm basally as the apices constrict would normally result in bending of the sheet (Lewis, 1947), but the bending sheet in this case is situated around the large, rigid endodermal core, and therefore the marginal zone is pushed outwardly and vegetally, such that it overrides the bottle cells, and the vegetal edge of the deep mesodermal torus is involuted (Fig. 27b). Such involution has several important functions: It forms the interface between preinvolution and postinvolution material and aids in giving the migratory head mesoderm access to its substratum—the roof of the blastocoel. Secondly, it begins the reorientation of the prospective notochord, just behind the head mesoderm, in the first half of gastrulation. Thus, the powerful convergent extension of this region in the second half of gastrulation will be directed inward rather than outward (see Fig. 14). The bottle cells may be assisted in this function by the animalward migration of head mesodermal cells (Fig. 27b). If prospective bottle cells are removed before formation of the definitive bottle cells, exogastrulation is more likely to occur (Hardin and Keller, 1986).

The initial formation of the bottle cells contributes to the depth of the archenteron in that it results in a shallow invagination—the blastoporal groove (Fig. 27b). The bulk of the depth of the archenteron is then formed by the involution associated with convergent extension of the involuting marginal zone. The bottle cells make a second contribution to the depth of the archenteron by respreading and advancing the domain of the endodermal epithelium by an amount equal to the maximum extent of the bottle cells, or about 150 μm (Keller, 1981).

7.4. Bottle Cell Function in Urodeles

The dorsal bottle cells in urodeles consist of prospective endoderm, and judging from Vogt's work (Vogt, 1929b), they appear to form and then respread to make the lining of the anterior archenteron. In this behavior they resemble the bottle cells of Xenopus. But the remaining bottle cells in urodeles are different from those of Xenopus. In the axolotl and in related Ambystoma species, bottle-shaped cells in the lateral and ventral sectors of the blastopore represent mesodermal cells in the process of ingression into the deep region (Holtfreter, 1944; Lundmark, 1986); such does not appear to be the case in Xenopus (Keller, 1975; Smith and Malacinski, 1983). Thus, the lateral and ventral bottle cells of the urodeles possibly have more in common with those of the primitive streak in the chick (see Chapter 12), as ingressing cells, than they do with their counterparts in Xenopus. Bottle or flask cells are found in association with many invaginations and involutions in metazoan development, but they do not always represent a common cellular behavior or function in morphogenesis.

8. Cell and Tissue Behavior as Related to Gastrulation

There have been many attempts to learn how cells bring about gastrulation by studying their behavior in culture. This method has the advantage of exposing cells of the opaque amphibian embryo to observation and the disadvantage of the uncertain relationship between the behavior observed in an artificial culture environment and that occurring *in vivo*.

8.1. The "Surface Coat" and Epithelial Cell Behavior

A number of early investigators thought that the amphibian embryo has a tough, elastic surface layer that gives it special mechanical properties (see Rhumbler, 1902; Holtfreter, 1943*a*). Holtfreter (1943*a,b*) investigated the physical properties of this "surface coat" and its function in gastrulation and concluded that it is elastic and has the capacity to contract. He guessed it to be proteinaceous and to possess a submicroscopic fibrillar structure. Holtfreter's view of its function was that "it amalgamates the peripheral cells into a plastic supracellular unit. Any displacement of individual cells will be reflected with the whole system." (Holtfreter, 1943*b*). According to Holtfreter, the surface coat has a non-adhesive outer surface, but cells such as the lateral mesoderm lose the surface coat and become sticky when they leave the superficial layer of the gastrula and migrate into the interior. Later they would regain the surface coat and form a mesentery around the gut (Holtfreter, 1943*a*). Surface cells could be made to lose the coat by culturing them in the tadpole coelom, and, conversely, deep cells could be made to form a coat by culture in dilute saline. The surface coat is broken down by pH >9.0, hypertonicity, monovalent cations (Na^+, K^+, and Li^+), and anoxia; it is stabilized by acidity, hypotonicity, and Ca^{2+} ions (Holtfreter, 1943*b*). Holtfreter argued that the surface coat, with its lack of adhesiveness, defines the polarity of the superficial cells, and he showed this polarity to be reversible: After a dilute alcohol or KCN treatment, ectoderm fragments will form vesicles with the coated surface inside rather than out; after a few days the polarity of the cells reverses and the exterior ends are again "coated" and nonadhesive.

The "surface coat" described by Holtfreter probably does not exist as a structural entity. Ultrastructural studies (Baker, 1965; Balinsky, 1961; Perry and Waddington, 1966; Luckenbill, 1971) have failed to show an extracellular, syncytial structure that might have the mechanical properties of the surface coat. A layer of material, lacking definite structure, was seen by Perry and Waddington (1966). This layer may be equivalent to the mucopolysaccharide coat found by Bell (1958, 1960) and the cell surface material seen in SEMs by Moran and Mouradian (1975), but it is not likely to be the "surface coat" described by Holtfreter.

The significance of Holtfreter's observations on the surface coat is not lessened by the failures to confirm its existence. Holtfreter's characterization of the "surface coat" makes it clear that he was describing the behavior and properties of the superficial epithelial sheet of cells. The mechanical properties of the surface coat are, in fact, due to the apical cytoskeleton and circumapical junctional complex characteristic of epithelia (see Section 2.2 and Chapter 4). His observations describe the loss and regaining of the epithelial character in the course of development, its acquisition by nonepithelial, deep cells, and change in the polarity of epithelial cells in culture (Holtfreter, 1943a,b). Further investigation of these changes in the embryonic amphibian epithelium would further our understanding of how epithelial sheets are made, how they are altered, and how they function in development.

8.2. Motile Behavior of Individual Cells in Culture

From the earliest days of experimental analysis of gastrulation, gastrula cells were cultured in the hope that they would reveal what could not be seen in the opaque embryo (Roux, 1894, 1896). Roux thought that gastrula cells cultured in egg white and saline showed positive and negative chemotactic responses, a conjecture that was not substantiated by timelapse cinemicrography (Kuhl, 1937; Voigtländer, 1932). Holtfreter (1943a,b, 1944) described several basic types of motile behavior of cultured gastrula cells. He used short treatments with alkaline (pH 10) or calcium-free medium and mechanical isolation to obtain individual cells. Many isolated, unattached cells showed "circus" or "limnicola" movement, named after the protozoan *Amoeba limnicola*, which shows this behavior (see Rhumbler, 1902). A hemispherical bleb forms on one side of a rotund cell and then proceeds around the cell on the order of a micrometer per second (Kageyama and Sirakami, 1976), broadening as it goes and usually disappearing into the cell profile after a good part of a full turn. The protrusions originate as hyaline structures, but yolky cytoplasm may intrude as the turn develops. Limnicola is shown by cultured early embryonic cells of amphibians (see Holtfreter, 1943b, 1944, Kageyama and Sirakami, 1976; Satoh et al., 1976; Johnson, 1976b; Johnson and Adelman, 1981) and fish (see Trinkaus, 1973; Fujinami, 1976). The frequency of cells expressing it increases during gastrulation (Satoh et al., 1976; Johnson and Adelman, 1981), and it occurs in fewer cells or not at all in arrested gastrulae of interspecific *Rana* hybrids (Johnson and Adelman, 1981).

There are several reasons for believing limnicola may have nothing to do with gastrulation. In culture, limnicola does not result in translocation beyond pushing packed cells apart. Holtfreter (1943a,b) interpreted it as artifact due to loss of contact with other cells. Indeed, morphological features characteristic of limnicola are not found *in vivo* (see Keller and Schoenwolf, 1977; Keller, 1980), but they can be seen after several minutes at freshly exposed edges of tissue in commonly used culture media (R. E. Keller, unpublished observations).

Gastrula and neurula cells also show several types of polarized, wormlike movements in which the cortical region of the cell forms a stable relationship to the substratum and the cell advances by cytoplasmic flow through the central core of the cell. These include the vermiform type of movement (see Holtfreter, 1943b; Satoh et al., 1976), a creeping locomotion of endodermal cells of *Cynops pyrrhogaster*, the Japanese newt (Kubota, 1981), and perhaps the "tug cells" described by Sirikami (1959). It is not known whether these movements occur *in vivo*, and if so, how they function in gastrulation.

Under the right conditions, cultured gastrula cells appear to show behavior that resembles what is thought to occur *in vivo*. Mesodermal cells of *Xenopus*, cultured on collagen substrata in a serum-containing medium, move at about 4 μm/min by extension of a large lamellipodium and continuous flow of the cell body into the lamellipodium (Nakatsuji and Johnson, 1982). The cells remain rotund during migration, and the lamellipodium is usually led by fine filopodia. They are similar in appearance to some of the mesodermal cells seen inside the gastrula by SEM (Nakatsuji, 1975*b*; Karfunkel, 1977; Keller and Schoenwolf, 1977) and show contact paralysis of protrusive activity (Nakatsuji and Johnson, 1982; see Johnson, 1976*a*). Leblanc and Brick (1981*a*) also made similar observations on the formation of filiform and lamelliform protrusions, adherance, and spreading of blastula and gastrula cells in culture (see also Stanistreet and Smith, 1978). External calcium is necessary in order for cells to make filiform and lamelliform protrusions (LeBlanc and Brick, 1981*b*), whereas it is not necessary for the blebbing and limnicola movements (Holtfreter, 1943*b*; M. S. Cooper and R. E. Keller, unpublished observations).

Cultured amphibian cells are capable of protrusive activity before gastrulation. The percentage of cells from the ectodermal animal cap and from the mesodermal marginal zone that show lobopodial activity in culture rises slowly at the onset of gastrulation to one-third or more of the cells by the late midgastrula stage (Satoh *et al.*, 1976). Cultured cells show surface activity as early as the mid-blastula stage (Stage 8) (Newport and Kirschner, 1982). Translocation and adhesiveness of cells to the substratum and to one another, in culture, increases near the onset of gastrulation (Johnson, 1970). However, there is no evidence that any of these behaviors occur *in vivo*; they may be products of the culture conditions. In the sea urchin (Gustafson and Wolpert, 1967) and in fish (Trinkaus, 1963, 1973), where cells can be seen *in vivo*, protrusive activity does occur before cells make stable contacts with the substratum and begin to translocate.

The onset of protrusive activity and translocation may be related to physical changes in the cell cortex or cell surface. Such changes may be indicated by alterations in lectin binding and cell surface charge. Johnson and Smith (1976) found that blastula cells would not show capping of fluorescein-labeled Con A, whereas gastrula cells would. Gadenne *et al.* (1984) have shown that an increased rate of patching and capping of Con A receptors at the onset of gastrulation is accompanied by an increase in the lateral mobility of the receptor molecule and also in the lateral mobility of the plasma membrane lipids. Con A receptors appear to be associated with cell–substrate adhesions in tissue culture cells (Chen and Singer, 1982), and they are found between amphibian blastomeres (O'Dell *et al.*, 1974). All blastula cells have the same cell surface charge, but surface charge increases, particularly in the dorsal lip region, with the onset of gastrulation (Schaeffer *et al.*, 1973*a,b*). These changes in physical properties may be related to the increased capacity of cells to make adhesions (Johnson, 1969, 1970, 1972) or to generate the protrusive activity involved in gastrulation. Fish gastrula cells, which translocate, are more deformable than blastula cells, which do not (Tickle and Trinkaus, 1973); corresponding measurements should be made on amphibians.

Relatively little is known of the cytoskeleton of amphibian gastrula cells despite the early interest in "submicroscopic fibrils" (Holtfreter, 1943a) and an oriented cytoskeleton (Waddington, 1942) as the basis of polarized cell behavior. The superficial epithelial cells have a subapical system of actin-containing microfilaments (see Perry and Waddington, 1966; Sanders and Zalik, 1972; Perry, 1975), and the deep mesodermal cells also have typical microfilaments (Nakatsuji, 1976). Cytochalasin B stops gastrulation, suggesting that proper microfilament organization is necessary for gastrulation (Nakatsuji, 1979); this is expected, considering the ubiquity of actomyosin systems in metazoan morphogenesis. As a direct effect or through its effect on the cytoskeleton, cytochalasin B appears to decrease gastrula cell adhesion and surface charge (Schaeffer et al., 1973c). Agents disrupting microtubules appear to have no immediate effect on gastrulation (Nakatsuji, 1979). Generation of force in amphibian gastrulation is undoubtedly based largely on an actomyosin system, but it remains to be seen whether or not the gastrular cytoskeleton shows subtleties of organization and regulation that are unique to the process of gastrulation. The gastrula cytoskeleton may impart a mechanical toughness not found in the blastula stage. Waddington (1942) found the former to be more resistant to disruption by surface tension than the latter. Sanders and Zalik (1972) noted that it is only at the gastrula stage that large tracts of microfilaments appear in thin sections. Much more is known about the cytoskeleton of earlier stages of development, both in terms of structure (Selman and Perry, 1970; Luckenbill, 1971; Singal and Sanders, 1974b; Perry, 1975; Franke et al., 1976; Campanella and Gabbiani, 1980; Columbo et al., 1980; Franz et al., 1983; Gall et al., 1983) and regulation (Gingell, 1970; Schroeder and Strickland, 1974; Hollinger and Schuetz, 1976; Clark and Merriam, 1978; Meeusen et al., 1980; Ezzel et al., 1984). Analyses of similar depth should be done on gastrulae.

8.3. Holtfreter's Concept of Tissue Affinity

Holtfreter (1939) explanted various tissues of the gastrula or neurula, alone or in combination with one another, and observed stereotyped responses that led him to the concept of tissue affinity as a force in morphogenesis. According to this idea, cells develop affinities for one another; these can be positive or negative and can change in sign during the course of development. They result in fusion of explanted tissue masses, the separation of tissue masses, and the spreading of tissues on one another to form specific configurations that, in some cases, mimic morphogenetic events in gastrulation, neurulation, and organogenesis. For example, fragments of endoderm and ectoderm of the early gastrula initially adhere to one another and form a single unit, but later they develop a negative affinity and separate from one another (self-isolation) (Fig. 28a). However, if mesoderm is added to the culture, ectoderm and endoderm minimize their contact with one another but both maintain it with the mesoderm; thus both retain positive affinity for mesoderm (Fig. 28b). If the initial conditions are suitable (an explant of ectoderm surrounding masses of endo-

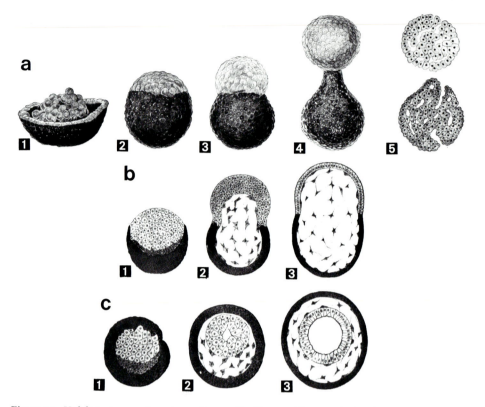

Figure 28. Holtfreter's experiments on tissue affinities are illustrated. Combination of early gastrula ectoderm and endoderm (a,1) is followed first by fusion of the tissue masses (a,2) and then self-isolation of the two from one another (a,3–5). Self-isolation of ectoderma and endoderm is prevented if mesoderm is included in the explants (b,1–3), and a tissue array mimicking that of the embryo is produced if the ectoderm is allowed to cover the mesoderm and endoderm (c,1–3). (From Holtfreter, 1939, with permission.)

derm and mesoderm), a concentric array reminiscent of the organization of the gastrula will result (Fig. 28c). Moreover, when dissociated, individual cells of two different origins would initially reaggregate, with the two populations intermingled, and later sort out according to tissue type (Holtfreter, 1943c; Townes and Holtfreter, 1955). Selective affinity may operate *in vivo*: [^{3}H]thymidine-labeled cells injected into the blastocoel of *Pleurodeles* embryos were later found preferentially in the tissue of their origin—ectoderm in ectodermal derivatives, endoderm in gut, and mesoderm in notochord and somites (Boucaut, 1974).

The specific behavioral responses observed by Holtfreter (1939) were interpreted to mean that cells could choose their associations, and this notion was called **tissue affinity.** The cellular behavior involved in making such choices was never observed, and in absence of facts, various ideas arose to explain what cellular properties could cause such tissue behavior. These ideas were narrower than the original concept of tissue affinity, which did not exclude any of the many cellular properties that could possibly stabilize one cell–cell association over another. The first of these was the notion of **selective adhesion** (Townes and Holtfreter, 1955), which held that cells make absolute or graded choices in making adhesions. Steinberg (1964, 1970) formulated the hypothesis that tissue combinations are driven by *differences* in adhesion between cells of the various types to form, at equilibrium, an array of homotypic and heterotypic adhesions that represents a minimum free-energy state. Steinberg and Kelland (1967) showed that combination of amphibian deep ectoderm, mesoderm, and endoderm will form concentric arrays in that order with ectoderm innermost. This is the opposite of the intact gastrula. But if "coated" (epithelial) ectoderm was used, the order of tissues in the concentric array was reversed, because of the nonadhesive outer surface of the ectoderm, and thus it resembled the gastrular pattern. Phillips and Davis (1978) argued that embryonic tissues could temporarily behave like immiscible liquids and flow between one another based on their surface tension relationships. Thus, lateral and ventral mesoderm would "flow" between the overlying ectoderm and the endoderm during gastrulation if the surface tension between ectoderm and endoderm is greater than the sum of that between mesoderm and ectoderm and mesoderm and endoderm (Phillips and Davis, 1978). Their measurements of aggregate surface tension with the medium are consistent with this hypothesis. Interestingly, the dorsal mesoderm of the gastrula does not appear to behave as a liquid. This fact argues that the convergent extension behavior characteristic of this region is driven by internal forces rather than forces generated at interfaces with other tissues (see Phillips and Davis, 1978; Davis, 1984; Phillips, 1984).

There is no direct evidence that sorting out and tissue rearrangements observed by Holtfreter and others are a function of selective or differential adhesion alone, or whether such behaviors result from more complex contact-induced reactions of the cell that might involve additional cellular systems. Differences in adhesion or the forces necessary for deadhesion have never been satisfactorily measured (see Chapter 5 in Trinkaus, 1984). Indeed, it seems difficult to define adhesion as an independent property of a cell. The notion that cells have cell surface components that behave like glues to yield differences in adhesion, which then result in some sort of self-assembly, is probably an oversimplification of a more complex process in which internal, regulatory processes and the cytoskeleton play a large role (see Harris, 1976; Curtis, 1978; Steinberg, 1978; Chapter 4 in Trinkaus, 1984).

8.4. Ionic Conditions and Gastrula Cell Behavior

Gastrula cell behavior in culture is difficult to interpret because it is a function of culture conditions. It would be useful to know what ionic conditions would allow normal gastrula cell behavior.

Development of intact gastrulae is sensitive to the ionic and osmotic conditions in the external medium (Jenkinson, 1906; Wilson, 1897; Morgan, 1906a), probably due, initially, to effects on the stability of the superficial epithelium. Intact embryos without their vitelline envelopes often gastrulate abnormally and usually exogastrulate when cultured in full to double-strength Holtfreter solution (Holtfreter, 1933b). Holtfreter (1943a,b) described the sensitivity of the surface coat (epithelium) to high salt and pH. Half-strength Ringer's solution or Holtfreter solution (which are roughly equivalent in tonicity at about 120 mOsM) result in violent contraction of the apices of the epithelial cells, ingression of these cells, destruction of the epithelium, and partial exogastrulation in *Xenopus* (Keller, unpublished data; see Section 9.3). In contrast, exposed deep cells require use of full-strength Holtfreter solution, or its equivalent (Holtfreter, 1943a,b, 1944; Shapiro, 1958). Thus, microsurgery must be done in standard solution, and the embryos must be returned to dilute (10–20% Holtfreter's) solution after healing (see Hamburger, 1960). The capsular fluid, lying beneath the vitelline envelope, is normally dilute saline of near-neutral pH (Richards, 1940; Stableford, 1949).

The blastocoelic conditions allowing gastrulation were investigated by Shapiro (1958), who perfused the blastocoel with solutions varied systematically in pH and tonicity. A saline with tonicity of 1.2–1.5 that of Holtfreter solution gave the best gastrulation. Low (6–7) and high (8–9) pH supported gastrulation much better than neutral pH (7–8). Holtfreter selected the tonicity of his "standard solution" at 0.38% salt solution on the basis of its ability to support cell division, movement, and differentiation (see Holtfreter, 1943a). Gastrula cells are sensitive to osmotic effects; Nakatsuji (1979) found that injected hypertonic (2.0 M) sorbitol solutions prevent gastrulation.

Another approach is to measure the ionic concentrations in the blastocoel. Blastocoel pH was measured at 8.7 (Buytendijk and Woerdeman, 1927), 8.3 to 8.7 (Stableford, 1949), and 8.4 (Turin and Warner, 1980; Gillespie, 1983). The calcium concentration was measured at 0.5 mM (Stableford, 1967) and 1.0 mM, using an ion-selective electrode in *Xenopus* (Rink *et al.*, 1980). Gillespie (1983), also using ion-selective electrodes on *Xenopus laevis* and *Ambystoma mexicanum* embryos cultured in Ringer's solution, measured the ionic composition of the blastocoel fluid as shown in Table I.

In *Xenopus*, the sodium concentration decreases from 91 in the midgastrula to 75 mM at the end of gastrulation, whereas potassium rises from 4 to 17 mM in the same period. Calcium falls from 1.5 mM in the blastula to 0.5 mM

Table I. Blastocoel Ion Concentrations

Ion	*Xenopus* (mM)	*Ambystoma* (mM)
Na^+	91–75	93
K^+	4–17	7–2
Cl^-	59	59
Ca^{2+}	1.5–0.5–1.8	—
pH	8.4	—

at the onset of gastrulation and then rises again to 1.8 mM in the late neurula. Tuft (1962) found the blastocoel fluid at the early blastula stage to be 190 mOsm, based on freezing point depression.

Stableford (1949) measured the pH and surface tension of the capsular fluid and blastocoel fluid in order to examine the contention of Rhumbler (1902) and Holtfreter (1943b) that bottle cells migrate toward the interior in response to a surface tension or pH gradient. A pH gradient of about 1 pH unit (inside higher) exists, but removing the blastocoel roof (presumably destroying the gradient) does not stop gastrulation (Morgan, 1906a; Stableford, 1949; Lewis, 1952; Holtfreter, 1933b).

The concentrations of common ions in media commonly used for culture of early amphibian cells differ greatly from one another as well as from the values for blastocoel fluid measured by Gillespie and others. Generally, media yielding the best movement of gastrula cells in culture have high pH and low calcium (Kubota and Durston, 1978; Nakatsuji and Johnson, 1982). The culture medium used in the open-faced explants approximates the composition of blastocoel fluid as measured by Gillespie, and it allows normal morphogenesis and differentiation of notochord and somites in culture (see Section 5.4) (Keller et al., 1985).

9. Analysis of Abnormal Gastrulation

Abnormal gastrulation occurs under a variety of conditions, and its analysis could contribute to our knowledge of the normal process.

9.1. Development of Interspecific Hybrids

Interspecific hybrids in the genus *Rana* show various degrees of normal development—from fertile hybrids to early lethality, depending on the species used in the cross and which species supply the maternal and paternal gametes (see Moore, 1955; Subtelny, 1974; Elinson, 1975). Hybrids that arrest early in gastrulation, or perhaps never really begin gastrulation, show abnormal cell spreading on substrata and reaggregation in culture, reduced intercellular contact in the embryo, increased rates of dissociation in the absence of cations, abnormal behavior of tissue explants, and reduced extracellular matrix material (see Johnson, 1969, 1970, 1972, 1977a–d, 1978, 1981; Jumah and Stanisstreet, 1980).

Moore (1946) made a detailed and valuable analysis of the *Rana pipiens* female × *Rana sylvatica* male cross, which attempts gastrulation, whereas the reciprocal cross fails to begin gastrulation and cytolyzes when normals have completed gastrulation (unilateral lethality). The blastopore of the hybrid forms 3.5 hr later and 30° farther vegetally than in normals. Thinning of the hybrid blastocoel roof is slower but appears equivalent in all other respects to that of normal gastrulae. The blastocoel appears to swell, and the hybrids show an abnormal increase in volume, but the large blastocoel is not invaded by involuted cells. The blastoporal lip stalls at about 150° from the animal pole when the normals have reached almost 180°, and the archenteron depth reaches about 35°, rather than nearly 180° at the end of gastrulation. Convergent extension does not occur, at least not to the degree that it does in normals; the dorsal side remains short, and the blastopore does not constrict as it does in normals. Moore's interpretation is that convergent extension is abnormal, whereas epiboly is normal; thus the increase in size of the embryo and lack of movement of material inside. But the uninvaded blastocoel might also suggest that the migration of the leading mesodermal cells is also abnormal. Alternatively, this migration may not proceed far unless the circumblastoporal convergent extension occurs. In any case, shortened axial structures (notochord, neural tube and archenteron) develop on the dorsal side.

Gregg and Klein (1955) explanted tissues of this hybrid in various combinations and compared their behavior to that of explants of normal embryos, in terms of the stereotyped tissue interactions expected from the work of Holtfreter (see Section 8.3). The dorsal (involuting) marginal zone of the hybrids failed to undergo the convergent extension normally characteristic of this region but instead showed spreading behavior. These authors conclude that the major defect in this hybrid is the lack of convergent extension of the dorsal, posterior (notochordal) mesoderm. Thus, the behavior seen in the hybrid gastrulae may reflect what happens when convergent extension fails. The initial invasion of the blastocoel, the short archenteron, and the short neural tube would be expected if bottle cell formation and migration of the leading mesodermal cells is normal. But these processes are unlikely to be sufficient to pull the material around the blastoporal lip if convergent extension fails. Temporary development of an annular groove, called the blastocoelic furrow, on the exterior of the embryo at the level of the leading edge of involuted material (Moore, 1946) may represent folding due to increased traction in absence or convergent extension.

Further analysis of this and other hybrids is complicated by the fact that the nature of the perturbation is not defined and is probably broader than one would wish. Hybrid embryos show some metabolic deficiencies (Gregg, 1957). The existence of regional differences in metabolism and sensitivity to metabolic inhibitors (see reviews by Boell, 1956 and Gregg, 1957) leave open the possibility that localized failure of morphogenesis in hybrids may be due to a general metabolic deficiency having a differential effect in specific regions of the embryo.

9.2. Development of Axis-Deficient Embryos

There is a critical period before first cleavage of *Xenopus* in which determination of the axial structures of the body can be disturbed by ultraviolet radiation of the vegetal pole (Grant and Wacaster, 1972; Malacinski *et al.*, 1977; Scharf and Gerhart, 1983) and by agents that affect the cytoskeleton, including cold and hydrostatic pressure (Scharf and Gerhart, 1983). These treatments result in a dose-dependant, graded series of increasing axis deficiency, beginning with the loss of head structures, followed by trunk and tail structures. The extreme case is a "barrel-shaped" embryo, radially symmetrical around the original animal–vegetal axis, and lacking all the dorsal, axial structures. These embryos gastrulate with the simultaneous formation of a symmetrical, circular blastopore (see Fig. 2, Scharf and Gerhart, 1983; Fig. 8, Gerhart *et al.*, 1983; Scharf *et al.*, 1984).

These embryos may be examples of gastrulation in absence of the convergent extension behavior of the dorsal, posterior mesoderm. Convergent extension of the dorsal, posterior mesoderm never occurs in these embryos, and the dorsal, axial structures normally derived from this mesoderm never form. Instead, all the mesoderm in these embryos resembles the nonconvergent lateral and ventral mesoderm of normal embryos. Their movements appear to be similar, though they do not move as far, and they differentiate into tissues similar to that seen in the "Bauchstück" (belly pieces) developing from isolated ventral blastomeres (Spemann, 1902) and halves of embryos isolated with ligatures (Fankhauser, 1948). In contrast, embryos exposed to D_2O before the critical period develop a precocious, circular blastoporal lip and gastrulate symmetrically, with excessive extension of the circumblastoporal region (see Fig. 8, Gerhart *et al.*, 1983). Thus, by manipulating the axis determination processes, whatever they are, before first cleavage, Scharf and Gerhart have been able to change the proportion of the mesodermal torus that participates in convergent extension from none to nearly all of it.

9.3. Exogastrulation and Pseudogastrulation

In several circumstances, the involuting marginal zone will not turn inside but move in the opposite direction, away from the animal region, and form a total exogastrula (Holtfreter, 1933b) (Fig. 29). Vital dye marks show that the normal movements of convergent extension occur but are directed externally instead of internally (Holtfreter, 1933b). Removal of the vitelline envelope increases the probability of exogastrulation, probably because it normally forms a mechanical barrier to extrusion of the vegetal region. Generally, eggs of urodeles (e.g., the *Ambystoma* species), are more prone to this behavior than those of *Xenopus* and anurans with similar eggs.

It is fashionable to call almost any gastrula with a wrinkled animal cap and protruding yolk plug an exogastrula, but this state can arise in numerous ways, and unless the embryos have been filmed, the degree to which they represent misdirected, normal gastrulation movements, as implied by the term "exogastrulation," and the degree to which they represent general derangement of cellular behavior, is a matter of speculation. For example, time-lapse cinemicrography of *Xenopus* "exogastrulae" produced by high salt concentrations shows powerful contractions of the apices of superficial cells at the animal pole, tearing of intercellular junctions, and ingression of the cells to form a

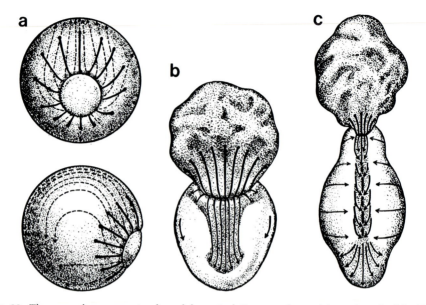

Figure 29. The normal movements of urodele gastrulation are shown (a), as described by Vogt (1929*b*). These movements are directed outward, rather than inward, and the endoderm and mesoderm (below) separate from the animal cap ectoderm (above) in the course of total exogastrulation (b,c). The arrows indicate the morphogenetic movements. (From Holtfreter, 1933*b*, with permission.)

thickened blastocoel roof (R. E. Keller, unpublished data; see Smith *et al.*, 1976). In these cases, the convergent extension machinery is not reoriented, and, if it functions at all, it pushes outward instead of inward and attempts to constrict the central yolk mass, usually without success.

Frog oocytes kept for several days in tap water or Holtfreter solution show expansion of the pigmented animal cap vegetally over the unpigmented hemisphere and what appears to be an invagination at the vegetal pole, complete with yolk plug (Holtfreter, 1943*a,b*; see Malacinski *et al.*, 1978). This "pseudogastrulation" suggested to Holtfreter that some aspects of the gastrulation movements are controlled by a cytomechanical system that is not dependent on segmentation of the egg into individual cells. Lillie (1902) described what may be a similar case of development of *Chaetopterus* larvae without cleavage.

Pseudogastrulation probably represents an abnormal activity of the cortical and subcortical cytoskeleton of the egg, perhaps due to its inability to regulate intracellular calcium levels. Pseudogastrulation is prevented by cytochalasin B, colchicine, and vinblastine and is dependent on the continuous presence of high NaCl or KCl but not Ca^{2+} (Malcinski *et al.*, 1978).

The idea that pseudogastrulation represents the activity of a preorganized cytoskeleton that is poised to carry out an orderly supracellular distortion, such as epiboly, pending some regulatory stimulus, deserves further study. Dorfman and Cherdantsev (1977*a,b*) contend that clinostat rotation of fertilized frog eggs results in abnormal distribution of local cytoplasms, each capable of specific morphogenetic movements, and that these form local, elementary patterns of morphogenesis. However, there is no direct evidence that the cytoskeleton is organized in a supracellular, global pattern. In addition, the movements of pseudogastrulation movements differ significantly from those of normal gastrulation (Holtfreter, 1943*a*).

10. The Diversity of Amphibian Gastrulation

The view of amphibian gastrulation described thus far is accurate for those few species of amphibians that are commonly available and popular with developmental biologists, but it is not representative of the Amphibia. The early morphogenetic movements of closely related organisms may differ much more than previously thought, and those of unrelated species may be very similar (Ballard, 1976, 1981; Ballard and Ginsberg, 1980). Among amphibians, species in the genera *Rana, Bufo, Bombina, Discoglossus,* and *Hyla* with small eggs similar to those of *Xenopus* also gastrulate in similar ways, except that in some mesoderm appears to occupy the superficial layer (see King, 1902*b*; Vogt, 1929*b*). Most of the commonly studied urodeles—including the California newt, *Taricha torosus,* the spotted newt, *Diemyctylus viridescens,* the European newt, *Triturus, Pleurodeles waltlii,* and most species of *Ambystoma*—appear to be much alike and differ little from the Mexican axolotl, *Ambystoma mexicanum.*

But even the well-known amphibians use several different mechanisms to bring their mesoderm into its definitive position and form notochord, and there are many unresolved questions about these differences. Also, there are many lesser known amphibians that gastrulate in ways that are different from these well-known species. Serious analysis of the cellular behaviors underlying these variants has only begun, but what has been done suggests something of significance for developmental and evolutionary biologists alike: There is stability in the basic macroscopic patterns of distortion and mechanical aspects of gastrulation, and there appears to be a common set of basic cellular, morphogenetic behaviors, but the particular combination of these cellular behaviors used in a given species varies greatly, even between closely related amphibians and at different axial levels of the same organism.

10.1. Variation in Notochord Formation

The notion considered above is supported by the variation in notochord formation. Sections 3.3 and 3.4 described two patterns of notochord formation. In *Xenopus,* the entire mesoderm lies in the deep region and the notochord forms as clefts appear between the notochordal and somitic mesoderm in the late gastrula stage. The cells of the notochord become flattened and closely applied to one another's surfaces and acquire lamelliform protrusions that distinguish them from the somitic mesodermal cells (Nieuwkoop and Florshutz, 1950; Youn *et al.*, 1980; Keller, 1984). In contrast, the notochord of the axolotl and similar urodeles forms by ingression of a single layer of epithelial cells to form a multilayered mesenchymal cell population in the deep region (King, 1903; Kingsbury, 1924; Ruffini, 1925; Vogt, 1929*a,b*; Lofberg, 1974; Brun and Garson, 1984; Lundmark, 1986). Moreover, in the anurans *Rana* and *Bufo*, a combination of these two patterns is used. The anterior end of the notochordal mesoderm lies in the deep region and the notochord forms as it does in *Xenopus*, but the posterior notochordal mesoderm lies on the surface, as well as in the deep region, and appears to be removed from the surface as it is in the axolotl (King, 1903). Species of anura other than *Xenopus* that have superficial mesoderm and appear to form the notochord in the *Bufo–Rana* fashion (see Vogt, 1929*b*) also show the powerful convergent extension behavior (see Holtfreter, 1938*b*).

In all these organisms, convergent extension occurs regardless of whether the cells are organized as an epithelium or mesenchyme, or whether they are deep or superficial. The evidence at hand supports the notion that convergent extension and definitive structure of the notochord are common and basic features of amphibian gastrulation, but that there are several patterns of cellular organization and behavior that have been used, in various combinations, in generating this pattern of morphogenesis.

10.2. Gastrulation in Amphibians with Yolky Eggs

Yolk content and egg size are parameters that vary greatly between species and even within them. Intrapopulation variation in egg size (presumably due mostly to differences in yolk content) is related to clutch size, developmental rate, time-to-feeding, and size of larvae at hatching in several *Ambystoma* species; these parameters have implications for offspring fitness and reproductive strategy (see Kaplan, 1980). Amphibians that have aquatic early embryonic stages usually lay large numbers of small eggs (1–3 mm diameter) with moderate yolk content; these are the animals that have been most commonly studied by developmental biologists. However, many other, less-studied amphibians lay a few very large, yolky eggs in damp terrestrial sites, brood the eggs, and may pass the larval stages in the egg membranes, or show direct development into the adult form (see del Pino and Humphries, 1978; del Pino and Escobar, 1981; Salthe and Duellman, 1973).

Analysis of gastrulation in several large-egged amphibians shows that the supposedly conservative morphogenetic movements of gastrulation display a remarkable agility in accommodating the changes in egg organization and yolk content that are part of the evolution of different reproductive strategies. With increasing yolk, the pattern of cleavage planes changes, and the vegetal hemisphere cleaves slowly and forms large blastomeres. The moderately yolked eggs of *Cryptobranchus allegheniensis* (Smith, 1912), the gigantic salamander *Megalobactrachus maximus* (Ishikawa, 1905), *Necturus*, and *Spelerpes bilineatus* (Goodale, 1911*b*), all show some accomodation to greater yolk content in their

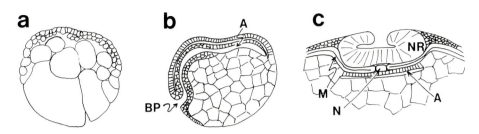

Figure 30. Early development of *Desmognathus fusca* involves formation of a field of micromeres lying above a mass of large endodermal cells during cleavage (a). During gastrulation, a blastopore (BP) forms within the micromeres, and from the micromeres a long, narrow archenteron and dorsal, axial structures form on one side of the endodermal mass (b,c). a, section of a blastula; b, saggital section of a gastrula; c, transverse section of neurula. A, archenteron; M, mesoderm; N, notochord; NR, forming neural tube. (Redrawn from Hilton, 1909, with permission.)

cleavage pattern. But their gastrulation differs little from that of small-egged amphibians except that the blastopore forms relatively high and is slow in closing. In the more heavily yolked, large egg (4 mm in diameter) of *Desmognathus fusca*, the gastrulation movements, as well as the cleavage pattern, are atypical (Wilder, 1899, 1904; Hilton, 1909). The animal half cleaves rapidly, and the vegetal half slowly. The blastocoel forms but is obliterated. The result is a sheet of small, animal micromeres forming a thin monolayer over a large mass of large endodermal cells (Fig. 30a). A small crescentic blastopore forms near one edge of the small-celled blastoderm, and a long, narrow archenteron and associated mesoderm involutes to form an embryonic shield, perched on one side of the large mass of endodermal cells, which do not participate in morphogenesis (Fig. 30b,c).

Similar gastrulation is found among the anurans with large, yolky eggs. Del Pino and Elinson (1983) studied the development of several egg-brooding hylid frogs—*Gastrotheca riobambae*, *Gastrotheca plumbea*, and *Eleutherodactylus coqui*. The egg of *G. riobambae* cleaves completely and forms a blastula consisting of large deep cells, surrounded by a thin superficial epithelium, and containing a blastocoel with a thin translucent roof. A blastopore forms and cells involute to form an embryonic disc, composed of small cells and comprising a small area on one side of the egg; involution and formation of the disk appear to be radially symmetrical. After closure of the blastopore, the part of the embryonic disk next to the blastocoel elongates to form the embryonic body. The embryo, perched on one side of the large egg, looks much like that of some teleost fish, reptiles, or chicks. Del Pino and Elinson point out that the chief difference between gastrulation in this animal and *Xenopus* is the fact that elongation of the embryo is coincident with gastrulation in *Xenopus*, whereas it follows gastrulation in *Gastrotheca*.

It is tempting to equate the late elongation movements in *Gastrotheca* to the exercise of convergent extension in *Xenopus*. But it seems unlikely that symmetrical, animalward migration could close the blastopore of *Gastrotheca*; the same applies to the axis-deficient embryos. Is convergent extension exercised in two spatial patterns, one symmetrical and the other biased to the dorsal sector? In any case, the modification of the timing and degree of convergent extension may be a major element in tailoring early morphogenetic movements to differences in yolk content that necessarily accompany changes in amphibian reproductive stratagies. But such tailoring is not obligatory in eggs of the size and yolk content of *G. riobambae*. *Eleutherodactylus coqui*, which has reproductive habits and an egg similar to that of *G. riobambae*, gastrulates without an embryonic disk in a manner similar to that found in *Xenopus* (del Pino and Elinson, 1983).

Comparative analysis of amphibian gastrulation presents a conflict with the basic assumptions inherent in attempting to homologize gastrulation movements among vertebrates (see Pasteels, 1940; Waddington, 1952). For the sake of argument, comparison of morphogenesis could be done in terms of the macroscopic tissue distortions or in terms of the basic cellular processes driving these distortions. Ballard (1976, 1981) has argued convincingly that there is little correlation of phylogenetic distance and similarity of morphogenetic movements among vertebrates. For example, the sturgeon—a ganoid fish—forms a blastula very similar to that of *Xenopus*, and there is much more similarity between the gastrulation of this fish and *Xenopus* than there is between *Xenopus* and the axolotl (see Ballard and Ginsberg, 1980). Among vertebrates, the "gastrulation" process is so diverse in what it accomplishes that it is definable only in general terms, so general that the definition is not very useful (see Ballard, 1976, 1981). If the macroscopic features vary greatly, what about the basic cellular processes driving them? One might argue that early, basic processes must be conservative; to alter them in the course of evolution would require an impossible cascade of changes at following stages. But what appears to be important (and conserved) in amphibians is a workable mechanical design. Perhaps in our analysis of morphogenesis, we have forgotten that the primary issue is the generation of a pattern of unbalanced forces that produce ordered tissue distortion. The little comparative analysis of amphibian gastrulation done thus far shows that any of several basic cellular processes can be recruited to accomplish a given function, such as convergent extension, involution, and notochord formation, and that various combinations of cellular processes will accomplish the same end, depending on the geometrical and mechanical context. The variation in cellular organization in the process of notochord formation and the execution of involution by using cellular rearrangement in *Xenopus* and ingression and rearrangement in the axolotl are cases supporting this point. In our investigations, the cellular processes that we can expect to function in a given morphogenetic movement will not follow some universal program or rationale; they will be combinations of processes that happen to work.

11. Prospectus

Review of the investigations of amphibian gastrulation carried out over the past 100 years brings several general points to the fore that might serve us well in future investigations.

First, there is no substitute for knowing what cells actually do to bring about their movement. In an integrated morphogenetic system there are any number of ways that a cell can be displaced, and we know far too little about the relationship between cell behavior and mechanical stresses. Much of the good work arising out of Holtfreter's concept of tissue affinity has been taken to mean that cells differentiate cell surface properties that result directly in the movement of germ layers to their proper position by some sort of vague self-assembly process in which motile behavior plays only a general role. Thus, there is a common perception that the major issue in morphogenesis is learning what molecular constituents change in the course of morphogenesis. In fact, we have neither a homogeneous cell population nor a specific and well-understood morphogenetic behavior to analyze at the molecular level. To reiterate, Holtfreter (1943b) pointed out that gastrulation is "mainly the result of the predisposed arrangement of cells with a locally different kinetic behavior." The major issue at hand is to identify and characterize local patterns of cellular behavior that can be demonstrated to produce the macroscopic distortion observed.

Second, analysis of gastrulation at the molecular level must be directed at these homogeneous cell populations of defined and specific morphogenetic function. The evidence thus far only crudely defines such populations and more must be done, but it is clear that traditional regions, such as the "dorsal lip," are heterogeneous with respect to morphogenetic function, and their analysis is likely to yield little that is useful. The profitable molecular analyses will be those that focus on specific patterns of cell behavior in regions that show discrete, powerful, and autonomous distortion patterns, rather than on undefined cellular processes and on the gastrula as a whole.

Third, the evidence suggests that the forceful narrowing (convergence) and lengthening (extension) of cell populations by active intercalation of cells plays a greater role in the gastrulation of amphibians and probably in the morphogenesis of other metazoa as well. Gastrulation of many metazoans essentially involves the transformation of a disc into a cylinder (Fig. 31). This transformation may be performed by involution, as in the amphibians, or by invagination, as in the echinoderms. In either case, it has generally been analyzed in sectional view with the intent of relating change in cell shape—such as bottle cell formation—to bending of the cell sheet (Fig. 31a). Several facts suggest an alternative way of perceiving the problem. The dramatic change in shape of the bottle cells, in their sectional aspect, has a limited role in gastrulation of amphibians (Keller, 1981). Moreover, in the sea urchin the changes in cell shape assumed to occur at the base of the primary invagination could not be demonstrated to occur (Ettensohn, 1984a,b). Could it be that the important events occur in the plane of the sheet? There are several reasons to think this is true. First, Ettensohn (1984b) has indirect evidence that cell rearrangement occurs during the secondary (elongation) phase of archenteron formation. Such repacking may also occur during primary invagination and function in bringing it about. Second, one of the best-analyzed examples of evagination—the *Drosophila* imaginal disc—involves cell repacking in a disc to form narrower circumferential aspect and longer radii (see Fristrom and Rickoll, 1983). This is essentially a convergent extension, and it occurs by cell repacking. Third, it is

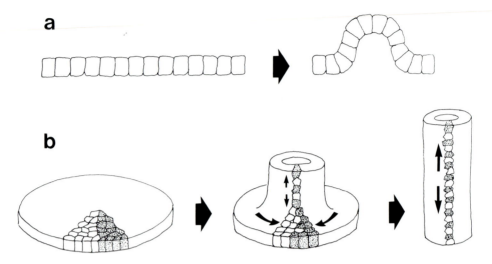

Figure 31. The traditional way of viewing formation of cavities by cell populations is in sectional view, wherein cell shape change is related to bending of the sheet (a). Analysis of amphibian gastrulation strongly suggests that a more appropriate view is the surface view, where cell repacking is the essential process in producing the deformation of the cell population (b). This holds true for epithelial sheets as well as for mesenchymal cell populations.

convergent extension by circumferential intercalation of cells in a plane tangential to the surface that has the major role in gastrulation of amphibians. Fourth, the sectional analysis ignores the fact that the principal distortion of the sheet involves progressive decrease in the circumference of the disc and an accompanying increase in radius—convergent extension (Fig. 31b). To some degree, all invaginations or evaginations in development, and many other distortions as well, involve convergent extension and most rely on it heavily. These observations argue for the abandonment of our preoccupation with the sectional view as the sole context in which to analyze gastrulation and suggest that planar tissue distortion by active intercalation of cells to form a longer, narrower array is perhaps more important.

ACKNOWLEDGMENTS.   I wish to thank Dave Erkenbrack, John Shih, and Paul Tibbits for their help in getting materials, the librarians at University of California for their diligent efforts in finding what was needed, and Cathy Lundmark, Stan Scharf, Mark Cooper, and especially Ann Sutherland for their corrections and comments.

References

Baker, P., 1965, Fine structure and morphogenetic movements in the gastrula of the treefrog, *Hyla regilla, J. Cell Biol.* **24:**95–116.

Balinsky, B. I., 1961, Ultrastructural mechanisms of gastrulation and neurulation, in: *Symposium on Germ Cells and Development*, pp. 550–563, Pallanza, Institut International d'Embryologie, and Fondazione, A. Baselli, Pallanza.

Ballard, W., 1955, Cortical ingression during cleavage of amphibian eggs, studied by means of vital dyes, *J. Exp. Zool.* **129:**77–97.

Ballard, W., 1976, Problems of gastrulation: Real and verbal, *Bioscience* **26:**36–39.

Ballard, W., 1981, Morphogenetic movements and fate maps of vertebrates, *Am. Zool.* **21:**391–399.

Ballard, W., and Ginsberg, A., 1980, Morphogenetic movements in acipenserid embryos, *J. Exp. Zool.* **213:**69–103.

Bautzmann, H., 1933, Über Determinationsgrad und Wirkungsbezichungen der Randzonenteilanlagen (Chorda, Ursegmente Seitenplatten und Kopfdarmanlage) bei Urodelen und Anuren, *Roux Arch.* **128:**666–765.

Bell, E., 1958, Removal of the surface coat with ultrasound, *Anat. Rec.* **131:**532.

Bell, E., 1960, Some observations on the surface coat and intercellular matrix material of the amphibian ectoderm, *Exp. Cell Res.* **20:**378–383.

Beloussov, L. V., 1978, Formation and cellular structure of cross-lines in axial rudiments of the amphibian embryo, *Soviet J. Dev. Biol.* **9:**105–109. [Transl. from *Ontogenez.* **9:**124–130.]

Beloussov, L. V., 1980, The role of tensile fields and contact cell polarization in the morphogenesis of amphibian axial rudiments, *Wilhelm Roux Arch.* **188:**1–7.

Beloussov, L. V., Dorfman, J. G., and Cherdantzev, V. G., 1975, Mechanical stresses and morphological patterns in amphibian embryos, *J. Embryol. Exp. Morphol.* **34:**559–574.

Bijtel, J. H., 1930, Beiträge zur Schwanzentwicklung der Amphibien, *Anat. Anz.* **71:**87–93.

Bijtel, J. H., 1931, Über die Entwicklung des Schwanzes bei Amphibien, *Wilhelm Roux Arch. Entwicklungsmech. Org.* **125:**448–486.

Bijtel, J. H., 1958, The mode of growth of the tail in urodele larvae, *J. Embryol. Exp. Morphol.* **6:**466–460.

Black, S., and Gerhart, J., 1985, Experimental control of the site of embryonic axis formation in *Xenopus Laevis* eggs centrifuged before first cleavage, *Dev. Biol.* **108:**310–324.

Black, S., and Gerhart, J., 1986, High-frequency twinning in double-centrifuged eggs of *Xenopus laevis, Dev. Biol.* (in press).

Bluemink, J. G., Tertoolen, L. G. J., Ververgaert, P. H. J., and Verkleij, A. J., 1976, Freeze-fracture electron microscopy of preexisting and nascent cell membrane in cleaving eggs of *Xenopus laevis, Biochim. Biophys. Acta* **443:**143–155.

Bluemink, J. G., and deLaat, S. W., 1973, New membrane formation during cytokinesis in normal and cytochalasin-B treated eggs of *Xenopus laevis.* I. Electron microscopic observations. *J. Cell Biol.* **59:**89–108.

Bluemink, J. G., and deLaat, S. W., 1977, Plasma membrane assembly as related to cell division, in: *The Synthesis, Assembly, and Turnover of Cell Surface Components* (G. Poste and G. L. Nicolson, eds.), pp. 403–461, Elsevier/North-Holland, Amsterdam.

Boell, E. J., 1956, Energy exchange and enzyme development during embryogenesis, in: *Analysis of Development* (B. H. Willier, P. Weiss, and V. Hamburger, eds.), pp. 520–555, W. B. Saunders, Philadelphia.

Boucaut, J.-C., 1974, Étude autoradiographique de la distribution de cellules embryonnaires isolées, transplantées dans le blastocoele chez *Pleurodeles waltlii* Michah (Amphibien, Urodele), *Ann. d'Embryol. Morphol.* **7:**7–50.

Boucaut, J.-C., and Darriberre, T., 1983, Presence of fibronectin during early embryogenesis in amphibian *Pleurodeles waltii, Cell Diff.* **12:**77–83.

Boucaut, J.-C., Darribere, T., Boulekbache, H. and Thierry, J.-P., 1984a, Antibodies to fibronectin prevent gastrulation but do not perturb neurulation in gastrulated amphibian embryos, *Nature* (Lond.) **307:**364–367.

Boucaut, J.-C., Darribere, T., Poole, T. J., Aoyama, H., Yamada, K. M., and Thiery, J. P., 1984b, Biologically active synthetic peptides as probes of embryonic development: A competitive peptide inhibitor of fibronectin functions inhibits gastrulation in amphibian embryos and neural crest cell migration in avian embryos, *J. Cell Biol.* **99:**1822–1830.

Brachet, A., 1903, Rechèrches sur l'ontogénèse des Amphibiens, Urodeles et Anoures (*Siredon piciformis, Rana temporaria*), *Arch. Biol.* **20:**1–243.

Bragg, A. N., 1938, The organization of the early embryo of *Bufo cognatus* as revealed especially by the mitotic index, *Z. Zellforsch. Mikrosk. Anat.* **28:**154–178.

Brick, I., Schaeffer, B. E., Schaeffer, H. E., and Gennaro, J. F., Jr., 1974, Electrokinetic properties and morphological characteristics of amphibian gastrula cells, *Ann. N.Y. Acad. Sci.* **238:**390–407.

Briggs, R., 1939, Changes in the density of the frog embryo (*Rana pipiens*) during development, *J. Cell Comp. Physiol.* **13:**77–98.

Brun, R., and Garson, J., 1984, Notochord formation in the Mexican salamander (*Ambystoma mexicanum*) is different from notochord formation in *Xenopus laevis*, *J. Exp. Zool.* **229:**235–240.

Buytendijk, F. J. J., and Woerdeman, M. W., 1927, Die physicochemischen Erscheinungen während der Entwicklung, *Arch. Entwicklungsmech. Org.* **112:**387–410.

Campanella, C., and Gabbiani, G., 1980, Cytoskeletal and contractile proteins in coelomic oocytes, unfertilized, and fertilized eggs of *Discoglossus pictus* (Anura), *Gamete Res.* **3:**99–114.

Chen, W.-T., and Singer, S. J., 1980, Fibronectin is not present in focal adhesions between normal and cultured fibroblasts and their substrata, *Proc. Natl. Acad. Sci. USA* **77:**7318–7322.

Chuang, H. H., 1947, Defekt- und Vitalfarbungsversuche zur Analyse der Entwicklung der kaudelen Rumpfabschnitte und des Schwanzes bei Urodelen, *Roux Arch.* **143:**19–125.

Clark, T. G., and Merriam, R. W., 1978, Actin in *Xenopus* oocytes. I. Polymerization and gelation *in vitro*, *J. Cell Biol.* **77:**427–438.

Columbo, R., Benedusi, P., and Valle, G., 1980, Actin in *Xenopus* development: Indirect immunofluorescence study of actin localization, *Differentiation* **20:**45–51.

Cooke, J., 1972a, Properties of the primary organization field in the embryo of *Xenopus laevis*. I. Autonomy of cell behavior at the site of initial organizer formation, *J. Embryol. Exp. Morphol.* **28:**13–26.

Cooke, J., 1972b, Properties of the primary organization field in the embryo of *Xenopus laevis*. II. Positional information for axial organization in embryos with two head organizers, *J. Embryol. Exp. Morphol.* **28:**27–46.

Cooke, J., 1972c, Properties of the primary organization field in the embryo of *Xenopus laevis*. III. Retention of polarity in cell groups excised from the region of the early organizer, *J. Embryol. Exp. Morphol.* **28:**47–56.

Cooke, J., 1973a, Properties of the primary organization field in the embryo of *Xenopus laevis*. IV. Pattern formation and regulation following early inhibition of mitosis, *J. Embryol. Exp. Morphol.* **30:**49–62.

Cooke, J., 1973b, Properties of the primary organization field in the embryo of *Xenopus laevis*. V. Regulation after removal of the head organizer in normal early gastrulae and in those already possessing a second implanted organizer, *J. Embryol. Exp. Morphol.* **30:**283–300.

Cooke, J., 1975, Local autonomy of gastrulation movements after dorsal lip removal in two anuran amphibians, *J. Embryol. Exp. Morphol.* **33:**147–157.

Cooke, J., 1979a, Cell number in relation to primary pattern formation in the embryo of *Xenopus laevis*. I. The cell cycle during new pattern formation in response to implanted organizers, *J. Embryol. Exp. Morphol.* **51:**165–182.

Cooke, J., 1979b, Cell number in relation to primary pattern formation in the embryo of *Xenopus laevis*. II. Sequential cell recruitment and control of the cell cycle, during mesoderm formation, *J. Embryol. Exp. Morphol.* **53:**269–289.

Curtis, A. S. G., 1978, Cell positioning, in: *Receptors and Recognition*, Vol. 4, Ser. B, *Specificity of Embryological Interactions* (D. R. Garrod, ed.) pp. 159–195, Chapman and Hall, London.

Daniel, J. F., and Yarwood, E., 1939, The early embryology of *Triturus torosus*, *Univ. Calif. Publ. Zool.* **43:**321–356.

Davis, G., 1984, Migration-directing liquid properties of embryonic amphibian tissues, *Am. Zool.* **24:**649–655.

Decker, R. S., 1981, Disassembly of the zonula occludens during amphibian neurulation, *Dev. Biol.* **81:**12–22.

Decker, R. S., and Friend, D. S., 1974, Assembly of gap junctions during amphibian neurulation, *J. Cell Biol.* **62:**32–47.

deLaat, S. W., and Barts, P. W. J. A., 1976, New membrane formation and intercellular communication in the early *Xenopus* embryo. II. Theoretical analysis, *J. Membr. Biol.* **27**:131–151.

deLaat, S. W., Barts, P. W. J. A., and Bakker, M. I., 1976, New membrane formation and intercellular communication in the early *Xenopus* embryo. I. Electrophysiological analysis, *J. Membr. Biol.* **27**:109–129.

del Pino, E., and Elinson, R., 1983, A novel development pattern for frogs: Gastrulation produces an embryonic disk, *Nature (Lond.)* **306**:589–591.

del Pino, E., and Escobar, B., 1981, Embryonic stages of *Gastrotheca riobambae* (Fowler) during maternal incubation and comparison of development with that of the egg-brooding hylid frogs, *J. Morphol.* **167**:277–295.

del Pino, E., and Humphries, A. A., Jr., 1978, Multiple nuclei during early oogenesis in *Flectonotus pygmaeus* and other marsupial frogs, *Biol. Bull.* **154**:198–212.

Detlaff, T. A., 1983, A study of the properties, morphogenetic potencies and prospective fate of outer and inner layers of ectodermal and chrodamesodermal regions during gastrulation, in various Anuran amphibians, *J. Embryol. Exp. Morphol.* **75**:67–86.

DiCaprio, R. A., French, A. S., and Sanders, E. J., 1974, Dynamic properties of electrotonic coupling between cells of early *Xenopus* embryos, *Biophys. J.* **14**:387–411.

Dictus, W. J. A. G., van Zoelen, E. J. J., Tetteroo, P. A. J., Tertoolen, L. G. J., DeLaat, S. W., and Bluemink, J. G., 1984, Lateral mobility of plasma membrane lipids in *Xenopus* eggs; regional differences related to animal/vegetal polarity become extreme upon fertilization, *Dev. Biol.* **101**:201–211.

Dorfman, Ya., and Cherdantsev, V. G., 1977a, Structure of morphogenetic movements of gastrulation in anuran amphibians. Communication I. Destabilization of ooplasmatic segregation and subdivision under the influence of clinostat rotation, *Soviet J. Dev. Biol.* **8**:201–210.

Dorfman, Ya., and Cherdantsev, V. G., 1977b, Structure of morphogenetic movements of gastrulation in anuran amphibians. Communication II. Elementary morphogenetic processes, *Soviet J. Dev. Biol.* **8**:211–219.

Doucet-de Bruine, M. H. M., 1973, Blastopore formation in *Ambystoma mexicanum*, *Wilhelm Roux Arch.* **173**:136–163.

Eakin, R., 1933, Regulatory development in *Triturus torosus* (Rathke), *Univ. Calif. Publ. Zool.* **39**:191–200.

Eakin, R., 1939, Further studies in regulatory development of *Triturus torosus*, *Univ. Calif. Publ. Zool.* **43**:185–209.

Elinson, R., 1975, Isozymes and morphology of five amphibian hybrid embryo combinations which develop beyond gastrulation, *Can. J. Zool.* **53**:993–1003.

Elinson, R., 1980, The amphibian egg cortex in fertilization and early development, *Symp. Soc. Dev. Biol.* **38**:217–234.

Ellinson, R., 1984, Cytoplasmic phases in the first cleavage cycle of the activated frog egg, *Dev. Biol.* **100**:440–451.

Ettensohn, C., 1984a, Primary invagination of the vegetal plate during sea urchin gastrulation, *Am. Zool.* **24**:571–588.

Ettensohn, C., 1984b, An analysis of invagination during sea urchin gastrulation, Ph.D. thesis, Yale University, New Haven, Connecticut.

Eycleshymer, A. C., 1895, The early development of *Ambystoma*, with observations on some other vertebrates, *J. Morphol.* **10**:343–418.

Ezzell, R., Brothers, A. J., and Cande, W. Z., 1984, Phosphorylation-dependent contraction of actomyosin gels from amphibian eggs, *Nature (Lond.)* **306**:620–622.

Fankhauser, G., 1948, The organization of the amphibian egg during fertilization and cleavage, *Ann. N.Y. Acad. Sci.* **49**:684–702.

Franke, W. W., Rathke, P. C., Sieb, E., Trendelenburg, M. F., Osborn, M., and Weber, K., 1976, Distribution and mode of arrangement of microfilamentous structures and actin in the cortex of the amphibian oocyte, *Cytobiologie* **14**:111–130.

Franz, J. K., Gall, L., Williams, M., Picheral, B., and Franke, W., 1983, Intermediate-size filaments in a germ cell: Expression of cytokeratins in oocytes and eggs of the frog *Xenopus*, *Proc. Natl. Acad. Sci. USA* **80**:6254–6258.

Fristrom, D., 1976, The mechanism of evagination of imaginal discs of Drosphila melanogaster. III. Evidence for cell rearrangement, Dev. Biol. **54**:163–171.

Fristrom, D., and Chihara, C., 1978, The mechanism of evagination of imaginal discs of Drosophila melanogaster. V. Evagination of disc fragments, Dev. Biol. **66**:564–570.

Fristrom, D., 1982, Septate junctions in imaginal disks of Drosophila: A model for the redistribution of septa during cell rearrangement, J. Cell Biol. **94**:77–87.

Fristrom, D., and Rickoll, W., 1983, Morphogenesis of imaginal discs of Drosophila, in: Insect Ultrastructure. I. (R. King and H. Akai, eds.), Plenum Press, New York.

Fujinami, N., 1976, Studies on the mechanism of circus movement in dissociated cells of a teleost, Oryzias latipes: Fine structural observations, J. Cell Sci. **22**:133–147.

Gadenne, M., van Zoelen, E. J. J., Tencer, R., and deLaat, S. W., 1984, Increased rate of capping of Concanavalin A receptors during early Xenopus development is related to changes in protein and lipid mobility, Dev. Biol. **104**:461–468.

Gall, L., Picheral, B., and Gounon, P., 1983, Cytochemical evidence for the presence of intermediate filaments and microfilaments in the egg of Xenopus laevis, Biol. Cell **47**:331–342.

Gerhart, J., 1980, Mechanisms regulating pattern formation in the amphibian egg and early embryo, in: Biological Regulation and Development, Vol. 2 (R. Goldberger, ed.), pp. 133–316, Plenum Press, New York.

Gerhart, J., Black, S., Gimlich, R., and Scharf, S., 1983a, Control of polarity in the amphibian egg, in: Time, Space, and Pattern in Embryonic Development (W. Jeffery and R. Raff, eds.), pp. 261–286, Alan Liss, New York.

Gerhart, J., Black, S., and Scharf, S., 1983b, Cellular and pancellular organization of the amphibian embryo, in: Modern Cell Biology, Vol. 2: Spatial Organization of Eukaryotic Cells (R. McIntosh, ed.), pp. 403–508, Alan Liss, New York.

Gillespie, J. I., 1983, The distribution of small ions during the early development of Xenopus laevis and Ambystoma mexicanum embryos, J. Physiol. (Lond.) **344**:359–377.

Gimlich, R., and Cooke, J., 1983, Cell lineage and the induction of second nervous systems in amphibian development, Nature (Lond.) **306**:471–473.

Gimlich, R., and Gerhart, J., 1984, Early cellular interactions promote embryonic axis formation in Xenopus laevis, Dev. Biol. **104**:117–130.

Gingell, D., 1970, Contractile responses at the surface of an amphibian egg, J. Embryol. Exp. Morphol. **23**:583–609.

Goerttler, K., 1925a, Die Formbildung der Medullaranlage bei Urodelen, Sitzgsber. Ges. Morphol. Physiol. Munch. **36**:57–66.

Goerttler, K., 1925b, Die Formbildung der Medullaranlage bei Urodelen, im Rahmen der Verscheiebungsvorgange von Keimbezirken wahrend der Gastrulation und als entwicklungsphysiologishes Problem, Roux Arch. **106**:503–541.

Goerttler, K., 1926, Experimentell erqeugte "Spina bifida" und "Ringembryonenbildungen" und ihre Bedeutung für die Entwicklungsphysiologie der Urodeleneier, Z. Anat. Anz. Erg. **63**:75–80.

Goodale, H. D., 1911a, On blastopore closure in amphibia, Anat. Anz. **38**:275–279.

Goodale, H. D., 1911b, The early development of Spelerpes bilineatus (Green), Am. J. Anat. **12**:173–247.

Graf, L., and Gierer, A., 1980, Size, shape and orientation of cells in budding hydra and regulation of regeneration in cell aggregates, Wilhelm Roux Arch. Dev. Biol. **188**:141–151.

Grant, P., and Wacaster, J. F., 1972, The amphibian grey crescent—A site of developmental information?, Dev. Biol. **28**:454–471.

Gregg, J. R., 1957, Morphogenesis and metabolism of gastrula-arrested embryos of the hybrid Rana pipiens O × Rana sylvatica O, in: The Beginnings of Embryonic Development (A. Tyler, A. C. von Borstel, and C. Metz, eds.) pp. 231–261, American Association for the Advancement of Science, Washington, D.C.

Gregg, J., and Klein, D., 1955, Morphogenetic movements of normal and gastrula-arrested hybrid amphibian tissues, Biol Bull. **109**:265–270.

Gustafson, T., and Wolpert, L., 1967, Cellular movement and contact in sea urchin morphogenesis, Biol. Rev. **42**:442–498.

Hall, E. K., 1937, Regional differences in the action of the organization center, *Wilhelm Roux Arch. Entwicklungsmech. Org.* **135**:671–688.

Hama, T., 1978, Dynamics of the organizer. B. New findings on the regionality and morphogenetic movement of the organizer, in: *Organizer—A Milestone of a Half-century from Spemann* (O. Nakamura and S. Toivonen, eds.), pp. 71–92, Elsevier/North-Holland, Amsterdam.

Hamburger, V., 1960, *A Manual of Experimental Embryology*, rev. ed., University of Chicago Press, Chicago.

Hara, K., 1971, Cinematographic observation of "surface contraction waves" (SCW) during the early cleavage of axolotl eggs, *Wilhelm Roux Arch.* **167**:183–186.

Hara, K., Tydeman, P., and Hengst, R. T. M., 1977, Cinematographic observation of "Post-Fertilization Waves" (PFW) on the zygote of *Xenopus laevis*, *Wilhelm Roux Arch.* **181**:189–192.

Hardin, J., and Keller, R. E., 1986, The role of bottle cells in gastrulation of *Xenopus laevis*, in preparation.

Harris, T. M., 1964, Pregastrular mechanisms in the morphogenesis of the salamander *Ambystoma maculatum*, *Dev. Biol.* **10**:247–268.

Harris, A., 1976, Is cell sorting caused by differences in the work of intercellular adhesion? A critique of the Steinberg hypothesis, *J. Theoret. Biol.* **61**:267–285.

Hatta, S., 1907, Gastrulation in *Petromyzon*, *J. Colloq. Sci. Imp. Univ. Tokyo* **21**:3–44.

Hertwig, O. 1892, Urmund und Spina bifida. Eine vergleichend morphologische teratologishe Studie an missgebildeten Froscheieren, *Arch. Mikrosk. Anat.* **39**:353–505.

Hilton, W. A., 1909, General features of the early development of *Desmognathus fusca*, *J. Morphol.* **20**:533–547.

Hirose, G., and Jacobson, M., 1979, Clonal organization of the central nervous system of the frog. I. Clones stemming from individual blastomeres of the 16 cell and earlier stages, *Dev. Biol.* **71**:191–202.

Hollinger, T. G., and Schuetz, A. W., 1976, "Cleavage" and cortical granule breakdown in *Rana pipiens* oocytes induced by direct microinjection of calcium, *J. Cell Biol.* **71**:395–401.

Holmdahl, D. E., 1947, Das Verhalten des Entoderms und Hautektoderms bei der sekundaren Korperentwicklung, *Anat. Anz.* **96**:56–69.

Holtfreter, J., 1929, Über die Aufzucht isolierter Teile des Amphibienkeimes, *Arch. Entwicklungsmech. Org.* **124**:404–466.

Holtfreter, J., 1933*a*, Nicht typische Gestaltungsbewegungen, sondern Induktions vorgange Bedingen die medullare Entwicklung von Gastrulaektoderm, *Arch. Entwicklungsmech. Org.* **127**:591–618.

Holtfreter, J., 1933*b*, Die totale Exogastrulation, eine Selbstablösung des Ektoderms von Entomesoderm, *Arch. Entwicklungsmech. Org.* **129**:669–793.

Holtfreter, J., 1934, Formative Reize in der Embryonalentwicklung der Amphibien, dargestellt an Explantationsversuchen, *Arch. Exp. Zellforsch.* **15**:281–301.

Holtfreter, J., 1938*a*, Differenzierungspotenzen isolierter Teile der Urodeleangastrula, *Wilhelm Roux Arch.* **138**:522–656.

Holtfreter, J., 1938*b*, Differenzierungspotenzen isolierter Teile der Anurengastrula, *Wilhelm Roux Arch.* **138**:657–738.

Holtfreter, J., 1939, Gewebeaffinität, ein Mittel der embryonalen Formbildung, *Arch. Exp. Zellforsch. Besonders Geweb.* **23**:169–209.

Holtfreter, J., 1943*a*, Properties and function of the surface coat in amphibian embryos, *J. Exp. Zool.* **93**:251–323.

Holtfreter, J., 1943*b*, A study of the mechanics of gastrulation. Part I, *J. Exp. Zool.* **94**:261–318.

Holtfreter, J., 1943*c*, Experimental studies on the development of the pronephros, *Rev. Can. Biol.* **3**:220–249.

Holtfreter, J., 1944, A study of the mechanics of gastrulation. Part II, *J. Exp. Zool.* **95**:171–212.

Holtfreter, J., 1946*a*, Structure, motility and locomotion in isolated embryonic amphibian cells, *J. Morphol.* **72**:27–62.

Holtfreter, J., 1946*b*, Experiments on the formed inclusions of the amphibian egg. I, *J. Exp. Zool.* **101**:355–405.

Holtfreter, J., 1946c, Experiments on the formed inclusions of the amphibian egg. II, *J. Exp. Zool.* **102**:51–108.

Holtfreter, J., 1946d, Experiments on the formed inclusions of the amphibian egg. III, *J. Exp. Zool.* **103**:81–112.

Holtfreter, J., 1947a, Observations on the migration, aggregation and phagocytosis of embryonic cells, *J. Morphol.* **80**:25–55.

Holtfreter, J., 1947b, Changes of structure and kinetics of differentiating embryonic cells, *J. Morphol.* **80**:57–92.

Holtfreter, J., 1948, The significance of the cell membrane in embryonic processes, *Ann. N.Y. Acad. Sci.* **49**:709–760.

Honda, H., Ogita, Y., Higuchi, S., and Kani, K., 1982, Cell movements in a living mammalian tissue: Long-term observation of individual cells in wounded corneal endothelia of cats, *J. Morphol.* **174**:25–39.

Horstadius, S., 1944, Über die Folge von Chordaextirpation an spaten Gastrulae und Neurulae von *Ambystoma punctatum, Acta Zool. Stockh.* **25**:75–88.

Ikedo, S., 1902, Contributions to the embryology of the amphibia: The mode of blastopore closure and the position of the embryonic body, *J. Colloq. Sci. Imp. Univ. Tokyo* **17**:82–90.

Ikushima, N., 1958, Kinetic properties of the marginal zone of the amphibian egg in relation to the histological differentiation, *Mem. Colloq. Sci. Univ. Kyoto* **25**:145–160.

Ikushima, N., 1959, The formation of two independent notochords in an explant taken from the dorsal blastoporal area of the early gastrula of Amphibia, *Experientia* **15**:475–476.

Ikushima, N., 1961, Formation of notochord in an explant derived from the dorsal marginal zone of the early gastrula of Amphibia, *Jpn. J. Zool.* **13**:117–140.

Ikushima, N., and Maruyama, S., 1971, Structure and developmental tendency of the dorsal marginal zone in the early amphibian gastrula, *J. Embryol. Exp. Morphol.* **25**:263–276.

Ishikawa, C., 1905, The gastrulation of the gigantic salamander, *Megalobatrachus maximus, Zool. Mag. Tokyo Zool. Soc.* **17**:26–28.

Jacobson, A., 1982, Morphogenesis of the neural plate and tube, in: *Morphogenesis and Pattern Formation* (T. G. Connelly, L. Brinkley, and B. Carlson, eds.), pp. 223–263, Raven Press, New York.

Jacobson, A., and Gordon, R., 1976, Changes in the shape of the developing vertebrate nervous system analyzed experimentally, mathematically, and by computer simulation, *J. Exp. Zool.* **197**:191–246.

Jenkinson, J. W., 1906, On the effect of certain solutions upon the development of the frog's egg, *Arch. Entwicklungsmech. Org.* **21**:367–460.

Johnson, K., 1969, Altered contact behavior of presumptive mesodermal cells from hybrid amphibian embryos arrested at gastrulation, *J. Exp. Zool.* **170**:325–332.

Johnson, K., 1970, The role of changes in cell contact behavior in amphibian gastrulation, *J. Exp. Zool.* **175**:391–428.

Johnson, K., 1972, The extent of cell contact and the relative frequency of small and large gaps between presumptive mesodermal cells in normal gastrulae of *Rana pipiens* and the arrested gastrulae of the *Rana pipiens* ♀ × *Rana catesbeiana* ♂ hybrid, *J. Exp. Zool.* **179**:227–238.

Johnson, K., 1976a, Ruffling and locomotion in *Rana pipiens* gastrula cells, *Exp. Cell Res.* **101**:71–77.

Johnson, K., 1976b, Circus movements and blebbing locomotion in dissociated embryonic cells of an amphibian, *Xenopus laevis, J. Cell Sci.* **22**:575–583.

Johnson, K., 1977a, Changes in the cell coat at the onset of gastrulation in *Xenopus laevis* embryos, *J. Exp. Zool.* **199**:137–142.

Johnson, K., 1977b, Extracellular matrix synthesis in blastula and gastrula stages of normal and hybrid frog embryos. I. Toluidine blue and lanthanum staining, *J. Cell Sci.* **25**:313–322.

Johnson, K., 1977c, Extracellular matrix synthesis in blastula and gastrula stages of normal and hybrid frog embryos. II. Autoradiographic observations on the sites of synthesis and mode of transport of galactose- and glucosamine-labelled materials, *J. Cell Sci.* **25**:323–334.

Johnson, K., 1977d, Extracellular matrix synthesis in blastula and gastrula stages of normal and hybrid frog embryos. III. Characterization of gaslactose- and glucosamine-labelled materials. *J. Cell Sci.* **25**:335–356.

Johnson, K., 1978, Extracellular matrix synthesis in blastula and gastrula stages of normal and hybrid frog embryos. IV. Biochemical and autoradiographic observations on fucose-, glucose-, and mannose-labelled materials, *J. Cell Sci.* **32**:109–136.

Johnson, K., 1981, Normal frog gastrula extracellular materials serve as a substratum for normal and hybrid cell adhesion when covalently coupled with CNBr-activated Sepharose beads, *Cell Diff.* **10**:47–55.

Johnson, K., and Adelman, M., 1981, Circus movements in dissociated cells in normal and hybrid frog embryos, *J. Cell Sci.* **49**:205–216.

Johnson, K., and Smith, E., 1976, The binding of concanavalin A to dissociated embryonic amphibian cells, *Exp. Cell Res.* **101**:63–70.

Johnson, K., and Smith, E., 1977, Lectin binding to dissociated cells from two species of *Xenopus* embryos, *Cell Diff.* **5**:301–309.

Jordan, E. O., 1893, The habits and development of the newt (*Diemyctylus viridescens*), *J. Morphol.* **8**:269–366.

Jumah, H., and Stanisstreet, M., 1980, Scanning electron microscopy of cells from *Xenopus* hybrid embryos, *Acta Embryol. Morphol. Exp.* **1**:129–135.

Kageura, H., and Yamana, K., 1983, Pattern regulation in isolated blastomeres of early *Xenopus laevis*, *J. Embryol. Exp. Morphol.* **74**:221–234.

Kageyama, T., 1982, Cellular basis of epiboly of the enveloping layer in the embryo of the medaka, *Oryzias latipes*. II. Evidence for cell rearrangement, *J. Exp. Zool.* **209**:241–256.

Kageyama, T., and Sirakami, K., 1976, Circus movement in dissociated embryonic cells of amphibia, with special reference to velocity, *Zoological Magazine (Dobutsugaku Zasshi)* **85**:169–172.

Kalt, M., 1971a, The relationship between cleavage and blastocoel formation in *Xenopus laevis*. I. Light microscopic observations, *J. Embryol. Exp. Morphol.* **26**:37–50.

Kalt, M., 1971b, The relationship between cleavage and blastocoel formation in *Xenopus laevis*. II. Electron microscopic observations, *J. Embryol. Exp. Morphol.* **26**:51–66.

Kaneda, T., and Hama, T., 1979, Studies on the formation and state of determination of the trunk organizer in the newt, *Cynops pyrrhogaster*, *Wilhelm Roux Arch.* **187**:25–34.

Kaplan, R. H., 1980, The implications of ovum size variability for offspring fitness and clutch size within several populations of salamanders (*Ambystoma*), *Evolution* **34**:51–64.

Karfunkel, P., 1977, SEM analysis of amphibian mesodermal migration, *Wilhelm Roux Arch.* **181**:31–40.

Keller, R. E., 1975, Vital dye mapping of the gastrula and neurula of *Xenopus laevis*. I. Prospective areas and morphogenetic movements of the superficial layer, *Dev. Biol.* **42**:222–241.

Keller, R. E., 1976, Vital dye mapping of the gastrula and neurula of *Xenopus laevis*. II. Prospective areas and morphogenetic movements in the deep region, *Dev. Biol.* **51**:118–137.

Keller, R. E., 1978, Time-lapse cinemicrographic analysis of superficial cell behavior during and prior to gastrulation in *Xenopus laevis*, *J. Morphol.* **157**:223–248.

Keller, R. E., 1980, The cellular basis of epiboly: An SEM study of deep-cell rearrangement during gastrulation in *Xenopus laevis*, *J. Embryol. Exp. Morphol.* **60**:201–234.

Keller, R. E., 1981, An experimental analysis of the role of bottle cells and the deep marginal zone in gastrulation of *Xenopus laevis*, *J. Exp. Zool.* **216**:81–101.

Keller, R. E., 1984, The cellular basis of gastrulation in *Xenopus laevis*: Active, postinvolution convergence and extension by mediolateral interdigitation, *Am. Zool.* **24**:589–603.

Keller, R., Danilchik, M., Gimlich, R., and Shih, J., 1985, Convergent extension by cell intercalation during gastrulation of *Xenopus laevis*, in: *Molecular Biology, New Series 31* (G. M. Edelman, ed.), Alan R. Liss, New York.

Keller, R. E., and Schoenwolf, G. C., 1977, An SEM study of cellular morphology, contact, and arrangement, as related to gastrulation in *Xenopus laevis*, *Wilhelm Roux Arch.* **182**:165–186.

Keller, R. E., and Trinkaus, J. P., 1982, Cell rearrangement in a tightly-joined epithelial layer during *Fundulus* epiboly, *J. Cell Biol.* **95**:325a.

King, H. D., 1902a, Experimental studies on the egg of *Bufo lentiginosus*, *Arch. Entwicklungsmech. Org.* **13**:545–564.

King, H. D., 1902b, The gastrulation of the egg of *Bufo lentiginosus*, *Am. Nat.* **36**:527–548.

King, H. D., 1903, The formation of the notochord in the amphibia, *Biol. Bull.* **4**:287–300.

Kingsbury, B. J., 1924, The developmental significance of the notochord (Chorda dorsalis), *Z. Morphol. Anthropol.* **24**:59–73.

Kirschner, M., Gerhart, J., Hara, K., and Ubbels, G., 1980, Initiation of the cell cycle and establishment of bilateral symmetry in *Xenopus* eggs, *Symp. Soc. Dev. Biol.* **38**:187.

Kitchen, J. C., 1938, The effects of extirpation of the notochord undertaken at the medullary plate stage in *Ambystoma mexicanum*, *Anat. Rec.* **72**:34a.

Kitchen, J. C., 1949, The effects of notochordectomy in *Ambystoma mexicanum*, *J. Exp. Zool.* **112**:393–415.

Klag, J. J., and Ubbels, G., 1975, Regional morphological and cytochemical differentiation in the fertilized egg of *Discoglossus pictus* (Anura), *Differentiation* **3**:15–20.

Kubota, H., 1981, Creeping locomotion of the endodermal cells dissociated from gastrulae of the Japanese newt, *Cynops pyrrhogaster*, *Exp. Cell Res.* **133**:137–148.

Kubota, H., and Durston, A. J., 1978, Cinematographical study of cell migration in the opened gastrula of *Ambystoma mexicanum*, *J. Embryol. Exp. Morphol.* **44**:71–80.

Kuhl, W., 1937, Untersuchungen über das Verhalten kunstlich getrennter Furchungszellen und Zellaggregate einiger Amphibienarten mit Hilfe des Zeitrafferfilms, *Arch. Entwicklungsmech. Org.* **136**:591–671.

LeBlanc, J., and Brick, I., 1981a, Morphological aspects of adhesion and spreading behavior of amphibian blastula and gastrula cells, *J. Embryol. Exp. Morphol.* **61**:145–163.

LeBlanc, J., and Brick, I., 1981b, Calcium and spreading behaviour of amphibian blastula and gastrula cells, *J. Embryol. Exp. Morphol.* **64**:149–168.

Lee, G., Hynes, R. O., and Kirschner, M., 1982, Temporal and spatial expression of fibronectin in early *Xenopus* development, *J. Cell Biol.* **95**:135a.

Lee, G., Hynes, R. O., and Kirschner, M., 1984, Temporal and spatial regulation of fibronectin in early *Xenopus* development, *Cell* **36**:729–740.

Lehman, F. E., 1932, Die Beteiligung von Implantats- und Wirtsgewebe bei der Gastrulation und Neurulation induzierter Embryonalanlagen, *Arch. Entwicklungsmech. Org.* **125**:566.

Lehman, F. E., 1937, Mesodermisierung des presumptiven Chorda-materials durch Einwirkung von Lithiumchlorid auf die Gastrula von *Triton alpestris*, *Wilhelm Roux Arch.* **136**:111–146.

Lehman, F. E., 1938, Regionale verschudenherten des Organizators von *Triton*, insbesondere in der vorderen und hintern Kopfregion, nachgewiesen durch phasen- Spezifishi Erzeugung von Lithiumbedingten und operativ Bewirkten Regionaldefekten, *Roux Arch.* **138**:106–158.

Lehman, F. E., and Ris, H., 1938, Weitere Untersuchungen über die Entwicklung der Achsenorgane bei partiell chordalosen Tritonlarven, *Rev. Suisse Zool.* **45**:419–424.

Lewis, W. H., 1947, Mechanics of invagination, *Anat. Rec.* **97**:139–156.

Lewis, W. H., 1948, Mechanics of *Ambystoma* gastrulation, *Anat. Rec.* **101**:700.

Lewis, W. H., 1952, Gastrulation of *Ambystoma punctatum*, *Anat. Rec.* **112**:473.

Lillie, F. R., 1902, Differentiation without cleavage in the egg of the annelid *Chaetopterus pergamentaceus*, *Arch. Entwicklungsmech. Org.* **14**:477–499.

Lofberg, J., 1974, Apical surface topography of invaginating and noninvaginating cells. A scanning-transmission study of amphibian neurulae, *Dev. Biol.* **36**:311–329.

Løvtrup, S., 1966, Cell type distribution, germ layers, and fate maps, *Acta Zool.* **47**:209–276.

Løvtrup, S., 1975, Fate maps and gastrulation in amphibia—A critique of current views, *Can. J. Zool.* **53**:473.

Luckenbill, L., 1971, Dense material associated with wound closure in the axolotl egg (*Ambystoma mexicanum*), *Exp. Cell Res.* **66**:263–267.

Luchinskaya, N. N., and Beloussov, L. V., 1977, Electron microscopic investigation of rapid morphogenetic processes in amphibian tissue explants, *Soviet J. Dev. Biol.* **8**:220–226.

Lundmark, C., 1986, Role of bilateral zones of superficial mesodermal cell ingression during gastrulation of *Ambystoma mexicanum*, submitted for publication.

Lundmark, C., Shih, J., Tibbetts, P., and Keller, R., 1984, Amphibian gastrulation as seen by scanning electron microscopy, *Scanning Electron Microscopy* **III**:1289–1300.

MacMurdo-Harris, H., and Zalik, S. E., 1970, Microelectrophoresis of early amphibian embryonic cells, *Dev. Biol.* **24**:335–347.

Malacinski, G. M., Brothers, A. J., and Chung, H.-M., 1977, Destruction of components of the neural induction system of the amphibian egg with ultraviolet irradiation, Dev. Biol. **56**:24–39.

Malacinski, G., Ryan, B., and Chung, H.-M., 1978, Surface coat movements in unfertilized amphibian eggs, Differentiation **10**:101–107.

Malacinski, G., and Youn, B.-W., 1981, Neural plate morphogenesis and axial stretching in "notochord defective" Xenopus laevis embryos, Dev. Biol. **88**:352–357.

Manes, E., and Elinson, R., 1980, Ultraviolet light inhibits grey crescent formation on the frog egg, Wilhelm Roux Arch. **189**:73–76.

Mangold, O., 1920, Fragen der Regulation und Determination an umgeordneten Furchungsstadien und verschmolzenen Keimen von Triton, Roux Arch. **47**:250–301.

Mangold, O., 1923, Transplantationsversuche zur Frage der Spezefitat und der Bildung der Keimblatter bei Triton, Arch. Mikrosk. Anat. Entwicklungsmech. Org. **100**:198–301.

Meeusen, R. L., Bennet, J., and Cande, W. Z., 1980, Effect of microinjected N-ethylmaleimide-modified heavy meromyosin on cell division in amphibian eggs, J. Cell Biol. **86**:858–865.

Meeusen, R. L., Bennet, J., and Cande, W. Z., 1983, J. Cell Biol. **97**:1062–1071.

Mookerjee, S., 1953, An experimental study of the development of the notochord, J. Embryol. Exp. Morphol. **1**:411–416.

Mookerjee, S., Deuchar, E., and Waddington, C. H., 1953, The morphogenesis of the notochord in Amphibia, J. Embryol. Exp. Morphol. **1**:399–409.

Monroy, A., Baccetti, B., and Denis-Donini, S., 1976, Morphological changes in the surface of the egg of Xenopus laevis in the course of development. III. Scanning electron microscopy of gastrulation, Dev. Biol. **59**:250–259.

Moore, J. A., 1946, Development of frog hybrids. I. Embryonic development in the cross Rana pipiens O × Rana sylvatica O, J. Exp. Zool. **101**:173–219.

Moore, J. A., 1955, Abnormal combinations of nuclear and cytoplasmic systems in frogs and toads, Adv. Genet. **7**:139–182.

Moran, D., and Mouradian, W. E., 1975, A scanning electron microscopic study of the appearance and localization of cell surface material during amphibian gastrulation, Dev. Biol. **46**:422–429.

Morgan, T. H., 1897, The Development of the Frog's Egg: An Introduction to Experimental Embryology, Macmillan, New York.

Morgan, T. H., 1906a, Experiments with frog's eggs, Biol. Bull. **11**:71–92.

Morgan, T. H., 1906b, The origin of the organ-forming materials in the frog's egg, Biol. Bull. **11**:124–136.

Morrill, G. A., Kostellow, A. B., and Murphy, J. B., 1975, Role of Na^+,K^+-ATPase in early embryonic development, Ann. N.Y. Acad. Sci. **242**:543–559.

Moscona, A., and Folkman, J., 1981, Role of cell shape in growth control, Nature (Lond.) **273**:345–349.

Muchmore, W. B., 1951, Differentiation of the trunk mesoderm in Ambystoma maculatum, J. Exp. Zool. **118**:137–186.

Nakamura, O., 1938, Tail formation in the urodele, Zool. Mag. Tokyo **50**:442–446.

Nakamura, O., 1942, Die Entwicklung der hinteren Korperhalfte bei Urodelen, Annot. Zool. Jpn. **21**:169–235.

Nakamura, O., 1947, Determination and differentiation in the development of the urodele tail, Exp. Morphol. **3**:169 (English summary).

Nakamura, O., and Kishiyama, K., 1971, Prospective fates of blastomeres at the 32 cell stage of Xenopus laevis embryos, Proc. Jpn. Acad. **47**:407–412.

Nakatsuji, N., 1974, Studies on the gastrulation of amphibian embryos; pseudopodia in the gastrula of Bufo bufo japonicus and their significance to gastrulation, J. Embryol. Exp. Morphol. **32**:795–804.

Natatsuji, N., 1975a, Studies on the gastrulation of amphibian embryos: Light and electron microscopic observations of a urodele Cynops pryyhogaster, J. Embryol. Exp. Morphol. **34**:669–685.

Nakatsuji, N., 1975b, Studies on the gastrulation of amphibian embryos: Cell movement during gastrulation in Xenopus laevis embryos, Wilhelm Roux Arch. **178**:1–14.

Nakatsuji, N., 1976, Studies on the gastrulation of amphibian embryos: Ultrastructure of the migrating cells of anurans, Wilhelm Roux Arch. **180**:229–240.

Nakatsuji, N., 1979, Effects of injected inhibitors of microfilament and microtubule function on the gastrulation movements in *Xenopus laevis, Dev. Biol.* **68:**140–150.

Nakatsuji, N., 1984, Cell locomotion and contact guidance in amphibian gastrulation, *Am. Zool.* **24:**615–627.

Nakatsuji, N., Gould, A. C., and Johnson, K., 1982, Movement and guidance of migrating mesodermal cells in *Ambystoma maculatum* gastrulae, *J. Cell Sci.* **56:**207–222.

Nakatsuji, N., and Johnson, K., 1982, Cell locomotion in vitro by *Xenopus laevis* gastrula mesodermal cells, *Cell Motil.* **2:**149–161.

Nakatsuji, N., and Johnson, K., 1983a, Conditioning of a culture substratum by the ectodermal layer promotes attachment and oriented locomotion by amphibian gastrula mesodermal cells, *J. Cell Sci.* **59:**43–60.

Nakatsuji, N., and Johnson, K., 1983b, Comparative study of extracellular fibrils on the ectodermal layer in gastrulae of five amphibian species, *J. Cell Sci.* **59:**61–70.

Nakatsuji, N., and Johnson, K., 1984, Experimental manipulation of a contact guidance system in amphibian gastrulation by mechanical tension, *Nature (Lond.)* **307:**453–455.

Newport, J., and Kirschner, M., 1982, A major developmental transition in early *Xenopus* embryos. I. Characterization and timing of cellular changes at the midblastula stage, *Cell* **30:**675–686.

Nicholas, J. S., 1945, Blastulation, its role in pregastrular organization in *Ambystoma punctatum, J. Exp. Zool.* **51:**159–184.

Nieuwkoop, P., 1947, Experimental investigations on the origin and determination of the germ cells, and on the development of the lateral plates and germ ridges in urodeles, *Arch. Neerl. Zool.* **8:**1–205.

Nieuwkoop, P., 1977, Origin and establishment of embryonic polar axes in amphibian development, *Curr. Top. Dev. Biol.* **11:**115.

Nieuwkoop, P., and Faber, J., 1967, *Normal Table of Xenopus laevis (Daudin)*, 2nd ed., North-Holland, Amsterdam.

Nieuwkoop, P., and Florshutz, P., 1950, Quelques caractères speciaux de la gastrulation et de la neurulation de l'oeuf de *Xenopus laevis*, Daud. et de quelques autres Anoures. 1'ère partie. Étude déscriptive, *Arch. Biol. (Liège)* **61:**113–150.

O'Dell, D., Tencer, R., Monroy, A., and Brachet, J., 1974, The pattern of concanavalin A-binding sites during the early development of *Xenopus laevis, Cell Diff.* **3:**193–198.

Palecek, J., Ubbels, G. A., and Rzehak, K., 1978, Changes of the external and internal pigment pattern upon fertilization in the egg of *Xenopus laevis, J. Embryol. Exp. Morphol.* **45:**203–214.

Pasteels, J., 1940, Un aperçu comparatif de la gastrulation chez les chordes, *Biol. Rev.* **15:**59–106.

Pasteels, J., 1942, New observations concerning the maps of presumptive areas of the young amphibian gastrula (*Ambystoma* and *Discoglossus*), *J. Exp. Zool.* **89:**255–281.

Perry, M., 1975, Microfilaments in the external surface layer of the early amphibian embryo, *J. Embryol. Exp. Morphol.* **32:**127–146.

Perry, M., and Waddington, C. H., 1966, Ultrastructure of the blastoporal cells in the newt, *J. Embryol. Exp. Morphol.* **15:**317–330.

Phillips, H., 1984, Physical analysis of tissue mechanisms in amphibian gastrulation, *Am. Zool.* **24:**657–672.

Phillips, H., and Davis, G. S., 1978, Liquid tissue mechanics in amphibian gastrulation: Germ-layer assembly in *Rana pipiens, Am. Zool.* **18:**81–93.

Poole, T., and Steinberg, M., 1981, Amphibian pronephric duct morphogenesis; segregation, cell rearrangement and directed migration of the *Ambystoma* duct rudiment, *J. Embryol. Exp. Morphol.* **63:**1–16.

Rhumbler, L., 1902, Zur Mechanik des Gastrulationsvorganges, insbesondere der Invagination. Eine entwicklungsmechanishe Studie, *Wilhelm Roux Arch. Entwickslungsmech. Org.* **14:**401–476.

Richards, O. W., 1940, The capsular fluid of *Amblystoma punctatum* eggs compared with Holtfreter's and Ringer's solutions, *J. Exp. Zool.* **83:**401–406.

Rink, T. J., Tsien, R. Y., and Warner, A. E., 1980, Free calcium in *Xenopus* embryos measured with ion-selective microelectrodes, *Nature (Lond.)* **283:**658–660.

Roberson, M., and Armstrong, P., 1980, Carbohydrate-binding component of amphibian embryo cell surfaces: Restriction to surface regions capable of cell adhesion, *Proc. Natl. Acad. Sci. USA* **77**:3460–3463.

Roberson, M., Armstrong, J., and Armstrong, P., 1980, Adhesive and nonadhesive membrane domains of amphibian embryo cells, *J. Cell Sci.* **44**:19–31.

Roux, W., 1894, Über den "Cytotropismus" der Furchungszellen des Grasfrosches (*Rana fusca*), *Arch. Entwickslungmech. Org.* **1**:43–68; 160–202.

Roux, W., 1896, Über die Selbstordung (Cytotaxis) sich "beruhrender" Furchungszellen des Froscheies durch Zellen zussamenfugung, Zellentrennung, und Zellengleiten, *Roux Arch.* **3**:387–468.

Ruffini, A., 1925, *Fisogenia*, Francesco Vallardi, Milan.

Salthe, S., and Duellman, W. E., 1973, *Evolutionary Biology of the Anurans* (J. L. Vial, ed.), pp. 229–249, University of Missouri Press, Columbia.

Sanders, E. J., and DiCaprio, R. A., 1976, A freeze-fracture and concanavalin A-binding study of the membrane of cleaving *Xenopus* embryos, *Differentiation* **7**:13–21.

Sanders, E. J., and Singal, P. K., 1975, Furrow formation in *Xenopus* embryos. Involvement of the Golgi body as revealed by ultrastructural localization of thiamine pyrophosphatase activity, *Exp. Cell Res.* **93**:219–224.

Sanders, E. J., and Zalik, S. E., 1972, The blastomere periphery of *Xenopus laevis* with special reference to intercellular relationships, *Wilhelm Roux Arch.* **171**:181–194.

Satoh, N., 1977, Metachronous cleavage and initiation of gastrulation in amphibian embryos, *Dev. Growth Diff.* **19**:111–117.

Satoh, N., Kageyama, T., and Sirakami, K.-I., 1976, Motility of dissociated embryonic cells in *Xenopus laevis*: Its significance to morphogenetic movements, *Dev. Growth Diff.* **18**:55–67.

Schaeffer, B., Schaeffer, H., and Brick, I., 1973a, Cell electrophoresis of amphibian blastula and gastrula cells; the relationship of surface charge and morphogenetic movement, *Dev. Biol.* **34**:66–76.

Schaeffer, H., Schaeffer, B., and Brick, I., 1973b, Effects of cytochalasin B on the adhesion and electrophoretic mobility of amphibian gastrula cells, *Dev. Biol.* **34**:163–166.

Schaeffer, H., Schaeffer, B., and Brick, I., 1973c, Electrophoretic mobility as a function of pH for disaggregated amphibian gastrula cells, *Dev. Biol.* **35**:376–381.

Scharf, S., and Gerhart, J., 1983, Axis determination in eggs of Xenopus laevis: A critical period before first cleavage, identified by the common effects of cold, pressure and ultraviolet irradiation, *Dev. Biol.* **99**:75–87.

Scharf, S. R., Vincent, J.-V., and Gerhart, J., 1984, Axis determination in the *Xenopus* egg, in: *Molecular Biology of Development* (E. Davidson and R. Firtel, eds.), pp. 51–73, Allan Liss, New York.

Schechtman, A. M., 1934, Unipolar ingression in Triturus torosus: A hitherto undescribed movement in the pregastrula stages of a urodele, *Univ. Calif. Publ. Zool.* **39**:303.

Schechtman, A. M., 1942, The mechanism of amphibian gastrulation. I. Gastrulation-promoting interactions between various regions of an anuran egg (*Hyla regilla*), *Univ. Calif. Publ. Zool.* **51**:1–39.

Schenk, R., 1952, Über quantitativ gestufte Defektversuche an der dorsalen Urmundlippe Jungen Gastrulae von *Triton alpestris*, *Roux Arch.* **145**:345–386.

Schroeder, T. E., and Strickland, D. L., 1974, Ionophore A23187, calcium, and contractility in frog eggs, *Exp. Cell Res.* **83**:139–148.

Selman, G., and Perry, M., 1970, Ultrastructural changes in the surface layers of the newt's egg in relation to the mechanism of its cleavage, *J. Cell Sci.* **6**:207–227.

Shapiro, B., 1958, Influences of the salinity and pH of blastocoelic perfusates on the initiation of amphibian gastrulation, *J. Exp. Zool.* **139**:381–396.

Singal, P. K., and Sanders, E. J., 1974a, Cytomembranes in first cleavage *Xenopus* embryos. Interrelationship between Golgi bodies, endoplasmic reticulum and lipid droplets, *Cell Tissue Res.* **154**:189–209.

Singal, P. K., and Sanders, E. J., 1974b, An ultrastructural study of the first cleavage of *Xenopus* embryos, *J. Ultrastruct. Res.* **47**:433–451.

Sirikami, K. I., 1959, Cyto-embryological studies of amphibians. II. On "tug cells" and their bearing upon the gastrulation and some other morphogenetic movements, *Mem. Fac. Liberal Arts Educ. Yamanashi Univ.* **10**:125–127.

Sirlin, J. L., 1956, Tracing morphogenetic movements by means of labeled cells, *Wilhelm Roux Arch.* **148**:489–493.

Slack, C., and Warner, A., 1973, Intracellular and intercellular potentials in the early amphibian embryo, *J. Physiol. (Lond.)* **232**:313–330.

Slack, C., Warner, A., and Warren, R. L., 1973, The distribution of sodium and potassium in amphibian embryos during early development, *J. Physiol. (Lond.)* **232**:297–312.

Smith, B. G., 1912, The embryology of *Cryptobranchus allegheniensis*, including comparisons with some other vertebrates. II. General embryonic and larval development, with special reference to external features, *J. Morphol.* **23**:455–579.

Smith, J. C., and Malacinski, G. M., 1983, The origin of the mesoderm in an Anuran, *Xenopus laevis*, and a Urodele, *Ambystoma mexicanum*, *Dev. Biol.* **98**:250–254.

Smith, J. L., Osborn, J. C., and Stanisstreet, M., 1976, Scanning electron microscopy of lithium-induced exogastrulae of *Xenopus laevis*, *J. Embryol. Exp. Morphol.* **36**:513–522.

Smithberg, M., 1954, The origin and development of the tail of the frog, *J. Exp. Zool.* **127**:397.

Spemann, H., 1902, Entwicklungsphysiologische Studien am Triton-Ei. II, *Arch. Entwicklungsmech. Org.* **15**:448–534.

Spemann, H., 1931, Über den Anteil von Implantat und Wirtskeim an der Orientierung und Beschaffenheit der induzierten Embryonalanlage, *Arch. Entwicklungsmech. Org.* **123**:390–517.

Spemann, H., 1938, *Embryonic Development and Induction*, Yale University Press, repr. 1962, Hafner, New York.

Spemann, H., and Mangold, H., 1924, Über Induction von Embryonalanlagen durch Implantation artfremder Organisatoren, *Wilhelm Roux Arch.* **100**:599–638.

Spofford, W., 1945, Observations on the posterior part of the neural plate in *Ambystoma*, *J. Exp. Zool.* **99**:35–53.

Spofford, W., 1948, Observations on the posterior parts of the neural plate in *Ambystoma*. II. The inductive effect of the intact posterior part of the chordamesodermal axis on competent prospective ectoderm, *J. Exp. Zool.* **107**:123–164.

Stableford, L. J., 1949, The blastocoel fluid in amphibian gastrulation, *J. Exp. Zool.* **112**:529–546.

Stableford, L. J., 1967, A study of calcium in the early development of the amphibian embryo, *Dev. Biol.* **16**:303–314.

Stanisstreet, M., and Smith, J., 1978, Scanning electron microscopy of cells isolated from amphibian early embryos, *J. Embryol. Exp. Morphol.* **48**:215–223.

Steinberg, M., 1964, The problem of adhesive selectivity in cellular interactions, in: *Cellular Membranes in Development* (M. Locke, ed.), *Symp. Soc. Study Dev. Growth* **22**:321–366.

Steinberg, M., 1970, Does differential adhesion govern self-assembly processes in histogenesis? Equilibrium configurations and the emergence of a hierarchy among populations of embryonic cells, *J. Exp. Zool.* **173**:395–434.

Steinberg, M., 1978, Cell–cell recognition in multicellular assembly: Levels of specificity, in: *Cell–Cell Recognition* (A. S. G. Curtis, ed.), *Soc. Exp. Biol. Symp.* **32**:25–49.

Steinberg, M., and Kelland, J. L., 1967, Cellular and adhesive differentials in the determination of the structure of the amphibian gastrula, in: *Control Mechanisms in Morphogenesis, American Association for the Advancement of Science, 134th Annual Meeting, New York, 1967.*

Stewart-Savage, J., and Grey, R. D., 1982, The temporal and spatial relationships between cortical contraction, sperm trail formation, and pronuclear migration in fertilized *Xenopus* eggs, *Wilhelm Roux Arch.* **191**:241–245.

Stopak, D., and Harris, A., 1982, Connective tissue morphogenesis by fibroblast traction. I. Tissue culture observations, *Dev. Biol.* **90**:383–398.

Subtelny, S., 1974, Nucleocytoplasmic interactions in the development of amphibian hybrids, *Int. Rev. Cytol.* **39**:35–88.

Tickle, C. A., and Trinkaus, J. P., 1973, Change in surface extensibility of *Fundulus* deep cells during early development, *J. Cell Sci.* **13**:721–726.

Todd, E. H., 1904, Results of injuries to the blastopore region of the frog's embryo, *Arch. Entwicklungsmech. Org.* **18**:489–506.

Tondury, G., 1936, Beitrage zum Problem der Regulation und Induktion, *Roux Arch.* **134**:1–111.

Townes, P. L., and Holtfreter, J., 1955, Directed movements and selective adhesion of embryonic amphibian cells, *J. Exp. Zool.* **128**:53–120.

Trinkaus, J. P., 1963, The cellular basis of *Fundulus* epiboly. Adhesivity of blastula and gastrula cells in culture, *Dev. Biol.* **1**:513–532.

Trinkaus, J. P., 1973, Surface activity and locomotion of *Fundulus* deep cells during blastula and gastrula stages, *Dev. Biol.* **30**:68–103.

Trinkaus, J. P., 1984, *Cells into Organs: Forces That Shape the Embryo*, 2nd ed., Prentice-Hall, Englewood Cliffs, New Jersey.

Tuft, P., 1957, The osmotic activity of the blastocoel and archenteron fluids, *Proc. R. Physiol. Soc. Edinb.* **26**:42–48.

Tuft, P., 1962, The uptake and distribution of water in the embryo of *Xenopus laevis* (Daudin), *J. Exp. Biol.* **39**:1–19.

Tuft, P., 1965, The uptake and distribution of water in the developing amphibian embryo, *Soc. Exp. Biol. Symp.* **19**:385–402.

Turin, L., and Warner, A. E., 1980, Intracellular pH in early *Xenopous* embryos: Its effect on current flow between blastomeres, *J. Physiol. (Lond.)* **300**:489–504.

Viamontes, G., Fochtmann, L., and Kirk, D., 1979, Morphogenesis in *Volvox*: Analysis of critical variables, *Cell* **17**:537–550.

Vogt, W., 1922a, Die Einrollung und Streckung der Urmundlippen bei Triton nach Versuchen mit einer neuen Methode embryonaler Transplantation, *Verh. Zool. Ges.* **27**:49–51.

Vogt, W., 1922b, Operativ bewirkte "Exogastrulation" bei *Triton* und ihre Bedeutung für die Theorie der Wirbeltiergastrulation, *Anat. Anz. Erg.* **55**:53–64.

Vogt, W., 1923a, Morphologische und physiologische Fragen der Primitiv Entwicklung, Versuche zu ihrer Lösung mittels vitaler Farbmarkierung, *Sitz. Ber. Ges. Morph. Physiol. Munch.* **35**:22–32.

Vogt, W., 1923b, Weitere Versuche mit vitaler Färbmarkierung von Triton, *Anat. Anz. Erg.* **57**:30–38.

Vogt, W., 1925, Gestaltungsanalyse am Amphibienkeim mit örtlicher Vitalfärbung. I Teil. Method und Wirkungsweise der örtlicher Vitalfärbung mit Agar als Farbtrager, *Roux Arch.* **106**:542–610.

Vogt, W., 1929a, Chorda, Hypochorda und Darmentoderm bei anuren Amphibien, *Verh. Anat. Ges. Tubingen. Anat. Anz. Erg.* **67**:153–163.

Vogt, W., 1929b, Gestaltungsanalyse am Amphibienkeim mit örtlicher Vitalfärbung. II Teil. Gastrulation und Mesodermbildung bei Urodelen and Anuren, *Wilhelm Roux Arch. Entwicklungsmech. Org.* **120**:384–706.

Vogt, W., 1939, Die Rumpfschwanzknopse bei Amphibien und die Theorie der sekundaren Körperentwicklung (Holmdahl), *Anat. Anz. Erg.* **88**:112–127.

Voigtländer, G., 1932, Untersuchungen über den "Cytotropismus" der Furchungszellen, *Arch. Entwicklungsmech. Org.* **127**:151.

Waddington, C. H., 1939, Order of magnitude of morphogenetic forces, *Nature (Lond.)* **144**:637.

Waddington, C. H., 1940, *Organizers and Genes*, Cambridge University Press, Cambridge.

Waddington, C. H., 1942, Observations on the forces of morphogenesis in the amphibian embryo, *J. Exp. Biol.* **19**:284–293.

Waddington, C. H., 1952, Modes of gastrulation in vertebrates, *Q. J. Microsc. Sci.* **93**:221–229.

Wilder, H. H., 1899, *Desmongnathus fusca* (Rafinesque) and *Spelerpes bilineatus* (Green), *Am. Nat.* **33**:231–246.

Wilder, H. H., 1904, The early development of *Desmognathus fusca*, *Am. Nat.* **38**:231–246.

Wilson, C. B., 1897, Experiments on the early development of the amphibian embryos under the influence of Ringer's and salt solutions, *Arch. Entwicklungsmech. Org.* **5**:615–648.

Youn, B.-W., Keller, R. E., and Malacinski, G. M., 1981, An atlas of notochord and somite morphogenesis in several Anuran and Urodelean amphibians, *J. Embryol. Exp. Morphol.* **59**:223–247.

Zotin, A. I., 1965, The uptake and movement of water in embryos, *Soc. Exp. Biol. Symp.* **19**:365–384.

Chapter 8

Cell–Cell Interactions in Mammalian Preimplantation Development

PATRICIA CALARCO-GILLAM

1. Introduction

Cells in tissues or in culture interact in a variety of ways. Communication between cells can occur by gap junctions, which ionically couple them, by receptor–ligand interactions, by direct inductive influences, or by indirect metabolic influences on the cellular microenvironment. Adhesion to other cells and to extracellular matrix components will influence both cell shape and cell function (i.e., the differentiated state of that cell) as well as help determine the cell's degree of motility. Nearly all these interactions are mediated, at least initially, by the cell surface.

This chapter examines cell–cell interactions in early mammalian development. Not surprisingly many of the interactions outlined above play a role in development. Thus, we shall see that cells of the embryo establish communication among themselves and create a microenvironment. They adhere to one another, interact with certain extracellular elements, and show limited motility before implantation. Of more interest are the cellular and developmental processes unique to developing systems that are mediated by cell–cell interactions. Although categories overlap, some of these interactions include cell recognition between unlike cells such as that occurring at fertilization or implantation, cell epithelialization (i.e., the series of compaction-related events that mold embryonic cells into a polarized epithelium), cell communication, cell differentiation from totipotential precursor cells, and cell motility as seen in the invasive trophoblast cell. For the understanding of these and like events, the mammalian embryo presents a useful model system.

Sections 2 and 3 of this chapter summarize our understanding to date of the initial arena for cell–cell interactions, at the morphological and molecular levels—the embryonic cell surface. Next we explore a problem inherent to many preimplantation mammals: the synthesis and translocation of proteins to

PATRICIA CALARCO-GILLAM • Department of Anatomy, School of Medicine, University of California, San Francisco, California 94143.

the surface in cells deficient in both rough endoplasmic reticulum (RER) and Golgi material. Finally, developmental processes and their regulation by cell–cell interactions and future directions for research are discussed. Unless noted otherwise, most information presented here has been derived from studies on the preimplantation mouse embryo. Therefore, a brief description of early mouse development is included.

After fertilization, the zygote progresses by a series of five cleavages to the blastocyst stage (see Fig. 1A–F). Relatively little change in embryo size occurs during this period, but an increasing number of metabolic and synthetic processes begin to function for the first time. During the 8-cell stage, a process termed **compaction** ensues, which is characterized by increasing adhesion between blastomeres and a general rounding of the embryo. After the development of tight junctions between external cells of the morula (approximately the 16-cell stage), the embryo accumulates fluid in the intercellular space and blastulates at approximately the 32-cell stage. The blastocyst consists of a spherical outer epithelial layer of trophectoderm cells surrounding the fluid-filled blastocoel cavity and a group of internal cells, the inner cell mass (ICM), localized at one end of the blastocyst. The fate of the polar trophectoderm cells, which overlie the ICM, differs from that of the mural trophectoderm cells, which bound the fluid-filled cavity. They can be distinguished by certain criteria, e.g., polar trophectoderm cells possess a centrosome during interphase (Calarco-Gillam et al., 1983) and a more rapid rate of division (Copp, 1979). In the ICM, the cells bordering the blastocoel cavity differentiate into the cells of the primitive endoderm, whereas the interior multipotential cells are termed the ectoderm. The primitive endoderm will form both the visceral endoderm, which remains in contact with the ectoderm, and the parietal endoderm, which migrates out along the inner aspect of the mural trophectoderm cells. After hatching from the zona pellucida, the embryo will implant in the uterus. There is general similarity in preimplantation development among all mammals, although there may be slight differences in detail, e.g., the appearance of the ICM as a dispersed monolayer in some bats (Rasweiler, 1979) and the partial externalization and exposure to the uterine medium of the embryonic disc in the ewe, sow, and rabbit just before implantation (Flechon, 1978).

2. Morphology of the Embryo Surface

Interactions between cells may depend on the extracellular matrix, on distinct features of the plasma membrane, on the cortical cytoskeleton of the cell, or on a combination of these features. All these aspects have been described morphologically and are considered here.

2.1. The Extracellular Matrix

The most prominent extracellular matrix observed in the mammalian embryo is the zona pellucida, an acellular coat deposited around the oocyte in the

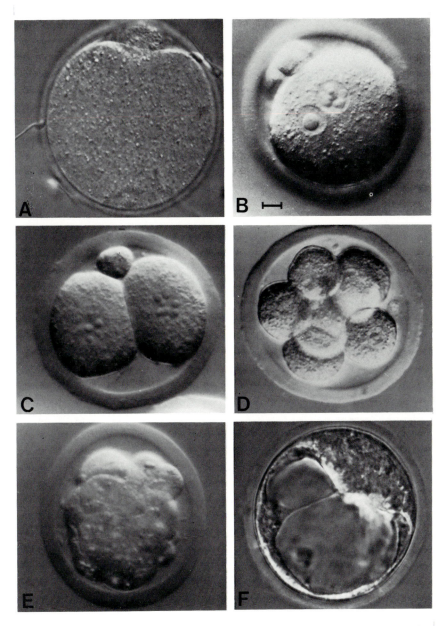

Figure 1. Micrographs of live preimplantation mouse embryos observed with Nomarski optics. Each embryo is surrounded by its acellular zona pellucida. Scale bar: 10 μm. (A) Early zygote. Egg has extruded its second polar body at top. A sperm can be seen adhering to the zona at left. (B) Zygote. Note the two pronuclei lying adjacent to one another. (C) Two-cell embryo. Nuclei with nucleoli are visible in each blastomere. (D) Eight-cell embryo before compaction. Individual blastomeres are rounded. (E) Morula of about 16 cells. Embryo has now compacted, and individual blastomeres are not readily distinguished. (F) Blastocyst. The inner cell mass and fluid-filled blastocoel are surrounded by the epithelial trophectoderm.

follicle and from which the embryo "hatches" before implantation. The dry mass of the zona of the mouse is approximately 30% protein, with the remaining 70% presumed to be carbohydrate (Loewenstein and Cohen, 1964; Dunbar et al., 1980). It can be removed by pronase digestion in the mouse (Mintz, 1962) or by trypsin digestion in the hamster or rat (Chang and Hunt, 1956) and binds a variety of lectins (Nicolson et al., 1975). Evidence suggests that specific zona proteins play a role in fertilization (Bleil and Wassarman, 1980). Because there is no evidence that the blastomeres of the embryo interact with the zona and because development occurs normally in its absence, it is not discussed further here.

The major components of the extracellular matrix (ECM) include collagenous glycoproteins, noncollagenous glycoproteins (e.g., fibronectin, entactin, laminin), and proteoglycans. For a more complete discussion of the ECM in early development, the reader is referred to the review by Leivo (1983). Conflicting results exist on the presence of collagens in preimplantation mouse embryos. Sherman and co-workers (1980) note the presence of collagens III, IV, and V and the absence of collagens I and II. Other researchers report the absence of collagens I, III, and IV from early cleavage stages but the presence of type IV both on the ICM and along the mural trophectoderm surfaces facing the blastocoel (Adamson and Ayers, 1979; Leivo et al., 1980; Wartiovaara and Leivo, 1982). Similarly, fibronectin does not appear until the blastocyst stage, where it is seen both on the ICM (Wartiovaara et al., 1978; Zetter and Martin, 1978) and along the blastocoelic margins of mural trophectoderm cells (Wartiovaara et al., 1979; Leivo et al., 1980). Another ECM molecule not seen until the late blastocyst stage is entactin, which is seen initially in hatched blastocysts between cells of the ICM and polar trophectoderm (Wu et al., 1983). Heparan sulfate proteoglycan (HSPG) has been reported in the ICM of blastocysts as well (Wartiovaara and Leivo, 1982) but may appear as early as the 8-cell stage (A. Sutherland and P. Calarco, unpublished observations).

By contrast, laminin appears earlier in development and has been visualized on the surface of the 2-cell mouse embryo (Dziadek and Timpl, 1985) and on the compacted 16-cell morula, where it outlines intercellular contours (Leivo et al., 1980). Its presence has been noted focally on the cell surface of the 8-cell embryo (Wu et al., 1983). Studies of laminin synthesis indicate that laminin polypeptides B1 and B2 are synthesized at the 4–8-cell stage, whereas laminin polypeptide A is not synthesized until the 16-cell stage (Cooper and MacQueen, 1983). Laminin continues to be present in the blastocyst on both the ICM and trophectoderm cells (Leivo, 1983), but trophoblast cells no longer produce laminin if permitted to attach and grow out in culture (Wu et al., 1983).

An 80,000-M_r extracellular glycoprotein, nidogen, was reported (Timpl et al., 1983), which is present after compaction on 8–16-cell mouse embryos, where it appears to codistribute with laminin (Dziadek and Timpl, 1985).

Thus, of the standard ECM components investigated to date, laminin, nidogen, and possibly HSPG are present before the blastocyst stage. Collagen IV is present at least by the blastocyst stage. Whether they are involved in cell interactions is unclear. It is clear, however, that the formation of the first polarized epithelium of the embryo at compaction is not correlated with the deposition of a defined basal lamina. After blastulation, other ECM components appear at approximately the same time as the differentiation of the endoderm cells from the ectoderm of the ICM. ECM components characterize the nascent basal lamina between these two cell types and may presage the appearance of Reichert's membrane between the parietal endoderm and the mural trophectoderm after implantation.

2.2. The Plasma Membrane

Interactions between cells may also depend on distinct features of the plasma membrane. Scanning electron microscopy (SEM) of preimplantation embryos shows a microvillous exterior surface on oocytes and cleavage-stage embryos (Calarco and Epstein, 1973; Calarco, 1975b; Phillips et al., 1978; Stastna and Prochazka, 1983). Studies of fertilization in the hamster demonstrate close contact between oocyte microvilli and the fusogenic equatorial segment of the sperm (Lin et al., 1975b). Microvilli are also observed in regions of cell contact in the early embryo (Calarco, 1975b). Treatments that lessen cell adhesion in compacted 8-cell embryos reveal a distinct ring of microvilli at the lateral margins between apposed cells (Johnson and Ziomek, 1982; Sutherland and Calarco, 1983). These lateral microvilli are strongly implicated as active mediators of the compaction process (Sutherland and Calarco, 1983). Transmission EM (TEM) studies suggest that tight junctions will form in this region.

Areas of nonmicrovillous or smooth membrane have been noted in at least four cases in the early oocyte and embryo: (1) on the unfertilized egg over the second metaphase spindle (Eager et al., 1976; Nicosia et al., 1977), (2) over the sperm incorporation cone of fertilization in the rat (Phillips et al., 1978), (3) along adjacent cell boundaries at the time of compaction and polarization in the 8-cell embryo (Calarco and Epstein, 1973; Ducibella et al., 1977; Johnson, 1981; Reeve and Ziomek, 1981; Sutherland and Calarco, 1983), and (4) on the portion of ICM and trophectoderm cells facing the blastocoel (Calarco and Epstein, 1973; Calarco, 1975).

In summary, microvilli may be important in establishing the close contact between sperm and egg required for gamete fusion and may also provide close focal apposition between cells at compaction. Although membrane apposition at compaction may also be enhanced by the regions of smooth membrane described, this could be functioning to limit cell contact to the regions of the lateral microvilli.

2.3. Intercellular Junctions

Junction formation does not begin until the 8-cell stage in the mouse. Before that time, blastomeres adhere but can be separated by pipetting and/or by light trypsin treatment. Gap junctions appear for the first time in the compacted 8-cell embryo and are present between all cells of the morula and blastocyst (Lo and Gilula, 1979). The components for gap junctions preexist in the embryo before their assembly, since inhibition of protein synthesis in 4-cell embryos by cycloheximide fails to block their appearance (McLachlin *et al.*, 1983). However, these preexisting components are not sufficient for further gap junction formation between one compacted embryo and another—a process that is cycloheximide sensitive (McLachlin *et al.*, 1983).

Macular tight junctions also appear at the 8-cell stage but do not become zonular until the morula stage (Magnuson *et al.*, 1977), a process that is calcium dependent (Ducibella and Anderson, 1979). After the development of zonular tight junctions and of a permeability seal between external cells destined to become the trophectoderm, blastocoel fluid accumulates. Interestingly, precocious blastulation with the accumulation of intercellular fluid can occur as early as the 2-cell stage, when the embryo is incubated in the lectin wheat germ agglutinin (Johnson, 1983; 1986). In these embryos, the formation of close areas of membrane apposition of sufficient strength provides a permeability seal. Although these precocious "junctions" may prove to be distinct from tight junctions, their formation within a few hours suggests they too may arise from preexisting components.

Desmosomes appear at the blastocyst stage between trophectoderm cells, proximal to tight junctions (Enders and Schlafke, 1965; Calarco and Brown, 1969; Magnuson *et al.*, 1977), where they frequently are associated with bands of filaments running parallel to the plasma membrane (P. Calarco, unpublished observations). By SEM, evidence of junction formation between trophectoderm cells is seen as epithelial ridges between cells of the late blastocyst (Calarco and Epstein, 1973).

In summary, junction formation seems to be requisite for normal mammalian development and represents one of the first examples of cell–cell interaction. Junctional elements may prove to be stockpiled by the maternal genome during oogenesis.

2.4. The Cortex and Cytoskeleton

The cortical region of the embryo must also function in an appropriate manner to permit cell–cell interactions, most likely *via* elements of the cytoskeleton. Here, one must consider the roles of microtubules, microfilaments, intermediate filaments, myosin, and spectrinlike molecules.

Since the initial observation of Ducibella *et al.* (1977) that microtubules are aligned parallel to the apposed cellular membranes of the compacting 8-cell embryo, studies using colchicine or colcemid to inhibit microtubule polymerization have provided conflicting data on the requirement of intact microtubules for maintenance of embryonic shape. Some investigators claim that colcemid does not interfere with compaction or the maintenance of the compacted state (Ducibella, 1982; Wiley and Eglitis, 1980), whereas others agree that compaction can occur in the presence of colcemid but state that it cannot be maintained (Surani et al., 1980; Sutherland and Calarco, 1983), although this may also be due to cells entering mitosis (Maro and Pickering, 1984). Interestingly, after pronase treatment, embryos are completely unable to compact in the presence of colcemid (Sutherland and Calarco, 1983).

A cortical network of thin filaments is present in the unfertilized egg and zygote (Szollosi, 1967; Nicosia et al., 1977) and in the preimplantation mouse embryo (Ducibella et al., 1977; Opas and Soltynska, 1978), possibly corresponding to a submembranous concentration of actin (Lehtonen and Badley, 1980). Microfilaments are also present within microvilli (Calarco and Brown, 1968). There may also be a significant pool of unpolymerized actin monomers in the cytoplasm (Lehtonen and Badley, 1980). When actin polymerization is blocked by cytochalasin, pronounced effects on compacted embryos are seen: Microvilli appear in patches, blastomeres round up, and embryos decompact. Continued incubation blocks cell division as well as blastulation (Surani et al., 1980; Pratt et al., 1981, 1982; Sutherland and Calarco, 1983). To a certain extent, these effects are reversible, and metabolic studies show that embryos continue to develop biochemically in the presence of cytochalasin (Surani et al., 1980). The ring of lateral microvilli between apposed cell surfaces seems to be less sensitive to cytochalasin, suggesting there are at least two populations of microfilaments in the compacted embryo (Sutherland and Calarco, 1983).

Intermediate filaments have not been observed until the late morula to blastocyst stage (Jackson et al., 1980; Lehtonen and Badley, 1980). However, immunoblotting has demonstrated the presence of cytokeratin polypeptides in oocytes and in all preimplantation stages of the mouse (Lehtonen et al., 1983). Because few 10–11-nm filaments are seen before trophoblast outgrowth and because the immunofluorescent staining is diffuse, most of the cytokeratin in cleavage-stage embryos may exist in a nonfilamentous pool (Lehtonen et al., 1983). Work by Oshima et al. (1983) detects the synthesis of two intermediate filament proteins, endo A and endo B, as early as the 4–8-cell stage. Antibodies to these two proteins react with intermediate filaments of the blastocyst by immunoelectron microscopy. Because inhibitors of intermediate filaments are not available, it has not been possible to determine whether their presence is required during early cell interaction events.

Myosin shows a very interesting pattern of localization in the preimplantation mouse embryo (Sobel, 1983). Blastomeres of 2- and 4-cell embryos exhibit a continuous band of cortical myosin in their rounded apical regions but none in flattened regions of cell apposition. This is seen also in later-stage embryos, in which inner cells of the morula fail to show cortical myosin. Thus, the polarized distribution of myosin distinguishes outer from inner regions of the embryo and may be functionally implicated in cell–cell interactions (Sobel, 1983).

Preliminary studies indicate that a spectrinlike molecule may be localized to the cell cortex of preimplantation mouse embryos beginning at the late 2-cell stage (Sobel and Alliegro, 1985, and personal communication). Analogous to the presence of spectrin molecules in erythroid and nonerythroid cells (Lazarides and Nelson, 1982), future work may demonstrate that embryo spectrin functions to bind actin to the plasma membrane as it does in erythrocytes, or in cortical stabilization of the plasma membrane. One might postulate that spectrinlike molecules would play a role in many cell–cell interactions during early development. α-Actinin, another protein that binds actin, is also present in the cortex of early mouse embryos (Lehtonen and Badley, 1980), where it may crosslink actin filaments to each other, as in other cells (Burridge *et al.*, 1982).

The ultrastructural visualization of many of these cortical cytoskeletal elements may soon be at hand. Preliminary work on membrane lawns by Sobel (1984) (see Fig. 2) reveals a cortical meshwork of filaments on the inner face of the plasma membrane. Immunoelectron microscopy to identify individual components of the cortical cytoskeleton is eagerly awaited.

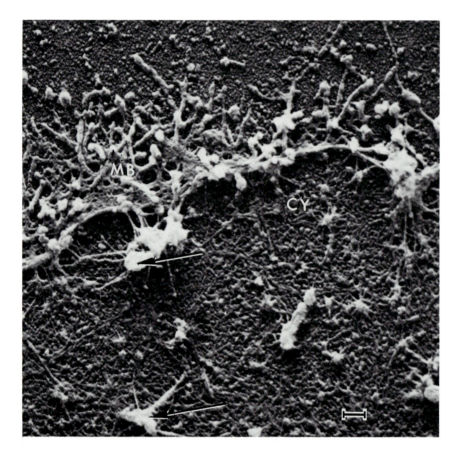

Figure 2. A membrane lawn preparation of a 2-cell mouse embryo attached to a Con A-coated substrate. A portion of the microvillar border (MB) and the cytoplasmic surface of the plasma membrane (CY) attached to the substrate are visible. The cytoplasmic face of the membrane is covered by a dense interlocking meshwork of fine filaments. Thicker filaments extend from centers composed of variable numbers of subunits that are scattered over the field (arrows). Some of the centers appear to serve as anchoring sites for the band of filaments at the cell border. Scale bar: 0.1 μm. (Figure courtesy of Dr. Sobel.)

The cortex may also show an enhanced complement of certain intracellular organelles, probably maintained there by the cytoskeleton. Mitochondria reportedly are localized to cortical regions of apposed membranes at compaction in the mouse (Ducibella, 1977; Wiley and Eglitis, 1981). Smooth membrane cisternae were frequently noted along apposed cell surfaces in the preimplantation sheep embryo (Calarco and McLaren, 1976). Although free polysomes appear to be evenly distributed in early mouse cleavage stages, some portion of them may be localized to the cortex (Calarco and Brown, 1969). Recent work also suggests cortical localization of a specific glycoprotein destined for surface expression in the mouse (see Section 4).

Because the outer cells of the embryo organize into an epithelium, it is not surprising that noncortical organellar localizations typical of epithelia have been observed as well. In the rat, Reeve (1981a) reports a polarized column of organelles between the nucleus and the periphery of the embryo. In the mouse, organelle polarization is less obvious, but a localization of horseradish peroxidase-containing endosomes can be seen in the apical cortex under the polar cap of microvilli (Reeve, 1981b). Nuclear migration away from the cortex toward the base of each cell has also been described in mouse 4- and 8-cell stages (Reeve and Kelly, 1983).

Thus, changes in cell shape that may be requisite for cell interactions in the embryo are likely to be mediated by elements of the cytoskeleton. Before the blastocyst stage, this probably does not involve intermediate filaments. In addition, the cortical localization of several subcellular organelles may play a role in the structuring of the membrane during preimplantation development.

In summary, morphological studies have revealed much about cell structure and cell interactions. Epithelialization at the 8-cell stage in the mouse is associated with changes in the plasma membrane, cytoskeleton, and the extracellular matrix. Although most morphological observations are static in nature, taken together they give insight into the dynamic events of early mammalian development.

3. Molecules at the Cell Surface

Considerable information is available on various components of the plasma membrane from metabolic-labeling and lectin-binding studies. Certain enzymes have been localized to the plasma membrane, transport systems have been characterized, and some surface antigens have been defined in terms of their molecular weights and carbohydrate moieties. Although the immunological characterization of the cell surface has been fruitful, very few of the detected antigens have a known function. This section briefly presents some of this information as an aid to understanding cell interactions.

3.1. Proteins

The first study of embryo surface proteins involved the analysis of glycopeptides cleaved by trypsin from mouse embryos labeled with glucosamine.

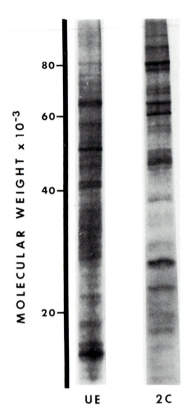

MOLECULAR WEIGHT × 10⁻³

UE   2 C

Figure 3. Marked differences in the pattern of surface proteins exist between mouse unfertilized eggs (UE) and 2-cell embryos (2C). Proteins were labeled by lactoperoxidase-catalyzed iodination and analyzed by SDS–polyacrylamide gel electrophoresis and autoradiography.

Glycopeptides of higher molecular weight characterized the morula to blastocyst stages as compared with the 4–8-cell stages (Pinsker and Mintz, 1973). Pronase digestion of embryos labeled with radioactive fucose for 48 hr also revealed large glycopeptides as the major labeled component of the morula/blastocyst stage (Jacob, 1979). When intact surface proteins on living embryos are iodinated and analyzed by one-dimensional gels, the protein patterns are seen to be nearly identical from the 2-cell through the morula stages with distinct changes in the surface protein pattern occurring after fertilization (see Fig. 3) and at blastulation (Johnson and Calarco, 1980a). These alterations in protein pattern at blastulation may correspond to the appearance of the large glycopeptides noted after proteolytic digesion of the surface, although the specific glycoproteins responsible have not been identified. Interestingly, large glycopeptides characterize both early embryos and undifferentiated teratocarcinoma cells and decrease markedly in size between days 7 and 10 of development (Jacob, 1979).

Several enzymes have been localized to the cell surface or to the immediately subjacent cortex. When morulae are incubated in [UDP]galactose, the substrate for galactosyltransferase, blastulation can be blocked, which suggests a function for this surface enzyme in early development (Shur et al., 1979). Six other membrane-bound enzymes have been shown to be active by the blastocyst stage: alkaline phosphatase, 5′-nucleotidase, Mg^{2+}-ATPase, transport Na^+,K^+-ATPase, cAMP-phosphodiesterase, and adenylate cyclase (Vorbrodt et al., 1977). Alkaline phosphatase, an enzyme often seen in undifferentiated cells, probably localizes to the ICM after blastulation (Johnson et al., 1977; but see Izquierdo and Marticorena, 1975).

Changes also occur during early development in membrane transport systems for carbohydrates, amino acids, and nucleic acid precursors (DiZio and Tasca, 1977; Powers and Tupper, 1977; Sherman, 1979). Briefly, these studies indicate the activation of new transport systems with increasing age and the segregation of certain activities to different regions of the membrane. For example, Na^+-, K^+-ATPases are localized to the blastocoelic surfaces of trophectoderm cells, whereas probable amino acid carriers (both sodium independent and sodium dependent) are located on the external blastocyst surface (DiZio and Tasca, 1977; Keefer and Tasca, 1984).

3.2. Carbohydrates

Carbohydrates exist on the cell surface as both glycoproteins and glycolipids. Much of the information we have about them during early development has been derived from lectin-binding studies. Receptors for the lectins concanavalin A (mannose, glucose), wheat germ agglutinin (N-acetylglucosamine, N-acetyl neuraminic acid), Ricinus communis agglutinin I (galactose) and II (galactose, galactosamine), peanut agglutinin (galactose), and isofucose-binding proteins from lotus (fucose) have been noted on early mouse embryos. For a summary of some of these earlier studies, the reader is directed to the reviews by Jacob (1979) and Johnson and Calarco (1980b). Receptors for Dolichos biflorus agglutinin (N-acetylgalactosamine) have also been described on cleavage-stage mouse embryos (Fujimoto et al., 1982). Briefly, lectin studies to date indicate the presence of a variety of sugars on the cell surface of preimplantation mammalian embryos, some of which show changes during development in distribution and in their ability to be agglutinated or participate in hemagglutination. Because lectins bind to many different molecules, information on the potential developmental function of a lectin receptor cannot be gathered. However, developmental changes in the Con A-binding surface proteins during mouse preimplantation development reported by Webb et al. (1977) and Konwinski et al. (1977) probably have their explanation in the stage-specific appearance and disappearance of a $65,000–70,000-M_r$ glycoprotein doublet on gels (Johnson and Calarco, 1980c,d; Magnuson and Epstein, 1981).

Several studies have explored the effects of glycosylation inhibitors on normal development. Using tunicamycin to block N-asparagine-linked glycosylation in the mouse embryo, Surani (1979) showed that the incorporation of mannose into glycoproteins is inhibited by 80%. A similar inhibition of the incorporation of labeled glucosamine is also seen (Maylie-Pfenniger, 1979). Under these conditions, compaction cannot be maintained, blastulation does not occur, trophoblast adhesion to a substratum is blocked, and giant cell outgrowth is inhibited (Surani, 1979), underscoring the importance of surface glycoproteins in developmental events. Furthermore, binding of both Con A and peanut lectin to the cell surface is reduced, and two proteins of approximately 68,000 M_r and 165,000 M_r fail to appear (Surani et al., 1981). The most striking surface changes visible on tunicamycin-treated embryos are the loss of cell–cell adhesion and the eventual loss of microvilli (Sutherland and Calarco, 1983).

Compactin, an inhibitor of HMG CoA reductase, which normally converts acetate to mevalonic acid, also causes decompaction of embryos and an inhibition of protein glycosylation—effects that are reversible by the addition of exogenous mevalonic acid (Surani et al., 1983). These effects were interpreted as resulting from a lack of dolichol intermediates required for glycosylation and further underscore the importance of glycosylation in preimplantation developmental interactions.

A recent study comparing the polysaccharides synthesized by preimplantation mouse morulae and F9 teratocarcinoma cells indicates that embryos contain fewer large and less acidic polysaccharides than do F9 cells (Iwakura, 1983). Interestingly, embryos have most of their mannose-labeled glycopeptides as the high mannose type with only a small proportion as the complex type, suggesting some immaturity in carbohydrate processing (Iwakura, 1983). Since most of the processing of the high mannose type to the complex type of oligosaccharide occurs in the Golgi apparatus, this observation may be explained by the small amounts of classical Golgi material present in preimplantation embryos (Enders and Schlafke, 1965; Calarco and Brown, 1969; Polak-Charcon et al., 1983a). This paucity of Golgi material might also explain the glycopeptide differences between early embryos and blastocysts noted in Section 3.1.

3.3. Glycolipids and Lipids

Although little is known about the total glycolipid composition of the cell surface of early embryos, work with several monoclonal antibodies has revealed specific information on several developmentally regulated glycolipid antigens. For a summary of this work, see Chapter 13 in Volume 4 of this series (Banka and Calarco, 1986). In brief, even the addition of a single terminal sugar can radically change the antigenicity of surface glycolipids, a mechanism that may be exploited in developmental interactions by the early embryo.

Exogenous fatty acids seem to be required for the development of early mouse embryos, at least up to the 8-cell stage (Quinn and Whittingham, 1982), as well as rabbit embryos (Kane, 1979). However, several studies report the ability of the preimplantation mouse embryo to synthesize major membrane lipids. Labeled glucose can be used in lipid synthesis in early embryos (Flynn and Hillman, 1978). Choline can be incorporated into lipid as early as the 2-cell stage and increases severalfold by the blastocyst stage (Pratt, 1980). Labeled palmitic acid is incorporated primarily into neutral glycerides at the 8-cell stage, although some polar lipids are synthesized, primarily choline phosphatides (Flynn and Hillman, 1980). Mevalonate can be converted into the membrane sterols lanosterol and cholesterol from the 2-cell stage on, with cholesterol synthesis more characteristic of the older, compacted embryo (Pratt, 1982). Incubation of mouse embryos in oxygenated sterols such as 7-ketocholesterol is capable of blocking compaction (Pratt et al., 1980) and supports the importance of sterol synthesis to normal development (Pratt, 1978).

Just when the embryo begins to modulate its own membrane composition is not clear, although available evidence indicates that it is capable of doing so by the 2-cell stage. Functionally, the presence of cholesterol probably regulates membrane fluidity by increasing the mechanical stability of the lipid bilayer. One might therefore predict that compaction would be accompanied by a localized decrease or loss of cholesterol from the basolateral regions of apposing cell membranes. Whether changes occur in the composition of the egg's membrane after fertilization is not known, but available evidence from membrane diffusion studies shows no generalizable effect on the diffusion of membrane proteins in the newly fertilized mouse egg, suggesting rather an alteration in the ensemble of lipid domains (D. E. Wolf et al., 1981; D. E. Wolf and Ziomek, 1983).

In summary, the presence of sterols and phospholipids in embryonic membranes and their interactions probably stabilizes and regulates membrane fluidity and enzyme activity as occurs in other cells. Changes in lipid composition appear likely to accompany major cell interaction events, such as compaction and blastulation.

3.4. Antigens

A large number of surface antigens expressed during early mouse development have been detected, but it is not my intention to describe all of them. Instead, the reader is directed to several reviews of this topic (Solter and Knowles, 1979; Wiley, 1979; Johnson and Calarco, 1980b; Heyner, 1980b, 1984; Banka and Calarco, 1986). Only those surface antigens that have been described biochemically, functionally, or both, are dealt with in this chapter.

3.4.1. Compaction-Related Antigens

3.4.1a. The F9 Antigen—Uvomorulin? This "antigen" has been characterized by syngeneic sera, heterosera, and monoclonal antibodies. It is clear that published results with these various sera have not all described the same molecular antigen. Early work with syngeneic sera described antigens present on sperm, but not oocytes, that appear on the zygote surface a few hours after fertilization, peak in expression at the 8–16-cell stage, and are present in reduced amounts on both the trophoblast and inner cell mass of the blastocyst

(Artzt *et al.*, 1973; Jacob, 1977). The small amounts of antibody available prohibited the biochemical characterization of the F9 antigen. At this point we cannot be sure which, if any, of the reported antibodies to F9 cells correspond to the original anti-F9 serum (Artzt *et al.*, 1973), so it is perhaps wiser not to designate an *F9 antigen*.

However, one antigen has been described functionally and biochemically, using primarily sera raised against F9 cells. Work with a rabbit heteroserum made against F9 cells showed that Fab fragments (Kemler *et al.*, 1977) or divalent antibodies (Ducibella, 1980) could cause the decompaction or prevent the compaction of mouse embryos at the 8-cell stage. Using detergent extraction and trypsin treatment of F9 or PCC4 teratocarcinoma cell membranes, Hyafil and co-workers (1980) extracted a Fab-target molecule that inhibits these anti-F9 effects on cells and embryos. This tryptic fragment (UMt) of a larger surface molecule (uvomorulin) is a glycoprotein of 84,000 M_r (Hyafil *et al.*, 1981) that, when added to the culture medium of early mouse embryos, will completely inhibit the decompaction effects of anti-F9 Fab, presumably by neutralizing the antibody fragment (Hyafil *et al.*, 1980). These workers conclude that either this 84,000-M_r UMt or a cross-reactive molecule is involved in mouse embryo compaction. A calcium-dependent equilibrium between two conformations of UMt has been proposed to explain the observed dependence of compaction on calcium (Hyafil *et al.*, 1981). According to this model, the presence of calcium results in a conformation of UMt that is resistant to trypsin digestion and that can be recognized by a conformation-specific monoclonal antibody (DE1); the absence of calcium results in a UMt conformation that is trypsin sensitive and that cannot be recognized by DE1 (Hyafil *et al.*, 1981). Later work reports that the intact 120,000-M_r uvomorulin exists as an exocellular molecule on teratocarcinoma cells and certain mouse organs; early embryos were not examined (Peyrieras *et al.*, 1983). A rabbit antibody similar to DE1 also detects the 123,000-M_r nonintegral glycoprotein uvomorulin characteristic of epithelial cells and implicated in adhesion (Vestweber and Kemler, 1984). Ultrastructural immunochemistry with this antibody shows a restricted localization of uvomorulin to the intermediate junctions of adult intestinal epithelial cells (Boller *et al.*, 1985).

Monoclonal antibodies to F9 cells include three IgMs reacting with preimplantation embryos (Kemler *et al.*, 1979): (1) DE1, which was actually prepared against PCC4 Aza teratocarcinoma cells and purified UMt (see preceding paragraph); (2) IIC3, which reacts with morula and trophoblast cells of blastocysts (Marticorena *et al.*, 1983); and (3) anti-SSEA-1, which is described below. All but DE1 detect molecules different from uvomorulin.

Although incubation in anti-F9 causes the loss of tight and gap junctions in confluent teratocarcinoma cells (Dunia *et al.*, 1979), it is not clear whether this is a primary or a secondary effect of the incubation in anti-F9. Any direct effects on junctions in the mouse embryo can only be inferred from its inhibition of compaction. Electron microscopic (EM) localization of uvomorulin in early embryos is needed to verify its association with junctions. With the advent of gel overlay techniques, it would be most interesting to see whether anti-F9 and DE1 recognize specific molecules or perhaps carbohydrate moieties on early embryos.

3.4.1b. Cadherin. Takeichi and co-workers (Yoshida-Noro et al., 1984) have described a monoclonal antibody raised against trypsin-treated F9 cells that detects a calcium-dependent adhesion system functioning at compaction in mouse embryos. The major antigen detected by this antibody has a molecular weight of 124,000 and is also detected by DE1 (Yoshida-Noro et al., 1984) and thus corresponds to the antigen "uvomorulin" described in Section 3.4.1a. A thorough discussion of cadherin and related antigens can be found in Chapter 9.

3.4.1c. Cell CAM 120/80. Two antisera were involved in the description of this antigen: One was prepared in goats against material shed into serum-free medium by MCF-7 human mammary carcinoma cells, and the second was prepared in rabbits against a glycoprotein of 80,000 M_r detected by the first antibody and eluted from gels (Damsky et al., 1983). Both reagents detect an antigen on mouse embryos that appears to be related to cell–cell adhesion, since incubation of precompacted embryos in the antibody blocks compaction. The antigen (which appears to be 120,000 M_r in detergent extracts of epithelial cells, but 80,000 M_r after shedding into the medium) is restricted to sites of cell–cell interaction in a variety of epithelial cells (Damsky et al., 1983).

3.4.1d. SFM-I-Defined Antigen. The anticompaction activity of the goat antibody SFM-I does not appear to be absorbed by purified GP80, suggesting that SFM-I may be detecting an antigen unrelated to cell CAM 120/80 described in Section 3.4.1c (Richa et al., 1985). Although other interpretations are possible, a second antigen may be involved in compaction, possibly a ligand recognized by uvomorulin–cadherin–CAM120/80.

3.4.2. Blood-Group Related Antigens

Several antigens have been defined in terms of their relatedness to specific blood group antigens.

3.4.2a. The Forssman Antigen. A monoclonal antibody prepared from a rat immunized with mouse spleen has been used to detect a glycolipid with the tissue and species distribution of the Forssman antigen (Stern et al., 1978). This antibody recognizes antigens on trophectoderm cells of blastocysts that decrease markedly after hatching; the ICM remains positive.

3.4.2b. Globotetraosylceramide (Human Blood Group P Antigen). Rabbit antibodies prepared against neutral glycolipids detect the presence of globotetraosylceramide beginning at the 2–4-cell stage and reaching a peak at the morula stage (Willison *et al.*, 1982). The data indicate that this antigen is carried predominantly by glycolipids.

     3.4.2c. Stage-Specific Embryonic Antigen-1. SSEA-1 is recognized by a monoclonal IgM antibody prepared against F9 teratocarcinoma cells (Solter and Knowles, 1978). The antigen appears first on cells of the 8-cell-stage mouse embryo, is maximally expressed at the morula stage, and decreases on trophoblast cells but remains high on inner cell masses. It is also present on sperm but not on oocytes (Solter and Knowles, 1979). Subsequent work has demonstrated that anti-SSEA-1 recognizes a terminal trisaccharide that is formed by the $\alpha 1$–3-fucosylation of blood group I or i antigens (Gooi *et al.*, 1981) and is carried by a set of lactoseries glycolipids (Kannagi *et al.*, 1982). This represents the most completely characterized of the mouse embryo surface glycolipids and suggests that glycosylation changes alone could be a basis for stage-specific expression of antigens. However, one of the limitations of monoclonal antibodies is also evident from the observed reactivity of anti-SSEA-1 with other tissues. Antigens detected by them may not be functionally similar; i.e., several antigens may possess the same terminal trisaccharide.

     3.4.2d. Stage-Specific Embryonic Antigen 3. SSEA-3, which is detected by a monoclonal IgM antibody made against 4–8-cell embryos, is present on mouse oocytes and early embryos and becomes restricted to the inner cell mass and eventually to the primitive endoderm as development progresses (Shevinsky *et al.*, 1982). Expression of this carbohydrate antigen appears to be inherited from the cytoplasm of the unfertilized ovum (McGrath and Solter, 1983) and may prove a useful marker for extraembryonic visceral endoderm (Fox *et al.*, 1984). Studies on human teratocarcinoma cells indicate that the SSEA-3 antigenic determinant may be carried by a set of globoseries glycolipids that bear some relatedness to blood group antigens (Kannagi *et al.*, 1983).

3.4.3. Stage-Specific Embryonic Antigen 2

     Anti-SSEA-2 is an IgM antibody prepared in Balb/c mice against a somatic cell hybrid containing the genome from a Balb/c mouse and human chromosome 7 with integrated SV40 (Shevinsky *et al.*, 1981). SSEA-2 is maximally expressed on 4–8-cell mouse embryos and decreases in amount on morulae and blastocysts. Shevinsky and co-workers suggest that the antigenic determinant may well be a carbohydrate. Most of the work in this field would suggest a glycolipid as the most likely carrier of the antigen; in fact, Shevinsky *et al.* (1981) were unable to detect a protein component to the antigen by immunoprecipitation.

3.4.4. OTT6050-Related Antigens

     Using a rabbit antiserum prepared against embryoid bodies of the teratocarcinoma line OTT6050, Webb (1980) described reactivity against several mouse embryo surface proteins that were iodinated by a lactoperoxidase-catalyzed reaction. Immunoprecipitation and gel analysis show proteins of 115,000, 70,000, and 48,000 M_r on cleavage-stage embryos, as well as the loss of the 70,000-M_r protein and the new appearance of 82,000- and 22,000-M_r proteins on the blastocyst (Webb, 1980). In the absence of complement, this antiserum did not have deleterious effects on *in vitro* development of preimplantation embryos.

3.4.5. BL Antigens

The first report of stage-specific glycoprotein surface antigens immunoprecipitated directly from preimplantation embryos was made using a rabbit antiserum prepared against mouse blastocysts mechanically freed of their zonae (Johnson and Calarco, 1980c,d). Their synthesis is confined to the period between the 2-cell and morula stages of mouse development, whereas their expression on the cell surface begins at the 4-cell stage, peaks at the 8-cell to morula stages, and declines thereafter. Biochemical characterization of these antigens has shown them to be a pair of glycoproteins with apparent molecular weights between 65,000 and 70,000, with a single protein species of 60,000 M_r detected when embryos are cultured in tunicamycin. Two-dimensional gel analysis shows the BL antigens to consist of at least three pairs of proteins showing charge heterogeneity on isoelectric focusing, the predominant pair at an isoelectric point between 6.5 and 7.0. A possible functional role of these surface antigens is suggested because incubation in anti-BL2 causes a concentration-dependent decrease in the proportion of embryos reaching the blastocyst stage (Johnson and Calarco, 1980c). These are the same molecules detected by con A-mediated precipitation (Magnuson and Epstein, 1981) and may be the same as the 70,000-M_r species reported by Webb *et al.* (1980). Their disappearance at the blastocyst stage could also explain the changes in surface binding of con A reported by Konwinski *et al.* (1977).

3.4.6. Intracisternal A Particle Antigens

The only viral antigens known to be expressed on the surface of the preimplantation embryo have been detected by an antiserum made in rabbits against intracisternal A particles (IAP) purified from the MOPC-104E cell line (Huang and Calarco, 1981a). These antigens appear first on zygotes, reach maximal expression on 2–8-cell embryos, and are absent from morulae and all later stages. Although immunoprecipitation from NP-40 lysates of embryos shows the presence of the 73,000-M_r core protein characteristic of IAP (Huang and Calarco, 1981b), this does not appear to be the protein precipitated from embryos surface-labeled with ^{125}I, where two proteins of 25,000 and 50,000 M_r are faintly detected (P. Calarco and M. C. Siebert, unpublished observations). By ultrastructural immunocytochemistry, this antibody also reacts with the viruses seen budding into ER cisternae during the 2–8-cell stage (Huang and Calarco, 1982), which have been termed ϵ particles (Yotsuyanagi and Szollosi,

1981). Whether ε particles and IAP share more than morphological and immunological similarities remains to be determined. Interestingly, incubation of the zygote in actinomycin D blocks the appearance of these ε particles at the 2-cell stage, but incubation in α-amanitin does not (Calarco, 1975a), the latter finding suggesting that the viral RNA is transcribed before this time.

3.4.7. gp67 Antigen

A rabbit heteroserum prepared against a nondifferentiating clonal teratocarcinoma stem cell line SCC-1 (Martin and Evans, 1975) has been used to define a glycoprotein of 67,000 M_r (gp67) with an approximate pI of 5.3 (Banka and Calarco, 1985). gp67 appears to be heavily glycosylated and may be located on the cell surface of early embryos. gp67 also appears on pluripotential PSA-1 teratocarcinoma cells but not on a variety of differentiated adult cells (Banka and Calarco, 1985). Interestingly, culture in this antibody interferes with normal development (Calarco and Banka, 1979).

More specific antibodies to this nondifferentiating stem cell line include (1) a syngeneic IgM that recognizes antigens appearing first on the 8-cell embryo and present, at least, on the ICM and 6½ day embryonic ectoderm (Hahnel and Eddy, 1983); (2) a monoclonal IgM antibody (5D4) that detects antigens of a probable carbohydrate nature that are present on unfertilized eggs and zygotes, decline in expression to the 8-cell stage, and are absent from trophectoderm cells (Stern et al., 1983); and (3) two monoclonal antibodies (C6 and A5) that recognize no antigens on intact early embryos, but do recognize a carbohydrate specificity after treatment of embryos with neuraminidase (Fenderson et al., 1983; see also review of carbohydrate-containing antigens in Banka and Calarco, 1986). None of these appears to detect gp67, nor do they appear to overlap in specificity with each other. This is not surprising in view of the complexity of the immunogen; similar results were obtained when F9 cells were used as an immunogen (see Section 3.4.1a).

3.4.8. Histocompatibility Antigens

$β_2$-Microglobulin ($β_2$m) is a 12,000-M_r protein noncovalently linked to the major and minor histocompatibility antigens on the plasma membrane. Sawicki et al. (1981) showed that $β_2$m synthesis begins at the 2-cell stage in the mouse embryo but that surface expression as assessed by immunofluorescence does not occur until the blastocyst stage. In the pig it has been shown by EM immunochemistry that $β_2$m is expressed first at the early blastocyst stage and is localized to the apical surface of the trophectoderm cells from that stage onward (Meziou et al., 1983). It is likely that surface expression is linked to embryonic expression of major and minor histocompatibility antigens (see below).

Many studies have examined the question of histocompatibility antigen expression on the preimplantation mouse embryo. To summarize briefly, major histocompatibility antigen, or H-2, expression is easily detected at the blastocyst stage, where expression predominates on the inner cell mass and its derivatives. Low levels of H-2 expression seen on trophectoderm cells apparently disappear before implantation (Hakansson *et al.*, 1975; Searle *et al.*, 1976). Low levels of H-2 antigen expression on some blastomeres of the 8-cell stage can be detected by ultrastructural immunochemistry (Warner and Spannaus, 1984). In contrast to the reports of early H-2 expression, synthesis of H-2 has not been reported before day 18 in the mouse (Sawicki, 1984). H-Y antigen expression begins at the 8-cell stage (Krco and Goldberg, 1976; Epstein *et al.*, 1980), and minor histocompatibility antigen expression has been detected as early as the 2-cell stage in the mouse (Palm *et al.*, 1971; Muggleton-Harris and Johnson, 1976). For more complete information on this topic, the reader is directed to Chapter 12 of Volume 4 of this series (Heyner, 1986).

Mta is a maternally transmitted cell surface antigen of the mouse that is detected by unrestricted cytotoxic T lymphocytes on cells of most strains of mice. Mta expression is either positive or negative and is not influenced by paternal Mta type. It is of interest for us to consider because as early as 10 hr after fertilization, Mta type becomes set and is not influenced by transfer to a uterine environment of the opposite type (Lindahl and Burki, 1982). Although Mta appears to be inherited cytoplasmically, it is not known when the genetic element specifying Mta is duplicated, transcribed, and translated or whether it is associated with a cytoplasmic organelle. Lindahl *et al.* (1983) suggest that the H-2 linked gene, *Hmt*, is the structural gene for Mta and that Mta belongs to class I of histocompatibility antigens, a 40,000–50,000-dalton heavy chain associated with β_2m. Hence, the expression of an adult surface antigen is shown to be dependent on a maternal cytoplasmic contribution, possibly mitochondrial in nature.

3.4.9. T-Complex Antigens

Numerous studies have been directed toward the surface expression of antigens specified by the T complex during early mouse development (for review, see Johnson and Calarco, 1980b). The T complex is an interesting region of chromosome 17 in which specific *t* mutants and all members of their complementation group show an association with a specific H-2 haplotype (Levinson and McDevitt, 1976). The T region, marked by crossover suppression, shows an altered transmission ratio and may eventually be explained as a physically altered region of the chromosome (Silver and Artzt, 1981; Shin *et al.*, 1983). Whatever the explanation for the altered T region of chromosome 17, research suggests that different surface antigens are expressed by each *t* mutant haplotype, as well as their counterpart wild-type haplotypes on testicular cells, sperm, and early embryos. The major antigenic determinants of these *t*-antigens are reported to be oligosaccharides (Cheng and Bennett, 1980). This observation has led to the isolation of the only *t*-related antigen described in molecular terms to date—an 87,000-M_r glycoprotein (gp87) identified by anti-t^{12} antisera (Cheng *et al.*, 1983). N-linked carbohydrates comprise about one-half the molecular weight of gp87, which shows a higher level of galactosylation on *t*-bearing testicular cells. In early wild-type embryos, gp87 synthesis peaks at the 4–

8-cell stage but peaks at the 8-cell to morula stage in litters consisting entirely of homozygous and heterozygous t^{12} embryos (Cheng et al., 1983). The gp87 molecule thus appears to be either specified by genes in the t^{12} haplotype or altered in some way by the presence of t^{12}. This appears to be a useful approach to untangling the various postulated t antigens and should be combined with studies to explore t antigen function in early development.

3.4.10. Substratum Adhesion Antigen (GP 140, or CSAT ag)

A goat antibody raised against glycoproteins isolated from BHK (baby hamster kidney) fibroblasts defines a set of three cell–substratum adhesion glycoproteins of 120,000–160,000 M_r (Knudsen et al., 1981). GP 140 also appears to be involved in trophoblast attachment of the mouse blastocyst during outgrowth in vitro (see Fig. 4) (Richa et al. 1985). Its function in vivo remains to be investigated. Studies on chick myogenic cells and fibroblasts show that CSAT ag co-aligns with portions of stress fibers and codistributes with fibronectin (Damsky et al., 1985).

3.4.11. Summary

To summarize, the early embryo expresses a variety of serologically detectable molecules on its cell surface, both glycoprotein and glycolipid in nature. Several of these antigens are expressed in a stage-specific manner. Three of these (uvomorulin, CAM 120/80, and cadherin) are implicated in the process of compaction in the mouse and may be different names for the same molecule. Another antigen (GP140) seems to mediate attachment to a substratum by the embryo during outgrowth in vitro. Much of the antigenic specificity described for the early embryo seems to reside in the carbohydrate portions of glycolipids. Although the complexing of certain antigens with their antibodies interferes with development, no specific function for most of these molecules is known. Future work should be directed toward elucidating antigen function(s) in early development.

4. Synthesis and Expression of Surface Glycoproteins

In the ovulated oocyte of the mouse, endoplasmic reticulum (ER) usually exists only as small saccules without associated ribosomes (Calarco and Brown, 1969; Szollosi, 1971). These ER saccules are associated with mitochondria, an association that arises during late oogenesis. After the onset of ribosomal RNA synthesis at the late 2-cell to early 4-cell stage, these saccules accumulate a few ribosomes, but elongated RER cisternae are not observed until the late morula to blastocyst stage (Enders and Schlafke, 1965; Calarco and Brown, 1969; Hillman and Tasca, 1969). In nine additional mammalian species, it has been noted that RER is scant, even at the blastocyst stage (Enders and Schlafke, 1965). Similarly, a characteristic supranuclear Golgi apparatus is not usually observed

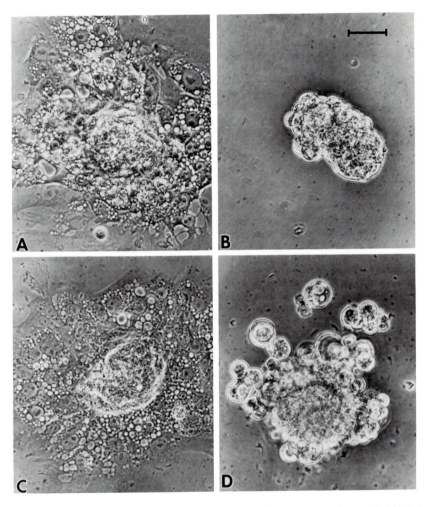

Figure 4. Effect of antiserum to GP 140 on attachment of blastocysts in culture. (A–C) Blastocyst cultured for 40 hr in control medium, DMA (A), in GP 140 antiserum (1 : 100) (B), or in a mixture of GP 140 antiserum (1 : 100) and GP 140 (1.5 μg) purified from BHK cells (C). (D) Blastocyst cultured for 48 hr in DMA, permitting attachment to the substratum, followed by 20-hr incubation in GP 140 antiserum (1 : 100). Scale bar: 100 μm. (After Richa *et al.*, 1985, with permission from Academic Press, Inc.)

before the blastocyst stage in the mouse, although variable amounts of Golgi material can be observed in early cleavage stages (Enders and Schlafke, 1965; Calarco and Brown, 1969). As noted earlier, this paucity of Golgi material (where processing of high mannose-type to complex-type oligosaccharides takes place) may explain the finding that mouse preimplantation embryos have most of their mannose-labeled glycopeptides as the high mannose type with only a small proportion as the complex type (Iwakura, 1983). Taken together, these observations suggest that the early mammalian embryo is deficient in organelles thought to be required for protein synthesis, glycosylation, and translocation to the cell surface (Sabatini *et al.*, 1982). Yet embryos do synthesize and express surface glycoproteins, as the research described above on surface antigens attests.

We have recently used ultrastructural immunocytochemistry to explore this seeming contradiction and have focused initially on the intracellular localization of antigens recognized by an anti-blastocyst serum (A-BL$_2$) made in rabbits against mouse blastocysts (see Section 3.4.5). These antigens (BL) can be labeled with [^{35}S]methionine and [^{3}H]glucosamine and are synthesized from the 2-cell stage through the early blastocyst stage. Thus, synthesis of the protein portion of BL at the 2-cell stage begins before the appearance of any RER in the embryo, and glycosylation of BL (which can be blocked by tunicamycin at the 2-cell stage) also occurs before the appearance of a classical Golgi apparatus. BL expression on the cell surface begins at the late 2-cell to early 4-cell stage, peaks at the 8–16-cell stage, and ends during the late blastocyst stage, demonstrating that translocation of BL antigens to the cell surface also begins before the appearance of a characteristic Golgi apparatus. BL antigens are a pair of glycoproteins with apparent molecular weights of 65,000–70,000, with a single protein species of 60,000 M_r detected when N-linked glycosylation is blocked by tunicamycin (Johnson and Calarco, 1980c,d).

The results of our studies to date show that BL antigens are localized exclusively to the cortical region of unfertilized eggs and zygotes, subjacent to the plasma membrane. Beginning with the first cleavage the distribution of BL in the cytoplasmic cortex is seen to be polarized, with expression confined to the region underlying the free or nonapposed plasma membrane of blastomeres and no antigens present at the region of cell contact. This pattern of BL antigen distribution is seen in all subsequent stages with antigen observed only in the cortex underlying the free or nonapposed surface of the embryo, whereas antigens detected by a rabbit antibody prepared against mouse L cells are distributed throughout the embryo (compare Fig. 5A and B). Immunocytochemistry at the EM level with a peroxidase (Fig. 5C) or colloidal gold (Fig. 5D) marker verifies the cortical localization of BL antigens. Apposed regions of membrane (Fig. 5C), inner cells in the morula, and the inner cell mass of the blastocyst are all negative for BL antigens. There were no other intracellular sites of antigen localization; RER and Golgi (normal sites for synthesis of surface glycoproteins), when observed, were always negative (Polak-Charcon *et al.*, 1983*a,b*, 1985).

Several interesting points have emerged from these studies. First, the presence of BL antigen in the unfertilized egg shown by EM immunocytochemistry and Western blots (Polak-Charcon *et al.*, 1985) indicates that early synthesis of BL antigen occurs on oocyte mRNA and then ceases. Embryonic synthesis of BL antigen does not begin until the 2-cell stage. Second, we see an example of membrane regionalization with the BL antigen at the 2-cell stage that exists before polarization or compaction (Handyside, 1980; M. H. Johnson *et al.*, 1981; Sutherland and Calarco, 1983) or the development of tight junctions (Calarco and Brown, 1969; Magnuson *et al.*, 1977; Ducibella and Anderson, 1979). Whether this represents nonassociation of BL antigens with regions of new

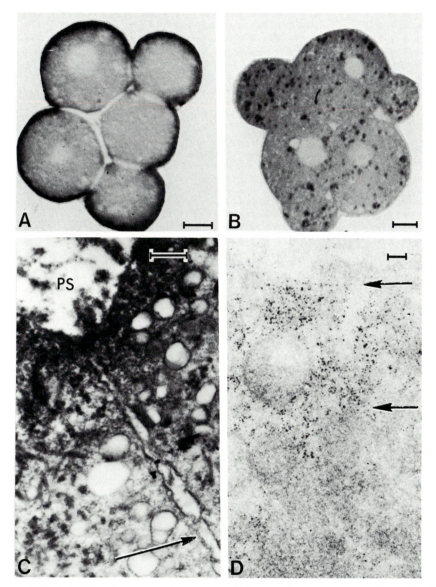

Figure 5. Unstained 8-cell mouse embryos permeabilized with saponin, exposed to antibodies to BL antigens (A, C, D) or mouse L-cell antigens (B) and processed immunocytochemically. (A) Light micrograph of 1–2 μm Epon section of an 8-cell embryo treated with antibodies to BL and labeled by the immunoperoxidase procedure. Labeling is confined to the free or apical cortex of cells. Scale bar: 10 μm. (B) As in A except 8-cell embryo was exposed to antibody to L cell antigens before immunoperoxidase labeling. Antigen is found throughout the cytoplasm in dark spots of varying sizes. Scale bar: 10 μm. (C) Electron micrograph of thin section of an 8-cell embryo exposed to antibodies to BL antigens and labeled by immunoperoxidase. Note concentration of label in cortex and lack of label along cell–cell boundary (arrow). PS, perivitelline space. Scale bar: 1 μm. (D) Electron micrograph of 8-cell embryo. Protein A–colloidal gold (5 nm) used subsequent to exposure of embryo to BL antibodies. Gold particles are concentrated in the cortex between the arrows. Scale bar: 0.1 μm.

membrane or co-localization of BL with elements of the cytoskeleton is not yet known. One might speculate that antigen distribution reflects an earlier event: the association of the mRNA for the BL antigen with some cortical element during oogenesis. In fact, work from several laboratories indicates that mRNA, and, secondarily, polysomes are associated with the cytoskeleton (Farmer et al., 1983; Kasamatsu et al., 1983; Cervera et al., 1981), and specific localization of mRNA to the cortical cytoskeleton has been reported in sea urchin (Moon et al., 1983) and Chaetopterus embryos (Jeffery, 1983) and in ascidian follicle cells (Jeffery, 1982).

However, the most interesting aspect of these studies is the suggestion that the localization of BL antigens in the cortex may reflect the site of antigen synthesis. This would further suggest an unusual pathway for the synthesis and translocation to the surface of certain membrane glycoproteins in early embryos. Although little RER is present, free polysomes are present abundantly throughout the cytoplasm and in the cortical region of early embryos (Calarco and Brown, 1969), as are a variety of small vesicles. Although several hypotheses allowing for the free polysome synthesis of the protein portion of an integral membrane protein are presented in the literature, and several cases of such free polysome-synthesized proteins destined for chloroplasts, mitochondria, or peroxisomes exist, Sabatini and co-workers (1982) conclude that there is no well-documented case for insertion of such a protein into the plasma membrane. Recent work on pp60 src, however, demonstrates that a free polysome-synthesized protein can insert into the cytoplasmic face of the plasma membrane (Courtneidge and Bishop, 1982). Alternative explanations are that small amounts of BL antigen synthesized in a different location move rapidly to the cell cortex, where they reach a concentration sufficient for immunocytochemical detection or that the conformation of BL antigens during early stages of synthesis may make them unrecognizable by the antibody.

An interesting working hypothesis, then, is that at least some portion of BL synthesis occurs in the cell cortex. The most likely vehicles for the glycosylation and translocation of BL to the surface are the small vesicles observed in the cortex, which may be Golgi derived. Future research will include the verification of the cortex as a site of BL antigen synthesis, the development of monospecific antibodies to the oligosaccharide portions of BL, and similar studies of other antigens synthesized by the early embryo, in order to delineate the features involved in surface glycoprotein expression in early mammalian development.

5. Developmental Processes Mediated by Cell–Cell Interaction

As we have seen in preceding sections, changes in the array of surface glycoproteins and glycolipids expressed after fertilization may occur by several mechanisms. Molecules synthesized before fertilization may be unmasked, either by delayed insertion into the plasma membrane or by slight changes (of even one sugar) in the oligosaccharide portion of those preexisting in the membrane. Some glycoproteins, dominated by forms with immature oligosaccharide chains, will be newly synthesized, secreted, or expressed on the surface, perhaps by a somewhat unusual route. Changes in the array of surface molecules that the embryo presents to its environment are mandated both by the developmental program of the embryo and the maternal oocyte cytoplasm. These changes undoubtedly regulate—in concert with the cytoskeleton—those processes involved in cell–cell interactions.

We know that several developmental events in the preimplantation mammalian embryo depend explicitly on cell–cell interaction. For the purposes of discussion, these have been divided into recognition, adhesion, motility, and communication, although there is significant overlap between some of these categories.

5.1. Cell Recognition

In early mammalian development, two prime examples of recognition between cells occur at fertilization and implantation. Both events bring into contact unlike cells—cells that usually differ genetically and whose surface molecular composition is quite different. These recognition events may well differ from the recognition events that occur between like cells of an embryo or between cells of a tissue. We may also expect that there is something unique about recognition between sperm and egg, since this results in membrane fusion, rather than the coexistence of two cells that are in contact and communication. Although one might predict a high degree of specificity for these two recognition events on which the future of a species rests, this does not seem to be the case at a cell–cell level.

A great deal is now known about mammalian fertilization, and the reader is directed to several excellent reviews on this topic (D. P. Wolf, 1981; Yanagimachi, 1981; Wassarman, 1983). Unfortunately, many questions remain on the specific topic of cell recognition under consideration here. It appears that the major recognition event in mammalian fertilization, the event that is responsible for most of the species specificity of fertilization, occurs between the sperm and the zona pellucida. In the mouse, Bleil and Wassarman (1980) identified a specific zona glycoprotein of 83,000 M_r, ZP3, which serves as a sperm receptor, with O-linked carbohydrates providing the specificity for the recognition (Florman and Wassarman, 1985). This zona–sperm recognition event may be mediated by lectins, for recent work with guinea pigs demonstrates that it can be blocked by fucoidin, a highly sulfated carbohydrate containing mostly L-fucose, which binds to the inner acrosome membrane (Huang et al., 1984).

Less is known about cell–cell recognition between egg and sperm. When the zona pellucida is removed from mammalian eggs, they frequently can be penetrated by heterologous sperm, although varying degrees of species specificity exist (Yanagimachi, 1981). For example, the mouse egg will bind only mouse sperm, but the hamster egg is more egalitarian and will bind a variety of foreign sperm, although still preferring its own (Yanagimachi, 1981). Morphologically, this binding event occurs between microvilli on the egg's surface and the sperm plasma membrane, although there is disagreement as to whether it involves the postacrosomal region or the equatorial segment of the sperm. Biochemically, the surface molecules of the egg that are involved in sperm recognition have not been identified in any species and may differ from one species to another. For example, protease treatment of mouse eggs decreases their ability to be fertilized but has no effect on hamster eggs (D. P. Wolf *et al.*, 1976). Although there are many difficulties in the identification of surface molecules (which are by definition transient) from material that is as limited as mammalian eggs, it is disappointing that work to date has not pinpointed the egg receptor. However, the sperm molecule(s) involved in recognition and binding to the egg plasma membrane may prove easier to characterize biochemically, since material is not as limiting. In fact, it has been shown that antibodies made to human sperm (Menge and Black, 1979) or hamster sperm (Yanagimachi *et al.*, 1981) can prevent sperm attachment and penetration of the zona-free egg. O'Rand and co-workers have made progress toward this goal with the identification of a low-molecular-weight family of glycoproteins (RSA) on rabbit and human sperm (O'Rand *et al.*, 1984; O'Rand and Irons, 1984); antibodies to RSA will block fertilization and also demonstrate RSA on the egg after fertilization. Needless to say, future work in this area is needed and would have profound medical implications, not only for its obvious contraceptive potential, but for understanding events permissive for cell fusion as well.

The second example of cell recognition mentioned above was that occurring between trophectoderm cells and the uterine epithelium at implantation. Some species exhibit surface changes in charge, lectin binding, and antigen expression of the embryo at blastulation, but these changes do not seem to be a universal feature of implantation. For instance, in the rat, no gross reduction in surface glycoproteins or change in charge at implantation was noted, and material of a possible secretory nature that could aid in adhesion was observed between the embryonic and maternal epithelia (Enders *et al.*, 1980). Initial adhesions might then be stabilized by the formation of junctional complexes: desmosomes and gap junctions, which have been observed in a variety of mammals, including rabbit, bat, ferret, mouse, rat, and human (Enders and Schlafke, 1967; Morris *et al.*, 1982; Tachi and Tachi, 1979). From this point on, implantation would progress *via* the displacement of uterine epithelial cells by the invasive trophectoderm cells (Schlafke and Enders, 1975; Glass *et al.*, 1979; Van Blerkom and Chavez, 1981), although phagocytic and enzymatic mechanisms also may play a part.

But what of the initial cell–cell recognition event? Morphologically, the electron microscope has chronicled implantation *in vivo* (Schlafke and Enders, 1975) and *in vitro*, where blastocysts invade uterine epithelial vesicles (Morris *et al.*, 1982) after making initial contacts *via* microvilli (Morris *et al.*, 1983). Biochemically, research is hindered by the small amount of material available and by the technical difficulties of reaching the implantation site. However, blastocysts will attach to substrata and grow out (Spindle and Pedersen, 1973); here it has been shown that trypsin will inhibit the attachment of the trophectoderm to glass, whereas neuraminidase has no effect and inhibition of glycoprotein synthesis has some effect (Glass et al., 1979). As discussed in Section 3.4.10, the GP140 antigens may also play a role in this attachment. Although useful in the study of differentiation, this in vitro model may be of limited use for understanding implantation at a biochemical level.

Surprisingly, the initial recognition and adhesion events of implantation will occur independently of genotype. For example, desmosomes and gap junctions will develop between rat trophectoderm and mouse uterine epithelium (Tachi and Tachi, 1979), but how broad a phylogenetic gap can be bridged in this manner is not known. However, successful implantation and survival to term has been shown to be dependent on genotype, requiring at least some of the cells of the trophectoderm to be of the same genotype as the mother (Rossant et al., 1982). Even sheep–goat chimeras have been achieved in this fashion (Fehilly et al., 1984; Meinecke-Tillman and Meinecke, 1984)! Rejection of an incompatible embryo is likely due to the action of cytotoxic T cells (Rossant et al., 1982). Thus, it appears there has been little selection for specificity in cell recognition at implantation, and the specificity of implantation has come to lie primarily with the maternal immune response and perhaps secondarily with later events such as invasion.

So far, the recognition events discussed have been between unlike cells. Cell recognition also occurs between the like cells of the early embryo but is poorly understood. For example, although trophectoderm cells form gap junctions with cells of the inner cell mass (Magnuson et al., 1977), they apparently will not form gap junctions or cooperate metabolically with some embryonal carcinoma cells (Gaunt and Papaioannou, 1979), a cell type often treated as equivalent to ICM cells. Similarly, cleavage-stage blastomeres will adhere to each other and to embryonal carcinoma cells (Fujii and Martin, 1980; Stewart, 1980, 1982), but not to cultured cells (Fujii and Martin, 1980). Trophoblast cells will not adhere to cultured cells or macrophages (Glass et al., 1980). Whether the two events of recognition and adhesion involve the same systems at the molecular level is unknown, but recognition is often defined by the ability of cells to adhere.

In summary, cell recognition in the early mammalian embryo is poorly understood and poorly characterized. Future work should be directed toward sperm–egg binding in a species that shows species specificity, such as the mouse. Since fertilization in mice can be deleteriously affected by protease treatment of the egg, recognition of a sperm receptor may be favored in this system. Another promising avenue would seem to be immunological, by developing first a heterospecific antiserum to sperm or to a tryptic digest of unfertilized eggs that blocks fertilization. Secondarily, more specific sera made against components immunoprecipitated from sperm or eggs could be developed. The technically difficult characterization of cell recognition at implantation may have to depend on ultrastructural immunocytochemistry for identification of developmentally significant molecules. For example, by using antibodies to surface molecules implicated in cell adhesion in other systems (CAMs, uvomorulin, CSAT ag) and ultrastructural immunocytochemistry, one could determine whether similar molecules were involved in the initial stages of implantation.

5.2. Cell Adhesion

Although cell adhesion ultimately occurs in the systems described above, I would like to focus this section on the process of adhesion between the blastomeres of the developing embryo.

As development progresses in the mouse, one observes that blastomeres of the zona-free 2-cell, 4-cell, and early 8-cell embryo have a tendency to adhere that increases with time. Thus, 2-cell blastomeres can generally be separated by repeated pipetting, but the 4-cell blastomeres are less easily dissociated. Brief exposure to trypsin or calcium, magnesium-free medium accompanied by pipetting is usually required to separate blastomeres at the 8-cell stage, which often separate first into couplets representing sister cells held together by their midbody. Some of this early tendency to adhere is probably due to physical reasons (midbodies, microvilli) but may also be enhanced by specific cytoskeletal elements. For instance, all blastomeres of these early stages are slightly flattened where they contact one another (Calarco, 1975b), and, as described earlier, spectrin is localized to this contact region (Sobel and Alliegro, 1985), whereas myosin is localized exclusively to the noncontact regions (Sobel, 1983). In addition, it is known that there are biochemical alterations to, and new expression of, cell surface moieties during these early cleavage stages, although no direct correlations have been made between these events. Interestingly, tunicamycin treatment, which eventually eliminates the close cellular adhesion of compacted embryos, does not interfere with the general tendency of embryo blastomeres to adhere and might provide a tool for the investigation of early cell adhesion.

Beginning with the mid- to late 8-cell embryo, a number of events occur whose end result is a marked increase in cell adhesion (compaction), a process that culminates in the formation of tight junctions and desmosomes between outer cells by the 32-cell stage. Here one has the chance to study the process of epithelialization as certain of the 8-cell blastomeres become polarized, develop apical and basal surfaces, and establish membrane domains. Research into these events is very active but is not the subject of this chapter; the reader is directed to a review by Johnson (1981) for more information.

What do we know about the molecules involved in compaction? Some evidence indicates the involvement of proteins. For instance, brief pronase treatment of precompacted 8-cell embryos to remove the zona does not block compaction, nor does incubation in colcemid, by itself, block compaction. However, if pronase treatment is followed immediately by incubation in colcemid, compaction is completely blocked (Sutherland and Calarco, 1983). This implicates a surface protein in the process of compaction and suggests, too, that the cytoskeleton is involved in the initial stages of this process. Other evidence indicates the involvement of oligosaccharides in compaction, for when N-linked glycosylation is blocked by tunicamycin, embryos eventually decompact (Surani, 1979; Sutherland and Calarco, 1983). (Alternatively, tunicamycin may only block events that stabilize the compacted state, since compaction will initially occur in tunicamycin but cannot be maintained.) Finally, evidence suggests that at least one newly synthesized compound is expressed on the cell surface at the 8-cell stage and is responsible for the increased adhesion observed at compaction. This is the molecule termed uvomorulin by one group (Hyafil et al., 1980, 1981), CAM 120/80 by another group (Damsky et al., 1983), and cadherin by yet another group (Yoshida-Noro et al., 1984). For a more complete description of these molecules and their calcium dependence the reader is referred to Chapter 9. Possibly, compaction will ultimately be explained in terms of a lectin–oligosaccharide receptor model although it is likely that uvomorulin would fit only one of these roles.

Future work should continue to probe possible mechanisms of action of uvomorulin (CAM 120/80, cadherin). In addition, future research into the possible involvement of extracellular matrix molecules in adhesion and compaction is definitely indicated.

5.3. Cell Motility

Cells of the preimplantation embryo do not appear to be actively motile before the blastocyst stage. Investigations into the amount of cell mixing show that the descendants of a single 8-cell blastomere will remain within one-quarter of the blastocyst (Kelly, 1979). In spite of this, chimeras between two or more embryos are readily formed, probably by cell adhesion coupled with limited blastomere motility. Even at the blastocyst stage, trophoblastic giant cells do not appear to be uniform in their migratory ability, with those of the mouse showing limited invasiveness whereas those of the vole may be found throughout the placenta (Copp, 1980). The parietal endoderm also begins its migration around the inner aspect of the trophectoderm cells at about the time of implantation and presents a very different phenotype from the static visceral endoderm cells (Enders et al., 1978). Because of their late appearance, endoderm motility and trophoblast motility will not be considered further here.

During the 8-cell stage, blastomeres develop the ability to spread on one another (Burgoyne and Ducibella, 1977; Kimber et al., 1982), an ability we consider here under cell "motility." However, in this case cells have only a leading edge; no trailing edge is apparent. This can be visualized in: the engulfment of older (Spindle, 1982) or earlier dividing cells by others (Kimber and Surani, 1981; Kimber et al., 1982); the reassortment of an inner or an outer cell to the same location in an aggregate (Surani and Handyside, 1983); the presence of long filopodia extending from one cell across another (Calarco, 1975b); and the interaction of 8-cell blastomeres with lectin-coated beads, where PNA, WGA, and Con A trigger calcium-insensitive spreading (Kimber and Surani,

1982). Although comparisons of spreading on beads with "compaction" are obvious, compaction appears to be a calcium-sensitive process. Perhaps this tells us that compaction consists of a process of adhesion, which is calcium sensitive and a separable process, cell spreading, which is not. (Alternatively, spreading on lectin beads may not be a suitable model for spreading at compaction.) When an 8-cell blastomere associated with a lectin bead divides, both daughter cells invariably spread on the lectin-coated bead rather than one cell encircling the other. Thus, the presentation of a lectin, such as WGA, to only one region of a cell triggers a markedly different cellular response (spreading) than when the lectin is presented to all surfaces. In the latter case, culture in WGA triggers precocious "blastulation" (i.e., close membrane apposition and accumulation of fluid in the intercellular space; Johnson, 1983).

Interestingly, although specific sugars block lectin-induced spreading on beads, they are ineffective at blocking compaction in the intact embryo (Kimber and Surani, 1982), again suggesting that a single lectin–oligosaccharide interaction may not be the only mechanism involved in compaction. By contrast, cytochalasin, which prevents microfilament polymerization, blocks both spreading on beads and compaction (Kimber and Surani, 1982), underscoring the importance of the cytoskeleton to motility. Whether the appearance of spreading indicates an increase in adhesiveness of inside cells and a decrease in adhesivity of outside cells (Kimber *et al.*, 1982), or is associated with junction formation (Burgoyne and Ducibella, 1977) or an increase in contractility of certain cells and/or differential activity of the cytoskeleton, is not yet clear. Obviously cell surface interactions are very important in determining embryo behavior, and manipulations of the sort described above will provide a powerful tool for deciphering the response of the embryo. Again, the known effects of the extracellular matrix on cell spreading warrant examination of the involvement of ECM molecules with compaction.

5.4. Cell Communication and Differentiation

How do cells of the preimplantation embryo interact? What mechanisms could possibly mediate the effect of one cell on another? Under the rubric of cell communication, we shall explore some of the possible ways this might occur. With terminology borrowed from the field of induction (Saxen, 1977), these might include instructive communications *via* gap junctions, ionic currents, receptor–ligand interactions, and/or the creation of a microenvironment. There is evidence that all of these play a role in early mammalian development. We must be cautious in our interpretations, however, for these cell interactions may only be permissive, merely permitting the normal expression of developmental potential. Thus, the time elapsed, or the number of divisions completed after fertilization, or the attainment of a certain cell mass may be the critical factor for some developmental events.

Gap junctions are known to appear at the 8-cell stage, preceding compaction (Lo and Gilula, 1979; Goodall and Johnson, 1982) and even form between trophectoderm and inner cell mass cells (Magnusen et al., 1977). By allowing the passage of ions and small molecules, one cell could influence or instruct another. This does not seem sufficient to explain differences in cell fate in early mammals, however, since all cells are in gap junctional communication. Work on ionic currents in early embryos is just beginning, but initial results are interesting and indicate polarization of cellular current concomitant with compaction (Nuccitelli and Wiley, 1983). Whether this is the cause or effect of the observed cellular polarization at this stage will be important to determine.

Receptor–ligand interactions could effect changes in gene expression as they do in other cells, but little evidence of this exists in preimplantation embryos. Embryonic differentiation, blastulation, and implantation in the hamster are reportedly dependent on estrogen (Sengupta et al., 1983). Diapause, an extended quiescent state of the blastocyst before implantation, can be terminated by injection of estrogen into the mother, triggering both translational and transciptional changes as embryonic development begins again (Weitlauf, 1984). Because preimplantation development can take place in a defined medium, we would have to conclude that any dependence of the embryo on receptor-mediated events would require synthesis of the appropriate ligand by the embryo itself.

Perhaps, as distinct from the above events, we should consider some events wherein cell "contact" (mechanism unknown) plays an instructive role. This would include the phenomenon of a 2-cell blastomere inducing polarity of an 8-cell blastomere, even though no gap junctions are formed between them (Goodall and Johnson, 1982) and the dependence of trophoblast invasiveness on a prior interaction with the ICM (Copp, 1980). Also giant cell transformation has been reported to correlate directly with substrate contact and inversely with homotypic cell contacts (Ilgren, 1981), which fits in well with evidence that trophoblast differentiation may not require cell interaction (Sherman and Atienza-Samols, 1979). For these phenomena, too little information is available to indicate whether a receptor–ligand interaction is occurring, but it would seem a likely alternative.

Lastly, we should consider the microenvironment that exists as a result of junction formation between outside cells of the preimplantation embryo. Cells of the inner cell mass are exposed to a modified environment when compared with the external cells of the embryo. Could this environment directly function in instructional cell communication? It is known that teratocarcinoma cells can be redirected into normal developmental pathways after being injected into the blastocoel (Brinster, 1974; Mintz and Illmensee, 1975) and that there are limitations on the number of cells that can be normalized, with only a slight effect on three cells and no regulation of four to five cells (Pierce et al., 1979). Conversely, ICM cells removed from the normalizing influence of the blastocoel can behave as though transformed (Evans and Kaufman, 1981; Martin, 1981). How normalization is achieved is not known, but work by Pierce et al. (1984) suggests that direct contact of the embryonal carcinoma cells with the blastocoelic surface of the trophectoderm is required. The blastocoel environment is unable to regulate certain tetracarcinoma cell lines (Papaioannou and Rossant, 1983) and certain other transformed cells, such as B-16 melanoma, L1210 leukemia, or the cell line CHO (Pierce *et al.*, 1979, 1984). However, this may reflect less the environment *per se* and more the ability of the injected cells to make cellular contact once inside the blastocyst. Eight-cell embryos injected into the cavities of large chimeric blastocysts appear to differ in developmental fate, depending on their ability to make contact with cells of the host blastocyst. When injected embryos are encased within an intact or torn zona pellucida, differentiation into a blastocyst occurs on schedule. When removed from their zona and injected as above, the embryo adheres to host trophoblast cells and forms a compact mass (Pedersen and Spindle, 1980). Although it is not yet clear that these masses are absolutely equivalent to the inner cell mass, it is suggestive that cellular contact rather than the microenvironment may play the dominant role in determining cell fate here as well.

6. Conclusions and Future Directions

The mammalian embryo presents a useful system for studying many developmental questions, but the limited quantities of embryos available limit what can be done. During the past few decades, researchers have scaled down biochemical, morphological, and immunological techniques to the level of a few embryos; much has been learned about the embryo and about cell–cell interactions. Future work will continue to advance this knowledge, with the techniques of molecular biology offering the opportunity for a closer look at gene expression in early development.

One of the more interesting problems facing the mammalian developmental biologist will be the investigation of maternal inheritance in mammals. There has been so much interest in the early involvement of the embryonic genome in development that the contributions of the oocyte have been overlooked. Research described in this chapter has indicated that certain surface antigens, retrovirus (ϵ), and possibly junctional elements may be preformed in the oocyte. Does the preprogramming of the oocyte include the stockpiling of proteins, their mRNA, or both? Will mechanisms for onset of translation of oocyte preformed message involve the cytoskeleton?

Another major question to be answered is the mechanism(s) for glycoprotein synthesis and expression in mammalian embryos, which are almost free of RER and frequently deficient in classical Golgi material. How the embryo has solved these problems should also be of interest to the cell biologist. Antibody probes and molecular probes are now available and should be used to investigate this unique system.

The increased adhesion and cell spreading that characterize compaction at the 8-cell stage deserve further attention. Despite the discovery of uvomorulin (CAM 120/80; cadherin), there is not sufficient information to explain the events of adhesion. Whether uvomorulin is extracellular or an integral membrane component, it may represent only one of the components in a receptor–

ligand interaction. Further research is needed, particularly on SFM-I-defined antigens (see Section 3.4.1d), to characterize possible interacting components. The involvement of extracellular matrix molecules in compaction and the possible interaction of surface receptors with any of them would seem a very exciting area of research. An anticipated corollary of this research would be the identification of transmembrane molecules, which could interact with the extracellular matrix and also interact with internal cytoskeletal structures. Already we know that there is a polarized distribution of two cytoskeletal molecules, myosin and spectrin. The spreading of cells, which also characterizes compaction, may involve an event separable from adhesion, and the interesting work begun on this system should continue.

Two other areas seem particularly worthy of attention. Evidence suggests that a sperm receptor exists on the egg cell surface, although it may not show much species specificity. If sperm–egg recognition depends on a unique molecule and not on one used in general cell recognition, it probably could be detected by current immunological methods. It would be exceedingly useful to have information on such a mammalian receptor and to determine what differences exist across species.

Another research goal of great potential significance is to determine why the developmental fate of cells placed in the blastocoel environment differs from that of other cells. Evidence suggests that potential fate may be mediated by a cell–cell interaction event. Not only are teratocarcinoma cells exposed to this environment "normalized," but normal ICM cells removed from it can be "carcinoma-ized." It is possible that ICM cells produce a growth factor that, at sufficient intracellular concentration, acts to trigger uncontrolled growth. The utilization of this factor would then have to be controlled either by the environment of the blastocoel or by cell contact with the trophectoderm. Interesting work is progressing to characterize such a factor and should reveal much about growth control in the early embryo.

ACKNOWLEDGMENTS. I would like to acknowledge gratefully the helpful discussions and valuable input into portions of this chapter from Dr. Karen Artzt, Dr. Verne Chapman, Dr. Allen Enders, Dr. Tom Huang, Dr. Roger Pedersen, Dr. Sylvie Polak-Charcon, Ann Poznanski, Dr. Sabina Sobel, Ann Sutherland, Dr. Richard Tasca, and Dr. Harry Weitlauf. Special thanks are due to Dr. Dan Szollosi, Dr. Linc Johnson, and Dr. Carole Banka for critical evaluation of the entire manuscript as well as many useful discussions. Thanks are also due to Sandy Tsarnas for the collating and typing of the many references.

References

Adamson, E. D., and Ayers, S. E., 1979, The localization and synthesis of some collagen types in developing mouse embryos, *Cell* **16**:953–965.

Artzt, K., Dubois, P., Bennett, D., Condamine, H., Babinet, C., and Jacob, F., 1973, Surface antigens common to mouse cleavage embryos and primitive teratocarcinoma cells in culture, *Proc. Natl. Acad. Sci. USA* **70**:2988–2992.

Banka, C. L., and Calarco, P. G., 1985, A new embryonic antigen p67, common to preimplantation embryos and pluripotent embryonic cells, *Gamete Res.* (in press).

Banka, C. L., and Calarco, P. G., 1986, The immunologic approach to the study of preimplantation mammalian development, in: *Developmental Biology, A Comprehensive Synthesis*, Vol. 4 (R. B. L. Gwatkin, ed.), pp. 353–381, Plenum Press, New York.

Biggers, J., and Borland, R., 1976, Physiological aspects of growth and development of the pre-implantation mammalian embryo, *Annu. Rev. Physiol.* **38**:95–119.

Bleil, J. D., and Wassarman, P. M., 1980, Mammalian sperm-egg interaction: Identification of a glycoprotein in mouse egg zonae pellucidae possessing receptor activity for sperm, *Cell* **20**:873–882.

Boller, K., Vestunelier, D., and Kemler, R., 1985, Cell-adhesion molecule uvomorulin is localized in the intermediate junctions of adult intestinal epithelial cells, *J. Cell Biol.* **100**:327–332.

Brinster, R. L., 1974, The effect of cells transferred into the mouse blastocyst on subsequent development, *J. Exp. Med.* **140**:1049–1056.

Burgoyne, P. S., and Ducibella, T., 1977, Changes in the properties of the developing trophoblast of preimplantation mouse embryos as revealed by aggregation studies, *J. Embryol. Exp. Morphol.* **40**:143–155.

Burridge, K., Kelly, T., and Connel, L., 1982, Proteins involved in the attachment of actin to the plasma membrane, *Phil. Trans. R. Soc. Lond. B.* **299**:291–299.

Calarco, P., 1975a, Intracisternal A particle formation and inhibition in preimplantation embryos, *Biol. Reprod.* **12**:448–454.

Calarco, P. G., 1975b, Cleavage (mouse), in: *Scanning Electron Microscope Atlas of Mammalian Reproduction* (E. S. E. Hafez, ed.), pp. 306–317, Igaku-Shoin, Tokyo.

Calarco, P. G., and Banka, C. L., 1979, Surface antigens of preimplantation mouse embryos, *Biol. Reprod.* **20**:699–704.

Calarco, P. G., and Brown, E. H., 1968, Cytological and ultrastructural comparisons of t^{12}/t^{12} and normal mouse morulae, *J. Exp. Zool.* **168**:169–186.

Calarco, P., and Brown, E., 1969, An ultrastructural and cytological study of preimplantation development in the mouse, *J. Exp. Zool.* **171**:253–284.

Calarco, P. G., and Epstein, C. J., 1973, Cell surface changes during preimplantation development in the mouse, *Dev. Biol.* **32**:208–213.

Calarco, P., and McLaren, A., 1976, Ultrastructural observations of preimplantation stages of the sheep, *J. Embryol. Exp. Morphol.* **36**:609–622.

Calarco-Gillam, P. D., Siebert, M. C., Hubble, R., Mitchison, T., and Kirshner, M., 1983, Centrosome development in early mouse embryos as defined by an autoantibody against pericentriolar material, *Cell* **35**:621–629.

Cervera, M., Dreyfuss, G., and Penman, S., 1981, Messenger RNA is translated when associated with the cytoskeletal framework in normal and VSV-infected HeLa cells, *Cell* **23**:113–120.

Chang, M., and Hunt, D., 1956, Effects of proteolytic enzymes on the zona pellucida of fertilized and unfertilized mammalian eggs, *Exp. Cell Res.* **11**:497–502.

Cheng, C., and Bennett, D., 1980, Nature of the antigenic determinants of T locus antigens, *Cell* **19**:537–543.

Cheng, C., Sege, K., Alton, A., Bennett, D., and Artzt, K., 1983, Characterization of an antigen present on testicular cells and preimplantation embryos whose expression is modified by the t^{12} haplotype, *J. Immunogen.* **10**:465–486.

Cooper, A. R., and MacQueen, H. A., 1983, Subunits of laminin are differentially synthesized in mouse eggs and early embryos, *Dev. Biol.* **96**:467–471.

Copp, A. J., 1979, Interaction between inner cell mass and trophectoderm of the mouse blastocyst. II. The fate of the polar trophectoderm, *J. Embryol. Exp. Morphol.* **51**:109–120.

Copp, A. J., 1980, The development of field vole (Microtus agrestis) and mouse blastocysts in vitro: A study of trophoblast cell migration, *Placenta* **1(1)**:47–59.

Courtneidge, S., and Bishop, M., 1982, Transit of pp60[v-src] to the plasma membrane, *Proc. Natl. Acad. Sci. USA* **79**:7117–7121.

Damsky, C., Richa, J., Solter, D., Knudsen, K., and Buck, C., 1983, Identification and purification of a cell surface glycoprotein mediating intercellular adhesion in embryonic and adult tissue, *Cell* **34**:455–466.

Damsky, C., Knudsen, K., Bradley, D., Buck, C., and Horwitz, A., 1985, Distribution of the cell substratum attachment (CSAT) antigen on myogenic and fibroblastic cells in culture, *J. Cell Biol.* **100**:1528–1539.

DiZio, S. M., and Tasca, R. J., 1977, Sodium-dependent amino acid transport in preimplantation mouse embryos. III. Na$^+$-K$^+$-ATPase-linked mechanism in blastocysts, *Dev. Biol.* **49**:198–205.

Ducibella, T., 1980, Divalent antibodies to mouse embryonal carcinoma cells inhibit compaction in the mouse embryos, *Dev. Biol.* **79**:356–366.

Ducibella, T., 1982, Depolymerization of microtubules prior to compaction, *Exp. Cell Res.* **138**:31–38.

Ducibella, T., and Anderson, E., 1979, The effects of calcium deficiency on the formation of the zonula occludens and blastocoel in the mouse embryo, *Dev. Biol.* **73(1)**:46–58.

Ducibella, T., Ukena, T., Karnovsky, M., and Anderson, E., 1977, Changes in cell surface and cortical cytoplasmic organization during early embryogenesis in the preimplantation mouse embryo, *J. Cell Biol.* **74(1)**:153–167.

Dunbar, B. S., Wardrip, N. J., and Hedrick, J. L., 1980, Isolation, physicochemical properties and macromolecular composition of zona pellucida from porcine oocytes, *Biochemistry* **19**:356–365.

Dunia, I., Nicolas, J., Jakob, H., Benedetti, E., and Jacob, F., 1979, Junctional modulation in mouse embryonal carcinoma cells by Fab fragments of rabbit anti-embryonal carcinoma cell serum, *Proc. Natl. Acad. Sci. USA* **76**:3387–3391.

Dziadek, M., and Timpl, R., 1985, Expression of nidogen and laminin in basement membranes during mouse embryogenesis and in teratocarcinoma cells, *Dev. Biol.* **111**:372–382.

Eager, D. D., Johnson, M., and Thurley, K., 1976, Ultrastructural studies on the surface membrane of the mouse egg, *J. Cell Sci.* **22**:345–353.

Enders, A. C., and Schlafke, S. J., 1965, The fine structure of the blastocyst: Some comparative studies, in: *Preimplantation Stages of Pregnancy* (G. E. W. Wolstenholme and M. O'Connor, eds.), pp. 29–54, Little, Brown, Boston.

Enders, A., and Schlafke, S., 1967, A morphological analysis of the early implantation stages in the rat, *J. Anat.* **120**:185.

Enders, A. C., Given, R. L., and Schlafke, S., 1978, Differentiation and migration of endoderm in the rat and mouse at implantation, *Anat. Rec.* **190(1)**:65–77.

Enders, A. C., Schlafke, S., and Welsh, A. O., 1980, Trophoblastic and uterine luminal epithelial surfaces at the time of blastocyst adhesion in the rat, *Am. J. Anat.* **159(1)**:59–72.

Epstein, C. J., Smith, S., and Travis, B., 1980, Expression of H-Y antigen on preimplantation mouse embryos, *Tissue Antigen* **15**:63–67.

Evans, M., and Kaufman, M., 1981, Establishment in culture of pluripotent cells from mouse embryos, *Nature (Lond.)* **292**:154–156.

Farmer, S., Wan, K., Ben Zeev, A., and Penman, S., 1983, Regulation of actin mRNA levels and translation responds to changes in cell configuration, *Mol. Cell Biol.* **3**:182–189.

Fehilly, C. B., Willadsen, S. M., and Tucker, E. M., 1984, Interspecific chimaerism between sheep and goat, *Nature (Lond.)* **307**:634–636.

Fenderson, B. A., Hahnel, A. C., and Eddy, E., 1983, Immunohistochemical localization of two monoclonal antibody-defined carbohydrate antigens during early murine embryogenesis, *Dev. Biol.* **100**:318–327.

Flechon, J. E., 1978, Morphological aspects of embryonic disc at the time of its appearance in the blastocyst of farm mammals, in: *Scanning Electron Microscopy*. Vol. II (R. P. Becker and O. Johari, eds.), pp. 541–547, Scanning Electron Microscopy, Inc., Elk Grove Village, Illinois.

Florman, H., and Wassarman, P., 1985, O-linked oligosaccharides of mouse egg ZP3 account for its sperm receptor activity, *Cell* **41**:313–324.

Flynn, T. J., and Hillman, N., 1978, Lipid synthesis from (U^{14}C) glucose in preimplantation mouse embryos in culture, *Biol. Reprod.* **19**:922–926.

Flynn, T. J., and Hillman, N., 1980, The metabolism of exogenous fatty acids by preimplantation mouse embryos developing *in vitro*, *J. Embryol. Exp. Morphol.* **56**:157–168.

Fox, H. W., Damjanov, I., Knowles, B. B., and Solter, D., 1984, Stage-specific embryonic antigens as a marker of visceral extraembryonic endoderm, *Dev. Biol.* **103**:263–266.

Fujimoto, H., Muramatsu, T., Urushihara, H., and Yanagisawa, K., 1982, Receptors to Dolichos biflorus agglutinin, *Differentiation* **22:**59–61.

Fujii, J. T., and Martin, G. R., 1980, Incorporation of teratocarcinoma stem cells into blastocysts by aggregation with cleavage-stage embryos, *Dev. Biol.* **74(1):**239–244.

Gaunt, S. J., and Papaioannou, V. E., 1979, Metabolic co-operation between embryonic and embryonal carcinoma cells of the mouse, *J. Embryol. Exp. Morphol.* **54:**263–275.

Glass, R. H., Spindle, A. I., and Pedersen, R. A., 1979, Mouse embryo attachment to substratum and interaction of trophoblast with cultured cells, *J. Exp. Zool.* **208(3):**327–336.

Glass, R. H., Spindle, A. I., Maglio, M., and Pedersen, R. A., 1980, The free surface of mouse trophoblast in culture is non-adhesive for other cells, *J. Reprod. Fertil.* **59(2):**403–407.

Goodall, H., and Johnson, M. H., 1982, Use of carboxyfluorescein diacetate to study formation of permeable channels between mouse blastomeres, *Nature (Lond.)* **295:**524–526.

Gooi, H., Feizi, T., Kapadia, A., Knowles, B., Solter, D., and Evans, M., 1981, Stage-specific embryonic antigen involves 1–3 fucosylated type 2 blood group chains, *Nature (Lond.)* **292:**156–158.

Hahnel, A. C., and Eddy, E. M., 1983, Syngeneic antiserum to Nulli SCC1 embryonal carcinoma cells recognizing surface antigens of embryonic cells, *J. Reprod. Immunol.* **5:**371–382.

Hakansson, S., Heyner, S., Sundqvist, K.-G., and Bergstrom, S., 1975, The presence of paternal H-antigens on hybrid mouse blastocysts during experimental delay of implantation and the disappearance of these antigens after onset of implantation, *Int. J. Fert.* 91. **20:**137–140.

Handyside, A., 1980a, Distribution of antibody- and lectin-binding sites on dissociated blastomeres from mouse morulae: Evidence for polarization at compaction, *J. Embryol. Exp. Morphol.* **60:**99–116.

Heyner, S., 1980b, Antigens of trophoblast and early embryo, in: *Immunological Aspects of Infertility and Fertility Regulation* (D. Dhindsa and J. Schumacher, eds.), pp. 183–203, Elsevier North-Holland, New York.

Heyner, S., 1986, Immunogenetic approaches to the analysis of mammalian development, in: *Developmental Biology: A Comprehensive Synthesis*, Vol. 4: *Manipulation of Mammalian Development* (R. B. L. Gwatkin, ed.), pp. 335–352, Plenum Press, New York.

Hillman, N., and Tasca, R., 1969, Ultrastructural and autoradiographic studies of mouse cleavage stages, *Am. J. Anat.* **126:**151–174.

Huang, T. T., and Calarco, P. G., 1981a, Evidence for the cell surface expression of intracisternal A particle associated antigens during early mouse development, *Dev. Biol.* **82:**388–392.

Huang, T. T., and Calarco, P. G., 1981b, Immunoprecipitation of intracisternal A particle-associated antigens from preimplantation mouse embryos, *J. Natl. Cancer Inst.* **67:**1129–1134.

Huang, T. T., Jr., and Calarco, P. G., 1982, Immunologic relatedness of intracisternal A-particles in mouse embryos and neoplastic cell lines, *J. Natl. Cancer Inst.* **68(4):**643–649.

Huang, T. T., and Yanagimachi, R., 1984, Fucoidin inhibits attachment of guinea pig spermatozoa to the zona pellucida through binding to the inner acrosomal membrane and equatorial domains, *Exp. Cell Res.* **153:**363–373.

Hyafil, F., Morello, D., Babinet, C., and Jacob, F., 1980, A cell surface glycoprotein involved in the compaction of embryonal carcinoma cells and cleavage stage embryos, *Cell* **21:**927–934.

Hyafil, F., Balinet, C., and Jacob, F., 1981, Cell–cell interactions in early embryogenesis: A molecular approach to the role of calcium, *Cell* **26:**447–454.

Ilgren, E. B., 1981, On the control of the trophoblastic giant-cell transformation in the mouse: Homotypic cellular interactions and polyploidy, *J. Embryol. Exp. Morphol.* **62:**183–202.

Iwakura, Y., 1983, Comparison of polysaccharide synthesis between preimplantation stage mouse embryos and F9 embryonal carcinoma cells, *Exp. Cell Res.* **146:**329–338.

Izquierdo, L., and Marticorena, P., 1975, Alkaline phosphatase in preimplantation mouse embryos, *Exp. Cell Res.* **92:**399–402.

Jackson, B. W., Grund, C., Schmid, E., Burki, K., Franke, W. W., and Illmensee, K., 1980, Formation of cytoskeletal elements during mouse embryogenesis. Intermediate filaments of the cytokeratin type and desmosomes in preimplantation embryos, *Differentiation* **17(3):**161–179.

Jacob, F., 1977, Mouse teratocarcinoma and embryonic antigens, *Immunol. Rev.* **33:**3–32.

Jacob, F., 1979, Cell surface and early stages of mouse embryogenesis, in: *Current Topics in Developmental Biology*, Vol. 13 (A. Moscona and A. Monroy, eds.), pp. 117–137, Academic Press, New York.

Jeffery, W., 1982, Messenger in the cytoskeletal framework: Analysis by *in situ* hybridization, *J. Cell Biol.* **95:**1–7.

Jeffery, W., 1983, Pattern and mechanism of maternal mRNA localization during early development of Chaetopterus, *J. Cell Biol.* **97:**34a.

Johnson, L. V., 1983, Wheat germ agglutinin induces compaction- and cavitation-like events in 2-cell mouse embryos, *J. Cell Biol.* **97:**36a.

Johnson, L. V., 1986, Wheat germ agglutinin induces compaction- and cavitation-like events in 2-cell mouse embryos, *Dev. Biol.* (in press).

Johnson, L. V., and Calarco, P. G., 1980a, Electrophoretic analysis of cell surface proteins of preimplantation mouse embryos, *Dev. Biol.* **77(1):**224–227.

Johnson, L. V., and Calarco, P. G., 1980b, Mammalian preimplantation development: The cell surface, *Anat. Rec.* **196:**202–219.

Johnson, L. V., and Calarco, P. G., 1980c, Immunological characterization of embryonic cell surface antigens recognized by anti-blastocyst serum, *Dev. Biol.* **79:**208–223.

Johnson, L. V., and Calarco, P. G., 1980d, Stage-specific embryonic antigens detected by an anti-serum against mouse blastocysts, *Dev. Biol.* **79:**224–231.

Johnson, L. V., Calarco, P., and Siebert, M., 1977, Alkaline phosphatase activity in the preimplantation mouse embryo, *J. Embryol. Exp. Morphol.* **40:**83–89.

Johnson, M. H., 1981, Membrane events associated with the generation of a blastocyst, *Int. Rev. Cytol. (Suppl.)* **12:**1–37.

Johnson, M. H., and Ziomek, C., 1982, Cell subpopulations in the late morula and early blastocyst of the mouse, *Dev. Biol.* **91:**431–439.

Johnson, M. H., Pratt, H., and Handyside, A., 1981, The generation and recognition of positional information in the preimplantation mouse embryo, in: *Cellular and Molecular Aspects of Implantation* (S. R. Glasser and D. W. Bullock, eds.), pp. 55–79, Plenum Press, New York.

Kane, M. T., 1979, Fatty acids as energy sources for culture of one-cell rabbit ova to viable morulae, *Biol. Reprod.* **20:**323–332.

Kannagi, R., Nudelman, E., Levery, S. B., and Hakomori, S., 1982, A series of human erythrocyte glycosphingolipids reacting to the monoclonal antibody directed to a developmentally regulated antigen, SSEA-1, *J. Biol. Chem.* **257:**14865–14874.

Kannagi, R., Levery, S. B., Ishigami, F., Hakamori, S-I., Shevinsky, L. H., Knowles, B. B., and Solter, D., 1983, New globoseries glycosphingolipids in human teratocarcinoma reactive with the monoclonal antibody directed to a developmentally regulated antigen, Stage-specific Embryonic Antigen 3, *J. Biol. Chem.* **258:**8934–8942.

Kasamatsu, H., Lin, W., Edens, J., and Revel, J. P., 1983, Visualization of antigens attached to cytoskeletal framework in animal cells: Colocalization of SV40 Vp1 polypeptide and actin in TC7 cells, *Proc. Natl. Acad. Sci. USA* **80:**4339–4343.

Keefer, C. L., and Tasca, R. J., 1984, Modulation of amino acid transport in preimplantation mouse embryos by low concentrations of non-ionic and zwitterionic detergents, *J. Reprod. Fertil.* **70:**399–407.

Kelly, S. J., 1979, Investigations into the degree of cell mixing that occurs between the 8-cell stage and the blastocyst stage of mouse development, *J. Exp. Zool.* **207(1):**121–130.

Kemler, R., Babinet, C., Eisen, H., and Jacob, F., 1977, Surface antigen in early differentiation, *Proc. Natl. Acad. Sci. USA* **74:**4449–4452.

Kemler, R., Morello, D., and Jacob, F., 1979, Properties of some monoclonal antibodies raised against mouse embryonal carcinoma cells, in: *Cell Lineage, Stem Cells and Cell Determination* (N. LeDouarin, ed.), pp. 101–113, North-Holland, Amsterdam.

Kimber, S. J., and Surani, M. A. H., 1981, Morphogenetic analysis of changing cell associations following release of 2-cell and 4-cell mouse embryos from cleavage arrest, *J. Embryol. Exp. Morphol.* **61:**331–345.

Kimber, S. J., and Surani, M. A., 1982, Spreading of blastomeres from eight-cell mouse embryos on lectin-coated beads, *J. Cell Sci.* **56:**191–206.

Kimber, S. J., Surani, M. A., and Barton, S. C., 1982, Interactions of blastomeres suggest changes in cell surface adhesiveness during the formation of inner cell mass and trophectoderm in the preimplantation mouse embryo, *J. Embryol. Exp. Morphol.* **70:**133–152.

Knudsen, K. A., Rao, P. E., Damsky, C. H. and Buck, C. A., 1981, Membrane glycoproteins involved in cell-substratum adhesion, *Proc. Natl. Acad. Sci. USA* **78**:6071–6075.

Konwinski, M., Vorbrodt, A., Solter, D., and Koprowski, H., 1977, Ultrastructural study of concanvalin-A binding to the surface of preimplantation mouse embryos, *J. Exp. Zool.* **200(3)**:311–323.

Krco, C. J., and Goldberg, E. H., 1976, H-Y (male) antigen: Detection on eight-cell mouse embryos, *Science* **193**:1134–1135.

Lazarides, E., and Nelson, W., 1982, Expression of spectrin in nonerythroid cells, *Cell* **31**:505–508.

Lehtonen, E., and Badley, R. A., 1980, Localization of cytoskeletal proteins in preimplantation mouse embryos, *J. Embryol. Exp. Morphol.* **55**:211–225.

Lehtonen, E., Lehto, V.-P., Vartio, T., Bodley, R., and Virtanen, I., 1983, Expression of cytokeratin polypeptides in mouse occytes and preimplantation embryos, *Dev. Biol.* **100**:158–165.

Leivo, I., 1983, Structure and composition of early basement membranes: Studies with early embryos and teratocarcinoma cells, *Med. Biol* **61(1)**:1–30.

Leivo, I., Vaheri, A., Timpland, R., and Wartiovaara, J., 1980, Appearance and distribution of collagens and laminin in the early mouse embryo, *Dev. Biol.* **76**:100–114.

Levinson, J., and McDevitt, H., 1976, Murine t factors: An association between alleles at t and at H-2, *J. Exp. Med.* **144**:834.

Lin, T. P., Glass, R., Bronson, R., Florence, J., and Maglio, M., 1975, Interspecies sperm-egg interaction, in: *SEM Atlas of Mammalian Reproduction* (E. Hafeg, ed.), pp. 300–305, lgaku-Shoin, Tokyo.

Lindahl, K., and Burki, K., 1982, Mta, a maternally inherited cell surface antigen of the mouse, is transmitted in the egg, *Proc. Natl. Acad. Sci. USA* **79**:5362–5366.

Lindahl, K. F., Hausmann, B., and Chapman, V. M., 1983, A new H-2-linked class I gene whose expression depends on a maternally inherited factor, *Nature (Lond.)* **306**:383–385.

Lo, C. W., and Gilula, N. B., 1979, Gap junctional communication in the preimplantation mouse embryo, *Cell* **18(2)**:399–409.

Loewenstein, J. E., and Cohen, A. I., 1964, Drymass, lipid content, and protein content of the intact and zona-free mouse ovum, *J. Embryol. Exp. Morphol.* **12**:113–121.

Magnuson, T., and Epstein, C., 1981, Characterization of con A precipitated proteins from early mouse embryos: A 2-dimensional gel electrophoresis study, *Dev. Biol.* **81**:193–199.

Magnuson, T., Demsey, A., and Stackpole, C. W., 1977, Characterization of intercellular junctions in the preimplantation mouse embryo by freeze-fracture and thin-section electron microscopy, *Dev. Biol.* **61(2)**:252–61.

Maro, B., and Pickering, S., 1984, Microtubules influence compaction in preimplantation mouse embryos, *J. Embryol. Exp. Morphol.* **84**:217–232.

Marticorena, P., Hogan, B., DiMeo, A., Artzt, K., and Bennett, D., 1983, Carbohydrate changes in pre- and peri-implantation mouse embryos as detected by a monoclonal antibody, *Cell Diff.* **12**:1–10.

Martin, G., 1981, Isolation of a pluripotent cell line from early mouse embryos cultured in medium conditioned by teratocarcinoma stem cells, *Proc. Natl. Acad. Sci. USA* **78**:7634–7638.

Martin, G., and Evans, M., 1975, The differentiation of clonal lines of teratocarcinoma cells: Formation of embryoid bodies in vitro, *Proc. Natl. Acad. Sci. USA* **72**:1441–1445.

Maylie-Pfenniger, M-F., 1979, The effect of tunicamycin on the development of mouse preimplantation embryos, *J. Cell Biol.* **83**:216a.

McLachlin, J., Caveney, S., and Kidder, G., 1983, Control of gap junction formation in early mouse embryos, *Dev. Biol.* **98**:155–164.

McGrath, J., and Solter, D., 1983, Nuclear transplantation in mouse embryos, *J. Exp. Zool.* **228**:355–362.

Meinecke-Tillman, S., and Meinecke, B., 1984, Experimental chimaeras—Removal of reproductive barriers between sheep and goat, *Nature (Lond.)* **307**:637–638.

Menge, A. C., and Black, C. B., 1979, Effects of antisera on human sperm penetration of zona-free hamster ova, *Fertil. Steril.* **32**:214–218.

Meziou, W., Chardon, P., Flechon, J., Kalil, J., and Vaiman, M., 1983, Expression of B-2-microglobulin on preimplantation pig embryos, *J. Reprod. Immunol.* **5**:73–80.

Mintz, B., 1962, Experimental study of the developing mammalian egg: Removal of the zona pellucida, *Science* **138**:594–595.

Mintz, B., and Illmensee, K., 1975, Normal genetically mosaic mice produced from malignant teratocarcinoma cells, *Proc. Natl. Acad. Sci. USA* **72**:3585–3589.

Moon, R., Nicosia, R., Olsen, C., Hille, M., and Jeffery, W., 1983, The cytoskeletal framework of sea urchin eggs and embryos: Developmental changes in the association of messenger RNA, *Dev. Biol.* **95**:447–458.

Morris, J. E., Potter, S. W., and Buckley, P. M., 1982, Mouse embryos and uterine epithelia show adhesive interactions in culture, *J. Exp. Zool.* **222**(2):195–198.

Morris, J. E., Potter, S. W., Rynd, L. S., and Buckley, P. M., 1983, Adhesion of mouse blastocysts to uterine epithelium in culture: A requirement for mutual surface interactions, *J. Exp. Zool.* **225**(3):467–479.

Muggleton-Harris, A., and Johnson, M., 1976, The nature and distribution of serologically detectable alloantigens on the preimplantation mouse embryo, *J. Embryol. Exp. Morphol.* **25**:52–72.

Nicolson, G. L., Yanagimachi, R., and Yanagimachi, H., 1975, Ultrastructural localization of lectin-binding sites on the zonae pellucida and plasma membranes of mammalian eggs, *J. Cell Biol.* **66**: 263–274.

Nicosia, S., Wolf, D., and Inoue, M., 1977, Cortical granule distribution and cell surface characteristics in mouse eggs, *Dev. Biol.* **57**:56–74.

Nuccitelli, R., and Wiley, L. M., 1983, Polarity of isolated blastomeres from mouse morulae: Detection of transcellular ion currents, *J. Cell Biol.* **97**:32a.

Opas, J., and Soltynska, M., 1978, Reorganization of the cortical layer during cytokinesis in mouse blastomeres, *Exp. Cell Res.* **113**:208–211.

O'Rand, M., and Irons, G., 1984, Monoclonal antibodies to rabbit sperm autoantigens. II. Inhibition of human sperm penetration of zona-free hamster eggs, *Biol. Reprod.* **30**:731–736.

O'Rand, M., Irons, G., and Porter, J., 1984, Monoclonal antibodies to rabbit sperm autoantigens. I. Inhibition of *in vitro* fertilization and localization on the egg, *Biol. Reprod.* **30**:721–729.

Oshima, R., Howe, W., Klier, G., Anderson, E., and Shevinsky, L., 1983, Intermediate filament protein synthesis in preimplantation murine embryos, *Dev. Biol.* **99**:447–455.

Palm, J., Heyner, S., and Brinster, R. L., 1971, Differential immunofluorescence of fertilized mouse eggs with H-2 and non H-2 antibody, *J. Exp. Med.* **133**:1282–1293.

Papaioannou, V., and Rossant, J., 1983, Effects of the embryonic environment on proliferation and differentiation of embryonal carcinoma cells, *J. Cancer Surveys* **2**:165–183.

Pedersen, R. A., and Spindle, A. I., 1980, Role of the blastocoele microenvironment in early mouse embryo differentiation, *Nature (Lond.)* **284**:550–552.

Peyrieras, N., Hyafil, F., Louvard, D., Ploegh, H. L., and Jacob, F., 1983, Uvomorulin: A nonintegral membrane protein of early mouse embryo, *Proc. Natl. Acad. Sci. USA* **80**:6274–6277.

Phillips, D., Shalgi, R., Kraicer, P., and Segal, S., 1978, The rat oocyte–cumulus complex during ovulation and fertilization as seen with the SEM, in: *Scanning Electron Microscopy*, Vol. II (R. P. Becker and O. Johari, eds.), pp. 1113–1122, Scanning Electron Microscopy Inc., Elk Grove Village, Illinois.

Pierce, G. B., Lewis, S. H., Miller, G. J., Moritz, E., and Miller, P., 1979, Tumorigenicity of embryonal carcinoma as an assay to study control of malignancy by the murine blastocyst, *Proc. Natl. Acad. Sci. USA* **76**:6649–6651.

Pierce, G. B., Aguilar, D., Hood, G., and Wells, R. S., 1984, Trophectoderm in control of murine embryonal carcinoma, *Cancer Res.* **44**:3987–3996.

Pinsker, M., and Mintz, B., 1973, Change in cell-surface glycoproteins of mouse embryos before implantation, *Proc. Natl. Acad. Sci. USA* **70**:1645–1648.

Polak-Charcon, S., Johnson, L., and Calarco, P., 1983a, Cortical synthesis of a stage-specific surface glycoprotein during early mammalian development, in: *Proceedings of the Forty-First Electron Microscopic Society of America*, pp. 508–509.

Polak-Charcon, S., Johnson, L., and Calarco, P., 1983b, Intracellular localization and surface expression of stage-specific glycoprotein antigens during early mouse development, *J. Cell Biol.* **97**:33a.

Polak-Charcon, S., Calarco-Gillam, P., and Johnson, L., 1985, Intracellular localization and surface expression of a stage-specific embryonic glycoprotein, *Gamete Res.* **12**:329–343.

Powers, R. D., and Tupper, J. T., 1977, Developmental changes in membrane transport and per-meability in the early mouse embryo, *Dev. Biol.* **56**:306–315.

Pratt, H. P., 1978, Lipids and transitions in embryos, in: *Development in Mammals*, Vol. 3 (M. Johnson, ed.), pp. 83–129, Elsevier, New York.

Pratt, H. P. M., 1980, Phospholipid synthesis in the preimplantation mouse embryo, *J. Reprod. Fertil.* **58**:237–248.

Pratt, H. P. M., 1982, Preimplantation mouse embryos synthesize membrane sterols, *Dev. Biol.* **89(1)**:101–110.

Pratt, H. P. M., Keith, J., and Chakraborty, J., 1980, Membrane sterols and the development of the preimplantation mouse embryo, *J. Embryol. Exp. Morphol.* **60**:303–319.

Pratt, H. P. M., Chakraborty, J., and Surani, M. A. H., 1981, Molecular and morphological differ-entiation of the mouse blastocyst after manipulations of compaction with cytochalasin D, *Cell* **26**:279–292.

Pratt, H. P. M., Ziomek, C. A., Reeve, W. J. R., and Johnson, M. W., 1982, Compaction of the mouse embryo: An analysis of its components, *J. Embryol. Exp. Morphol.* **70**:113–132.

Quinn, P., and Whittingham, D. G., 1982, Effect of fatty acids on fertilization and development of mouse embryos in vitro, *J. Androl.* **3**:440–444.

Rasweiler, J. J., IV, 1979, Early embryonic development and implantation in bats, *J. Reprod. Fertil.* **56(1)**:403–416.

Reeve, W. J., 1981a, Cytoplasmic polarity develops at compaction in rat and mouse embryos, *J. Embryol. Exp. Morphol.* **62**:351–367.

Reeve, W. J., 1981b, The distribution of ingested horseradish peroxidase in the 16-cell mouse embryo, *J. Embryol. Exp. Morphol.* **66**:191–207.

Reeve, W. J., and Kelly, F. P., 1983, Nuclear position in the cells of the mouse early embryo, *J. Embryol. Exp. Morphol.* **75**:117–139.

Reeve, W. J., and Ziomek, C. A., 1981, Distribution of microvilli on dissociated blastomeres from mouse embryos: Evidence for surface polarization at compaction, *J. Embryol. Exp. Morphol.* **62**:339–350.

Richa, J., Damsky, C. H., Buck, C. A., Knowles, B. B., and Solter, D., 1985, Cell surface glycoproteins mediate compaction, trophoblast attachment and endoderm formation during early mouse development, *Dev. Biol.* **108**:513–521.

Rossant, J., Mauro, V. M., and Croy, B. A., 1982, Importance of trophoblast genotype for survival of interspecific murine chimaeras, *J. Embryol. Exp. Morphol.* **69**:141–149.

Sabatini, D., Kreibich, G., Morimoko, T., and Adesnik, M., 1982, Mechanisms for the incorporation of proteins in membranes and organelles, *J. Cell Biol.* **92**:1–22.

Sawicki, J., 1984, Developmental expression of H-2, class I genes, *J. Cell Biochem. (Suppl.)* **8B**:78.

Sawicki, J., Magnusen, T., and Epstein, C., 1981, Evidence for expression of the paternal genome in the two-cell mouse embryo, *Nature (Lond.)* **294**:450–451.

Saxen, L., 1977, Directives. Permissive induction: A working hypothesis, in: *Cell and Tissue Interactions*, Vol. 32 (J. Lash and M. Burger, eds.), pp. 1–9, Society of General Physiologists Series, New York.

Schlafke, S., and Enders, A., 1975, Cellular basis of interaction between trophoblast and uterus at implantation, *Biol. Reprod.* **12**:41–65.

Searle, R. F., Sellens, M. H., Elson, J., Jenkinson, E. J., and Billington, W. D., 1976, Detection of alloantigens during preimplantation. Development and early trophoblast differentiation in the mouse by immunoperoxidase labeling, *J. Exp. Med.* **143**:348–359.

Sengupta, J., Paria, B. C., and Manchanda, S. K., 1983, Effect of an estrogen antagonist on develop-ment of blastocysts and implantation in the hamster, *J. Exp. Zool.* **225**:119–122.

Sherman, M. I., 1979, Developmental biochemistry of preimplantation mammalian embryos, *Annu. Rev. Biochem.* **48**:443–470.

Sherman, M. I., and Atienza-Samols, S. B., 1979, Differentiation of mouse trophoblast does not require cell–cell interaction, *Exp. Cell Res.* **123**:73–77.

Sherman, M. I., Gay, R., Gay, S., and Miller, E., 1980, Association of collagen with preimplantation and peri-implantation mouse embryos, *Dev. Biol.* **74**:470–478.

Shevinsky, L. H., Knowles, B. B., Howe, C., Aden, D. P., and Solter, D., 1981, A murine stage-specific embryonic antigen (SSEA-2) is expressed on some murine SV40-transformed cells, *J. Immunol.* **127**:632–636.

Shevinsky, L. H., Knowles, B. B., Damijanov, I., and Solter, D., 1982, Monoclonal antibody to murine embryos defines a stage-specific embryonic antigen expressed on mouse embryos and human teratocarcinoma cells, *Cell* **30**:697–705.

Shin, H. S., Flaherty, L., Artzt, K., Bennett, D., and Ravetch, J., 1983, Inversion in the H-2 complex of t-haplotypes in mice, *Nature (Lond.)* **306**:380–383.

Shur, B., Oettgen, P., and Bennett, D., 1979, UDP galactose inhibits blastocyst formation in the mouse, *Dev. Biol.* **73**:78–181.

Silver, L. M., and Artzt, K., 1981, Recombination suppression of mouse 5-haplotypes due to chromatin mismatching, *Nature (Lond.)* **290**:68–70.

Sobel, J. S., 1983, Localization of myosin in the preimplantation mouse embryo, *Dev. Biol.* **95(1)**:227–231.

Sobel, J. S., 1984, Organization of plasma membrane-associated filaments during development of cell contacts in early mouse embryos, *J. Cell Biol.* **99**:36a.

Sobel, J. S., and Alliegro, M. A., 1985, Changes in the distribution of a spectrin-like protein during development of the preimplantation mouse embryo, *J. Cell Biol.* **100**:333–336.

Solter, D., and Knowles, B., 1978, Monoclonal antibody defining a stage-specific mouse embryonic antigen (SSEA-1), *Proc. Natl. Acad. Sci. USA* **75**:5565–5569.

Solter, D., and Knowles, B., 1979, Developmental stage-specific antigens during mouse embryogenesis, in: *Current Topics in Developmental Biology*, Vol. 13 (A. Moscona and A. Monroy, eds.), pp. 139–165, Academic Press, New York.

Spindle, A., 1982, Cell allocation in preimplantation mouse chimeras, *J. Exp. Zool.* **209**:361–367.

Spindle, A. I., and Pedersen, R. A., 1973, Hatching, attachment, and outgrowth of mouse blastocysts in vitro: Fixed nitrogen requirements, *J. Exp. Zool.* **186**:305–318.

Stastna, J., and Prochazka, V., 1983, Scanning electron microscopy of cleaving mouse eggs, *Fol. Morphol.* **31**:212–217.

Stern, P., Willson, I., Lennox, E., Galfre, G., Milstein, C., Secher, D., and Sziegler, A., 1978, Monoclonal antibodies as probes for differentiation and tumor-associated antigens: A Forssman specificity on teratocarcinoma stem cells, *Cell* **14**:775–783.

Stern, P. L., Gilbert, P., Heath, J. K., and Furth, M., 1983, A monoclonal antibody which detects a cell surface antigen on murine embryonal carcinoma and early mouse embryo stages may recognize a carbohydrate determinant involving alpha-linked galactose, *J. Reprod. Immunol.* **5**:145–160.

Stewart, C., 1980, Aggregation between teratocarcinoma cells and preimplantation mouse embryos, *J. Embryol. Exp. Morphol.* **58**:298–302.

Stewart, C. L., 1982, Formation of viable chimeras by aggregation between teratocarcinomas and preimplantation mouse embryos, *J. Embryol. Exp. Morphol.* **67**:167–179.

Surani, M. A., 1979, Glycoprotein synthesis and inhibition of glycosylation by tunicamycin in preimplantation mouse embryos: Compaction and trophoblast adhesion, *Cell* **18(1)**:217–227.

Surani, M. A. H., and Handyside, A. H., 1983, Reassortment of cells according to position in mouse morulae, *J. Exp. Zool.* **225**:505–511.

Surani, M. A. H., Barton, S. C., and Burling, A., 1980, Differentiation of 2-cell and 8-cell mouse embryos arrested by cytoskeletal inhibitors, *Exp. Cell Res.* **125**:275–286.

Surani, M. A., Kimber, S. J., and Handyside, A. H., 1981, Synthesis and role of cell surface glycoproteins in preimplantation mouse development, *Exp. Cell Res.* **133(2)**:331–339.

Surani, M. A. H., Kimber, S. J., and Osborn, J. C., 1983, Mevalonate reverses the developmental arrest of preimplantation mouse embryos by compactin, an inhibitor of HMG Co A reductase, *J. Embryol. Exp. Morphol.* **75**:205–223.

Sutherland, A. E., and Calarco, P., 1983, Analysis of compaction in the preimplantation mouse embryo, *Dev. Biol.* **100**:328–338.

Szollosi, D., 1967, Development of cortical granules and the cortical reaction in rat and hamster eggs, *Anat. Rec.* **159**:431–446.

Szollosi, D., 1971, Nucleoli and ribonucleoprotein particles in the preimplantation conceptus of the rat and mouse, in: *The Biology of the Blastocyst* (R. Blandau, ed.), pp. 95–113, University of Chicago Press, Chicago.

Tachi, S., and Tachi, C., 1979, Ultrastructural studies on maternal–embryonic cell interaction during experimentally induced implantation of rat blastocysts to the endometrium of the mouse, *Dev. Biol.* **68**:203–223.

Timpl, R., Dziadek, M., Fugiwara, S., Nowack, H., and Wick, G., 1983, Nidogen: A new, self-aggregating basement membrane protein, *Eur. J. Biochem.* **137**:455–465.

Van Blerkom, J., and Chavez, D. J., 1981, Morphodynamics of outgrowths of mouse trophoblast in the presence and absence of a monolayer of uterine epithelium, *Am. J. Anat.* **162**:143–155.

Vestweber, D., and Kemler, R., 1984, Rabbit antiserum against a purified surface glycoprotein decompacts mouse preimplantation embryos and reacts with specific adult tissues, *Exp. Cell Res.* **152**:169–178.

Vorbrodt, A., Konwinski, M., Solter, D., and Koprowski, H., 1977, Ultrastructural cytochemistry of membrane-bound phosphatases in preimplantation mouse embryos, *Dev. Biol.* **55(1)**:117–134.

Warner, C., and Spannaus, D., 1984, Demonstration of H-2 antigens on preimplantation mouse embryos using conventional antisera and monoclonal antibody, *J. Exp. Zool.* **230**:37–52.

Wartiovaara, J., and Leivo, I., 1982, Basement membrane matrices and early mouse development, in: *New Trends in Basement Membrane Research* (K. Kuhn, H. Schone and R. Timpl, eds.), pp. 239–246, Raven Press, New York.

Wartiovaara, J., Leivo, I., and Vaheri, A., 1979, Expression of the cell-surface associated glycoprotein, fibronectin, in the early mouse embryo, *Dev. Biol.* **69**:274–257.

Wartiovaara, J., Leivo, I., Virtanen, I., Vaheri, A., and Graham, C. P., 1978, Appearance of fibronectin during differentiation of mouse teratocarcinoma in vitro, *Nature (Lond.)* **272**:355–356.

Wassarman, P. M., 1983, Fertilization, in: *Cell Interactions and Development* (K. Yamada, ed.), pp. 1–27, Wiley, New York.

Webb, C., 1980, Characterization of antisera against mouse teratocarcinoma OTT 6050: Molecular species recognized on embryoid bodies, preimplantation embryos and sperm, *Dev. Biol.* **76**:203–214.

Webb, C., Gall, W., and Edelman, G., 1977, Synthesis and distribution of H-2 antigens in pre-implantation mouse embryos, *J. Exp. Med.* **146**:923–932.

Weitlauf, H. M., 1984, Changes in synthesis of RNA and protein during reactivation of delayed implanting mouse blastocysts, in: *Molecular Aspects of Early Development* (G. Malacinski and W. Klein, eds.), pp. 289–307, Plenum Press, New York.

Wiley, L., 1979, Early embryonic cell surface antigens as developmental probes, in: *Current Topics in Developmental Biology*, Vol. 13 (A. Moscona and A. Monroy, eds.), pp. 167–197, Academic Press, New York.

Wiley, L., and Eglitis, M., 1980, Effects of colcemid on cavitation during mouse blastocoele formation, *Exp. Cell Res.* **127**:89–101.

Wiley, L. M., and Eglitis, M., 1981, Cell surface and cytoskeletal elements: Cavitation in the mouse preimplantation embryo, *Dev. Biol.* **86**:493–501.

Willison, K., Karol, R., Suzuki, A., Kundu, S., and Marcus, D., 1982, Neutral glycolipid antigens as developmental markers of mouse teratocarcinoma and early embryos: An immunologic and chemical analysis, *J. Immunol.* **129**:603–609.

Wolf, D. E., Edidin, M., and Handyside, A. H., 1981, Changes in the organization of the mouse egg plasma membrane upon fertilization and first cleavage: Indications from the lateral diffusion rates of fluorescent lipid analogs, *Dev. Biol.* **15(1)**:195–198.

Wolf, D. E., and Ziomek, C. A., 1983, Regionalization and lateral diffusion of membrane proteins in unfertilized and fertilized mouse eggs, *J. Cell Biol.* **96**:1786–1790.

Wolf, D. P., 1981, The mammalian egg's block to polyspermy, in: *Fertilization and Embryonic Development in Vitro* (L. Mastroianni and I. Biggers, eds.), pp. 183–197, Plenum Press, New York.

Wolf, D. P., Inoue, M., and Stark, R. A., 1976, Penetration of zona-free mouse ova, *Biol. Reprod.* **15**:215–221.

Wu, J-C., Wan, Y-J., Chung, A., and Damjanov, I., 1983, Immunohistochemical localization of entactin and laminin in mouse embryos and fetuses, *Dev. Biol.* **100**:496–505.

Yanagimachi, R., 1981, Mechanisms of fertilization in mammals, in: *Fertilization and Embryonic Development in Vitro* (L. Mastroianni and J. Biggers, eds.), pp. 81–182, Plenum Press, New York.

Yanagimachi, R., Okada, A., and Tung, K. S. K., 1981, Sperm autoantigens and fertilization. II. Effects of anti-guinea pig autoantibodies on sperm-ovum interactions, *Biol. Reprod.* **24:**512–518.

Yoshida-Noro, C., Suzuki, N., and Takeichi, M., 1984, Molecular nature of the calcium-dependent cell–cell adhesion system in mouse teratocarcinoma and embryonic cells. Studies with a monoclonal antibody, *Dev. Biol.* **101:**19–27.

Yotsuyanagi, Y., and Szollosi, D., 1981, Early mouse embryo intracisternal particle: Fourth type of retrovirus-like particle associated with the mouse, *J. Natl. Cancer Inst.* **67:**677–683.

Zetter, B. R., and Martin, G. R., 1978, Expression of a high molecular weight cell surface glycoprotein (LETS protein) by preimplantation mouse embryos and teratocarcinoma stem cells, *Proc. Natl. Acad. Sci. USA* **75(5):**2324–2328.

Chapter 9

Molecular Basis for Teratocarcinoma Cell–Cell Adhesion

MASATOSHI TAKEICHI

1. Introduction

Cell–cell interactions are essential for many developmental systems. Molecules responsible for joining cells presumably play important roles in cellular interaction—not only by providing bridges between cells but also by functioning as a key factor for cell–cell recognition and for other cellular processes. Characterizing cell adhesion molecules is therefore of great interest in elucidating the molecular nature of various cell–cell interacting systems.

Teratocarcinoma is a tumor that expresses properties of early embryonic cells, such as pluripotency for differentiation and embryonic surface antigens. From this property, the stem cells of teratocarcinoma (referred to as "teratocarcinoma cells" in this chapter for simplicity) have been used extensively as an ideal model system for studying early embryonic cells. It is expected that the cell adhesion molecules in teratocarcinoma cells can be related directly to those molecules in early embryonic cells that may be involved actively in cellular interactions in embryonic development. This chapter discusses recent progress on the characterization of cell–cell adhesion molecules in mouse teratocarcinoma cells and their possible roles in morphogenesis of early mouse embryos.

2. Recent Progress in Studies on Cell Adhesion Molecules

Earlier studies on identification of cell adhesion molecules focused primarily on substances showing activity in promoting cell aggregation. Such substances were detected in conditioned media of cell cultures (e.g., Lilien et al., 1978) or in the supernatant of suspensions of cells after dissociating tissues (e.g., Turner, 1978); these substances were regarded as the molecules involved

MASATOSHI TAKEICHI • Department of Biophysics, Faculty of Science, Kyoto University, Kyoto 606, Japan.

in cell–cell adhesions. In this type of approach, however, a couple of weak points should be noted. For example, there is no assurance that cell adhesion molecules are released from cell surfaces in an active form either spontaneously or by artificial means. Usually, no evidence is available to support the hypothesis that substances with cell aggregation-promoting activity are really involved in the natural states of cell adhesion. It is possible that many classes of aggregation-promoting substances so far detected are extracellular adhesive molecules, such as fibronectin, but are not the molecules involved directly in cell–cell adhesion.

More successful results have been achieved by immunological approaches. Antibodies recognizing cell adhesion molecules—if such can be obtained—are expected to block cell adhesion. Spiegel (1954a,b) first described the aggregation-inhibitory effect of antibodies, which were raised against sponge and amphibian embryonic cells. Once the antibodies displaying this activity are obtained (usually used as Fab fragments), it becomes possible to search for the substances capable of neutralizing their aggregation-inhibitory effect from cell lysates or some fractions containing cell surface components. One can regard the substances showing this activity, which are the targets for the aggregation-inhibitory antibodies, as candidates for cell adhesion molecules. Applying this method (denoted the **Fab strategy**), Müller and Gerisch (1978) first succeeded in identifying cell–cell adhesion molecules in slime molds (contact site A). In vertebrate systems, Edelman and associates (see Edelman, 1983, for review) approached the identification of neural cell adhesion molecules by the Fab strategy. These investigators prepared antibodies against cell surface components of chicken neural retina, the Fab fragments of which can affect cell–cell adhesion of this tissue. The glycoproteins that neutralize the adhesion-inhibitory effect of these antibodies were identified and termed N-CAM. With the aid of adhesion-inhibitory antibodies, molecules relevant to **cell–substrate** adhesion have also been identified (Damsky et al., 1981; Knudsen et al., 1981; Oesch and Birchmeier, 1982).

There is another type of approach to studying cell adhesion molecules. According to this strategy, tissues and cultured cells are disaggregated into single cells by various treatments. These cells can then be reaggregated under proper culture conditions. In this experimental system, it is possible to determine which cell surface components must be removed to disaggregate cells and which components are required for reaggregation of these cells. Takeichi (1977) adopted this approach to investigate the adhesive properties of V79 fibroblast cells and found that these cells are provided with two classes of cell–cell adhesion systems, **Ca^{2+}-dependent systems (CDS)** and **Ca^{2+}-independent systems (CIDS)**. Further studies clarified that these systems can be distinguished in immunological and functional specificities and also in trypsin sensitivities (Urushihara et al., 1979; Takeichi et al., 1979). The presence of these two distinct cell–cell adhesion systems has been confirmed in a variety of cell types (see Takeichi et al., 1982, for review). By combining these findings with the Fab strategy, Urushihara and Takeichi (1980) succeeded in identifying a glycoprotein involved in CIDS of V79 cells. N-CAM belongs to the CIDS group as well (Brackenbury et al., 1981).

3. Cell Adhesion Molecules of Teratocarcinoma

Adhesive properties of teratocarcinoma cells were studied in comparison with those of fibroblastic cells by Takeichi et al. (1981). Teratocarcinoma cells were found to have the two distinct cell–cell adhesion systems described above, as do other cell types. An interesting observation made during this study was that teratocarcinoma and fibroblastic cells show preferential adhesiveness to their own cell types when they are artificially mixed. It was found that this selective adhesion occurs when CDS are active on the cell surface. This observation suggested that CDS can be grouped into at least two types—**teratocarcinoma type** and **fibroblast type**—which are distinct in their functional specificity.

The Fab strategy was applied to the identification of molecules involved in teratocarcinoma CDS (**t-CDS**). Fab fragments of antibodies raised against teratocarcinoma F9 cells (anti-TC-F9) inhibited Ca^{2+}-dependent aggregation of teratocarcinoma cells of various lines but had no effect on fibroblastic cells, indicating that CDS in teratocarcinoma and fibroblasts are immunologically distinct (Takeichi et al., 1981).

In identifying the targets for the Fab in inhibiting cell aggregation, Yoshida and Takeichi (1982) employed the following strategy. CDS show a unique trypsin sensitivity; i.e., they are destroyed by trypsin treatment of cells in the absence of Ca^{2+} but not by the same treatment in the presence of Ca^{2+} (Takeichi, 1977). In other words, CDS are left intact on the cells treated with trypsin plus Ca^{2+} (TC) but not on the cells treated with trypsin plus EGTA (TE), a calcium chelator. Therefore, among cell surface antigens reacting with anti-TC-F9, those detected on the surfaces of TC-treated cells but not of TE-treated cells can be regarded as the targets for the adhesion-inhibiting Fab. Yoshida and Takeichi (1982) found that one cell surface protein with a molecular weight of 124,000 is present only in TC-treated cells (this value was originally reported to be 140,000 but was corrected to 124,000 in recent experiments).

To obtain evidence that the above TC-treated cell-specific protein is really a component of t-CDS, the following experiment was performed. Since the TE treatment of cells inactivates CDS, it is expected that t-CDS is released as tryptic fragments from cell surfaces after this treatment. Actually, Yoshida and Takeichi (1982) found that a substance with molecular weight of 34,000, which can adsorb the adhesion-inhibitory effect of anti-TC-F9 Fab, is released from cells by TE treatment. When anti-TC-F9 was absorbed with this 34,000-M_r protein, its reactivity to the 124,000-M_r protein was also lost, suggesting that the 34,000-M_r protein is a fragment of the 124,000-M_r protein. These results strongly supported the hypothesis that the 124,000-M_r protein is a target for the Fab in inhibiting t-CDS.

3.1. E-Cadherin: A Ca^{2+}-Dependent Cell Adhesion Molecule

Conclusive evidence in identifying cell adhesion molecules by immunological methods can be provided only by the use of monospecific antibodies. Toward this aim, Yoshida-Noro *et al.* (1984) generated a monoclonal antibody (ECCD-1) recognizing t-CDS. This antibody not only inhibits aggregation of teratocarcinoma cells involving CDS but also induces disruption of cell–cell adhesion of teratocarcinoma cells in monolayer cultures (Fig. 1a,b).

The target molecules for ECCD-1 have been determined by Western blot analysis: Antigen sources were collected from nontrypsinized F9 cells, separated by electrophoresis, and blotted onto nitrocellulose. The blot was then incubated with ECCD-1. The antibody reacted with a major band with a molecular weight of 124,000 and several other relatively minor bands with molecular weights of 104,000, 97,000, 87,000, 85,000, and 73,000, respectively (Fig. 2, left lane). When antigens were collected from TC-treated cells, the 124,000-M_r band was again a major component reacting with ECCD-1, but in this case another major band (100,000 M_r) was also detected (results not

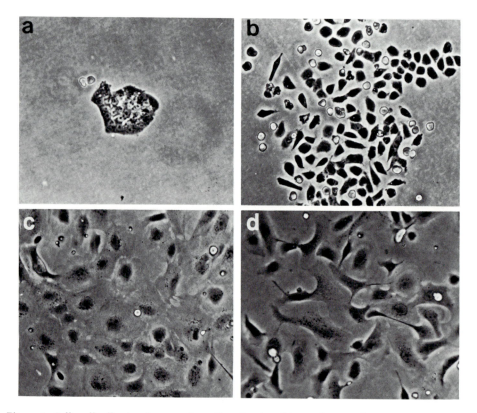

Figure 1. Cell–cell adhesion in normal medium (a, c) and in medium with ECCD-1 (b, d). (a, b) Teratocarcinoma PCC3 cells. (c, d) Epidermal cells derived from mouse fetus. Magnification is the same for a and b and for c and d, respectively.

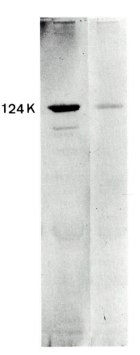

124 K

Figure 2. Detection of antigens to ECCD-1 by Western blot analysis. In the right lane, the blot was rinsed with EGTA after incubation with ECCD-1. See Yoshida-Noro et al. (1984) for technical details.

shown). None of these components was detected in TE-treated F9 cells, in which CDS are inactive.

The electrophoretic mobility and trypsin sensitivity of the major 124,000-M_r polypeptide were exactly the same as the protein detected as a component of t-CDS by the use of anti-TC-F9 in the experiment described in the previous section. Thus, different approaches gave the same result in identification of the target molecules for the antibodies that inhibit the activity of t-CDS.

Yoshida-Noro et al. (1984) proposed to call the cell adhesion molecules recognized by ECCD-1 **cadherin.** Since our recent study suggested that non-epithelial tissues also contain cadherinlike molecules (Hatta et al., 1985), we will call the ECCD-1 targets specifically **E-cadherin.**

3.2. Cell-Type Specificity of E-Cadherin

Tissue specificity or cell-type specificity of E-cadherin was examined by Yoshida-Noro et al. (1984). These workers found that ECCD-1 binds specifically to epithelial components of various tissues of fetal and adult mice and actively disrupts their cell–cell adhesions. For example, when ECCD-1 was added to epithelial colonies derived from hepatocytes or epidermal cells, the epithelia were broken into smaller cell clusters or individual cells (see Fig. 1c,d). ECCD-1 also caused rounding up of blastomeres of compacted mouse embryos (see Section 5). Therefore, it appears certain that the ECCD-1 targets play an active role in binding cells in various tissues.

Molecular species reacting with ECCD-1 in nonteratocarcinoma cells were examined using liver cells as a representative of differentiated tissues (Ogou *et al.*, 1983). The 124,000- and 104,000-M_r components were detected as the major antigens to ECCD-1 in liver cells. Therefore, the molecular constitution of CDS in teratocarcinoma and liver cells is essentially the same. Although the identity of molecular species reacting with ECCD-1 in other tissues has not been determined, the molecules detected in teratocarcinoma and liver cells are presumably of general importance for cell–cell binding in a wide variety of tissues.

3.3. Relationship of E-Cadherin to the Molecules Identified in Other Laboratories

Polyclonal antibodies raised against teratocarcinoma cells are known to inhibit compaction of mouse embryos (Kemler *et al.*, 1977; Johnson *et al.*, 1979; Ducibella, 1980; Nicolas *et al.*, 1981; see Chapter 8). Such antibodies also cause rounding up of teratocarcinoma cells (decompaction) in monolayer cultures (Hyafil *et al.*, 1980; Nicolas *et al.*, 1981; Yoshida and Takeichi, 1982). These effects are apparently the same as that of ECCD-1.

Hyafil *et al.* (1980, 1981) attempted to identify a substance that can neutralize the activity of such antibodies in inducing decompaction of teratocarcinoma cells. These investigators found that a glycoprotein with a molecular weight of 84,000 (gp 84), which is released from the membrane fraction of teratocarcinoma cells by treatment with trypsin (in the presence of Ca^{2+}), neutralizes the decompacting activity of the antibodies. Hyafil *et al.* (1981) then generated monoclonal antibodies against gp84. One of these (DE1) showed an interesting property in that it binds to gp84 only in the presence of Ca^{2+}. This binding feature is quite similar to that shown by ECCD-1 (see Section 4). Although DE1 by itself showed no effect on cell–cell adhesion or teratocarcinoma cells when added to their cultures, this monoclonal antibody was capable of inhibiting the decompacting effect of the polyclonal antibodies against teratocarcinoma cell surfaces. All data described above suggest that ECCD-1 and DE1 may recognize common target molecules. Indeed, the search for the native target molecules for DE1 revealed that this antibody also recognizes the 124,000-M_r protein as the major antigen (Yoshida-Noro *et al.* 1984). It is therefore likely that gp84 is a tryptic fragment of the 124,000-M_r protein, a possibility that has also been supported by Peyrieras *et al.* (1983).

Other groups working with nonteratocarcinoma cells also identified E-cadherinlike molecules. Gallin *et al.* (1983) isolated a tryptic fragment (81,000 M_r) of some cell surface component that is released by treating the membrane of chicken liver cells with trypsin (in the present of Ca^{2+}). Fab of antibodies raised against this tryptic fragment was effective in inhibiting liver cell–cell adhesion. Gallin and co-workers generated monoclonal antibodies against this protein and detected a predominant 124,000-M_r antigen to these antibodies from liver cell lysate. The tissue distribution of this adhesion molecule (called

L-CAM) determined by Edelman *et al.* (1983) is quite similar to that of E-cadherin. These findings suggest that L-CAM belongs to the same molecular species as E-cadherin. Damsky *et al.* (1983), Imhof *et al.* (1983), and Vestweber and Kemler (1984) also reported cell adhesion molecules similar in molecular weight and cell-type specificity to E-cadherin.

3.4. Other Molecules Implicated in Teratocarcinoma Adhesion

Grabel *et al.* (1979) presented evidence that teratocarcinoma cell surfaces express a lectinlike component. Further studies by this group found that Ca^{2+}-independent aggregation of teratocarcinoma cells is inhibited by certain classes of polysaccharides such as fucans (fucoidan) and mannans, suggesting that a fucan/mannan-specific lectin is involved in this adhesion system (Grabel *et al*, 1983). These workers also reported that the aggregation of teratocarcinoma cells harvested by deleting divalent cations is promoted both by Ca^{2+} and Mg^{2+}. Since this divalent cation-dependent aggregation was not completely inhibited by the above polysaccharides, it was concluded that this aggregation was mediated by an adhesion system separable from the Ca^{2+}-independent one. This divalent cation-dependent aggregation system is also distinct from CDS, because CDS-dependent aggregation is not promoted by Mg^{2+}. It is known that Mg^{2+} is an essential ion for cells to attach to noncellular substrata. Ca^{2+} is also effective in promoting the cell–substrate adhesion, although it is usually less effective than Mg^{2+}. Therefore, it is probable that the divalent cation-dependent aggregation observed by these investigators depends on the same mechanism as for cell–substrate adhesion. It is also possible that CDS cooperates with the Mg^{2+}-dependent adhesion system in the observed aggregation. In any case, a third group of cell adhesion mechanism other than CDS and CIDS would be involved in teratocarcinoma cell adhesion.

Oppenheimer and Humphreys (1971) and Oppenheimer (1975) reported that mouse ascites fluid in which teratocarcinoma cells had grown contains a macromolecular factor promoting the aggregation of teratocarcinoma cells. The role of this factor in cell adhesion remains to be elucidated.

4. Molecular Mechanisms of Cell–Cell Adhesion by E-Cadherin

In monolayers of cells of various origins including teratocarcinoma cells treated with EGTA or EDTA, the cell layers dissociate and cells are rounded up. Treatment of compacted mouse embryos by these chelators exhibits the same effect. On the basis of our current knowledge of CDS, this effect is assumed to result from the inactivation of CDS.

How is Ca^{2+} involved in the function of CDS? An interesting feature of the response of ECCD-1 to cells is that the binding of ECCD-1 to cell surfaces requires Ca^{2+} (Yoshida-Noro *et al.*, 1984). Other divalent cations are ineffective for this binding, although Mn^{2+} is slightly effective. This observation

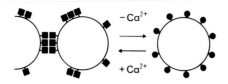

Figure 3. Model of the Ca^{2+}-dependent binding of cells with E-cadherin. (■) Active form; (●) inactive form of E-cadherin molecules.

suggests that the antigen molecules must bind Ca^{2+} to be recognized by ECCD-1. This Ca^{2+}-dependent binding of ECCD-1 to the antigens was observed even *in vitro*. In the Western blot experiments described in Section 3.1, the blot of antigens was incubated with ECCD-1 in the presence of Ca^{2+}. However, when Ca^{2+} was eliminated in the incubation medium, ECCD-1 did not bind to the antigens. Alternatively, when the blot was rinsed with EGTA, after incubation with ECCD-1 in the presence of Ca^{2+}, the antibody once bound was readily released (Fig. 2, right lane). Under the assumption that Ca^{2+} has no effect on antibody structure, these results indicate that Ca^{2+} has an active role in conformational changes of E-cadherin. This conclusion was also suggested from the observation that trypsin sensitivity of CDS is altered with Ca^{2+} (Takeichi, 1977).

We assume that this Ca^{2+}-sensitive property of E-cadherin must be related directly with the Ca^{2+}-dependent function in binding cells. It is most likely that the role of Ca^{2+} in cell adhesion is to convert the E-cadherin molecule into an active form, as illustrated in Figure 3. However, we have no further idea based on experimental data to explain the molecular mechanisms underlying cellular adhesion mediated with E-cadherin.

The mode of action of E-cadherin and related molecules such as L-CAM in binding cells could be deduced from a structural analysis of these molecules. Thus far, we have observed that when intact cells are trypsinized in the presence of Ca^{2+}, no proteolysis of E-cadherin seems to occur. However, when isolated cell membranes are subjected to the same treatment, tryptic fragments with an approximate molecular weight of 80,000 are produced (Hyafil et al., 1980, 1981; Gallin et al., 1983). It remains uncertain as to how the molecules become susceptible to trypsin/Ca^{2+} when the isolated membranes are trypsinized. It has also been observed that substances with this molecular size are spontaneously released into the medium in serum-free culture of neural retina cells (Grunwald et al., 1982) and mammary tumor cells (Damsky et al., 1983) and that they are capable of neutralizing the aggregation-inhibitory effect of antibodies raised against the cell surfaces of each cell type. This phenomenon could be attributed to cleavage of CDS by cellular intrinsic proteases, which might occur under particular culture conditions (such as in serum-free media).

The above tryptic fragments retain the property of reacting with Ca^{2+}, as shown by Hyafil et al. (1981). It seems, therefore, that as long as Ca^{2+} is present, the Ca^{2+}-binding site of E-cadherin is not destroyed by proteolysis. In the absence of Ca^{2+}, however, the tryptic fragments are further degraded by proteolysis into smaller components even when intact cells were treated. One such component was detected as a protein with a molecular weight of 34,000 in the case of E-cadherin (Yoshida and Takeichi, 1982) and of 40,000 in the case of chicken L-CAM (Gallin *et al.*, 1983). The tryptic fragments produced in the absence of Ca^{2+} do not react with ECCD-1 (M. Takeichi, unpublished data), suggesting that they lose their original molecular conformation.

It is interesting to speculate how ECCD-1 disrupts cell–cell adhesion. One possibility is that the antibody directly blocks the functional sites of adhesion molecules. Cell–cell adhesion may be under the control of a sort of dynamic equilibrium between the connection and disconnection of the apposed cell surfaces. If this were the case, the antibodies would be able to reach and bind to the functional site of the adhesion molecules when they are in an unbound state in their connection–disconnection cycle. When most molecules are masked with the antibodies in this way, the cell–cell adhesions would then be blocked.

However, if the molecular bridges once established between cell surfaces are structurally stable, the above mechanism may not work. In this case, the functional sites of adhesion molecules engaged in connecting cells may not be exposed for attack by antibodies. If this is so, the antibodies must bind to other portions of the molecule but still affect the functions of the molecule, leading to disruption of molecular connections of cell surfaces. It is highly possible that the effect of binding of antibodies to the nonactive sites of a protein is transmitted to other portions of this protein, affecting its function. None of the monoclonal antibodies raised against gp84 by Hyafil *et al.* (1981) actively inhibits cell–cell adhesion. This suggests that the binding of antibodies to the nonactive sites of adhesion molecules does not always cause inhibition of their function. Further studies will be required to determine which mechanism is correct.

Figure 4 shows ECCD-1 bound to F9 cells in monolayer cultures by fluorescence staining. The stain was concentrated in the cell–cell contact portions of

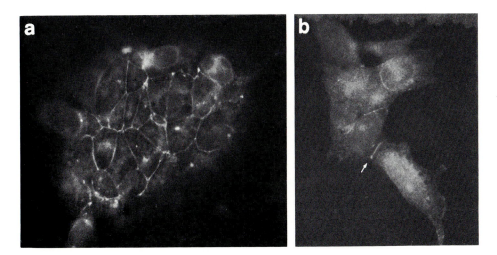

Figure 4. Immunofluorescence staining of ECCD-1 bound to F9 cells in a monolayer culture. (a) Dense cell colony. (b) Sparse cell cluster. Note that the stain is detected only at the cell–cell contact portions of cell surfaces (b, arrow).

cell surfaces but not clearly detected in portions of cell surfaces not engaged in cell–cell contact. Furthermore, careful microscopic observation of these specimens led to the impression that the stain was detected only when the focus was adjusted to the upper portions of cell layers, suggesting that E-cadherin may be located at a particular upper region of cell surfaces. The boundary of cells traced by the stain of ECCD-1 usually did not match the overall boundary of cells detected by phase contrast microscopy. This strongly suggests a special structural organization of a portion of cell surfaces carrying E-cadherin.

5. The Role of E-Cadherin in Early Embryonic Development

Teratocarcinoma cells resemble early embryonic cells in various respects. Ogou et al. (1982) showed that mouse embryos at cleavage stage are provided with CDS with the same functional and immunological specificities as those in teratocarcinoma cells. ECCD-1 induces rounding up of blastomeres of the compacted embryos (see Section 3.2). This antibody should therefore be an ideal tool to investigate the role of cell–cell contact mediated by E-cadherin in embryonic development. An account of our recent experiments on the effect of ECCD-1 on the development of mouse embryos follows (Shirayoshi et al., 1983).

Mouse embryos at the 4-cell stage were cultured in vitro in the absence or presence of ECCD-1 (Fig. 5). In normal medium, embryos undergo compaction at the late 8-cell stage and then develop to the blastocyst stage. However, embryos cultured in ECCD-1 showed a distinct developmental profile. Cell proliferation was not affected with ECCD-1, so that cell number in the embryos increased at a normal rate in the presence of this antibody. When they reached the late 8-cell stage, however, compaction did not occur, and the grapelike morphology remained unchanged up to the 16-cell stage. (Although compaction was inhibited, cells were still adherent, probably because of the presence of Ca^{2+}-independent cell–cell adhesion systems.) At the late 16-cell stage, however, the surface of embryos became smooth, suggesting the increase in adhesiveness between cells. After approximately the 32-cell stage, the embryos developed to the blastocyst stage with an apparently normal external morphology. Analysis of these blastocysts by morphological and histochemical techniques showed that they do not contain an inner cell mass (ICM), however, (Fig. 6).

We may now ask: How does the block of CDS-dependent cell–cell contact in morula cause the inhibition of ICM formation? A current hypothesis, the so-called "inside-outside hypothesis," proposes that cells positioned on the inside of embryos become ICM and that cells positioned on the periphery of embryos become trophectoderm as a consequence of their locations (Mintz, 1965; Tarkowski and Wroblewska, 1967; Hillman et al., 1972). In normal mouse embryos, determination of these two groups of cells occurs at the 16-cell stage. In embryos treated with ECCD-1, it is likely that all cells at this stage are forced to face the same environment—presumably the outer environment—

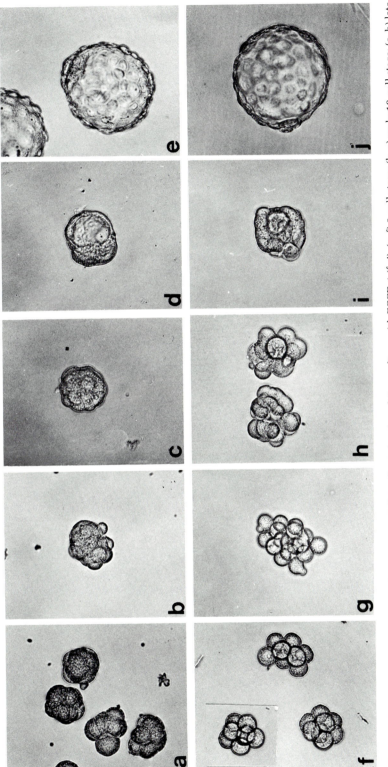

Figure 5. Development of mouse embryos in normal medium (a–e) and in medium with ECCD-1 (f–j). (a, f) 8-cell stage; (b, g) early 16-cell stage; (c, h) late 16-cell stage; (d, i) approximately the 32-cell stage; (e, j) blastocyst stage.

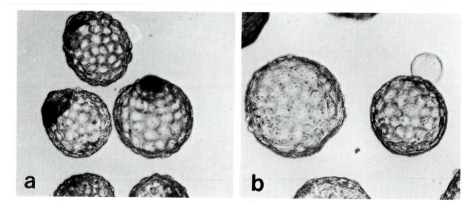

Figure 6. Inhibition of ICM formation with ECCD-1. (a) Control embryos. (b) Experimental embryos. To detect the ICM, alkaline phosphatase was stained.

due to the lack of a cell layer partitioning the inside and outside of embryos. This environmental effect could drive all cells in an embryo to the trophectoderm lineage. Leaving this idea as one possibility, we shall now discuss another possible mechanism for the inhibition of ICM formation by ECCD-1.

There have been reports suggesting that cell–cell contacts in early mouse embryos are essential for the formation of polarity of blastomeres. Handyside (1980) reported that when cleaving embryos at various stages were stained with FITC-conjugated concanavalin A (FITC-Con A) or FITC-conjugated antibodies reacting with cell surface materials, the stain was observed evenly on cell surfaces until the early 8-cell stage, but after compaction the stain became concentrated at one portion of the surface of a cell; this was called **polarization.** Ziomek and Johnson (1980) confirmed Handyside's observation and suggested that polarization is a cell–cell contact-dependent process. In this context, it should be stressed that the position of the pole is always located away from the plane of cell–cell contact (Johnson and Ziomek, 1981a).

Additional work from Johnson's group (Johnson and Ziomek, 1981b) suggested that polarization is implicated in differentiation of blastomeres. According to their studies, when polarized 8-cell–stage blastomeres cleave, many of them divide asymmetrically; i.e., one inherits the pole and the other the opposite part. The former, usually larger than the other, maintains the polar surface, but the latter loses the polarity. Ziomek and Johnson (1982) observed a tendency after this division for the larger polarized cells to enter the trophectoderm lineage and the smaller unpolarized cells to enter the ICM lineage, as long as they are in their original normal positions in an embryo. Thus, the formation of polarity in a cell appears to be somehow connected to the mechanisms for differential division of a cell to generate multiple cell lineages.

These results suggested the possibility that the cell–cell contact by CDS is required primarily for the induction of polarity on cell surfaces. Shirayoshi et al. (1983) examined whether ECCD-1 actually affects the polarization of blastomeres by conducting the following experiments. Single blastomeres were

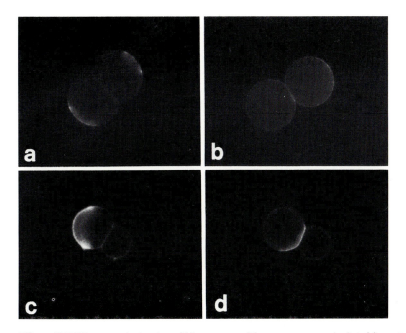

Figure 7. Effect of ECCD-1 on polarization of blastomeres. Blastomeres were isolated from 4-cell (a, b) or 8-cell (c, d) embryos and incubated in the presence or absence of ECCD-1. Several hours after the division of each blastomere, pairs were stained with FITC-conjugated concanavalin A. (a, c) Control pairs. (b, d) Experimental pairs. Note the difference in intensity and position of the pole between the control and experimental pairs.

isolated from embryos at various stages and were cultured in the absence or presence of ECCD-1. At proper incubation periods after the first division, each cell pair was stained with FITC-Con A to observe the distribution of receptors for Con A.

The results obtained were as expected. When blastomeres isolated from 4-cell–stage embryos were cultured in ECCD-1 until they reached the stage corresponding to the late 8-cell stage (compaction stage), polarization was not detected (Fig. 7a,b). When blastomeres isolated from 8-cell-stage embryos were cultured in ECCD-1, many of them divided to a pair consisting of a larger and a smaller cell. The pole was not observed, however, in many of these pairs and, if observed, the position of the pole was abnormal. The pole was always detected apart from the positions of cell–cell contact in normal embryos. In ECCD-1 treated embryos, the pole tended to be located close to, or associated with, the cell–cell contact planes (Fig. 7c,d). These observations suggest that polarization is regulated by the cell–cell contact mediated by E-cadherin.

If the hypothesis that polarization is essential for differential division of a cell to produce multiple cell lineages is correct, the results described above suggest that ECCD-1 suppresses formation of the ICM by perturbing polarization. Even if this is the case, it remains to be resolved how the trophectoderm lineage is exclusively produced in the ECCD-1-treated embryos. One possible explanation is that perhaps the cells capable of differentiating into ICM can be generated solely from blastomeres that have once been polarized in a normal pattern. Under this assumption, an ICM may not be formed in embryos in which the polarization pattern has been affected with ECCD-1.

To identify the factors that are most essential for the suppression of formation of ICM by ECCD-1, further studies are needed. In any case, compaction of embryos mediated by E-cadherin is thus important for regulation of differentiation of blastomeres.

6. Concluding Remarks

The strong effect of the monoclonal antibody ECCD-1 in disrupting cell–cell adhesion of teratocarcinoma and early embryonic cells suggests that E-cadherin plays the major role in joining these cells. We have asked whether this adhesion molecule is implicated in any dynamic processes of embryonic morphogenesis. The experiments described here indicate that E-cadherin is required not only for mechanical binding of cells but also for some functional communication between cells, which appears essential for regulation of cell differentiation. So far, we have studied the role of this adhesion molecule only at very early developmental stage. Since expression of E-cadherin persists in a certain group of cells until the maturation of the organism, this adhesion molecule must play active roles in various steps of embryogenesis. We intend to investigate such possible roles of E-cadherin in future studies.

References

Brackenbury, R., Rutishauser, U., and Edelman, G. M., 1981, Distinct calcium-independent and dependent adhesion systems of chicken embryo cells, *Proc. Natl. Acad. Sci. USA* **78**:387–391.

Damsky, C. H., Knudsen, K. A., Dorio, R. J., and Buck, C. A., 1981, Manipulation of cell–cell and cell–substratum interactions in mouse mammary tumor epithelial cells using broad spectrum antisera, *J. Cell Biol.* **89**:173–184.

Damsky, C. H., Richa, C., Solter, D., Knudsen, K., and Buck, C. A., 1983, Identification and purification of a cell surface glycoprotein mediating intercellular adhesion in embryonic and adult tissues, *Cell* **34**:455–456.

Ducibella, T., 1980 Divalent antibodies to mouse embryonal carcinoma cells inhibit compaction in the mouse embryos, *Dev. Biol.* **79**:356–366.

Edelman, G. M., 1983, Cell adhesion molecules, *Science* **219**:450–457.

Edelman, G. M., Gallin, W. J., Delouvee, A., Cunningham, B. A., and Thiery, J.-P., 1983, Early epochal maps of two different cell adhesion molecules, *Proc. Natl. Acad. Sci. USA* **80**:4384–4388.

Gallin, W. J., Edelman, G. M., and Cunningham, B. A., 1983, Characterization of L-CAM, a major cell adhesion molecule from embryonic liver cells, *Proc. Natl. Acad. Sci. USA* **80**:1038–1042.

Grabel, L. B., Rosen, S., and Martin, G. R., 1979, Teratocarcinoma stem cells have a cell surface carbohydrate-binding component implicated in cell–cell adhesion, *Cell* **17**:477–483.

Grabel, L. B., Singer, M. S., Martin, G. R., and Rosen, S. D., 1983, Teratocarcinoma stem cell adhesion: The role of divalent cations and a cell surface lectin, *J. Cell Biol.* **96**:1532–1537.

Grunwald, G. B., Pratt, R. S., and Lilien, J. 1982, Enzymic dissection of embryonic cell adhesive mechanisms. III. Immunological identification of a component of the calcium-dependent adhesive system of embryonic chick neural retina cells, *J. Cell Sci.* **55:**69–83.

Handyside, A. H., 1980, Distribution of antibody- and lectin-binding sites on dissociated blastomeres from mouse morulae: Evidence for polarization at compaction, *J. Embryol. Exp. Morphol.* **60:**99–116.

Hatta, K., Okada, T. S., and Takeichi, M., 1985, A monoclonal antibody disrupting calcium-dependent cell–cell adhesion of brain tissues: Possible role of its target antigen in animal pattern formation, *Proc. Natl. Acad. Sci. USA* **82:**2789–2793.

Hillman, N., Sherman, M. I., and Graham, C. F., 1972, The effect of spatial arrangement on cell determination during mouse development, *J. Embryol. Exp. Morphol.* **28:**263–278.

Hyafil, F., Morello, D., Babinet, C., and Jacob, F., 1980, A cell surface glycoprotein involved in the compaction of embryonal carcinoma cells and cleavage stage embryos, *Cell* **21:**927–934.

Hyafil, F., Babinet, C., and Jacob, F., 1981, Cell–cell interactions in early embryogenesis: A molecular approach to the role of calcium, *Cell* **26:**447–454.

Imhof, B. A., Vollmers, H. P., Goodman, S. L., and Birchmeier, W., 1983, Cell–cell interaction and polarity of epithelial cells: Specific perturbation using a monoclonal antibody, *Cell* **35:**667–675.

Johnson, M. H ., Chakraborty, J., Handyside, A. H., Willison, K., and Stern, P., 1979, The effect of prolonged decompaction on the development of the preimplantation mouse embryo, *J. Embryol. Exp. Morphol.* **54:**241–261.

Johnson, M. H., and Ziomek, C. A., 1981a, Induction of polarity in mouse 8-cell blastomeres: Specificity, geometry, and stability, *J. Cell Biol.* **91:**303–308.

Johnson, M. H., and Ziomek, C. A., 1981b, The foundation of two distinct cell lineages within the mouse morula, *Cell* **24:**71–80.

Kemler, R., Babinet, C., Eisen, H., and Jacob, F., 1977, Surface antigen in early differentiation, *Proc. Natl. Acad. Sci. USA* **74:**4449–4452.

Knudsen, K. A., Rao, P. E., Damsky, C. H., and Buck, C. A., 1981, Membrane glycoprotein involved in cell–substratum adhesion, *Proc. Natl. Acad. Sci. USA* **78:**6071–6075.

Lilien, J., Hermolin, J., and Lipke, P., 1978, Molecular interactions in specific cell adhesion, in: *Specificity of Embryological Interactions* (D. R. Garrod, ed.), pp. 133–155, Chapman and Hall, London.

Mintz, B., 1965, Experimental genetic mosaicism in the mouse, in: *Preimplantation Stages of Pregnancy* (G. W. Wolstenholme and M. O'Connor, eds.), pp. 194–207, J and A Churchill, London.

Müller, K., and Gerisch, G., 1978, A specific glycoprotein as the target site of adhesion blocking Fab in aggregating Dictyostelium cells, *Nature (Lond.)* **274:**445–449.

Nicolas, J., Kemler, R., and Jacob, F., 1981, Effects of antiembryonal carcinoma serum on aggregation and metabolic cooperation between teratocarcinoma cells, *Dev. Biol.* **81:**127–132.

Oesch, B., and Birchmeier, W., 1982, New surface component of fibroblast's focal contacts identified by a monoclonal antibody, *Cell* **31:**671–679.

Ogou, S., Okada, T. S., and Takeichi, M., 1982, Cleavage stage mouse embryos share a common cell adhesion system with teratocarcinoma cells, *Dev. Biol.* **92:**521–528.

Ogou, S., Yoshida-Noro, C., and Takeichi, M., 1983, Calcium-dependent cell–cell adhesion molecules common to hepatocytes and teratocarcinoma stem cells, *J. Cell Biol.* **97:**944–948.

Oppenheimer, S. B., 1975, Functional involvement of specific carbohydrate in teratocarcinoma cell adhesion factor, *Exp. Cell Res.* **92:**122–126.

Oppenheimer, S. B., and Humphreys, T., 1971, Isolation of specific macromolecules required for adhesion of mouse tumor cells, *Nature (Lond.)* **232:**125–127.

Peyrieras, N., Hyafil, F., Louvard, D., Ploegh, H. L., and Jacob, F., 1983, Uvomorulin: A nonintegral membrane protein of early mouse embryo, *Proc. Natl. Acad. Sci. USA* **80:**6274–6277.

Shirayoshi, Y., Okada, T. S., and Okada, T. S., 1983, The calcium-dependent cell–cell adhesion system regulates inner cell mass formation and cell surface polarization in early mouse development, *Cell* **35:**631–638.

Spiegel, M., 1954a, The role of specific surface antigens in cell adhesion. Part I. The reaggregation of sponge cells, *Biol. Bull.* **107:**130–148.

Spiegel, M., 1954b, The role of specific surface antigens in cell adhesion. Part II. Studies on embryonic amphibian cells, *Biol. Bull.* **107**:149–155.

Takeichi, M., 1977, Functional correlation between cell adhesive properties and some cell surface proteins, *J. Cell Biol.* **75**:464–474.

Takeichi, M., Ozaki, H. S., Tokunaga, K., and Okada, T. S., 1979, Experimental manipulation of cell surface to affect cellular recognition mechanisms, *Dev. Biol.* **70**:195–205.

Takeichi, M., Atsumi, T., Yoshida, C., Uno, K., and Okada, T. S., 1981, Selective adhesion of embryonal carcinoma cells and differentiated cells by Ca^{2+}-dependent sites, *Dev. Biol.* **87**:340–350.

Takeichi, M., Atsumi, T., Yoshida, C., and Ogou, S., 1982, Molecular approaches to cell–cell recognition mechanisms in mammalian embryos, in: *Teratocarcinoma and Embryonic Cell Interactions* (T. Muramatsu, G. Gachelin, A. A. Mossona, and Y. Ikawa, eds.), pp. 283–293, Japan Scientific Societies Press, Tokyo.

Tarkowski, A. K., and Wroblewska, J., 1967, Development of blastomeres of mouse eggs isolated at the 4- and 8-cell stage, *J. Embryol. Exp. Morphol.* **18**:155–180.

Turner, R. S., 1978, Sponge cell adhesions, in: *Specificity of Embryological Interactions* (D. R. Garrod, ed.), pp. 202–231, Chapman and Hall, London.

Urushihara, H., and Takeichi, M., 1980, Cell–cell adhesion molecule: Identification of a glycoprotein relevant to the Ca^{2+}-independent aggregation of chinese hamster fibroblasts, *Cell* **20**:363–371.

Urushihara, H., Ozaki, H. S., and Takeichi, M., 1979, Immunological detection of cell surface components related with aggregation of Chinese hamster and chick embryonic cells, *Dev. Biol.* **70**:206–216.

Vestweber, D., and Kemler, R., 1984, Rabbit antiserum against a purified surface glycoprotein decompacts mouse preimplantation embryos and react with specific adult tissues, *Exp. Cell Res.* **152**:169–178.

Yoshida, C., and Takeichi, M., 1982, Teratocarcinoma cell adhesion: Identification of a cell surface protein involved in calcium-dependent cell aggregation, *Cell* **28**:217–224.

Yoshida-Noro, C., Suzuki, N., and Takeichi, M., 1984, Molecular nature of the calcium-dependent cell–cell adhesion system in mouse teratocarcinoma and embryonic cells studied with a monoclonal antibody, *Dev. Biol.* **101**:19–27.

Ziomek, C. A., and Johnson, M. H., 1980, Cell surface interaction induces polarization of mouse 8-cell blastomeres at compaction, *Cell* **21**:935–942.

Ziomek, C. A., and Johnson, M. H., 1982, The roles of phenotype and position in guiding the fate of 16-cell mouse blastomeres, *Dev. Biol.* **91**:440–447.

III

Cell Migration

Chapter 10

Migration of Sea Urchin Primary Mesenchyme Cells

MICHAEL SOLURSH

1. Introduction

1.1. Basic Morphogenetic Processes

In considering the cellular basis of morphogenesis during development among various organisms, one is immediately struck by the fact that a small number of basic mechanisms are of widespread usage. Although minor modifications can be observed, it is clear that one obtains a broad understanding of development by understanding the underlying mechanism of a process in any one system.

Embryonic tissues are organized as either epithelia or mesenchyme. Recognizing this organization is essential to any consideration of cellular morphogenesis (Hay, 1968). Briefly, epithelia are sheets of cells. The cells are joined together by characteristic junctions and are polarized so that there is an apical and a basal side. The basal side normally rests on an extracellular matrix, the basal lamina. Mesenchyme consists of a three-dimensional network of cells that are associated by punctate junctions and separated by an extensive extracellular matrix.

Basic morphogenetic processes can be considered in terms of the tissue organization of the cells involved. Epithelia can fold or spread or give rise to mesenchyme cells. Mesenchyme can migrate as single cells, expand as a group of cells, or reform an epithelium.

The sea urchin embryo has been extremely useful as a model system for studying such processes as epithelial folding and mesenchyme cell behavior. Epithelial folding is first involved in invagination during formation of the archenteron, and primary mesenchyme formation is a prototype for mesenchyme morphogenesis. The sea urchin provided a basis for many of the early ideas concerning these two types of morphogenetic activities (see Dan, 1960; Gustafson and Wolpert, 1963).

Studies on echinoid embryos have a long and exciting history. This might

MICHAEL SOLURSH • Department of Biology, University of Iowa, Iowa City, Iowa 52242.

be in part because of the exquisite transparency of the embryos. One can directly observe morphogenesis *in situ* in a manner that is not possible in most embryos. Sea urchins have the additional advantage that large numbers of synchronously-staged embryos can be obtained readily for biochemical and molecular approaches. In spite of such advantages and early successes, other systems have received more attention during the past 20 years. It is hoped that this chapter will serve to stimulate more interest in the sea urchin as a model system for cellular morphogenesis.

This chapter focuses specifically on the formation and subsequent behavior of the primary mesenchyme cells in the sea urchin embryo. As considered in some detail below, these cells are derived from the vegetal plate epithelium by a process (**ingression**) that appears similar to gastrulation in amniote embryos (Solursh and Revel, 1978), including humans. Other migratory cells, such as neural crest (Tosney, 1978) (see Chapter 13), and perhaps even invasive carcinoma cells during early steps in metastasis (Hart and Fidler, 1980), are probably released by similar mechanisms. Once ingressed, the primary mesenchyme cells (so-called because they are the first mesenchyme formed in the embryo) migrate as individual cells a short distance, where they aggregate at specific locations and produce the larval skeleton. The migratory phase serves as a model for cell–matrix interactions, the regulation of migration, and cell patterning *in situ*.

1.2. Normal Development: An Overview

Before considering the primary mesenchyme cells in some detail, it is useful to review briefly normal early echinoid development. There are numerous descriptions in the literature to which the reader can refer for additional details. Some of these are by Dan (1960), Gustafson and Wolpert (1963), Hinegardner (1967), and Kumé and Dan (1968). (See also Chapter 6 of this volume for a discussion of early echinoid development.)

After fertilization and formation of the hyaline layer, the embryos undergo synchronous cleavages. With each cleavage, the central blastocoel enlarges. The first two cleavages are meridional (Fig. 1), and the third is equatorial forming eight equal-size blastomeres. The fourth cleavage is unequal (Fig. 1d). The upper four blastomeres divide equally and meridionally to form a ring of eight mesomeres. The lower four blastomeres divide unequally and at right angles to form four large, upper macromeres and four small, lower micromeres. According to Endo's time-lapse movies (see Okazaki, 1975), the micromeres again divide unequally. The progeny of the upper, larger micromeres later give rise to the primary mesenchyme cells. The lower, small blastomeres divide once more and contribute to the endoderm. By the end of the tenth division, septate desmosomes are fully formed between adjacent blastomeres (Balinsky, 1959) (see also Chapter 6), so that the blastocoel is no longer open to the outside (Dan, 1960). With the blastula stage (Fig. 1e), divisions become less frequent, and the blastula hatches from the fertilization envelope. The blastocoel continues to enlarge, and the vegetal region of the blastula becomes thickened and flattened, forming the vegetal plate. The presumptive primary mesenchyme cells then ingress from the vegetal plate to form a mass of about thirty spherical primary mesenchyme cells within the blastocoel, although the number varies with species. After a lag period, these cells begin to migrate on the inner wall of the blastocoel (Fig. 1f) to form a ring of primary mesenchyme cells. Many then move back toward the vegetal end, converging into two ventrolateral aggregates. The remaining mesenchyme cells align in other chains of cells that extend from the aggregates. Cell processes fuse (Gibbins et al., 1969) to form cables, and the skeleton begins to be produced. The skeleton starts as two triradiate crystals (spicules). The radii grow and branch as the primary mesenchyme cells move apart to form the definitive skeletal elements (Fig. 1h).

After primary mesenchyme cell formation—the timing depends on the species—invagination begins at the vegetal plate so that the archenteron is formed. In species with large blastocoels, the secondary mesenchyme cells, which are located at the tip of the archenteron, send out long cell processes that attach to the animal pole and pull the archenteron in further. The tripartite gut becomes apparent with distinct esophagus, stomach, and intestine. It bends ventrally and fuses with the stomodeal invagination at the animal pole. The ventral side of the larval ectoderm flattens, and ciliated bands develop. As the skeleton develops, pairs of oral and anal arms form. Within, the coelomic rudiments extend laterally from the gut to form two sacs that will produce the coelom and pore canal of the larva.

2. Origins of the Primary Mesenchyme

2.1. The Micromeres

2.1.1. Cytoplasmic Determinants

Since the work of Driesch (1896), it has been well known that isolated sea urchin blastomeres of the two- and four-cell stages form small but morphologically normal larvae. In spite of the regulative capacity of echinoid embryos, the animal–vegetal axis is established during oogenesis (Schroeder, 1980; Chapter 3). The expression of this organization is not visible in most species, however, until the unequal, fourth cleavage, when the micromeres are produced. The basis for the organization that results in this unequal cleavage is unknown but appears to be fixed firmly in the embryo's cytoarchitecture. This is suggested by the inability of centrifugation of Arbacia eggs to disrupt the site of micromere formation (Morgan and Spooner, 1909). That the animal–vegetal polarity may well be fixed in the organization of the cytoskeleton needs to be examined.

The asymmetry of the fourth cleavage can be traced to the movement of the four vegetal nuclei to the vegetal pole in the embryo just before this cleavage (Dan, 1979). This asymmetry of nuclear location then results in the formation of asymmetrical mitotic apparati. This unequal cleavage can be made to occur

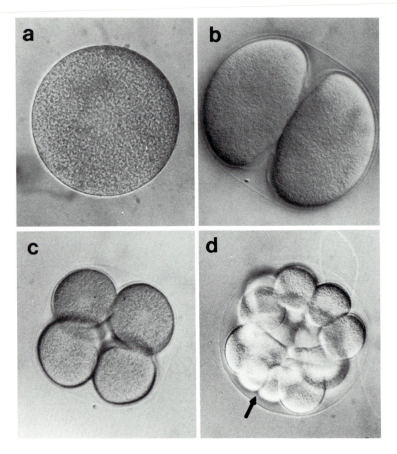

Figure 1. Photomicrographs of a sequence of stages during the normal development of *Lytechinus pictus*. (a) Zygote. (b) Two blastomeres. (c) Four blastomeres. (d) Sixteen blastomeres. Note the presence of mesomeres (eight in one row), macromeres (four), and micromeres at the vegetal pole (arrow). (e) Hatched blastula. (f) Mesenchyme blastula. Note primary mesenchyme cells on the inner wall of the blastula. (g) Midgastrula. Note the secondary mesenchyme cells extending from the tip of the archenteron. The primary mesenchyme cells have largely moved back toward the vegetal pole and are forming the spicules. (h) Prism-stage larva. Note the pair of well-formed spicules, the thickened apical tuft at the animal pole and the gut. Scale bar: 50 μm. (f, g, and h are reprinted from Karp and Solursh, 1974, with permission.)

equally, although apparently normally in other respects, by treating embryos with detergents at the proper time (Tanaka, 1976). It is of particular interest that many of these treated embryos fail to form primary mesenchyme cells and therefore lack spicules. However, other defects, including impaired gastrulation, are also produced. In such embryos the resulting blastomeres are of equal size. The vegetal cytoplasm is still present, but in normal embryos the micromeres contain solely vegetal cytoplasm. Dan (1979) suggests that the segregation of vegetal cytoplasm is required for the formation of primary mesenchyme. Langelan and Whiteley (1985) have shown recently that neither orientation of

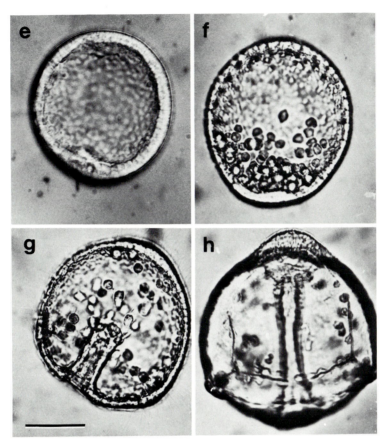

Figure 1. (*continued*)

the fourth cleavage spindle nor the unequal fourth cleavage is required for the determination of spicule forming cells.

2.1.2. Determination

It is clear that normally at the 16-cell stage the micromeres are determined in their capacity to give rise to the primary mesenchyme. This property is best demonstrated by the elegant studies of Okazaki (1975) on the *in vitro* behavior of isolated micromeres. The fertilization envelope is not permitted to form, and at the 16-cell stage the blastomeres are dissociated in Ca^{2+},Mg^{2+}-free seawater. The micromeres are isolated from the other, larger blastomeres by their slower rate of settling. The isolated micromeres from some species when cultured in sea water supplemented with horse serum will form spicules. In some cases, the micromeres undergo cleavages that resemble those occurring *in situ*. The cells then form tight cell aggregates that have indistinct cell borders. At the time when ingression is occurring *in situ*, pulsatile activity is displayed by some of the blastomeres, and the aggregates sometimes move. The cell contours again become clear. Some of the aggregates flatten on the culture substratum, but these never form spicules. Others do not flatten but begin to produce skeletal rods. The serum appears to be required to promote the formation of pseudopodial networks between primary mesenchyme cells (McCarthy and Spiegel, 1983). Although the shape of the spicules that are formed *in vitro* is not quite normal (Harkey and Whiteley, 1980), these observations demonstrate that the micromeres are already determined as spicule-forming cells, even at the 16-cell stage.

2.1.3. Early Changes in Micromere Derivatives

Because of their early determination and because they can be isolated in large quantities (Harkey and Whiteley, 1985), micromeres provide a potentially very useful system in which to study cell determination. Qualitative differences in protein synthesis among the early blastomeres have been difficult to detect. Slightly higher rates of glycoprotein synthesis in the micromeres have been reported, but qualitative differences were not detected (Brown and Bosmann, 1978). Only recently have reports of differences in the labeled protein profiles on two-dimensional gel electrophoretograms appeared (Harkey and Whiteley, 1982). Ingression appears to be the time when the major change—increased or decreased synthesis—in protein synthetic pattern occurs (Harkey and Whiteley, 1983). Approaches utilizing *in situ* molecular hybridization might be especially useful in the future for detecting local differences in the distribution of messenger RNA (e.g., Lynn *et al.*, 1983; Angerer and Davidson, 1984).

It would be especially important to find molecular differences that could provide the basis for the special migratory capacity of the micromere derivatives. Some interesting differences in the surface of micromeres and their derivatives have been detected. Micromeres can sort out from other blastomeres (Spiegel and Spiegel, 1978). Carbohydrate-binding proteins such as concanavalin A (Con A) have been used as probes to compare exposed carbohydrates on blastomeres. This lectin binds α-mannopyranosyl, α-glucopyranosyl, and sterically related residues. The Con A dimer contains two binding sites, permitting it to agglutinate some cells. Krach *et al.* (1974) found that Con A can agglutinate dissociated blastomeres and that this capacity decreases with stage of development. It is perhaps more significant that the micromere derivatives in the 32- to 64-cell stage are more agglutinable with Con A than are other blastomeres (Roberson and Oppenheimer, 1975; Sano, 1980). Because these effects are inhibited by α-mannoside, they are thought to reflect the presence or distribution of specific sugar residues. Indeed, Neri *et al.* (1975) found Con A binding to be quantitatively the same in micromeres, macromeres, and mesomeres but micromeres to have clustered binding sites. Increased mobility of the Con A binding sites and agglutinability are correlated for the micromere derivatives (Roberson *et al.*, 1975). Sano (1977) found that cultured micromere-derived cells acquire an increased negative surface charge at the time when the primary mesenchyme cells are forming *in vivo*. At the same time, the negative surface charge of mesomeres actually decreases slightly. In addition, the mesenchyme cells lack the microvilli found on epithelial cells (Sano and Usui, 1980). Recently, a micromere specific cell surface protein has been identified (DeSimone and Spiegel, 1985). The protein has a molecular weight of 133,000 M_r and binds wheat germ agglutinin but not Con A.

More clearly related to the migratory behavior of the primary mesenchyme cells are developmentally regulated changes in their ability to bind specific extracellular matrix components, as considered in Sections 2 and 3. Forming primary mesenchyme cells show decreased affinity to hyaline (Fink and McClay, 1985) and an increased affinity for fibronectin (Venkatasubramanian and Solursh, 1984; Fink and McClay, 1985). Studies on changes in cell surface receptors for extracellular matrix components would be very useful. Clearly, this system has the potential to provide information concerning the molecular nature of the migratory phenotype.

2.2. Formation of the Primary Mesenchyme Cells

2.2.1. Description of Ingression

As already considered, the primary mesenchyme cells are derived from the vegetal plate epithelium by the process of ingression. This process has been described at the ultrastructural level in most detail by Gibbins *et al.* (1969) and Katow and Solursh (1980). Viewed from the outside of the late hatched blastula, a depression is visible in the vegetal plate (Fig. 2). This transient depression disappears after the primary mesenchyme cells are shed (Kinnander and Gustafson, 1960). In the early mesenchyme blastula stage, the location of the presumptive primary mesenchyme cells is visible from the outside (Fig. 3). There are eight central, ciliated cells that do not ingress, presumably derivates of the small division products of the micromeres. These cells are surrounded by a ring of about 30 unciliated, presumptive primary mesenchyme cells. The area of the apical surfaces of these cells can be observed to narrow as they ingress (Fig. 3a,b), as a consequence of the narrowing of the apical ends of these cells as they elongate toward the blastocoel. From the blastocoel side, the early presumptive primary mesenchyme cells can be seen to protrude from the vegetal plate into the blastocoel (Fig. 4). The basal cell surfaces appear smooth and bulbous in the scanning electron microscope (SEM). Transmission electron microscopy (TEM) shows that the apical side these cells still have well-formed septate desmosomes (Gibbins *et al.*, 1969; Katow and Solursh, 1980) (see Fig. 5) and are intimately associated with the hyaline layer by short cell processes. On the basal side, the basal lamina, which was laid down at the blastula stage (Endo and Uno, 1960; Wolpert and Mercer, 1963), has disappeared in the vegetal plate region (Fig. 5). At this stage, time-lapse movies show that the presumptive primary mesenchyme cells exhibit waves of pulsatile activity (Gustafson and Kinnander, 1956; Kinnander and Gustafson, 1960), forming bulbous protrusions into the blastocoel. As pointed out by Gibbins *et al.* (1969), these basal lobes lack most observable cytoplasmic components but are rich in ribosomes (Fig. 5c).

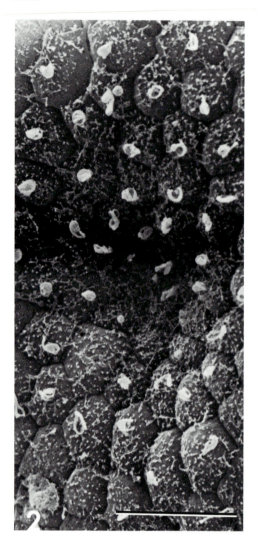

Figure 2. Scanning electron micrograph of apical (outer) surface of the vegetal pole of the late-hatched blastula. The vegetal plate is composed of ciliated cells. The depression is transient, since it is gone after ingression. Scale bar: 10 μm. (From Katow and Solursh, 1980, with permission.)

As ingression proceeds, the presumptive primary mesenchyme cells elongate further (Figs. 5 and 10). The cytoplasm in the narrow neck region contains microtubules oriented along the long axes of the cells (Gibbins *et al.*, 1969; Katow and Solursh, 1980). After additional elongation, changes can be observed at the apical ends. Numerous cell processes form in association with the hyaline (Fig. 6), and the septate desmosomes disappear rapidly. The long neck region of the primary mesenchyme cells then shortens (Fig. 8), as the apical end detaches from the hyaline and adjacent cells. At the same time, apical cell processes of neighboring cells fill in the forming vacant spaces and make contact with their new neighbors (Figs. 7 and 8). Once detached, the primary mesenchyme cells are round in shape and have few cell processes (Fig. 9). Pulsatile activity ceases for awhile (Kinnander and Gustafson, 1960). These

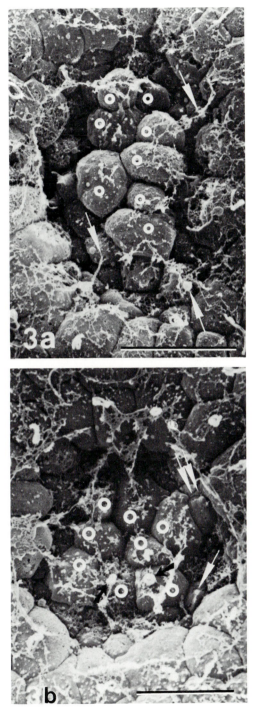

Figure 3. Scanning electron micrograph of the vegetal pole of the mesenchyme blastula. The apex of the vegetal pole is slightly depressed. The depressed area is composed of eight centrally located cells (circles) and about 30 peripheral cells. Several ciliated cells are observed among the peripheral cells (a, arrows). However, the deeper the depression, the fewer are the number of ciliated cells (b, white arrows). An apical cell process (b, double arrow) extends from a neighboring blastomere toward a central cell. Cilia of central cells are indicated by black arrows in b. Scale bars: 10 μm. (From Katow and Solursh, 1980, with permission.)

Figure 4. Scanning electron micrographic view of the inner surface of the depression at the vegetal pole. Some cells have extended into the blastocoel. Note their smooth, bulbous basal surface. Scale bar: 10 μm. (H. Katow and M. Solursh, unpublished data.)

cells remain at the vegetal plate during a lag period before the onset of cell migration. The major descriptive features of ingression are summarized in Figure 10.

2.2.2. Mechanism of Ingression

The process of ingression is clearly a complex, integrated sequence of events leading to the translocation of epithelial cells into the blastocoel. One has to wonder what special properties endow the micromere derivatives with this capacity. Although little is known concerning this important question, a number of separate activities appear to contribute to the ingression process.

One early change is the localized loss of the basal lamina. Although the mechanism of basal lamina removal is not known, it is noteworthy that some other epithelia form basal blebs after enzymatic removal of the basal lamina (Sugrue and Hay, 1981). These appear much like those seen during the normal development of the primary mesenchyme.

This blebbing activity appears to lead to further changes in cell shape. The presumptive primary mesenchyme cells elongate and extend further into the blastocoel. There is a shift of their cytoplasm toward the broadening basal side, as the apical end narrows. A role of the oriented microtubules that are seen in

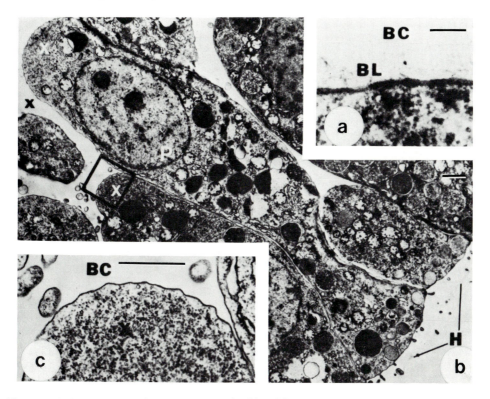

Figure 5. (a) Transmission electron micrograph of basal lamina (BL) in the area adjacent to the vegetal plate in the mesenchyme blastula. BC, blastocoel. Scale bar: 0.1 μm. (b) TEM of elongated presumptive primary mesenchyme cells (P). The basal side of these cells forms a large bulbous structure (white crosses) and is not covered by the basal lamina. Black cross, blastocoel; H, hyaline layer. Scale bar: 1 μm. (c) TEM of basal surface of presumptive primary mesenchyme cell squared in b. The cytoplasm in the bulbous structure contains few visible organelles but is rich in ribosomes. The basal lamina has disappeared from the surface. BC, blastocoel; cross, bulbous structure. Scale bar: 0.5 μm. (From Katow and Solursh, 1980, with permission.)

these elongating cells (Gibbins *et al.*, 1969; Katow and Solursh, 1980) in causing these cell shape changes has been emphasized (Tilney and Gibbins, 1969). However, the roles of other cytoskeletal components in the formation and translocation of these bottle-shaped cells have not yet been determined. Considering the numerous advances in our knowledge of the cytoskeleton, other cytoskeletal components may indeed play important roles in bottle cell formation and translation.

In addition to changes in the shapes of individual presumptive primary mesenchyme cells, cell–cell interactions are thought to play a major role in ingression. One important change in cell–cell interactions involves the localized, rapid loss of apical cell junctions. This is obviously an essential step that finally permits detachment of the forming mesenchyme cells from their epithelial neighbors. Little is known about this fundamental morphogenetic process.

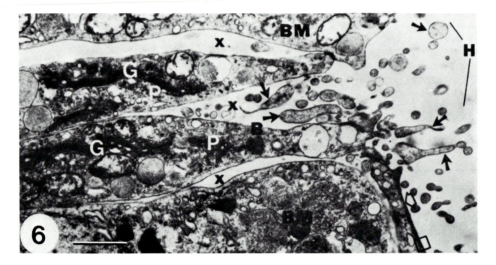

Figure 6. Transmission electron micrograph of longitudinal section through the vegetal plate. Presumptive primary mesenchyme cells (P) are elongated toward the basal side (left hand side of the picture) and have formed convoluted apical cell processes (arrows). These cell processes, however, do not penetrate the hyaline layer (H). Among the presumptive primary mesenchyme cells and adjacent blastomeres (BM) are large intercellular spaces (crosses). The basal body (B) of the presumptive primary mesenchyme cell is located on the apical side of the Golgi cisternae (G). The open arrow indicates a cell process, which extends from a neighboring blastomere to a presumptive primary mesenchyme cell. Scale bar: 1 μm. (From Katow and Solursh, 1980, with permission.)

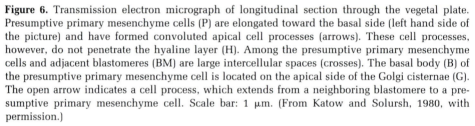

Figure 7. Transmission electron micrograph of longitudinal section through the apical cell process formed by the blastomere squared in Fig. 6. Microfilaments (arrows) are oriented along the long axis of the cell process. Scale bar: 0.1 μm. (From Katow and Solursh, 1980, with permission.)

Figure 9. Transmission electron micrograph of the vegetal plate. The shed primary mesenchyme cells (P) are now round in shape within the blastocoel (BC). A large intercellular space is observed where a presumptive primary mesenchyme cell is apparently detaching (X) as suggested by the numerous cell processes. H, hyaline layer. The arrow indicates the limit of the basal lamina at the vegetal plate. Scale bar: 2 μm. (H. Katow and M. Solursh, unpublished data.)

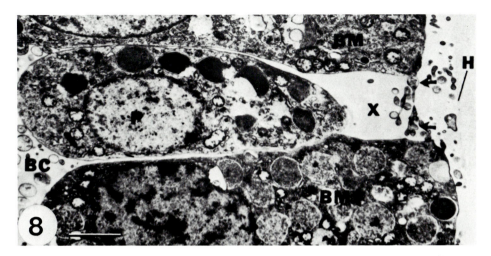

Figure 8. Transmission electron micrograph of longitudinal section through the vegetal plate. The primary mesenchyme cells (P) have separated from neighboring blastomeres (BM) and have left large intercellular spaces in the vegetal plate (cross). Apical cell processes (arrows) extend from neighboring blastomeres to occlude the spaces. H, hyaline layer. Scale bar: 1 μm. (From Katow and Solursh, 1980, with permission.)

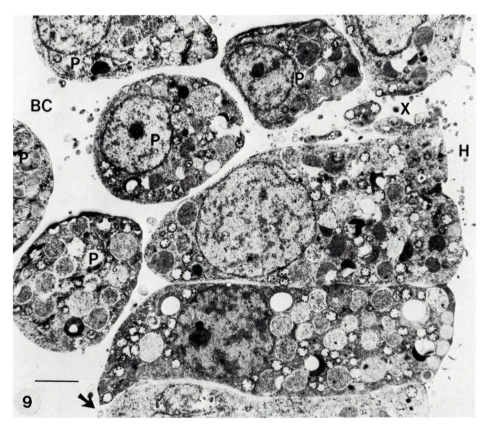

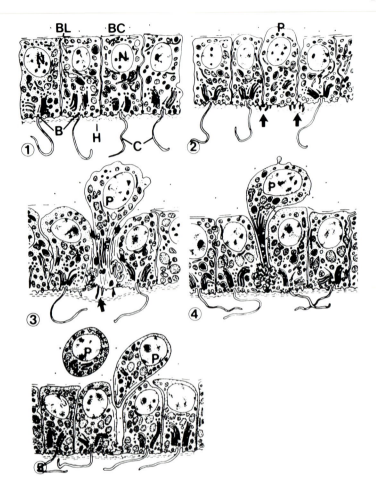

Figure 10. Diagram showing primary mesenchyme shedding. Crosses, blastocoel material. (1) The initial morphology of presumptive primary mesenchyme cells and the neighboring blastomeres is identical. They have cilia (C) and are covered by the hyaline layer (H) on the apical surface. The basal surface is covered with a thin, patchy basal lamina (BL). The basal body (B) is surrounded by Golgi cisternae (G). Nuclei (N) are located on the basal side. (2) Elongation of presumptive primary mesenchyme cell (P) into the blastocoel. Presumptive primary mesenchyme cells and the neighboring blastomeres form cell processes on the apical side (arrows). The basal lamina has disappeared from this region. (3) Apical detachment of the primary mesenchyme cell. The presumptive primary mesenchyme cell elongates further (P), and the apical region of the cell forms convoluted cell processes (large arrow). Desmosomes between presumptive primary mesenchyme cells and neighboring blastomeres have disappeared. Microtubules (small arrows) are found in the peripheral cytoplasm of the neck region in the presumptive primary mesenchyme cells and in the cytoplasm of neighboring blastomeres preferentially on the sides adjacent to presumptive primary mesenchyme cells. (4) Separation of the primary mesenchyme cell (P). Primary mesenchyme cells separate from the vegetal plate after the disappearance of desmosomes. The neighboring blastomeres extend apical cell processes toward each other, resulting in the occlusion of the intercellular spaces. Also, blastocoel material is observed in the intercellular spaces. (5) Rounding of the primary mesenchyme cells. The neck region of the primary mesenchyme cells shortens toward the basal side, and the mesenchyme cells become round in shape (P). (From Katow and Solursh, 1980, with permission.)

There are indications that another type of cell–cell interaction is involved in ingression, in the sense that ingression represents the outcome of the concerted efforts of both epithelial cells and the presumptive primary mesenchyme cells. Gustafson and Kinnander (1956) suggested that adjacent nonmesenchymal cells might squeeze the presumptive primary mesenchyme cells out of the vegetal plate. This hypothesis is supported by the presence of oriented microtubules in epithelial cells adjacent to the presumptive primary mesenchyme cells (Katow and Solursh, 1980). Such microtubules have only been seen on the sides of the cells that face the presumptive primary mesenchyme cells. In addition, the forming mesenchyme cells appear to be compressed by their neighboring cells (Fig. 11), since depressions can be observed on the surface of the presumptive primary mesenchyme cells. How these lateral forces

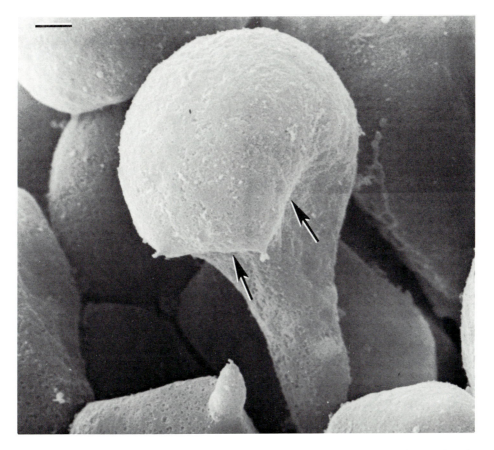

Figure 11. Scanning electron micrograph of bottle-shaped cell viewed from the blastocoel at the vegetal plate. The surfaces of the basal and neck regions of this cell are smooth and have no cell processes. Note the spherical shape of the basal end of this cell and the depressions (arrows) in the neck region, which are presumably impressions of the adjacent blastomeres, formerly pressed closely against the cell. The vegetal plate was broken open to observe the bottle-shaped cell, so the intercellular spaces (dark background) are artifcially enlarged. Scale bar: 1 μm. (From Katow and Solursh, 1980, with permission.)

act selectively on the presumptive primary mesenchyme cells is an important, but unanswered question.

Another type of activity that appears to be involved with the last steps of ingression concerns cell–hyaline layer interactions. Interactions between the apical cell surfaces and the hyaline have been considered to play a central role in morphogenetic process in echinoid embryos (Dan, 1960; Gustafson and Wolpert, 1963). In particular, reduced cell–hyaline binding at the vegetal plate has been considered to be involved in both ingression and invagination. Consistent with such a role, Dan (1960) noted that one can observe a blister of the hyaline associated with the forming primary mesenchyme cells. McClay and Fink (1982) obtained direct evidence for a reduced affinity of the primary mesenchyme cells for hyaline. In contrast epithelial cells do not change affinity for hyaline.

With the combination of several activities, one can begin to understand the multiple forces that contribute to the ingression process. These activities include basal blebbing and elongation of the presumptive primary mesenchyme cells, crowding by adjacent cells, dissolution of apical cell junctions, and detachment from the hyaline, followed by cell shortening. The net result is the formation of single, round cells lying within the blastocoel.

3. Migration of the Primary Mesenchyme

3.1. The Normal Process

The primary mesenchyme cells in the sea urchin embryo represent one of the few instances in which it is possible to observe cell migration *in situ* during normal development. In species with large blastocoels, the primary mesenchyme cells undergo a remarkable migration. Many cells migrate more than halfway up the wall of the blastocoel and then converge into two ventrolateral clumps (Fig. 12). The events leading up to this stage will now be considered in some detail.

Gustafson and Kinnander (1956) made time-lapse movies of this process and were able to analyze the dynamics of migration. More recent work in the author's laboratory is used here for illustration. In *Lytechinus pictus*, a 2–4-hr lag occurs after ingression but before the onset of migration (Katow and Solursh, 1980). During the lag period, the cells are round but do move over each other. It is not known what is delaying the onset of migratory behavior, but one possibility is that the mesenchyme cells must first synthesize some surface component that is required for attachment to the migratory substratum. It is noteworthy that migrating primary mesenchyme cells stain intensely by indirect immunofluorescence with antibody directed against human plasma fibronectin (Katow et al., 1982) (see Fig. 13). The staining is particularly intense on what appear to be short migratory cell processes. The cross-reacting antigen has not yet been characterized, but vertebrate fibronectin is well known for its role in cell attachment (Hynes and Yamada, 1982). It will be important to identify

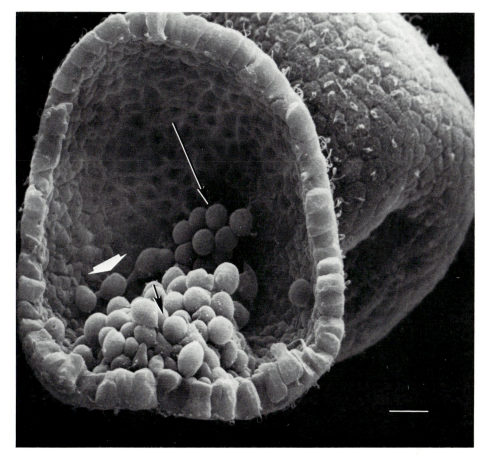

Figure 12. Scanning electron micrograph of *Lytechinus pictus* early gastrulae. The embryo in the forefront has been fractured so that the area of invagination (short black arrow) is seen. Behind, one can observe the embryo's left ventrolateral clump of primary mesenchyme cells (long arrow) connected to the ring of primary mesenchyme cells (thick, white arrow). The invagination is seen from the outside in the embryo in the background. Scale bar: 10 μm. (H. Katow and M. Solursh, unpublished data.)

surface components of the primary mesenchyme cells that are involved with the onset of their migratory behavior.

As migration begins, cells at the edge of the vegetal plate move onto the inner wall of the blastocoel (see Fig. 1f). As documented by Gustafson's films (Gustafson and Kinnander, 1956; Gustafson and Wolpert, 1963), pseudopodia mediate migration. The pseudopodia are about 0.5 μm in diameter and can be 30–40 μm long. They can form rapidly, extending a considerable distance from the cell in minutes. They appear to be stiff or bristlelike and can be observed to explore the substratum. The thicker processes contain microtubules while the thinner ones contain microfilaments that are oriented along their long axis. If a pseudopod fails to attach to the substratum, it is resorbed. However, if it forms

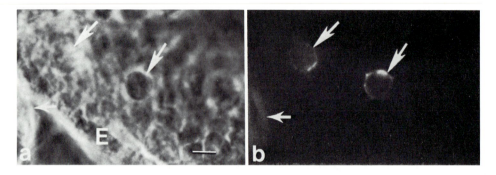

Figure 13. Ingressed primary mesenchyme cells in cut mid-mesenchyme blastulae are stained with FITC-IgG after incubation with polyclonal antibody directed against human plasma fibronectin (a,b, large arrows). Most of the ingressed primary mesenchyme cells are stained, but those out of the plane of focus do not photograph well. The apical surface of the ectoderm is occasionally stained with FTIC-IgG nonspecifically (a,b, small arrow), (a) Phase-contrast micrograph; (b) fluorescence micrograph of the same embryo. Scale bar: 10 μm. (From Katow *et al.*, 1982, with permission.)

a stable attachment, the cell moves forward as the cell body detaches. In this manner, the mesenchyme cells can move up the wall of the blastocoel.

Through SEM used in conjunction with time-lapse cinematography, the types of processes on the primary mesenchyme cells have been described in *L. pictus* (Katow and Solursh, 1981). In this species, the cell processes utilized for migration include long pseudopodia (Fig. 14a), as well as short cell processes (Fig. 14b). The use of the short process for migration is illustrated in the sequence shown in Figure 15. In either case, it is clear that the attachment of the migratory cell process to the substratum can be central in determining the rate and direction of movement.

Isolated primary mesenchyme cells will also migrate *in vitro*. Migration is dependent upon the presence of fibronectin on the substratum (Venkatasubramanian and Solursh, 1984). Under these conditions, the primary mesenchyme cells form thin filopodia resembling those seen *in vivo* (Karp and Solursh, 1985). If the filopodia contact blastocoel extracellular matrix that had been deposited on the substratum, a response is initiated which causes the cell to move in a directed manner toward the extracellular matrix. Thus, cell behavior can be altered by contact with specific matrix components in the substratum.

The substratum *in vivo* has been shown to be the basal lamina that lines the wall of the blastocoel (Gibbins *et al.*, 1969; Katow and Solursh, 1981). As shown in Figure 16, the cell processes are intimately associated with components of the blastocoel but appear to be anchored to the basal lamina. Gustafson's earlier light microscopic observations on compressed embryos (1963) indicated that the cell processes are preferentially located between ectodermal cells. However, the higher resolution obtained with SEM indicates no such preferential localization. Thus, the molecular nature of the interaction between the pseudopodial surface and components of the blastocoel, particularly the basal lamina, must be fundamental to the whole migratory process. In this

Figure 14. Scanning electron micrograph of two different types of kinetic cell processes extended from primary mesenchyme cells. (a) These cell processes are unbranched long cell processes (arrows). Occasional terminal branches are present. (b) This type of process is a short and unbranched cell process, seen simultaneously at several sites on migrating cells (small arrows) and extend toward the presumed direction of migration (long arrows). The cell on the lower right in this figure has one retraction process (arrow on lower right). Scale bars: 10 μm. (From Katow and Solursh, 1981, with permission.)

regard, Fink and McClay (1985), using a cell adhesion assay, find that as the primary mesenchyme cells have reduced affinity for the hyaline and other cells, they acquire an increased affinity for basal lamina and fibronectin. The components of the blastocoel with which the primary mesenchyme cells might interact are considered in the next section.

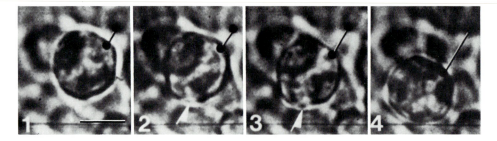

Figure 15. Time-1apse cinematography of short cell processes. Two black dots and line in (1) at 0 sec show original location of the migrating cell. (2) After 22.5 sec, the cell extends a short cell process (arrow). (3) After 37.5 sec, the cell process tip (arrow) became sharper during the 15-sec interval, and the cell body has moved a little (line). (4) At 90 sec, the cell has migrated about 6 μm during the last 52.5 sec interval, and the short cell process has disappeared. Scale bar: 10 μm. (From Katow and Solursh, 1981, with permission.)

3.2. The Extracellular Matrix of the Blastocoel

3.2.1. Morphology

The contents of the blastocoel have been of interest for some time, but only recently has its ultrastructure been studied. Using SEM, Endo and Noda (1977) observed the presence of a fine fibrous matrix in the blastocoel of *Mespilia* at morula, blastula, and gastrula stages. In *L. pictus* blastulae (Katow and Solursh, 1979), the matrix consists of fine fibers (Fig. 17) of variable diameter. The material fills the blastocoel and is attached to the basal surface of the blastula wall. Transmission electron microscopic examination indicates that the fibrous structure consists of long chains of approximately 30-nm-diameter granules (inset). In the mesenchyme blastula the primary mesenchyme cells are embedded in the blastocoel matrix (Fig. 18). By this time in *L. pictus* and earlier in S. *purpuratus*, a fine fibrillar component (7–8 nm) has been added to the matrix. This component can be observed by TEM (Fig. 19). As indicated in Figure 16c, these structures are continuous with the basal lamina. At least on morphological grounds, the blastocoel resembles a bag of basal lamina material.

3.2.2. Composition

Histochemical methods have demonstrated the presence of sulfated glycans in the blastocoel as well as in the hyaline layer (Monné and Slautterback, 1950; Monné and Härde, 1951; Immers, 1956, 1961). Such stained, as well as [^{35}S]labeled, material is particularly intense in association with the primary mesenchyme cells (Immers, 1961; Sugiyama, 1972). Indeed, at the light microscopic level, the primary mesenchyme cells appear to be embedded in sulfated glycans. Sugiyama (1972) found that the vegetal plate stained selectively for acidic glycans and selectively labeled autoradiographically with $^{35}SO_4^{2-}$ even before ingression. Karp and Solursh (1974) noted that sulfate incorporation into glycans was maximal at the time of primary mesenchyme formation. In addi-

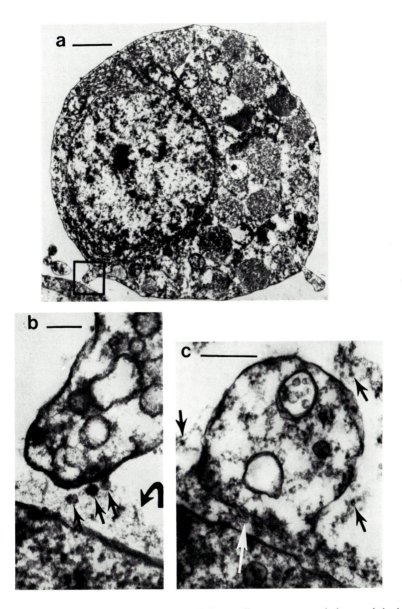

Figure 16. (a) Transmission electron micrograph of short cell process, extended toward the basal lamina. (b) High magnification of square in (a). At the tip of the cell process, the basal lamina extends up toward the cell process (curved arrow), and the 30-nm-diameter granules (arrows) in the basal lamina remain associated with the tip of the cell process. This specimen was fixed in Millonig's phosphate buffer. (c) Short cell process, which is attached to the basal lamina (white arrow). The process is surrounded by 30-nm-diameter granules present in both the basal lamina and in the blastocoel (arrows). This specimen was fixed in s-collidine buffer for better preservation of the 30-nm-diameter granules. Scale bars: 1 μm (a), 0.1 μm (b), and 2 μm (c). (From Katow and Solursh, 1981, with permission.)

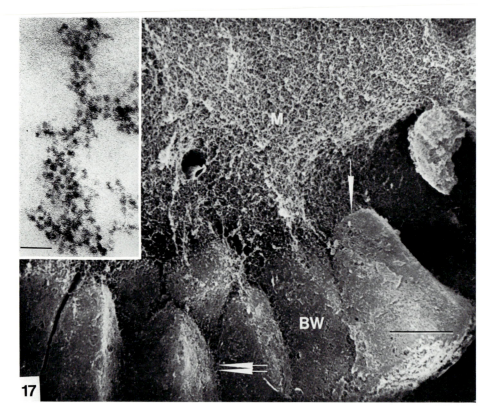

Figure 17. In the normal blastula of *Lytechinus pictus*, the blastocoel material (M) fills the entire blastocoel and the intercellular spaces (double arrow) on the basal side of the blastular wall (BW). The material has a complex, intertwined, fine fiberlike structure bound to the basal surface of the blastular wall (single arrow). Scale bar: 5 µm. By TEM (inset), however, the blastocoel material appears to be composed of aggregates of approximately 30-nm-diameter granules. The blastocoel material appears similar in *Strongylocentrotus purpuratus*, except that thin fibers are also present at this early stage. Scale bar: 0.1 µm. (H. Katow and M. Solursh, unpublished data.)

tion, as is considered in the next section, sulfate deprivation prevents primary mesenchyme cell migration. Thus, the sulfated macromolecules are of considerable interest in relationship to primary mesenchyme cell migration.

Biochemical studies of sulfated components from whole embryos have indicated considerable heterogeneity in the classes of sulfated macromolecules (Karp and Solursh, 1974). These include heparinlike (Kinoshita, 1971) and dermatan sulfate proteoglycans (Yamagata and Okazaki, 1974; Oguri and Yamagata, 1978), a sulfated fucogalactan (Akasaka and Terayama, 1980), as well as sulfated glycoproteins (Heifetz and Lennarz, 1979).

The intact blastocoel can actually be isolated in large quantities for biochemical studies. The procedures were first worked out by Okazaki and Niijima (1964) and later refined by Harkey and Whiteley (1980). The epithelial cells are dissociated and separated from the blastocoel "bag" by differential

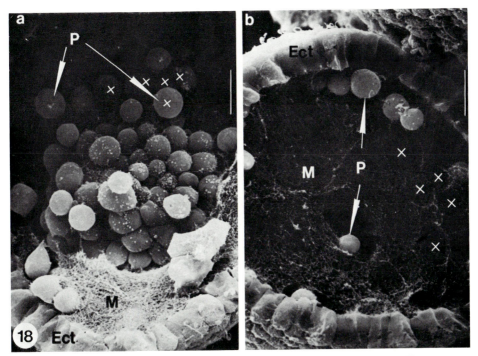

VEGETAL HALF   **ANIMAL HALF**

Figure 18. In the mesenchyme blastula the blastocoel material (M) appears more loosely woven, and the material is integrated into a more organized fiberlike structure than that at the blastula stage. The blastocoel material tends to bind to the basal surface of the animal half ectoderm (Ect in b) stronger than to that of the vegetal half ectoderm (Ect in a) and the surface of the primary mesenchyme cells (P). Depressions in the blastocoel material in the animal half (crosses in b) form molds of the primary mesenchyme cells in the vegetal half (crosses in a). By TEM, the blastocoel material contains thin fibers as a minor component in addition to the major granular component mentioned above (not shown). Scale bar: 10 μm. (H. Katow and M. Solursh, unpublished data.)

centrifugation. As shown in Figure 20, the bag includes the basal lamina and blastocoelic matrix. Trapped primary mesenchyme cells are often present as well but can be removed after breaking the bag. Gel electrophoresis of the solubilized bags (Fig. 21) demonstrates the presence of at least 38 polypeptides after silver staining. About 14 bands become labeled with $^{35}SO_4{}^{2-}$, and about 28 bind Con A (B. A. Norbeck and M. Solursh, unpublished). In spite of the large number of components, comparisons between different species and developmental stages indicate impressive similarities, suggesting that many of these polypeptides might have important structural functions in the blastocoel.

The sulfated components of the isolated blastocoels are largely in the form of very high molecular weight proteoglycans. Most of the $^{35}SO_4{}^{2-}$ in radiolabeled extracts does not even enter 7% polyacrylamide gels (Solursh and Katow, 1982). If the extracts are dissolved in a dissociative solvent like 4 M

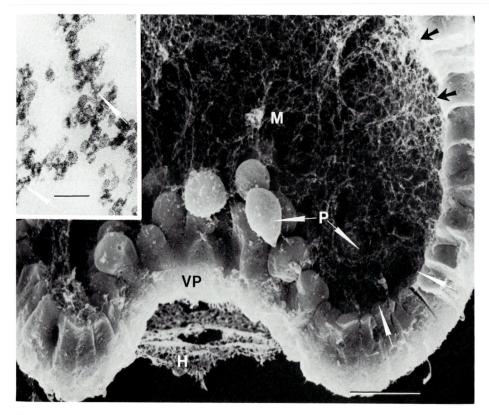

Figure 19. In the early gastrula, the blastocoel material (M) has a thicker, rough-surfaced, fibrous appearance. The blastocoel material forms a looser network in the vegetal half (white arrows) than in the animal half (black arrows). The fibrous structure is composed of the thin fibers that are first observed in mesenchyme blastulae (inset). These fibers have become more extensive than in the mesenchyme blastula and are associated with 30 nm diameter granules (inset, arrows). H, hyaline layer; P, primary mesenchyme cell; VP, vegetal pole. Scale bars: 10 μm and 0.1 μm (inset). (H. Katow and M. Solursh, unpublished data.)

guanidine HCl and are chromatographed on a Sepharose CL2B column, most of the label is in the excluded fraction, suggesting that the material is of a very large size ($>20 \times 10^6$ daltons) (see Fig. 22). The size is drastically reduced by pretreatment with chondroitinase ABC, which degrades dermatan sulfate and chondroitin sulfates -4 and -6. Additional analyses indicate the presence of largely dermatan sulfate, chondroitin-6-SO_4, heparan sulfate, and other still uncharacterized components (Table I). How these are associated in macromolecular structures and organized within the blastocoel remains to be studied. It will be important to relate biochemical information with morphologically recognizable components.

Considering their widespread occurrence and their importance in vertebrate development (Hay, 1981), it is surprising that collagens have not yet been identified in echinoid embryos, although collagen has been studied in the adults (Matsumura *et al.*, 1979). Hydroxyproline (a major amino acid in col-

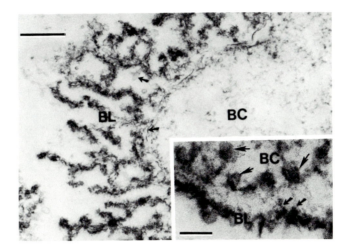

Figure 20. Transmission electron micrograph of isolated bag. The basal lamina (BL) consists of an undulating complex and surrounds the blastocoelic matrix (BC). The membranelike components associated with the basal lamina (arrows) are probably cell-derived contaminants. The blastocoel contains numerous 25–30-nm-diameter granules (large arrows, inset) and fibers (small arrows) connecting to the basal lamina (BL). Scale bars: 0.5 μm and 0.1 μm (inset). (From Solursh and Katow, 1982, with permission.)

lagens) can be detected at gastrulation, when it increases progressively in association with spicule formation (Golob *et al.*, 1974). Proline hydroxylase is present as early as the blastula stage (Benson and Sessions, 1980), and several inhibitors of proline hydroxylation inhibit gastrulation (Mizoguchi and Yasumasu, 1983). However, some noncollagenous proteins also contain hydroxyproline. Striated and nonstriated fibers have been described in prism-stage larvae and are purported to be collagen (Crise-Benson and Benson, 1979; Pucci-Minafra *et al.*, 1972). Extracts studied so far have a rather low hydroxyproline content for collagen (Pucci-Minafra *et al.*, 1972) and appear to be resistant to collagenase (Mintz *et al.*, 1981). On the basis of morphological criteria, which are not diagnostic for collagen, Spiegel and Spiegel (1979) claimed that collagen is present in the hyaline layer (see Chapter 6). By contrast, using biochemical approaches, Hall and Vacquier (1982) found no evidence for the presence of collagen in the hyaline. By indirect immunofluorescence with unpurified antisera, antigens that cross-react with vertebrate types I, III, and IV collagens have been detected on the inside and outside of the blastocoel wall, as well as in granules in the uncleaved egg (Wessel *et al.*, 1984). The nature of the cross-reacting antigens is unknown. It is interesting that vertebrate collagen promotes spicule formation *in vitro* (Blankenship and Benson, 1984). Although it is certainly likely that collagenous proteins will eventually be found in echinoid embryos, we shall have to wait until appropriate approaches have been used before recognizing its developmental significance.

Antibodies raised against some noncollagenous extracellular matrix components from vertebrates have been used to detect some additional cross-react-

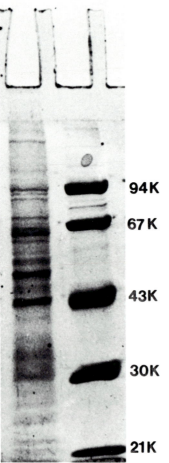

Figure 21. Coomassie blue-stained SDS–polyacrylamide gel electrophoretogram of reduced blastocoel extracts prepared from *L. pictus* gastrulae (left lane). The positions of molecular weight markers in the right lane are indicated on the right by the molecular weight values (in kilodaltons). Note that a number of major polypeptides are present in the extract. Based on autofluorography, some of these are sulfated. In addition, many bind concanavalin A. (B. A. Norbeck and M. Solursh, unpublished data.)

ing antigens in sea urchin embryos. An antigen that cross-reacts with antibody raised against vertebrate laminin, a large glycoprotein in vertebrate basal laminae, was detected in the extracellular matrix and on the inner and outer surfaces of the blastula wall (Spiegel *et al.*, 1983). No biochemical information is yet available. In addition, antigen(s) that cross-react with antibodies raised against human plasma fibronectin have been detected between blastomeres and on the outer (Spiegel *et al.*, 1980) and inner surfaces (Spiegel *et al.*, 1983; Wessel *et al.*, 1984) of the embryo. Using a different antiserum, a different species, and a different method of sample preparation, however, Katow *et al.* (1982) detected only nonspecific staining at these locations with specific staining on the migrating primary mesenchyme. The reason for these different results is not known, but they could reflect a higher antigen concentration on the mesenchyme cells, distinct antigens, or, most likely, the masking of antigen. Masking could result from antigen compartmentalization or its association with other components of the extracellular matrix. There are some suggestions that fibronectinlike antigens are stored within the embryo (Wessel *et al.*, 1984;

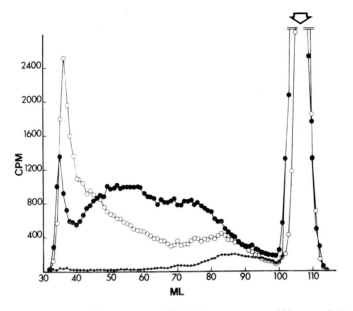

Figure 22. Characterization of the major [³⁵S]labeled components of blastocoel. Distribution of [³⁵S]labeled materials from blastocoels of mid-mesenchyme blastulae of *S. purpuratus*. The ³⁵S in the untreated extract (o———o) is largely voided from a Sepharose CL-2B column, except for material that co-elutes with free $^{35}SO_4^{2-}$ (arrow). Pretreatment of the extract with chondroitinase ABC (•———•) results in the breakdown of the labeled materials, indicating that the integrity of the ³⁵S-labeled matrix is sensitive to chondroitinase ABC. Extract from embryos grown in the presence of β-D-xyloside contains a smaller proportion of [³⁵S]labeled material in the voided fractions than extract from controls (•———•). (From Solursh and Katow, 1982, with permission.)

Iwata and Nakano, 1984). Altering the availability of fibronectin to migrating cells could play a role in regulating cell movement (Katow *et al.*, 1982).

These different results emphasize the importance of biochemical identification of the antigens. A fibronectinlike protein has been extracted from sea urchin ovaries (Iwata and Nakano, 1981) and embryos (DeSimone *et al.*, 1985). It will be interesting from a comparative standpoint to examine echinoid extracellular matrix components biochemically and functionally. Of particular interest will be the molecular interactions that occur between the surface of mesenchyme cell pseudopodia and the basal lamina, to which these cell processes attach.

Table I. Summary of Biochemical Analysis
of [³⁵S]Labeled Materials

	S. purpuratus	L. pictus
Sulfated glycoprotein	20	24
Heparan sulfate	13	13
Chondroitin-6-SO₄	16	4
Dermatan SO₄	37	25
Uncharacterized	14	34
Total	100%	100%

[a]From Solursh and Katow (1982).

3.3. Environmental Disruption of Migration

Several related treatments interfere with normal primary mesenchyme cell migration. With all these treatments, ingression occurs, but the primary mesenchyme cells either remain at the vegetal plate or migrate abnormally. Although none of these treatments has much specificity on the biochemical processes they influence, their effects are still of interest, because they point the way to future, more specific approaches.

The most widely studied such treatment is sulfate deprivation, first reported by Herbst (1904). The physiological effects were considered by Lindahl (1942) and then by Immers (1956) and Immers and Runnström (1965) in *Paracentrotus lividus*. These latter investigators noted that the vegetal half of the embryo is more affected than the animal half in this species, since sulfate deprivation produced animalized embryos with inhibited primary mesenchyme formation and inhibited gastrulation but with an overdeveloped apical tuft. Their studies emphasized the animal–vegetal gradient and demonstrated the multiple effects of sulfate deprivation. Runnström *et al.* (1964) found that sulfate deprivation even reduced [³H]thymidine and [³H]uridine incorporation, especially after gastrulation (Gezelius, 1974). Sulfate deficiency also results in the nuclear retention of newly synthesized RNA after the late mesenchyme blastula stage (Gezelius, 1976). This multiplicity of effects clearly indicates the use of caution in considering the role of sulfate in cell migration.

One conspicuous effect of sulfate deprivation is the reduction in sulfated glycans in the blastocoel (Immers, 1956; Sugiyama, 1972; Karp and Solursh, 1974). Immers and Runnström (1965) hypothesized that the mobility of the primary mesenchyme cells requires the presence of sulfated glycans. In the case of *L. pictus*, there is no indication of animalization, but the first known effect of sulfate deprivation is to prevent primary mesenchyme cell migration, more specifically than in some species (Karp and Solursh, 1974) (see Fig. 23). Ultrastructural examination of sulfate-deprived embryos demonstrates the selective reduction of the 30-nm-diameter granules within the blastocoel and basal lamina (Katow and Solursh, 1979) (see Fig. 24). As indicated in Figure 16 (Katow and Solursh, 1981), kinetic cell processes are intimately associated with such granules in normal embryos. It is therefore of some interest that time-lapse films of sulfate-deprived embryos (Katow and Solursh, 1981) demonstrate that migratory cell processes form normally but fail to attach to the migratory substratum. These observations are consistent with the hypothesis that sulfate-deprived embryos lack a specific component of the basal lamina that is required for cell attachment and migration. Consistent with such a defect, Akasaka *et al.* (1980) noted that even the secondary mesenchyme cells lack normal, stably attached cell processes in sulfate-deprived embryos.

The nature of the affected component is difficult to ascertain by this sort of approach, as even the specific alteration of one component of a highly structured extracellular matrix can secondarily affect other components of the matrix. In addition, a single treatment can affect more than one matrix component. Even collagen could be involved in the effects of sulfate deprivation, as sug-

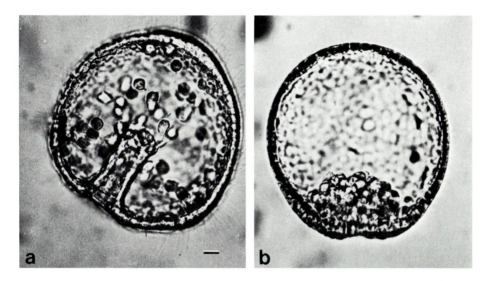

Figure 23. Photomicrographs of living *L. pictus* embryos 38 hr after fertilization in normal seawater (a) or sulfate-free seawater (b). The normal embryo is a mid-gastrula, and the primary mesenchyme cells are localized as ventrolateral clumps. The embryo in sulfate-free seawater is still blocked as an early mesenchyme blastula. Scale bar: 10 μm. (From Karp and Solursh, 1974, with permission.)

gested by Mizoguchi and Yasumasu (1982). These workers found that exposure of sulfate-deprived *Hemicentrotus pulcherrius* embryos to ascorbate increased hydroxyproline and reversed the inhibition of gastrulation. However, they did not examine the effect of ascorbate on sulfated proteoglycan synthesis, which is stimulated in some systems (e.g., Hajek and Solursh, 1977). Also, ascorbate does not improve primary mesenchyme cell migration in sulfate-deprived *L. pictus* (M. Solursh, unpublished).

To make matters more difficult for further analysis of the mechanism of action of sulfate deprivation is the probability that several unrelated sulfated components of the extracellular matrix are affected. Akasaka and Terayama (1983) have found that a sulfated fucan, which is located largely outside of the blastocoel, and which stimulates cellular reaggregation (Akasaka and Terayama, 1984), is reduced by sulfate deprivation. Also, sulfate-deprivation inhibits the synthesis of dermatan sulfate, which is found in the blastocoel (Yamaguchi and Kinochita, 1985).

In addition to effects of sulfate deprivation on the extracellular matrix, there are indications that the primary mesenchyme cells themselves are abnormal. Karp and Solursh (1974) showed by SEM that while normal primary mesenchyme cells have a rough surface, such cells in sulfate-deprived embryos are smooth. Furthermore, normal, isolated primary mesenchyme cells spread and migrate extensively *in vitro* on a substratum of human plasma fibronectin (Venkatasubramanian and Solursh, 1984). On the other hand, cells from sulfate-deprived embryos fail to spread or migrate on the same substratum. The

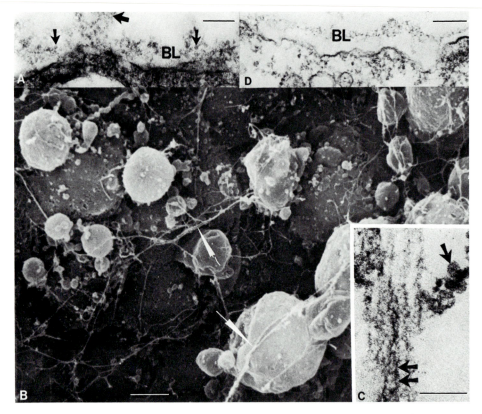

Figure 24. In the mid-mesenchyme blastula of *L. pictus* the blastocoelic material (thick arrow in A) appears to be continuous with the basal lamina (BL in A) in normal embryos. Similar granules are also found among the fibrous material (small arrows in A), forming the basal lamina. On the other hand, the blastocoelic material of *L. pictus* embryos that had been cultured in sulfate-deficient seawater until the control embryos cultured in complete artificial seawater had reached the early gastrula stage consists of thinner and smoother surfaced fibers, as seen by SEM (arrows in B). These fibers (double arrow in C) are associated with a conspicuously decreased number of the granules (single arrow in C) and combine with each other to form thicker fibers (B, C). The basal lamina (BL in D) also lacks the granules, but the fibrous component appears morphologically intact. Scale bars: 0.2 μm (A), 2 μm (B), 0.1 μm (C), and 0.2 μm (D).

defect can be reversed *in vitro* within 6 hr after the addition of normal sea water to the isolated cells. These results support the hypothesis that the primary mesenchyme cells produce a sulfate-dependent cell surface component that is required for cell spreading and migration. The production of monoclonal antibodies directed against antigens that first appear on ingressed primary mesenchyme cells and that subsequently line the blastocoel wall (Wessel *et al.*, 1984) indicates that the primary mesenchyme cells themselves contribute to the extracellular matrix. Such antibodies might provide a means of identifying components required for cell migration.

A related treatment that has been studied is the exposure of embryos to sodium selenate, which presumably acts by inhibiting sulfate activation, as in yeast (Wilson and Bandurski, 1958). Selenate does reduce sulfate incorporation into total glycosaminoglycans in sea urchin embryos (Kinoshita and Saiga, 1979). Sugiyama (1972) found, however, that in the presence of selenate, the primary mesenchyme cells ingressed and even migrated. Their distribution, however, was abnormal, and they were often in clumps.

Another treatment that might have a more specific action on proteoglycans than sulfate deprivation is exposure of embryos to β-xylosides. Earlier, Kinoshita and Saiga (1979) found that β-, but not α-xylosides block gastrulation and inhibit proteoglycan synthesis. Akasaka et al. (1980) found β-xyloside treatment to permit ingression but to block migration of primary mesenchyme cells. Confirming this effect in S. purpuratus, Solursh et al., 1986) found xyloside-treated embryos to resemble sulfate-deprived L. pictus (see Fig. 23). TEM observations demonstrate a drastic reduction in 30-nm granules in the blastocoel as well.

Proteoglycans consist of a core protein to which glycosaminoglycans and other carbohydrate chains are covalently linked. In other systems, β-xyloside derivatives substitute in vivo for the proteoglycan core proteins as initiators for the synthesis of glycosaminoglycans that are linked by xylose to the core protein. This is true for chondroitin sulfate synthesis (Schwartz et al., 1974; Fukunaga et al., 1975; Gibson and Segen, 1977) and heparan sulfate synthesis (Hart and Lennarz, 1978). Dermatan sulfate, the major glycosaminoglycan in the blastocoel (Solursh and Katow, 1982), is also normally xyloside linked (Akiyama and Seno, 1981). As shown in Figure 22, xyloside treatment reduces the size of sulfated components in the blastocoel. Much of the labeled material now present in the retained fractions consists of free chondroitin-6-sulfate and dermatan sulfate chains (Solursh et al., 1986). These results are consistent with a role for intact chondroitin sulfate/dermatan sulfate proteoglycans in primary mesenchyme cell migration.

The effects of tunicamycin, an inhibitor of protein glycosylation, on primary mesenchyme cell behavior have also been reported. Schneider et al. (1978) observed that the primary mesenchyme cells ingress in Arbacia after tunicamycin treatment. These investigators did not comment on mesenchyme cell migration but found that gastrulation was blocked. In S. purpuratus, Heifetz and Lennarz (1979) reported that migration was blocked by the drug under conditions where glycosaminoglycan synthesis was unimpaired. The treatment did, however, depress total protein synthesis slightly in addition to inhibiting glycosylation. They suggested that sulfated glycoproteins might play an important role in morphogenesis. More recently, using much lower concentrations of tunicamycin, Akasaka et al. (1980) found that gastrulation was inhibited, whereas primary mesenchyme cell migration occurred (although the pattern was abnormal).

All these treatments lack the specificity needed to implicate in any rigorous sense a particular class of macromolecule in primary mesenchyme cell migration. Sulfated proteoglycans deserve careful attention, however, as do other components of the basal lamina. One, more direct, approach might involve studies of mesenchyme cell–matrix interactions in vitro (Karp and Solursh, 1986). In this way, abnormal or reconstituted matrices can be used as substrata.

4. Patterning of the Primary Mesenchyme Cells

4.1. Pattern of Migration

Weiss (1947) pointed out that three distinct mechanisms can be involved in the patterning of cell distributions. **Selective conduction** involves the influence of a nonrandom environment on the direction of migration. **Selective elimination** involves differential survival of cells at certain sites. **Selective fixation** involves the random movement of cells until they are trapped at certain locations. On the basis of time-lapse movies (Gustafson and Kinnander, 1956; Gustafson and Wolpert, 1961), it is apparent that the primary mesenchyme cells move up the blastula wall in a random fashion. Each migrating cell moves differently from the others. During migration, pseudopodial contacts are continually being made and broken. Even after most of the mesenchyme cells aggregate in the two ventrolateral clumps near the vegetal pole, pseudopodial processes are still unstable; there is just no net cell movement. Clearly, selective fixation is involved in the patterning process. As Gustafson and Wolpert (1963) proposed, patterning of these cells must involve gradual and localized changes in adhesive properties of the substratum. The final positions are those determined by the formation of the most stable cell attachments.

The underlying basis for selective fixation of cells in a pattern like that shown in Figure 12 is not known. Historically, the ectoderm has been thought of as a template. Presumably, it would have to provide a substratum with regionalized, differential adhesiveness. If the mesenchyme cells are experimentally displaced (e.g., by centrifugation), they will move back to their normal position (Driesch, 1896; Hörstadius, 1928). In vegetalized larvae, the primary mesenchyme cells shift toward the animal pole (Herbst, 1896; Runnström, 1917, 1928; Gustafson and Wolpert, 1961). At low magnification, Okazaki et al. (1962) noted an optical pattern in the ectoderm that resembles two fans, coinciding with the two ventrolateral clumps of mesenchyme. They also shifted appropriately in vegetalized embryos. Okazaki and co-workers suggested that the patterns are reflections of ectodermal cell shape. Gustafson (1963) has emphasized that the primary mesenchyme cells are localized in regions of ectodermal curvature and thickening. At the light microscopic level, there appear to be spaces between the ectodermal cells at such locations (Gustafson and Wolpert, 1963). Gustafson (1963) suggested that at regions of curvature there could be more contacts per cell between a particular mesenchyme cell and the substratum, thereby providing a more stable attachment. However, examination of specimens like that shown in Figure 12 provides no morphological basis for such a hypothesis.

It is also possible that localized specializations in the ectoderm and its underlying basal lamina, to which the mesenchyme cells attach, develop with

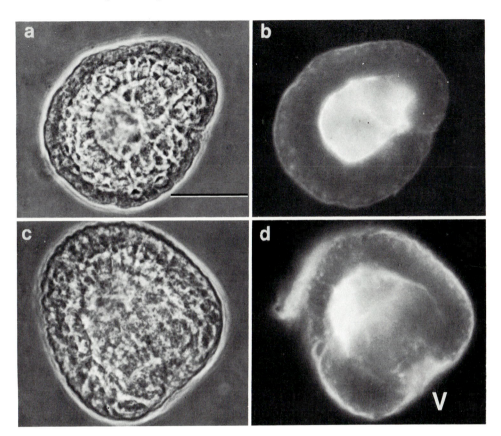

Figure 25. (a, c) Phase-contrast micrographs of mesenchyme blastula- and early gastrula-stage *S. purpuratus* embryos that have been cut with a razor blade and exposed to fluorescein-tagged concanavalin A(Con A) as described by Katow and Solursh (1982). (b, d) Fluorescence micrographs of the same embryos, showing the distribution of FITC-Con A. Note that in the mesenchyme blastula, binding occurs around the entire basal surface of the ectoderm and in the blastocoel. In the early gastrula, binding is restricted to the animal half ectoderm and is not on the vegetal half ectoderm (V). The binding is completely blocked in the presence of α-D-mannoside, but not β-D-mannoside (not shown) (Stanford and M. Solursh, unpublished data). Similar results have been reported for *L. pictus* (Katow and Solursh, 1982), suggesting that the change in pattern occurs widely among different echinoid species. Scale bar: 100 μm.

time. Although there is still no direct evidence for this hypothesis, Katow and Solursh (1982) have described temporal changes in the spatial distribution of Con A binding sites on the basal surface of the ectoderm (Fig. 25). At the mesenchyme blastula stage, these binding sites are distributed around the entire blastocoel. However, in early gastrulae, the Con A binding sites are lost in the vegetal half ectoderm and associated primary mesenchyme cells. Moreover, vegetalized embryos retain the uniform distribution, and the primary mesenchyme cells reach the animal pole, where they remain. Thus, there is a correlation between the spatial distribution of Con A binding sites and the patterning of the primary mesenchyme cells. It will be important to examine the cell attachment properties of specific glycoproteins in the blastocoel, as well as modifications of such glycoproteins during development. It is noteworthy that Con A injected into the blastocoel can cause detachment of filopodia from the secondary mesenchyme (Spiegel and Burger, 1982). However, if Con A-binding glycoproteins are involved in pseudopod attachment, one must eventually be able to explain how the primary mesenchyme cells attach near the vegetal plate and the secondary mesenchyme to the dorsal ectoderm and later to the oral zone (Gustafson, 1963). Clearly, the echinoid system has considerable potential in permitting one to approach this fundamental process of developmental biology.

4.2. Spicule Pattern

The continued patterning of the primary mesenchyme as the larval skeleton is laid down becomes spatially more complex. However, the underlying basis for the pattern may be similar in principle to that involved in the earlier stages of mesenchyme cell localization.

The descriptive details of spicule morphogenesis have been presented in a number of reviews (e.g., Dan, 1960; Gustafson and Wolpert, 1963; Okazaki, 1960, 1975) and are highlighted here only briefly for perspective. There is a long history of studies on spicule formation dating from Selenka (1879), and there are extensive species variations. The two ventrolateral clumps of primary mesenchyme cells become connected by dorsal and ventral circumferential chains of mesenchyme cells, resulting in the formation of a ring of cells around the blastocoel. Remaining mesenchyme cells align anteriorly in two longitudinal cords from the ventrolateral clumps. These longitudinal cords are later joined together more anteriorly by an anterior–dorsal chain of cells. Pseudopodia from the cells actually fuse to form a cable syncytium (Gibbins *et al.*, 1969), which lies close to the basal lamina under the ectoderm. The primary mesenchyme cells apparently have an intrinsic tendency to fuse with each other and even to other cell types, if given the opportunity (Hagström and Lönning, 1969). The skeleton is formed within the cable and consists of calcite (calcium and some magnesium carbonate) in an organic matrix of unknown composition. It first appears as a triradiate crystal within each triangular ventrolateral clump of primary mesenchyme (Fig. 26a). Each crystal then enlarges by the addition of stacks of microcrystals (Okazaki and Inoué, 1976).

In the present context, the important question is what factors determine the form of the spicules (Fig. 26b). As emphasized by Von Ubisch (1937) and Gustafson and Wolpert (1963), factors both intrinsic and extrinsic to the primary mesenchyme must be involved. The importance of factors intrinsic to the mesenchyme is indicated by the observation that in chimeras produced by implantation of micromeres of one species into the blastocoel of another, the spicules tend to resemble the species that provided the mesenchyme cells (e.g., whether fenestrated or simple) (Von Ubisch, 1939; Hörstadius, 1973). In addi-

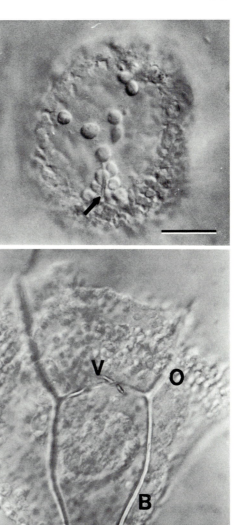

Figure 26. (a) Photomicrograph of a side view gastrula of *L. pictus*, focused to show a ventrolateral clump of primary mesenchyme cells. Within the clump the triradiate spicule can be observed (arrow). (b) Photomicrograph of a crushed *L. pictus* pluteus larva to illustrate the more advanced spicules. In this ventral view, three radii corresponding to the oral rod (0), mid-ventral rod (V), and body rod (B) are visible. The mid-ventral rods from the two sides have fused medially. The region of intersection of these three rods corresponds to the earlier rudiment shown in (a). The anal rods, which form as new branches, are behind the plane of focus. Scale bars: 50 μm.

tion, Okazaki (1975) showed isolated primary mesenchyme cells to have considerable autonomous capacity to form spicules *in vitro*. In culture, spicule formation is a community effort of many primary mesenchyme cells (Okazaki, 1965). In this case, the pattern of the spicules formed might be determined more by crystal properties of the spicule itself. Unlike spicules formed *in situ*, those formed *in vitro* tend to be straight, uniformly thick parallel rods, often connected by short perpendicular bridges (Okazaki, 1975).

As for primary mesenchyme cell migration, the extrinsic influence is thought to be provided by the ectoderm (Von Ubisch, 1937; Gustafson and Wolpert, 1963). Gustafson and Wolpert (1963) suggest that the spicules will tend to be straight (determined by factors intrinsic to the mesenchyme) unless altered by extrinsic factors provided by the ectoderm. Thus, in chimeras the position of the spicules is like the species supplying the ectoderm. They emphasize that the pattern of spicule elongation is determined by the distribution of the primary mesenchyme cells, in particular the pseudopodial syncytia, which are attached to the basal lamina. The pattern of the pseudopodia is constantly changing, apparently by alterations in cell attachments, like those involved in the early patterning of the primary mesenchyme. The ectodermal influence is apparently not stabilized in the extracellular matrix of the blastocoel, since the spicule pattern is not normal if the primary mesenchyme cells are cultured within intact blastocoelic matrices after removal of the epithelial cells (Harkey and Whiteley, 1980). There is some improvement, however, in spicule formation if clumps of ectodermal cells are also present. Harkey and Whiteley (1980) suggested that the ectoderm must be closely associated with the mesenchyme in order to influence spicule morphogenesis. The nature of the ectodermal influence and the manner in which it is integrated with the developing larva, as a whole, are major challenges for future analysis. Studies of such phenomena may provide important information concerning the mechanisms of pattern formation.

5. Conclusions and Summary

The echinoid primary mesenchyme cells provide a model system for analysis of the process of ingression and cell–matrix interaction during cell migration and cell patterning. The embryos are transparent and readily obtained in large quantities as synchronous cultures. The origin of the primary mesenchyme cells, which have special migratory properties, can be traced to the 16-cell stage and even to the animal–vegetal polarity of the ovarian egg.

The primary mesenchyme cells form as a result of localized loss of the basal lamina in the blastocoel, cell shape changes, compression from adjacent blastomeres, and changes in cell junctions and cell–hyaline layer interactions. The primary mesenchyme cells migrate by attaching motile cell processes to the basal lamina, which lines the blastocoel. The cells move randomly and become selectively fixed in patterns that influence the shape of the developing calcareous skeleton. The interactions between the migratory cell processes and the substratum must provide the basis for migration and patterning.

Components associated with cell processes and the substratum used for migration are only just being described. These components appear to be numerous and quite heterogeneous. Ultrastructural observations suggest that these components are highly organized.

The pattern of primary mesenchyme cell distribution is thought to result from localized changes in cell–matrix interactions during development. The cells stop at sites of most stable attachment. Localized modifications of the extracellular matrix by epithelial products is a likely mechanism for cell patterning in this system.

ACKNOWLEDGMENT.   This work was supported by grant HD16549 from the National Institutes of Health.

References

Akasaka, K., and Terayama, H., 1980, General pattern, $^{35}SO_4$-incorporation and intracellular localization of glycans in developing sea urchin (Anthocidaris) embryos, Dev. Growth Diff. **22**:749–762.

Akasaka, K., and Terayama, H., 1983, Sulfated glycan present in the EDTA extract of Hemicentrotus embryos (mid-gastrula), Exp. Cell Res. **146**:177–185.

Akasaka, K., and Terayama, H., 1984, A proteoglycan fraction isolated from the EDTA extract of sea urchin (Hemicentrotus pulcherrimus) gastrulae stimulates reaggregation of dissociated embryonic cells, Exp. Cell Res. **150**:226–233.

Akasaka, K., Amemiya, S., and Terayama, H., 1980, Scanning electron microscopical study of the inside of sea urchin embryos (Pseudocentrotus depressus), Exp. Cell Res. **129**:1–13.

Akiyama, F., and Seno, N., 1981, Linkage regions between dermatan polysulfates and peptides, Biochim. Biophys. Acta **674**:280–296.

Angerer, R. C., and Davidson, E. H., 1984, Molecular indices of cell lineage specifications in sea urchin embryos, Science **226**:1153–1160.

Balinsky, B. I., 1959, An electron microscopic investigation of the mechanism of adhesions of the cells in a sea urchin blastula and gastrula, Exp. Cell Res. **16**:429–433.

Benson, S. C., and Sessions, A., 1980, Prolyl hydroxylase activity during sea urchin development, Exp. Cell Res. **130**:467–470.

Blankenship, J., and Benson, S., 1984, Collagen metabolism and spicule formation in sea urchin micromeres, Exp. Cell Res. **152**:98–104.

Brown, A. E., and Bosmann, H. B., 1978, Glycoprotein synthesis in developing sea urchin embryos, Biochem. Biophys. Res. Commun. **80**:833–840.

Crise-Benson, N., and Benson, S. C., 1979, Ultrastructure of collagen in sea urchin embryos, W. Roux Arch. **186**:65–70.

Dan, K., 1960, Cyto-embryology of echinoderm and amphibia, in: International Review of Cytology, Vol. 9 (G. H. Bourne and J. K. Danielli, eds.), pp. 321–367, Academic Press, New York.

Dan, K., 1979, Studies on unequal cleavage in sea urchins. I. Migration of the nuclei to the vegetal pole, Dev. Growth Diff. **21**:527–535.

DeSimone, D., and Spiegel, M., 1985, Micromere-specific cell surface proteins of 16-cell stage sea urchin embryos, Exp. Cell Res. **156**:7–14.

DeSimone, D. W., Spiegel, E., and Spiegel, M. 1986, The biochemical identification of fibronectin in the sea urchin embryo, Biochem. Biophys. Res. Commun. **133**:183–188.

Driesch, M., 1896, Die taktische Reizbarken der Mesenchymzellen von Echinus microtuberculatus, W. Roux Arch. **3**:362–380.

Endo, Y., and Noda, Y. D., 1977, Ultrastructure of blastocoel of sea urchin embryos, Zool. Mag. **86**:309.

Endo, Y., and Uno, N., 1960, Intercellular bridges in sea urchin blastula, Zool. Mag. **69**:8.

Fink, R. D., and McClay, D. R., 1985, Three cell recognition changes accompany in ingression of sea urchin primary mesenchyme cells, Dev. Biol. **107**:66–74.

Fukunaga, Y., Sobue, M., Suzuki, N., Kushida, H., and Suzuki, S., 1975, Synthesis of a fluorogenic mucopolysaccharide by chondrocytes in cell culture with 4-methylumbelliferyl β-D-xyloside, Biochim. Biophys. Acta **381**:443–447.

Gezelius, G., 1974, Aspects on the role of sulphate in sea urchin development, Acta Univ. Upsal. Diss. Fac. Sci. 318.

Gezelius, G., 1976, Effects of sulphate deficiency on the RNA synthesis of sea urchin larvae, *Zoon* **4**:43–46.

Gibbins, J. R., Tilney, L. G., and Porter, K. R., 1969, Microtubules in the formation and development of the primary mesenchyme in *Arbacia punctulata*. I. The distribution of microtubules, *J. Cell Biol.* **41**:201–226.

Gibson, K. D., and Segin, B. J., 1977, The mode of action of 4-methyl-umbelliferyl β-D-xyloside on the synthesis of chondroitin sulphate in embryonic-chicken sternum, *Biochem J.* **168**:65–79.

Golob, K., Chetsanga, C. J., and Doty, P., 1974, The onset of collagen synthesis in sea urchin embryos, *Biochim. Biophys. Acta* **349**:135–141.

Gustafson, T., 1963, Cellular mechanisms in the morphogenesis of the sea urchin embryo. Cell contacts within the ectoderm and between mesenchyme and ectoderm cells, *Exp. Cell Res.* **32**:570–589.

Gustafson, T., and Kinnander, H., 1956, Microaquaria for time-lapse cinematographic studies of morphogenesis in swimming larvae and observations on sea urchin gastrulation, *Exp. Cell Res.* **11**:36–51.

Gustafson, T., and Wolpert, L., 1961, Studies on the cellular basis of morphogenesis in the sea urchin embryo. Directed movements of primary mesenchyme cells in normal and vegetalized larvae, *Exp. Cell Res.* **24**:64–79.

Gustafson, T., and Wolpert, L., 1963, The cellular basis of morphogenesis and sea urchin development, in: *International Review of Cytology*, Vol. 15 (G. H. Bourne and J. K. Danielli, eds.), pp. 139–214, Academic Press, New York.

Hagström, B. E., and Lönning, S., 1969, Time-lapse and electron microscopic studies of sea urchin micromeres, *Protoplasma* **68**:271–288.

Hajek, A. S., and Solursh, M., 1977, The effect of ascorbic acid on growth and synthesis of matrix components by cultured chick embryo chondrocytes, *J. Exp. Zool.* **200**:377–388.

Hall, H. G., and Vacquier, V. D., 1982, The apical lamina of the sea urchin embryo: Major glycoproteins associated with the hyaline layer, *Dev. Biol.* **89**:168–178.

Harkey, M. A., and Whiteley, A. H., 1980, Isolation, culture and differentiation of echinoid primary mesenchyme cells, *W. Roux. Arch.* **189**:111–122.

Harkey, M. A., and Whiteley, A. H., 1982, Cell-specific regulation of protein synthesis in the sea urchin gastrula: A two dimensional electrophoretic study, *Dev. Biol.* **93**:453–462.

Harkey, M. A., and Whiteley, A. H., 1983, The program of protein synthesis during the development of the micromere-primary mesenchyme cell line in the sea urchin embryo, *Dev. Biol.* **100**:12–28.

Harkey, M. A., and Whiteley, A. H., 1985, Mass isolation and culture of sea urchin micromeres, *In Vitro* **21**:108–113.

Hart, G. W., and Lennarz, W. J., 1978, Effects of tunicamycin on the biosynthesis of glycosaminoglycans by embryonic chick cornea, *J. Biol. Chem.* **253**:5795–5801.

Hart, I. R., and Fidler, I. J., 1980, Cancer invasion and metastasis, *Q. Rev. Biol.* **55**:121–142.

Hay, E. D., 1968, Organization and fine structure of epithelium and mesenchyme in the developing chick embryo, in: *Epithelial–Mesenchymal Interactions* (R. Fleischmajer and R. E. Billingham, eds.), pp. 31–55, Williams & Wilkins, Baltimore.

Hay, E. D., 1981, Collagen and embryonic development, in: *Cell Biology of Extracellular Matrix* (E. D. Hay, ed.), pp. 379–409, Plenum Press, New York.

Heifetz, A., and Lennarz, W. J., 1979, Biosynthesis of N-glycosidically linked glycoproteins during gastrulation of sea urchin embryos, *J. Biol. Chem.* **254**:6119–6127.

Herbst, C., 1896, Experimentelle Untersuchungen über den Einfluss der Veränderten chemischen Zusammensetzung des umgebenden Mediums auf die Entwicklung der Thiere, *Arch. Entwicklungsmech. Org.* **2**:455–516.

Herbst, C., 1904, Über die zur Entwicklung des seeigellarven notwendigen anorganischen Stoffe, ihre Rolle und Vertretbarkeit. II Teil. Die Rolle der notwendigen anorganischen Stoffe, *W. Roux Arch.* **17**:306–520.

Hinegardner, R. T., 1967, Echinoderms, in: *Methods in Developmental Biology* (F. H. Wilt and N. K. Wessells, eds.), pp. 139–155, T. Y. Crowell, New York.

Hörstadius, S., 1928, Über die Determination deskeimes bei Echinodermen, *Acta Zool. (Stockh.)* **9**:1–191.

Hörstadius, S., 1973, *Experimental Embryology of Echinoderms*, Clarendon Press, Oxford.

Hynes, R. O., and Yamada, K. M., 1982, Fibronectins: Multifunctional modular glycoproteins, *J. Cell Biol.* **95:**369–377.

Immers, J., 1956, Changes in acid mucopolysaccharides attending the fertilization and development of the sea urchin, *Arch. Zool.* **9:**367–375.

Immers, J., 1961, Comparative study of the localization of incorporated ^{14}C-labeled amino acids and ^{35}SO$_4$ in the sea urchin ovary, egg and embryo, *Exp Cell Res.* **24:**356–378.

Immers, J., and Runnström, J., 1965, Further studies of the effects of deprivation of sulfate on the early development of the sea urchin *Paracentrotus lividus*, *J. Embryol. Exp. Morphol.* **14:**289–305.

Iwata, M., and Nakano, E., 1981, Fibronectin from the ovary of the sea urchin, *Pseudocentrotus depressus*, *W. Roux Arch.* **190:**83–86.

Iwata, M., and Nakano, E., 1984, Cell-to-substratum adhesion of dissociated embryonic cells of the sea urchin, *Pseudocentrotus depressus*, *W. Roux Arch.* **193:**71–77.

Karp, G. C., and Solursh, M., 1974, Acid mucopolysaccharide metabolism, the cell surface and primary mesenchyme cell activity in the sea urchin embryo, *Dev. Biol.* **41:**110–123.

Karp, G. C., and Solursh, M., 1985, Dynamic activity of the filopodia of sea urchin embryonic cells and their role in exploratory behavior of the primary mesenchyme *in vitro*, *Dev. Biol.* (in press).

Katow, H., and Solursh, M., 1979, Ultrastructure of blastocoel material in blastulae and gastrulae of the sea urchin, *Lytechinus pictus*, *J. Exp. Zool.* **210:**561–567.

Katow, H., and Solursh, M., 1980, Ultrastructure of primary mesenchyme cell ingression in the sea urchin *Lytechinus pictus*, *J. Exp. Zool.* **213:**231–246.

Katow, H., and Solursh, M., 1981, Ultrastructural and time-lapse studies of primary mesenchyme cell behavior in normal and sulfate deprived sea urchin embryos, *Exp. Cell Res.* **136:**233–245.

Katow, H., and Solursh, M., 1982, *In situ* distribution of Con A binding sites in mesenchyme blastulae and gastrulae of the sea urchin *Lytechinus pictus*, *Exp. Cell Res.* **139:**171–180.

Katow, H., Yamada, K. M., and Solursh, M., 1982, Occurrence of fibronectin on the primary mesenchyme cell surface during migration in the sea urchin embryo, *Differentiation* **22:**120–124.

Kinnander, H., and Gustafson, T., 1960, Further studies on the cellular basis of gastrulation in the sea urchin larva, *Exp. Cell Res.* **19:**278–290.

Kinoshita, S., 1971, Heparin as a possible initiator of genomic RNA synthesis in early development of sea urchin embryos, *Exp. Cell Res.* **64:**403–411.

Kinoshita, S., and Saiga, H., 1979, The role of proteoglycan in the development of sea urchins. I. Abnormal development of sea urchin embryos caused by the disturbance of proteoglycan synthesis, *Exp. Cell Res.* **123:**229–236.

Krach, S. W., Green, A., Nicolson, G. L., Oppenheimer, S. B., 1974, Cell surface changes occurring during sea urchin embryonic development monitored by quantitative agglutination with plant lectins, *Exp. Cell Res.* **84:**191–198.

Kumé, M., and Dan, K., 1968, *Invertebrate Embryology*, Nolit, Belgrade, Yugoslavia.

Langelan, R. E., and Whiteley, A. H., 1985, Unequal cleavage and the differentiation of echinoid primary mesenchyme, *Dev. Biol.* **109:**464–475.

Lindahl, P. E., 1942, Contribution to the physiology of form generation in sea urchin development, *Q. Rev. Biol.* **17:**213–227.

Lynn, D. A., Angerer, L. M., Brusken, A. M., Klein, W. H., and Angerer, R. C., 1983, Localization of a family of mRNAs in a single cell type and its precursors in sea urchin embryos, *Proc. Natl. Acad. Sci. USA* **80:**2656–2660.

Matsumura, T., Hasegawa, M., and Shigei, M., 1979, Collagen biochemistry and phylogeny of echinoderms, *Comp. Biochem. Physiol.* **62B:**101–105.

McCarthy, R. A., and Spiegel, M., 1983, Serum effects on the *in vitro* differentiation of sea urchin micromeres, *Exp. Cell Res.* **149:**433–441.

McClay, D. R., and Fink, R. D., 1982, Sea urchin hyalin: Appearance and function in development, *Dev. Biol.* **92:**285–293.

Mintz, G. R., DeFrancesco, S., and Lennarz, W. J., 1981, Spicule formation by cultured embryonic cells from the sea urchin, *J. Biol. Chem.* **256:**13105–13111.

Mizoguchi, H., and Yasumasu, I., 1982, Archenteron formation induced by ascorbate and α-ket-oglutarate in sea urchin embryos kept in SO_4^{2-}-free artificial sea water, *Dev. Biol.* **93**:119–125.

Mizoguchi, H., and Yasumasu, I., 1983, Inhibition of archenteron formation by the inhibitors of prolyl-hydroxylase in sea urchin embryos, *Cell Diff.* **12**:225–231.

Monné, L., and Härde, S., 1951, On the formation of the blastocoel and similar embryonic cavities, *Ark. Zool.* **1**:463–469.

Monné, L., and Slautterback, D. B., 1950, Differential staining of various polysaccharides in sea urchin, *Exp. Cell Res.* **1**:477–491.

Morgan, T. H., and Spooner, G. B., 1909, The polarity of the centrifuged egg, *Arch. Entwichlungs-mech. Org.* **28**:104–117.

Neri, A., Roberson, M., Connolly, D. T., and Oppenheimer, S. B., 1975, Quantitative evaluation of concanavalin A receptor site distribution on the surface of specific populations of embryonic cells, *Nature (Lond.)* **258**:342–344.

Oguri, K., and Yamagata, T., 1978, Appearance of a proteoglycan in developing sea urchin embryos, *Biochim. Biophys. Acta* **541**:385–393.

Okazaki, K., 1960, Skeleton formation of sea urchin larvae. II. Organic matrix of the spicules, *Embryologia* **5**:283–320.

Okazaki, K., 1965, Skeleton formation of sea urchin larvae. V. Continuous observation of the process of matrix formation, *Exp. Cell Res.* **40**:585–596.

Okazaki, K., 1975, Spicule formation by isolated micromeres of the sea urchin embryo, *Am. Zool.* **15**:567–581.

Okazaki, K., and Inoué, S., 1976, Crystal property of the larval sea urchin spicule, *Dev. Growth Diff.* **18**:413–434.

Okazaki, K., and Niijima, I., 1964, Basement membrane in sea urchin larvae, *Embryologia* **8**:89–100.

Okazaki, K., Fukuski, T., and Dan, K., 1962, Cyto-embryological studies of sea urchins. IV. Correlation between the shape of the ectodermal cells and the arrangement of the primary mesenchyme cells in sea urchin larvae, *Acta Embryol. Morphol. Exp.* **5**:17–31.

Pucci-Minafra, I., Casana, C., and LaRosa, C., 1972, Collagen synthesis and spicule formation in sea urchin embryos, *Cell Diff.* **1**:157–165.

Roberson, M., and Oppenheimer, S. B., 1975, Quantitative agglutination of specific populations of sea urchin embryo cells with concanavalin A, *Exp. Cell Res.* **91**:263–268.

Roberson, M., Neri, A., and Oppenheimer, S. B., 1975, Distribution of concanavalin A receptor sites on specific populations of embryonic cells, *Science* **189**:639–640.

Runnström, J., 1917, Analytische Studien über die Seeigelentwicklung. III, *Arch. Entwicklungsmech. Org.* **43**:223–328.

Runnström, J., 1928, Zur Experimentellen Analyse der Wirkung des Lithiums auf den Seeigelkeim, *Acta Zool. (Stockh.)* **9**:365–424.

Runnström, J., Hörstadius, S., Immers, J., and Fudge-Mastrengelo, M., 1964, An analysis of the role of sulfate in the embryonic differentiation of the sea urchin (*Paracentrotus lividus*), *Rev. Suisse Zool.* **71**:21–54.

Sano, K., 1977, Changes in cell surface charges during differentiation of isolated micromeres and mesomeres from sea urchin embryos, *Dev. Biol.* **60**:404–415.

Sano, K., 1980, Changes in concanavalin A, mediated cell agglutinability during differentiation of micromere- and mesomere-derived cells of the sea urchin embryo, *Zool. Mag.* **89**:321–325.

Sano, K., and Usui, N., 1980, Changes in cell surface morphology during differentiation of micromere and mesomere-derived cells in the sea urchin embryo, *Dev. Growth Diff.* **22**:179–185.

Schneider, E. G., Nguyen, H. T., and Lennarz, W. J., 1978, The effect of tunicamycin, an inhibitor of protein glycosylation on embryonic development in the sea urchin, *J. Biol. Chem.* **253**:2348–2355.

Schroeder, T. E., 1980, Expression of the prefertilization polar axis in sea urchin eggs, *Dev. Biol.* **79**:428–443.

Schwartz, N. B., Galligani, L., Ho, P.-L., and Dorfman, A., 1974, Stimulation of synthesis of free chondroitin sulfate chains by β-D-xylosides in cultured cells, *Proc. Natl. Acad. Sci. USA* **71**:4047–4051.

Selenka, E., 1879, Keimblätter and Organanlage der Echiniden, *Z. Wiss. Zool.* **33**:39–54.

Solursh, M., and Katow, H., 1982, Initial characterization of sulfated macromolecules in the blastocoels of mesenchyme blastulae of *Strongylocentrotus purpuratus* and *Lytechinus pictus*, *Dev. Biol.* **94**:326–336.

Solursh, M., and Revel, J.-P., 1978, A scanning electron microscope study of cell shape and cell appendages in the primitive streak region of chick and rat embryos, *Differentiation* **11**:185–190.

Solursh, M., Mitchell, S. L., and Katow, H., 1986, Inhibition of cell migration in sea urchin embryos by β-D-xyloside, *Dev. Biol.* (submitted).

Spiegel, M., and Burger, M. M., 1982, Cell adhesion during gastrulation, *Exp. Cell Res.* **139**:377–382.

Spiegel, M., and Spiegel, E., 1978, Sorting out of sea urchin embryonic cells according to cell type, *Exp. Cell Res.* **117**:269–271.

Spiegel, E., and Spiegel, M., 1979, The hyaline layer as a collagen-containing extracellular matrix in sea urchin embryos and reaggregating cells, *Exp. Cell Res.* **123**:434–441.

Spiegel, E., Burger, M., and Spiegel, M., 1980, Fibronectin in the developing sea urchin embryo, *J. Cell Biol.* **87**:309–313.

Spiegel, E., Burger, M. M., and Spiegel, M., 1983, Fibronectin and laminin in the extracellular matrix and basement membrane of sea urchin embryos, *Exp. Cell Res.* **144**:47–55.

Sugiyama, K., 1972, Occurrence of mucopolysaccharides in the early development of the sea urchin embryo and its role in gastrulation, *Dev. Growth Diff.* **14**:63–73.

Sugrue, S. P., and Hay, E. D., 1981, Response of basal epithelial cell surface and cytoskeleton to solubilized extracellular matrix molecules, *J. Cell Biol.* **91**:45–54.

Tanaka, Y., 1976, Effects of the surfactants on the cleavage and further development of the sea urchin embryos. I. The inhibition of micromere formation at the fourth cleavage, *Dev. Growth Diff.* **18**:113–122.

Tilney, L. G., and Gibbins, J. R., 1969, Microtubules in the formation and development of the primary mesenchyme in *Arbacia puntulata*. II. An experimental analysis of their role in development and maintenance of cell shape, *J. Cell Biol.* **41**:227–250.

Tosney, K. W., 1978, The early migration of neural crest cells in the trunk region of the avian embryo: An electron microscopic study, *Dev. Biol.* **62**:317–333.

Venkatasubramanian, K., and Solursh, M., 1984, Adhesive and migratory behavior of normal and sulfate-deficient sea urchin cells *in vitro*, *Exp. Cell Res.* **154**:421–431.

Von Ubisch, L., 1937, Die normale Skelettbuldung bei *Echinocyamus pusillus* und *Psammechinus miliaris* und die Bedeutung dieser Vorgänge für die Analyse der Skelette von Keimblattchimären, *Z. Wiss. Zool.* **149**:402–476.

Von Ubisch, L., 1939, Keimblattchimärenforschung an Seeigellarven, *Biol. Rev.* **14**:88–103.

Weiss, P., 1947, The problem of specificity in growth and development, *Yale J. Biol. Med.* **19**:233–278.

Wessel, C. M., Marchese, R. B., and McClay, D. R., 1984, Ontogeny of the basal lamina in the sea urchin embryo, *Dev. Biol.* **103**:235–245.

Wilson, L. G., and Bandurski, R. S., 1958, Enzymatic reactions involving sulfate, sulfite, selenate and molybdate, *J. Biol. Chem.* **233**:975–981.

Wolpert, L., and Mercer, E. H., 1963, An electron microscopic study of the development of the blastula of the sea urchin embryo and its radial polarity, *Exp. Cell Res.* **30**:280–300.

Yamagata, T., and Okazaki, K., 1974, Occurrence of a dermatan sulfate isomer in sea urchin larvae, *Biochim. Biophys. Acta* **372**:469–473.

Yamaguchi, M., and Kinoshita, S., 1985, Polysaccharides sulfated at the time of gastrulation in embryos of the sea urchin *Clypeaster japonicus*, *Exp. Cell Res.* **159**:353–365.

Chapter 11
Primordial Germ Cell Migration

C. C. WYLIE, D. STOTT, and P. J. DONOVAN

1. Introduction—PGCs and Their Migratory Properties

Many, if not all, cells in the body can show motile activity when placed in a permissive environment. Cells held in an apparently immutable array in the intact body will nonetheless exhibit some sort of motile behavior when disaggregated and placed in culture. It is clear, therefore, that all cells possess or can make the machinery for cell movement. However, few cells actively move in the adult body; the phase of the life cycle during which large scale cell and tissue movements take place is during embryonic development. These movements result in the correct tissue architecture of the embryonic body becoming established.

During this period, the differentiation of several cell lineages involves a stage when the cells migrate from the site at which they first become determined to the site at which further cell differentiation will take place. Examples are early germ-line cells (the primordial germ cells), neurons, mesenchyme cells, somite-derived cells that will form bone and muscle, blood-forming cells, developing glands, neural crest, and many others. In each of these cases, the movements are fairly precise and the targeting highly accurate.

These cell migrations in the embryo raise fascinating questions for the cell biologist. For a cell to become migratory at a particular period in embryogenesis and to move relative to the cells around it, many changes in its surface and cytoskeletal properties must be involved. These must be the result of expression of a particular set of genes and can loosely be defined as the **migratory phenotype.** The cell properties included in a migratory phenotype would be the ability to make transient cell contacts with cells and matrix encountered on the migratory route, the ability to recognize the correct migratory route (and the target), and the many different factors that result in actual propulsion of the cell.

We know very little about the detailed cellular and molecular mechanisms involved in the migratory phenotype, and we know even less about the trigger-

C. C. WYLIE, D. STOTT, and P. J. DONOVAN • Department of Anatomy, St. George's Hospital Medical School, London SW17 ORE, England.

ing mechanisms that start and stop it. It is important to remember that migration is only a transient property of many differentiating cell types and is switched off once the target is reached, to be replaced by another, often quite different phenotype.

This chapter concentrates on the migration of primordial germ cells (PGCs) of vertebrate embryos. PGCs have been an attractive model for cell studies for a number of reasons: (1) they can be identified easily in histological sections, particularly in anuran amphibian species such as *Xenopus laevis* and *Rana pipiens*; (2) they have only one target, the developing gonad, whereas other migrating populations such as neurons or neural crest cells have many different targets; and (3) it is easy in some species to isolate a pure population of primordial germ cells for study. For these reasons, PGCs have been used extensively in studies of embryonic cell movement and its guidance.

2. The Migratory Route

This discussion is confined to those vertebrates most studied, since there are many features common to PGC migration in even the most disparate vertebrate groups. Much less is known about PGCs in invertebrates (see Nieuwkoop and Sutasurya, 1979, 1981, for more extensive reviews).

2.1. PGC Migration in Amphibians

In the two amphibian groups studied, the Urodeles and the Anurans, fundamental differences in the origin and migratory route have been found, perhaps reflecting the long time period since their divergence in evolution (Nieuwkoop and Sutasurya, 1983). In both groups, the gonads form as ridgelike thickenings in the epithelium lining the abdominal cavity. These ridges lie on either side of the root of the gut mesentery, on the dorsal abdominal wall. The ridges, however, provide only somatic tissue of the adult gonad; the primordial germ cells, progenitors of the gametes, are determined much earlier than this stage and in another part of the embryo. During the cleavage and blastula stages of anuran embryos, a small number of blastomeres at the vegetal pole are found to contain aggregates of basophilic cytoplasm (the **germinal cytoplasm**), which they have inherited from the vegetal pole cytoplasm of the egg. Some of the descendants of these cells will eventually become the primordial germ cells. The normal fate of cells at the vegetal pole of the anuran blastula is to become incorporated by the morphogenetic movements of gastrulation into the endodermal gut tube. The germinal cytoplasm-containing cells follow this fate and end up in the wall of the developing gut. We do not know whether they can still enter other cell lineages at this stage, since they are only recognizable by their small masses of basophilic cytoplasm. For this reason, the term **presumptive PGCs (pPGCs)** was coined by Nieuwkoop to describe the germinal cytoplasm-containing cells found from blastula to early larval stages.

During early larval stages, the PGCs migrate dorsally and caudally in the gut tube until they lie in a dorsal crest over the hindgut. Profound changes in the lateral plate mesoderm then occur that result in formation of the abdominal cavity and the mesentery that suspends the gut in this cavity from the abdominal wall. When the mesentery of the gut forms, it does so in an oblique fashion compared with the conventional axes of the embryo (craniocaudal or longitudinal axis and dorsoventral or transverse axis). For this reason, a rather confusing picture of PGCs is often seen in transverse sections, as they migrate from the dorsal crest of the gut into the mesentery (see Heasman and Wylie, 1981). The PGCs migrate obliquely along the mesentery to the dorsal body wall and then laterally for a variable distance before stopping beneath the strip of coelomic epithelium that is destined to thicken and become the gonadal ridge. Although the gonadal thickening does not form before the PGCs arise, they probably do not induce it to do so; the thickening can occur in embryos with no recognizable PGCs at this stage, after irradiation of the fertilized egg with ultraviolet (UV) light to destroy the germinal cytoplasm. Once the gonadal ridge is formed, the PGCs enter a series of mitotic divisions and rapidly become smaller. This loss in relative size is at least partly due to loss of the large yolk platelets, which are a characteristic feature of their cytoplasm during migration along the dorsal mesentery.

In urodele amphibians the migratory pathway is altered by a different origin of the PGCs (Nieuwkoop and Sutasurya, 1979), which are derived from blastomeres in the marginal zone of the blastula. Instead of becoming part of the gut, this region becomes incorporated into the lateral plate mesoderm, and it is from here that the PGCs migrate to the developing gonad. The pathway of their migration is less well documented than in anurans, largely due to difficulty of identification of the PGCs. No large basophilic masses of cytoplasm can be seen by light microscopy.

2.2. PGC Migration in Birds

There is still some uncertainty over the origin of PGCs in avian embryos. PGCs first appear at primitive streak stages, between the epiblast and hypoblast in a crescent-shaped region cranial and lateral to the primitive streak, at the border between area pellucida and area opaca. The uncertainty is over which germ layer gives rise to the PGCs. Morphological evidence suggests that they separate from the hypoblast layer. However, interspecific grafting of chick and quail epiblast and hypoblast suggests that their origin is from the epiblast (Eyal-Giladi et al., 1981), although this interpretation has been questioned (see England, 1983, for review). Whichever layer of the embryo gives rise to the PGCs, they gather between the epiblast and hypoblast and migrate caudally alongside the developing embryonic body. It is in this region that blood islands first differentiate to form the vitelline circulation, shuttling blood back and forth between embryo and yolk sac. PGCs enter these blood vessels and are carried for part of their migratory route in the vasculature. They leave the blood vessels in various regions near the developing gonads, into which they migrate. The progress of PGCs can be followed in fixed material because their high levels of periodic acid-Schiff (PAS)-positive material permit accurate identification.

2.3. PGC Migration in Mammals

There is also uncertainty about the site of origin of PGCs in mammalian embryos. PGCs become identifiable at postgastrula stages in some mammals because of their high levels of alkaline phosphatase. These alkaline phosphatase-positive cells are found at the base of the allantois, near the endoderm that will give rise to the hindgut. As the hindgut forms, first by development of the tail folds, and then by the lateral body folds, these cells are found embedded in its wall. They then leave the hindgut wall and migrate along its dorsal mesentery to the dorsal body wall and laterally to the gonadal ridges. This is similar to the situation in anuran amphibians. Since mammalian PGCs cannot be identified before their expression of alkaline phosphatase at the base of the developing allantois, it is not possible to study their movements before this time or even to know whether there is a population of cells determined only to enter the germ line before this. The only evidence that bears on their origin comes from single-cell labeling studies using chimeric embryos and from extirpation studies. In the former technique, donor cells (whose progeny can be distinguished genetically from those of the host) are introduced into host embryos at the blastula stage. Using donor cells from morulae, blastulae, primary ectoderm, and primary endoderm, it seems that all except primary endoderm can give rise to PGCs, but also to other lineages of the same host embryo. This is interpreted to mean that potential PGCs are pluripotent at least until they are part of the primary ectoderm of the gastrula (for review, see Eddy and Hahnel, 1983). Extirpation experiments show that removal of a small part of the mouse gastrula including the posterior end of the primitive streak dramatically reduces the number of PGCs in the operated embryo. Furthermore, these same fragments of the embryo, when cultured *in vitro*, produce, among other things, cells that become alkaline phosphatase positive (reviewed in Snow and Monk, 1983). Thus, most of the evidence points to an origin of PGCs in the primary ectoderm of the gastrula stage, near the posterior end of the primitive streak.

3. Germ-Line Markers in the Study of PGC Migration

The above descriptions show that the precise origin of PGCs in all vertebrate groups studied (except perhaps in the anuran amphibian) is still something of a mystery. This is because there are no features of these cells that distinguish them—either in the living embryo or in histological sections—from cells around them. Nor is it certain exactly when these cells cease to be pluripotent and become determined only to form germ-line cells. It is often said that PGCs are determined very early in anuran amphibian embryos, perhaps even at the blastula stage. This is more a statement of convenience than of strict accuracy. What is meant is that certain blastomeres can be recognized to contain a histological structure that can be followed into the germ line as development proceeds. It is perfectly feasible to assume, however, that these blastomeres divide so that the germinal cytoplasm is retained in only one progeny cell, and thus cells can be given off into other lineages for an unspecified length of time. This process has been suggested to occur, at least up to the blastula stage, by Dixon and colleagues (Whitington and Dixon, 1975). Only the kind of single-cell labeling done by making embryo chimeras, as is currently done with mammalian embryos, will solve this central issue.

PGCs become recognizable in different species according to different criteria. The basophilic "germinal cytoplasm" permits the early part of PGC migration to be followed in anuran amphibians, albeit with difficulty. Later, when the PGCs leave the developing gut and enter its mesentery, they are easy to identify due to their very large size, content of large yolk platelets, and multi-lobed nuclei (see Wylie and Heasman, 1976). In fact, the PGCs are so large while in the mesentery that they can be picked out under a dissecting microscope, either for quantitation by direct counting or for collection and study *in vitro*.

In other species, no natural markers are known that unequivocally identify the early part of PGC migration, and the initial identification of PGCs is attributable to the appearance of such markers at comparatively late stages of development, i.e., gastrula in the chick and postgastrula in the mouse. Chick PGCs become recognizable due to their high staining affinity for the PAS reaction, whereas mouse and other mammalian PGCs express surface alkaline phosphatase activtiy. One natural marker present in early germ-like cells of many species, from nematodes to mammals, is **nuage** (see reviews by Strome and Wood, 1983; Eddy, 1975; Mahowald, 1977). This term refers to small discrete clumps of electron-dense material. In electron micrographs (EMs) they are similar in appearance to electron-dense material in insect eggs (the **polar granules**), in later germ-line cells (oocytes and developing sperm of many species), as well as apparently non-germ-line-related cells of early embryos. Since nuage can only be recognized under the electron microscope, its usefulness is rather limited in tracing PGCs. Furthermore, since its biochemical nature is unknown, no direct relationship can be inferred from either the rather similar structures found in PGCs of diverse species or between these and similar-looking structures seen in cells that apparently do not enter the germ line. Much work needs to be done on this intriguing structure.

Considerable recent effort has been made to identify germ-line-specific antigenic markers in or on the surface of PGCs. This work has been done particularly with respect to mouse and rat PGCs, since large numbers of them can be isolated easily from the early gonad. The rationale here is (1) to establish when PGC-specific molecules appear, (2) to establish their function in *in vitro* studies, and (3) to trace the origin of PGCs at earlier stages than can be done using the conventional histological markers already described.

One fact established by these studies is the close relationship between some teratocarcinoma cells and migrating PGCs. Several monoclonal anti-bodies raised against F9 teratocarcinoma and embryonal carcinoma cells are found to stain PGCs (reviewed by Stern, 1983; Eddy and Hahnel, 1983). In general it seems that several cell types in very early embryos react with most of these antibodies, whereas PGCs do so later on, during their migratory phase. No antibodies have yet been found that exclusively label PGCs in mouse embryos before they are found at the base of the allantois in postgastrulation stages. Whether such antigens exist is unknown.

4. The Migratory Phenotype

We shall now turn to the rather fragmentary data suggesting that PGCs have behavioral characteristics and surface molecules that are uniquely expressed during the migratory phase and switched off in the later postmigratory stages of germ cell differentiation after arrival at the gonad. These data fall into three groups: cell morphology and cytoskeleton, adhesive properties, and surface antigens.

4.1. PGC Morphology and Cytoskeleton

In histological sections during their migratory phase, PGCs of *Xenopus laevis* show a remarkable plasticity of shape, sometimes highly elongated, other times almost spherical. They conform to the shape of the space available to them and in particular to that of the cells over which they migrate (Heasman and Wylie, 1981). In some PGCs, fine processes can be seen, each of which has a core of microfilaments. Since these processes have been observed to be ex-truded from isolated PGCs *in vitro*, they are assumed to be filopodial-type extensions characteristic of motile cells. These extensions are complex in nature, often a combination of flattened lamellar areas and small filopodia. In time-lapse films of PGCs *in vitro*, they are constantly extruded and retracted (Heasman *et al.*, 1977). Postmigratory PGCs have none of these characteristics, being rounded in the developing gonad (and certainly in the female once meiosis has started), remaining rounded *in vitro*, and showing no evidence of locomotory activity at all.

Transmission electron microscopic (TEM) evaluation of *Xenopus* PGCs during migration either *in vitro* or *in vivo* shows specialized junctional con-tacts thought to be characteristic of moving cells. They are known as **focal adhesion sites** and consist of small electron-dense regions of apposed plasma membranes of both PGCs and adjacent somatic cells. The intercellular space narrows to 0–20 nm, and small bundles of microfilaments are inserted into the membrane plaque, particularly on the somatic cell side. That these are sites of adhesion is shown by treatment with EDTA, after which retraction of somatic cells from over the PGCs leaves long retraction fibers ending at these junctions (Heasman and Wylie, 1981). Focal adhesion sites of this type have been observed in fibroblasts crawling *in vitro* as well as in other motile cells. They have not been reported in later germ-line cells.

The cytoskeleton of migrating PGCs is characteristic of moving cells. There are no obvious microfilament or intermediate filament bundles, and the most obvious cytoskeletal element is a thin cortical network of microfilaments; filopodia also have a core of microfilaments. These results are supported by immunocytochemistry of *Xenopus* PGCs *in vitro* (Heasman and Wylie, 1983). Actin staining is diffuse, with a more brightly fluorescent cortical shell; myosin staining is diffuse and continues into all processes. No evidence of organized bundles of cytoskeletal material has been reported. By contrast, later stages of germ cell differentiation are accompanied by dramatic accumulations of highly organized cytoskeletal material. During oogenesis, for example, organized arrays of intermediate filaments are formed (Godsave *et al.*, 1984a,b). The migratory phenotype is therefore transient and is replaced by quite different morphology and cytoskeletal arrangements.

4.2. Adhesive Behavior

Attempts to culture isolated PGCs from both *Xenopus* and mouse were frustrated at first by the fact that they are nonadhesive on most artificial substrates. However, the discovery that *Xenopus* PGCs adhere to a feeder layer of cells derived from adult mesentery permitted the study of their behavior *in vitro* (Heasman *et al.*, 1977). PGCs from the migratory phase will adhere to, and spread on, cellular substrates.

Similarly, mouse PGCs isolated as single cells will not stick to glass or plastic substrata (Heath, 1978), nor will they adhere to substrates coated with purified extracellular matrix components (D. Stott and P. J. Donovan, unpublished observations). However, when fragments of genital ridge are grown as explant cultures, PGCs do adhere to the somatic cells of the explant outgrowth (de Felici and McLaren, 1983). Fibroblastic cell lines derived from mouse embryos also provide a suitable substrate for germ cell adhesion. These cell lines provide a more reproducible substrate for adhesion than do explant cultures. Such a system should therefore enable assays to be carried out on changes in germ cell adhesivity that occur during the period of their migration and establishment in the gonad anlagen. Initial results indicate a decrease in adhesivity of PGCs once they have reached the genital ridge (P. J. Donovan and L. Cairns, unpublished observation).

4.3. Surface Antigens

Several antibodies have been described that react with surface antigens expressed by mouse PGCs in a temporally regulated fashion. Some of these reagents were raised against teratocarcinoma cell lines, and some were produced against early mouse embryos (reviewed by Eddy and Hahnel, 1983;

Stern, 1983). To date, only one antibody, **PG1,** has been raised against embryonic germ cells (Heath, 1978). All these antibodies react with other tissues, or with cell lines, as well as PGCs, the general pattern being that they are present on many cells in the early embryo, are lost during the postimplantation stage, and are re-expressed on PGCs during their migration.

The detailed evidence concerning the timing of appearance and disappearance of these antigens on identified PGCs is mostly fragmentary (Eddy and Hahnel, 1983). Thus, we have undertaken a survey of the temporal pattern of staining of migratory PGCs with six serological reagents. In all experiments, histological sections were double stained for alkaline phosphatase activity and specific antibody reactivity; hence, specific recognition of is PGCs shown by the codistribution of the two stains. One of these antibodies, TG1, has not previously been reported to recognize mouse PGCs. TG1 is a mouse monoclonal antibody raised against a glycoprotein fraction of human thymocytes (Beverley *et al.,* 1980) that has diverse specificities, including human peripheral blood neutrophils, some murine teratocarcinoma cells, and migratory PGCs. It stains no other embryonic cells than PGCs at the same sagittal level during PGC migration.

Figure 1 shows histological sections from a developmental series stained with both alkaline phosphatase and TG1. Sections from embryos at 8.5, 10.5, 12.5, and 15.5 days of development are shown. At 8.5 days (Fig. 1a,b), the PGCs do not express TG1. At 9.5 days (Fig. 1c,d) and 10.5 days (Fig. 1e,f), the PGCs show staining of their surface membrane with TG1, which then disappears at the end of the migratory phase when the PGCs have settled in the gonad (Fig. 1g,h). The results from this and similar surveys with different monoclonal antibodies (all of which have been shown to stain PGCs) are shown in Table I. The temporal patterns of staining are similar with all these antibodies. In each case they appear around the time of onset of PGC migration and are lost when migration has ceased. This finding suggests that migration is accompanied by several changes in the surface of the PGC, presumably enabling it to interact in an appropriate manner with the changing environment encountered during migration (see below). No experimental situation has yet defined the role of any of these antigens. As soon as a suitable experimental system for the study of PGC migration is established, it will be essential to test the role of these surface groups in PGC locomotion.

All these diverse and fragmentary facts are undoubtedly linked. The morphology, behavior, and surface antigen spectrum characteristic of the migratory phase of PGCs are probably easily detected products of a coordinately expressed group of genes responsible for the locomotion and guidance of PGCs. The main interest in the future will be to characterize the gene products, identify their role in PGC migration, and find out what factors control their appearance at the beginning and disappearance at the end of migration.

One interesting fact to emerge from immunological studies of PGCs is that at least one surface antigen expressed by migratory mouse PGCs is also expressed by those of *Xenopus laevis* (Heath and Wylie, 1981). This finding confirms the suspicion that many molecules involved in germ-line formation should be highly conserved in evolution.

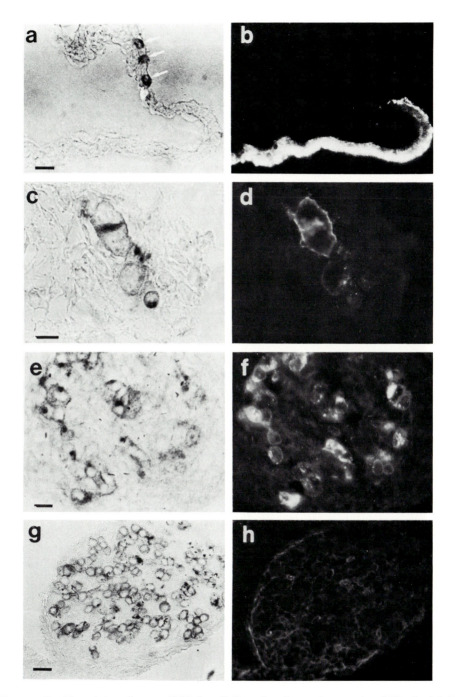

Figure 1. Double staining of mouse PGCs for alkaline phosphatase (a, c, e, g) and TG1 (b, d, f, h). Each horizontal pair of pictures represents a phase-contrast and immunofluorescent micrograph of the same field of view. (a, b) 8.5-day embryo. TG1 stains extraembryonic endoderm but not alkaline phosphatase-positive PGCs. Scale bar: 20 μm. (c, d) 10.5-day embryo. High-power micrograph of double-labeled PGCs in hindgut mesentery. Scale bar: 10 μm. (e, f) 12.5-day embryo. Double-labeled PGCs in developing gonad. Scale bar: 15 μm. (g, h) 15.5-day embryo. TG1 no longer stains alkaline phosphatase-positive PGCs in gonad. Scale bar: 20 μm.

Table I. Presence of Specific Markers on Mouse PGCs during Embryogenesis[a,b]

Age (dpc)	Alkaline phosphatase	PG1[c]	Anti-Forssman[c]	SSEA-1[d]	SSEA-3[d]	TG1[d]	M1 22/25[e]
7.5	−	nd	nd	nd	nd	−	nd
8.5	+	−	−	−	−	−	−
9.5	+	−	−	+	−	+	−
10.5	+	+	+	+	−	+	+
11.5	+	+	+	+	−	+	+
12.5	+	+	+	+	+	+	+
13.5	+	+	+	+	+	+	+
14.5	+	+	+	+	−	+	+
15.5 (F)	+ (weak)	−	−	−	−	−	−
15.5 (M)	+	+	−	−	−	−	−
17.5 (F)	−	−	−	nd	nd	nd	−
17.5 (M)	−	−	−	nd	nd	nd	−

[a]Sections containing PGCs were stained for alkaline phosphatase and double stained with one of six antibodies.
[b]+, positive staining of PGCs; −, negative staining of PGCs; nd, not done; dpc, days postcoitus.
[c]PG1 (Heath, 1977) and anti-Forssman (Karol et al., 1981) are rabbit polyclonal antisera.
[d]SSEA-1 (Solter and Knowles, 1978), SSEA-3 (Shevinsky et al., 1982), and TG1 (Beverley et al., 1980) are mouse monoclonal antibodies.
[e]M1 22/25 (Willison et al., 1982) is a rat monoclonal directed against the Forssman antigen.

5. Guidance of PGC Migration

The appearance of gene products that endow PGCs with migratory properties is not sufficient to get them to the gonad. In addition to motility, PGCs must possess the ability to recognize guidance signals and to react to them such that they will eventually move to their target in the embryonic body. The nature of both of these signals and of the molecules on the PGCs surface that respond to them has long been the subject of experimentation and discussion. First, we shall consider the nature of the substrate on which PGCs move, next, the way in which that substrate may guide migration, and, lastly, the evidence for chemotaxis.

5.1. Substrate for PGC Migration

Evidence from both in situ and in vitro studies strongly suggests that PGCs interact directly with other cells along their migratory route. Electron micrographs of PGCs migrating in Xenopus gut mesentery show very close apposition of their surfaces and of those of the coelomic epithelial cells. In places, focal adhesion sites can be seen between the two cell types (Heasman and Wylie, 1981). The same is seen in vitro when isolated PGCs are seeded onto cellular feeder layers. PGCs isolated from their migratory path in Xenopus or mouse embryos will not stick to conventional tissue culture surfaces, nor will mouse PGCs stick to surfaces covered with purified extracellular matrix components (laminin, fibronectin, and collagen). This finding suggests that PGCs use other factors associated with the surrounding cells as a substrate for migration.

However, several observations point to the involvement of extracellular matrix molecules in some way in PGC migration. Electron microscopic examination indicates the presence of bundles of matrix fibrils around the leading processes of PGCs in Xenopus. Furthermore, the addition of antifibronectin to Xenopus PGCs during their adhesion to cellular feeder layers shows that this antibody, but not nonimmune IgG, inhibits their adhesion (Heasman et al., 1981). In the mouse embryo, immunohistochemistry demonstrates that the distribution of matrix components around mouse germ cells differs between migratory and postmigratory phases (D. Stott and P. J. Donovan, unpublished observations).

Figure 2 shows that fibronectin and laminin are present interstitially between cells of the mesentery and dorsal body wall through which the PGCs migrate at 9½ and 10½ days of development (Fig. 2a–d). In the developing gonad as well, interstitial staining for these components is seen when the germ cells first arrive at 11.5 days of development. However, later in development, this interstitial staining is lost, and the matrix components become restricted to the basal lamina of the tubules or follicles of the gonad (Fig. 2e,f), so that by 15.5 days of development no interstitial staining is detected. The same pattern is found with type IV collagen (not shown). It may also be significant that this loss of interstitial matrix distribution corresponds approximately to the loss of several antigens from the PGC surface (see Table I).

Although PGCs have not been shown to adhere to purified matrix molecules in vitro, it may be that a particular combination or orientation of the components is required for an interaction to occur. We do not yet know what mediates the interaction between PGCs and feeder cells. At the moment it seems likely that extracellular matrix molecules may be involved in PGC adhesion during their migration.

5.2. Substrate Guidance of PGCs

There are several possible ways in which a cell might be directed by the substrate. We discuss these mechanisms below.

5.2.1. Physical Shape

Fibroblasts in culture are unable to migrate over angles sharper than a certain limit (see Dunn, 1982, for review). Situations can thus be envisaged in the embryo where narrow linear channels exist, making an exit impossible due to their angle of curvature.

5.2.2. Adhesive Gradients

There is good evidence to show that certain cells in vitro will accumulate in areas of high adhesiveness and that a gradient of adhesiveness will thus have

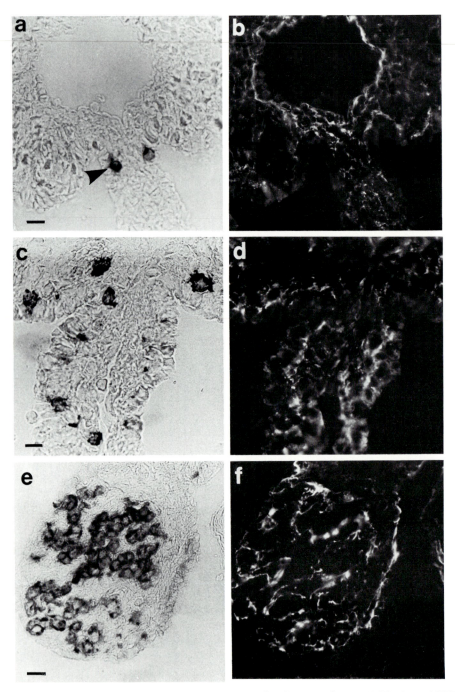

Figure 2. Immunolocalization of extracellular matrix molecules around mouse PGCs. (a, c, e) PGCs (arrows) are identified by alkaline phosphatase staining. (b, d, f) Same field of view stained with antifibronectin (b) or antilaminin (d, f). (a, b) From a 9.5-day embryo; fibronectin is present interstitially around PGCs in the hindgut mesentery. (c, d) From a 10.5-day embryo. Laminin is present interstitially around PGCs in the hindgut mesentery and in the dorsal body wall. (e, f) From a 13.5-day embryo. Laminin is now restricted to basal laminae surrounding groups of PGCs. Scale bar: 20 μm for all micrographs.

an orientating effect (Carter, 1967; see Chapter 13). Concentration gradients of adhesive molecules have not been detected along the path of PGCs in situ, although this remains a plausible hypothesis.

5.2.3. Linear Arrays of Adhesion Sites

If migrating cells are presented with a linear sequence of binding sites—either on extracellular matrix molecules or cell surfaces—a chemical substrate guidance becomes possible that does not depend on visible inhomogeneities in the substrate. No evidence exists for this in the case of PGCs, as the binding sites on their migratory path are as yet unidentified.

5.2.4. Opportunistic Channels

This is a rather more nebulous concept, and it is extremely difficult to test. If a migrating cell is unable to penetrate the three-dimensional arrangement of cells and matrix unless a permissive channel is temporarily opened, such channels will act as guidance cues in that the cell can only move when they are open and in a direction imposed by them. Such permissive channels could include physical spaces, temporary adhesion sites for the migratory cell, or temporary appearance of a particular matrix molecule.

Many of these theoretical possibilities cannot be tested in the living embryo and can only be tested in vitro. However, some circumstantial evidence is accumulating concerning the relationship between Xenopus PGCs and their substrate. Scanning electron microscopic (SEM) studies show that the living cells of the gut mesentery, beneath which PGCs migrate to the dorsal body wall, are highly polarized. Moreover, PGCs conform closely to their orientation (Wylie et al., 1979; Heasman and Wylie, 1981). When isolated Xenopus PGCs are seeded onto orientated feeder layers, they spread on their surface in the same orientation. Xenopus PGCs are clearly capable of responding to guidance cues related to the cells around them. What these cues are is not known.

PGCs of chick embryos are also confronted by an orientated substrate at the earliest stage of their migration, before they enter the vascular system. Between the epiblast and hypoblast of the early embryo in the germinal crescent is an aligned group of extracellular matrix fibers derived from the basal lamina of the epiblast. SEM shows the PGCs to associate with this band of fibrils. In twinned embryos in which two primitive streaks are present, the germinal crescents, and thus the fibrous bands, meet between the two embryonic axes. The presence at this point of large numbers of PGCs suggests that PGCs do move along the fibrous band. However, its precise role in the migration process of PGCs is not clear (for review, see England, 1983).

5.3. Chemotaxis

The orientation of the substrate becomes immaterial (provided that the PGCs can penetrate it) if guidance is furnished by a concentration gradient of a diffusable molecule produced by the target. This mechanism of guidance has been suggested to occur both in chick (Dubois, 1968) and anuran (Gipouloux, 1970; Giorgi, 1974) embryos.

In anurans these conclusions are based on grafting experiments. In order to show that dorsal components attract the PGCs from the more ventral endoderm, dorsal components of donor embryos have been grafted into the most ventral regions of host embryos. In these experimental embryos, some PGCs migrated to the host dorsal regions as they normally do, some remained in the endoderm, and some migrated to the grafted dorsal region. However, it is difficult to exclude the fact that such grafted tissue may "attract" PGCs in other ways, e.g., by causing new orientation of the lateral plate mesoderm so that new substrate guidance pathways are established. For example, it has been noted that more PGCs accumulate next to dorsal structures derived from the graft when a new dorsal mesentery forms connecting the grafted dorsal region to the host endoderm. In addition, there is no evidence as to which cells are from the donor and which from the host in these experiments. More rigorous evidence is required from *in vitro* experiments in which the only variable is the concentration gradient of the putative chemotactic agent. This type of experiment has been elegantly done with respect to neutrophils (Zigmund, 1978). Chemotaxis remains a perfectly reasonable mechanism for imposing directionality, as well as polarity, on migrating anuran PGCs.

In chick embryos, in which PGCs leave the vasculature only in the gonadal region, chemotaxis is a plausible hypothesis. Evidence to support this possibility comes from experiments in which tissues containing chick (Dubois, 1968) or mouse (Rogulska *et al.*, 1971) germ cells have been grafted into the coelomic cavity next to the developing gonad, or in which PGCs have been injected into the blood vascular system of a host whose own PGCs have been previously destroyed (Raynaud, 1969). In both cases, the host gonadal epithelium "attracts" the donor PGCs. The mechanism behind these phenomena remains unclear. Selective trapping in the region of the gonad or release of a humoral chemotactic agent would both lead to the same observed results. Once again, *in vitro* culture of migrating PGCs, where rigorous testing of chemotaxis is possible, will be of enormous benefit.

6. Conclusions

Despite a great deal of work, our knowledge about the mechanism of PGC locomotion, the surface molecules involved, the substrate, and the mechanism of guidance is largely fragmentary and circumstantial. The future seems to lie in two areas. First, a molecular analysis is required of the surface molecules that confer the migratory phenotype. It will be important to establish the roles of these molecules and to find out whether they are expressed in other migrating cells and in malignant cells. It would appear unlikely that cells of different lineages expressing a migratory phenotype at some stage of differentiation use different sets of genes to do so. More likely is the existence in all cells of a coordinated set of genes responsible for migratory activity and that this particular combination of genes is expressed in different cell lineages if and when cell movement is required for their development. Metastasis is apparently the inappropriate expression of these genes in certain cells of a malignant tumor. Second, purified populations of migrating PGCs are required, so that more rigorous testing is possible of their mode of locomotion and their guidance.

ACKNOWLEDGMENTS. Work in the authors' laboratory is supported by the Cancer Research Campaign and the Medical Research Council, whose support to Janet Heasman and C. C. Wylie is gratefully acknowledged. We thank Linda Cairns for her help with the work on mouse PGCs and Melanie Coulton for typing the manuscript.

References

Beverley, P. C. L., Linch, D., and Delia, D., 1980, Isolation of human haematopoietic progenitor cells using monoclonal antibodies, *Nature (Lond.)* **287**:332–333.

Carter, S. B., 1967, Haptotaxis and the mechanisms of cell motility, *Nature (Lond.)* **215**:256–260.

de Felici, M., and McLaren, A., 1983, *In vitro* culture of mouse primordial germ cells, *Exp. Cell Res.* **144**:417–427.

Dubois, R., 1968, La colonisation des étranches gonadiques par les cellules germinales de l'embryon de poulet, en culture *in vitro*, *J. Embryol. Exp. Morphol.* **20**:189–213.

Eddy, E. M., 1975, Germ plasm and the differentiation of the germ cell line, *Int. Rev. Cytol.* **43**:229–280.

Eddy, E. M., and Hahnel, A. C., 1983, Establishment of the germ line in mammals, in: *Current Problems in Germ Cell Differentiation* (A. McLaren and C. C. Wylie, eds.), pp. 41–70, Cambridge University Press, Cambridge.

England, M. A., 1983, The migration of primordial germ cells in avian embryos, in: *Current Problems in Germ Cell Differentiation* (A. McLaren and C. C. Wylie, eds.), pp. 91–114, Cambridge University Press, Cambridge.

Eyal-Giladi, H., Ginsburg, M., and Farborov, A., 1981, Avian primordial germ cells are of epiblastic origin, *J. Embryol. Exp. Morphol.* **65**:139–147.

Giorgi, P. P., 1974, Germ cell migration in toad (*Bufo bufo*): Effect of ventral grafting on embryonic dorsal regions, *J. Embryol. Exp. Morphol.* **31**:75–87.

Gipouloux, J. D., 1970, Recherches experimentales sur l'origine de la migration des cellules germinales, et l'édification des crêtes génitales chez les Amphibiens Anoures, *Bull. Biol. Fr. Belg.* **104**:22–93.

Godsave, S. F., Anderton, B. H., Heasman, J., and Wylie, C. C., 1984a, Oocytes and early embryos of *Xenopus laevis* contain intermediate filaments which react with anti-mammalian vimentin antibodies, *J. Embryol. Exp. Morphol.* **83**:169–187.

Godsave, S. F., Wylie, C. C., Lane, E. B., and Anderton, B. H., 1984b, Intermediate filaments in the *Xenopus* oocyte: The appearance and distribution of cytokeratin-containing filaments, *J. Embryol. Exp. Morphol.* **83**:157–167.

Heasen, J., and Wylie, C. C., 1981, Contact relations and guidance of primordial germ cells on their migratory route in embryos of *Xenopus laevis*, *Proc. R. Soc. Lond. B* **213**:41–58.

Heasman, J., Wylie, C. C., 1983, Amphibian primordial germ cells—What can they tell us about directed cell migration?, in: *Current Problems in Germ Cell Differentiation* (A. McLaren and C. C. Wylie, eds.), pp. 73–90, Cambridge University Press, Cambridge.

Heasman, J., Mohun, T. J., and Wylie, C. C., 1977, Studies on the locomotion of primordial germ cells from *Xenopus laevis* in vitro, *J. Embryol. Exp. Morphol.* **42**:149–161.

Heasman, J., Hynes, R. O., Swan, A. P., Thomas, V. A., and Wylie, C. C., 1981, Primordial germ cells of *Xenopus* embryos; the role of fibronectin in their adhesion during migration, *Cell* **27**:437–447.

Heath, J., 1977, Characterization of a Xenogeneic antiserum raised against the fetal germ cells of the mouse: Cross-reactivity with embryonal carcinoma cells, *Cell* **15**:299–306.

Heath, J., 1978, Mammalian primordial germ cells, in: *Development in Mammals*, Vol. 3 (M. H. Johnson, ed.), pp. 267–298, Elsevier/North-Holland, New York.

Heath, J., and Wylie, C. C., 1981, Cell surface molecules of mammalian foetal germ cells, in: *Development and Function of Reproductive Organs* (A. G. Byskov and H. Peters, eds.), pp. 83–92, Excerpta Medica, Amsterdam.

Karol, R. A., Kundu, S. K., and Marcus, D. M., 1981, Immunochemical relationship between Forssman and globoside glycolipid antigens, *Immunol. Commun.* **10**:137–250.

Mahowald, A. P., 1977, The germ plasm of Drosophila: A model system for the study of embryonic determination, *Am. Zool.* **17**:551–563.

Nieuwkoop, P. D., and Sutasurya, L. A., 1979, *Primordial Germ Cells in the Chordates*, Cambridge University Press, Cambridge.

Nieuwkoop, P. D., and Sutasurya, L. A., 1981, *Primordial Germ Cells in the Invertebrates*, Cambridge University Press, Cambridge.

Nieuwkoop, P. D., and Sutasurya, L. A., 1983, Some problems in the development and evolution of the chordates, in: *Development and Evolution* (B. C. Goodwin, N. Holder, and C. C. Wylie, eds.), pp. 123–136, Cambridge University Press, Cambridge.

Reynaud, G., 1969, Transfert de cellules germinales primordiales de dindon à l'embryon de poulet par injection intravasculaire, *J. Embryol. Exp. Morphol.* **21**:485–507.

Rogulska, T. R., Ozdzenski, L., and Komar, A., 1971, Behaviour of mouse primordial germ cells in the chick embryo, *J. Embryol. Exp. Morphol.* **25**:155–164.

Shevinsky, L. H., Knowles, B. B., Damjanov, I., and Solter, D., 1982, Monoclonal antibody to murine embryos defines a stage-specific embryonic antigen expressed on mouse embryos and human teratocarcinoma cells, *Cell* **30**:697–705.

Snow, M. H. L., and Monk, M., 1983, Emergence and migration of mouse primordial germ cells, in: *Current Problems in Germ Cell Differentiation* (A. McLaren and C. C. Wylie, eds.), pp. 115–136, Cambridge University Press, Cambridge.

Solter, D., and Knowles, D. P., 1978, Monoclonal antibody defining a stage specific mouse embryonic antigen (SSEA-1), *Proc. Natl. Acad. Sci. USA* **75**:5565–5569.

Stern, P., 1983, Serological and cell-mediated immune recognition of teratocarcinomas, in: *Current Problems in Germ Cell Differentiation* (A. McLaren and C. C. Wylie, eds.), pp. 157–174, Cambridge University Press, Cambridge.

Strome, S., and Wood, W. B., 1983, Generation of asymmetry and segregation of germ line granules in early *C. elegans* embryos, *Cell* **35**:15–25.

Whitington, P. McD., and Dixon, K. E., 1975, Quantitative studies of germ plasm and germ cells during early embryogenesis of *Xenopus laevis*, *J. Embryol. Exp. Morphol.* **33**:57–74.

Willison, K. R., Karol, R. A., Suzuki, A., Kundu, S. K., and Marcus, D. M., 1982, Neutral glycolipid antigens as developmental markers of mouse teratocarcinoma and early embryos: An immunologic and chemical analysis, *J. Immunol.* **129(2)**:603–609.

Wylie, C. C., and Heasman, J., 1976, The formation of the gonadal ridge in *Xenopus laevis*. I. A light and transmission electron microscope study, *J. Embryol. Exp. Morphol.* **35**:125–138.

Wylie, C. C., Heasman, J., Swan, A. P., and Anderton, B. H., 1979, Evidence for substrate guidance of primordial germ cells, *Exp. Cell Res.* **121**:315–324.

Zigmund, S. H., 1978, Chemotaxis by polymorphonuclear leucocytes, *J. Cell Biol.* **77**:269–287.

Chapter 12

Mesoderm Migration in the Early Chick Embryo

ESMOND J. SANDERS

1. Introduction

The idea that mesoderm cells migrate during the course of early morphogenesis was either implicit or specifically expressed in some of the earliest works of the nineteenth century embryologists. The origin of these moving cells during gastrulation by a process of de-epithelialization of the hitherto coherent epiblast was finally settled and fully appreciated by Abercrombie (1937), Pasteels (1937), Waddington and Taylor (1937), and Jacobson (1938). The contributions of the many investigators whose work led to the current concepts in this area of developmental biology are thoroughly covered in the remarkable review by Rudnick (1944), which must be considered obligatory reading for students of early embryology.

This chapter follows the characteristics of mesoderm cell movement from the time of the first emergence of this tissue from the primitive streak, through the continuous process of differentiation leading to the early limb bud stage of development. This period embraces the outward migration of primary mesenchyme and its division into segmental plate and lateral plate populations, somitogenesis and the subsequent formation of dermomyotome and sclerotome, and the differentiation of intermediate mesoderm and area vasculosa mesoderm. The sclerotome and the mesenchymal derivatives of the nephrotome and lateral plate constitute the "secondary mesenchyme" (Hay, 1968). Some of these cell populations are clearly migratory in the sense that individual cells translocate from one region of the embryo to another, whereas other populations are not. By emphasizing the continuity in the process of mesoderm differentiation, however, it is clear that cell populations pass through both migratory and nonmigratory phases. The migratory sclerotome cells, for example, are ultimately derived from migratory primary mesenchyme but pass through segmental plate and early somitic phases, in which cell movement is restricted.

ESMOND J. SANDERS • Department of Physiology, University of Alberta, Edmonton, Alberta T6G 2H7, Canada.

What is meant by cell migration? Clearly there is a difference between the mass movements of cells from one defined location in the embryo to another and the movements or rearrangements of cells that occur in solid tissues. Our knowledge of both of these phenomena is far from complete, but both are perceived to rely on similar principles involving, for example, changes in cell–cell and cell–substratum adhesiveness and cytoskeletal changes. Both types of behavior are encountered in mesoderm morphogenesis and are considered here. Mass migrations and translocations commonly occur either by the coordinated movement of individual cells or by the flowing of coherent cell sheets. Both processes also occur during early mesoderm morphogenesis, the latter being exemplified by the expansion of sheets of area vasculosa mesoderm. Trinkaus (1976, 1982) has given detailed consideration to distinctions between various modes of cell movement and has provided examples from many morphogenetic situations.

Paradoxically, our understanding of these particular dynamic processes of cell movement and interactions in the chick embryo comes almost entirely from the study of fixed material using scanning and transmission electron microscopy (SEM and TEM). It is important to realize that the impressions of cell migration are built up by analyzing the relationships between cells in chronological series of static images. Whereas some other morphogenetic processes are observable directly in vivo using time-lapse film or video techniques, current technology does not permit detailed examination of the migration of individual mesoderm cells in the chick embryo. Although considerable success has been achieved in the direct filming of the gross tissue movements during gastrulation, the mesoderm is the most difficult to study even at this level due to the opacity of the material (Vakaet, 1970). For most of the tissues considered here, the sole recourse for investigators wishing to study cell movements directly is the use of in vitro techniques with the attendant difficulties and limitations. Behavioral characteristics of these cells may be compared using conventional culture techniques (Bellairs et al., 1980; Sanders, 1980) or the three-dimensional collagen gel system (Sanders and Prasad, 1983). In either case, interpretation and extrapolation to in vivo circumstances can only be made with great caution.

2. Gastrulation

Gastrulation in birds is generally thought to comprise the complex series of events associated with the invagination of the epiblast through the primitive streak. The result of the process is that epiblast cells emerge from the primitive streak transformed into endoblast or mesoderm cells. The former contribute to the lower layer of the embryo, replacing the primitive hypoblast (Sanders et al., 1978; Stern and Ireland, 1981; Bellairs, 1982), while the latter migrate laterally as primary mesenchyme (Fig. 1). Details of the painstaking grafting and labeling work that established the origin and fate of various regions of the mesoderm are outside the scope of this chapter and are dealt with elsewhere (Rosenquist, 1966; Bellairs, 1971; Nicolet, 1971; Hara, 1978). The remarkable change from

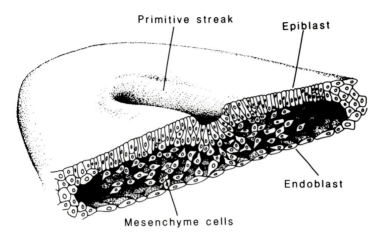

Figure 1. Diagrammatic representation (not to scale) of the area pellucida of a gastrulating chick embryo at stage 4 described by Hamburger and Hamilton (1951). Epiblast cells ingress through the primitive streak and give rise to endoblast and mesenchyme cells. (Redrawn from Balinsky, 1975, *An Introduction to Embryology* by copyright permission, 1975, of W. B. Saunders Company. Reprinted by permission of CBS College Publishing.)

epithelial to mesenchymal cell morphology that coincides with passage through the primitive streak has naturally attracted much interest, yet we still have only general ideas as to the environmental cues and intrinsic factors that trigger the change. Although the work considered below is extensive, much more needs to be done to understand this transformation. Changes in cell morphology such as this occur throughout nature in both normal and pathological situations. Gastrulation provides an example of cell transformation that is both well defined and accessible. An understanding of its underlying mechanisms may well be of widespread importance.

2.1. Migration of Cells through the Primitive Streak

On the basis of results from SEM of the dorsal surface of epiblast cells, it has been possible to identify cells destined for invagination by their characteristic surface topography (Bancroft and Bellairs, 1974; H. J. Jacob *et al.*, 1974). This has led to speculation that epiblast cells are at least partially "preprogrammed" for invagination (Bellairs, 1982), but no morphological criteria can be established to distinguish cells destined for either the endoblast or mesoderm (Wakely and England, 1977).

The morphology and apparent behavior of the cells passing through the primitive streak is well known (Balinsky and Walther, 1961; Granholm and Baker, 1970; Revel, 1974; Bancroft and Bellairs, 1975; Wakely and England, 1977, 1979a; Solursh and Revel, 1978). The cells migrate through the streak (Fig. 2) individually with their apical epithelial contacts persisting, while the body of the cell elongates ventrally. The result of this manoeuver is the forma-

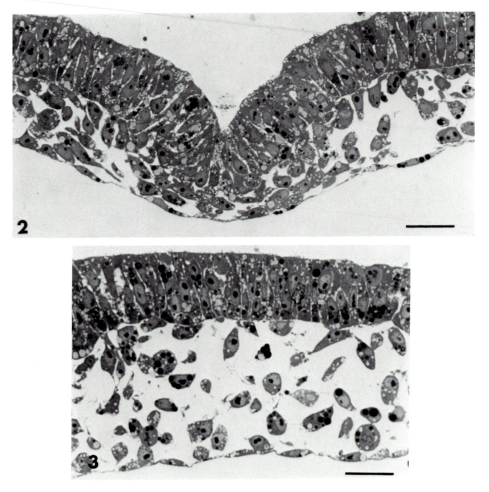

Figure 2. Section through the primitive streak at stage 4. Scale bar: 50 μm.

Figure 3. Section taken lateral to the primitive streak at stage 4. Note that some mesenchyme cells are attached to either the epiblast or endoblast, but many are not. Scale bar: 50 μm.

tion of flask-shaped ("bottle") cells with thin, attenuated apical necks and broad basal regions. The basolateral cell surfaces presumably disrupt their contacts with adjacent cells to facilitate this shape change. Freeze-fracture images of the lateral membranes of primitive streak cells appear to confirm this (Revel et al., 1973; Revel, 1974). Small accumulations of intramembranous particles are found between these cells and have been interpreted to be the remnants of larger gap junctions that were present earlier in the epiblast. Whether these particle accumulations are functional gap junctions during this phase of migration is not known. The apparent remnants of the apical tight junctions are also found between streak cells, consisting of isolated strings of particles. Presumably these strands are unable to perform the sealing function of the original tight junctions. The ventral extension, which is accompanied by the formation of both lamellipodia and filopodia (Vakaet et al., 1980), is followed by the release of the narrow apical contact. Thus freed, the cells flow basally into the outwardly migrating stream of mesenchyme, where the abundant lamellipodia and filopodia presumably effect locomotion.

Cytoplasmic microtubules are conspicuous in primitive streak cells and are aligned parallel to the long axis of the cells (Granholm and Baker, 1970; Sanders and Zalik, 1972). Although present in cells both before (Sanders and Zalik, 1970) and after (Trelstad et al., 1967) passage through the streak, they are nowhere as abundant or organized as in the streak cells themselves. This suggests their participation in the elongation of the flask cells, but, whereas much has been learned about microtubules since their demonstration in these cells, the subsequent information still does not provide a comprehensive explanation of their function. Little can be added to the conjecture of Granholm and Baker (1970) that the microtubules are involved in the cell shape change and the basal redistribution of cytoplasm. Microfilaments appear to be particularly concentrated, or perhaps more highly ordered, in the filopodia (but not the cell bodies) of the streak cells (Bancroft and Bellairs, 1975).

So, although the nature of the factors initiating these changes in the epiblast cells eludes us, the physiology of invagination itself can be seen to depend on sequential regional changes in adhesiveness, which are coincident with cytoskeletal reorganization. This pattern of events repeats itself in other morphogenetic rearrangements considered here and elsewhere. Changes in the composition of both cell surfaces and the extracellular matrix in the primitive streak, which accompany the adhesive changes, are considered in Section 2.3.

2.2. Lateral Migration of Primary Mesenchyme

Once invaginated, the mesoderm cells turn through 90° and migrate out away from the primitive streak. Whether all the cells migrate, or whether some remain close to the primitive streak (Bancroft and Bellairs, 1975) is uncertain. Both TEM and SEM studies have shown that although the mesoderm of this layer is a loose association of stellate cells, migration is not characterized by individual cell movement. Except at the leading edge of the advancing mesoderm where some cells are completely isolated from their neighbors, the migration is a coordinated mass movement (Trelstad et al., 1967; Ebendal, 1976; England and Wakely, 1977; Triplett and Meier, 1982). The cells contact one another and their substrata by means of their filopodia and lamellipodia (Fig. 3). The latter presumably correspond to the "ruffles" of moving cells in vitro (Abercrombie et al., 1970), but their presence on the migrating mesenchyme cells is a matter of debate. Whereas Ebendal (1976) and Revel (1974) claim their existence, others have failed to demonstrate them (England and Wakely, 1977; Wakely and England, 1979a), finding only filopodia. It is important to resolve this contradiction so that we know whether the ruffles characteristic of in vitro movement occur also in vivo. Aside from the possible differences in tissue preservation, the disagreement may arise from the definition of what constitutes a ruffle in a static image. These cells certainly do show ruffles, or at least undulations of the leading edge, when cultured *in vitro* (see Section 2.4), but there is some suggestion that the three-dimensional *in vivo* configuration suppresses the formation of large ruffles. In this connection, it is interesting to note that at least two investigators have commented that the mesenchyme cells migrating on the undersurface of the epiblast present a surprisingly two-dimensional appearance (Revel, 1974; Ebendal, 1976). That the mesenchyme cells use the surfaces of both epiblast and endoblast as substrata can be inferred from Figure 3 and from all the TEM and SEM studies quoted here. The influence of these surfaces on migration is considered later.

The intercellular contacts made by the filopodia of the migrating cells have been studied both by conventional TEM and freeze fracture with a view to elucidating their possible role in the migratory process (Trelstad *et al.*, 1967; Hay, 1968; Revel *et al.*, 1973; Revel, 1974). Earlier TEM studies had shown the presence of small "focal junctions" among mesoderm cells, but these were later shown by freeze-fracture studies to correspond to small gap junctions and rare tight junction remnants. The latter are more scarce in this location than in the primitive streak, supporting the contention that they are the remnants of apical junctions from the epiblast. Unlike the gap junctions, which occur both on filopodia and cell bodies, the tight junction strands are restricted to the latter location. It could be argued that the gap junctions are relics of larger structures from the epiblast, but there are reasons—admittedly still unsubstantiated—for suspecting them to be functional as channels of intercellular coupling (Revel *et al.*, 1973).

Much has been written about the attractive idea that gap junction-mediated intercellular coupling during development could influence morphogenesis (Bennett, 1973; Sheridan, 1976; Powers and Tupper, 1977; Wolpert, 1978; Gilula, 1980; Lo, 1980). However, little progress has been made in recent years in demonstrating such an influence in a substantial way, although this does not diminish the tenability of the concept. The early work of Sheridan (1968) remains the only intracellular electrode study on the early chick embryo. The results of that study suggest that intercellular coupling is widespread among all tissues and also between cells of different tissues. There appears ample scope for confirmation and extension of this work, but it has not been done.

The conceivable roles of intercellular coupling in influencing development have been amply discussed in the reviews quoted above. One possibility that warrants further consideration is that contact inhibition of locomotion in the mesenchyme might be mediated by coupling (Harris, 1974) and provide a mechanism for the outward migration. Contact inhibition of locomotion is defined as "the stopping of the continued locomotion of a cell in the direction which has produced a collision with another cell or, alternatively, the prohibition, when contact between cells has occurred, of continued movement such as would carry one cell across the surface of another" (Heaysman, 1978). The result of this behavior in culture is to encourage cells to move away from their neighbors, forming a monolayer. A clear analogy can be drawn between these *in vitro* events and the outgrowth of mesenchyme from the primitive streak, making contact inhibition an attractive control mechanism for this situation (Ebendal, 1976; Vakaet *et al.*, 1980). However, whether contact inhibition occurs *in vivo*, or whether it even *needs* to occur to explain outgrowth, has not been fully determined. Trinkaus (1976, 1982) has reviewed this subject comprehensively, citing a number of developmental situations in which contact inhibition clearly operates and expressing the view that the phenomenon could be of major importance in morphogenesis. Whether gap junctions transmit the required inhibitory signals between cells is a matter of conjecture. Harris (1974) is skeptical and critically evaluates the hypothesis.

A question often asked but seldom investigated is: To what extent does cell division contribute to primitive streak formation and mesoderm expansion? That cell division does occur is beyond doubt (Bellairs, 1957; Stern, 1979). What has been difficult to determine, however, are the temporal and regional variations in the occurrence of mitoses during this period. The general conclusion that can be drawn (Bellairs, 1982) is that there is a relatively high mitotic rate in the fully grown primitive streak in comparison with the regions to either side. The significance of this finding for morphogenesis in the streak region is not clear.

It has generally been considered that the primary mesenchyme cells eventually aggregate and organize into an epithelial configuration, forming the segmental plate and lateral plate mesoderm (Trelstad *et al.*, 1967). The former, by segmentation, gives rise to somites. Recent SEM studies have suggested, however, that the pattern of segmentation in the mesoderm is already organized in the primitive streak stages (Meier, 1979, 1981; Meier and Jacobson, 1982; Triplett and Meier, 1982). Examination of stereopairs of micrographs makes it possible to discern circular domains or "somitomeres" in the anterior regions of the very early mesoderm during gastrulation, which suggests the prepatterning of segmentation. This apparently occurs even before the primitive streak has reached its full length. The mesoderm cells migrate into these domains in a swirled pattern but remain stellate and loosely associated until the segmentation process generates clear somites.

2.3. The Cell Surface and Extracellular Matrix

It is difficult to draw a distinction between cell surface material and extracellular matrix. The continuity between these two glycoprotein-containing domains blurs the cell boundary and makes the older concept of a "surface coat" (Martinez-Palomo, 1970) hard to define. Nevertheless, the first part of this section discusses characteristics of mesoderm cells pertaining to what has traditionally been considered the cell surface.

2.3.1. The Cell Surface

Determination of whether differences in cell surface coat thickness or distribution correlate with the degree of cell motility has relied on ultrastructural stains for carbohydrate-containing surface components. Lanthanum ions, which may also demonstrate calcium-binding sites, result in exceptionally heavy staining of the surfaces of cells in passage through the primitive streak but very little stain on mesoderm cells themselves (Sanders and Zalik, 1972). Interstitial bodies also stain by this technique. By contrast, in the case of ruthenium red staining, mesodermal cells not only stain appreciably, but this staining increases as the cells migrate away from the streak (Martinez-Palomo, 1970; Mestres and Hinrichsen, 1974). Although these techniques do not permit any hard-and-fast correlations, they indicate that changes in adhesive behavior are accompanied by changes at the cell surface.

A similar conclusion may be drawn from examination of cationized ferritin binding to anionic sites on the free surface of epiblast cells in the primitive streak and lateral to it. Whereas cells in the latter location bind ample ferritin, those in the former do not (MacLean and Sanders, 1983). This correlates well with the finding that as more cells become committed to the primitive streak, the average electrophoretic mobility of cells of the blastoderm decreases (Zalik et al., 1972). In comparison with older embryonic or adult tissue, the ionogenic groups on cell surfaces at the primitive streak stages show a high affinity for calcium, magnesium, manganese, and lanthanum ions. The latter, in contrast to the others, are irreversibly bound, however. The high calcium and magnesium binding may well be relevant to adhesive changes occurring at this time in development (Harris and Zalik, 1974, 1977). Edelman et al. (1983) have identified two cell adhesion molecules (CAMs) in the chick embryo whose sequential appearance and disappearance from various cell surfaces during early development suggests their participation in some of the first differentiative events. For example, these molecules, isolated from neural tissue (N-CAM) and liver (L-CAM), are distributed uniformly throughout the embryo before gastrulation, but cells leaving the streak and migrating as mesoderm appear to lose L-CAM. By contrast, cells leaving the streak and entering the endoblast still express L-CAM. As development proceeds, N-CAM predominates in the neural plate and transiently in the somitic mesoderm, whereas L-CAM becomes associated with endodermally derived structures. The indications are, therefore, that these cell surface molecules influence cell adhesion and movement as well as cell–cell recognition from the earliest phases of development.

Evidence for cell surface changes during gastrulation also comes from examination of the lectin-binding properties of the cells (Hook and Sanders, 1977; Sanders and Anderson, 1979). Wheat germ agglutinin, which binds specifically to N-acetylglucosamine, binds to cell surfaces before and after—but not during—passage through the streak. The mesenchyme cells bind this lectin with increasing avidity as they move away from the streak, again showing that cell surface changes are temporally associated with altered adhesive and migratory characteristics. The mesoderm cells appear to be undergoing progressive modification of their surface after invagination. This idea is supported by the detection of active glycosyltransferases on migrating chick embryo cell surfaces, including those of the primary mesenchyme (Shur, 1977a,b). The observations suggest that the cells interact with and modify the substrata on which they move by the continual incorporation into the substrata of sugar residues. Similarly, the basal lamina of the epiblast may be modified as the mesoderm cells migrate over it (Harrisson *et al.*, 1984). These types of changes indicate that both the cell surface and extracellular matrix may play controlling roles in cell migration and differentiation.

2.3.2. The Extracellular Matrix

The extracellular matrix is a complex association of collagen, glycosamino-glycans, proteoglycans, and glycoproteins. A large recent review literature on various aspects of this matrix includes several that highlight its developmental significance as a factor influencing morphogenesis and differentiation (Manasek, 1975; Wartiovaara *et al.*, 1980; Hay, 1981a,b; Toole, 1981). In essence, the developmental implications rely on demonstrations of cell–matrix interactions occurring spatially and temporally to coincide with important morphogenetic events. These interactions may take the form of the modification of the chemical composition of the matrix by the cells or, conversely, the modification of cell behavior by the matrix. In either case it is necessary to consider some genetically determined predisposition of cells to modify, or to be modified by, the extracellular matrix. Whereas the cell–matrix interactions that occur are observable with varying degrees of difficulty, the contribution of the intrinsic genetic differentiative "program" is often an imponderable. Changes in extracellular matrix composition at key points in development always remain circumstantial evidence for matrix involvement in influencing behavioral events. Much of our understanding of the controlling role played by the extracellular matrix is based on an accumulation of such largely circumstantial correlations.

When one examines the tissue space surrounding the primary mesenchyme cells during gastrulation, one is usually disappointed. There is relatively little matrix to be seen, at least by current techniques of fixation and staining (Fig. 4). This is true whether the cells are viewed in fractured specimens by SEM (Revel, 1974; Bancroft and Bellairs, 1975; England and Wakely, 1977) or by TEM after application of contrast-enhancing procedures such as ruthenium red (Mestres and Hinrichsen, 1974) or tannic acid (Sanders, 1979). The major accumulations of extracellular materials occur in the form of the basal lamina of the epiblast (see Section 2.3.3) and a mat on the dorsal surface of the endoblast. The latter is not organized into a basal lamina (Sanders, 1979). The remainder of the space at this time contains clumps of amorphous material attached to the basal lamina (Fig. 5) termed **interstitial bodies** (Low, 1970), and unstriated fibrils (Low, 1968; Sanders, 1979), the latter being sparse but particularly apparent near the primitive streak (Fig. 6). The impression gained from the SEM and TEM images is that the mesenchyme cells are able to support themselves by their own stiffness and by filopodial contact. Yet these tissue spaces are almost certainly filled with hydrated glycosaminoglycans (Toole, 1981), in particular hyaluronic acid, which we are unable to fix adequately and visualize ultrastructurally.

Hyaluronic acid accounts for 80–90% of the glycosaminoglycan synthesized at this time in development (Manasek, 1975; Solursh, 1976; Toole, 1981).

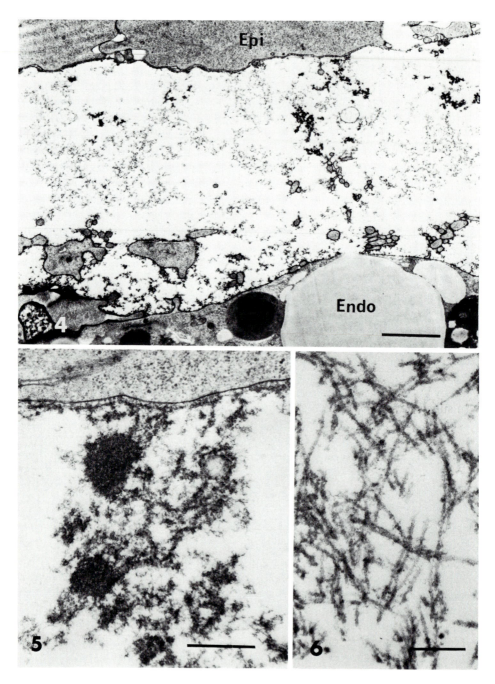

Figure 4. Transmission electron micrograph of the space between the epiblast (Epi) and endoblast (Endo). Note the apparent sparsity of extracellular materials. Tannic acid staining. Scale bar: 2 μm.

Figure 5. Transmission electron micrograph of the basal lamina of the epiblast and an interstitial body. Tannic acid staining. Scale bar: 0.5 μm.

Figure 6. Transmission electron micrograph of extracellular unstriated fibrils in the primitive streak region of a stage 5 embryo. Tannic acid staining. Scale bar: 0.25 μm.

Its distribution in the embryo has been determined by autoradiography of [³H]glucosamine incorporation, Alcian blue staining, and hyaluronidase digestion. Such studies show that hyaluronic acid or hyaluronidase-sensitive material is abundant between the epiblast cells, surrounding the mesenchyme cells, and in the basal lamina (Manasek, 1975; Fisher and Solursh, 1977; Sanders, 1979; Sanders and Anderson, 1979; Solursh et al., 1979a; Vanroelen et al., 1980a,b,c). Although the attractive idea of hyaluronic acid influencing development by forming a gradient (Manasek, 1975) has not been substantiated, there are other proposed modes of action for this substance. Various lines of evidence suggest that hyaluronic acid may be important in keeping cells separated and moving, thereby discouraging premature aggregation, adhesion, and differentiation (Toole, 1973, 1981). The early chick embryo shows indications that the highly hydrated hyaluronic acid serves to open up and maintain tissue spaces into which mesenchyme cells are able to migrate (Fisher and Solursh, 1977). This suggestion has been applied to a number of similar developmental situations (Toole, 1981) (see Chapter 13).

Despite the paucity of observable fibrillar elements at primitive streak stages, collagen of several types is being synthesized at this time (Manasek, 1975; Hay, 1981b). With the possible exception of type IV (basal lamina) collagen, it is unlikely that collagen provides any significant substratum for the movement of primary mesenchyme cells. It appears in fact as though these cells are only able to attach poorly to type I collagen, even when it is provided as the sole substratum (Sanders and Prasad, 1983).

The glycoprotein fibronectin is the matrix molecule with the greatest potential for influencing morphogenesis thus far investigated. Since its primary location during gastrulation is the basal lamina of the epiblast, it is considered in the following section.

2.3.3. The Basal Lamina

The basal lamina of the epiblast, a specialized region of extracellular matrix, has been examined in some detail by both TEM and SEM (Low, 1967; Sanders, 1979; Wakely and England, 1979b). It is found to be present without interstitial bodies and in a somewhat disorganized form at the time of laying, becoming complete by the time of gastrulation. Its principal features are a lucent *lamina interna* (*lamina lucida*) traversed by hyaluronidase-resistant fibrils and a filamentous *lamina densa* in which some periodic densities may be discerned (Sanders, 1979). The face of the basal lamina in contact with the mesenchyme cells is characterized by the presence of interstitial bodies (Low, 1970). By contrast, the extracellular material on the surface of the endoblast is not organized into a basal lamina. The epiblast retains the ability to produce a basal lamina when explanted *in vitro*, where it lines the roofs of characteristic "domes" that appear in the culture (Sanders and Dickau, 1981).

The fact that the basal lamina is formed well in advance of the appearance of the mesoderm (Sanders, 1979) suggests that the primary purpose of the basal lamina proper is to maintain the integrity of the epiblast against the stresses of gastrulation. The later appearance of the fibronectin-rich interstitial bodies places them ideally, both spatially and temporally, to influence the mesoderm cells. How this influence would be exerted by these extracellular clumps of material is difficult to imagine, and it may be that the interstitial bodies merely reflect the turnover of basal lamina constituents (Sanders, 1984). The continual reorganization of the basal lamina as the overlying ectoderm migrates medially may give rise to these bodies. Evidence supporting this view comes from experiments in which the basal lamina of the epiblast of living chick embryos is labeled with ultrastructural markers (Sanders, 1984). Results show that as the epiblast migrates, basal lamina material accumulates in the region of the primitive streak, suggesting that the basal lamina moves medially with the epiblast cells and becomes sloughed off as the latter ingress into the primitive streak.

Basal laminae have become the object of much interest as possible modulators of morphogenesis in a variety of situations (reviewed by Sanders, 1983). With respect to gastrulation, the speculation is twofold: (1) the presence of fibronectin in the basal lamina suggests that it may provide an attractive substratum for mesoderm migration; and (2) the basal lamina surface topography may provide tracks that direct the migrating mesoderm cells along a specified route by contact guidance (Section 2.3.4). These tracks may themselves contain fibronectin. The evidence for the first possibility is circumstantial but persuasive, and for the second possibility the evidence is equivocal.

Fibronectin (Hynes, 1981) has been localized using immunocytochemical techniques on both whole mounts (Critchley et al., 1979; Wakely and England, 1979b) and sections (Mitrani and Farberov, 1982; Duband and Thiery, 1982; Harrison et al., 1984) of gastrulating embryos. These studies, as well as investigations using ultrastructural immunocytochemistry (Sanders, 1982a), show that fibronectin is present—mainly in the basal lamina—from the time of laying (Figs. 7 and 8). Ultrastructural examination shows that the interstitial bodies are fibronectin-rich and also that the basal lamina adjacent to the primitive streak is unaccountably fibronectin poor. It is in this region that the basal lamina breaks up to allow the inward migration of presumptive mesoderm. This disruption of the basal lamina is accompanied by blebbing of the ventral surface of the epiblast cells (Vakaet et al., 1980). Whole-mount preparations show that the fibronectin is not uniformly distributed in the basal lamina throughout the embryo but appears concentrated in certain regions. In the anterior germinal crescent, for example, the fibronectin is associated with prominent bands of parallel fibrils observable both by immunofluorescence and SEM.

The potential importance of finding fibronectin in this location at this time is due to the well-known fact that it is able to affect cell adhesion and spreading in vitro (Hynes, 1981). This is true for primary mesenchyme cells (Sanders, 1980), which show an enhanced rate of spreading and changed morphology in the presence of fibronectin in vitro. The same result is obtained whether the substratum is coated with fibronectin itself or whether the fibronectin is present in the form of cell exudate derived from cultured endoblast. The mesenchyme cells themselves appear to produce relatively little fibronectin (Sanders, 1980). Fibronectin-containing extracellular material from the embryo at the

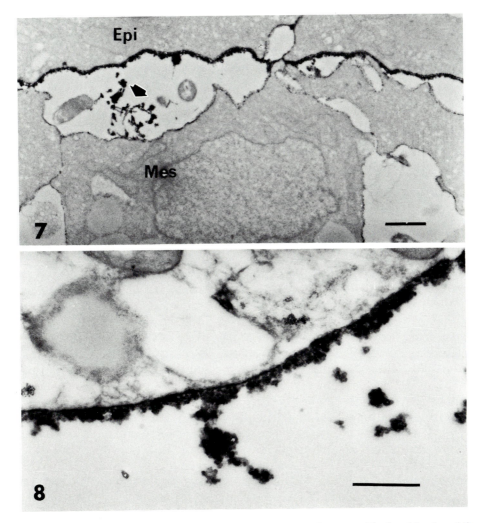

Figure 7. Fibronectin localized by ultrastructural immunocytochemistry. The basal lamina of the epiblast (Epi) is rich in fibronectin, but the mesenchyme cell surfaces (Mes) are not. The arrow indicates fibronectin associated with a group of extracellular fibrils. PAP technique. Scale bar: 1.0 μm.

Figure 8. An interstitial body staining heavily for fibronectin. PAP technique. Scale bar: 0.5 μm.

time of gastrulation can therefore influence the spreading of the mesenchyme. Whether it performs this function in the chick embryo is unknown, although experiments on amphibia have shown that the introduction of antibodies to fibronectin into the embryo prevents mesodermal migration (Boucaut *et al.*, 1984). Since no gradients of fibronectin have been discerned, it is hard to explain how directionality could be imparted to the mesenchyme by this means. It seems most likely that the cells tend to migrate away from the primitive streak, perhaps due to contact inhibition of locomotion or the high cell density, and that the fibronectin provides a hospitable surface for the attachment of filopodia or for cell spreading. Such problems of directional cell movements have recently been explored in detail (Trinkaus, 1982). Despite the unexplained problems, the presence of fibronectin represents our best clue so far as to the factors influencing this migration. The glycoprotein laminin has also been localized in the basal lamina and found to co-distribute with fibronectin (Mitrani, 1982). The significance of this substance to gastrulation is unclear and represents an important area for future work.

2.3.4. Contact Guidance

The idea that a cell's migration may be directed by the oriented topography of its substratum is a long-standing one (Weiss, 1958; Dunn, 1982) and has been shown to be effective in a number of *in vivo* situations (Trinkaus, 1976, 1982). The search for aligned extracellular matrix fibrils that could direct primary mesenchyme cell migration has been largely unsuccessful or equivocal at best (Revel, 1974; Bancroft and Bellairs, 1975; Ebendal, 1976; England and Wakely, 1977; England, 1981). The fibronectin-rich anterior band of fibrils (Critchley *et al.*, 1979) might be well placed to direct the movements of primordial germ cells, but it is not oriented appropriately to influence the mesenchyme. Radially oriented fibrils containing fibronectin (Wakely and England, 1979b) could be effective, however, but appear too sparse. The case for contact guidance as a major factor influencing primary mesenchyme directionality is not strong, especially in view of labeling studies, which show that the basal lamina is moving medially with the epiblast at this time—in the opposite direction to the outwardly migrating mesenchyme (Sanders, 1984). Examination of TEMs and SEMs shows that the mesenchyme migrates on the surfaces of both the epiblast and endoblast, but although the latter possesses fibronectin, it does not bear a basal lamina or fibrils. The cells moving on the endoblast—or those otherwise remote from the epiblast—could therefore not benefit directly from any putative contact guidance. It is conceivable that guidance imparted by the basal lamina to the more remote cells is indirect, such that the movement of cells in contact with the basal lamina is influenced, in turn influencing the remaining mesenchyme cells.

2.4. *In Vitro* Studies

The only way to observe living mesoderm cells actually moving is by means of *in vitro* culture techniques. When explanted onto glass or plastic substrata (Sanders, 1980), the primary mesenchyme cells migrate out from the explant mass with a "fibroblastic" appearance (Fig. 9). Ruffling of the leading edge is prominent in moving cells observed by time-lapse cinemicrography. Maximum outgrowth occurs by 16–24 hr, the cells having migrated radially from the explant mass forming a monolayer of largely nonoverlapping cells. When fibronectin is present on the substratum (see Section 2.3.3), the cells are able to spread within 4 hr of explanting and with a significantly different morphology (Fig. 10). The cells on such a substratum take on an epithelial appearance, forming a coherent sheet (Fig. 11). The reverse transformation, from epithelial to fibroblastic, is seen when epiblast cells are explanted (Sanders and Dickau, 1981). In this case, the epiblast cells form characteristic "domes," and fibroblastic cells resembling mesenchyme appear to migrate from the dome margins, hence there is plasticity in cell morphology both before and after passage through the primtive streak.

Cells in embryos move on other cells or on the extracellular matrix. Attempts to provide mesenchyme cells in culture with endoblast cell sheets as substrata (Sanders, 1982b) result in the penetration of the sheet by mesenchyme cells seeded from the top. The mesenchyme will not spread or locomote on the dorsal surface of the cultured endoblast as it does in vivo. Consequently, this technique cannot be used as a model for in vivo events. This result is presumably due to the difference between the dorsal surface of the endoblast sheet in vitro and its counterpart in vivo. This difference must reflect a different cell surface organization in the two situations. Indeed, endoblast in vitro shows considerable extracellular material on its ventral surface (Sanders, 1980), which is the opposite polarity from that seen in vivo.

In an effort to approach the spatial characteristics of physiological substrata, mesoderm cells have been cultured within gels of hydrated collagen (Sanders and Prasad, 1983). The optical qualities of such gels permit phase-contrast microscopy and time-lapse filming with considerable clarity. The gels consist of type I collagen, facilitating the incorporation of glycoproteins and glycosaminoglycans, which allows the investigator to reproduce some features of the in vivo matrix. In comparison with mesoderm from later stages of development, primary mesenchyme grows very poorly under these conditions, indicating a limited capacity to bind to the collagen. The addition of neither fibronectin nor hyaluronic acid improves outgrowth. As with somitic mesoderm, these cells show some suppression of the large ruffles during movement in this three-dimensional situation but retain undulations of the leading edge of the lamellipodia. Such observations tend to confirm the view that the features of cell movement in vitro (e.g., ruffling) on two-dimensional substrata may be exaggerations of normal in vivo phenomena (England and Wakely, 1979).

2.5. Galvanotaxis

Young chick embryos are able to generate extracellular pathways of electrical current by an ion-pumping mechanism located in the epiblast (Jaffe and Stern, 1979). These fields, centered near Hensen's node, leave the primitive streak from the dorsal surface of the embryo. Although in vitro attempts to demonstrate movement of mesoderm cells in response to such currents have not always been successful (Stern, 1981), more recent results indicate that galvanotaxic movements can be demonstrated using several types of embryonic cells (Nuccitelli and Erickson, 1983; Stump and Robinson, 1983; Cooper and

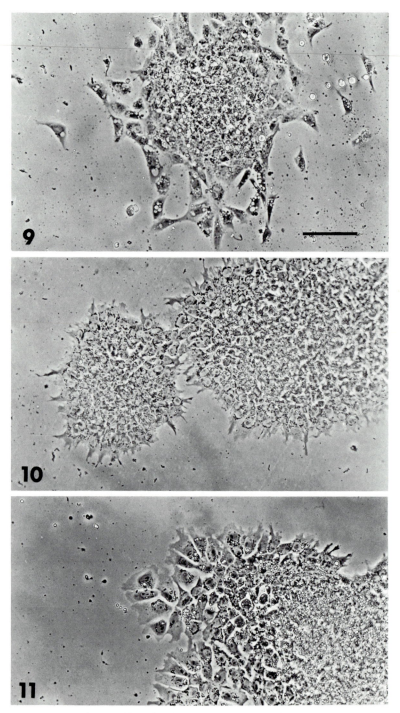

Figure 9. Stage 5 mesenchyme cells 24 hr after explanting onto a glass substratum. Scale bar: 100 μm.

Figure 10. Stage 5 mesenchyme cells 4 hr after explanting onto a fibronectin-containing substratum.

Keller, 1984; Erickson and Nuccitelli, 1984; see also Chapter 13). Evidence that mesoderm cells actually move under such influences in the embryo is still needed, however.

3. Intermediate Mesoderm and Area Vasculosa

Lateral migration of the primary mesenchyme is followed by condensation of the mesoderm into epithelial elements. Segmental plate is laid down on either side of the notochord and will give rise to somites (Section 4) of epithelial nature. Lateral to the segmental plate, cells of the intermediate mesoderm come to lie and give rise to the pronephric duct and elements of the vasculature. Farthest from the embryonic axis—and possibly the first to have invaginated—are cells of the lateral plate mesoderm, which organize into somatic and splanchnic epithelial layers enclosing the coelom.

Relatively little is known of the cell movements involved in the organization of these epithelial tissues. The migratory pathway of the nephrogenic cells has been traced by means of autoradiographic mapping (Rosenquist, 1970). In their passage through the primitive streak, these cells maintain their position between the presomitic mesoderm and lateral plate. After entering the mesoderm, they migrate anteriorly and laterally from the streak, where many participate in the development of the tubular pronephric duct.

The relationships between these tissues and the disposition of cells within them can best be appreciated by SEM (Meier, 1980); this method represents our principal source of knowledge concerning the overall reorganizations that occur at this time. By this means it has been possible to gain some feeling for the cell movement and shape changes that accompany differentiation as well as the extent of extracellular matrix development. What is most interesting, but not understood, is the manner in which the stellate mesenchyme cells transform into the frank epithelia of the organized pronephric duct and lateral plate layers. Soon after formation of the former, polarization of the cells occurs with the development of apical cell junctions and a basal lamina at the basal surface. The lateral plate cells form apical junctional complexes and interstitial bodies basally (Meier, 1980).

With the establishment of the nephrotome, gross cell migrations cease in this tissue for the time being, but cell movements probably do not. Evidence from the study of nephrogenic mesoderm *in vitro* suggests continued mobility of cells within this tissue (Armstrong and Armstrong, 1973). In fused fragments of this tissue, cells were shown to move across the interface between the fragments, resulting in a mixing of the cells. The same type of cells showed contact inhibition of locomotion in monolayer culture, but this phenomenon was clearly not fully operative in the solid tissue.

Figure 11. Stage 5 mesenchyme cells 24 hr after explanting onto a fibronectin-containing substratum. Note the epithelial appearance in comparison with that in Figure 9.

Anteriorly, mesoderm of the intermediate and lateral plate regions contributes to the precardiac mesenchyme, which undergoes elaborate migration to form the primitive tubular heart. The migrations of this mesoderm have been given careful attention by both *in vivo* time-lapse cinemicrography (DeHaan, 1963) and autoradiographic mapping (Rosenquist and DeHaan, 1966). These cells move as a coherent sheet, and the mechanism of the translocation has two components: (1) the cells are actively mobile on their endoderm substratum; and (2) they are passively carried by that substratum during its own movements in forming the foregut. The latter appears to be a unique observation and may prove to be a widespread, although neglected, mechanism of movement. De-Haan (1963) also considered that the elongated endodermal cells might influence the active migration of the precardiac mesoderm by contact guidance (see DeHaan, 1964, and Manasek, 1976, for in-depth discussion).

The splanchnic layer of the lateral plate mesoderm gives rise to the extensive vasculature of the area vasculosa and, together with the overlying somatic layer, expands radially as the embryo grows. The outward migration of the epithelial splanchnic layer and its blood vessels appears to rely partly on its interaction with the underlying endoderm, which provides a substratum for its attachment and maintains the mesoderm cells as a monolayer (Augustine, 1978, 1981). Mesodermal cells at the advancing edge of the area vasculosa, where the somatic and splanchnic layers fuse to form edge cells, are in contact with the basal lamina of the overlying epiblast and its associated interstitial bodies (Augustine, 1970, 1977). These extracellular matrix elements are rich in both glycosaminoglycans (Mayer and Packard, 1978) and fibronectin (Mayer *et al.*, 1981). Similar arguments as those used for gastrulation suggest the significance to this mesodermal migration of the presence of fibronectin in this basal lamina at this time.

4. Somitogenesis

The early cell movements associated with somite formation were broached earlier (see Section 2.2) with consideration of the organization of "somitomeres" (Meier, 1979, 1981; Meier and Jacobson, 1982). Details of the possible mechanisms of the subsequent segmentation process, which is responsible for somite formation, are discussed in Chapter 14. However, there are questions of cell movement and adhesiveness appropriate to the topic at hand.

The probability that segmentation is the result of adhesive changes among cells of the segmental plate was suggested by Rudnick (1944) but was not demonstrated until much more recently (Bellairs *et al.*, 1978a; Cheney and Lash, 1984). The finding was that cells from somites are more adhesive to one another than are cells from the unsegmented region. This difference is probably reflected in the distinct behavior of cells from these two tissues in culture (Bellairs and Portch, 1977; Bellairs *et al.*, 1980). It is not known with certainty whether the entire cell surface is involved in the adhesive change or whether, more probably, the cells become polarized in this respect. The possibility of a nonuniform adhesive change has been incorporated into a proposal that explains the cell movements or rearrangements occurring at segmentation (Bellairs, 1979). This involves the attachment of the somite-forming cells to one of the embryonic epithelia at one pole *via* extracellular materials, and their strong adhesion to each other at the luminal pole, followed by cell elongation. More recent evidence shows the importance of cell movement (or at least contraction) in this process (Chernoff and Lash, 1981; Chernoff and Hilfer, 1982; Beloussov and Naumidi, 1983; Ostrovsky et al., 1983a). Fibronectin is implicated in the adhesive changes by studies on somites *in situ* and somite cells *in vitro* (Ostrovsky et al., 1983b; Sanders and Prasad, 1983).

In vitro examination of presomitic, somitic, and other mesoderm cells has shown that each type of cell exhibits a characteristic pattern of behavior that may manifest itself as a difference in rate of spreading and locomotion or of cell shape (Bellairs and Portch, 1977; Bellairs et al., 1980). Cells from unsegmented mesoderm appear to be less adhesive to the substratum than somitic cells and show correspondingly poorer spreading and slower rate of movement. However, these characteristics may be modified by changes in the nature of the substratum. In conventional two-dimensional cultures, spreading and movement of somite cells are enhanced on basal lamina material but inhibited on hyaluronic acid-containing substrata (Fisher and Solursh, 1979a). When incorporated into three-dimensional gels of hydrated type I collagen, the spreading of cells from both segmented and unsegmented mesoderm is similar, with the latter being somewhat slower to attach (Sanders and Prasad, 1983). Under these conditions, fibronectin is necessary for optimum spreading of both types of cell, and hyaluronic acid promotes spreading of segmented but not unsegmented cells. These results show that the process of segmentation is accompanied by complex changes affecting the cell surface and cell behavior.

In some respects, somite cells behave in an unexpected manner in culture. Although they can be considered epithelial cells complete with lumenal surface, apical junctional complexes, and basement membrane (Hay, 1968), in culture they migrate as fibroblastlike cells (Bellairs et al., 1980). This is in clear distinction to other epithelia in the embryo (Sanders et al., 1978; Bellairs et al., 1978b) and even the dermomyotome (Bellairs et al., 1980). These other epithelia exhibit the characteristic coherent sheet of cells in culture. What this means is unclear; the somite cells are presumably not fully committed epithelial cells, which is highly appropriate since many of them disperse in a fibroblastic manner with the differentiation of the sclerotome.

5. Sclerotome

The sclerotome (Fig. 12) arises as a result of the disaggregation and dispersal of cells from the ventromedial side of the somites, followed by apparent migration of cells toward the notochord (Trelstad et al., 1967; Mestres and Hinrichsen, 1976). In this region they participate in chondrogenesis and the formation of the axial skeleton. The dorsal side of the somites gives rise to the epithelial layers of dermatome and myotome. Whether or not cells from the sclerotome migrate to the apical surface of the dermatome, reaggregate, and

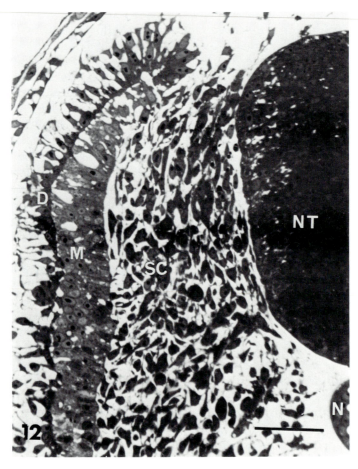

Figure 12. Section through the anterior trunk region of a stage 18 embryo. The somites have differentiated into dermatome (D), myotome (M), and sclerotome (SC). The latter occupies the space around the neural tube (NT) and notochord (N). Scale bar: 50 μm.

form myotome as suggested by Mestres and Hinrichsen (1976) is a matter of debate (Christ *et al.*, 1978). The factors responsible for the initiation of sclerotome dispersal are largely unknown but may be related to increased hyaluronic acid synthesis (Pintar, 1978).

Some of the following discussion concerns work carried out on rodent sclerotome. However, avian and rodent embryos appear very similar in this aspect of their development, and to omit the latter would be inappropriate.

5.1. Cell Movement

Before proceeding, it is necessary to consider to what extent the common assumption of sclerotome cell migration is justified. Gasser (1979) presented evidence that casts doubts on the generally accepted view. Measurements made on serial sections show that the migration of sclerotome cells from the somite to the notochord may be more apparent than real and may in fact be an illusion created by the relative movements of the surrounding tissues. After the ventromedial side of the somite breaks open, the dermomyotome moves laterally and dorsally in order to maintain its position relative to the overlying ectoderm. It is Gasser's firm conviction that the sclerotome cells are merely left behind as a trail resulting from the dermomyotome movement. Cell proliferation among the trailing cells accounts for the increasing cell density of the perinotochordal tissue. Gasser suggests that the presence of tight junctions between the sclerotome cells (Trelstad *et al.*, 1967; Flint and Ede, 1978) supports the view that migration is limited. To what extent this notion is valid is uncertain because, if the sclerotome cells do move, they do so in clusters that may not be hindered by the remnants of tight junctions from the somite stage. Other evidence shows that sclerotome dispersal is inhibited by cytochalasin D and is therefore a result of active cell movement (Chernoff and Lash, 1981; see Chapter 14). The careful examination of Gasser (1979) leads one to agree that growth movements of surrounding tissues do contribute to the appearance of sclerotome migration, but the weight of all other evidence does not permit the complete exclusion of active cell locomotion.

The dispersal of the cells from the somite may be viewed not necessarily as a mass directed migration, but as the gradual influx of cells by individual movement into areas of low cell density (Flint, 1977; Flint and Ede, 1978). The cells appear to be randomly oriented before emigration (Flint and Ede, 1978), but, by the time they reach the region of the notochord, a distinct orientation is discernable, reflecting periodic condensations of the developing vertebral bodies (Trelstad, 1977). The orientation involves both cell alignment parallel to the notochord axis as well as polarity with respect to the distribution of cell organelles. The notochord itself, by its proximity, appears to be able to influence the synthetic and differentiative potential of the sclerotome cells (Cheney and Lash, 1981). There is little overall agreement, however, as to the precise mode of migration of these mesenchymal cells *in vivo*. Some results, for example, even indicate that the sclerotome might move as a cohesive unit of cells similar to the spreading of an epithelial sheet (Fisher and Solursh, 1979b; Solursh *et al.*, 1979b). The reason for this diversity of opinion on the migratory pattern of the sclerotome is that our knowledge comes entirely from static microscopical images. The controversy will not be resolved until we can observe the cells moving *in vivo*.

In vitro investigation has shown that sclerotome cells are more successful at substratum attachment and migration than are cells from earlier mesodermal populations (Bellairs *et al.*, 1980; Sanders and Prasad, 1983). Sclerotome produces large outgrowths of individual cells in collagen gels (Figs. 13 and 15), which are able to attach to and pull the collagen fibers into alignment (Figs. 14 and 16)—a characteristic not observed with primary mesenchyme cells, which attach only very sparsely. It is not known whether the acquisition of this affinity for collagen reflects the development of a specific attachment mechanism or is simply due to an increase in general adhesiveness. The addition of fibronectin or hyaluronic acid has no detectable effect on sclerotome outgrowth on

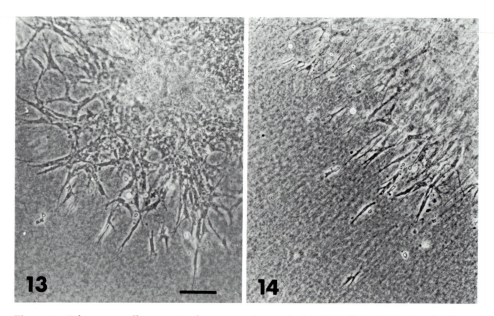

Figure 13. Sclerotome cells migrating from an explant embedded in a three-dimensional collagen gel. Scale bar: 100 μm.

Figure 14. Similar to Figure 13. Alignment by the explant of the fibrils of the gel is shown. Note the diagonal direction, from bottom left to top right. The cells are elongated and aligned in the direction of the collagen fibrils.

this substratum. Direct time-lapse observation of individual sclerotome cells indicates that cell-to-cell contact in this situation is not accompanied by contact paralysis of the undulating leading edge.

5.2. The Extracellular Matrix

The extracellular matrix surrounding sclerotome cells is of interest for two reasons: (1) since sclerotome is a population of moving cells, the matrix is of importance as a substratum and is even able to encourage migration in otherwise nonmigratory cells (Fisher and Solursh, 1979b); and (2) the matrix laid down by the notochord, neural tube, and sclerotome in this region is a prelude to the major process of chondrogenesis that follows. Indeed, the presence of certain of the matrix components promotes the differentiation of chondrocytes.

Matrix fibrils and the basal laminae of the notochord and neural tube are apparently substrata for the movement of sclerotome cells (M. Jacob *et al.*, 1975; Bancroft and Bellairs, 1976); in at least one study, aligned fibrils have been implicated in determining sclerotome directionality by contact guidance (Ebendal, 1977). Glycosaminoglycans are abundant in the axial region at this time (Kvist and Finnegan, 1970; O'Hare, 1973; Abrahamsohn *et al.*, 1975) and,

Figure 15. Scanning electron micrograph of sclerotome cells embedded in a collagen gel. The cells were exposed by breaking open the gel. Scale bar : 10 μm.

Figure 16. Alignment of collagen fibrils (arrow) by a sclerotome cell (see also Fig. 14). Scale bar: 10 μm.

as with gastrulation and initiation of neural crest cell migration (see Chapter 13), the presence of hyaluronic acid has been implicated in the facilitation of sclerotome dispersal by opening up tissue spaces (Solursh et al., 1979b). This glycosaminoglycan has been suggested to mediate the initial migration of the sclerotome (Pintar, 1978). The relatively nonspecific staining of ruthenium red also suggests the widespread occurrence of glycosaminoglycans in the form of proteoglycan granules both in the matrix and attached to the cell surfaces (Hay, 1978). Associated with the proteoglycans, in both interstitial bodies and extracellular fibrils, is fibronectin (Mayer et al., 1981). This glycoprotein is found in the basal laminae of the notochord and neural tube and is also distributed throughout the sclerotome at this time (Newgreen and Thiery, 1980; Thiery et al., 1982). Its function here may be to mediate the attachment of sclerotome and/or neural crest cells to collagen fibrils. As with gastrulation, however, this functional association is a circumstantial one based on its temporal and spatial relationship to sclerotome migration and on its known in vitro effects on cell spreading.

The greatest difference between the extracellular matrices of the primary mesenchyme and the sclerotome is in the increased appearance of collagen. This is reflected in the relative ease with which fibrillar elements may be demonstrated by SEM and TEM in the region of the dispersing sclerotome cells (Frederickson and Low, 1971; Cohen and Hay, 1971; Bancroft and Bellairs, 1976; Frederickson et al., 1977). Type I collagen is present in the perinotochordal region, and type II (cartilage) collagen is synthesized by the sclerotome cells themselves during their dispersal (von der Mark et al., 1976). This topic has been covered in detail by Hay (1981b), including consideration of collagen types present, their role in stimulating chondrogenesis, and the interactions of collagen with migrating cells. Although direct attachment of the cells to collagen might be possible, adhesion is usually thought to be mediated via an intermediate glycoprotein such as fibronectin, laminin, or chondronectin.

At this stage the importance of the matrix goes beyond that of supplying a substratum for cell movement. Its widespread involvement in regulating differentiation is nowhere as clear as in the interaction between sclerotome cells and the perinotochordal matrix. The result of the interaction is the stimulation of chondrogenesis in cells that are now called chondrocytes (Toole, 1972; Minor, 1973; Hall, 1977; Lash and Vasan, 1977; 1978; Belsky et al., 1980).

6. Somitic Contribution to Limb Bud Mesenchyme

At about the same time that the somite breaks open ventromedially to release sclerotome cells, certain other cells of the somite begin to migrate as well. At the levels of the prospective wing and leg buds, relatively undifferentiated cells leave the region of the dermomyotome and migrate to the mesenchymal region of the swellings that constitute the limb buds. A great deal of ablation and transplantation work has been done to show this contribution from the somitic mesoderm to the limb musculature (see, e.g., Chevallier et al.,

1977; Christ et al., 1977; Beresford, 1983). Determination of somitic cells as either chondrogenic (sclerotome) or myogenic appears to occur in the segmental plate before the formation of the somites, although the prospective sclerotomes also retain the ability to give striated muscle for some time (Wachtler et al., 1982).

The migratory pathway of the myogenic stem cells from the dermomyotome into the limb region is in the space between the somite and the somatopleure of the lateral plate. Here, the cells of somitic origin join and intermingle with those of lateral plate origin to constitute the limb bud mesenchyme. A great deal is known about the cell interactions and movements within the mesenchyme of the developing limb bud itself (Ede, 1976), but these are outside the scope of this chapter.

The cells of somitic origin apparently migrate singly and in groups and are identifiable by their lamellipodia and other processes (M. Jacob et al., 1978). Extracellular fibrils are identifiable in the pathway, as are glycosaminoglycans (Kvist and Finnegan, 1970). The former are oriented in a mediolateral direction, as are the cells themselves, and they may thus guide the cells (M. Jacob et al., 1978).

Although less attention has been given to the mechanism and pathway of this migration, the situation is analogous to those detailed earlier. It is, however, more complex, since the pathway is less well defined, and the migrating cells intermix with other populations. Much of the effort of investigators working on this aspect of mesoderm differentiation has thus far gone into the difficult task of proving the existence of this migration. We can now expect that the details of the process will be explored.

7. Concluding Remarks

During the past 15 years, considerable information has been accumulated, which, taken together, represents significant advances in defining our concept of the factors influencing cell migrations. It is nevertheless important to reiterate the caveat expressed in the Introduction with regard to the limitations of the techniques currently at our disposal. Scanning electron microscopy and immunocytochemistry are undoubtedly powerful and persuasive techniques, and for these very reasons it is vital to maintain a balanced view of the results and avoid overinterpretation. Scanning electron microscopy cannot reveal anything directly about cell movement. Immunocytochemistry alone cannot reveal anything about cause and effect relationships. In vitro studies are limited by the artificiality of the system and the destruction of the integrity of the tissue. To point this out is not to dampen enthusiasm but to maintain a realistic perspective. On the positive side, the combined results from these and other techniques have provided valuable insights into the mechanisms controlling various aspects of cell movement. Thus we know a great deal about cell–cell and cell–substratum interactions, cell surface and adhesive changes, and the composition of the extracellular matrix.

We still do not know the nature of the factors responsible for initiating and stopping migration or how our current knowledge can be used to explain the directionality of migration, but we do have some testable ideas. Most importantly, we need techniques that facilitate direct examination and modification of cell movements *in vivo*. It may be that this will only be possible in a few selected tissues, of which the chick embryo may not be one. Even if this is so, the work described in this chapter will provide a firm base from which to proceed to any other system of migrating cells.

ACKNOWLEDGMENT.   The author's work cited in this chapter was supported by grants from the Medical Research Council of Canada.

References

Abercrombie, M., 1937, The behaviour of epiblast grafts beneath the primitive streak of the chick, *J. Exp. Biol.* **14:**302–318.

Abercrombie, M., Heaysman, J. E. M., and Pegrum, S. M., 1970, The locomotion of fibroblasts in culture. II. "Ruffling," *Exp. Cell Res.* **60:**437–444.

Abrahamsohn, P. A., Lash, J. W., Kosher, R. A., and Minor, R. R., 1975, The ubiquitous occurrence of chondroitin sulfates in chick embryos, *J. Exp. Zool.* **194:**511–518.

Armstrong, P. B., and Armstrong, M. T., 1973, Are cells in solid tissues immobile? Mesonephric mesenchyme studied *in vitro*, *Dev. Biol.* **35:**187–209.

Augustine, J. M., 1970, Expansion of the area vasculosa of the chick after removal of the ectoderm, *J. Embryol. Exp. Morphol.* **24:**95–108.

Augustine, J. M., 1977, Mesodermal expansion after arrest of the edge in the area vasculosa of the chick, *J. Embryol. Exp. Morphol.* **41:**175–188.

Augustine, J. M., 1978, Changes in area and thickness of mesodermal explants from the area vasculosa of the chick on three substrata, *Anat. Rec.* **190:**329.

Augustine, J. M., 1981, Influence of the entoderm on mesodermal expansion in the area vasculosa of the chick, *J. Embryol. Exp. Morphol.* **65:**89–103.

Balinsky, B. I., 1975, *An Introduction to Embryology*, 4th. ed., W. B. Saunders, Philadelphia.

Balinsky, B. I., and Walther, H., 1961, The immigration of presumptive mesoblast from the primitive streak in the chick as studied with the electron microscope, *Acta Embryol. Morphol. Exp.* **4:**261–283.

Bancroft, M., and Bellairs, R., 1974, The onset of differentiation in the epiblast of the chick blastoderm (SEM and TEM), *Cell Tissue Res.* **155:**399–418.

Bancroft, M., and Bellairs, R., 1975, Differentiation of the neural plate and neural tube in the young chick embryo, *Anat. Embryol.* **147:**309–335.

Bancroft, M., and Bellairs, R., 1976, The development of the notochord in the chick embryo, studied by scanning and transmission electron microscopy, *J. Embryol. Exp. Morphol.* **35:**383–401.

Bellairs, R., 1957, Studies on the development of the foregut in the chick embryo. IV. Mesodermal induction and mitosis, *J. Embryol. Exp. Morphol.* **5:**340–350.

Bellairs, R., 1971, *Developmental Processes in Higher Vertebrates*, Logos Press, London.

Bellairs, R., 1979, The mechanism of somite segmentation in the chick embryo, *J. Embryol. Exp. Morphol.* **51:**227–243.

Bellairs, R., 1982, Gastrulation processes in the chick embryo, in: *Cell Behaviour* (R. Bellairs, A. Curtis, and G. Dunn, eds.), pp. 395–427, Cambridge University Press, Cambridge.

Bellairs, R., and Portch, P. A., 1977, Somite formation in the chick embryo, in: *Vertebrate Limb and Somite Morphogenesis* (D. A. Ede, J. R. Hinchliffe, and M. Balls, eds.), pp. 449–463, Cambridge University Press, Cambridge.

Bellairs, R., Curtis, A. S. G., and Sanders, E. J., 1978a, Cell adhesiveness and embryonic differentiation, *J. Embryol. Exp. Morphol.* **46:**207–213.

Bellairs, R., Sanders, E. J., and Portch, P. A., 1978b, In vitro studies on the development of neural and ectodermal cells from young chick embryos, *Zoon* **6:**39–50.

Bellairs, R., Sanders, E. J., and Portch, P. A., 1980. Behavioural properties of chick somitic mesoderm and lateral plate when explanted in vitro, *J. Embryol. Exp. Morphol.* **56:**41–58.

Beloussov, L. V., and Naumidi, I. I., 1983, Cell contacts and rearrangements preceding somitogenesis in chick embryo, *Cell Diff.* **12:**191–204.

Belsky, E., Vasan, N. S., and Lash, J. W., 1980, Extracellular matrix components and somite chondrogenesis: A microscopic analysis, *Dev. Biol.* **79:**159–180.

Bennett, M. V. L., 1973, Function of electrotonic junctions in embryonic and adult tissues, *Fed. Proc.* **32:**65–75.

Beresford, B., 1983, Brachial muscles in the chick embryo: The fate of individual somites, *J. Embryol. Exp. Morphol.* **77:**99–116.

Boucaut, J. C., Darribère, T., Boulekbache, H., and Thiery, J. P., 1984, Prevention of gastrulation but not neurulation by antibodies to fibronectin in amphibian embryos, *Nature (Lond.)* **307:**364–367.

Cheney, C. M., and Lash, J. W., 1981, Diversification within embryonic chick somites: Differential response to notochord, *Dev. Biol.* **81:**288–298.

Cheney, C. M., and Lash, J. W., 1984, An increase in cell–cell adhesion in the chick segmental plate results in a meristic pattern, *J. Embryol. Exp. Morphol.* **79:**1–10.

Chernoff, E. A. G., and Hilfer, S. R., 1982, Calcium dependence and contraction in somite formation, *Tissue Cell* **14:**435–449.

Chernoff, E. A. G., and Lash, J. W., 1981, Cell movement in somite formation and development in the chick: Inhibition of segmentation, *Dev. Biol.* **87:**212–219.

Chevallier, A., Kieny, M., Mauger, A., and Sengel, P., 1977, Developmental fate of the somitic mesoderm in the chick embryo, in: *Vertebrate Limb and Somite Morphogenesis* (D. A. Ede, J. R. Hinchliffe, and M. Balls, eds.), pp. 421–432, Cambridge University Press, Cambridge.

Christ, B., Jacob, H. J., and Jacob, M., 1977, Experimental analysis of the origin of the wing musculature in avian embryos, *Anat. Embryol.* **150:**171–186.

Christ, B., Jacob, H. J., and Jacob, M., 1978, On the formation of the myotomes in avian embryos. An experimental and scanning electron microscope study, *Experientia* **34:**514–516.

Cohen, A. M., and Hay, E. D., 1971, Secretion of collagen by embryonic neuroepithelium at the time of spinal cord–somite interaction, *Dev. Biol.* **26:**578–605.

Cooper, M. S., and Keller, R. E., 1984, Perpendicular orientation and directional migration of amphibian neural crest cells in d.c. electrical fields, *Proc. Natl. Acad. Sci. USA* **81:**160–164.

Critchley, D. R., England, M. A., Wakely, J., and Hynes, R. O., 1979, Distribution of fibronectin in the ectoderm of gastrulating chick embryos, *Nature (Lond.)* **280:**498–500.

DeHaan, R. L., 1963, Migration patterns of the precardiac mesoderm in the early chick embryo, *Exp. Cell Res.* **29:**544–560.

DeHaan, R. L., 1964, Cell interactions and oriented movements during development, *J. Exp. Zool.* **157:**127–138.

Duband, J. L., and Thiery, J. P., 1982, Appearance and distribution of fibronectin during chick embryo gastrulation and neurulation, *Dev. Biol.* **94:**337–350.

Dunn, G. A., 1982, Contact guidance of cultured tissue cells: a survey of potentially relevant properties of the substratum, in: *Cell Behaviour* (R. Bellairs, A. Curtis, and G. Dunn, eds.), pp. 247–280, Cambridge University Press, Cambridge.

Ebendal, T., 1976, Migratory mesoblast cells in the young chick embryo examined by scanning electron microscopy, *Zoon* **4:**101–108.

Ebendal, T., 1977, Extracellular matrix fibrils and cell contacts in the chick embryo, *Cell Tissue Res.* **175:**439–458.

Ede, D. A., 1976, Cell interactions in vertebrate limb development, in: *The Cell Surface in Animal Embryogenesis and Development* (G. Poste and G. L. Nicolson, eds.), pp. 495–543, Elsevier/North-Holland, Amsterdam.

Edelman, G. M., Gallin, W. J., Delouvée, A., Cunningham, B. A., and Thiery, J. P., 1983, Early epochal maps of two different cell adhesion molecules, *Proc. Natl. Acad. Sci. USA* **80:**4384–4388.

England, M. A., 1981, Applications of the SEM to the analysis of morphogenetic events, *J. Microsc.* **123:**133–146.

England, M. A., and Wakely, J., 1977, Scanning electron microscopy of the development of the mesoderm layer in chick embryos, *Anat. Embryol.* **150:**291–300.

England, M. A., and Wakely, J., 1979, Evidence for changes in cell shape from a 2-dimensional to a 3-dimensional substrata, *Experientia* **35:**664–666.

Erickson, C. A., and Nuccitelli, R., 1984, Embryonic fibroblast motility and orientation can be influenced by physiological electric fields, *J. Cell Biol.* **98:**296–307.

Fisher, M., and Solursh, M., 1977, Glycosaminoglycan localization and role in maintenance of tissue spaces in the early chick embryo, *J. Embryol. Exp. Morphol.* **42:**195–207.

Fisher, M., and Solursh, M., 1979a, The influence of the substratum on mesenchyme spreading *in vitro*, *Exp. Cell Res.* **123:**1–14.

Fisher, M., and Solursh, M., 1979b, The influence of local environment on the organization of mesenchyme cells, *J. Embryol. Exp. Morphol.* **49:**295–306.

Flint, O. P., 1977, Cell interactions in the developing axial skeleton in normal and mutant mouse embryos, in: *Vertebrate Limb and Somite Morphogenesis* (D. A. Ede, J. R. Hinchliffe, and M. Balls, eds.), pp. 465–484, Cambridge University Press, Cambridge.

Flint, O. P., and Ede, D. A., 1978, Cell interactions in the developing somite. *In vivo* comparisons between amputated (am/am) and normal mouse embryos, *J. Embryol. Exp. Morphol.* **31:**275–291.

Frederickson, R. G., and Low, F. N., 1971, The fine structure of perinotochordal microfibrils in control and enzyme-treated chick embryos, *Am. J. Anat.* **130:**347–376.

Frederickson, R. G., Morse, D. E., and Low, F. N., 1977, High-voltage electron microscopy of extracellular fibrillogenesis, *Am. J. Anat.* **150:**1–34.

Gasser, R. F., 1979, Evidence that sclerotomal cells do not migrate medially during normal embryonic development of the rat, *Am. J. Anat.* **154:**509–524.

Gilula, N. B., 1980, Cell-to-cell communication and development, in: *The Cell Surface: Mediator of Developmental Processes* (S. Subtelny and N. K. Wessels, eds.), pp. 23–41, Academic Press, New York.

Granholm, N. H., and Baker, J. R., 1970, Cytoplasmic microtubules and the mechanism of avian gastrulation, *Dev. Biol.* **23:**563–584.

Hall, B. K., 1977, Chondrogenesis of the somitic mesoderm, *Adv. Anat.* **53(4):**1–47.

Hamburger, V., and Hamilton, H., 1951, A series of normal stages in the development of the chick embryo, *J. Morphol.* **88:**49–92.

Hara, K., 1978, "Spemann's organizer" in birds, in: *Organizer—A Milestone of a Half-Century from Spemann* (O. Nakamura and S. Toivonen, eds.), pp. 221–265, Elsevier/North-Holland, Amsterdam.

Harris, A. K., 1974, Contact inhibition of cell locomotion, in: *Cell Communication* (R. P. Cox, ed.), pp. 147–185, Wiley, New York.

Harris, H. L., and Zalik, S. E., 1974, Studies on the surface of chick blastoderm cells. III. Calcium binding to ionogenic sites, *J. Cell Physiol.* **83:**359–368.

Harris, H. L., and Zalik, S. E., 1977, The binding of cations to the surfaces of cells from early chick blastoderms, *Differentiation* **7:**83–88.

Harrisson, F., Vanroelen, C., Foidart, J-M., and Vakaet, L., 1984, Expression of different regional patterns of fibronectin immunoreactivity during mesoblast formation in the chick blastoderm, *Dev. Biol.* **101:**373–381.

Hay, E. D., 1968, Organization and fine structure of epithelium and mesenchyme in the developing chick embryo, in: *Epithelial-Mesenchymal Interactions* (R. Fleischmajer and R. E. Billingham, eds.), pp. 31–55, Williams & Wilkins, Baltimore.

Hay, E. D., 1978, Fine structure of embryonic matrices and their relation to the cell surface in ruthenium red-fixed tissues, *Growth* **42:**399–423.

Hay, E. D., 1981a, Extracellular matrix, *J. Cell Biol.* **91:**205s–223s.

Hay, E. D., 1981b, Collagen and embryonic development, in: *Cell Biology of Extracellular Matrix* (E. D. Hay, ed.), pp. 379–409, Plenum Press, New York.

Heaysman, J. E. M., 1978, Contact inhibition of locomotion: A reappraisal, *Int. Rev. Cytol.* **55**:49–66.

Hook, S. L., and Sanders, E. J., 1977, Concanavalin A-binding by cells of the early chick embryo, *J. Cell Physiol.* **93**:57–68.

Hynes, R. O., 1981, Fibronectin and its relation to cellular structure and behaviour, in: *Cell Biology of Extracellular Matrix* (E. D. Hay, ed.), pp. 295–334, Plenum Press, New York.

Jacob, H. J., Christ, B., Jacob, M., and Bijvank, G. J., 1974, Scanning electron microscope (SEM) studies on the epiblast of young chick embryos, *Z. Anat. Entwickl.* **143**:205–214.

Jacob, M., Jacob, H. J., and Christ, B., 1975, Die frühe Differenzierung des chordanahen Bindegewebes. Raster—und transmissionselektronenmikroskopische Untersuchungen an Hühnerembryonen, *Experientia* **31**:1083–1086.

Jacob, M., Christ, B., and Jacob, H. J., 1978, On the migration of myogenic stem cells into the prospective wing region of chick embryos, *Anat. Embryol.* **153**:179–193.

Jacobson, W., 1938, The early development of the avian embryo. II. Mesoderm formation and the distribution of presumptive embryonic material, *J. Morphol.* **62**:445–502.

Jaffe, L. F., and Stern, C. D., 1979, Strong electrical currents leave the primitive streak of chick embryos, *Science* **206**:569–571.

Kvist, T. N., and Finnegan, C. V., 1970, The distribution of glycosaminoglycans in the axial region of the developing chick embryo. I. Histochemical analysis, *J. Exp. Zool.* **175**:221–240.

Lash, J. W., and Vasan, N. S., 1977, Tissue interactions and extracellular matrix components, in: *Cell and Tissue Interactions* (J. W. Lash and M. M. Burger, eds.), pp. 101–113, Raven Press, New York.

Lash, J. W., and Vasan, N. S., 1978, Somite chondrogenesis *in vitro*. Stimulation by exogenous extracellular matrix components, *Dev. Biol.* **66**:151–171.

Lo, C. W., 1980, Gap junctions and development, in: *Development in Mammals*, Vol. 4 (M. H. Johnson, ed.), pp. 39–80. Elsevier/North-Holland, Amsterdam.

Low, F. N., 1967, Developing boundary (basement) membranes in the chick embryo, *Anat. Rec.* **159**:231–238.

Low, F. N., 1968, Extracellular connective tissue fibrils in the chick embryo, *Anat. Rec.* **160**:93–108.

Low, F. N., 1970, Interstitial bodies in the early chick embryo, *Am. J. Anat.* **128**:45–56.

MacLean, I. M., and Sanders, E. J., 1983, Cationized ferritin and phosvitin uptake by coated vesicles of the early chick embryo, *Anat. Embryol.* **166**:385–397.

Manasek, F. J., 1975, The extracellular matrix: A dynamic component of the developing embryo, *Curr. Top. Dev. Biol.* **10**:35–102.

Manasek, F. J., 1976, Heart development: interactions involved in cardiac morphogenesis, in: *The Cell Surface in Animal Embryogenesis and Development* (G. Poste and G. L. Nicolson, eds.), pp. 545–598, Elsevier/North-Holland, Amsterdam.

Martinez-Palomo, A., 1970, The surface coats of animal cells, *Int. Rev. Cytol.* **29**:29–75.

Mayer, B. W., and Packard, D. S., 1978, A study of the expansion of the chick area vasculosa, *Dev. Biol.* **63**:335–351.

Mayer, B. W., Hay, E. D., and Hynes, R. O., 1981, Immunocytochemical localization of fibronectin in embryonic chick trunk and area vasculosa, *Dev. Biol.* **82**:267–286.

Meier, S., 1979, Development of the chick embryo mesoblast. Formation of the embryonic axis and establishment of the metameric pattern, *Dev. Biol.* **73**:25–45.

Meier, S., 1980, Development of the chick embryo mesoblast: Pronephros, lateral plate, and early vasculature, *J. Embryol. Exp. Morphol.* **55**:291–306.

Meier, S., 1981, Development of the chick embryo mesoblast: Morphogenesis of the prechordal plate and cranial segments, *Dev. Biol.* **83**:49–61.

Meier, S., and Jacobson, A. G., 1982, Experimental studies of the origin and expression of metameric pattern in the chick embryo, *J. Exp. Zool.* **219**:217–232.

Mestres, P., and Hinrichsen, K., 1974, The cell coat in the early chick embryo, *Anat. Embryol.* **146**:181–192.

Mestres, P., and Hinrichsen, K., 1976, Zur Histogenese des Somiten beim Hühnchen, *J. Embryol. Exp. Morphol.* **36**:669–683.

Minor, R. R., 1973, Somite chondrogenesis. A structural analysis, *J. Cell Biol.* **56**:27–50.

Mitrani, E., 1982, Primitive streak forming cells of the chick invaginate through a basement membrane, *Wilhelm Roux Arch.* **191**:320–324.

Mitrani, E., and Farberov, A., 1982, Fibronectin expression during the processes leading to axis formation in the chick embryo, *Dev. Biol.* **91**:197–201.

Newgreen, D., and Thiery, J. P., 1980, Fibronectin in early avian embryos: Synthesis and distribution along the migration pathways of neural crest cells, *Cell Tissue Res.* **211**:269–291.

Nicolet, G., 1971, Avian gastrulation, *Adv. Morphogen.* **9**:231–262.

Nuccitelli, R., and Erickson, C. A., 1983, Embryonic cell motility can be guided by physiological electric fields, *Exp. Cell Res.* **147**:195–201.

O'Hare, M. J., 1973, A histochemical study of sulphated glycosaminoglycans associated with the somites of the chick embryo, *J. Embryol. Exp. Morphol.* **29**:197–208.

Ostrovsky, D., Sanger, J. W., and Lash, J. W., 1983a, Light microscopic observations on actin distribution during morphogenetic movements in the chick embryo, *J. Embryol. Exp. Morphol.* **78**:23–32.

Ostrovsky, D., Cheney, C. M., Seitz, A. W., and Lash, J. W., 1983b, Fibronectin distribution during somitogenesis in the chick embryo, *Cell Diff.* **13**:217–223.

Pasteels, J., 1937, Études sur la gastrulation des Vertébrés méroblastiques. III. Oiseaux. IV. Conclusions générales, *Arch. Biol. (Paris)* **48**:381–488.

Pintar, J. E., 1978, Distribution and synthesis of glycosaminoglycans during quail neural crest morphogenesis, *Dev. Biol.* **67**:444–464.

Powers, R. D., and Tupper, J. T., 1977, Intercellular communication in the early embryo, in: *Intercellular Communication* (W. C. DeMello, ed.), pp. 231–251, Plenum Press, New York.

Revel, J.-P., 1974, Some aspects of cellular interactions in development, in: *The Cell Surface in Development* (A. A. Moscona, ed.), pp. 51–65, Wiley, New York.

Revel, J.-P., Yip, P., and Chang, L.L., 1973, Cell junctions in the early chick embryo—A freeze etch study, *Dev. Biol.* **35**:302–317.

Rosenquist, G. C., 1966, A radioautographic study of labeled grafts in the chick blastoderm. Development from primitive-streak stages to stage 12, *Carnegie Inst. Wash. Publ. 625 Contrib. Embryol.* **38**:71–110.

Rosenquist, G. C., 1970, The origin and movement of nephrogenic cells in the chick embryo as determined by radioautographic mapping, *J. Embryol. Exp. Morphol.* **24**:367–380.

Rosenquist, G. C., and DeHaan, R. L., 1966, Migration of precardiac cells in the chick embryo: A radioautographic study. *Carnegie Inst. Wash. Publ. 625 Contrib. Embryol.* **38**:111–121.

Rudnick, D., 1944, Early history and mechanics of the chick blastoderm, *Q. Rev. Biol.* **19**:187–212.

Sanders, E. J., 1979, Development of the basal lamina and extracellular materials in the early chick embryo, *Cell Tissue Res.* **198**:527–537.

Sanders, E. J., 1980, The effect of fibronectin and substratum-attached material on the spreading of chick embryo mesoderm cells in vitro, *J. Cell Sci.* **44**:225–242.

Sanders, E. J., 1982a, Ultrastructural immunocytochemical localization of fibronectin in the early chick embryo. *J. Embryol. Exp. Morphol.* **71**:155–170.

Sanders, E. J., 1982b, Inability of mesoderm cells to locomote on the modified free surface of epithelial cell sheets in vitro, *In Vitro* **18**:71–78.

Sanders, E. J., 1983, Recent progress towards understanding the roles of the basement membrane in development, *Can. J. Biochem. Cell Biol.* **61**:949–956.

Sanders, E. J., 1984, Labelling of basement membrane constituents in the living chick embryo during gastrulation, *J. Embryol. Exp. Morphol.* **79**:113–123.

Sanders, E. J., and Anderson, A. R., 1979, Ultrastructural localization of wheat germ agglutinin-binding sites on surfaces of chick embryo cells during early differentiation, *J. Cell Physiol.* **99**:107–124.

Sanders, E. J., and Dickau, J. E., 1981, Morphological differentiation of an embryonic epithelium in culture, *Cell Tissue Res.* **220**:539–548.

Sanders, E. J., and Prasad, S., 1983, The culture of chick embryo mesoderm cells in hydrated collagen gels, *J. Exp. Zool.* **226**:81–92.

Sanders, E. J., and Zalik, S. E., 1970, The occurrence of microtubules in the pre-streak chick embryo, *Protoplasma* **71**:203–208.

Sanders, E. J., and Zalik, S. E., 1972, Studies on the surface of chick blastoderm cells. II. Electron microscopy of surface binding characteristics, *J. Cell Physiol.* **79**:235–248.

Sanders, E. J., Bellairs, R., and Portch, P. A., 1978, *In vivo* and *in vitro* studies on the hypoblast and definitive endoblast of avian embryos, *J. Embryol. Exp. Morphol.* **46**:187–205.

Sheridan, J. D., 1968, Electrophysiological evidence for low-resistance intercellular junctions in the early chick embryo, *J. Cell Biol.* **37**:650–659.

Sheridan, J. D., 1976, Cell coupling and cell communication during embryogenesis, in: *The Cell Surface in Animal Embryogenesis and Development* (G. Poste and G. L. Nicolson, eds.), pp. 409–447, Elsevier/North-Holland, Amsterdam.

Shur, B. D., 1977a, Cell-surface glycosyltransferases in gastrulating chick embryos. I. Temporally and spacially specific patterns of four endogenous glycosyltransferase activities, *Dev. Biol.* **58**:23–39.

Shur, B. D., 1977b, Cell-surface glycosyltransferases in gastrulating chick embryos. II. Biochemical evidence for a surface localization of endogenous glycosyltransferase activities, *Dev. Biol.* **58**:40–55.

Solursh, M., 1976, Glycosaminoglycan synthesis in the chick gastrula, *Dev. Biol.* **50**:525–530.

Solursh, M., and Revel, J-P., 1978, A scanning electron microscope study of cell shape and cell appendages in the primitive streak region of the rat and chick embryo, *Differentiation* **11**:185–190.

Solursh, M., Fisher, M., and Singley, C. T., 1979a, The synthesis of hyaluronic acid by ectoderm during early organogenesis in the chick embryo, *Differentiation* **14**:77–85.

Solursh, M., Fisher, M., Meier, S., and Singley, C. T., 1979b, The role of extracellular matrix in the formation of the sclerotome, *J. Embryol. Exp. Morphol.* **54**:75–98.

Stern, C. D., 1979, A re-examination of mitotic activity in the early chick embryo, *Anat. Embryol.* **156**:319–329.

Stern, C. D., 1981, Behaviour and motility of cultured chick mesoderm cells in steady electrical fields, *Exp. Cell Res.* **136**:343–350.

Stern, C. D., and Ireland, G. W., 1981, An integrated experimental study of endoderm formation in avian embryos, *Anat. Embryol.* **163**:245–263.

Stump, R. F., and Robinson, K. R., 1983, *Xenopus* neural crest cell migration in an applied electrical field, *J. Cell Biol.* **97**:1226–1233.

Thiery, J. P., Duband, J. L., and Delouvée, A., 1982, Pathways and mechanisms of avian trunk neural crest cell migration and localization, *Dev. Biol.* **93**:324–343.

Toole, B. P., 1972, Hyaluronate turnover during chondrogenesis in the developing chick limb and axial skeleton, *Dev. Biol.* **29**:321–329.

Toole, B. P., 1973, Hyaluronate and hyaluronidase in morphogenesis and differentiation, *Am. Zool.* **13**:1061–1065.

Toole, B. P., 1981, Glycosaminoglycans in morphogenesis, in: *Cell Biology of Extracellular Matrix*, (E. D. Hay, ed.), pp. 259–294, Plenum Press, New York.

Trelstad, R. L., 1977, Mesenchymal cell polarity and morphogenesis of chick cartilage, *Dev. Biol.* **59**:153–163.

Trelstad, R. L., Hay, E. D., and Revel, J.-P., 1967, Cell contact during early morphogenesis in the chick embryo, *Dev. Biol.* **16**:78–106.

Trinkaus, J. P., 1976, On the mechanism of metazoan cell movements, in: *The Cell Surface in Animal Embryogenesis and Development* (G. Poste and G. L. Nicolson, eds.), pp. 225–329, Elsevier/North-Holland, Amsterdam.

Trinkaus, J. P., 1982, Some thoughts on directional cell movement during morphogenesis, in: *Cell Behaviour* (R. Bellairs, A. Curtis, and G. Dunn, eds.), pp. 471–498, Cambridge University Press, Cambridge.

Triplett, R. L., and Meier, S., 1982, Morphological analysis of the development of the primary organizer in avian embryos, *J. Exp. Zool.* **220**:191–206.

Vakaet, L., 1970, Cinephotomicrographic investigations of gastrulation in the chick blastoderm, *Arch. Biol. (Liège)* **81**:387–426.

Vakaet, L., Vanroelen, C., and Andries, L., 1980, An embryological model of non-malignant invasion or ingression, in: *Cell Movement and Neoplasia* (M. de Brabander, M. Mareel, and L. de Ridder, eds.), pp. 65–75, Pergamon Press, Oxford.

Vanroelen, C., Vakaet, L., and Andries, L., 1980a, Localization and characterization of acid mucopolysaccharides in the early chick blastoderm, *J. Embryol. Exp. Morphol.* **56:**169–178.

Vanroelen, C., Vakaet, L., and Andries, L., 1980b, Distribution and turnover of testicular hyaluronidase sensitive macromolecules in the primitive streak stage chick blastoderm as revealed by autoradiograph, *Anat. Embryol.* **159:**361–367.

Vanroelen, C., Vakaet, L., and Andries, L., 1980c, Alcian blue staining during the formation of mesoblast in the primitive streak stage chick blastoderm, *Anat. Embryol.* **160:**361–367.

von der Mark, H., von der Mark, K., and Gay, S., 1976, Study of differential collagen synthesis during development of the chick embryo by immunofluorescence. I. Preparation of collagen type I and type II specific antibodies and their application to early stages of the chick embryo, *Dev. Biol.* **48:**237–249.

Wachtler, F., Christ, B., and Jacob, H. J., 1982, Grafting experiments on determination and migratory behaviour of presomitic, somitic and somatopleural cells in avian embryos, *Anat. Embryol.* **164:**369–378.

Waddington, C. H., and Taylor, J., 1937, Conversion of presumptive ectoderm to mesoderm in the chick, *J. Exp. Biol.* **14:**335–339.

Wakely, J., and England, M. A., 1977, Scanning electron microscopy (SEM) of the chick embryo primitive streak, *Differentiation* **7:**181–186.

Wakely, J., and England, M. A., 1979a, The chick embryo late primitive streak and head process studied by scanning electron microscopy, *J. Anat.* **129:**615–622.

Wakely, J., and England, M. A., 1979b, Scanning electron microscopical and histochemical study of the structure and function of basement membranes in the early chick embryo, *Proc. R. Soc. Lond. B* **206:**329–352.

Wartiovaara, J., Leivo, I., and Vaheri, A., 1980, Matrix glycoproteins in early mouse development and in differentiation of teratocarcinoma cells, in: *The Cell Surface: Mediator of Developmental Processes* (S. Subtelny and N. K. Wessels, eds.), pp. 305–324, Academic Press, New York.

Weiss, P., 1958, Cell contact, *Int. Rev. Cytol.* **7:**391–423.

Wolpert, L., 1978, Gap junctions: channels for communication in development, in: *Intercellular Junctions and Synapses* (J. Feldman, N. B. Gilula, and J. D. Pitts, eds.), pp. 81–96, Chapman and Hall, London.

Zalik, S. E., Sanders, E. J., and Tilley, C., 1972, Studies on the surface of chick blastoderm cells. I. Electrophoretic mobility and pH-mobility relationships, *J. Cell Physiol.* **79:**225–234.

Chapter 13

Morphogenesis of the Neural Crest

CAROL A. ERICKSON

1. Introduction

The vertebrate neural crest has been, and continues to be, the subject of active investigation, because it offers developmental biologists a unique model system with which to study control of morphogenesis as well as cellular differentiation. Neural crest cells migrate great distances in the embryo and move along many diverse pathways. Then, after arriving at their destination or while migrating, they differentiate into a dazzling array of cell types (for review, see Weston, 1970, 1982; Noden, 1973; Le Douarin, 1980, 1982). Presented with such diversity of behavior and phenotype within one population of cells, we are in an excellent position to exploit these differences and to use a comparative approach to investigate the control of these major developmental processes. One hopes that understanding morphogenesis and differentiation of the crest will contribute to our knowledge of these fundamental processes in other aspects of embryogenesis as well.

The neural crest has been studied for many years; indeed, one classic review was published more than 30 years ago (Hörstadius, 1950). Since that time, our understanding of the neural crest has expanded rapidly, primarily due to the development of several important techniques in two general categories. First, a variety of cell-marking techniques now enable us to distinguish crest cells from surrounding embryonic tissues and so to trace accurately their movements. These include [^{3}H]thymidine labeling (Weston, 1963; Chibon, 1967; Johnston, 1966; Noden, 1975), permanent cell markers revealed by Feulgen staining in chick–quail chimaeras (Le Douarin, 1973; Noden, 1978b), and, recently, acetylcholinesterase (AChE) staining (Kussäther et al., 1967; Cochard and Coltey, 1983) and monoclonal antibodies that recognize migrating crest cells (Vincent et al., 1983; Vincent and Thiery, 1984; Tucker et al., 1984). These techniques represent a significant improvement over simple morphological studies or the excision and cauterization methods used in earlier investigations (e.g., Weston, 1970, discusses limitations of the latter tech-

CAROL A. ERICKSON • Department of Zoology, University of California—Davis, Davis, California 95616.

niques). Second, modern biochemical tools and immunocytochemistry now permit more precise determination of the composition of the matrix through which cells migrate (see Weston, 1980), as well as identification of differentiated cell phenotypes (see Le Douarin, 1982).

The focus of this chapter is the control of neural crest cell morphogenesis. The regulation of neural crest differentiation, itself an exciting and rapidly progressing field, now represents too vast a literature to be discussed in this chapter, except for instances in which predetermination of cell phenotypes may dictate morphogenetic behavior. Fortunately, this subject was recently reviewed; the reader is directed to several excellent papers and monographs (Le Douarin, 1980, 1982; Weston, 1982; Johnston et al., 1973; Noden, 1973, 1980; Patterson, 1978; Black, 1982).

The following questions regarding the control of neural crest morphogenesis are addressed, although without pretense of being able to answer them yet: What are the events that control when and whence the neural crest initiates its migration? What factors—both extrinsic and intrinsic to the crest cells themselves—determine the pathways these cells will follow in the embryo? Are these pathways specific to the crest? What directs the dispersion of the crest away from the neural tube? Are the crest cells intrinsically more motile than other embryonic cells? If so, what determines this capability? And, finally, what directs crest cells to the positions where migration ceases?

It should be noted that because most research concerning the neural crest has employed chick or quail embryos, the papers discussed in this chapter, of necessity, concentrate on these species. Recently, however, studies of mouse and rat have added to our understanding of mammalian crest (e.g., Nichols, 1981; Erickson and Weston, 1983). There is also an old and very extensive literature concerning amphibian neural crest, thoroughly reviewed by Hörstadius (1950) and Weston (1970). Presently there is a resurgence of interest in amphibians, and these more recent studies, which approach an understanding of the mechanisms controlling crest morphogenesis, are addressed here. Unfortunately, teleosts and reptiles have been exploited very little, but they represent a potentially fertile line of investigation (see Ferguson, 1981).

2. Appearance of Crest Population and Initiation of Migration

2.1. Morphological Appearance

The neural crest arises as a distinct population of cells from the neural folds as the folds fuse to form the brain and spinal cord. Although this is the classic description, the precise origin of the crest continues to be debated, because the crest cells usually are not morphologically distinct from the adjacent neural epithelium and the epidermal ectoderm flanking them. For this reason, the crest has been thought by some to arise from the epidermis as well as from the neural epithelium itself (see the discussion in Hörstadius, 1950). Whatever their precise origin, the appearance of crest cells (as well as their dispersal away from the neural tube) generally occurs in an anterior-to-posterior wave along the neural axis. The mode of separation from the neural epithelium, as well as the timing of this event in relationship to neural fold fusion, varies enormously, both at different levels in the same animal and from species to species. The following sections consider the initiation of migration in chick, mouse, and the axolotl.

2.1.1. Chick Crest

In the chick, the neural crest separates from the neural epithelium and begins its migration first at the level of the mesencephalon (DiVirgilio *et al.*, 1967; Noden, 1975, 1978*a*; Duband and Thiery, 1982). Thereafter, the crest arises from the prosencephalon (specifically the diencephalon) and simultaneously from the rhombencephalon (metencephalon and preotic myelencephalon). Few crest cells are produced by the telencephalon (Johnston, 1966; Noden, 1975, 1978*a*) or at the metencephalic–myelencephalic junction (Noden, 1980). This sequence of appearance of the cranial crest is particularly well demonstrated in the scanning electron micrographs prepared by Anderson and Meier (1981) and has been precisely correlated with developmental stages (Tosney, 1982). The postotic rhombencephalic crest (vagal crest) appears after the preotic crest, followed by the trunk crest in a sequential wave from the anterior to posterior ends of the neural tube (Bancroft and Bellairs, 1976; Tosney, 1978; Thiery *et al.*, 1982*a*) (see Fig. 1). This process of "individualization" or separation of the crest from the neural tube differs in several respects from one axial level to another. In particular, the crest cells from different regions vary in relationship to the timing of neural fold fusion and the disruption of the basal lamina.

In the head region, crest cells leave the neural tube at different times relative to neural fold fusion. For example, crest cells in the mesencephalon have begun to separate from the neural tube as the folds become apposed, whereas in the metencephalon, neural tube fusion is complete before cells migrate out of the tube (see Tosney, 1982, for detailed description). Regardless of these differences in timing, several consistent changes in embryonic architecture accompany cranial crest migration:

1. A large extracellular space, which is filled with extracellular matrix materials, appears lateral to the presumptive cranial crest cells (Pratt *et al.*, 1975; Bolender *et al.*, 1980; Tosney, 1982; Duband and Thiery, 1982). The volume of this space appears to be proportional to the number of crest cells that will appear at a particular axial level (Tosney, 1982).

2. The first suggestion of imminent crest migration is the rounding up of the presumptive crest cells and the extension of filopodial processes into the spaces between the cells. There is no obvious collection of matrix materials among neural tube cells (Tosney, 1982), but this may merely be a reflection of our inability to retain these materials during fixation.

Figure 1. The neural crest cells appear and then migrate in a wave from the anterior to posterior end of the trunk. This chick embryo (12+ stage) has had its ectoderm removed, revealing the underlying crest cells. In the older regions (anterior, right), crest migration has reached the apex of the somites, whereas at the posterior end (left), fusion of the neural tube has just occurred, and crest cells have not yet separated from the neural epithelium. ×135. (From Tosney, 1978.)

3. Before migration, the basal lamina* overlying the presumptive crest appears discontinuous, and the cells of the crest extend processes through this patchy basal lamina into the extracellular matrix above (Tosney, 1982). Interestingly, the basal lamina of the adjacent neural epithelium and contiguous epidermal ectoderm remains intact, and the basal lamina over the dorsal neural tube becomes complete once separation of the neural crest from the neural epithelium has ceased.

The post-otic cranial crest and trunk crest appear only after the neural folds have fused and the epidermal ectoderm has separated from the neural epithelium (Bancroft and Bellairs, 1976; Thiery et al., 1982a; Tosney, 1978). At this time, the overlying ectoderm is underlaid by a complete basal lamina (Tosney, 1978; Martins-Green and Erickson, 1985). Before crest migration begins, the dorsal neural tube is covered by a discontinuous basal lamina, whereas the lateral and ventral regions of the neural tube are invested with a complete basal lamina (Tosney, 1978; Mayer et al., 1981; Newgreen and Gibbins, 1982; Martins-Green and Erickson, 1985). Just before the onset of migration, the crest cells begin to extend processes dorsally through the discontinuous basal lamina and into the overlying space; they also become rounded and irregular in shape as compared with the neural tube cells. Eventually these cells lose their terminal bar junctions (Newgreen and Gibbins, 1982) and move onto the surface of the neural tube. Both Bancroft and Bellairs (1976) and Tosney (1978) noted that once the crest cells have left the neural tube and flattened on its surface, they become randomly oriented. However, shortly after their appearance as individual cells, they become tangentially oriented (i.e., their long axes are perpendicular to the long axis of the neural tube) and begin to move in unison toward the apex of the somites.

---

*The term **basal lamina** refers in this chapter to the extracellular matrix associated with epithelial layers; it is a structure visible only at the electron microscopic level, generally visualized as three layers: (1) an inner electron-lucent lamina rara interna, (2) a middle electron-dense lamina densa, and (3) an outer electron-lucent lamina rara externa. **Basement membrane** is the term used by light microscopists to refer to extracellular matrix materials associated with epithelia; it includes the basal lamina as well as their adjacent **reticular fibers** and **ground substance**.

2.1.2. Mouse Crest

As in the chick, the appearance of mouse crest begins in the cranial region and subsequently propagates in a rostral–caudal wave. Crest emigration from the cranial folds in the mouse and rat begins long before neural fold fusion (Johnston and Listgarten, 1972; Waterman, 1976; Nichols, 1981; Morriss and Thorogood, 1978), whereas in the trunk the neural tube and epidermal ectoderm separate before crest emergence (Erickson and Weston, 1983). The complete sequence of closure of the neural folds is fairly complicated in the mouse and rat, but details can be found in Waterman (1976), Morriss and New (1979) and Morriss-Kay (1981). As in the chick, however, there appears to be no correlation between stage of neural fold closure and appearance of the crest.

The most detailed study of the appearance of mouse cranial crest is by Nichols (1981), who exploits a toluidine blue staining technique that appears to distinguish presumptive crest cells from the neural and epidermal ectoderm by a difference in staining intensity. The first indication of crest formation using this staining method is seen in the midbrain and rostral hindbrain of the 3–4-somite embryo (Nichols, 1981). Here, a portion of the epithelial cells lose their attachment to each other at the apical surface and accumulate at the basal surface of the epidermal ectoderm. The cells have not yet breached the basement membrane, however, as evidenced by the smooth contour of the basal cell surface. These first crest cells actually form from the epidermis contiguous with the neural folds and not from the neural epithelium itself; they are released at a stage two somites later, leaving behind the apical cells to form the definitive surface ectoderm. At later stages at this axial level, crest cells are derived exclusively from the neural plate. By the 5–7-somite stage, crest cells emerge from the neural plate of the forebrain adjacent to the optic pits as well as from the mid- and hindbrain. Unlike the chick, a few neural crest cells even leave the rostral margin of the forebrain. Crest emigration from the mid- and hindbrain continues beyond the 8–16-somite stage, at which time the neural folds have fused.

Mouse cranial crest morphogenesis has not been studied at the transmission electron microscopic (TEM) level as extensively as has chick morphogenesis, but light microscopic sections and scanning electron microscopic (SEM) images suggest that the presumptive crest cells round up before the basement membrane loses its integrity (Nichols, 1981). Morriss and Thorogood (1978) stated that the basal lamina also is discontinuous, but provide no micrographs.

In the trunk, neural fold fusion and epidermal ectoderm separation from the neural epithelium occur before the appearance of the crest (Derby, 1978; Erickson and Weston, 1983). As in the chick, the overlying ectoderm and the lateral and ventral neural tube are covered by a complete basal lamina, while the basal surface of the dorsal neural tube contains only flocculent material (Erickson and Weston, 1983; Martins-Green and Erickson, 1985). Shortly after fusion of the neural folds, presumptive crest cells begin to bleb and then to extend filopodia and lamellipodia into the overlying extracellular space (Fig.

2), but they are delimited in this extension by the laterally intact basal lamina. It is worth noting that despite a glycosaminoglycan-filled space above the neural tube (Derby, 1978; Erickson and Weston, 1983), it is not of the dimension seen in the chick (see Johnston *et al.*, 1981). In any event, the crest cells then crawl out of the neural tube, presumably while breaking (or after losing) their adhesions with the luminal surface. As emigration continues, the basal lamina becomes continuous medially, and crest migration is progressively restricted within the borders of the complete basal lamina. Once the crest cells delaminate from the neural tube, they begin their lateral migration on its surface, although the flattening and tangential arrangement of cells observed in the chick is not evident in the mouse.

2.1.3. Axolotl Crest

As in the mouse and chick, axolotl crest cells first arise from the neural tube in the cranial region and later appear progressively in an anterior–posterior wave in the trunk (Löfberg and Ahlfors, 1978; Löfberg *et al.*, 1980). The details of their appearance at the TEM level are not available, but at the light microscopic (Löfberg *et al.*, 1980) and SEM (Spieth and Keller, 1984) levels important differences are observed between amphibians and chick.

First, well before neural fold fusion, the trunk crest can be recognized within the neural folds as rounded cells that contrast with the adjacent elongated cells of the neural tube. Second, rather than migrating as single cells out of the neural tube epithelium, these crest cells appear to be segregated as a population by exclusion during neural fold fusion. Spieth and Keller (1984) observe that the neural folds are first apposed at their apical edge (i.e., luminal surface of the forming neural tube), while the basal ends are separated by a wedge of neural crest cells. After the apical ends of the neural tube cells contact each other, their adjacent surfaces become progressively apposed by "zipping-up", and the crest cells are gradually excluded from the neural tube. Eventually, the neural tube forms a continuous columnar epithelium, with the crest cells sitting as a ridge on its dorsal surface (Fig. 3). A similar sequence of events is seen in *Xenopus* (Schroeder, 1970). Although there is no doubt that most of the crest is formed in the manner just described, it is apparently an unresolved matter as to whether some neural crest cells in amphibians may actually be derived from the thinner contiguous epidermal ectoderm as well (see Hörstadius, 1950, p. 5 for further discussion). Cytochemical markers for premigratory crest (such as AChE in chick or crest-specific antibodies) will presumably enable us to settle this question.

Third, once situated on top of the neural tube, there is a considerable lag before the crest cells begin their migration. This contrasts with the chick, in which lateral migration follows fairly quickly after individualization, and with the mouse, in which the appearance and migration of the crest are virtually simultaneous events. Once segregated, the cells become spindle shaped and aligned tangentially to the long axis of the neural tube. Only after alignment is complete does migration begin (Fig. 3). Furthermore, only the cells resting on

Figure 2. Scanning electron micrograph of a cross section of a neural tube two somite lengths posterior to the last somite in a 9.5 day mouse embryo. One neural crest cell has extended a process onto the surface of the dorsal neural tube, while its trailing edge (arrowheads) remains within the neural epithelium. ×3290. (From Erickson and Weston, 1983.)

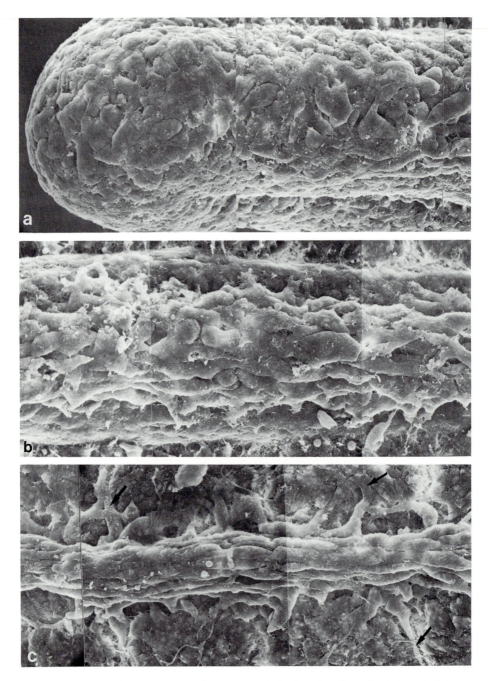

Figure 3. Scanning electron micrograph montages of the dorsal surface of a dark axolotl larva at stage 32 with the epidermis removed. (Top) Neural crest population immediately after segregation from the tube. (Middle) Farther anteriorly, between somites 15 and 20, alignment and orientation of crest cells has begun. (Bottom) At more advanced stages, between somite levels 9 and 11, alignment and orientation are complete, and lateral migration from the base of the neural crest is under way (arrows). ×300. (From Spieth and Keller, 1984.)

the neural tube surface begin to migrate (Löfberg *et al.*, 1980; Spieth and Keller, 1984), with the more dorsal cells in the pile eventually dispersing to the dorsal fin (Raven, 1931; DuShane, 1935; Twitty and Bodenstein, 1941).

2.2. Control of Initiation of Migration

Although the events surrounding the separation of the neural crest cells from the neural epithelium have been described to some extent, the underlying controls of this emigration are unknown. The following regulatory events have been suggested to initiate neural crest migration; at the very least, they provide a conceptual framework around which further experiments can be designed. These proposed regulators of initiation include (1) liberation of the crest from the neural tube due to degradation or disruption of the basal lamina, (2) changes in neural crest cell surface components that specifically alter adhesiveness, (3) creation of an adjacent extracellular space into which the crest can move; and (4) acquisition of motile capabilities. There is at least correlative evidence that each of these four may be involved under different conditions and, indeed, several may be necessary to permit migration.

2.2.1. Basal Lamina Disruption

The correlative observations just discussed provide the strongest evidence in support of the proposal that breakdown of the basal lamina permits release of the crest. That is, the basal lamina is discontinuous before the first signs of migration, and the regions of discontinuity coincide with those from which the neural crest will soon migrate.

Several pieces of evidence suggest that the crest is unable to penetrate an intact basal lamina and that an intact basal lamina therefore could indeed trap the crest within the neural tube. First, during early migration in the embryo, crest cells do not breach the basal lamina covering the neural tube, ectoderm, or blood vessel endothelia (Bancroft and Bellairs, 1976; Newgreen *et al.*, 1982; Tosney, 1978, 1982; Erickson and Weston, 1983)—or, for that matter, penetrate a basement membrane (Thiery *et al.*, 1982a; Duband and Thiery, 1982). Occasionally, however, they have been seen "probing" breaks in the basal lamina surrounding the sclerotome (Bancroft and Bellairs, 1976). Second, when neural crest cells are confronted with a basal lamina isolated from human amnion, they spread on its surface but do not invade the underlying stroma (C. A. Erickson, unpublished results), as tumor cells do (Liotta *et al.*, 1980; Russo *et al.*, 1982). Finally, if neural crest cells are grafted into the lumen of a stage 14 (approx. 48–52-hr) chick neural tube, some cells will migrate through the lateral walls until they reach the basal surface, where EM preparations show that they spread on, but do not penetrate, the basal lamina of the neural tube (C. A. Erickson, unpublished results) (see Fig. 4).

The events that produce a discontinuous basal lamina over the presumptive neural crest cells are unknown (see Tosney, 1982, for detailed description), but a number of possibilities exist. First, the crest may degrade the

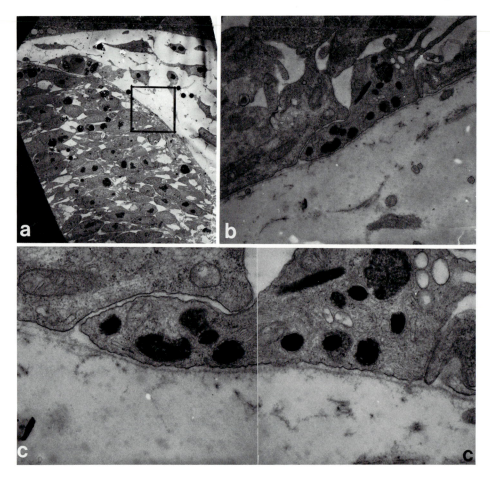

Figure 4. (a) Transmission electron micrograph of thin section through a chick neural tube fixed 24 hr after grafting pigmented neural crest cells into the lumen of the tube. ×720. (b) Higher magnification of the boxed area in (a), showing that a pigmented neural crest cell has reached the basal surface of the neural tube. ×9,600. (c) High magnification of (b), demonstrating that the crest cell has spread on the basal lamina of the neural tube but has not penetrated it. ×24,800.

basal lamina or prevent its initial formation through the release of proteolytic enzymes. Recently Valinsky and Le Douarin (1985) have demonstrated that cephalic crest cells produce the serine protease, plasminogen activator (PA), and R. R. Isseroff and C. A. Erickson (unpublished results) have similarly found PA activity in trunk crest cells. As yet, there is no proven role for PA in crest morphogenesis. However, several examples in developing systems of invasive processes have been associated with proteolytic activity. In particular, the colonization of the epithelium of the bursa of Fabricius in avian embryos by hematopoietic cells is accompanied by proteolysis generated by production of plasminogen activator (Valinsky *et al.*, 1981); in this example, the invasive hematopoietic cells are the source of the enzyme. Antibodies against chicken

PA and inhibitors of PA activity should allow us to test the role of PA in turnover or degradation of the basal lamina.

A second possibility is that the extracellular matrix produced by the premigratory crest cells accumulates between them, expanding the intercellular spaces and preventing the formation of a complete basal lamina. This is a particularly attractive hypothesis because the crest produces hyaluronate (Pintar, 1978), which has a space-expanding capability (see Toole, 1981, for review of this evidence). However, accumulation of hyaluronate or other matrix molecules is not revealed either by Alcian blue staining of glycosaminoglycans in mouse embryo sections (Erickson and Weston, 1983) or in thin sections of chick neural tubes (Newgreen and Gibbins, 1982; Tosney, 1982; C. A. Erickson, unpublished results). Of course, this may only reflect our inability to retain these molecules with present fixation methods.

A third possibility begins with the observation that during early stages of neural fold elevation the ectoderm and neural folds are in close apposition and share a common basal lamina (Tosney, 1982). The proposal is that perhaps the simple pulling apart of these two epithelia as neural fold elevation proceeds is sufficient to disrupt the lamina. There is no simple explanation, however, for why the epidermal ectoderm stays intact while the neural epithelium is discontinuous.

Fourth, in a number of instances in which epithelia fuse, it has been suggested that adjacent mesenchyme is needed to maintain the basal lamina (Waterman, 1977). Specifically, studies of the basal lamina of embryonic mouse tooth epithelium suggest that fibronectin derived from the adjacent mesenchyme is necessary for the supramolecular organization of the epithelium-derived basal lamina (Brownell et al., 1981). As neural folds elevate and the epidermal and neural ectoderms become apposed, they might effectively exclude the influence of the adjacent mesoderm. This might, in fact, be possible in the trunk region, but in the head region of the chick the changes in the basal lamina are far too complicated to be accounted for in this way (see discussion in Tosney, 1982).

Finally, it is unlikely that motile activity of the crest cells boring their way through the basal lamina produces these disruptions. The basal laminae, at least in the trunk regions of the chick and mouse, are discontinuous hours before migration begins (Fig. 5) (Tosney, 1978; Newgreen and Gibbins, 1982; Erickson and Weston, 1983). Indeed recent TEM studies show that during neurulation in the trunk of chick and mouse the basal lamina is never complete over the presumptive neural crest cells (Martins-Green and Erickson, 1985). It should be remembered, however, that disruption of the basal lamina in the cranial regions seems to be the last event before production of lamellipodia; in this case, the generation of tears in the basal lamina by moving crest cells (Tosney, 1982; Nichols, 1981) is at least consistent with morphological events.

This problem clearly awaits detailed analysis of the structural and biochemical changes in the basal lamina during this critical developmental period. Even if the basal lamina is not instrumental in triggering the migration of the crest from the neural epithelium, it seems clear that these changes must

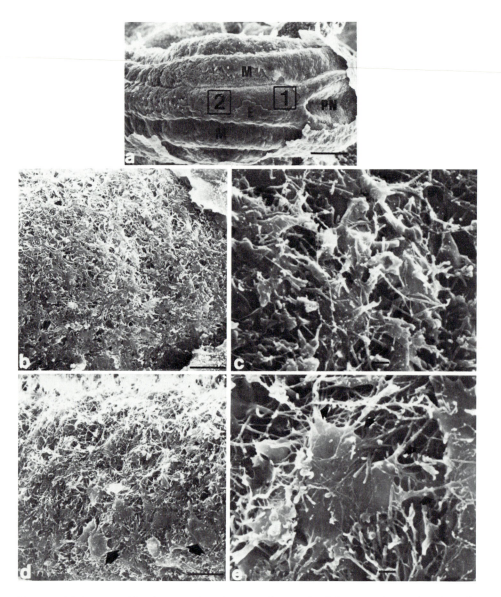

Figure 5. (a) Low-magnification scanning electron micrograph of the posterior trunk region of a 9.5-day mouse embryo, from which the ectoderm (E) has been partially removed. The posterior neuropore (PN) is visible to the right, and the neural tube at this axial level is bordered by unsegmented mesoderm (M). (b) High magnification of box 1 in Fig. 5a. Soon after neural tube fusion, neural crest cells are not distinguishable as a separate cell population. Note the distinct cell outlines of the dorsal neural tube cells and the large spaces between them. (c) Detail from Fig. 5b. Although some extracellular material is seen, there is clearly no complete basement membrane, since cell outlines are so distinct. (d) High magnification of box 2 in Fig. 5a. At this more anterior axial region, broad, flattened lamellipodia now extend from the neural tube, while their trailing ends are still contained within the tube. Some cells have completely separated from the neural tube and are observed as single cells (black arrowheads). (e) Detail of cell marked with white arrowhead from Fig. 5d, showing a lamellipodium with filopodial-like processes (arrowheads) extending from it. Some extracellular matrix is also observed which is frequently hard to distinguish from the filopodia. Scale bars: 100 μm (a); 10 μm (b), (d); 1 μm, (c), (e). (From Erickson and Weston, 1983.)

have important consequences for the function and integrity of the neural epithelium (e.g., see Banerjee *et al.*, 1977; Bernfield and Banerjee, 1982; Ormerod *et al.*, 1983; Martinez-Hernandez and Amenta, 1983).

2.2.2. Changes in Crest Adhesiveness

Despite the previous correlative evidence, it could equally be true that lack of a complete basal lamina is simply an outcome of other changes in the epithelium more basic to the process of crest delamination. A likely alternative hypothesis is that changes in the adhesiveness of presumptive neural crest cells permit them to migrate. Tissue culture studies have demonstrated that crest cells leave the explanted neural tube on a schedule similar to one they would have followed in the embryo (Newgreen *et al.*, 1979; Newgreen and Gibbins, 1982; Tucker and Erickson, 1984), suggesting that endogenous changes occur in the presumptive crest. Newgreen and Gibbins (1982) have evidence that these changes are adhesive in nature. In their experiments, neural tubes were isolated from the trunk region of chick embryos and divided into premigratory pieces from which crest cells had not yet migrated and into migratory segments from axial levels at which separation of crest cells from the neural tube had just started. The tubes were cultured on a substratum of fibronectin, to which they adhered tenaciously. The tubes were then detached with a stream of culture medium 0.5–2.0 hr after culturing. The migratory pieces left behind a ridge of neural crest cells, whereas the premigratory tubes did not. Because the basal lamina was discontinuous in both cases, some additional change must have preceded migration. The authors suggest that the neural crest cells become less adhesive to the other neural tube cells. [Another possibility not addressed by Newgreen and Gibbins is that the crest cells may have acquired motile capabilities; see Section 2.2.4.] Recent experiments are consistent with the notion that the onset of neural crest migration is controlled by the inactivation of Ca^{2+}-dependent cell adhesions (Newgreen and Gooday, 1985). In the cranial regions of chick and mouse, however, morphological studies suggest that cell–cell adhesiveness is lost before the basal lamina disintegrates (Tosney, 1982; Nichols, 1981) (see Section 2.1.1). In these regions, then, loss of adhesion may not be the final trigger involved in emigration.

Recently, Thiery *et al.* (1982b) identified at least one known adhesion molecule that disappears at the time of neural crest migration. N-CAM, a molecule thought to mediate neuronal cell adhesion (Rutishauser *et al.*, 1978), can be detected in the early chick neural tube with monoclonal antibodies. After the crest cells have detached from the neural tube, N-CAM antibodies no longer bind to their surface. Unfortunately, we do not know from this study whether loss of N-CAM precedes migration or whether it disappears once migration is underway. Furthermore, N-CAM may still be present but may no longer be accessible to the antibody. Other cell surface changes have been identified at the time of migration, and these may reflect adhesive modifications. Recently Currie *et al.* (1984) reported that there are alterations in lectin-binding patterns to crest cells concomitant with their migration in the trunk of rat embryos. In these studies only the apical surface (luminal surface of the neural tube) was exposed to the lectins and not the basal surface. Because crest cells extend lamellipodia from their basal surface, the significance of these findings is unclear.

Several other studies show physical evidence for changes in association of neural crest cells with the flanking neural and epidermal ectoderm. In freeze-fracture studies Revel and Brown (1975) demonstrated that cell junctions change during development in a variety of tissues, including the neural tube. In particular, there is evidence of disruption of preexisting tight junctions after neural fold fusion in the chick. Warner (1973) further showed in the axolotl embryo that gap junctions are lost between the neural epithelium and the epidermal ectoderm at the time of closure of the neural folds.

2.2.3. Appearance of Extracellular Spaces

Extracellular spaces appear above or lateral to the neural crest cells before their emergence from the dorsal neural tube (Pratt *et al.*, 1975; Derby, 1978; Tosney, 1982; Newgreen and Gibbins, 1982; Löfberg *et al.*, 1980; Erickson and Weston, 1983) (see Fig. 6). Not only could these spaces provide pathways of presumed lower resistance through which the cells might migrate more easily for mechanical reasons, but they also contain extracellular matrix molecules, including fibronectin, hyaluronate, and chondroitin sulfate (see Section 2.2.4 for detailed description), which could provide a preferred substratum on which to adhere. The extracellular matrix is produced by the overlying ectoderm and the neural tube (Cohen and Hay, 1971; Hay and Meier, 1974; Pintar, 1978), as well as by the crest cells themselves (Pintar, 1978; Greenberg *et al.*, 1980). Even though these initial spaces, primarily opened by hyaluronate (see Fisher and Solursh, 1977; Pratt *et al.*, 1975), may be important in defining the pathways of crest migration, they are probably not essential to, and not the trigger for, migration from the neural tube. Although there is not much evidence on the subject, this conclusion is based on the observation that reduction or elimination of these spaces by hyaluronidase treatment does not interfere with the normal migration of cranial crest cells (Anderson and Meier, 1982).

The extracellular matrix material produced by the ectoderm might well control the timing and onset of crest migration in amphibians, however (Löfberg *et al.*, 1985). If ectoderm from a migratory level is grafted to a more posterior region (where the crest has not begun to disperse), a premature stimulation of migration is observed, apparently due to some extracellular matrix molecule produced by the ectoderm, which can be collected on nucleopore filters and when grafted beneath the ectoderm in living embryos produces premature migration. It should be recalled that initiation of crest migration in amphibians is different from either chick or mouse in that the cells are excluded from the neural tube by fusion of the neural folds and are subsequently delayed in their lateral migration. Nevertheless, these experiments demonstrate the importance of the extracellular matrix in controlling the onset of neural crest migration in at least one group of animals.

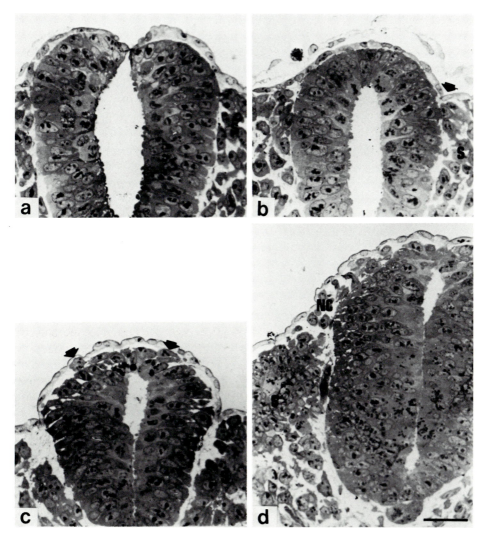

Figure 6. Micrographs of toluidine blue-stained Epon thick sections of a day 9.5 mouse neural tube. (a) Cross section of neural folds at the moment they meet and fuse. The epithelium and somites are closely apposed to the neural tube. (b) Cross section of a neural tube just after fusion of the neural folds. The neural tube cells are arranged in a tight pseudostratified epithelium. Note that the somites (S) and epithelium are tightly apposed to the neural tube, with little intervening extracellular matrix. (c) Cross section two somite lengths posterior to the last somite. Extracellular spaces now separate the dorsal neural tube cells, and the dorsal neural tube has widened. The neural tube is separated from the overlying ectoderm and surrounding somites by extracellular matrix-filled spaces. A few neural crest cells have now emigrated from the neural tube and appear as a distinct population of cells in the space above the neural tube (arrows). Note that the epithelium is still tightly apposed to the somites in a "cleft", and no neural crest cells have moved laterally over the somites. (d) Cross section four somites anterior to the last somite. Neural crest cells (nc) have spread ventrally in the large extracellular space between the neural tube and somites (S) and have also collected in the cleft between adjacent somites. None have spread between the ectoderm and somites. Scale bar: 50 μm. (From Erickson and Weston, 1983.)

2.2.4. Endogenous Change in Motile Capacity

In a number of reported instances, endogenous cues appear to dictate the point at which embryonic cells begin migration. For example, *Fundulus* deep cells suddenly begin blebbing and then translocating at a precise time in the late blastula, even when isolated in tissue culture (Trinkaus, 1963, 1973). Onset of motility is correlated with an increase in cellular deformability (Tickle and Trinkaus, 1973), suggesting that cytoskeletal or cortical changes may accompany the onset of migration. Likewise, the separation of primary mesenchyme cells from the vegetal plate in sea urchins appears to be an intrinsic property (Kinnander and Gustafson, 1960; Gustafson and Wolpert, 1961) and one that also is preceded by bleb formation. In this regard, it is interesting to note that neural crest cells of the mouse trunk begin blebbing just before emergence from the neural tube (Erickson and Weston, 1983). Acquisition of motile capacity will be very difficult to distinguish from changes in adhesiveness, however, because motility is certainly dependent on the ability of a cell to stick to a substratum.

3. Pathways of Migration

3.1. Patterns of Migration

Once the neural crest cells have separated from the neural folds, they generally move laterally and ventrally in the embryo along defined pathways. The precise routes followed by the crest cells and the timing of their migration are strikingly different at different axial levels.

Patterns of neural crest migration have been studied in most detail in the chick embryo. These embryos are used primarily because of the relative ease with which they can be manipulated (unlike the mouse, for example) and the ability to discern the migrating crest from the surrounding mesodermally derived mesenchyme, even during later stages of migration, thanks to the powerful tool of the chick–quail chimera pioneered by Le Douarin (1973). The discussion of migratory pathways therefore centers around this species, and our meager knowledge of mouse and amphibian pathways is only briefly mentioned. The routes taken by the crest at each axial level are summarized and reference made to specific papers for detailed descriptions.

3.1.1. Chick Pathways

Cranial crest cells follow different migratory routes in the prosencephalon, mesencephalon, and rhombencephalon. This pathway has been determined by Johnston (1966) and Noden (1975), who grafted [³H]thymidine-labeled crest-containing neural tubes into unlabeled host embryos and then followed their migration. These findings were later substantiated by grafting quail neural crest into chick hosts and identifying the donor cells using Feulgen stain (e.g., see

Noden, 1978b; Le Lièvre and Le Douarin, 1975). Cranial crest pathways are diagrammatically summarized in Figure 7.

3.1.1a. Mesencephalon. The mesencephalic crest cells are the first to appear and to begin migrating (Johnston, 1966; Noden, 1975, 1978b; Anderson and Meier, 1981; Duband and Thiery, 1982; Cochard and Coltey, 1983). These cells move laterally in a cell-free space between the epidermal ectoderm and underlying mesoderm, directing most of their locomotory processes toward the ectoderm (Tosney, 1982). There is no evidence that they significantly invade the mesoderm adjacent to the neural tube, although Johnston (1966) did observe a few labeled crest cells within this mesoderm—perhaps an artifactual result due to experimental trauma. No crest cells move ventrally between the mesencephalon and mesoderm. As the population continues to expand laterally and ventrally beneath the ectoderm, the crest cells eventually vacate the top of the neural tube and the dorsal region of the embryo and move into the first visceral arch, where they contribute to the mesenchyme of the maxillary process and partially to the mandibular process. Later, when the oculomotor and ophthalmic nerves grow out toward the eye, the crest cells from this region undergo a secondary migration medially along these nerves (Noden, 1975, 1978a). Some crest cells also collect at their intersection, forming the ciliary ganglion.

3.1.1b. Prosencephalon. After the mesencephalic crest cells have begun migration, crest cells rostral and caudal to this population appear. Crest cells arise only from the caudal portion of the prosencephalon (future diencephalon) and, along with some anterior mesencephalic crest, begin to migrate rostrally rather than laterally (Noden, 1975). In so doing they come to lie dorsal to the expanding optic stalks. They then stream caudally and later rostrally around the surface of the optic stalk as it continues to invaginate. The optic cup at its distal end obviously acts as an obstruction to any further lateral migration, and, as the crest cells move around the optic stalk, they collect at the margins of the optic cup (Noden, 1975; Johnston, 1966). Later, they will undergo a secondary migration between the edges of the optic cup and lens (Nelson and Revel, 1975; Bard et al., 1975), ultimately forming the endothelium and corneal matrix of the eye (Noden, 1978b; Johnston et al., 1979). In addition, crest cells continue to move rostrally over the prosencephalon to form the frontonasal process (Johnston, 1966). This area will later fuse with the maxillary process as torsion and flexure of facial components bring the two regions together. Apparently, no crest cells arise from the most anterior portion of the prosencephalon in the chick.

3.1.1c. Rhombencephalon. Crest migratory patterns in this region deviate significantly from those in the mesencephalon in that the population rapidly becomes segmented, rather than migrating strictly as an unsegmented sheet. As the crest begins its migration, some cells enter the extracellular matrix lateral to the neural tube and then extend under the ectoderm; the rest move

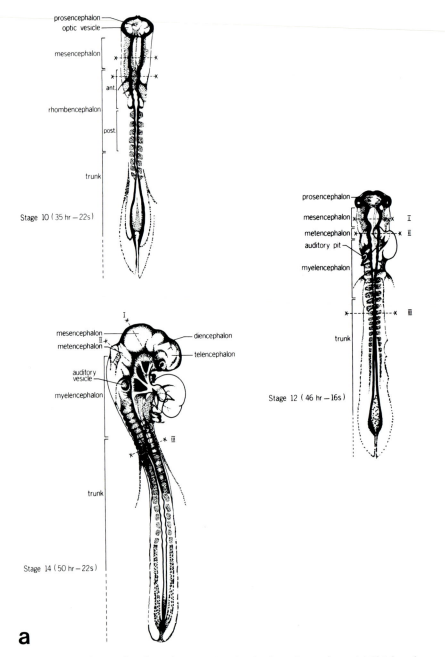

Figure 7. Pathways of cranial and trunk crest migration in the avian embryo. (a) Chick embryos at stages 10, 12, and 14–15, respectively. (b) Axial levels (X- - - -X) redrawn in cross section. From left to right, the patterns of crest migration are shown diagrammatically at the level of the anterior mesencephalon, posterior mesencephalon, metencephalon, and trunk, respectively. (■) Crest pop-

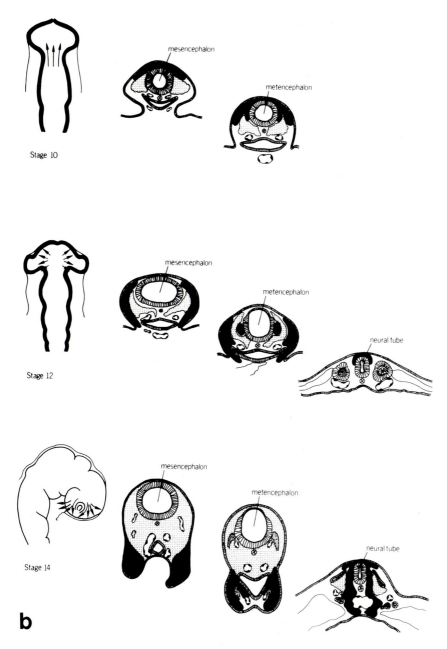

ulation, (::::::) mesodermal mesenchyme, (⊛) pharynx (ventral) and neural tube (dorsal). Note the unique timing and distribution of crest cells at each axial level. A description of migration is given in the text. (Redrawn according to data from Noden, 1975, Thiery *et al.*, 1982, and Duband and Thiery, 1982.)

ventrally between the neural tube and mesoderm. At the level of the metencephalon, the crest cells beneath the ectoderm spread into visceral arch I to
form the mandibular arch mesenchyme (Noden, 1978b; but see Le Lièvre,
1974). The ventrally-moving crest cells at this level aggregate adjacent to the
brain vesicle and contribute to the proximal trigeminal ganglion (V) (Noden,
1975; D'Amico-Martel and Noden, 1983). At the level of the rostral (preotic)
myelencephalon, the laterally-moving crest cells populate the hyoid arch (arch
II), while the rest form the root ganglion of nerve VII adjacent to the neural tube.
The postotic crest (posterior myelencephalon) fills the third and fourth visceral
arches with mesenchyme and also aggregates near the lateral margin of the
brain to form the neurons and glia of the superior (IX) and jugular (X) cranial
sensory ganglia. In addition, as the crest cells reach the level of the pharynx,
they disperse through the splanchnic mesoderm along the gut, giving rise to the
enteric ganglia (Le Douarin and Teillet, 1973). Interestingly, at the level of the
otic placode, the crest is divided into an anterior and posterior stream as it
moves laterally. Apparently the placode obstructs its pathway (Anderson and
Meier, 1981; Noden, 1975, 1978a).

3.1.1d. Trunk Crest.   At somite level 6 and farther posterior, the chick
trunk crest follows two major extracellular pathways. These are extracellular
spaces over the somites and also between the neural tube and somites (Weston,
1963). Movement along the former pathway proceeds a few hours after the
ventral migration has begun (Thiery et al., 1982a). These cells eventually disperse into the dermis and epidermis and become pigment cells (Teillet and Le
Douarin, 1970; Teillet, 1971). Those crest cells moving ventrally split into two
streams. One extends a short distance between the neural tube and somite,
where it gives rise to the sensory ganglia; the other rapidly invades the large
extracellular space between adjacent somites and produces the sympathetic
chain ganglia, the nerve plexus around the dorsal aorta, and the adrenal me-
dulla (Le Douarin and Teillet, 1974; Thiery et al., 1982a). These pathways
explain the staggered arrangement of the sensory and sympathetic ganglia.

Generally, trunk crest cells do not extensively invade other embryonic
tissues during their early stages of migration, aside from their movement
through trunk splanchnic mesoderm to form enteric ganglia (but see Section
4.3). Until recently it was reported that only a few crest cells extend into the
somite by migrating along the basement membrane of the dermatome–myotome after the eventual dispersion of the sclerotome (Le Douarin, 1982; Teillet
and Le Douarin, 1983). In several studies, neural crest derivatives such as
sensory and autonomic ganglia have been inserted into the early migratory
pathway, where they reinitiate migration (Ayer-Le Lièvre and Le Douarin,
1982; Le Lièvre et al., 1980). Crest cells under these circumstances are dispersed throughout the sclerotome 48 hr postsurgery. However, since the grafted
crest cells are considerably delayed in their migration by the time they dissociate from the grafted tissue, this probably does not represent the usual migratory
sequence. The precise distribution of crest cells with relationship to the trunk
mesoderm can now be answered definitively with monoclonal antibodies spe-

cific to neural crest cells (Vincent *et al.*, 1983; Tucker *et al.*, 1984), because the normal embryonic events need not be perturbed as they might with grafting studies. Preliminary results from several laboratories using the NC-1 antibody that recognizes neural crest cells suggest that many more crest cells migrate under the anterior portion of the dermatome–myotome than had been previously observed. Some crest cells are also found in the sclerotome, but these are aligned on the motor fibers from the ventral root (Thiery *et al.*, 1985) and may therefore use these nerve fibers as the substratum for their migration (J. Loring, J. L. Duband, C. A. Erickson, and J. P. Thiery, unpublished results). These issues clearly need more investigation in order to be resolved.

3.1.2. Mouse Crest Pathways

Crest pathways in the mouse and rat have not been studied in the same detail as in the chick, because permanent, crest-specific markers are not yet available. Grafting of [³H]thymidine-labeled neural tubes into unlabeled rat hosts has recently been attempted, but such embryos do not survive well *in vitro* and only preliminary results have been published (Johnston *et al.*, 1981). Cephalic crest cells apparently can be differentially stained with toluidine blue after fixation with cetylpyridinium chloride (Nichols, 1981), but this procedure identifies crest cells only in their earliest stages of appearance and is somewhat nonspecific. Consequently, detailed, long-term studies have not been accomplished. Likewise, because trunk crest cells cannot be easily distinguished once they have migrated past the dorsal margin of the sclerotome, little mention is made of these pathways in the mouse. Nonetheless, during their initial stages of migration, when crest cells can be identified by virtue of their obvious position in the matrix-filled spaces, the migratory routes seem to be substantially similar to those in the chick. In the trunk, for example, crest cells extend first between the neural tube and somites (Derby, 1978; Erickson and Weston, 1983). However, because no cell marking experiments have been done, it is not certain as to whether the crest cells invade the somitic mesoderm. As in the chick, migration laterally over the somites is delayed, although in the mouse the delay is as long as several days (Derby, 1978; Mayer and Oddis, 1977; Erickson and Weston, 1983).

In the head, crest migratory routes are similar to those of the chick, at least as far as can be determined (Johnston *et al.*, 1981). The timing of this migration is somewhat different, however, presumably because of the variation in fusion of the neural folds (see Waterman, 1976; Morriss and Thorogood, 1978; Nichols, 1981).

3.1.2. Axolotl Pathways

The crest migratory pathways in amphibians, as in mammals, have not been elucidated due to a lack of critical experiments with permanently marked neural crest cells. Although extirpation and ablation experiments and xeno-plastic grafts have provided a great deal of information regarding derivatives of the crest, they have not been appropriate for studying the pathways themselves. Some pertinent information is available, however. The early trunk crest migratory patterns have been examined in the most detail, and axolotls also have a lateral pathway that is invaded first (Löfberg and Ahlfors, 1978; Löfberg et al., 1980) and is subsequently followed by movement into a ventral pathway. In addition, crest cells that remain on the surface of the tube contribute to the mesenchyme of the dorsal fin (Raven, 1931; DuShane, 1935).

3.2. Determination of Pathways

Our hypotheses concerning the determination of patterns of crest migration are dependent on our knowledge of the developmental limitations already placed on the neural crest cells before they have quit the neural tube. If we assume that the crest cells are restricted in their developmental capabilities early in the process of migration, their patterns of migration could depend largely on the cells themselves. Different "determined" phenotypes would respond selectively to migratory cues in the environment and, in effect, "sort out" in the embryonic milieu (see Weston et al., 1984). Alternatively, if crest cells are pluripotential, the control for their patterns of migration must lie external to the cells, in the extracellular environment.

3.2.1. Environmental Control

The diversity of patterns of crest migration is determined primarily by the environment through which the cells move. This is best demonstrated by heterotopic transplant experiments, which exploit the differences in crest cell behavior at different axial levels. For example, Noden (1975) replaced the chick mesencephalic crest with labeled metencephalic crest. The mesencephalic crest normally disperses as a uniform mesenchyme, whereas the metencephalic crest contributes to the visceral arch mesenchyme but also coalesces to form ganglia adjacent to the neural tube. In this experiment, not only do the metencephalic crest cells migrate in the pathways appropriate to their new position, but they differentiate according to their new axial position as well (Noden, 1978b). Likewise, mesencephalic crest cells grafted into the hindbrain will migrate between the neural tube and mesoderm to form cranial ganglia (Noden, 1978c).

The differences between cranial and trunk patterns of migration are even more pronounced. In such exchanges, crest cells still generally behave according to directions from their new environment. As one of many examples, Le Douarin and Teillet (1974) grafted a segment of quail thoracic spinal cord (from somite level 18–24) into the position where a vagal level neural tube (somites 1–7) had been removed. The donor crest cells migrated to the endoderm and moved caudally to form normal enteric ganglia, using a pathway they normally would not have followed from their original position. However, the reciprocal graft of vagal-derived crest replacing thoracic crest yielded some confounding results. For the most part, the donor cranial crest cells dispersed along the pathways usually followed by the thoracic crest, giving rise to sensory and sympathetic ganglia, adrenomedullary tissue, and pigment cells. Some donor crest cells, however, produced connective tissue derivatives and, surprisingly, also entered the intestine to form enteric ganglia. This latter experiment, which suggests that some subset of the vagal population is determined to be enteric ganglia, is discussed in more detail in Section 5.2.

Grafts of whole cell populations cannot exclude the possibility that individuals within that population are predetermined in their migratory patterns and developmental fates (see Noden, 1980, for further details). Certainly the question of whether the phenotypes of some of these cells are predetermined lies beyond the scope of this chapter. However, the preponderance of evidence from such grafting studies clearly shows that the environment is primarily responsible for directing at least the patterns of migration (see also Le Douarin, 1982; Noden, 1980; Weston, 1980).

The environment not only dictates the pathways of migration but in some cases also determines the extent of that movement. Weston and Butler (1966) found that if an older neural tube was replaced with a [³H]thymidine-labeled "young" segment from which the crest had not yet migrated, the grafted cells, which presumably contain complete migratory capacity, were found only in the more dorsal structures. Similar results were obtained by Bronner-Fraser and Cohen (1980), when they injected crest cells into progressively older somites. Thus, the older environment does not provide the same migratory information as the younger environment.

Experimental evidence strongly suggests that the complex environment through which the crest moves is the primary control of its behavior. The elements in this embryonic milieu that play a role in directing the crest cells along their migratory pathways undoubtedly are many; each of these components, as well as its temporal and spatial distribution, is described below.

3.2.2. Extracellular Spaces

Neural crest cells initially do not invade other embryonic tissues, but rather move between them. Consequently, it seems clear that extracellular spaces, in part, determine the pattern of migration of neural crest cells, because the timing and direction of migration are coordinated with the appearance of an appropriate space filled with extracellular matrix. This is especially obvious when one compares the variable behavior of the neural crest at different axial levels. In this regard, the crest cells are like many other cell types that undergo morphogenetic movement in precise pathways correlated with the appearance of space, e.g., sternal primordium (Fell, 1939), optic nerves (Silver and Robb, 1979), and endocardial cushion cells (Markwald et al., 1978). There are numerous examples of barriers deflecting or delimiting crest cell movements, and a few of these are discussed here.

The anterior mesencephalic crest initiates its migration in a rostral direction along the dorsal midline and eventually joins the diencephalic crest dorsal to the optic stalk (Noden, 1975). In this instance, lateral migration is apparently inhibited by the close apposition of the epidermal ectoderm with the lateral margin of the brain (see Noden, 1975, 1978a). This tight contact does not appear to be an artifact of fixation, because, according to Noden, the neural epithelium in this region becomes distorted as the ectoderm is pulled on during surgical removal; furthermore, an indentation of the epidermal ectoderm can be found in living, obliquely illuminated embryos (Tosney, 1982). Once the prosencephalic and anterior mesencephalic crest cells reach the level of the optic stalk, they move ventrally, first around its caudal surface and then around the rostral surface as the stalk further constricts; they do not, however, spread laterally over the stalk's dorsal surface, presumably because of the close apposition of the overlying ectoderm (Duband and Thiery, 1982). Likewise, the advancing crest splits into pre- and postotic streams when it is obstructed by the invaginating otic placode (Anderson and Meier, 1981).

Other variations in crest behavior may also be dictated by available space, although these are more subtle and therefore controversial. At the level of the posterior mesencephalon the crest cells move only laterally over the mesoderm, possibly because the ventral space next to the brain vesicle is too small (Duband and Thiery, 1982). This point, however, is contradicted by Tosney's (1982) demonstration of a large space between the neural tube and mesoderm with her fixation techniques for TEM (Fig. 8). Tosney contends that the exclusive lateral migration is dictated by contact guidance due to the shape of the mesencephalon. In the rhombencephalon there does appear to be a space be-

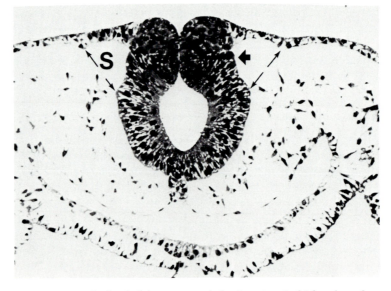

Figure 8. Cross section at the level of the mesencephalon in a stage 9-chick embryo demonstrating the cell-free spaces (S) that appear before emergence of the crest. The size of this space varies axially with the size of the presumptive crest population. Here, the presumptive crest is apparent as closely packed cells in the medial neural folds (large arrow). The dorsolateral edge of the neural tube may direct crest cell dispersion laterally. ×300. (From Tosney, 1982.)

tween the brain and mesoderm; and crest cells aggregate in this region to form some of the cranial sensory ganglia (Duband and Thiery, 1982).

In the chick and mouse trunk, the space between the somites and the neural tube is the first to appear and, consequently, first to be invaded by crest, whereas the lateral pathway is used only later. Indeed, the ectoderm and somites are so closely apposed in the chick that their basal laminae appear to be fused (Newgreen and Gibbins, 1982). Lateral migration of the mouse crest provides another example of the apparent role of space. Migration here is associated with the breakdown of the dermatome into a mesenchyme (Derby, 1978). In *Ambystoma*, however, Löfberg *et al.* (1980) documented the opposite sequence of trunk migration. They found that the first space to open up is in fact over the somites, rather than between the somites and neural tube.

Spaces are important in later (secondary) migratory events as well. Hay and co-workers (Bard and Hay, 1975; Bard *et al.*, 1975) described the invasive behavior of crest cells into the cornea, where they first form the endothelium and later the corneal stroma (Noden, 1978*b*; Johnston *et al.*, 1979); in the case of both structures, a space opens up just before the invasion of the crest. The following sequence of events has been documented: Just after their initial migration, cranial crest cells are found at the periphery of the optic cup as the lens invaginates. Then, at a later stage (stage 22, day 3), a space filled with an acellular collagenous stroma appears between the lens and corneal epithelium. Crest cells, however, do not invade this region immediately, probably because a dense fibrous matrix around the lips of the optic vesicle precludes movement. On day 4.5 (stage 23) this matrix swells, and crest cells invade the space by using both the surface of the lens capsule and the apposed stroma as substrata for migration (Bard *et al.*, 1975). The crest cells form a monolayer, which is apparently dictated by the size of the space; if the lens is removed and the anterior eye is cultured on a nylon raft, crest cells migrate on the stromal surface and pile up to form a multilayer of cells (Figs. 38 and 39 in Bard *et al.*, 1975). By day 5.5, the stroma swells to >50 μm in thickness (Toole and Trelstad, 1971), permitting a second wave of crest cells to colonize the collagenous stroma. Eventually, the stroma begins to collapse and, concomitant with this event, the crest cells cease migration (Toole and Trelstad, 1971).

The eye can be manipulated experimentally to demonstrate further the role of spaces in crest migration. If the anterior eye is grown in culture before endothelial migration begins, the lens is sometimes inadvertently pulled away from the edges of the optic cup, creating a small gap. Some crest cells will spread along the underside of the lens, rather than bridging the gap and moving onto the upper lens surface (Bard *et al.*, 1975). Thus, in this case it seems irrefutable that a simple opening of space can direct migration, provided that a suitable substratum for cell attachment is present as well.

The spaces in almost all these situations have been shown to expand due to the presence of the hydrophilic glycosaminoglycan, hyaluronic acid. This has been demonstrated in several ways. First, hyaluronate is produced at the same time that spaces appear (Toole and Trelstad, 1971; Pratt *et al.*, 1975; Derby, 1978) and disappears when spaces collapse (Toole and Trelstad, 1971). Second, when hyaluronate is removed experimentally by enzymatic digestion with hyaluronidase, these spaces collapse (Anderson and Meier, 1982; Fisher and Solursh, 1977). In most cases, the actual cells producing the hyaluronate that appears in the crest pathway have not been precisely determined. However, it has been demonstrated from tissue culture studies that ectoderm, somite and neural tube all can produce hyaluronate (Pintar, 1978; Hay and Meier, 1974), as can the neural crest cells themselves (Pintar, 1978; Glimelius and Pintar, 1981). It is interesting to note that in the early stages of migration, as the crest cells move on the dorsal surface of the neural tube, the spaces appear to be opening just in advance of their migration (Duband and Thiery, 1982; Thiery et al., 1982a; Löfberg et al., 1980; Derby, 1978). Thus, the crest may be opening its own path. As a particular example, the crest-derived endothelium of the eye has been shown to produce the hyaluronate responsible for corneal matrix swelling (Meier and Hay, 1973). Other tissues in addition to the crest may be responsible for creating spaces, because at times crest cells are delayed for several days before a pathway opens that they can exploit, e.g., over the somites in mouse (Derby, 1978).

Although the availablity of space seems to be an important parameter in directing the dispersion of the crest, several unsubstantiated points should be kept in mind:

1. It is not clear how large a space is necessary to permit a crest cell to pass. Anderson and Meier (1982) removed most, if not all, hyaluronate in chick embryos by enzyme digestion and even though the spaces collapsed, the cranial crest migrated as usual. In fact, in some cases the block to migration may not be the actual size of the space itself, but rather the structure of the space; fused basal laminae or dense collections of extracellular matrix may delimit the migration (see Newgreen and Gibbins, 1982; Thiery et al., 1982a; Tucker and Erickson, 1984). For example, in the study by Anderson and Meier (1982), hyaluronate-induced swelling may have already disrupted such blockages by the time the enzyme was added, so that the spaces themselves, beyond some minimum, may be trivial in later stages of migration. The precise condition of the extracellular matrix in the examples discussed has not yet been examined with high-resolution TEM.

2. As most of the studies cited failed to undertake a scrupulous examination of the matrix molecules present in these occluded spaces, we cannot really eliminate the possibility that molecules are present that themselves prevent migration.

3. It is possible that spaces appear concomitant with matrix molecules that are necessary for crest adhesion to a substratum. Thus, the patterns of migration may not be determined by passive spreading into spaces but rather by active movement on specific matrix components.

3.2.3. Basal Lamina Barriers

The initial migratory pathways of the neural crest are in many cases bounded by epithelia resting on basal laminae. It has been suggested that these basal laminae and the epithelia with their tightly adhering junctions may well delimit early crest migration by acting as an impenetrable barrier (e.g., see Weston, 1982). In the chick and mouse trunk, the lateral and ventral neural tube, the epidermal ectoderm, and the dorsal and lateral portions of the somites are all bordered by intact basal laminae (e.g., see Hay, 1978; Tosney, 1978, 1982; Erickson and Weston, 1983). Because crest cells can locomote within the neural epithelium (C. A. Erickson, unpublished results), the epidermal ecto-derm (Teillet and LeDouarin, 1971), and the somites (Erickson et al., 1980; Bronner-Fraser and Cohen, 1980)—if grafted there—the basal lamina may well be the barrier that keeps crest cells out during early stages of migration. There is little direct evidence that this is the case, other than the fact that neural crest cells grown on isolated basal laminae from human amnions (see Liotta et al., 1980) do not breach them (C. A. Erickson, unpublished data). Even so, there is no guarantee that an embryonic basal lamina and the amnionic basal lamina are identical in structure and therefore would be expected to elicit the same behav-ior from the neural crest (but see Fig. 4).

The basal lamina could also separate the crest from other tissues by provid-ing a substratum that is particularly adhesive and on which crest cells would prefer to flatten and spread. Indeed, one component of the basal lamina, lami-nin, reported to be highly adhesive for the crest (Rovasio et al., 1983; New-green, 1984), has been demonstrated to direct Schwann cell movement on the endoneurium of sciatic nerves by haptotaxis (McCarthy et al., 1983). It should be noted, however, that although some neural crest cells have been observed contacting basal laminae (Tosney, 1982; Bancroft and Bellairs, 1976), most of the crest population is multilayered and enveloped by the three-dimensional matrix of the pathways. Therefore, if the basal lamina is a barrier to migration, it may not function alone in that capacity. Perhaps the extracellular matrix material associated with the lamina (i.e., the basement membrane) may be a deterrent to invasion as well. For example, in the early developing limb of Xenopus (stage 44), the mesenchyme cells are separated from the ectoderm by a thick basement membrane consisting of highly organized collagen fibrils asso-ciated with the basal lamina. By stage 47, this collagen becomes disorganized; subsequently, mesenchyme filopodia are able to penetrate it to contact the basal lamina (Kelley and Bluemink, 1974).

3.2.4. Extracellular Matrix

Although the spaces and basal laminae may be important and even neces-sary in defining the crest pathway, they are not sufficient, because the cells must also be able to adhere to a substratum in order to translocate. A variety of macromolecules have been localized in the crest migratory pathway that may be important in crest morphogenesis. Indeed, these may be crucial in defining the pathways if very adhesive molecules are found only in the extracellular spaces and are absent from adjacent embryonic tissues. One approach to deter-mining which factors in the extracellular matrix are responsible for this control is to examine the relationship between the timely appearance and distribution of matrix components and changes in crest morphogenesis. Although this exercise can provide correlative evidence, it does not test crest–matrix interactions directly. A second, and more direct, approach is to confront migrating neural crest cells with isolated matrix materials and to observe their response to these molecules in tissue culture. This approach suffers from the possibility that molecules important to crest migration may not have been included or that the configuration and binding characteristics of known components *in vivo* may be different in culture. Taken together, however, these two experimental approaches can provide a more detailed understanding of the controls of crest morphogenesis. The individual molecules presently thought to determine crest migratory pathways are discussed below.

Fibronectin (FN) has been identified in the crest pathways and basement membranes bordering these pathways using antibody localization techniques (Fig. 9), but it is relatively scarce in the areas adjacent to the pathways (Newgreen and Thiery, 1980; Duband and Thiery, 1982; Thiery *et al.*, 1982a). Fibronectin has also been localized by TEM using antibodies conjugated to ferritin or peroxidase and appears as 5–10-nm microfibrils (Mayer *et al.*, 1981); these fibrils are similar in size to those produced in culture by chick embryo fibroblasts (Chen *et al.*, 1978). Fibronectin has also been found in the amorphous interstitial bodies first described by Low (1970). Furthermore, fibronectin greatly stimulates flattening and speed of movement of crest cells when used as a substratum *in vitro* (Newgreen *et al.*, 1982; Rovasio *et al.*, 1983; Erickson and Turley, 1983; Tucker and Erickson, 1984) (see Fig. 10). Finally antibodies to fibronectin stop crest cell movement on fibronectin *in vitro* (Rovasio *et al.*, 1983). Thus, because of its distribution in the embryo and its dramatic stimulation of motility in culture, fibronectin is considered one of the primary matrix molecules used as a substratum by the crest.

Recently two studies that perturb neural crest cell migration *in vivo* have corroborated the conclusions based on tissue culture studies. When decapeptides that competitively inhibit fibronectin function are injected into the mesencephalon at the time neural crest cells initiate their migration, normal morphogenetic movement ceases (Boucaut *et al.*, 1984). Also, when antibodies that recognize a 140,000 M_r protein complex that is thought to be a fibronectin receptor are similarly injected, a marked reduction in cranial crest migration is observed (Bronner-Fraser, 1985b).

Collagen is also present in the initial crest pathway (Frederickson and Low, 1971), although the type, amount, and distribution have not been characterized in detail. Type I collagen is found in the trunk at slightly later stages than when the crest is dispersing (von der Mark *et al.*, 1976), and in some instances striated fibrils characteristic of collagen have been seen coincident with crest dispersal (Hay, 1978; Frederickson and Low, 1971; Löfberg *et al.*, 1980). Most of the collagen, however, if it is present, may be in the form of procollagen. As for the mobility of crest cells on collagen, *in vitro* studies show that they do not migrate on denatured collagen (Newgreen, 1982b; Erickson and Turley, 1983) unless fibronectin is also present. However, crest cells do move quite readily through a three-dimensional collagen gel in the absence of fibronectin (Davis, 1980; Tucker and Erickson, 1984) (see Fig. 11). Until the distribu-

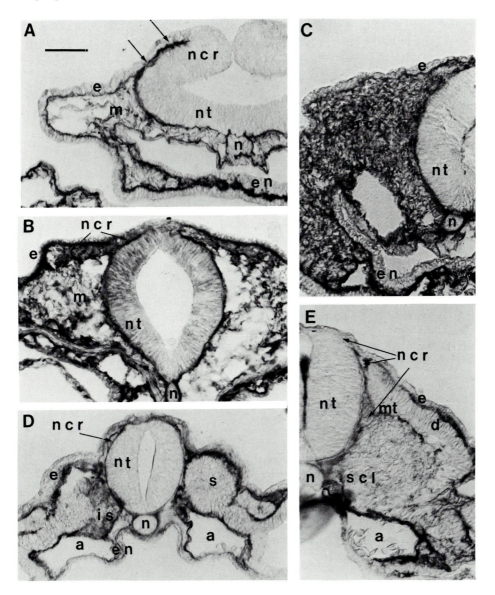

Figure 9. Transverse sections of chick embryos after immunoperoxidase labeling for fibronectin. (A) Mesencephalic level before crest cell migration (5-somite stage). Note the fused stained basement membrane (arrow) lateral to the neural crest. (B) Initial crest migration at the mesencephalic level (8-somite stage). Crest cells move under the ectoderm. (C) Mesencephalic level after invasion of the paraneural space by crest cells (16-somite stage). Note the heavy staining among the crest cells. (D) Trunk level at the onset of crest cell migration (somite 15, 18 somite stage). Note the stained perisomitic band. (E) Trunk level during later crest cell migration (somite 20, 30-somite embryo). Crest cells are surrounded by stained extracellular matrix. Note the decrease in perisomitic stain near the sclerotome and the stain extending under the myotome. ncr, neural crest; nt, neural tube; n, notochord; s, somite; is, intersomitic cleft; scl, sclerotome; d, dermatome; mt, myotome, e, ectoderm; en, endoderm; a, aorta; m, cranial endomesenchyme. Scale bar: 50 μm. (From Newgreen and Thiery, 1980.)

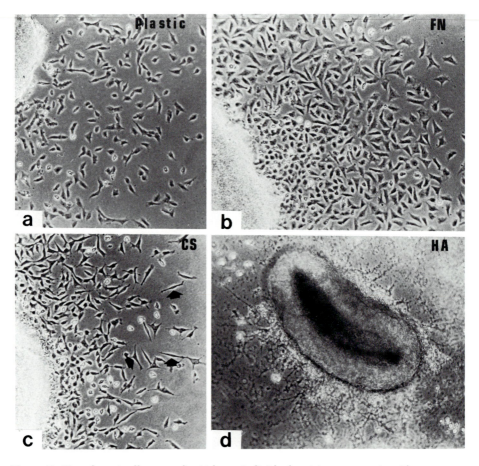

Figure 10. Neural crest cells grown for 18 hr on individual matrix components with no serum or embryo extract in the medium. The neural tube is visible at the margin of each micrograph. (a) Neural crest cells on tissue culture plastic. Most cells are relatively rounded. (b) Neural crest cells grown in medium containing 25 μg fibronectin. The cells are considerably flatter and at higher density than on plastic alone. (c) Neural crest cells on chondroitin sulfate derivatized to tissue culture plastic. Cell morphology is similar to that on plastic alone. (d) Neural tube attached to hyaluronate derivatized to tissue culture plastic. Only a few neural crest cells have been able to migrate on this substratum. ×990 (From Erickson and Turley, 1983.)

tion and organization of collagen are known for certain, it remains possible that the crest cells use collagen as a substratum as well as fibronectin.

Glycosaminoglycans (GAG) have been localized in the trunk crest pathway of mouse and chick embryos (Kvist and Finnegan, 1970a; Strudel, 1976; Derby, 1978; Pintar, 1978) and in the cranial crest pathway (Pratt *et al.*, 1975; Bolender *et al.*, 1980; Brauer *et al.*, 1985), using a variety of staining techniques. The embryonic GAG appear to be primarily hyaluronate and chondroitin sulfate, although smaller quantities of other types may not have been detected at the level of resolution provided by these techniques.

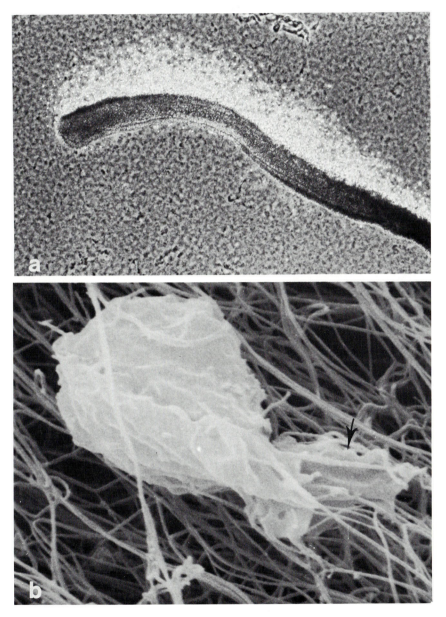

Figure 11. (a) Neural crest cells have migrated away from this neural tube, which was embedded in a three-dimensional hydrated collagen lattice, to form a halo of cells. ×400. (b) Scanning electron micrograph of a neural crest cell migrating through a collagen matrix. Note the rounded cell body being pulled by a ruffling lamellipodium (arrow). ×9000. (R. P. Tucker and C. A. Erickson, unpublished results.)

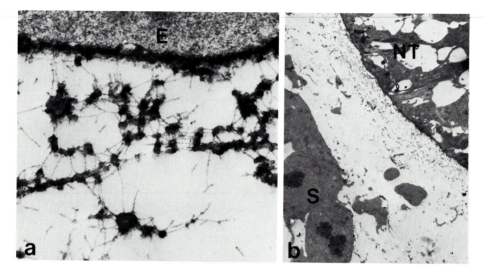

Figure 12. (a) Extracellular matrix beneath the ectoderm (E), showing ruthenium red-stained granules (chondroitin sulfate) linked by stained microfibrils that are probably hyaluronate. Posterior somite level of a 25-somite chick embryo. ×30,000. (b) Ruthenium red-stained extracellular matrix between the neural tube (NT) and the somites (S) is more dense adjacent to the neural tube. Posterior somite level of a 24-somite chick embryo. ×2500. (From Newgreen *et al.*, 1982.)

Hyaluronate and chondroitin sulfate have been localized with higher-resolution TEM, although to a limited extent. Cetylpyridinium chloride or ruthenium red both retain GAG in fixed material. Using such fixation techniques, 1–3-nm fibrils have been preserved in the matrix of avian neural crest pathways and have been identified as hyaluronate (Fig. 12a) because they are removed by hyaluronidase digestion (Frederickson and Low, 1971; Hay, 1978). These fibrils appear scattered sparsely throughout the extracellular spaces that will later be occupied by crest cells. Similar microfibrils have been observed in cardiac jelly (Markwald *et al.*, 1978) and in the chick limb (Singley and Solursh, 1980) and have been identified as hyaluronate. Chondroitin sulfate is also retained using cetylpyridinium chloride or ruthenium red; it appears as 40-nm granules (Fig. 12b), which are especially dense near the ectoderm, basal lamina of the neural tube, and around the notochord (Newgreen *et al.*, 1982; Hay, 1978). Because these granules are sensitive to digestion by both chondroitinase ABC digestion and trypsin (Hay, 1978), they probably contain chondroitin sulfate proteoglycan.

Hyaluronate is present in high concentrations (estimated to be as high as 20 mg/ml; Derby, 1978) in regions in which crest cells are migrating, but it is greatly reduced where crest migration has ceased, as in the sensory ganglia (Derby, 1978). Similarly, the colonization of the corneal stroma occurs concomitant with hyaluronate synthesis and swelling of the stroma, whereas cessation of migration is associated with hyaluronidase production and collapse of this space (Toole and Trelstad, 1971). Thus, appearance of hyaluronate is correlated with crest migration, as it is with migration in other systems (see Toole, 1981). Hyaluronate is not, however, an adhesive molecule capable of sustaining crest migration (Fisher and Solursh, 1979a; Erickson and Turley, 1983). Its function in crest morphogenesis is most likely to expand the extracellular spaces, as has been suggested by many other investigators. Indeed, recent experiments from our laboratory demonstrate that speed of crest locomotion in collagen gels is increased by the addition of low concentrations of hyaluronate (250 μg/ml) that expand the space between collagen fibrils, further suggesting a space-opening function (Tucker and Erickson, 1984). Likewise, similar stimulation by hyaluronate of endocardial cushion cell migration into collagen gels has been reported (Bernanke and Markwald, 1979). One curious observation is that hyaluronate concentrations of >1 mg/ml in collagen gels, or in the medium when crest cells are spread on fibronectin, greatly reduce cell motility (Erickson and Turley, 1983; Tucker and Erickson, 1984). This concentration is considerably lower than the estimated concentration of hyaluronate in the crest pathway employing densitometry techniques (Derby, 1978). Either the concentration of hyaluronate is considerably lower than estimated, and indeed the biochemical analysis of whole trunk shows hyaluronate concentrations at 0.18 mg/ml (Kvist and Finnegan, 1970b), or hyaluronate *in vivo* is complexed with other binding proteins in such a way as not to interfere with crest migration while still performing its space-expanding function. Recently a hyaluronate-binding protein, hyaluronectin, was localized in the crest migratory pathways of the rat just before the onset of migration (Delpech and Delpech, 1984).

Using Alcian blue staining, chondroitin sulfate has been identified in the crest pathway, although in relatively lower concentrations than hyaluronate (Derby, 1978; Pintar, 1978); using the higher-resolution TEM, chondroitin sulfate proteoglycan granules can be localized more precisely. Crest cells do not normally move in areas containing chondroitin sulfate proteoglycan granules (e.g., around the notochord, initially over the somites, or along the lateral surface of the neural tube; Newgreen *et al.*, 1982) (see Fig. 12b). Furthermore, chondroitin sulfate and chondroitin sulfate proteoglycan reduce crest cell adhesiveness to a fibronectin substratum (Newgreen *et al.*, 1982; Newgreen, 1982a; Erickson and Turley, 1983) or to collagen in a gel (Tucker and Erickson, 1984). Therefore chondroitin sulfate may act to exclude crest cells from certain areas, and such a mechanism of exclusion of migration would be as effective as stimulation of migration in determining the pathways of migration (see Morris, 1979).

4. Crest Migratory Ability

4.1. Determination of Migration Capacity

The crest pathway is similar in its composition to a variety of other extracellular pathways utilized by migrating embryonic cells. For example, many cell types invade matrices expanded by hyaluronate, such as cardiac cushion cells (Markwald *et al.*, 1978); furthermore, most cells in culture and some embryonic cells such as primordial germ cells (Heasman *et al.*, 1981; see Chapter 11) move on a substratum of fibronectin (for review, see Yamada, 1980,

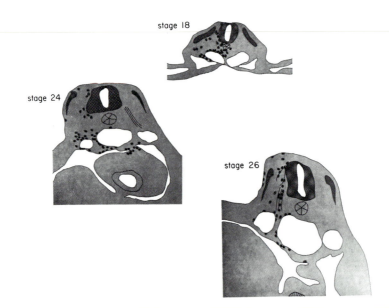

Figure 13. Pigmented crest in crest pathway. Camera lucida drawings of compilations of five representative sections of chick embryos showing distributions of pigmented neural crest cells 24 hr (stage 18), 48 hr (stage 24), and 72 hr (stage 26) after grafting cells into the anterior trunk crest pathway (at level of last somite of 15-somite hosts). The distribution of pigmented crest cells is identical to the usual distribution of the neural crest, except that few or no pigment cells are incorporated into the dorsal root ganglia. Note also the extensive migration of pigment cells into the dorsal mesenteries. (From Erickson et al., 1980.)

1983). It was our original conjecture, as well as that of others, that the crest pathway is relatively nonspecific and would sustain migration of other cell types introduced into it. If freshly isolated crest cells (Erickson et al., 1980; Bronner and Cohen, 1979; Bronner-Fraser and Cohen, 1980) and crest derivatives, e.g., pigment cells (Le Douarin et al., 1978; Erickson et al., 1980; Bronner-Fraser and Cohen, 1980; Le Lièvre et al., 1980; Ayer-Le Lièvre and Le Douarin, 1982) are introduced into the crest pathway (see Fig. 13), they will resume migration along the appropriate pathways. By contrast, however, fibroblasts of various embryonic origins, e.g., somite, heart, and limb bud, will remain at the graft site (Erickson et al., 1980; see also Fisher and Solursh, 1979b; Noden, 1978b; Nakamura and Ayer-Le Lièvre, 1982). Indeed, the only noncrest cell type demonstrated to use this pathway as a substratum for migration is the mouse sarcoma cell line S180 (Erickson et al., 1980) (see Fig. 14).

These results suggest that most embryonic cell types cannot migrate on the crest pathway, even though they are fully capable of motility at certain times during embryogenesis and in tissue culture. However, one tumor cell known to be invasive in other assays (e.g., Tickle et al., 1978) can utilize the crest pathway. This finding suggests (1) that the crest pathway is not specific only for newly migrating crest; and (2) perhaps more interesting and surprising, that the

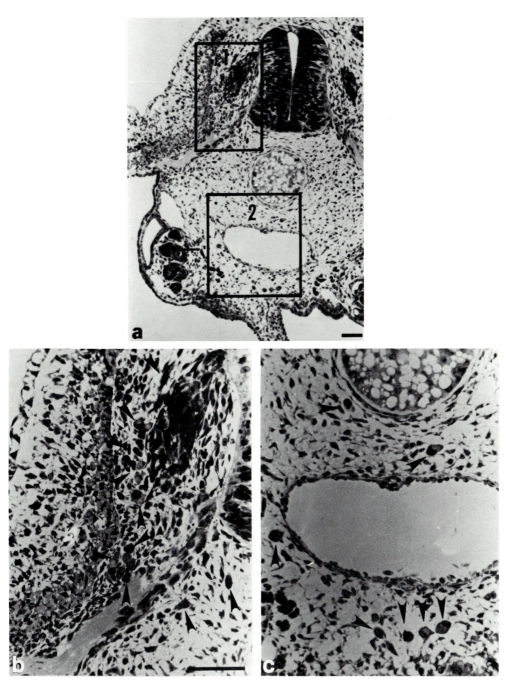

Figure 14. (a) Low magnification of a stage 24 chick embryo fixed 48 hr after a pellet of sarcoma 180 cells was grafted into the crest pathway at level of somite 26 of 27-somite host. (b) High magnification of boxed area 1 in (a). The sarcoma 180 cells (arrowheads), which are characterized by large, deep-staining nuclei, are distributed beside the dorsal root ganglion and along the ventral nerve root. (c) High magnification of boxed area 2 in (a). Sarcoma 180 cells (arrowheads) have distributed themselves in the crest pathway around the dorsal aorta and into the dorsal mesentery. Scale bars: 25 μm; b and c are at the same magnification. (From Erickson *et al.*, 1980.)

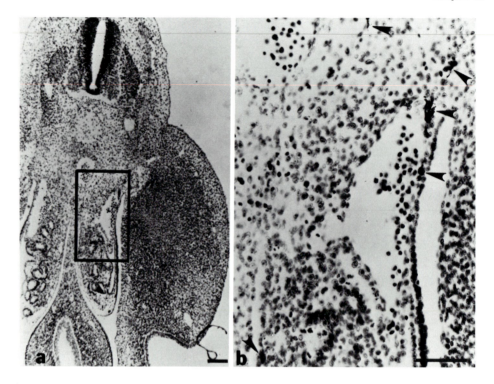

Figure 15. (a) Low magnification of a chick embryo fixed 48 hr after pigmented crest cells were grafted into a stage 21 hind limb bud. (b) High magnification of boxed area in (a). Arrowheads indicate pigmented crest cells that have migrated from the limb into the ventral crest pathway and are now distributed in the dorsal mesentery, region of the adrenal cortex, and around the kidney tubules. Scale bars: 25 μm. (From Erickson *et al.*, 1980.)

crest cells display unusual migratory properties, apart from other embryonic cells, similar to some of the invasive properties of tumor cells.

Not only are crest cells motile in their normal embryonic milieu, but they show extraordinary migratory capabilties when grafted into abnormal embryonic sites. When 2-day-old cultured neural crest or pigmented crest are grafted into a 3-day chick limb, the crest cells not only disperse through the mesenchyme and reach the ectoderm, but they also move medially and then ventrally to localize near the developing kidney and dorsal aorta (Erickson *et al.*, 1980) (see Fig. 15). Because they are found medial to the limb, we presume this is not merely passive movement caused by the growth and expansion of the limb mesenchyme. Interestingly, the crest cells do not normally invade the limb except along the basal surface of the ectoderm. They also seem unable to do so if co-cultured with chick limb mesenchyme on the chick chorioallantoic membrane (Le Douarin, 1982) or if grafted between the paraxial mesoderm and the base of the limb bud (Erickson, 1985; Erickson *et al.*, 1980). Thus, although the crest cells can disperse within the limb tissue and can even escape when grafted into it, they do not seem to be able to invade it during the usual course of development.

Crest cells have also been grafted into the somite, which represents another structure not extensively invaded during normal migration (but see Section 3.1.1d). The crest cells disperse ventrally from this site (Bronner and Cohen, 1979; Bronner-Fraser and Cohen, 1980; Erickson, 1985; Erickson et al., 1980) (see Fig. 16), although some crest cells remain in a dorsad position, but appear to be trapped by the basement membrane of the dermatome–myotome (Erickson et al., 1980). It is likely that some movement from the interior of the somite is passive, since Latex beads injected into the lumen of the somites are later found in ventral portions of the embryo, presumably pushed there by the dispersing sclerotome (Bronner-Fraser, 1982). Also, Weston (1963) found that neural crest cells will disperse through unsegmented lateral plate mesoderm when neural tubes are grafted in that atypical position.

As we have already noted, embryonic fibroblasts grafted into some of these aberrant positions are not motile if grafted into the dorsal crest pathway. Likewise, even limb mesenchyme will not mix with host mesenchyme if positioned in the limb; the cells remain fixed at the site of graft placement (Erickson et al., 1980; Fisher and Solursh, 1979b). Such experiments suggest that crest cells have motile capabilities beyond those of the usual fibroblast and that they are capable of locomotion even in abnormal embryonic sites.

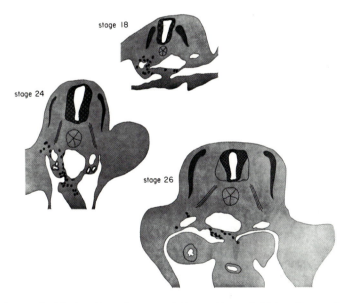

Figure 16. Camera lucida drawings of representative sections of chick embryos, showing distribution of pigmented crest cells 24 hr (stage 18; graft at somite 16 of 17-somite host), 48 hr (stage 24; graft at somite 17 of 17-somite host), and 72 hr (stage 26; graft at somite 17 of 17-somite host), after grafting. Crest cells are seen only in the ventral pathway, primarily above and lateral to the dorsal aorta after 24 hr, but eventually ventral to the aorta and in the dorsal mesenteries. (From Erickson et al., 1980.)

4.2. Crest "Invasive" Behavior

4.2.1. Absence of Fibronectin Production by Neural Crest Cells

What accounts for the unusual migratory ability of neural crest cells is unknown, but several points that may be related are worth noting. Neural crest cells in general do not make fibronectin, nor is it found at their surface (Loring *et al.*, 1977; Newgreen and Thiery, 1980; Sieber-Blum *et al.*, 1981), although some cephalic crest cells at the periphery of an outgrowth *in vitro* or *in vivo* appear to produce small amounts (Newgreen and Thiery, 1980). Tumor cells and transformed cell lines known to be invasive in a variety of model systems also have reduced or no fibronectin on their cell surface (see Hynes, 1976, for review). Because cells without fibronectin on their surface seem to attach to, and spread more avidly on, a fibronectin substratum (Ali and Hynes, 1978), it is possible that such cells could better use the fibronectin-filled pathways of the embryo. Alternatively, cell surface fibronectin may act as a stop signal to prevent migration of noncrest cells, perhaps by gluing the cells in place. For example, Bronner-Fraser (1982, 1985a) reported that Latex beads coated with fibronectin remain in the somites after being injected there, whereas uncoated beads are passively dispersed by the sclerotome. This hypothesis cannot explain, however, how some cephalic crest cells that do have cell surface fibronectin are also able to migrate.

4.2.2. Proteolytic Enzymes

A variety of tumor cells produce proteolytic enzymes such as plasminogen activator or collagenase (see Ossowski *et al.*, 1975), which have been thought to aid their invasion through tissues and basal laminae by degrading the matrix. Recently cephalic crest cells (Valinsky and Le Douarin, 1985) and trunk neural crest cells (R. R. Isseroff and C. A. Erickson, unpublished results) have been shown to produce the serine protease plasminogen activator. Thus far direct evidence for the involvement of proteases in permitting the crest to tunnel through the extracellular matrix has not been obtained.

4.2.3. Reduced Tractional Force

Recently, Harris and co-workers (1980, 1981) showed that fibroblasts exert a tractional force as they locomote of sufficient strength to deform either a silicone rubber sheet or a low concentration collagen gel. These investigations have pointed out that the tractional force measured far exceeds that necessary for locomotion and thus may have a more important role in tissue modeling by stretching and aligning the collagen (Harris *et al.*, 1981). Other cell types (i.e., those that move rather rapidly and appear weakly adherent to their substratum, such as nerve growth cones and lymphocytes), do not deform such substrata and indeed move extensively in embryonic and adult tissues and spaces.

Although crest cells can deform very low-concentration collagen gels, i.e., only those at ≤ 100 μg/ml (Davis and Trinkaus, 1981), we have detected no such deformation in collagen gels exceeding a concentration of 250 µg/ml (Tucker *et al.*, 1985). In addition, crest cells are less flattened and are probably less adhesive than most fibroblasts (Newgreen *et al*, 1979), further suggesting that the former might exert only a slight tractional force. On the other hand fibroblasts that have been grafted into the embryo and do not move are known to deform collagen gels or silicone sheets (Tucker *et al.*, 1985). They are also known to tear low-concentration collagen gels in which they are immobile. One interesting possibility is that neural crest cells are more migratory in the early embryonic spaces because they are less likely to deform or tear the matrices on and through which they move. Although this is an intriguing idea, it remains very difficult to test.

4.3. Passive Dispersion

There is bountiful evidence that crest cells actively migrate in the embryo. Furthermore, in the only time-lapse study of early crest cell migration *in vivo*, Keller and Spieth (1984) clearly show that the crest cells that move over the somites to populate the epidermis do so by lamellipodium-directed motility (Fig. 17). Recently, however, a number of studies have suggested that some of the remarkable distances traversed by the crest may be due in part to passive movement directed by growth or dispersion of the surrounding ectoderm or mesoderm.

In a recent review, Noden (1984) summarized three different studies in which host chick ectoderm, mesoderm, or neural crest at the level of the mesen-cephalon were replaced with equivalent quail tissue. When the grafted cells were followed in successive stages, the lateral expansion of the ectoderm, underlying mesoderm, and the crest progressed at the same rate. In the head region, at least, once the crest cells have managed to spread over the head mesoderm, they may attach and continue their ventral progression more or less by riding along. In an even more remarkable study in which chick splanchno-pleure was replaced at the level of somites 1–6 with equivalent quail tissue, Noden (1984) found that the mesoderm of the entire foregut and part of the midgut is derived from this grafted tissue in older embryos. Thus, the crest cells that reach the level of the pharynx may merely passively ride the endoderm (or mesoderm) as it grows. This cannot explain, however, how vagal crest cells transplanted to the thoracic region are able to invade the gut and migrate cranially and caudally (Le Douarin and Teillet, 1974). Crest cells are clearly able to migrate under their own power, but passive movements may account for part of their extensive dispersal. Although still far from substantiated, the possibility of passive movements of crest deserves serious consideration and investigation.

Bronner-Fraser (1982) showed that passive movements may also occur in the trunk region of the chick. If neural crest cells or Latex beads are injected into the somites, they are found in the ventral region of the embryo after the sclerotome disperses. The dispersal of the sclerotome may occur in part by active motility (see Chapter 12), but a large component of dispersal is undoubt-

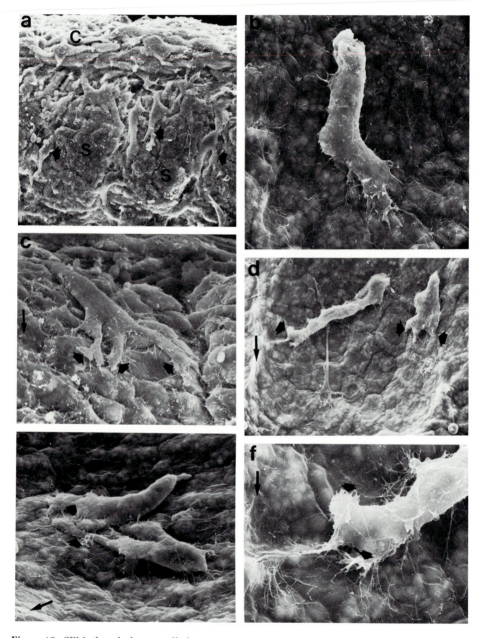

Figure 17. SEM of axolotl crest cells between stages 32–34. The pointers in (a) indicate cells that have left the neural crest (C) and are moving laterally and ventrally on the outer surface of the somites (S). Others are attached to the inner surface of the epidermis (b, d, e, f). Cells moving on the somite (c) or on the epidermis have one to three large protrusions at their ventral (leading) edges (arrowheads in c, d, e). These end in fine, filiform extensions (f). a: ×190; b: ×640; c: ×500; d: ×285; e: ×500; f: ×1140. (From Spieth and Keller, 1984.)

edly passive due to swelling by hyaluronate (Solursh *et al.*, 1979) and by growth (Gasser, 1979). It should come as no surprise, then, that beads in the sclerotome might be moved along as well. However, those crest cells that do migrate through the somites follow highly patterned routes whereas the early pattern of bead dispersion is not clear. Furthermore, many trunk crest cells have found their way to the dorsal aorta and other ventral structures well in advance of extensive sclerotome dispersion (cf. Thiery *et al.*, 1982a; Newgreen, 1982a). Thus it seems unlikely that the bulk of the trunk crest dispersion is passive. Nevertheless, the migration over the somites and into the skin is delayed with respect to ventral dispersion between the neural tube and somites; some of this later movement into the dermis and epidermis may well result from the underlying growth of the dermatome–myotome. This process clearly needs to be examined further.

5. Directionality of Migration

Neural crest cells must receive some directional cues to control their migration, because the crest pathways as we have described them in Section 4 define orientation only. In other words, some stimulus must direct the neural crest cells to disperse from their point of origin, rather than just to mill about randomly in place; furthermore, once cells enter a pathway, they must know whether to turn right or left [for lucid discussions of orientation versus directionality, see Trinkaus (1976, 1982)]. Numerous hypotheses for the directional migration of the neural crest, as well as other cell types, have been advanced. Contact guidance could provide orientation, whereas chemotaxis, negative chemotaxis, contact inhibition, haptotaxis, and galvanotaxis could ensure directionality away from the neural tube. Evidence in support of and contrary to each is presented in the following section.

5.1. Contact Guidance

Weiss (1945) conducted an extensive investigation of the phenomenon of contact guidance, first described by Harrison (1910), which is simply that cells can be guided by discontinuities in their substrata. For example, cells will align on scratches on a glass petri dish or on oriented native collagen fibrils (Dunn and Ebendal, 1978). The mechanisms that control contact guidance are complicated and beyond the scope of this chapter (for a detailed discussion, see Dunn, 1982).

Löfberg *et al.* (1980) demonstrated fibrils along the neural tube that are aligned parallel to the direction of crest migration; these workers hypothesized a role for these fibrils in orientation of the crest. It is unlikely, however, that contact guidance plays much of a role in crest directional movement for a variety of reasons. Aside from Löfberg's observation in amphibians, there is no evidence in other species, including chick (Bancroft and Bellairs, 1976;

Tosney, 1978, 1982; Ebendal, 1977) or mouse (Erickson and Weston, 1983), that fibrils are aligned in the direction of crest cell migration. In fact, Tosney (1982) showed that fibrils in the mesencephalon of the chick span the crest pathway from the mesenchyme to the ectoderm; thus, these fibrils are oriented normal to the direction of head crest migration (Fig. 18). In some instances, fibrils are slightly aligned at the lamellipodium of a crest cell (Hay, 1978; Spieth and Keller, 1984); similarly aligned fibrils are seen radiating from the leading edge of endocardial cushion cells (Markwald et al., 1979). These more likely represent transitory tractional alignment at points of adhesion to the substratum (see Harris et al., 1981), however, and although this might aid in persistence of directional movement, it is unlikely to account for initial orientation of the cells.

5.2. Chemotaxis

A variety of cell types such as the polymorphonuclear leukocyte (Zigmond, 1982) orient their locomotion along a gradient in concentration of certain diffusible molecules by extending lamellipodia toward the source. It has been proposed that a chemotactic source may be present in the embryo to direct ventral and lateral movements. Using a Boyden chamber assay, Greenberg and co-workers (1981) demonstrated that fibronectin not only stimulates motility, but appears to produce a chemotactic response as well. Both the intact molecule and the cell-binding fragment elicit similar behavior. In this assay, the substratum filter was first coated with collagen, to which intact fibronectin can bind. It remains to be shown that fibronectin and the fibronectin fragments do not bind to the filter and that the cells are not responding to substratum-bound material (see McCarthy et al., 1983). Thus, there is little evidence to support the notion of chemotaxis, and no gradient has been identified in the embryo. Although this lack of evidence certainly does not constitute proof that such a mechanism does not exist, several pieces of evidence lobby against chemotaxis in the case of the neural crest. Weston (1963) showed that inverting the neural tube caused the crest to move dorsally with respect to the embryo (i.e., normal in relationship to the neural tube and reversed in relationship to the usual situation). Likewise, crest cells grafted deep into the ventral pathway before host crest migration begins can also move dorsad along both sides of the neural tube (Erickson, 1985; Erickson et al., 1980) (see Fig. 19). These results would not be expected if a chemotactic source were located in the ventral portion of the embryo.

One instance is noteworthy when considering chemotaxis, however. Vagal crest cells migrate into the gut, whereas crest cells from other axial levels stop short of the endoderm. If vagal crest cells are transplanted into the thoracic level, however, they will enter the gut as well, even though they are in a region that usually does not support migration to that extent (Le Douarin and Teillet, 1974). Likewise, crest derived from nodose and ciliary ganglia will similarly invade the gut when grafted into the thoracic level (Ayer-Le Lièvre and Le

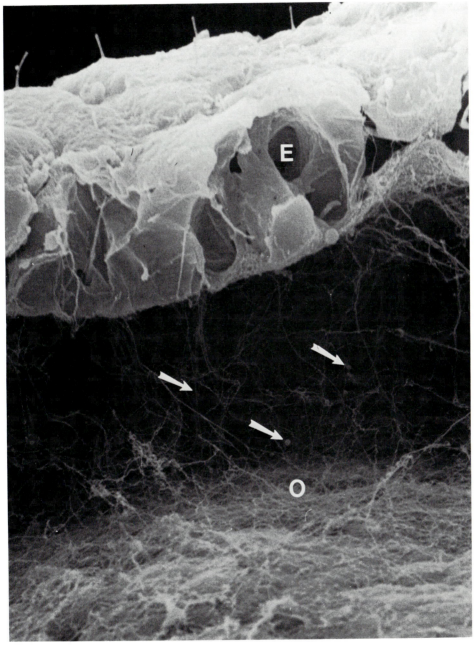

Figure 18. SEM of the extracellular matrix along the anterior mesencephalic pathway in a stage 10 chick embryo. Many fibrils extend between the ectoderm (E) and optic epithelium (O). Since crest cells will enter this space from a plane perpendicular to the micrograph, the orientation of the matrix fibers is normal to the direction of crest movement. Arrows indicate interstitial bodies. ×5800. (From Tosney, 1982.)

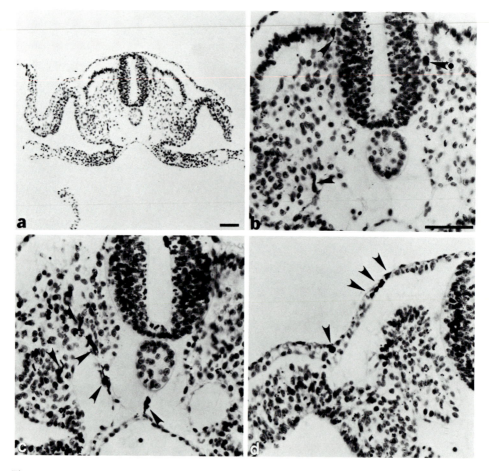

Figure 19. (a) Low magnification of a chick embryo fixed 24 hr after pigmented crest cells were grafted into the segmental plate mesenchyme approximately 10 somite lengths posterior to the last somite of a 19-somite host. (b) High magnification of section in (a), showing crest cells (arrowheads) beside the neural tube and near the dorsal aorta on the side that received the graft, as well as another quail crest cell, which is located contralaterally beside the neural tube (arrowhead). (c) An adjacent section from the same embryo showing crest cells dispersed throughout the sclerotome. (d) Another section from the same embryo, showing that when grafted into very young axial levels, crest cells can migrate dorsad and localize in the ectoderm. Scale bars: 25 μm; b–d are at the same magnification. (From Erickson *et al.*, 1980.)

Douarin, 1982). This finding prompted Le Douarin and Teillet (1974) to predict that the gut releases a chemotactic factor to which some predetermined crest cells respond (see also Le Douarin, 1982). D. Newgreen (unpublished results) was unable to demonstrate chemotaxis when he confronted either chick vagal crest or sacral crest with isolated endoderm in culture. Likewise, Epperlein (1974) could find no evidence for chemotaxis in similar *Xenopus* cultures. Newgreen (1982a) and Noden (1978a) suggest that the vagal crest represents a larger population than arises at thoracic or sacral levels, and perhaps population pressure simply drives these cells more ventrad into the gut. Both sides have evidence in support of their opinions, but this issue is yet to be resolved.

5.3. Negative Chemotaxis

Negative chemotaxis (the repulsion of cells away from a source) has been suggested as a mechanism that drives neural crest cells away from the neural tube. In fact, Twitty and Niu (1948, 1954) showed that neural crest isolated from the newt *Taricha* and grown beneath a coverslip appeared to be more dispersed (i.e., the cells were at lower density) than were those that escaped and were covered by medium only. Furthermore, crest cells drawn into narrow-bore capillary tubes dispersed farther than did those in wide-bore tubes. On the basis of these observations, Twitty and Niu (1948, 1954) asserted that crest cells produce a factor that repels other crest cells when it accumulates in sufficiently high concentrations, such as beneath a coverslip or in the narrow confines of a capillary tube. Presumably, then, crest cells *in vivo* would disperse from a point at which they are in the highest concentration, which is the dorsal surface of the neural tube. Recent attempts to repeat these experiments with either quail neural crest cells (Erickson and Olivier, 1983) or amphibian crest (R. E. Keller, personal communication) have not proved successful; in fact, Erickson and Olivier's observations suggest that Twitty and Niu's results could be accounted for by artifacts of their culture system and by contact inhibition (see Section 5.5).

5.4. Haptotaxis

Tissue culture studies have demonstrated the ability of cells to move up an adhesive gradient (Carter, 1967), primarily because cell processes detach more readily from the less adhesive end than from the more adhesive end (Harris, 1973). Furthermore, neurons have been observed with time-lapse cinemicrography to move preferentially onto a more adhesive substratum from a less adhesive one (Letourneau, 1975). Crest cells *in vitro* are known to adhere differentially to different substrata, and their migratory behavior is subsequently affected (Erickson and Turley, 1983; Rovasio *et al.*, 1983; Newgreen *et al.*, 1982). Therefore, haptotaxis, or movement up a gradient in adhesiveness, provides an attractive model for the directional movement of embryonic cells, including the neural crest. No such established gradients have been demonstrated directly in the embryo. Furthermore the movement of grafted crest cells from a ventral position in the crest pathway to the top of the neural tube makes such a mechanism unlikely (Erickson, 1985).

Weston (1982) suggested a unique means by which such an adhesive gradient might be generated as the crest cells move. However, crest cells produce hyaluronate (see Pintar, 1978), which is relatively less adhesive than fibronectin or native collagen (see Newgreen, 1982b; Erickson and Turley, 1983). The concentration of hyaluronate would be expected to be greatest within the crest population and less at the leading edge of the outgrowth. This dilution of adhesiveness by hyaluronate might therefore drive the crest away from the highest concentration of hyaluronate, which presumably would be a less adhesive substratum. The crest pathways are known to change in their ability to sustain crest cell migration, which at least supports the idea that some condition in the pathway is altered (Weston and Butler, 1966; Erickson et al., 1980). The possible role of a chondroitin sulfate gradient remains untested, although crest cells also produce this sulfated glycosaminoglycan, which is lowly adhesive. In this regard, we do know that endocardial cushion cells similarly leave behind chondroitin sulfate as they migrate into the cardiac jelly (Markwald et al., 1978). If the jelly is removed intact after this migration, however, and is reseeded with fresh cushion cells, the latter are able to migrate in either direction into the jelly and seem to ignore the gradient of glycosaminoglycan. At least in this case, cell conditioning of the environment does not seem to influence directional cell migration (Kinsella and Fitzharris, 1982).

We should keep in mind that there is an obvious asymmetry of chondroitin sulfate proteoglycan distribution in the embryo and that the crest cells appear to avoid areas rich in this proteoglycan (see Section 3.2.4). Thus, matrix may control directionality, or at the very least, the pathway—not so much by haptotaxis, but simply by exclusion from these nonadhesive areas.

5.5. Contact Inhibition

The simplest explanation for the directional dispersion of the crest is contact inhibition of movement. Contact inhibition, as defined and analyzed in detail by Abercrombie and associates, is the paralysis of lamellipodium extension, and thus cessation of locomotion after two cells contact, with subsequent redirection of movement away from the point of contact (for review, see Abercrombie, 1970, 1982). This phenomenon has two consequences in tissue culture: Cells will (1) move away from a region of high concentration to one of low concentration and (2) be radially aligned with each other because any deviation to the side would result in contact with a neighbor. Crest cells display similar behavior in the embryo. For example, crest cells clearly move away from regions of highest density on the dorsal neural tube to areas free of crest and are generally aligned in their direction of migration (Bancroft and Bellairs, 1976; Tosney, 1978). In addition, when crest cells have begun to deplete the neural tube, and are therefore reduced in density, their orientation becomes more random, a phenomenon also predicted by contact inhibition. Furthermore, crest cells from any one axial level do not generally migrate anteriorly to any great extent. This is predicted by contact inhibition, because crest cells could move freely in a posterior direction where other crest cells have not yet appeared, but they would be restricted in anterior movement by high concentrations. For example, Teillet and Le Douarin (1970) observed that exchanging small segments of chick neural tube for segments of quail neural tube caused few grafted crest cells to move anterior to the graft site, while many more dispersed posteriorly (see Fig. 5.4 in Le Douarin, 1982). In addition, several investigators have found that deleting a portion of the population causes crest cells from other axial levels to move in to fill the gap (e.g., Yntema and Hammond, 1945; Hammond and Yntema, 1947; Twitty, 1949; Twitty and Bodenstein, 1944).

There is some dispute, however, as to whether contact inhibition has been directly demonstrated. It has been suggested that because crest cells on plastic seem to criss-cross more readily, they are less contact inhibited than are fibroblasts (Le Douarin, 1982). This increased criss-crossing seems to be identical to that observed in many transformed cell cultures. At least for transformed cells, this culture morphology is generated by the cells underlapping (i.e., one cell moving under a second cell by using the culture dish as the substratum) each other in regions in which there are few adhesions to the substratum, thereby avoiding contact altogether (Bell, 1977; Erickson, 1978). Crest cells, like many transformed cells, have long thin processes with few adhesions to the substratum and may similarly underlap each other more readily. Davis and Trinkaus (1981) also suggest that crest cells in collagen gels are not contact inhibited. This conclusion was based on the speed of movement of a cell compared with the number of contacts it had with a neighbor. It must be noted, however, that such contacts are very difficult to resolve in gels, and there are many more opportunities for underlapping, and consequently for avoiding contact, in a three-dimensional lattice than on a planar substratum. By contrast, in the embryo crest cells are tightly associated in high-density outgrowths or in narrow spaces in which fewer opportunities for underlapping would be expected.

Newgreen *et al.* (1979) and Rovasio *et al.* (1983) clearly showed that crest cells move directionally from high to low cell concentrations in culture, although neither demonstrates contact paralysis. Time-lapse cinemicrography has permitted observation of such paralysis, especially when the cells are on substrata such as collagen or laminin, where they produce unusually large lamellipodia (Erickson, 1985). In addition to these *in vitro* observations, Bard and Hay (1975) recorded contact inhibition *in vivo* by photographing a contact event between two neural crest cells in the corneal stroma, after which movement ceased. The difficult task of filming chick or mouse crest cells in their early migration *in vivo* has not been accomplished. However, Keller and Spieth (1984) managed to film pigmented crest cells in the axolotl as the cells migrated over the somites beneath the ectoderm. The cells are highly directional in their movement, and yet they are found in low density, moving without contacting each other. In this instance, contact inhibition cannot be invoked because the cells never touch.

Contact inhibition cannot account for the fact that all crest cells eventually vacate the dorsal surface of the neural tube, except in the prosencephalon, where the crest forms the leptomeninges (Noden, 1978b, 1980). A contact inhibition model would predict that some cells should remain behind. Perhaps further displacement away from their original dorsal position is passive due to the rapid growth of the adjacent ectoderm and mesoderm (see Noden, 1984). In any event, contact inhibition remains our best explanation of the control of neural crest dispersion.

Persistence of directional movement could also be accounted for by the related phenomenon of "nudging," which was first noted in *Fundulus* deep cells (Tickle and Trinkaus, 1976), in which touching the side of a cell induces bleb formation on the opposite side. Thus, crest migration away from the neural tube could be due to the cells closest to the tube contacting the cells in front of them, thereby inducing lamellipodium formation away from the neural tube.

5.6. Galvanotaxis

Since the turn of the century, it has been documented that the directional migration of some cells in culture can be controlled by electrical fields. The direction of growth and motility of a variety of embryonic cells was recently shown to be quite sensitive to an imposed direct current. Such cells include neurons (Hinkle *et al.*, 1981; Jaffe and Poo, 1979; Patel and Poo, 1982), myoblasts (Hinkle *et al.*, 1981), somitic fibroblasts (Nuccitelli and Erickson, 1983; Erickson and Nuccitelli, 1984), and neural crest cells (Cooper and Keller, 1984; Stump and Robinson, 1983; Nuccitelli and Erickson, 1983). Indeed, somite fibroblasts will translocate to the negative pole in a voltage gradient as small as 10 mV/mm (Erickson and Nuccitelli, 1984), and electrical currents comparable to those shown to influence direction of cell migration *in vitro* have been measured in embryos (Robinson and Stump, 1984; Jaffe and Stern, 1979; Lindemann and Voute, 1976; Barker *et al.*, 1982). The currents are generated by epithelia pumping ions across themselves; the ions then leak out at points of least resistance and create an ion-carrying current. Preliminary findings suggest that the neural tube, which is also epithelium, is generating a field around itself (R. Nuccitelli and C. A. Erickson, unpublished results). Although certainly intriguing, it remains to be tested whether any current that might be generated by the neural tube *in vivo* does in fact drive crest cells away from their origin.

Such currents might be more important in initiating migration. Specifically, because currents are generated through "leaky" points in epithelium, currents may arise when junctions disintegrate between the future epidermal ectoderm and neural epithelium (Revel and Brown, 1975; Warner, 1973). Indeed, it is shortly after electrical contact between the neural epithelium and epidermal ectoderm is lost that the crest cells begin their migration (see Warner, 1973).

6. Cell Localization

The crest may ultimately localize in appropriate regions of the embryo by at least two alternative means. According to one hypothesis, pluripotential crest cells may move randomly along the pathways and stop in specific regions according to environmental cues; they would then differentiate according to directives from the local environment. Alternatively, neural crest cells may be predetermined, and certain phenotypes "know" where to cease migration. Both hypotheses have data to support them, although it should be noted that this aspect of crest morphogenesis is particularly obscure.

6.1. Cessation of Cell Migration by the Environment

6.1.1. Barriers to Migration

Crest cells may simply stop migrating when they run into a barrier. For example, trunk crest cells migrate between the somites and neural tube and appear to cease migrating when the sclerotome disperses and fills in this pathway (Thiery *et al.*, 1982a). The role of the somite in sensory gangliogenesis is also supported by Detwiler's (1937) observation in amphibians that removal of somites resulted in no ganglion development at that axial level; substituting several smaller somites for a larger one produced a corresponding change in the pattern of ganglion segmentation. Furthermore, Teillet and Le Douarin (1983) demonstrated that survival of somite cells is essential for sensory and sympathetic gangliogenesis. It should be said, however, that although obliteration of a space occurs concomitant with cessation of migration, there is no direct evidence that the crest cells are unable to invade the sclerotome when it spreads around the neural tube. Indeed, some grafting studies suggest that at late stages of sclerotome dispersion, crest cells are dispersed throughout the sclerotome (Weston, 1963; Teillet and Le Douarin, 1983). Furthermore, when the neural tube is grafted into an embryo upside down (Weston, 1963) or into the limb bud, where the mesenchyme is unsegmented (Fisher and Solursh, 1979b), the sensory ganglia form in each case in their appropriate relationship to the neural tube. These latter studies suggest that cessation of migration is independent of the structure of the surrounding medium. These data are difficult to rationalize in light of Detwiler's (1937) results, however.

Hyaluronate is a space-expanding molecule associated with regions in which crest cells are migrating; conversely, the loss of hyaluronate is associated with cessation of migration (see Toole, 1981). This space-expanding role of hyaluronate is particularly dramatic in the development of the corneal stroma (Toole and Trelstad, 1971), but there is also a decrease in the amount of hyaluronate in the region of the coalescing sensory ganglion (Derby, 1978). In this latter case, however, it has not yet been determined whether removal of hyaluronate results in cessation of migration or is removed after the ganglion begins to develop, perhaps to enable the cells to associate with each other more closely.

6.1.2. Extracellular Matrix

Besides simply running into physical barriers, cessation of migration may occur by trapping action. That is, the cells may contact extracellular matrix that cannot sustain their migration, causing them to stop. Alternatively, the cells may encounter a particularly adhesive region from which they cannot detach (see Trinkaus, 1982). With regard to adhesion, it is interesting to note that there is little or no fibronectin in the region in which sympathetic ganglia begin to condense and, concomitant with sensory ganglion development, there is a loss of fibronectin (Thiery *et al.*, 1982*a*).

The crest cells appear to associate with at least two embryonic structures. Crest cells frequently coalesce near embryonic blood vessels, e.g., sympathetic ganglia and aortic plexus near the dorsal aorta (Brauer, 1932). In addition, many crest cells adhere to, or coalesce on, nerves, e.g., enteric ganglia along vagus nerves and ciliary ganglion at the juncture of oculomotor and ophthalmic nerves (Noden, 1975). Of course, we have no evidence as to whether crest cells are using these tissues as substrata for migration or are, instead, using the same cues as angiogenic cords and growth cones as pathways for patterned distribution. It would be interesting to explore these two embryonic structures for associated molecules that might be instrumental in adhesive recognition.

Changes in the crest cell surface may influence interactions of the cells with each other and with their environment. Thiery *et al.* (1982*b*) showed that the putative adhesive molecule N-CAM is not associated with migratory crest cells, but rather that it appears at the time crest cells begin to aggregate to form sensory ganglia. A similar sequence of events apparently occurs *in vitro*, when isolated crest cells after several days in culture begin to aggregate concomitant with the appearance of N-CAM (J. P. Thiery, unpublished data; Le Douarin, 1982). The appearance of N-CAM or of other surface molecules may therefore be instrumental in gangliogenesis. Whether cessation of migration is due to environmental influences triggering the appearance of N-CAM and subsequent aggregation, or whether N-CAM appearance is endogenously timed, has not yet been determined.

6.1.3. Passive Localization

It has been suggested that, at least for trunk crest, the ultimate distribution is determined not by active cell locomotion to that area but rather by passive dispersion directed by environmental controls. This idea is based on the observation that immobile Latex beads are found ventrally in the embryo after being injected into the somites (Bronner-Fraser, 1982). Presumably, the dispersing sclerotome is responsible for this movement, although this is not proven. At least for the trunk crest, this passive mechanism seems unlikely for several reasons: (1) Many crest cells migrate to their ventral destination before the sclerotome disperses substantially. (2) Crest cells enter the somites along the basal lamina of the dermatome–myotome (see Fig. 5 in Teillet and Le Douarin, 1983), but few seem to be dispersed at early stages in the sclerotome where the Latex beads are found. (3) If the neural tube is grafted upside down, the sensory ganglia develop in their proper orientation relative to the neural tube, and not to the dispersing sclerotome (Weston, 1963). (4) Crest migration continues in the absence of a somite (see Lewis *et al.*, 1981). This is not to say that the sclerotome does not alter the distribution of the cells. For example, the sclerotome may very well be responsible for the so-called secondary migration of the sympathetic ganglia, which originally form near the dorsal aorta and may be moved to their more dorsal position by the sclerotome. Clearly, these passive movements deserve more attention.

The final position of a crest-derived structure in the older embryo could also be a result of asymmetrical growth occurring long after the cells cease migrating. Such displacement by differential growth occurs and can be seen most readily in the head, where the squamosal bone and the cartilaginous elements of the hypobranchial grow dorsally into mesoderm (Noden, 1980).

6.2. Movements of Predetermined Phenotypes

Crest phenotypes in some instances may be predetermined, and may localize and differentiate according to prepattern in the population (see Weston et al., 1984). We have already examined the possibility that the future enteric ganglia of the vagal crest population are predetermined. Some evidence suggests that there are premigratory populations determined to give rise to the ciliary ganglion (Barald, 1982) and perhaps to the sensory ganglion (Ciment and Weston, 1982). In addition, monoclonal antibodies that recognize different gangliosides stain various subpopulations of newly migrating neural crest cells in culture (Girdlestone and Weston, 1985). A dramatic example of a population that is clearly predetermined in some unknown manner is that of the mesencephalic crest (Noden, 1983). In a series of experiments, Noden grafted the crest from the level of the first visceral arch into the region of the second and third arches. Not only did the donor cells give rise to the normal second arch derivatives, but a duplicate set of the first arch skeletal elements was formed as well, including the squamosal, quadrate, pterygoid, Meckel's, and angular elements; externally, an ectopic beak and a supernumerary auditory depression were formed. The simplest explanation for these observations is that the crest mesenchyme that fills the first arch is prepatterned; if directed into another arch, it will produce the appropriate arrangement of skeletal elements. Thus, it is possible that these cells localize and differentiate in an appropriate relationship to each other without many further cues from their environment, although they undoubtedly use environmental cues to find their way to the arch. Similar sorts of patterns may exist elsewhere in the embryo but have not yet been elucidated.

The control of crest distribution has been studied most extensively with pigmentation patterns. A fascinating series of reports demonstrated that pigment patterns of three species of the California newt, *Taricha*, are different, and, furthermore, that these patterns are dependent not on the embryonic environment in each species but on some intrinsic cues within the crest cells themselves (Twitty, 1936, 1949; Twitty and Bodenstein, 1939). For example, in *T. rivularis* the pigment cells are evenly distributed over the flank, whereas in *T. torosa* the pigment is compacted in a broad band along the dorsal margin of the somites. If neural folds are grafted from one species to the other, the pigment patterns are consistent with the donor crest cells rather than with the host environment. Even in tissue culture, *Taricha rivularis* crest remains dispersed, whereas *T. torosa* crest forms aggregates after several days (Twitty, 1945). The basis of this extraordinary behavior has not been further examined.

Although Twitty and associates clearly showed that intrinsic properties of *Taricha* melanoblasts determine pigmentation patterns, the majority of studies suggest that differences between pigmentation patterns in most species are directly related to their environment. An excellent example is the recently examined difference in crest behavior between the white and dark embryos of *Ambystoma mexicanum* (also reported by Dalton, 1950). Keller *et al.* (1982) demonstrated that the failure of melanophores in the white strain to migrate as far as the dark melanophores is attributable to the overlying white epidermis, which either inhibits or simply fails to provide an appropriate substratum for migration. Grafting experiments, in which neural folds, epidermis, or meso- derm were exchanged between embryos of the two species, showed that the mutant white pigmentation pattern is not corrected by dark mesoderm or neu- ral crest. Just what the white epidermis releases into the underlying space is not known, although Spieth and Keller (1984) observed substantially fewer spher- ical bodies (interstitial bodies?) (Low, 1970) in the subepidermal space of white as compared with the dark embryos. The composition of these bodies has not been determined in the present case, but in chick they are composed of chon- droitin sulfate proteoglycan and fibronectin (Mayer *et al.*, 1981).

Pigmentation patterns have been studied extensively in an attempt to an- swer questions related to cessation of migration because the pigmented crest derivatives contain an obvious cytological marker. In addition, numerous pig- ment mutants in frogs (e.g., Browder, 1975; Bagnara, 1982), salamanders (Frost and Malacinski, 1980), and mice (reviewed by Weston, 1970; Le Douarin, 1982) will permit crucial comparative studies. One must use caution, however, in interpreting pigmentation patterns in terms of arrest of crest cell movement; some of these patterns actually may reflect differences in differentiation or selective cell death, rather than in movement. As an example, Ohsugi and Ide (1983) examined the dorsoventral pigment pattern of *Xenopus*, the embryos of which are darkly pigmented dorsally, whereas the belly is devoid of pigment. It was formerly believed that presumptive pigment cells only migrate to the dor- sal skin and are inhibited from invading the ventral region. However, embryos incubated in 3,4-dihydroxyphenylalanine (DOPA) in an effort to stimulate melanogenesis become pigmented equally in the dorsal and ventral region. Thus, crest cells exist in the belly and contain premelanosomes, but apparently they do not receive instructions to complete differentiation. The differences in cues between the dorsal and ventral skin are not known, but the example should suffice to suggest caution in the interpretation of similar studies.

7. Conclusions

Neural crest cells escape from the neural tube by unknown means and disperse through the embryo along cell-free pathways filled with extracellular matrix material. Antibody-binding and specific staining techniques have identified and localized materials in these pathways, and tissue culture studies have enabled us to study directly the interactions between neural crest cells and various matrix components. Our best evidence suggests that crest cells use fibronectin as a substratum for migration, whereas hyaluronate helps open up spaces and chondroitin sulfate may inhibit migration. Dispersal of neural crest cells through their otherwise cell-free pathways is probably driven by contact inhibition, but other mechanisms have not yet been ruled out. Grafting experiments could be fruitful in testing what directional cues, if any, are established by the pathways. Although the extraordinary migratory capabilities of the neural crest have been well characterized, they are not well understood. Monoclonal antibodies directed against surface components of neural crest cells and comparisons between crest cells and invasive tumor and blood cells may clarify the way in which their exceptional motile capacity is controlled. Cessation of migration is perhaps the least well understood of all these morphogenetic phenomena. Experiments in which grafted crest cells are confronted by different embryonic structures, as well as by perturbation of the environment in vivo, may provide us with direct evidence of components that facilitate migration or that instruct cells to stop. In the future, the isolation of predetermined subpopulations of the crest will permit us to test whether the patterns of migration and localization of these particular cells are also predetermined by reinserting them into the early crest pathways.

Specific perturbation of the usual neural crest cell behavior is another means of understanding normal crest morphogenesis. In this regard, embryonic mutants of the neural crest are potentially of great value. For example, two mutants that affect the neural crest in the mouse are Splotch (Auerbach, 1954), in which the crest cells fail to migrate from the neural tube in regions of spina bifida (Erickson, unpublished results) and of Patch, in which crest cells enter abnormal pathways (Erickson and Weston, 1983). Weston (1970) has summarized the various mouse mutants that appear to perturb crest morphogenesis. Recent studies of mutant mice demonstrate the power of such mutants. Mutant recessive genes, such as Piebald-lethal (s¹) (Webster, 1973) and lethal-spotting (ls), produce severe megacolon in homozygotes due to the absence of enteric ganglia in the 2–3-mm terminal segment of the bowel. Recent experiments in which the guts from normal and lethal-spotting (ls/ls) mouse embryos were cultured in vitro suggest that the environment of the mutant's terminal gut cannot support invasion of crest cells (Rothman et al., 1982; Rothman and Gershon, 1984). If this is true, the ls/ls mutant may be instructive in identifying what extracellular matrix components sustain neural crest cell migration (Gershon and Rothman, 1984).

In conclusion, the migratory behavior of neural crest cells has been well characterized, and some of the controls of their behavior are now understood. The advent of powerful biochemical tools and tissue culture, combined with more traditional embryological techniques, have begun to illuminate essential mechanisms. As we continue to exploit these techniques, and as new methodology becomes available, the neural crest should become an increasingly important model system with which to study morphogenesis.

ACKNOWLEDGMENTS.   I would like to thank Dr. David W. Phillips, Dr. Jeanne F. Loring, Dr. James A. Weston, Dr. J. P. Trinkaus, and Dr. Drew N. Noden for critical reading of this manuscript and Dr. Christopher Womersley for the artwork in Figure 7. The research reported in this chapter was supported by NIH grant PHS DE05630 and NSF grant PCM 8004524. The author's work has also been supported by a Research Career Development Award from the National Institutes of Health.

References

Abercrombie, M., 1970, Contact inhibition in tissue culture, *In Vitro* **6:**128–142.

Abercrombie, M., 1982, The crawling movement of metazoan cells, in: *Cell Behaviour* (R. Bellairs, A. Curtis, and G. Dunn, eds.), pp. 19–48, Cambridge University Press, Cambridge.

Ali, I. U., and Hynes, R. O., 1978, Effects of LETS glycoprotein on cell motility, *Cell* **14:**439–446.

Allan, I. J., and Newgreen, D. F., 1980, The origin and differentiation of enteric neurons of the intestine of the fowl embryo, *Am. J. Anat.* **157:**137–154.

Anderson, C. B., and Meier, S., 1981, The influence of the metameric pattern in the mesoderm on migration of cranial neural crest cells in the chick embryo, *Dev. Biol.* **85:**385–402.

Anderson, C. B., and Meier, S., 1982, Effect of hyaluronidase treatment on the distribution of cranial neural crest cells in the chick embryo, *J. Exp. Zool.* **221:**329–335.

Auerbach, R., 1954, Analysis of the developmental effects of a lethal mutation in the house mouse, *J. Exp. Zool.* **127:**305–327.

Ayer-Le Lièvre, C. S., and Le Douarin, N. M., 1982, The early development of cranial sensory ganglia and the potentialities of their component cells studied in quail–chick chimeras, *Dev. Biol.* **94:**291–310.

Bagnara, J. T., 1982, Development of the spot pattern in the leopard frog, *J. Exp. Zool.* **224:**283–287.

Bancroft, M., and Bellairs, R., 1976, The neural crest cells of the trunk region of the chick embryo studied by SEM and TEM, *Zoon* **4:**73–85.

Banerjee, S. D., Cohn, R. H., and Bernfield, M. R., 1977, Basal lamina of embryonic salivary epithelia. Production of the epithelium and role in maintaining lobular morphology, *J. Cell Biol.* **73:**445–463.

Barald, K. F., 1982, Monoclonal antibodies to embryonic neurons, in: *Neuronal Development* (N. E. Spitzer, ed.), pp. 101–119, Plenum Press, New York.

Bard, J. B. L., and Hay, E. D., 1975, The behavior of fibroblasts from the developing avian cornea. Morphology and movement *in situ* and *in vitro, J. Cell Biol.* **67:**400–418.

Bard, J. B. L., Hay, E. D., and Meller, S. M., 1975, Formation of the endothelium of the avian cornea: A study of cell movement *in vivo, Dev. Biol.* **42:**334–361.

Barker, A. T., Jaffe, L. F., and Vanable, J. W., Jr., 1982, The glabrous epidermis of cavies contains a powerful battery, *Am. J. Physiol.* **242:**R358–R366.

Bell, P. B., Jr., 1977, Locomotory behavior, contact inhibition, and pattern formation of 3T3 and polyoma virus-transformed 3T3 cells in culture, *J. Cell Biol.* **74:**963–982.

Bernanke, D. H., and Markwald, R. R., 1979, Effects of hyaluronic acid on cardiac cushion tissue cells in collagen matrix cultures, *Tex. Rep. Biol. Med.* **39:**271–285.

Bernfield, M., and Banerjee, S. D., 1982, The turnover of basal lamina glycosaminoglycan correlates with epithelial morphogenesis, *Dev. Biol.* **90:**291–305.

Black, I. B., 1982, Stages of neurotransmitter development in autonomic neurons, *Science* **215:**1198–1204.

Bolender, D. L., Seliger, W. G., and Markwald, R. R., 1980, A histochemical analysis of polyanionic compounds found in the extracellular matrix encountered by migrating cephalic neural crest cells, *Anat. Rec.* **196:**401–412.

Boucaut, J.-C., Darribère, T., Poole, T. J., Aoyama, H., Yamada, K. M., and Thiery, J. P., 1984,

Biologically active synthetic peptides as probes of embryonic development: A competitive peptide inhibitor of fibronectin function inhibits gastrulation in amphibian embryos and neural crest cell migration in avian embryos, *J. Cell Biol.* **99**:1822–1830.

Brauer, A., 1932, A topographical and cytological study of the sympathetic nervous components of the suprarenal of the chick embryo, *J. Morphol.* **53**:277–325.

Brauer, P. R., Bolender, D. L., and Markwald, R. R., 1985, The distribution and spatial organization of the extracellular matrix encountered by mesencephalic neural crest cells, *Anat. Rec.* **211**:57–68.

Bronner, M. E., and Cohen, A. M., 1979, Migratory patterns of cloned neural crest melanocytes injected into host chicken embryos, *Proc. Natl. Acad. Sci. USA* **76**:1843–1848.

Bronner-Fraser, M., 1982, Distribution of latex beads and retinal pigment epithelial cells along the ventral neural crest pathway, *Dev. Biol.* **91**:50–63.

Bronner-Fraser, M., 1985a, Effects of different fragments of the fibronectin molecule on latex bead translocation along neural crest migratory pathways, *Dev. Biol.* **108**:131–145.

Bronner-Fraser, M., 1985b, Alterations in neural crest migration by a monoclonal antibody that affects cell adhesion, *J. Cell Biol.* **101**:610–617.

Bronner-Fraser, M. E., and Cohen, A. M., 1980, Analysis of the neural crest ventral pathway using injected tracer cells, *Dev. Biol.* **77**:130–141.

Browder, L. W., 1975, Frogs of the genus *Rana*, in: *Handbook of Genetics*, Vol. 4 (R. C. King, ed.), pp. 19–33, Plenum Press, New York.

Brownell, A. G., Bessem, C. C., and Slavkin, H. C., 1981, Possible functioning of mesenchyme cell-derived fibronectin during formation of basal lamina, *Proc. Natl. Acad. Sci. USA* **78**:3711–3715.

Carter, S., 1967, Haptotaxis and the mechanisms of cell motility, *Nature (Lond.)* **213**:256–260.

Chen, L. B., Murray, A., Segal, R. A., Bushnell, A., and Walsh, M. L., 1978, Studies on intercellular LETS glycoprotein matrices, *Cell* **14**:377–391.

Chibon, P., 1967, Marquage nucléaire par la thymidine tritiée des dérivés de la crête neurale chez l'Amphibien Urodèle *Pleurodeles waltlii* Michah, *J. Embryol. Exp. Morphol.* **18**:343–358.

Ciment, G., and Weston, J. A., 1982, Early appearance in neural crest and crest-derived cells of an antigenic determinant present in avian neurons, *Dev. Biol.* **93**:355–367.

Cochard, P., and Coltey, P., 1983, Cholinergic traits in the neural crest: Acetyl cholinesterase in crest cells of the chick embryo, *Dev. Biol.* **98**:221–238.

Cohen, A. M., and Hay, E. D., 1971, Secretion of collagen by embryonic neuroepithelium at the time of spinal cord–somite interaction, *Dev. Biol.* **26**:578–605.

Cooper, M. S., and Keller, R. E., 1984, Perpendicular orientation and directional migration of amphibian neural crest cells in dc electrical fields, *Proc. Natl. Acad. Sci. USA* **81**:160–164.

Currie, J. R., Maylié-Pfenninger, M.-F., and Pfenninger, K. H., 1984, Developmentally regulated plasmalemmal glycoconjugates of the surface and neural ectoderm, *Dev. Biol.* **106**:109–120.

Dalton, H. C., 1950, Inhibition of chromatoblast migration as a factor in the development of genetic differences in white and black axolotls, *J. Exp. Zool.* **115**:151–170.

D'Amico-Martel, A., and Noden, D. M., 1983, Contributions of placodal and neural crest cells to avian cranial peripheral ganglia, *Am. J. Anat.* **166**:445–468.

Davis, E. M., 1980, Translocation of neural crest cells within a hydrated collagen lattice, *J. Embryol. Exp. Morphol.* **55**:17–31.

Davis, E. M., and Trinkaus, J. P., 1981, Significance of cell-to-cell contacts for the directional movement of neural crest cells within a hydrated collagen lattice, *J. Embryol. Exp. Morphol.* **63**:29–51.

Delpech, A., and Delpech, B., 1984, Expression of hyaluronic acid-binding glycoprotein, hyaluronectin, in the developing rat embryo, *Dev. Biol.* **101**:391–400.

Derby, M. A., 1978, Analysis of glycosaminoglycans within the extracellular environments encountered by migrating neural crest cells, *Dev. Biol.* **66**:321–336.

Detwiler, S. R., 1937, Observations upon the migration of neural crest cells, and upon the development of the spinal ganglia and vertebral arches in *Ambystoma*, *Am. J. Anat.* **61**:63–94.

DiVirgilio, G., Lavenda, N., and Worden, J. L., 1967, Sequence of events in neural tube closure and the formation of neural crest in the chick embryo, *Acta Anat.* **68**:127–146.

Duband, J. L., and Thiery, J. P., 1982, Distribution of fibronectin in the early phase of avian cephalic neural crest cell migration, *Dev. Biol.* **93**:308–323.

Dunn, G. A., 1982, Contact guidance of cultured tissue cells: A survey of potentially relevant properties of the substratum, in: *Cell Behaviour* (R. Bellairs, A. Curtis, and G. Dunn, eds.), pp. 247–280, Cambridge University Press, Cambridge.

Dunn, G. A., and Ebendal, T., 1978, Contact guidance on oriented collagen gels, *Exp. Cell Res.* **111**:475–479.

DuShane, G. P., 1935, An experimental study of the origin of pigment cells in Amphibia, *J. Exp. Zool.* **72**:1–31.

Ebendal, T., 1977, Extracellular matrix fibrils and cell contacts in the chick embryo. The possible role in the orientation of cell migration and axon extension, *Cell Tissue Res.* **175**:439–458.

Epperlein, H. H., 1974, The ectomesenchymal–endodermal interaction system (EEIS) of *Triturus alpestris* in tissue culture. I. Observations on attachment, migration, and differentiation of neural crest cells, *Differentiation* **2**:151–168.

Erickson, C. A., 1978, Contact behaviour and pattern formation of BHK and polyoma virus-transformed BHK fibroblasts in culture, *J. Cell Sci.* **33**:53–84.

Erickson, C. A., 1985, Control of neural crest cell dispersion in the trunk of the avian embryo, *Dev. Biol.* **111**:138–157.

Erickson, C. A., and Nuccitelli, R., 1984, Embryonic fibroblast orientation and motility can be influenced by physiological electric fields, *J. Cell Biol.* **98**:296–307.

Erickson, C. A., and Olivier, K. R., 1983, Negative chemotaxis does not control quail neural crest cell dispersion, *Dev. Biol.* **96**:542–551.

Erickson, C. A., and Turley, E. A., 1983, Substrata formed by combinations of extracellular matrix components alter neural crest cell motility *in vitro*, *J. Cell Sci.* **61**:299–323.

Erickson, C. A., and Weston, J. A., 1983, An SEM analysis of neural crest cell migration in the mouse, *J. Embryol. Exp. Morphol.* **74**:97–118.

Erickson, C. A., Tosney, K. W., and Weston, J. A., 1980, Analysis of migratory behavior of neural crest and fibroblastic cells in embryonic tissues, *Dev. Biol.* **77**:142–156.

Fell, H. B., 1939, The origin and developmental mechanics of the avian sternum, *Philos. Trans. R. Soc. Lond. B* **229**:406–463.

Ferguson, M. W. J., 1981, Review: The value of the American alligator (*Alligator mississippiensis*) as a model for research in craniofacial development, *J. Craniofacial Genet. Dev. Biol.* **1**:123–144.

Fisher, M., and Solursh, M., 1977, Glycosaminoglycan localization and role in maintenance of tissue spaces in the early chick embryo, *J. Embryol. Exp. Morphol.* **42**:195–207.

Fisher, M., and Solursh, M., 1979a, The influence of the substratum on mesenchyme spreading *in vitro*, *Exp. Cell Res.* **123**:1–14.

Fisher, M., and Solursh, M., 1979b, The influence of local environments on the organization of mesenchyme cells, *J. Embryol. Exp. Morphol.* **49**:295–306.

Frederickson, R. G., and Low, F. N., 1971, The fine structure of perinotochordal microfibrils in control and enzyme-treated chick embryos, *Am. J. Anat.* **130**:347–376.

Frost, S. K., and Malacinski, G. M., 1980, The developmental genetics of pigment mutants in the Mexican axolotl, *Dev. Genet.* **1**:271–294.

Gasser, R. F., 1979, Evidence that sclerotomal cells do not migrate medially during normal development of the rat, *Am. J. Anat.* **154**:509–524.

Gershon, M. D., and Rothman, T. P., 1984, Experimental and genetic approaches to the study of the development of the enteric nervous system, *Trends Neurosci.* **7**:150–155.

Girdlestone, J., and Weston, J. A., 1985, Identification of early neuronal subpopulations in avian neural crest cell cultures, *Dev. Biol.* **109**:274–287.

Glimelius, B., and Pintar, J. E., 1981, Analysis of developmentally homogeneous neural crest cell populations in vitro. IV. Cell proliferation and synthesis of glycosaminoglycans, *Cell Diff.* **10**:173–182.

Greenberg, J. H., Foidart, J.-M., and Greene, R. M., 1980, Collagen synthesis in cultures of differentiating neural crest cells, *Cell Diff.* **9**:153–163.

Greenberg, J. H., Seppä, S., Seppä, H., and Hewitt, T., 1981, Role of collagen and fibronectin in neural crest cell adhesion and migration, *Dev. Biol.* **87**:259–266.

Gustafson, T., and Wolpert, L., 1961, Studies on the cellular basis of morphogenesis in the sea urchin embryo. Directed movements of primary mesenchyme cells in normal and vegetalized larvae, *Exp. Cell Res.* **24**:64–79.

Hammond, W. S., and Yntema, C. L., 1947, Depletions of the thoraco-lumbar sympathetic system following removal of neural crest in the chick, *J. Comp. Neurol.* **86**:237–266.

Harris, A. K., 1973, Behaviour of cultured cells on substrata of variable adhesiveness, *Exp. Cell Res.* **77**:285–297.

Harris, A. K., Wild, P., and Stopak, D., 1980, Silicone rubber substrata: A new wrinkle in the study of cell locomotion, *Science* **208**:177–179.

Harris, A. K., Stopak, D., and Wild, P., 1981, Fibroblast traction as a mechanism for collagen morphogenesis, *Nature (Lond.)* **290**:249–251.

Harrison, R. G., 1910, The outgrowth of the nerve fiber as a mode of protoplasmic movement, *J. Exp. Zool.* **9**:787–848.

Hay, E. D., 1978, Fine structure of embryonic matrices and their relation to the cell surface in ruthenium-red fixed tissue, *Growth* **42**:399–423.

Hay, E. D., and Meier, S., 1974, Glycosaminoglycan synthesis by embryonic inductors: Neural tube, notochord and lens, *J. Cell Biol.* **62**:889–898.

Heasman, J., Hynes, R. O., Swan, A. P., Thomas, V., and Wylie, C. C., 1981, Primordial germ cells of *Xenopus* embryos: The role of fibronectin in their adhesion during migration, *Cell* **27**:437–447.

Hinkle, L., McCaig, C. D., and Robinson, K. R., 1981, The direction of growth of differentiating neurones and myoblasts from frog embryos in an applied electric field, *J. Physiol. (Lond.)* **314**:121–135.

Hörstadius, S., 1950, *The Neural Crest: Its Properties and Derivatives in the Light of Experimental Research*, Oxford University Press, London.

Hynes, R. O., 1976, Cell surface proteins and malignant transformation, *Biochim. Biophys. Acta* **458**:73–107.

Jaffe, L. F., and Poo, M.-M., 1979, Neurites grow faster towards the cathode than the anode in a steady field, *J. Exp. Zool.* **209**:115–128.

Jaffe, L. F., and Stern, C. D., 1979, Strong electrical currents leave the primitive streak of chick embryos, *Science* **206**:569–571.

Johnston, M. C., 1966, A radioautographic study of the migration and fate of cranial neural crest cells in the chick embryo, *Anat. Rec.* **156**:143–156.

Johnston, M. C., and Listgarten, M. A., 1972, Observations on the migration, interaction, and early differentiation of orofacial tissues, in: *Developmental Aspects of Oral Biology* (H. C. Slavkin and L. A. Baretta, eds.), pp. 53–79, Academic Press, New York.

Johnston, M. C., Bhakdinaronk, A., and Reid, Y. C., 1973, An expanded role of the neural crest in oral and pharyngeal development, in: *Development in the Fetus and Infant* (J. Bosma, ed.), pp. 37–52, Fourth Symposium on Oral Sensation and Perception, DHEW Publ. No. NIH 73-549, Department of Health and Human Services, Washington, D.C.

Johnston, M. C., Noden, D. M., Hazelton, R. D., Coulombre, J. L., and Coulombre, A. J., 1979, Origins of avian ocular and periocular tissues, *Exp. Eye Res.* **29**:27–43.

Johnston, M. C., Vig, K. W. L., and Ambrose, L. J. H., 1981, Neurocristopathy as a unifying concept: Clinical correlations, in: *Advances in Neurology, Vol. 29: Neurofibromatosis* (V. M. Riccardi and J. J. Mulvihill, eds.), pp. 97–104, Raven Press, New York.

Kefalides, N. A., 1978, Current status of chemistry and structure of basement membranes, in: *Biology and Chemistry of Basement Membranes* (N. A. Kefalides, ed.), pp. 215–228, Academic Press, New York.

Keller, R. E., and Schoenwolf, G. C., 1977, An SEM study of cellular morphology, contact, and arrangement, as related to gastrulation in *Xenopus laevis*, *Wilhelm Roux Arch.* **182**:165–186.

Keller, R. E., and Spieth, J., 1984, Neural crest cell behavior in white and dark larvae of *Ambystoma mexicanum*: Time-lapse cinemicrographic analysis of pigment cell movement *in vivo* and in culture, *J. Exp. Zool.* **229**:109–126.

Keller, R. E., Löfberg, J., and Spieth, J., 1982, Neural crest cells behavior in white and dark embryos of *Ambystoma mexicanum*: Epidermal inhibition of pigment cell migration in the white axolotl, *Dev. Biol.* **89**:179–195.

Kelley, R. O., and Bluemink, J. G., 1974, An ultrastructural analysis of cell and matrix differentiation during early limb development in *Xenopus laevis*, *Dev. Biol.* **37**:1–17.

Kinnander, H., and Gustafson, T., 1960, Further studies on the cellular basis of gastrulation in the sea urchin larva, *Exp. Cell Res.* **19**:278–290.

Kinsella, M. G., and Fitzharris, T. P., 1982, Control of cell migration in atrio-ventricular pads during chick early heart development: Analysis of cushion tissue migration *in vitro*, *Dev. Biol.* **91**:1–10.

Kussäther, E., Usadel, K. H., and Drews, U., 1967, Cholinesterase-Aktivität beider Entstehung und Auflösung der Somiten des Hühnchens, *Histochemie* **8**:237–247.

Kvist, T. N., and Finnegan, C. V., 1970a, The distribution of glycosaminoglycans in the axial region of the developing chick embryo. I. Histochemical analysis, *J. Exp. Zool.* **175**:221–240.

Kvist, T. N., and Finnegan, C. V., 1970b, The distribution of glycosaminoglycans in the axial region of the developing chick embryo. II. Biochemical analysis, *J. Exp. Zool.* **175**:241–258.

Le Douarin, N., 1973, A biological cell labelling technique and its use in experimental embryology, *Dev. Biol.* **30**:217–222.

Le Douarin, N. M., 1980, The ontogeny of the neural crest in avian embryo chimaeras, *Nature (Lond.)* **286**:663–669.

Le Douarin, N. M., 1982, *The Neural Crest*, Cambridge University Press, Cambridge.

Le Douarin, N. M., and Teillet, M.-A., 1973, The migration of neural crest cells to the wall of the digestive tract in avian embryos, *J. Embryol. Exp. Morphol.* **30**:31–48.

Le Douarin, N. M., and Teillet, M.-A., 1974, Experimental analysis of the migration and differentiation of neuroblasts of the autonomic nervous system and of neurectodermal mesenchymal derivatives, using a biological cell marking technique, *Dev. Biol.* **41**:162–184.

Le Douarin, N. M., Teillet, M. A., Ziller, C., and Smith, J., 1978, Adrenergic differentiation of cells of the cholinergic ciliary and Remak ganglia in avian embryo following *in vivo* transplantation, *Proc. Natl. Acad. Sci. USA* **75**:2030–2034.

Leivo, I., Vaheri, A., Timpl, R., and Wartiovaara, J., 1980, Appearance and distribution of collagens and laminin in the early mouse embryo, *Dev. Biol.* **76**:100–114.

Le Lièvre, C., 1974, Rôle des cellules mesectodermiques issues des crêtes neurales céphaliques dans la formation des arcs branchiaux et du skelette viscéral, *J. Embryol. Exp. Morphol.* **31**:453–577.

Le Lièvre, C. S., and Le Douarin, N. M., 1975, Mesenchymal derivatives of the neural crest: Analysis of chimaeric quail and chick embryos, *J. Embryol. Exp. Morphol.* **34**:125–154.

Le Lièvre, C. S., Schweizer, G. G., Ziller, C. M., and Le Douarin, N. M., 1980, Restrictions of developmental capabilities in neural crest cell derivatives as tested by *in vivo* transplantation experiments, *Dev. Biol.* **77**:362–378.

Letourneau, P. C., 1975, Cell-to-substratum adhesion and guidance of axonal elongation, *Dev. Biol.* **44**:92–101.

Lewis, J., Chevallier, A., Kieny, M., and Wolpert, L., 1981, Muscle nerve branches do not develop in chick wings devoid of muscle, *J. Embryol. Exp. Morphol.* **64**:211–232.

Lindemann, B., and Voute, C., 1976, Structure and function of the epidermis, in: *Frog Neurobiology* (R. Llinas and W. Precht, eds.), pp. 169–210, Springer-Verlag, New York.

Liotta, L. A., Lee, C. W., and Morakis, D. J., 1980, New method for preparing large surfaces of intact human basement membrane for tumor invasion studies, *Cancer Lett.* **11**:1411–1152.

Löfberg, J., and Ahlfors, K., 1978, Extracellular matrix organization and early neural crest cell migration in the axolotl embryo, *Zoon* **6**:87–101.

Löfberg, J., Ahlfors, K., and Fällström, C., 1980, Neural crest cell migration in relation to extracellular matrix organization in the embryonic axolotl trunk, *Dev. Biol.* **75**:148–167.

Löfberg, J., Nynäs-McCoy, A., Olsson, C., Jönsson, L., and Perris, R., 1985, Stimulation of initial neural crest cell migration in the axolotl embryo by tissue grafts and extracellular matrix transplanted on microcarriers, *Dev. Biol.* **107**:442–459.

Loring, J., Erickson, C., and Weston, J. A., 1977, Surface proteins of neural crest, crest-derived and somite cells *in vitro*, *J. Cell Biol.* **75**:71a.

Low, F. N., 1970, Interstitial bodies in the early chick embryo, *Am. J. Anat.* **128**:45–56.

Markwald, R. R., Fitzharris, T. P., Bank, H., and Bernanke, D. H., 1978, Structural analysis on the matrical organization of glycosaminoglycans in developing endocardial cushions, *Dev. Biol.* **62:**292–316.

Markwald, R. R., Fitzharris, T. P., Bolender, D. L., and Bernanke, D. H., 1979, Structural analyses of cell:matrix association during morphogenesis of atrioventricular cushion tissue, *Dev. Biol.* **69:**634–654.

Martinez-Hernandez, A., and Amenta, P. S., 1983, The basement membrane in pathology, *Lab. Invest.* **48:**656–677.

Martins-Green, M., and Erickson, C. A., 1985, Neural crest cells: Initiation of migration is not triggered by disruption of basal lamina, *Cell Differ.* (in press).

Mayer, B. W., Jr., Hay, E. D., and Hynes, R. O., 1981, Immunocytochemical localization of fibronectin in embryonic chick trunk and area vasculosa, *Dev. Biol.* **82:**267–286.

Mayer, T. C., and Oddis, L., 1977, Pigment cell differentiation in embryonic mouse skin and isolated epidermis: An *in vitro* study, *J. Exp. Zool.* **202:**415–424.

McCarthy, J. B., Palm, S. L., and Furcht, L. T., 1983, Migration by haptotaxis of a Schwann cell tumor line to the basement membrane glycoprotein laminin, *J. Cell Biol.* **97:**772–777.

Meier, S., and Hay, E. D., 1973, Synthesis of sulfated glycosaminoglycans by embryonic corneal epithelium, *Dev. Biol.* **35:**318–331.

Morris, J. E., 1979, Steric exclusion of cells. A mechanism of glycosaminoglycan-induced cell aggregation, *Exp. Cell Res.* **120:**141–153.

Morriss, G. M., and New, D. A. T., 1979, Effect of oxygen concentration on morphogenesis of cranial neural folds and neural crest in cultured rat embryos, *J. Embryol. Exp. Morphol.* **54:**17–35.

Morriss, G. M., and Thorogood, P. V., 1978, An approach to cranial neural crest cell migration and differentiation in mammalian embryos, in: *Development in Mammals,* Vol. 3, (M. H. Johnson, ed.), pp. 363–412, North-Holland, Amsterdam.

Morriss-Kay, G. M., 1981, Growth and development of pattern in the cranial neural epithelium of rat embryos during neurulation, *J. Embryol. Exp. Morphol.* **65:**225–241.

Nakamura, H., and Ayer-Le Lièvre, C. S., 1982, Mesectodermal capabilities of the trunk neural crest of birds, *J. Embryol. Exp. Morphol.* **70:**1–18.

Nelson, G. A., and Revel, J.-P., 1975, Scanning electron microscopic study of cell movements in the corneal endothelium of the avian embryo, *Dev. Biol.* **42:**315–333.

Newgreen, D., 1982a, Role of the extracellular matrix in the control of neural crest cell migration, in: *Extracellular Matrix* (S. Hawkes and J. L. Wang, eds.), pp. 141–146, Academic Press, New York.

Newgreen, D. F., 1982b, Adhesion to extracellular materials by neural crest cells at the stage of initial migration, *Cell Tissue Res.* **227:**297–317.

Newgreen, D., 1984, Spreading of explants of embyronic chick mesenchymes and epithelia on fibronectin and laminin, *Cell Tissue Res.* **236:**265–277.

Newgreen, D., and Gibbins, I., 1982, Factors controlling the time of onset of the migration of neural crest cells in the fowl embryo, *Cell Tissue Res.* **224:**145–160.

Newgreen, D. F., and Gooday, D., 1985, Control of the onset of migration of neural crest cells in avian embryos. Role of Ca^{++}-dependent cell adhesion, *Cell Tissue Res.* **239:**329–336.

Newgreen, D., and Theiry, J.-P., 1980, Fibronectin in early avian embryos: Synthesis and distribution along migration pathways of neural crest cells, *Cell Tissue Res.* **211:**269–291.

Newgreen, D. F., Ritterman, M., and Peters, E. A., 1979, Morphology and behaviour of neural crest cells of chick embryo *in vitro, Cell Tissue Res.* **203:**115–140.

Newgreen, D. F., Gibbins, I. L., Sauter, J., Wallenfels, B., and Wütz, R., 1982, Ultrastructural and tissue culture studies on the role of fibronectin, collagen and glycosaminoglycans in the migration of neural crest cells in the fowl embryo, *Cell Tissue Res.* **221:**521–549.

Nichols, D. H., 1981, Neural crest formation in the head of the mouse embryo as observed using a new histological technique, *J. Embryol. Exp. Morphol.* **64:**105–120.

Noden, D. M., 1973, The migratory behavior of neural crest cells, in: *Fourth Symposium on Oral Sensation and Perception* (J. Bosma, ed.), pp. 9–35, DHEW Publ. No. NIH 73-546, Department of Health and Human Services, Washington, D.C.

Noden, D. M., 1975, An analysis of the migratory behavior of avian cephalic neural crest cells, *Dev. Biol.* **42:**106–130.

Noden, D. M., 1978a, Interactions directing the migration and cytodifferentiation of avian neural crest cells, in: *The Specificity of Embryonic Interactions* (D. Garrod, ed.), pp. 3–49, Chapman and Hall, London.

Noden, D. M., 1978b, The control of avian cephalic neural crest cytodifferentiation. I. Skeletal and connective tissues, *Dev. Biol.* **67**:296–312.

Noden, D. M., 1978c, The control of avian cephalic neural crest cytodifferentiation. II. Neural tissues, *Dev. Biol.* **67**:313–329.

Noden, D. M., 1980, The migration and cytodifferentiation of cranial neural crest cells, in: *Current Research Trends in Prenatal Craniofacial Development* (R. M. Pratt and R. L. Christiansen, eds.), pp. 3–25, Elsevier/North-Holland, New York.

Noden, D. M., 1983, The role of the neural crest in patterning of avian cranial skeletal, connective, and muscle tissues, *Dev. Biol.* **96**:144–165.

Noden, D. M., 1984, Neural crest development: New views on old problems, *Anat. Rec.* **208**:1–13.

Nuccitelli, R., and Erickson, C. A., 1983, Embryonic cell motility can be guided by physiological electric fields, *Exp. Cell Res.* **147**:195–201.

Ohsugi, K., and Ide, H., 1983, Melanophore differentiation in *Xenopus laevis,* with special reference to dorsoventral pigment pattern formation, *J. Embryol. Exp. Morphol.* **75**:141–150.

Ormerod, E. J., Warburton, M. J., Hughes, C., and Rudland, P. S., 1983, Synthesis of basement membrane proteins by rat mammary epithelial cells, *Dev. Biol,* **96**:269–275.

Ossowski, L., Quigley, J. P., and Reich, E., 1975, Plasminogen, a necessary factor for cell migration in vitro, in: *Proteases and Biological Control* (E. Riech, D. Rifkin, and E. Shaw, eds.), pp. 901–914, Cold Spring Harbor Laboratory, Cold Spring Harbor, New York.

Patel, N., and Poo, M.-M., 1982, Orientation of neurite growth by extracellular electric fields, *J. Neurosci.* **2**:483–496.

Patterson, P. H., 1978, Environmental determination of autonomic neurotransmitter functions, *Annu. Rev. Neurosci.* **1**:1–17.

Pintar, J. E., 1978, Distribution and synthesis of glycosaminoglycans during quail neural crest morphogenesis, *Dev. Biol.* **67**:444–464.

Poole, T. J., and Steinberg, M. S., 1982, Evidence for the guidance of pronephric duct migration by a cranio-caudally traveling adhesion gradient, *Dev. Biol.* **92**:144–158.

Pratt, R. M., Larsen, M. A., and Johnston, M. C., 1975, Migration of cranial neural crest cells in a cell-free hyaluronate-rich matrix, *Dev. Biol.* **44**:298–305.

Raven, C. P., 1931, Zur Entwicklung der Ganglienleiste. I. Die Kinematik der Ganglienleisten Entwicklung bei den Urodelen, *Wilhelm Roux Arch. Entwicklungsmech. Org.* **125**:210–293.

Revel, J.-P., and Brown, S. S., 1975, Cell junctions in development, with particular reference to the neural tube, *Cold Spring Harbor Symp. Quant. Biol.* **40**:443–455.

Rich, A. M., Pearlstein, E., Weissmann, G., and Hoffstein, S. T., 1981, Cartilage proteoglycans inhibit fibronectin-mediated adhesion, *Nature (Lond.)* **293**:224–226.

Robinson, K. R., and Stump, R. F., 1984, Self-generated electrical currents through neurulae, *J. Physiol. (Lond.)* **352**:339–352.

Rothman, T. P., and Gershon, M. D., 1984, Regionally defective colonization of the terminal bowel by the precursors of enteric neurons in lethal spotted mutant mice, *Neuroscience* **12**:1293–1311.

Rothman, T. P., Nilaver, G., and Gershon, M. D., 1982, Neural crest–microenivronment interactions in the formation of enteric ganglia: An analysis of normal and lethal spotted mutant mice, *Soc. Neurosci. Abstr.* **8**:6.

Rovasio, R. A., Delouvée, A., Yamada, K. M., Timpl, R., and Thiery, J. P., 1983, Neural crest cell migration: Requirements for exogenous fibronectin and high cell density, *J. Cell Biol.* **96**:462–473.

Russo, R. G., Thorgeirsson, U., and Liotta, L. A., 1982, In vitro quantitative assay of invasion using human ammion, in: *Tumor Invasion and Metastasis* (L. A. Liotta and I. R. Hart, eds.), pp. 173–187, Martinus Nijhoff, The Netherlands.

Rutishauser, U., Thiery, J.-P., Brackenbury, R., and Edelman, G. M., 1978, Adhesion among neural cells of the chick embryo. Relationship of the surface molecule CAM to cell adhesion and the development of histotypic patterns, *J. Cell Biol.* **79**:371–381.

Schroeder, T., 1970, Neurulation in *Xenopus laevis*. An analysis and model based on light and electron microscopy, *J. Embryol. Exp. Morphol.* **23**:427–462.

Sieber-Blum, M., Sieber, F., and Yamada, K. H., 1981, Cellular fibronectin promotes adrenergic differentiation of quail neural crest cells in vitro, *Exp. Cell Res.* **133**:285–295.

Silver, J., and Robb, R. M., 1979, Studies on the development of the eye cup and optic nerve in normal mice and in mutants with congenital optic nerve aplasia, *Dev. Biol.* **68**:175–190.

Singley, C. T., and Solursh, M., 1980, The use of tannic acid for the ultrastructural visualization of hyaluronic acid, *Histochemistry* **65**:93–102.

Solursh, M., Fisher, M., Meier, S., and Singley, C. T., 1979, The role of extracellular matrix in the formation of the sclerotome, *J. Embryol. Exp. Morphol.* **54**:75–98.

Spieth, J., and Keller, R. E., 1984, Neural crest cell behavior in white and dark larvae of *Ambystoma mexicanum*: Differences in cell morphology, arrangement and extracellular matrix as related to migration, *J. Exp. Zool.* **229**:91–107.

Strudel, G., 1976, The primary connective tissue matrix of the bird embryo, *Front. Matrix Biol.* **3**:77–99.

Stump, R. F., and Robinson, K. R., 1983, *Xenopus* neural crest cell migration in an applied electrical field, *J. Cell Biol.* **97**:1226–1233.

Teillet, M.-A., 1971, Recherches sur le mode de migration et la différenciation des mélanocytes cutanés chez l'embryon d'Oiseau: Étude expérimentale par la methode des greffes hétérospécifiques entre embryons de Caille et de Poulet, *Ann. Embryol. Morphog.* **4**:125–135.

Teillet, M.-A., and Le Douarin, N., 1970, La migration des cellules pigmentaires étudiée par la methode des greffes hétérospécifiques de tube nerveux chez l'embryon d'Oiseau, *C. R. Acad. Sci.* **270**:3095–3098.

Teillet, M.-A., and Le Douarin, N. M., 1983, Consequences of neural tube and notochord excision on the development of the peripheral nervous system in the chick embryo, *Dev. Biol.* **98**:192–211.

Thiery, J. P., Duband, J. L., and Delouvée, A., 1982a, Pathways and mechanisms of avian trunk neural crest cell migration and localization, *Dev. Biol.* **93**:324–343.

Thiery, J. P., Duband, J. L., Rutishauser, U., and Edelman, G. M., 1982b, Cell adhesion molecules in early chicken morphogenesis, *Proc. Natl. Acad. Sci. USA* **79**:6737–6741.

Thiery, J. P., Delouvèe, A., Grumet, M., and Edelman, G. M., 1985, Initial appearance and regional distribution of the neuron-glia cell adhesion molecule in the chick embryo, *J. Cell Biol.* **100**:442–456.

Tickle, C. A., and Trinkaus, J. P., 1973, Changes in surface extensibility of *Fundulus* deep cells during early development, *J. Cell Sci.* **13**:721–726.

Tickle, C. A., and Trinkaus, J. P., 1976, Observations on nudging cells in culture, *Nature (Lond.)* **261**:413.

Tickle, C., Crawley, A., and Goodman, M., 1978, Cell movement and the mechanism of invasiveness: A survey of the behavior of some normal and malignant cells implanted into the developing chick wing bud, *J. Cell Sci.* **31**:293–322.

Toole, B. P., 1981, Glycosaminoglycans in morphogenesis, in: *Cell Biology of Extracellular Matrix* (E. D. Hay, ed.), pp. 259–294, Plenum Press, New York.

Toole, B. P., and Trelstad, R. L., 1971, Hyaluronate production and removal during corneal development in the chick, *Dev. Biol.* **26**:28–35.

Tosney, K. W., 1978, The early migration of neural crest cells in the trunk region of the avian embryo: An electron microscopic study, *Dev. Biol.* **62**:317–333.

Tosney, K. W., 1982, The segregation and early migration of cranial neural crest cells in the avian embryo, *Dev. Biol.* **89**:13–24.

Trinkaus, J. P., 1963, The cellular basis of *Fundulus* epiboly. Adhesivity of blastula and gastrula cells in culture, *Dev. Biol.* **7**:513–532.

Trinkaus, J. P., 1973, Surface activity and locomotion of *Fundulus* deep cells during blastula and gastrula stages, *Dev. Biol.* **30**:68–103.

Trinkaus, J. P., 1976, On the mechanism of metazoan cell movements, in: *The Cell Surface in Animal Embryogenesis and Development* (G. Poste and G. L. Nicolson, eds.), pp. 225–329, Elsevier Biomedical Press, New York.

Trinkaus, J. P., 1982, Some thoughts on directional cell movement during morphogenesis, in: *Cell Behavior* (R. Bellairs, A. Curtis, and G. Dunn, eds.), pp. 471–498, Cambridge University Press, Cambridge.

Tucker, G. C., Aoyama, H., Lipinski, M., Tursz, T., and Thiery, J. P., 1984, Identical reactivity of monoclonal antibodies HNK-1 and NC-1: Conservation in vertebrates on cells derived from the neural primordium and on some leukocytes, *Cell Differ.* **14:**223–230.

Tucker, R. P., and Erickson, C. A., 1984, Morphology and behavior of quail neural crest cells in artificial three-dimensional extracellular matrices, *Dev. Biol.* **104:**390–405.

Tucker, R. P., Edwards, B. F., and Erickson, C. A., 1985, Tension in the culture dish: Microfilament organization and migratory behavior of quail neural crest cells, *Cell Motil.* **5:**225–237.

Turley, E. A., Erickson, C. A., and Tucker, R. P., 1985, The retention and ultrastructural appearances of various extracellular matrix molecules incorporated into three-dimensional hydrated collagen lattices, *Dev. Biol.* **109:**347–369.

Twitty, V. C., 1936, Correlated genetic and embryological experiments on *Triturus.* I and II, *J. Exp. Zool.* **74:**239–302.

Twitty, V. C., 1945, The developmental analysis of specific pigment patterns, *J. Exp. Zool.* **100:**141–178.

Twitty, V. C., 1949, Developmental analysis of amphibian pigmentation, *Growth* (Suppl. 9) **13:**133–161.

Twitty, V. C., and Bodenstein, D., 1939, Correlated genetic and embryological experiments of *Triturus.* III and IV, *J. Exp. Zool.* **95:**357–398.

Twitty, V. C., and Bodenstein, D., 1941, Experiments on the determination problem. I and II, *J. Exp. Zool.* **86:**343–380.

Twitty, V. C., and Bodenstein, D., 1944, The effect of temporal and regional differentials on the development of grafted chromatophores, *J. Exp. Zool.* **95:**213–231.

Twitty, V. C., and Niu, M. C., 1948, Causal analysis of chromatophore migration, *J. Exp. Zool.* **108:**405–437.

Twitty, V. C., and Niu, M. C., 1954, The motivation of cell migration studied by isolation of embryonic pigment cells singly and in smaller groups *in vitro, J. Exp. Zool.* **125:**541–574.

Valinsky, J. E., and Le Douarin, N. M., 1985, Production of plasminogen activator by migrating cephalic neural crest cells, *EMBO J.* **4:**1403–1406.

Valinsky, J. E., Reich, E., and Le Douarin, N. M., 1981, Plasminogen activator in the bursa of Fabricius: Correlations with morphogenetic remodeling and cell migration, *Cell* **25:**471–476.

Vincent, M., and Thiery, J.-P., 1984, A cell surface marker for neural crest and placodal cells: Further evolution in peripheral and central nervous system, *Dev. Biol.* **103:**468–481.

Vincent, M., Duband, J.-L., and Thiery, J.-P., 1983, A cell surface determinant expressed early on migrating neural crest cells, *Dev. Brain Res.* **9:**235–238.

von der Mark, H., von der Mark, K., and Gay, S., 1976, Study of differential collagen synthesis during development of the chick embyro by immunofluorescence, *Dev. Biol.* **48:**237–249.

Warner, A. E., 1973, The electrical properties of the ectoderm in the amphibian embryo during induction and early development of the nervous system, *J. Physiol. (Lond.)* **235:**267–286.

Waterman, R. E., 1976, Topographical changes along the neural fold associated with neurulation in the hamster and mouse, *Am. J. Anat.* **146:**151–172.

Waterman, R. E., 1977, Ultrastructure of oral (buccopharyngeal) membrane formation and rupture in the hamster embryo, *Dev. Biol.* **58:**219–229.

Webster, W., 1973, Embryogenesis of the enteric ganglia in normal mice and in mice that develop congenital aganglionic megacolon, *J. Embryol. Exp. Morphol.* **30:**573–585.

Weiss, P., 1945, Experiments on cell and axon orientation *in vitro:* The role of colloidal exudates in tissue organization, *J. Exp. Zool.* **100:**353–386.

Weston, J. A., 1963, A radioautographic analysis of the migration and localization of trunk neural crest cells in the chick, *Dev. Biol.* **6:**279–310.

Weston, J. A., 1970, The migration and differentiation of neural crest cells, *Adv. Morphogen.* **8:**41–114.

Weston, J. A., 1980, Role of the embryonic environment in neural crest morphogenesis, in: *Current*

Research Trends in Prenatal Craniofacial Development (R. M. Pratt and R. L. Christiansen, eds.), pp. 27–45, Elsevier/North-Holland, New York.

Weston, J. A., 1982, Motile and social behavior of neural crest cells, in: *Cell Behaviour* (R. Bellairs, A. Curtis, and G. Dunn, eds.), pp. 429–470, Cambridge University Press, Cambridge.

Weston, J. A., and Butler, S. L., 1966, Temporal factors affecting localization of neural crest cells in the chicken embryo, *Dev. Biol.* **14:**246–266.

Weston, J. A., Ciment, G., and Girdlestone, J., 1984, The role of extracellular matrix in neural crest development: A reevaluation, in: *The Role of Extracellular Matrix in Development* (R. L. Trelstad, ed.), pp. 433–460, Alan R. Liss, New York.

Yamada, K., 1980, Structure and function of fibronectin in cellular and developmental events, in: *Current Research Trends in Prenatal Craniofacial Development* (R. M. Pratt and R. L. Christiansen, eds.), pp. 297–314, Elsevier/North-Holland, New York.

Yamada, K. M., 1983, Cell surface interactions with extracellular materials, *Annu. Rev. Biochem.* **52:**761–799.

Yntema, C. L., and Hammond, W. S., 1945, Depletions and abnormalities in the cervical sympathetic system of the chick following extirpation of neural crest, *J. Exp. Zool.* **100:**237–263.

Zigmond, S. H., 1982, Polymorphonuclear leucocyte response to chemotactic gradients, in: *Cell Behavior* (R. Bellairs, A. Curtis, and G. Dunn, eds.), pp. 183–202, Cambridge University Press, Cambridge.

IV

Cellular Dynamics in Morphogenesis

Chapter 14

On The Formation of Somites

JAMES W. LASH and DAVID OSTROVSKY

1. History

How do animals wiggle and bend? Throughout the animal kingdom may be found a persistent body organization of repeated metameric segments. In practically all metazoans, this meristic body pattern provides an efficient arrangement for body movement and locomotion. This is particularly striking in the annelids, in which this meristic pattern is seen in body segments and organs within the segments. Naturalists as far back as Aristotle (1910, translated by Thompson) have noted the common feature of segmentation in animals as diverse as worms, maggots, and snakes and in chicken embryos. Because of the ready availability of chicken eggs and the accessibility of the developing embryo, the segmentation of the vertebrate embryo was noted as early as the sixteenth century. This segmentation was alluded to as the "division of the embryo," which could seldom be seen in the early stages of development without aids for magnification. Although lenses for magnification were noted as early as Aristophanes (fifth century BC), Roger Bacon (thirteenth century) is thought to have been the first scientist–naturalist to use lenses for investigative purposes (Needham, 1959). Embryologists were slow to use microscopes in their observations, although occasional use of the "perspicilli" was reported, and undoubtedly this instrument helped Harvey in his pioneer observations (1651) on early chick development (Meyer, 1936). Others in the seventeenth century also represented somites in their drawings of chicken embryos, although they did not remark upon them. Malpighi in 1673 portrayed with great clarity 11 pairs of somites in a chicken embryo, and in an older ("38-hr") embryo he portrayed 10 pairs of somites (Meyer, 1939).

During the 1830s, microscopes became generally available, and embryologists began to use them in their studies. One of the earliest portrayals of somite development in the chick embryo appeared in an 1834 monograph by Coste (see Fig. 1). What Coste called the "première apparition des vertébrés"

JAMES W. LASH • Department of Anatomy, School of Medicine, University of Pennsylvania, Philadelphia, Pennsylvania 19104.   DAVID OSTROVSKY • Department of Biology, Millersville University of Pennsylvania, Millersville, Pennsylvania 17551.

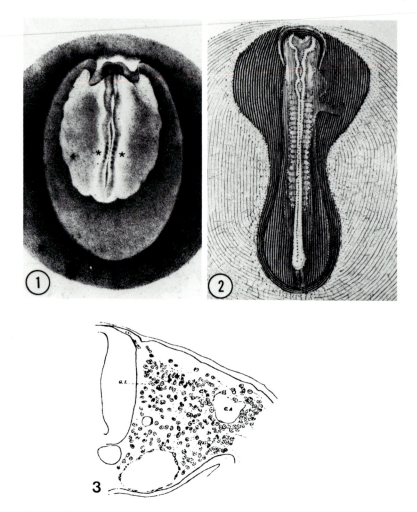

Figure 1. Drawing from Coste (1834), showing a chick embryo with six pairs of somites at the asterisks (*).

Figure 2. One of the "Hunterian drawings" of a chicken embryo, published in 1837. It closely approximates a Hamilton–Hamburger stage 12 embryo. (From Palmer, 1837.)

Figure 3. A drawing of a chick somite from Williams (1910), clearly showing the sclerotome and the myotome forming from the dermamyotome.

are clearly six pairs of somites. Apparently he did not place high importance on them, since in the next frame (not shown), when the embryo had developed from approximately 22–24 hours to 26–29 hr, the number of somite pairs had decreased to five. John Hunter, in the "Hunterian drawings" done between 1762 and 1798, drew a remarkably accurate representation of what today would be considered a stage 12 embryo (Hamburger and Hamilton, 1951) (see Fig. 2). These blocks of tissue, or soma, were realized by the early embryologists (although somewhat dimly) to be related to vertebral development and, in the mid-nineteenth century, were called provertebrae. The suffix *-ite* was added to soma to give these structures the name "somite" in the 1860s; during the 1890s, many papers appeared describing the development of somites. One of the earliest detailed descriptions of somite formation is that of Williams (1910), who traced the development of chick somites in a series of elegant drawings (Fig. 3). In spite of the very active period of descriptive embryology and descriptions of somites forming (Assheton, 1896; Peebles, 1898), the relationship of these meristic structures to the eventual segmental pattern remained elusive. The noted biologist, E. G. Conklin, described vertebral segmentation in Mason (1928) in the following manner: "His backbone begins as a notochord, is next a segmented cartilaginous rod, then each segment or vertebra consists of five separate bones, and finally fuses into a single bone." Clearly, the relationship between somites and vertebral segmentation had not yet become part of the mainstream of embryology.

Modern studies of somitogenesis began during the early 1950s. Fraser (1954) and Spratt (1955) were among the first to seek the origin or somite pattern formation.

2. Somite Origin and Sequence of Formation

A variety of techniques have been used to delineate presumptive somite mesoderm in the chick embryo. These methods have ranged from vital staining to stereoanalysis of scanning electron micrographs (SEMs). An example of an early study is that by Spratt and Condon (1947), who marked several hundred chick blastoderms with carbon particles and concluded that at the primitive streak stage, at least 5 hr before the appearance of the first somite, presumptive somites could be mapped in the epiblast behind the primitive pit. This region was 0.1–0.3 mm long and extended 0.1 mm from the edge of the primitive streak. The width of the region was later extended to 0.2–0.3 mm on the basis of vital dye-marking experiments (Spratt, 1955). A similar narrow band of presumptive somite mesoderm in the epiblast was previously identified in the vital staining experiments conducted by Pasteels (1937).

There are drawbacks, however, to both carbon marking and vital staining. Spratt (1955) pointed out that carbon particles were often brushed off the surface of invaginating cells, and it is well known that vital dyes can diffuse from their point of application. Nicolet (1970, 1971a) avoided these problems by using radioactively labeled tissue. A portion of an unlabeled embryo was removed and replaced by a similar portion from another embryo labeled with tritiated thymidine. These experiments confirmed the presumptive mesoderm fate maps prepared by Pasteels and Spratt.

Once presumptive somite mesoderm ingresses, time-lapse film studies indicate that this tissue lengthens and narrows to form the segmental plates, which first appear anterior to the regressing primitive pit (Lipton and Jacobson, 1974a). The somites, except for the first three, which "are not formed from segmental plate but from a direct aggregation of mesoderm cells" (Bellairs, 1979), will eventually develop from the anterior end of the segmental plate. Lipton and Jacobson (1974a) believe the boundaries of the somite-forming region of the segmental plate are set by the overlying neural plate. The only tissues that adhere to the basal lamina of the overlying epiblast are the somite mesoderm and the lateral plate. They point out that the first few somites from the segmental plate are significantly wider than later somites because of subsequent narrowing of segmental plate by the closure of the neural tube. Eventually the somite mesoderm ceases to adhere to the neural ectoderm but remains attached by fibrous (collagenous) material.

An interesting observation on segmental plate structure was made by Meier (1979). Meier agrees that the boundaries of the neural plate define the boundaries of the somite mesoderm; he also states that stereoanalysis of more than 2400 SEMs shows a meristic pattern in the segmental plate hours before the first somite would normally appear. The segmental plates contain tandem whorls of mesoderm, which he calls **somitomeres** (Fig. 4). This early pattern had not been noted previously, because somitomeres can only be seen by

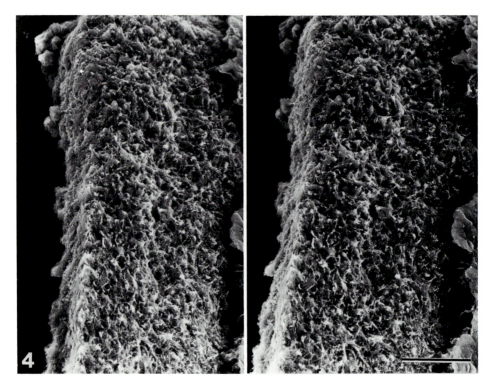

Figure 4. Stereo-scanning electron micrographs (stereo-SEMs) (tilt angle 10°) of the left segmental plate of a stage 12 chick embryo, from which the surface and neural ectoderm have been removed. A stereoscopic viewer permits visualization of three somitomeres. The upper part of the photograph is the posterior end of the segmental plate. Scale bar: 50 μm. (Unmarked photographs were kindly provided by Dr. Stephen Meier.)

stereoanalysis of SEM photographs. The segmental plate is consistently reported to have 10 or 11 somitomeres, which eventually convert into somites (Packard, 1978; Packard and Meier, 1983).

Until the turn of the century, new somites were believed to arise anterior to already formed somites. Disputes arose over just how many somites developed anterior to the first pair (Patterson, 1907). A study of sagittal sections of chick embryos, for example, led Platt (1898) to conclude that "two provertebrae" (or, more correctly, one and a half) "are slowly formed anterior to the first mesodermic cleft, in the time occupied by the formation of six or seven provertebrae posterior to that cleft." Experiments were carried out by Patterson (1907), in which the anterior boundary of the first somite was marked by cautery. No somites formed anterior to this mark. He concluded that somitogenesis proceeds only from the anterior to the posterior end of the embryo, a conclusion supported by contemporary observations (Bellairs, 1963).

3. The Stages of Somite Formation

Anyone who has tried to count the number of somite pairs in a chick embryo will appreciate the difficulty in determining just what to consider the last pair. In our laboratory we have tried multiple observer test counts to see how much agreement, or disagreement, would be found between somite counts made by different observers. Invariably, until we agreed on a visual definition of a somite, individual counts varied by as many as three somite pairs. The variance was consistent in that one observer's counts were always high, whereas another observer's counts tended to be low. It became obvious that we were using different end points in our visual definition of a somite. Careful and repeated observations on Hamburger and Hamilton's (1951) stages 12–15 indicate the following five discrete stages of somite formation (Fig. 5):

Stage I: The segmental plate appears to be homogeneous under the dissecting microscope. There is no discernible morphological distinction between the anterior region, where the next somite will form, and the more posterior regions. This stage might have been called stage 0, but for the observations of Meier (1979), who reported a meristic pattern visible only with stereoscopic examination of SEM photographs. This stage is somewhat ambiguous and may eventually have to be subdivided, since the meristic pattern ("somitomeres") extends throughout the segmental plate and the somitomeres undergo "maturation" (Packard and Meier, 1983). Stage I is a clearly recognizable stage that can be considered the stage of commitment and determination of the eventual somites.

Stage II: A subtle darkening of the anterior end of the segmental plate is seen during this stage. Histological preparations indicate that this period is marked by cellular condensations. The compaction of the presomite cells is caused by cell aggregation, which appears to be the result of increased cell–cell adhesion accompanying cellular distribution and localization of

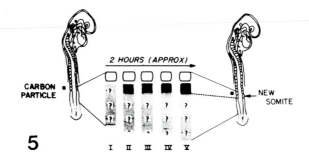

Figure 5. Stages I–V of somite formation. *Stage I:* The segmental plate appears to be homogeneous, but stereo-SEM analysis indicates the presence of meristic whorls of cells, the somitomeres described by Meier (1979). *Stage II:* The presomitic mass of the segmental plate undergoes compaction, apparently as the result of an increase in cell–cell adhesion in this region. *Stage III:* The anterior border of the presomitic mass becomes distinct. *Stage IV:* The lateral borders of the future somite become distinct. *Stage V:* The posterior border of the presomitic mass becomes distinct. As the nascent somite separates from the segmental plate, stage I begins.

actin filaments and as yet unknown interactions with extracellular matrix fibers surrounding the segmental plate (see Section 4). Stage II is thus one of compaction and aggregation of presomite cells.

Stage III: This stage marks the first indication of segmentation, or the separation of the presomitic mass. The anterior border of the segmental plate becomes clearly delineated, and a definite interface appears. Before this, the anterior border is not clearly defined. The cause of this delineation is thought to be increased compaction and the beginning of epithelialization (Ostrovsky *et al.*, 1983; Cheney and Lash, 1984). Stage III marks the beginning of true segmentation as opposed to the meristic pattern of somite determination that existed before this.

Stage IV: This stage gradually develops from the previous one and is easily recognized by the clear delineation of the lateral borders of the presomitic mass. This is caused by a continuation of the processes initiating stage III and marks the further epithelialization of the future somite.

Stage V: The posterior border appears. The somite now has distinct borders around its entire periphery and separates from the segmental plate. This stage can be considered that of the **nascent somite**—a term used here to describe the somite as it begins its existence as a separate mass of tissue and begins to grow and develop. When a space appears between the nascent somite and the segmental plate, stage V becomes stage I.

The rate of *in ovo* somite formation (i.e., from stage I–V) in the chick embryo has been reported to be one pair every 75 min (Elias and Sandor, 1965). In our studies the embryo was placed on a raft over nutrient medium (cf. Chernoff and Lash, 1981); one somite pair formed in 2 hr. This finding agrees with the *in vivo* rate of one pair every 115 min reported by Menkes *et al.* (1961, 1969).

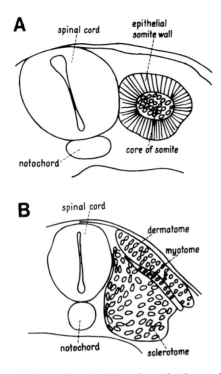

Figure 6. Drawings of a recently formed somite (A) and one that has undergone further diversification into sclerotome and dermamyotome (B). (Drawings modified from Lash and Cheney, 1982.)

Once formed, the nascent somite rapidly undergoes diversification. The epithelial vesicle contains cells in its interior, the **core cells.** Williams (1910) describes these cells as increasing in number in the more posterior somites. The core cells and the ventromedial wall of the somite become the mesenchymal **sclerotome,** which begins to migrate toward the notochord. The stratified dorsolateral portion of the somite separates from the sclerotomal cells and forms a double-layered structure, the **dermamyotome** (Fig. 6).

It is important to keep in mind that the somites are probably interacting significantly at this time with neighboring cells and structures. As we shall discuss in Section 4, extracellular matrix components have been implicated in somite compaction. Minor (1971) suggested a possible interaction between neural crest cells and somite diversification. He observed that diversification coincides with neural crest cells contacting the dorsal and medial surfaces of the dorsomedial angle of the somite (Fig. 7). Coincident with this event is a burst of mitotic activity in the presumptive myotome region of the dermamyotome; these cells then move ventrolaterally beneath the layer of tissue that will become the dermatome (Minor, 1971 and personal communication). Williams (1910) also noted mitotic activity in the presumptive myotomal cells and suggested that this activity generates the myotomal layer (cf. Fig. 3). Contrary

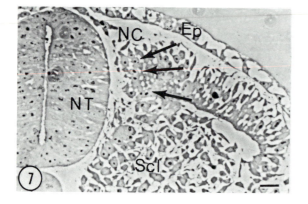

Figure 7. Cross section of a plastic-embedded chick embryo showing mitotic figures (arrows) in the region wherein the myotome is forming, NT, neural tube; NC, neural crest cells; Ep, epidermis; Scl, sclerotome. Scale bar: 10 μm. (Photograph kindly provided by Dr. Ronald R. Minor.)

evidence was reported by Langman and Nelson (1968), who treated embyros with tritiated thymidine to label dividing cells and then treated the embryos with vincristine sulfate to arrest mitoses in metaphase. These workers did not find mitotic figures in the presumptive myotomal layer and concluded that the myotome originated from the overlying dermatome.

Fitting together Williams's observations in 1910 with those of Minor in 1971, it appears that the myotome forms from the dorsomedial border of the dermamyotome, possibly stimulated to mitotic activity by impinging neural crest cells, and then migrates beneath the future dermatome. The treatment with vincristine sulfate by Langman and Nelson (1968) probably prevented cell migration by interfering with microtubule organization. Indeed, using quail grafts in chick embryos Christ et al. (1978) concluded that the myotomal cells did come from the edges of the dermatome and not from a separation of the dermamyotome into two layers.

4. The Mechanism of Somite Formation

The mechanism of somite formation has long intrigued embryologists. Recent work and theories are summarized in the multiauthored volume, *Vertebrate Limb and Somite Morphogenesis*, edited by Ede et al. (1977) and in Meier and Jacobson (1982). It has been difficult to test experimentally the factors that have been proposed to regulate somite formation. There are probably two reasons for this. One is the smallness of the embryo, which makes some experimentation difficult. Another is that some researchers have not yet recognized the complexity of the process.

Efforts to investigate the initiation of the somite forming process began with experiments indicating that Hensen's node (Fraser, 1954; Spratt, 1955), or its derivative, the notochord (Nicolet, 1971a), acts as an "inducer" of somite formation. Sandor and Amels (1970) reported that *in situ* segmental plates, plus spinal cord and notochord with ectoderm and endoderm removed, underwent normal segmentation. This research showed a requirement for axial structures (spinal cord and notochord) to ensure normal somite formation. Nicolet (1971a) even reported evidence that the somite-promoting effect of the notochord could be transmitted through a Millipore filter. Contrary evidence was presented by Bellairs (1963) and Lipton and Jacobson (1974b), who demonstrated that isolated segmental plates can segment in the absence of the notochord. It could be argued, of course, that the pattern of somite determination had already been set by Hensen's node or the notochord since, as we have already seen, there may be a meristic pattern of whorls throughout the segmental plate (i.e., the "somitomeres" described by Meier, 1979).

Butros (1967) and Lipton and Jacobson (1974b) implicated the neural tube in the stimulation of somitogenesis, even though somites can form from isolated segmental plates. The actual process of segmentation (i.e., separation of the nascent somite from the segmental plate) is only one of the processes that occur during somitogenesis. The process of segmentation may be temporally and even spatially removed from the agent or agents involved in initiating the process of somitogenesis. The earlier work clearly implicated the axial structures as being important for somitogenesis, but no one structure has yet been shown to be the essential component of somite formation. Lipton and Jacobson (1974b) attempted to resolve these conflicting results and concluded that Hensen's node and the notochord "shear" the prospective somite mesoderm and "release somite-forming capabilities already present." This hypothesis is difficult to test by experimentation.

In attempts to isolate the putative factor(s) responsible for somite formation, attention was brought to bear on the issue of determination: When does the segmental plate become committed to form somites? Lanot (1971) presented evidence that the nascent somites influence segmentation in the segmental plate just posterior to it, clearly suggesting that something from the nascent somite acts on the adjacent segmental plate. Packard and Jacobson (1976) and later Packard (1978) presented convincing evidence that the segmental plate becomes determined as it forms and that segmentation in the posterior portion of the segmental plate could occur in the absence of either the nascent somite or anterior segmental plate tissue. This again indicates that the pattern of somite formation is set very early, even though its initiation may not occur until a later time. Lipton and Jacobson (1974b) suggested that the mesoderm is imprinted with a "prepattern of segmentation," possibly as a result of interacting with the neural tube or notochord, or both. This "prepattern" is reminiscent of the meristic pattern visualized with stereo-SEM by Meier (1979, 1981), which seems to herald the eventual metameric somite pattern.

Interest in the earliest events of somitogenesis has been revived since the description of somitomeres by Meier (1979). Earlier workers had related the somite pattern to the events surrounding Hensen's node and suspected that Hensen's node might in some way impart the pattern to the segmental plate. The process of determination or commitment results in a "prepattern" of somites; i.e., the pattern of repetitive somites is already present in the segmental plate, waiting to unfold. Experiments reported earlier, e.g., those by Bellairs (1963) and Packard and Jacobson (1976), clearly showed that isolated segmental plates will attempt to segment, proving that the segmentation pattern is present within the apparently homogeneous mesenchyme. The segmentation of the isolated segmental plate is incomplete, however, and only one fully formed somite appears. This suggests that the microenvironment in the embryo provides additional factors for complete somitogenesis (Lash *et al.*, 1984; R. Bellairs, personal communication).

The first visual indication of this pattern may be the cell whorls signifying the somitomeres. Until recently, it was uncertain as to whether the "prepattern" was an outmoded abstract expression or was visible as somitomeres. Bellairs and Veini (1984) clearly showed recently that the stage 4 (Hamburger and Hamilton staging series, 1951) chick embryo contains preprogrammed cells in the presumptive somite area. Their evidence suggests that the programming occurs as the cells migrate out of the primitive streak. Despite uncertainty about the precise nature of the role played by Hensen's node and the process of ingression, the pattern itself seems to be established at this time. Thus, these fragile meristic whorls called somitomeres, which are morphologically indistinct without special methodology, may be the first visual indication of the eventual segmentation pattern, which was established at the time of ingression.

Mechanical disruption of the segmental plate with a glass needle (Menkes and Sandor, 1969, 1977), presumably disrupting the somitomeric pattern, had little or no effect on subsequent segmentation depending on age. After mechanical disruption of the segmental plate, 30-hr embryos showed some abnormalities, whereas 36-hr embryos were reported to show normal segmentation. So either the somitomeric pattern is not essential for segmentation, or it forms again soon after being disturbed. The fact that segmentation occurs in a nearly normal pattern after disruption suggests that the neighboring tissues and/or matrix components of the embryo can still impose a segmentation pattern on the segmental mesenchyme. Furthermore, this pattern seems to agree with what would normally be present in the embryo at that time. Of course, mechanical disruption may leave vestiges of the "prepattern," and total cell disaggregation of the segmental plate might give more definite information on both the stability of the commitment and the control of the somite pattern by the surrounding microenvironment.

Most of the work on mechanisms of somite formation has been on the chick embryo, undoubtedly because of its accessibility and relative ease of perturbation. Hypotheses have also been derived from experimentation with amphibian and mammalian somite formation. The detailed work of Hamilton (1969) on amphibian somite formation has led to interesting proposals of control mechanisms derived from the mathematics of topology and the catastrophe theory. This resulted in a clock and wavefront model by Cooke and Zeeman (1976) and Cooke (1977). (An earlier version of this model, as applied to avian embryos, was proposed by Zeeman in 1974.) The wavefront model is envisioned as a smooth progression of a morphogenetic process (similar to a biorhythm) that imparts a periodicity to the morphogenetic events as they proceed in the conventional cranial–caudal direction. It is periodically interrupted, which results in the repetitive somite pattern. Pearson and Elsdale (1979) showed that heat-shock treatment of amphibian embryos at selected times of early development results in somite anomalies that can be interpreted as being compatible with the clock and wavefront model. The effect of heat shock is not on the region of somite formation but on more posterior regions in the segmental plate, where the presumed interaction between the clock and the wavefront is taking place. There is no comparable supportive evidence from work on intact avian embryos. This model can also be considered an extension of the spatial pattern theories proposed by Goodwin and Cohen (1969) and Wolpert (1969, 1982). These theories propose that information can reside in the relative position of a cell to its neighbors; it is this positional information that plays a determining role in pattern formation and either programmed growth or differentiation, or both. This theory has been propounded most thoroughly for limb morphogenesis (e.g., see Archer *et al.*, 1983) but as yet remains only a brief theoretical proposal for somite formation.

All the above observations support the contention that somite formation is not the result of any one factor and strongly suggest that the determination, or commitment, of the segmental plate occurs as it is formed from the lateral mesoderm, probably as a result of influences of axial structures and the surrounding extracellular matrix.

Whereas theories have abounded, very little experimentation has been done on the mechanisms of somite formation. Although isolated segmental plates will attempt to form somites if the tissue does not stick and spread over the culture dish, differentiation and maintenance of somites is best when the neighboring tissues are present. There is no reported examination of the stages of segmentation in isolated segmental plates to see whether they follow the same regimen as segmental plates *in situ*. The work of Menkes and Sandor (1969, 1977) clearly indicates that the *in vivo* environment of the segmental plate plays an important role in regulating pattern formation. The somites that formed again after disruption did not form at random but attempted to form a pattern appropriate for that place and time in the embryo.

As discussed in Chapter 12, Bellairs *et al.* (1978) suggested that compaction during somite formation may be due to increased adhesion among the cells in the segmental plate. In addition to increased adhesion, Bellairs (1979) proposed that collagen fibers have a possible regulatory role. These practical observations place the theory of somite formation on a level that can be tested by experimentation. Do segmental plate cells that are forming somites adhere to their nearest neighbors more closely during somite formation than before, and do identifiable extracellular components have an effect on somitogenesis? Various other molecules have been proposed as playing an important role in somitogenesis. Cheney *et al.* (1980) and Lash *et al.* (1984) suggested a role for fibronectin. Bolzern *et al.* (1979) reported that chick somite formation can be induced by a 26S mRNA from muscle as well as by myosin. Although provocative, this latter report is too preliminary to evaluate properly.

Although the work uncovering a prepattern of the somites and establishing the controlling influence of the segmental plate on the eventual somite pattern is of great significance, we still require an understanding of the mechanisms involved in the transition from segmental plate to the eventual metameric somites.

Flint and Ede (1982), from their work with the mouse mutant *amputated*,

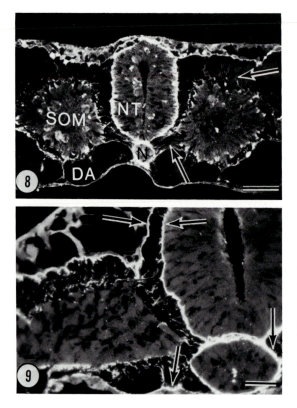

Figure 8. Immunohistological preparation localizing collagen type I in a cross section of a stage 12 chick embryo. The somites appear to be suspended in the body by a fine meshwork of collagen fibers (arrows). SOM, somites; DA, dorsal aorta; NT, neural tube; N, notochord. Scale bar: 50 μm.

Figure 9. Immunohistological preparation localizing cellular fibronectin (cFN) in the segmental plate of a stage 12 chick embryo. cFN is prominent in the regions containing basal lamina (arrows) but is sparse around the segmental plate. As somitogenesis proceeds, the amount of fibronectin around the segmental plate increases. Scale bar: 10 μm. (Compare with Ostrovsky et al., 1984; Lash, 1984.)

implicated cell motility as playing a role in the developing somite. Chernoff and Lash (1981) showed with chick embryos that cytochalasin D, which disrupts the integrity of the cell's contractile apparatus, inhibits somite formation. Chernoff and Hilfer (1982) reported the existence of a calcium-dependent contractile process that is involved in somite formation. Further indication that contractility is an important factor comes from the studies of Ostrovsky et al. (1983), in which the pattern of cellular actin localization was followed during the early stages of somite formation.

The segmental plate is suspended in the embryo by a fine network of extracellular matrix fibers, which are primarily (if not solely) collagen fibers (Fig. 8) (see Section 2). The distribution of cell-associated fibronectin during the early stages of somitogenesis (Fig. 9) strongly suggests that it plays a role in the compaction or aggregation of presomite cells. Further experiments involving the addition of exogenous fibronectin (cell-associated) to either intact embryos, isolated segmental plates, or dissociated segmental plate cells, show that fibronectin has a pronounced promotional effect on the rate of somite formation and on the aggregation competence of segmental plate cells (Cheney et al., 1980; Lash et al., 1984).

Thus, as far as mechanisms of somitogenesis are concerned, obviously more than one mechanism is involved in the multistage process of somite formation. The imprinting of the pattern is clearly related to the events surrounding Hensen's node, although the actual mechanism of the event is still elusive. The transition from stage I to stage II certainly seems to involve increased cell–cell adhesion and redistribution of intracellular actin (Bellairs *et al.*, 1978; Ostrovsky *et al.*, 1983; Cheney and Lash, 1984). Little is known of the influence of the extracellular matrix. The work of Meier and Jacobson (1982) and Packard and Meier (1983), as well as earlier studies by Menkes and Sandor (1969, 1977), suggests that the embryonic space containing the segmental plate has considerable influence on the expression of the somitic pattern. Obvious candidates for involvement with this regulation are the extracellular matrix components.

Other than the investigations discussed above on the mechanism of somite formation, no detailed biochemical studies of somitogenesis have been conducted. Shortly after the somite forms, it rapidly begins to diversify; this diversification has been analyzed biochemically. Since vertebral cartilages are one of the major products of somites, research has concentrated primarily on chondrogenesis. During chondrogenesis, however, the somite becomes a complicated structure composed of heterogeneous tissues undergoing intricate morphogenetic movements no less intriguing than the initial movements involved in the formation of the somite. These movements are discussed in more detail in Chapter 12; suffice it to say that problems still abound in the differentiation of the somite into its constituent subpopulation of tissues. The work of Cheney and Lash has addressed the question of biochemical changes related to chondrogenesis during diversification (Cheney and Lash, 1981; Lash and Cheney, 1982). These investigators found that the sclerotome and dermamyotome rapidly acquire different metabolic properties attributable to the chondrogenic properties of the sclerotome. Reviews on the biochemistry of somite chondrogenesis can be found in Lash and Cheney (1982), Lash and Vasan (1983), and Hall (1983).

5. The Fate of the Somite

This chapter emphasizes the processes of somite formation, but the rapid change from the nascent somite to its subsequent differentiation warrants brief mention of the somite's fate. The early changes in the somite are well known and have been frequently described in the literature (Williams, 1910; Nelson, 1953; Romer and Parsons, 1977; Solursh *et al.* 1979; Meier, 1980; Chernoff and Lash, 1981). The ventromedial portion of the somite becomes the sclerotome, while the dorsolateral portion differentiates into the dermamyotome (Fig. 6). Although Gasser (1979) contends that the sclerotomal cells do not migrate, the evidence seems uncontrovertible that the sclerotome cells do migrate toward the notochordal region (see Section 3). Whereas the dermatome migrates to the epithelium, where it contributes to the dermal portion of the skin, the sclerotome and myotome undergo more complicated morphogenetic changes (Nelson, 1953). The original segmentation pattern of the somites is disrupted

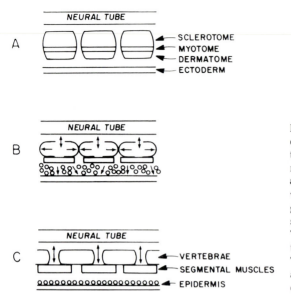

Figure 10. Diagram showing the developmental fate of somites. The segmental sclerotome (A), as the result of anterior and posterior migration (B, lateral arrows), becomes the intersegmental vertebrae (C). Neural crest cells migrate, to contribute to the segmental spinal nerves (B, C, double arrows). The dermatomal cells migrate to the ectoderm (B), to form the epidermis (C). The myotome maintains its segmental arrangement as the segmental muscles of the back (C).

and reorganized so that the anterior portion of one somite fuses with the posterior portion of the adjacent somite to form the muscular metameric structures that characterize vertebrates. These muscles connect the vertebral segments, which are derived from the original segmental pattern of the somites (Fig. 10).

Although the precise origin of the somite's influence on limb development is not comletely known, there is strong evidence not only that somites influence limb development (cf. Lash and Saxén, 1972; Stephens, 1983) but that they actually contribute the cells that eventually differentiate into the limb's skeletal muscles (Chevallier, 1978; Christ *et al.*, 1983; Kieny, 1983). Thus, a relatively simple vesicular structure, the embryonic somite, unfolds its differentiative program by contibuting to, and influencing, a myriad of other differentiative processes.

6. Projections

It is frequently very difficult to predict how a research field will develop. Saltatory advances frequently make difficult techniques and procedures readily available to most laboratories, as evidenced by the widespread use of monoclonal antibody production and genetic engineering. Considering the large number of reports implicating the axial structures in the imprinting of the meristic pattern on the segmental plate and the influence of the space surrounding the segmental plate on the unfolding of the segmental pattern, future work will undoubtedly center on identifying, isolating, and characterizing substances and molecules that impinge on the segmental plate. Thus far, two candidates have been investigated in a preliminary manner: fibronectin and collagen. The use of highly specific antisera and monoclonal antibodies will be invaluable in the characterization and localization of these extracellular matrix components.

References

Archer, C. W., Rooney, P., and Wolpert, L., 1983, The early growth and morphogenesis of limb cartilage, in: *Limb Development and Regeneration*, Part A (J. F. Fallon, and A. I. Caplan, eds.), pp. 267–278, Alan R. Liss, New York.

Aristotle, 1910, *Historia Animalium* (A. W. Thompson, tr. and ed.), Oxford University Press, Oxford.

Assheton, R., 1896, An experimental examination into the growth of the blastoderm of the chick, *Proc. R. Soc. Lond. B* **60**:349–356.

Bellairs, R., 1963, The development of somites in the chick embryo, *J. Embryol. Exp. Morphol.* **2**:697–714.

Bellairs, R., 1979, The mechanism of somite segmentation in the chick embryo, *J. Embryol. Exp. Morphol.* **51**:227–243.

Bellairs, R., and Veini, M., 1980, An experimental analysis of somite segmentation in the chick embryo, *J. Embryol. Exp. Morphol.* **55**:93–108.

Bellairs, R., and Veini, M., 1984, Experimental analysis of control mechanisms in somite segmentation in avian embryos. II. Reduction of material in the gastrula stages of the chick, *J. Embryol. Exp. Morphol.* **79**:183–200.

Bellairs, R., Curtis, A. S. G., and Sanders, E. J., 1978, Cell adhesiveness and embryonic differentiation, *J. Embryol. Exp. Zool.* **46**:207–213.

Bolzern, A., Leonardi, M. C., DeBernardi, F., Maci, R., and Ranzi, S., 1979, On the induction of somites in chick embryo, *Rend. Acc. Naz. Lincei* **64**:621–625.

Butros, J., 1967, Limited axial structures in nodeless chick blastoderm, *J. Embryol. Exp. Morphol.* **17**:119–130.

Cheney, C. M., and Lash, J. W., 1981, Diversification within embryonic chick somites: Differential response to notochord, *Dev. Biol.* **81**:288–298.

Cheney, C. M., and Lash, J. W., 1984, An increase in cell–cell adhesion in the segmental plate results in a meristic pattern, *J. Embryol. Exp. Morphol.* **79**:1–10.

Cheney, C. M., Seitz, A. W., and Lash, J. W., 1980, Fibronectin, cell adhesion and somite formation, *J. Cell Biol.* **87**:94a.

Chernoff, E. A. G., and Hilfer, S. R., 1982, Calcium dependence and contraction in somite formation, *Tissue Cell* **14**:435–449.

Chernoff, E. A. G., and Lash, J. W., 1981, Cell movement in somite formation and development in the chick: Inhibition of segmentation, *Dev. Biol.* **87**:212–219.

Chevallier, A., 1978, Étude de la migration des cellules somitiques dans le mesoderme somatopleural de l'ébauche de l'aile, *Wilhelm Roux Arch Entwicklungsmech.* **184**:57–73.

Christ, B., Jacob, H. J., and Jacob, M., 1978, On the formation of the myotomes in avian embryos. An experimental and scanning electron microscope study, *Experientia* **34**:514–516.

Christ, B., Jacob, H. J., Jacob, M., and Wachtler, F., 1983, On the origin, distribution and determination of avian limb mesenchymal cells, in: *Limb Development and Regeneration*, Part B (R. O. Kelley, P. F. Goetinck, and J. A. MacCabe, eds.), pp. 281–291, Alan R. Liss, New York.

Cooke, J., 1977, The control of somite number during amphibian development: Models and experiments, in: *Vertebrate Limb and Somite Morphogenesis* (D. A. Ede, J. R. Hinchliffe, and M. Balls, eds.), pp. 433–448, Cambridge University Press, London.

Cooke, J., and Zeeman, E. C., 1976, A clock and wavefront model for control of the number of repeated structures during animal morphogenesis, *J. Theoret. Biol.* **58**:455–476.

Coste, J. J. C. V., 1834, in: *Recherches sur la génération des mammifères*, Planche IV, pp. 47–185, Librairie des Sciences Medicales, Paris.

Ede, D. A., Hinchliffe, J. R., and Balls, M., eds., 1977, *Vertebrate Limb and Somite Morphogenesis*, Cambridge University Press, London.

Elias, S., and Sandor, S., 1965, A biostatistical study of the early morphogenesis of the chick embryo and its appendages, *Rev. Roum. d'Embryol. Cytol.* **2**:115–159.

Flint, O. P., and Ede, D. A., 1982, Cell interactions in the developing somite: *In vitro* comparisons between amputated (am/am) and normal mouse embryos, *J. Embryol. Exp. Morphol.* **67:**113–125.

Fraser, R. C., 1954, Studies on the hypoblast of the young chick embryo, *J. Exp. Zool.* **126:**349–400.

Gasser, R. F., 1979, Evidence that sclerotomal cells do not migrate medially during normal embryonic development of the rat, *Am. J. Anat.* **154:**509–524.

Goodwin, B. C., and Cohen, M. H., 1969, A phase-shift model for the spatial and temporal organization of developing systems, *J. Theoret. Biol.* **25:**49–107.

Hall, B. K., 1983, *Cartilage*, Vol. 2, Academic Press, New York.

Hamburger, V., and Hamilton, H. L., 1951, A series of normal stages in development of the chick embryo, *J. Morphol.* **88:**49–92.

Hamilton, L., 1969, The formation of somites in *Xenopus*, *J. Embryol. Exp. Morphol.* **22:**253–264.

Kieny, M., 1983, Cell and tissue interactions in the organogenesis of the avian limb musculature, in: *Limb Development and Regeneration*, Part B (R. O. Kelley, P. F. Goetinck, and J. A. MacCade, eds.), pp. 293–302, Alan R. Liss, New York.

Langman, J., and Nelson, G. R., 1968, A radioautographic study of the development of the somite in the chick embryo, *J. Embryol. Exp. Morphol.* **19:**217–226.

Lanot, R., 1971, La formation des somites chez l'embryon d'oiseau: Etude experimentale, *J. Embryol. Exp. Morphol.* **26:**1–20.

Lash, J. W., 1985, The analysis of immunohistological preparations using computer-assisted image processing, in: *Prevention of Physical and Mental Congenital Defects*, Part C: Basic and Medical Science, Education, and Future Strategies (M. Marois, ed.), pp. 255–257, Alan R. Liss, New York.

Lash, J. W., and Cheney, C. M., 1982, Diversification within embryonic chick somites: *In vitro* analysis of proteoglycan *in vitro*, in: *Differentiation in Vitro* (M. M. Yeoman and D. E. S. Truman, eds.), pp. 193–206, Cambridge University Press, London.

Lash, J. W., and Saxén, L., 1972, Human teratogenesis: *In vitro* studies on thalidomide inhibited chondrogenesis, *Dev. Biol.* **28:**61–70.

Lash, J. W., and Vasan, N. S., 1983, Glycosaminoglycans of cartilage, in: *Cartilage*, Vol. 1 (B. K. Hall, ed.), pp. 215–251, Academic Press, New York.

Lash, J. W., Seitz, A. W., Cheney, C. M., and Ostrovsky, D., 1984, On the role of fibronectin during the compaction stage of somitogenesis in the chick embryo, *J. Exp. Zool.* **232:**197–206.

Lipton, B. H., and Jacobson, A. G., 1974a, Analysis of normal somite development, *Dev. Biol.* **38:**73–90.

Lipton, B. H., and Jacobson, A. G., 1974b, Experimental analysis of the mechanisms of somite morphogenesis, *Dev. Biol.* **38:**91–103.

Mason, F., 1928, *Creation by Evolution*, Macmillan, New York.

Meier, S., 1979, Development of the chick embryo mesoblast. Formation of the embryonic axis and establishment of the metameric pattern, *Dev. Biol.* **73:**25–45.

Meier, S., 1980, Development of the chick embryo mesoblast: Pronephros, lateral plate, and early vasculature, *J. Embryol. Exp. Morphol.* **55:**291–306.

Meier, S., 1981, Development of the chick embryo mesoblast: Morphogenesis of the prechordal plate and cranial segments, *Dev. Biol.* **83:**49–61.

Meier, S., and Jacobson, A. G., 1982, Experimental studies of the origin and expression of metameric pattern in the chick embryo, *J. Exp. Zool.* **219:**217–232.

Menkes, B., and Sandor, S., 1969, Researches on the formation of axial organs, V. *Rev. Roum. Embryol. Cytol.* **6:**65–72.

Menkes, B., and Sandor, S., 1977, Somitogenesis: Regulation potencies, sequence determination and primordial interactions, in: *Vertebrate Limb and Somite Morphogenesis* (D. A. Ede, J. R. Hinchliffe, and M. Balls, eds.), pp. 405–419, Cambridge University Press, London.

Menkes, B., Miclea, C., Elias, S., and Deleanu, M., 1961, Researches on the formation of axial organs. I. Study on the differentiation of somites, *Acad. RPR Baza Timisoara Stud. Cerc. St. Med.* **8:**7–33 (in Rumanian).

Menkes, B., Sandor, S., Elias, S., and Deleanu, M., 1969, Contributions to the problems of somitogenesis, *Rev. Roum. Embryol. Cytol.* **6:**149–158.

Meyer, A. W., 1936, *An Analysis of De Generations Animalium of William Harvey*, Stanford University Press, Stanford, California.

Meyer, A. W., 1939, *The Rise of Embryology*, Stanford University Press, Stanford, California.

Minor, R., 1971, Somite differentiation and vertebral chondrogenesis: A structural study of *in vivo* and *in vitro* chondrogenic differentiation, Doctoral dissertation, University of Pennsylvania, Philadelphia.

Needham, J., 1959, *A History of Embryology*, Cambridge University Press, London.

Nelson, O. E., 1953, *Comparative Embryology of the Vertebrates*, Blakiston, New York.

Nicolet, G., 1970, Is the presumptive notochord responsible for somite genesis in the chick?, *J. Embryol. Exp. Morphol.* **24**:467–468.

Nicolet, G., 1971, Avian gastrulation, *Adv. Morphogen.* **9**:231–262.

Ostrovsky, D., Sanger, J. W., and Lash, J. W., 1983, Light microscope observations on actin distribution during morphogenetic movements in the chick embryo, *J. Embryol. Exp. Morphol.* **78**:23–32.

Ostrovsky, D., Cheney, C. M., Seitz, A. W., and Lash, J. W., 1984, Fibronectin distribution during somitogenesis in the chick embryo, *Cell Diff.* **13**:217–223.

Packard, D. S., 1978, Chick somite determination: The role of factors in young somites and the segmental plate, *J. Exp. Zool.* **203**:295–306.

Packard, D. S., and Jacobson, A. G., 1976, The influence of axial structures on chick somite formation, *Dev. Biol.* **53**:36–48.

Packard, D. S., and Meier, S., 1983, An experimental study of the somitomeric organization of the avian segmental plate, *Dev. Biol.* **97**:191–202.

Palmer, J. F., 1837, *The works of John Hunter*, F.R.S. Longman, Rees, Orme, Brown, Green and Longman, London.

Pasteels, J., 1937, Études sur la gastrulation des vertebrés méroblastiques. III. Oiseaux, *Arch. Biol.* **48**:381–463.

Patterson, J. T., 1907, The order of appearance of the anterior somites in the chick, *Biol. Bull.* **13**:121–133.

Pearson, M., and Elsdale, T., 1979, Somitogenesis in amphibian embryos. I. Experimental evidence for an interaction between two temporal factors in the specification of somite pattern. *J. Embryol. Exp. Morphol.* **51**:27–50.

Peebles, F., 1898, Some experiments on the primitive streak of the chick, *Arch. Entwicklungsmech. Org.* **7**:405–429.

Platt, J. B., 1898, Studies on the primitive axial segmentation of the chick, *Bull. Mus. Comp. Zool. Harvard Coll.* **17(4)**:171–190.

Romer, A. S., and Parsons, T. S., 1977, *The Vertebrate Body*, W. B. Saunders, Philadelphia.

Sandor, S., and Amels, D., 1970, Researches on the development of axial organs. VI. The influence of the neural tube on somitogenesis, *Rev. Roum. Embryol. Cytol.* **7**:49–57.

Solursh, M., Fisher, M., Meier, S., and Singley, C. T., 1979, The role of extracellular matrix in the formation of the sclerotome, *J. Embryol. Exp. Morphol.* **54**:75–98.

Spratt, N. T., 1955, Analysis of the organizer center in the early chick embryo. I. Localization of prospective notochord and somite cells, *J. Exp. Zool.* **128**:121–163.

Spratt, N. T., and Condon, L., 1947, Localization of prospective chorda and somite mesoderm during regression of the primitive streak in the chick blastoderm, *Anat. Rec.* **99**:53a.

Stephens, T., 1983, Parameters establishing the location and nature of the tetrapod limb, in: *Limb Development and Regeneration*, Part A (J. F. Fallon, and A. I. Caplan, eds.), pp. 3–12, Alan R. Liss, New York.

Williams, L. W., 1910, The somites of the chick, *Am. J. Anat.* **2**:55–100.

Wolpert, L., 1969, Positional information and the spatial pattern of differentiation, *J. Theoret. Biol.* **25**:1–47.

Wolpert, L., 1982, Cartilage morphogenesis in the limb, in: *Cell Behaviour* (R. Bellairs, A. Curtis, and G. Dunn, eds.), pp. 359–372, Cambridge University Press, London.

Zeeman, E. C., 1974, Primary and secondary waves in developmental biology. Lectures on Mathematics in the Life Sciences, *American Mathematical Society*, Providence (USA) **7**:69–161.

Chapter 15

Mechanisms of Axonal Guidance
The Problem of Intersecting Fiber Systems

MARK HARRIS HANKIN and JERRY SILVER

1. Introduction

The complex patterns exhibited by axon fiber tracts in animals present an intriguing challenge in the study of morphogenesis. How are these patterns established? What mechanisms determine the routes that particular axon growth cones will follow during development? This chapter first outlines the major historical theories of axonal guidance. These classic theories and their modern counterparts are then discussed in relationship to the development of intersecting fibers systems. Can these theories explain intersection? The second part of this chapter discusses an *in vivo* paradigm in which it is possible to study the ontogeny of an intersecting system of fibers in the mammalian central nervous system. A possible mechanism for axonal guidance in this system is discussed.

2. Historical Perspectives and Guidance Theories

2.1. The Discovery of Axons

For nearly a century, a focal point of neurobiological research has been the elucidation of the mechanisms by which outgrowing nerve fibers are guided during neurogenesis and regeneration. Investigations into this subject began almost immediately after the discovery of the mechanism of axonal elongation. During the latter half of the nineteenth century, a debate raged regarding the very origin of nerve cells and their processes. On the one hand, Schwann (1839), Hensen (1864), and others (Balfour, 1876; von Bünger, 1891; Bethe,

---

MARK HARRIS HANKIN • Department of Anatomy and Cell Biology, Center for Neuroscience, University of Pittsburgh, School of Medicine, Pittsburgh, Pennsylvania 15261. JERRY SILVER • Department of Developmental Genetics and Anatomy, School of Medicine, Case Western Reserve University, Cleveland, Ohio 44106.

1901) believed that nerve cells and their processes (at that time called "intermediate bridges") were initially formed by the sculpting away of surrounding cells, subsequently followed by the stretching and anastomosing of the remaining continuous cytoplasmic bridge (**catenary theory**). The alternative hyopthesis (**monogenetic**), championed by Küpffer (Bidder and Küpffer, 1857), His (1879, 1890), and Kölliker (1879, referenced in Elliot Smith, 1897), stated that primitive nerve cells, or neuroblasts, produced their processes by the budding and subsequent extension of cytoplasmic processes. In 1890, Ramon y Cajal shed light on this issue by describing for the first time in fixed tissue the **axonal growth cone** at the ends of regenerating axons in the central (proximal) stumps of transected sciatic nerves (Ramon y Cajal, 1937). It was left to Harrison (1907) to provide definitive proof that axons were active extensions of neurblasts when he showed, using explanted embryonic frog neural tubes, that "the nerve fiber develops by the outflowing of protoplasm from the central cells."

2.2. The Chemotactic Hypothesis

The discovery of growth cones and their significance for axonal elongation immediately raised further questions for Ramon y Cajal regarding the formation of fiber patterns in the nervous system. In order to explain the axonal patterns observed in normal embryonic and regenerating nerves, Ramon y Cajal, in 1892, proposed the theory of neurotropism, currently known as **chemotropism** or **chemotaxis,** as a mechanism for axonal guidance. The neurotropic hypothesis stated that the guidance of axons was "brought about by an orienting stimulus from attracting or neurotropic substances" (Ramon y Cajal, 1928).

The current understanding of the neurotropic hypothesis posits the existence of diffusible chemical substances that act over relatively large distances to influence the direction of axon growth. Ramon y Cajal's studies on the regeneration of transected sciatic nerves, as well as his observations of nerve fibers at the junction of the central and peripheral stumps, were especially interesting in this regard. Ramon y Cajal believed that neurotropic substances are produced specifically by the Schwann cells, which compose the neurolemma of the peripheral stump (Ramon y Cajal, 1928). If this were, in fact, the case, the regenerating fibers from a transected nerve, the ends of which had been brought to lie next to one another, would have to make a U-turn upon leaving the proximal stump in order to reach the distal stump. The regrowing fibers were observed to do this and—since no other means of guidance was apparent (e.g., physical)—diffusible chemical factors were thought to be responsible (Ramon y Cajal, 1928).

It should be pointed out that although neurotropism was, for Ramon y Cajal, the best explanation for the data at the time, it was not the only guidance mechanism he considered. Ramon y Cajal was also aware of the potential guidance role of the physical environment for growing axons; he made many observations about the surrounding cellular milieu in regenerating axonal pathways (Ramon y Cajal, 1928). In his observations, Ramon y Cajal noted a relationship between regenerating axons and the old Schwann cell tubes (Ramon y Cajal, 1928), wherein the Schwann cells in the latter formed longitudinal striations, i.e., *Zellenbänder*, or bands of Büngner (Büngner, 1891) after 8–10 days of regeneration, thereby giving the appearance of a preestablished, oriented ultrastructural pathway within the Schwann cell sheath that patterned the course of axonal growth. However, although Ramon y Cajal recognized these Schwann cells as a source of neurotropic substances, he never fully committed himself to the possibility that the geometric nature of the Schwann cell tube could also play a role in guiding axons.

In order to explain the stereotypic nature of axon tracts as well as the topographic synaptic mapping of nervous systems, Ramon y Cajal (1928) separately defined pathway selection and target matching:

> The action exercised by the *peripheral stump*—the proliferated cells of Schwann, etc., on the growth of the young fibers is not individual and specific, . . . but is *general and collective* . . . near the *terminal structures* . . . [However] the terminal clubs are influenced by another neurotropic influence which now has an *individual* and *specific* character. . . . (italics added)

This important distinction has remained relevant, as it has become increasingly apparent that the factors governing pathway guidance and synaptogenesis are certainly different (Constantine-Paton *et al.*, 1983; Reh *et al.*, 1983; Whitelaw and Hollyday, 1983*a*–*c*). Ramon y Cajal's studies of the developing cochlear apparatus revealed this dichotomy most clearly (Ramon y Cajal, 1908, 1960). Within the inner ear the peripheral branches of the spiral ganglion will ultimately innervate the organ of Corti. During early development, these axons grow along the inferior plane of the cochlear epithelium in a small compact bundle of fibers. Ramon y Cajal observed that these fibers initially pause for a while without penetrating the cochlear epithelium. Ramon y Cajal believed that the initial attraction to the otocyst is the result of a neurotropic substance of diffuse character. Later, when the fibers synapse specifically on the hair cells of the organ of Corti, this behavior was attributed to a different, selective neurotropic agent emanating from the target cells themselves (but, see Section 3.4). In this way, Ramon y Cajal believed that axons grow into the periphery and are directed to innervate appropriate targets.

Despite some ambiguity as to how neurotropic molecules might function *in vivo* and various criticisms of the neurotropic hypothesis (see Section 4.1), research in this area has continued. Perhaps most significant was the identification and purification of the protein **nerve growth factor** (**NGF**) during the 1950s (Levi-Montalcini, 1952, 1962) and the subsequent investigations of its properties (Letourneau, 1978; Gundersen and Barrett, 1979). NGF has been shown to have tropic influences on sensory and sympathetic ganglion cells in the chick embryo and on sympathetic ganglia only in newborn and adult mammals (for review, see Levi-Montalcini, 1982). Recently, NGF was also demonstrated to enhance neurite outgrowth in dissociated chick parasympathetic ciliary ganglion neurons (Collins and Dawson, 1983).

One aspect of NGF behavior that has been difficult to clarify, however, has been whether NGF directs axon outgrowth by affecting the survival of the axons (**trophism**) or whether NGF directs axonal migration *per se* (**tropism**). Various studies have been undertaken to distinguish the tropic versus trophic effects of NGF on sympathetic ganglia. An important first step was the demonstration *in vitro* that NGF acts locally at the growth cone (Campenot, 1977). At the same time, Ebendal and Jacobson (1977) showed that a selection of organ explants from chick embryos differentially affected the growth of axons from chick embryo ganglia *in vitro:* When a sympathetic ganglion was challenged with different target explants simultaneously, the extent of axon outgrowth was greatest in the direction of those organs that were characteristically innervated by that ganglion (e.g., heart > skin > spinal cord). Thus, *in vitro,* NGF clearly can influence the direction-finding apparatus of the axon—the growth cone— and could therefore potentially direct axons during development.

One of the most interesting and convincing demonstrations of the chemotactic properties of NGF *in vivo* was the fact that sympathetic nerve fibers would grow ectopically within the central nervous system (CNS) of neonatal rats and mice if NGF were injected intracerebrally (Menesini-Chen *et al.,* 1978). Thus only in the presence of NGF will sympathetic fibers innervate an abnormal target—the CNS. Further support for the chemotactic effect of NGF was the observation that dorsal root ganglion (DRG) axons continually readjust their direction of growth toward an *in vitro* NGF source whose position was being periodically changed (Gundersen and Barrett, 1979). One caveat, however, is that, although these experiments suggest that NGF acts in a diffusible form, it is possible that it elicits these behaviors as a substrate-bound molecule. Indeed, recent studies suggest that NGF (R. W. Gundersen, personal communication) and perhaps other molecules (Collins and Lee, 1984) can act haptotactically, i.e., as a substrate-bound molecule.

Is NGF the only potential chemotactic factor? Recent evidence clearly suggests the existence of other neurotropic substances and that NGF, in some systems, may exert only a trophic effect. Lumsden and Davies (1983) examined the relative *in vitro* influence of target and inappropriate tissues on trigeminal sensory neurite outgrowth at periods preceding the normal innervation in the embryo, embryonic days (E) 10–12 in the mouse. Neurites in E10 and E11 cultures showed an exclusive directional growth preference for their normal target that was *not* inhibited by addition of anti-NGF. This finding suggested the existence of another chemotactic substance, immunochemically unrelated to NGF, and was referred to as **trigeminal neurotropic factor** (**TNF**). By contrast, outgrowing fibers from E12 cultures (corresponding to the period of actual target cell innervation) exhibited less directional specificity than did neurites from earlier cultures and were very sensitive to anti-NGF treatment. Since E10–11 trigeminal neurites *in vitro* did respond to exogenous NGF, it was suggested that the behavior of E12 neurites did not represent the inability of the neurites to respond to NGF but rather that NGF production by target tissues was only initiated on E12. Is TNF an example of a "general and collective" tropic factor, as proposed by Cajal (see previous discussion), to which axons respond opportunistically by invading a developing area? Is TNF then superseded by a more specific factor (NGF?) on E12 when the fibers reach their targets, and does the specific factor then act trophically until the final synaptic arrangements have been made? Although NGF does not appear to play a neurotropic role in some *in vitro* paradigms of the developing nervous system, this does not rule out a possible function for it in others (e.g., the sympathetic system).

2.3. Contact Guidance Theory

The classic definition of **contact guidance** as a mechanism for axonal guidance was formalized by Paul Weiss (1955): "nerve fiber tips [growth cones] are guided in their course by contact with surrounding structures. . . ." Weiss, however, was not the first to posit the existence of "oriented ultrastructures" (preexisting pathways) as a mechanism for axonal guidance. As early as 1887, Wilhelm His suggested that the pattern of branching of sensory roots upon reaching the dorsal columns of the spinal cord may be determined by a "system of passages formed by the trabeculae of the spongiosa".

Ramon y Cajal too, although a proponent of neurotropism, was aware of the role of the environment in guiding axons: "The sprouts [of recurrent regenerating axons] lean on the sheaths of Schwann . . . and in that way they reach the ganglion cells" (Ramon y Cajal, 1928).

R. G. Harrison was the first to demonstrate experimentally that axons interact with, and could be guided by, the underlying substrate. Harrison showed that nerve fibers, *in vitro*, would only grow on a solid support such as a fibrin network from clotted amphibian lymph, spider webs, or the etched surface of a coverglass. A purely fluid medium, however, would not support growth of nerve fibers (Harrison, 1910, 1914). Later, Speidel (1932, 1933) made exquisitely detailed observations of the activities of living growth cones in the tail fin of frog tadpoles.

Paul Weiss was perhaps the most outspoken proponent of contact guidance during this century. Weiss's early experiments (1934), like Harrison's two decades earlier, also showed that axons *in vitro* would grow along highly oriented preformed substrates. The latter were created in two ways: (1) by placing fibrin networks under mechanical tension, and (2) as a result of the *in vitro* "two-center effect," whereby the intervening fibrous (and cellular) matrix between two foci for nerve outgrowth (e.g., two spinal ganglia) became stretched to form a "bridge." In both cases, but especially in experiments using the two-center effect, it was noted that only those fibers within close proximity to the connecting bridge were able to grow in the direction established by the oriented substrate. Fibers extending from more distant portions of the ganglionic circumference did not use the bridge and exhibited only radial outgrowth. Because all outgrowing fibers would have been exposed to diffusing neurotropic substances, and therefore should have grown in the direction of the tropic source (i.e., the facing ganglion), Weiss concluded that long-distance chemotaxis was not operant in this paradigm.

In another of Weiss's early experiments (1934), two different concentrations of plasma were used to form fibrin networks of different densities, thereby establishing a mechanical barrier at the interface between the clots of differing constitution. Fibers growing from a central ganglion encountered such a barrier and would alter their original radial growth pattern to a tangential one. This behavior was especially evident if the angle of approach was not orthogonal.

Weiss repeatedly saw such mechanisms of axonal guidance as an exclusive dichotomy—either neurotropism or contact guidance, never both. In fact, Weiss consistently and insistently scoffed at the very idea of neurotropism. The experiments of Weiss and Taylor (1944) were performed for the purpose of refuting neurotropism. In these long-term *in vivo* experiments, regenerating sciatic nerves were confronted with a Y-maze (formed by using the bifurcating abdominal aorta). The open ends, or "test channels," were (1) left open, (2) ligated, or (3) grafted with either a piece of a degenerating nerve (a potential neurotropic source) or a tendon. Significantly, regenerating fibers grew into ligated or open channels as readily as into those with a distal stump. Particularly instructive were cases in which the opening of the aorta was ligated and the degenerating nerve graft was placed in the recurrent (iliac) channel. Outgrowing nerve fibers from the other iliac channel again exhibited no preference for the channel with the graft. Both channels were filled, although the blind one apparently contained the greater proportion of the fibers. The results could not be explained satisfactorily by the action of tropic substances, and contact guidance was again invoked.

2.4. The Chemoaffinity Hypothesis

Contact guidance, in its extreme form with the physically oriented surface the only guidance parameter, could not explain the establishment of the more baroque fiber patterns found in the CNS. Although Weiss (1955) did realize this possibility during the latter part of his experimental career, it was most fully explored by his student, Roger Sperry. Sperry (1944) proposed the **chemoaffinity hypothesis** as an explanation for the formation of complex CNS fiber patterns and topographic target relationships. According to this theory, growing fibers are guided by an exquisite chemical matching system between individual growing neurites and individually appropriate pathways and targets. Sperry's early experiments demonstrated that optic nerve regeneration in anuran amphibians, after rough surgical sectioning of the optic nerve (just posterior to the globe) combined with 180° rotation of the eye, nevertheless resulted in the reestablishment of appropriate, although inverted, visual–behavioral associations. After such an experimental manipulation, which was believed to have disrupted any organization of the environment that might play a role in axonal guidance, the visual tracking and reflex behavior responses of the animal were evaluated: "When the lure was held above the head and a little caudad to the eye, the animals struck downward in front of them and got a mouthful of mud and moss" (Sperry, 1944). Additional tests of visual function (e.g., optokinetic response) were routinely performed and always yielded similar results. The resultant visual behavior was always appropriate for the part of the retinal field in the normal orientation, even though that hemiretina was now rotated 180°. The frogs would sense the lure above them but would respond to it as if it were below them. Sperry concluded from these experiments that the regenerating optic axons were in topographic disarray within the optic nerve and tract and that the "appropriate" visual behavior observed was a reflection of the orderly target synapsing that could only be a result of the sorting out due to specific chemoaffinities between individual retinal axons and tectal cells.

In later experiments on goldfish (Attardi and Sperry, 1963), specific retinal areas were surgically ablated, and the consequent fiber course to terminal tectal fields was anatomically mapped after periods of up to 8 weeks. As a result, some areas of the tectum were left partially denervated and, although potentially available for the formation of new synapses, these areas were by-passed by ingrowing fibers en route to their normal synaptic targets (Attardi and Sperry, 1963; Sperry, 1963). These results supported Sperry's hypothesis that there were matching sets of locus-specific chemoaffinities between retina and tectum.

More recently, Fujisawa et al. (1982) examined the anatomical projection of individual horseradish peroxidase (HRP)-filled regenerating optic axons to the tectum after 10 weeks of regeneration. In addition to confirming Sperry's anatomical results, several significant observations were made: (1) random branching of optic axons in the tectum was observed for all regenerating retinal axons but was especially vigorous for those that were misrouted, (2) the number of branches observed on individual fibers decreased as regeneration progressed, (3) branches directed toward ectopic tectal areas degenerated, and (4) those branches oriented toward normal sites of innervation had a more developed morphology and were maintained. These data support the notion that tectal target matching by growing axons results from a selective pruning process carried out at the tip of each growing fiber.

Although many investigators have confirmed the main aspects of Sperry's original results (Meyer, 1980; Fujisawa et al., 1982; Constantine-Paton et al., 1983; Stuermer and Easter, 1984), many of these recent investigations employing more sensitive methods have demonstrated that there are, indeed, fibers that do not reinnervate the appropriate tectal target areas (reviewed in Horder and Martin, 1978; Fraser and Hunt, 1980; Meyer, 1980). Although these fibers appear to be in the minority, their existence cannot be accommodated by Sperry's original model.

During the past two decades, evidence has also been presented that the regenerated retinotectal map may not be quite as stable as originally indicated by Sperry's results (reviewed in Levine, 1983). These observations have included that (1) the intact retina can compress its tectal map onto one-half of a tectum (Gaze and Sharma, 1970; Yoon, 1972, 1975), (2) a half-retina could expand onto a whole tectum (Horder, 1971; Cook and Horder, 1977), and (3) optic fibers could be rerouted to the ipsilateral tectum after their normal contralateral tectal targets were deleted and form a map there with a reversed polarity (Meyer, 1979). These results were not seen earlier, in part, because the regeneration times examined were not long enough. In these cases, the topography of the map was maintained, but its dimensions were correspondingly compressed, expanded, or translated. These observations point to an inherent plasticity in the retinotectal system and illustrate the necessity for postulating guidance factors other than (or in addition to) rigidly specific chemical labels on target and pathway cells (Constantine-Paton and Law, 1978; Fraser and Hunt, 1980; Reh et al., 1983).

Topographic ordering of retinal ganglion cell axons with respect to near-neighbors (e.g., fiber–fiber interactions) has recently been demonstrated within the optic nerve and tract of normally developing vertebrates (Horder and Martin, 1978; Bunt and Horder, 1983; Reh et al., 1983). It has been suggested that during development the positional information of topographically ordered incoming ganglion cell axons can be transcribed onto the tectum to form an orderly synaptic map. If topographic ordering is restored within the optic pathway proximal to the lesion during optic nerve regeneration, order might conceivably play a role in reestablishing tectal cues (Horder and Martin, 1978). However, recent exquisite anatomical studies have shown convincingly that the regenerating fiber bundle is scrambled (as originally suggested by Attardi and Sperry, 1963); thus, fiber order alone cannot account for the reesetablishment of the *regenerated* retinotectal map (Fawcett and Gaze, 1981; Fujisawa et al., 1982; Stuermer and Easter, 1984).

Can fiber order alone account for the *normal development* of the retinotectal system? A recent study of the abnormally developing retinotectal projection in chick demonstrates that fiber–fiber topography alone cannot account for the establishment of the normal retinotectal map (Thanos et al., 1984). Because **N-CAM (neural cell adhesion molecule)** [Thiery et al., 1977, 1982] had been implicated in the formation of neurite bundles *in vitro* (Rutishauser et al., 1978), Thanos and colleagues made intraocular injections of an anti-N-CAM Fab' to inhibit fiber–fiber interactions between optic axons *in vivo*. Although approximately 30–40% of the ganglion cell axons were located ectopically within the optic nerve and tract, some of the pioneering contingent of misrouted fibers could still, upon reaching the tectal surface, correct their ectopic positions to find appropriate innervation sites. However, the later-arriving population of axons were unable to rectify their aberrant courses and consequently, bypassed their usual sites of innervation. It has also been demonstrated recently that the fiber order in the normally developing optic pathway of embryonic mice is not nearly precise enough to create an effective synaptic map (Silver, 1984). Taken together, the *in vivo* and *in vitro* data suggest that—although fiber–fiber interactions are important for maintaining fiber topography within the optic nerve and tract and may play a role in directing fasciculating axons to their targets—"chemospecific" mechanisms may be operant at the tectum that help guide "pioneer" fibers to their correct synaptic destination.

How many "identification tags" would be needed to specify the many sensory projections present in the brain? Sperry (1963) himself realized that the number of distinct chemical labels necessary to direct all such pathways and synaptic events during neurogenesis was staggering (possibly $>10^9$) and almost certainly beyond the genetic capacity of the embryo, one estimate of which puts the coding capacity of the neuronal genome at approximately 150,000 average-size proteins (Van Ness *et al.*, 1979). This rather limited number of genomic products would not even be enough to specify chemoaffinity labels for a single axon tract, the corpus callosum in the mouse, which has approximately 300,000 fibers, not to mention the corpus callosum in the human, which has approximately 3×10^9.

One modification of the chemoaffinity hypothesis that circumvents this statistical improbability has been implicitly made by Silver and Rutishauser (1984), who have demonstrated that some stereotyped axon tracts are delineated not by labels for individual fibers, but by patterned accumulations of substrate molecules (e.g., N-CAM), which may generate constraints on growth of the entire axonal population by providing a preformed "adhesive" pathway.

3. Modern Synthetic Guidance Theories

3.1. Introduction

One conclusion that may be drawn from the results of the classic experiments is that no single mechanism (neurotropism, contact guidance, or chemoaffinity) can adequately explain all aspects of pattern formation in the nervous system. It is also unlikely that the forces governing the choice of pathway are the same as for synaptic targeting. As a consequence, modern investigations have tended to define the scope of the experiments more carefully and have dealt either with pathway formation or synapse specificity.

One result of the aforementioned shortcomings of the original chemoaffinity hypothesis has been the proposal of multiple mechanisms to explain axon tract patterning and synaptic matching. An example of such a "synthetic" model has been proposed by Fraser and Hunt (1980). In essence, this model proposes the existence of a number of "forces" or labels distributed over the entire retinotectal pathway, which are ordered (hierarchically) from strongest to weakest: (1) a general eye-to-tectum adhesive force; (2) a general fiber-to-fiber repulsive force; and (3) a number of graded labels, or (quantitative adhesive gradients)—the anteroposterior (AP) and the dorsoventral (DV) gradients (these gradients are not necessarily of equal strength).

These three forces were postulated to interact in such a way that axons would seek maximally energetic (i.e., strongest available) adhesive contacts. This interesting scheme has the unique quality of postulating a number of organizing factors dispersed over the length of the pathway and are potentially interactive. The behavior of a given axon during its development would then be a function of a summation of a number of unequal forces, some of which are expressed between axons and others between axons and their environment.

3.2. Adhesive Hierarchies in Axon Guidance

During the past decade, the role of adhesivity in promoting and directing growing axons has been tested. Letourneau (1975) characterized a series of substrates and established a hierarchy of growth cone–substratum adhesivity: glia ≈ polyornithine > collagen > tissue culture plastic = Pd > petri dish plastic. A general rule for axon growth in culture has emerged; that is, when given a choice, axons always grow on the surface with the highest adhesivity. How might the growth-promoting adhesive interaction provide guidance cues for growing axons? Conceivably, two possible mechanisms could generate directionality for axonal growth: (1) The adhesive surfaces *in vivo* could be arranged in a logical, prepatterned, spatial array (Singer *et al.*, 1979); and/or (2) the molecules responsible for the adhesion could be distributed as a gradient or hierarchy on the "adhesive" surfaces of the pathway (Nardi, 1983; Berlot and Goodman, 1984). The neuroepithelial endfeet (to which axons adhere preferentially) in the disc region of the preaxonal retina of mice and probably other animals are, indeed, arranged with an ordered, one-way radial geometry (Silver and Sidman, 1980) (see Section 3.4). Developmental studies of the *ocular retardation* mutant mouse (*or^j*) suggest that this channel system provides a contact-guidance "highway" that guides ganglion cell axons out of the eye (Silver and Robb, 1979).

What are the non-neuronal elements to which "pioneer" axons adhere and receive guidance cues *in vivo*? In order to address this *in vivo* question in *in vitro* experiments, Noble *et al.* (1984) showed that CNS and spinal cord neurons extend neurites when grown on both glial and nonglial monolayers with stereotyped behaviors. However, neurites grown on highly enriched glia (type I protoplasmic astrocytes) were characteristically not fasciculated, and they grew faster and farther on a glial terrain. Although the paths of the neurites were largely parallel, the fibers frequently crossed. By contrast, neurites grown on nonglia were always found to be fasciculated, and their growth was observed to be less robust than that of neurites grown on glia. These stereotyped behaviors were not altered by either (1) addition of conditioned medium from the heterotypic cultures (i.e., astrocyte monolayer with nonglia-conditioned medium and vice-versa), or (2) heat fixation of the monolayers. These data suggested that the control of neurite outgrowth on monolayers is regulated by intimate membrane–membrane interactions between the growth cone and the cellular substrate.

One possible mechanism for the regulation suggested by these results is that axons will follow specific migrating routes according to the relative adhesivity (or affinity) between neurite–neurite on the one hand and neurite–substrate on the other. Therefore, neurons grown on glia, for which they have an extremely high affinity, will extend neurites that preferentially adhere to the glial surface rather than to other neurites (to generate a nonfasciculated pattern). On the other hand, neurons grown on nonglia demonstrate an adhesivity for neighboring neurites that is greater than for the underlying substrate; consequently, the fasciculated pattern results. A question that is beginning to be resolved is whether adhesive hierarchies between growing axons and their glial environment play a role *in vivo* (Silver and Rutishauser, 1984).

3.3. The Blueprint Hypothesis

Within the past decade, increasing emphasis has been placed on the role played by the glial environment during the formation of nerve pathways. Although a large dose of classic chemoaffinity is still present in some glial theories, in many systems it has become clear that specific cells, groups of cells, or even tissues may play a primary (structural) role in guiding axons as well. Investigators at one extreme have proposed that there are specific chemical tracts on glia deposited for the purpose of guiding axons (Singer *et al.*, 1979; Katz and Lasek, 1979, 1981); those at the other extreme have demonstrated very little axon tract chemical specificity (Constantine-Paton, 1983). However, nearly all agree that any chemical guidance cues are contained on or near the surfaces of certain carefully positioned cells along the presumptive pathways; in the vertebrate, such pathway cells are usually glia or presumptive glia.

In 1979, Singer *et al.* made a proposal to explain spinal cord regeneration in the tail of the adult lizard and newt. They observed that the germinal ependyma extended radially oriented cytoplasmic processes, which were separated at the pial margin by long extracellular spaces. These spaces were found to be in register, suggesting that they formed a continuous longitudinal channel, and regenerating neurites were always found to grow along the walls that bordered the preformed channels. Later regenerating axons also grew within the channels, but always adjacent to glial membranes eventually filling them to capacity. The same situation was also found during embryogenesis. The **blueprint hypothesis** borrowed an element of Sperry's chemoaffinity hypothesis and suggested that axon-specific "physicochemical stripes" directing such growth might be contained on or near the surfaces of the crossing glial channels. Although this demonstration of a preformed pathway in the spinal cord suggested a mechanism by which axons might be guided in a relatively simple unidirectional system, the question still remained as to what directed the growth of more complicated intersecting axon tracts.

At the same time, Silver and Robb (1979) demonstrated the existence of preformed glial channels within the mouse retina on embryonic day 10 (E10), prior to the outgrowth of optic axons on E12–13. In the *ocular retardation* mutant mouse (*or^J*), with optic nerve aplasia, the spaces between glial processes are reduced by almost 90% compared with the normal condition. Ganglion cell axons in the *or^J* mouse never leave the eyeball and instead form large whorls, or neuromas, at the interface forced between the retina and the pigmented epithelium. These experimental data suggested for the first time that the ultimate failure of the optic fibers to exit from the eye in the mutant was a result of a drastic change in the non-neuronal environment. Later it was confirmed that the glial cells outlining the channels are in register and form a series of tunnels oriented toward the optic stalk (Silver and Sidman, 1980). At the rim of the eye, the channels end blindly, thereby creating a unidirectional tunnel system for the optic fibers to follow toward the optic disc. Importantly, the only region in which aligned channels have been found is in the optic disc itself, the nerve and tract being totally anastomotic.

3.4. Structured Environments

The channels described in the tail cord and optic system may be thought of as structured environments in the sense that certain non-neuronal elements, primitive ependyma, or glia arrange themselves into geometrically meaningful patterns before the arrival of the axons. Are there other such organized glial substrates in the CNS, and if so, what are their forms? An important common feature of these organized cellular structures may be the fact that a surface— usually, although perhaps not always, composed of glial cell processes rather than cell bodies—is available at the proper time for the axons to follow. It is also proposed that such surfaces must also (1) provide a general adhesivity for the advancing growth cones, and (2) be positioned at the right place and have the proper configuration before axon outgrowth.

Recent evidence from our laboratory has demonstrated additional examples of preformed glial guidance pathways. In the embryonic mouse, a preformed, cellular "funnel" is established along the route taken during development by the vestibular fibers from the eighth ganglion to the otocyst (Carney and Silver, 1983) and from that ganglion to the CNS (P. R. Carney and J. Silver, unpublished results). It has been proposed that the axons use the cells of the "funnel" (a population of primitive Schwann cells) as selective guides during their migration toward the otocyst and central target. Although Ramon y Cajal was unable to observe this structure in his studies of the otocyst (Ramon y Cajal, 1960), the existence of this structure does not preclude the need for a neurotropic substance on the surface of these cells to facilitate the directed growth of the vestibular fibers.

Another example of a preformed glial structure can be seen during the development of the corpus callosum. During the ontogeny of this commissure its growing fibers must make numerous direction decisions and, ultimately, synapsing decisions. One important direction decision made by the callosum is the hairpin turn it makes as it grows toward the midline, dorsal to the lateral ventricles, at the corticoseptal interface. In so doing, the callosum passes up a potential target, the septum, and terminates on the opposite side of the brain (this particular problem is discussed in Sections 4 and 5). In order to help overcome these obstacles, a special glial structure develops at the midline at the point of crossing of the first callosal fibers (approximately 200–250 μm rostral to the lamina terminalis). This structure, which is composed partly of radial glia and partly of a dense mat of primitive astrocytes, has been called the "sling", and it is on the dorsal surface of this structure that the "pioneer" callosal axons migrate across the cerebral midline (Silver *et al.*, 1982).

3.4. Axonal Guidance in Insects

The earliest insects predated the first mammals by 100 million years and arose approximately 300 million years ago during the carboniferous period of the pre-Cambrian era (Wigglesworth, 1964). It is therefore natural to wonder how axons are guided in the embryonic insect and whether there are any similarities to the guidance strategies found in the vertebrates. Developing insect nervous systems are markedly different from those found in mammals in one especially important respect. Whereas the mammalian pioneer axons usually exist as a large population of fibers, those in insects have remained small in number and in some cases consist of single neurites (for a recent review, see Goodman *et al.*, 1982). Are the guidance mechanisms for small versus large numbers of pioneer fibers the same or different?

The nervous system of the grasshopper has been studied extensively, and individual identified neurons and their processes can now be followed throughout the development of the organism. In the embryonic grasshopper, it has been shown that pioneer fibers from identified peripheral and central neurons selectively adhere to, and migrate upon, specific cells and processes with stereotypically fixed positions (Taghert *et al.*, 1982; Keshishian and Bentley, 1983*a–c*). Growth cones move consecutively along a chain of specific **landmark cells** (Taghert *et al.*, 1982) in the preaxonal environment, each successive cell landing being determined by the growth cone filopodial palpation. The matching of **guidepost cells** (Keshishian and Bentley, 1983*a*) for "pioneer" axons is not entirely specific, however. If the pioneer neuron is destroyed with UV irradiation, it can be shown that later forming axons (which normally fasciculate on the pioneer) can "pathfind" using the same guidepost cells that the pioneer normally would (Keshishian and Bentley, 1983*c*). As expected, if one of the guidepost cells is destroyed, the pioneer fiber becomes lost, and normal axon pathway formation is aborted (Bentley and Caudy, 1983).

Is there any similarity between the guidepost system of invertebrates and the structured non-neuronal pathways of vertebrates? Is it conceivable that the mammalian structured pathways evolved from a more simple system akin to the guidepost type? Could structured environments with axon-adhesive properties have evolved from a guidepost-type situation in response to the vast increase in neuron number found in the organisms with larger brains? Although these questions may never be answered conclusively, we may get some clues by gaining an understanding of some of the possible common features of these two mechanisms in extant species. Clearly, "guidepost cells," "slings," and "channels" may all provide an adherant cell surface for the axons to grow on and, because of their spatial organization, possess potential guidance cues. However, there is still considerable debate as to the level of specificity between axons and their pathway cells in vertebrates (Rager, 1980; Lance-Jones and Landmesser, 1981a,b; Silver et al., 1982; Levine, 1983).

4. Intersecting Fiber Systems

4.1. The General Problem

One of the hallmarks of nervous systems is the intricate intersection of nerve fiber tracts. Intersections are places at which two or more fiber tracts meet during development; the critical directional "choices" made by migrating growth cones at points of intersection will determine the final disposition of diversely targeted fiber tracts. However, except for occasional mention of intersection under in vitro conditions (Wessells et al., 1980), in analysis of peripheral nerve innervation of the limb (Lance-Jones and Landmesser, 1981a), and an in vivo study (Roberts and Taylor, 1982), there has been little systematic investigation of an in vivo intersecting fiber system and its development within the CNS of higher vertebrates. What are the potential forces acting at the site of intersection and over the pathway as a whole that govern the selection of one path over another?

Can the earlier theories of axonal guidance explain crossing fiber tracts? The classic theories, in their original forms, are all hard pressed to account for intersection. Chemotactic factors, which by definition act by diffusing over great distances, must be molded by physical constraints in the cellular environment in order to explain fiber tract intersection, the sharp turns made by growing fibers, and the confined nature of the tracts themselves. Candidates for such structures may be the "channels" (Silver and Robb, 1979) and "tunnels" described by the blueprint hypothesis (Singer et al., 1979), although the actual presence of diffusible chemicals within those spaces has not been demonstrated.

Contact guidance structures on the other hand can provide the physical substrate for sharp bending axon tracts, but—in situations in which fibers of two systems arrive at points of intersection simultaneously—these lack the specificity necessary for the formation of separate fiber projections. Thus as Weiss pointed out (1934), contact guidance could explain fiber crossing if the configuration of the substrate changed and if the fibers arrived at different times. In order to establish an intersection, the physical substrate must be oriented in the proper configuration to accommodate pathways arising at different times and with different trajectories. A puzzling example is found in the system described in detail in Section 5, in which the two tracts in question are in fact parallel to each other just before they intersect. At the point of intersection one fiber tract, the corpus callosum, makes a sudden 90° turn to insersect a corticoseptal pathway.

The chemoaffinity hypothesis (Sperry, 1963) in its original statement specified that:

> The cells and fibers . . . carry some kind of individual identification tags, presumably
> cytochemical in nature, by which they are distinguished one from another almost . . .
> to the level of the single neuron. . . .

In addition to the earlier criticism of the genetic impracticality and the statistical improbability of such an extreme version of the chemoaffinity principle, recent experimental evidence demonstrates that growing axons can use pathways other than those taken during normal development (Hibbard, 1959; Constantine-Paton and Law, 1978; Udin, 1978; Kormer *et al.*, 1981; Silver and Ogawa, 1983). In addition, we have shown *in vivo* that, given an equal opportunity to grow along an aberrant and a normal route, callosal axons will make the inc ect choice as readily as the correct one (see Section 5.5).

4.2. Corpus Callosum and Perforating Fibers: An *in Vivo* Paradigm

The intersecting system that is described in detail in Section 5 consists of the corpus callosum and its perforating fibers (PF). These two telencephalic fiber tracts, which are present in all placental mammals (Kappers *et al.*, 1936), intersect each other orthogonally just lateral to the cerebral mid-sagittal plane (Figs. 1 and 2). Although Malphigi recorded the existence of nerve fibers within the septum pellucidum in the latter half of the seventeenth century, it was not until 1845 that Arnold described a fiber system that connected the cingulum (*fornix periphericus*) with the septum (*fornix internus*) by running through the corpus callosum (cited in Elliot Smith, 1897) (Fig. 3). The fibers perforating the corpus callosum were variously described after Arnold's work (Kölliker, 1894) until Sir Grafton Elliot Smith (1897) published his studies on the development of the corpus callosum and fornix at the turn of the century. He termed them the *fornix superior* and described them in some detail in the ox:

> All the longitudinal uncrossed fibers . . . which break through a commissure of non-
> hippocampal, or a mixture of the latter, and hippocampal fibers [i.e., corpus callosum
> and its splenium].

This paradigm offers some advantages over the classic *in vitro* and regeneration models, foremost being the ease with which it may be followed during development and its accessibility to surgical manipulation. As will be seen, the ontogeny of the two fiber systems can be analyzed using standard microscopic techniques and fiber labeling methods.

5. Corticoseptal Barricade and Intersection

The intersection of the corpus callosum and of its perforating fibers is brought about by two carefully orchestrated morphogenetic processes: (1) the two fiber projections are formed at different times, and (2) the configuration of the non-neuronal environment changes at the point of intersection (Hankin, 1984).

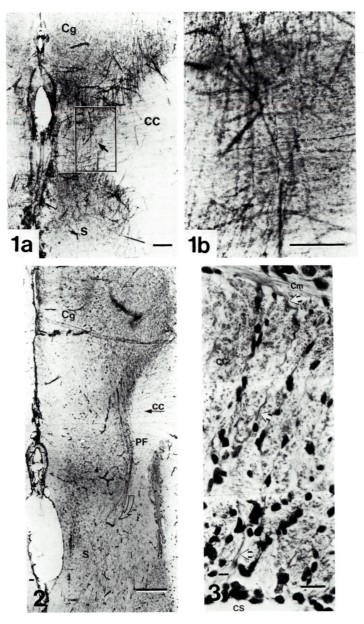

Figure 1. Coronal section just rostral to the genu of the corpus callosum (CC) in an adult mouse (C57BL/6J) that had received a horseradish peroxidase (HRP) injection in the cingulate cortex (Cg). (a) Note the orthogonal crossing (arrow) of the horizontally oriented callosal fiber with the vertically aligned perforating fiber. (b) The point of crossing shown at a higher magnification. Other PFs can be seen coursing ventrally to the septum(s). It is interesting to note the paucity of labeled callosal fibers compared with the number of labeled ones when dorsomedial or lateral cortex is injected with HRP (See Figs. 21 and 22). Scale bars: 100 μm (a) and 50 μm (b).

Figure 2. Coronal section 200 μm rostral to the genu of the corpus callosum (CC) in an adult mouse injected as in Figure 1. Note that the perforating fibers (PF) turn medially (open arrow) when they enter the septum toward the region of the medial septal nucleus. Scale bar: 200 μm.

5.1. Embryonic Day 15

Embryonic day 15 (E15) in the mouse (C57BL/6J) forebrain is a stage in which PF growth cones first appear but is preaxonal for the corpus callosum (Fig. 4). Although silver-stained sections at this stage fail to resolve labeled PF, transmission electron microscopy (TEM) demonstrates the presence of large growth cones located between the developing cortical plate and the intermediate zone (Fig. 5b). Four basic regions in the early developing cortex may be defined from the ventricle outward (Fig. 4a): (1) a thick, dense ventricular zone; (2) a slightly thinner and sparser intermediate zone subjacent to the ventricular zone; (3) a thin and variously cell body-sparse zone, the cortical subplate (Kostovič and Molliver, 1974) traversed by radial glial processes between the cortical plate and underlying intermediate zone; and (4) a very thin neuron-containing cortical plate. The intermediate layer, also called the subependymal zone (SE) (Smart, 1961; Privat and Leblond, 1972), has been recognized as a transient structure in mice (Smart, 1961), as well as in humans (Globus and Kuhlenbeck, 1944), that is present in embryonic and early postnatal animals but not in adults. TEM studies of the SE (Privat and Leblond, 1972) demonstrate that many of the cells in these areas are immature non-neuronal elements that differentiate primarily into astrocytes, but apparently not into neurons. Microglia have also been observed in these regions.

The subplate is open ventrally to the septum (Fig. 4a,c) but appears to lose patency at the dorsomedial aspect of the hemispheres, approximately at the junction of the archi- and neocortex (Fig. 4a,b). TEM reveals that this dorsal region actually contains many migrating cells (Fig. 5a) on their way from the intermediate layer to the cortical plate. The fact that there are so many cell bodies in this region compared with the more ventral area, where the PF grow (Fig. 5b) gives one the impression that the dorsal subplate is a closed interface. In this context, it is interesting to note that the PF growth cones are not seen in TEM in the dorsal region, where migrating cells are found (Fig. 5a). This observation raises the question of how the initial direction taken by a perforant fiber is determined. One possibility is that growth cones (e.g., PF) are generated and grow in regions in which cell body migration is minimal. Another possibility is that the growth cones themselves (or something that they produce) retard the migration of cell bodies into the pathway. Why growth cones are generated in these regions in the first place and how the timing of such events is regulated is unknown. An additional and especially exciting observation concerning this interface [confirming an earlier report (Nakanishi, 1983)] is that it contains an extracellular matrix material (Fig. 5c). Extracellular matrix fibrils have also been observed in the developing optic tectum of the chick (E10–11) ahead of the front of growing optic fibers (Krayanek, 1980). The matrix material reported

---

Figure 3. Parasagittal hematoxylin-stained section through the thickness of the corpus callosum in a P1 mouse, 70 μm lateral to the mid-sagittal plane. Note the fascicle of perforating fibers (white arrows) as it leaves the cingulate bundle (cingulum) and then traverses the corpus callosum. Scale bar: 10 μm. CS, cavum septum pellucidum.

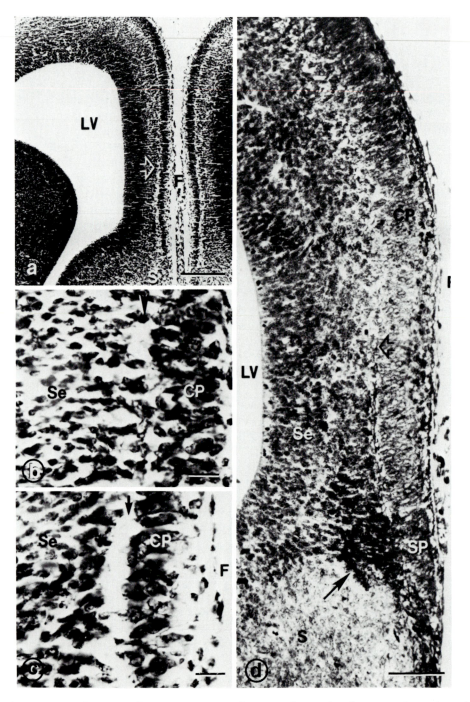

Figure 4. Coronal sections taken 280 μm (a–c) and 350 μm (d) rostral to the ventricular surface of the third ventricle (lamina terminalis). (a–c) Hematoxylin; (d) silver stain. Note the aligned inter-face or cortical subplate (arrow) between the cortical plate (CP) and subependymal zone (Se) in (c),

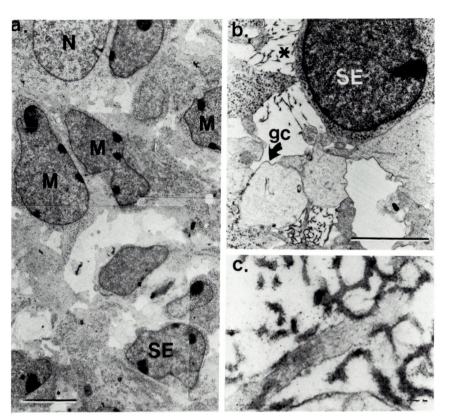

Figure 5. Transmission electron micrograph of (a) dorsal and (b) ventral cortical subplates. Note the numerous migrating cells between the neurons of the cortical plate and the cells of the subependymal zone in (a). There are many fewer growth cones in the dorsal subplate (isocortex in a) on E16 than in the ventral region (mesocortex, arrow in b), and there is more extracellular space in the latter. In addition to the presence of PF growth cones in the mesocortical subplate, an extracellular matrix appears to be more abundant in this region and is frequently seen to be intimately associated with the growth cones (b,c). The extracellular matrix (ECM) is composed of fibers of approximately 4 nm diameter (c) that commonly contain three fibrils. Collectively, these three fibrils constitute a trilamina, which does not exhibit any microalignment and usually forms anastomotic networks. Scale bars: 5 μm (a,b), 100 nm (c). gc, growth cone; SE, subependymal zone.

which is a high magnification view of the area ventral to the white arrow in a (mesocortex). A high magnification of the region dorsal to the white arrow in (a) is shown in (b) (isocortex). Note especially the numerous cell bodies that clutter the isocortical subplate (b) as opposed to the relatively patent interface of the mesocortical subplate (c). In the latter there are many radial glial processes. The PF grow ventrally within the mesocortical subplate (c) on E16 and can easily be demonstrated with silver stain at that time (d). The solid arrow in d refers to the cell dense corticoseptal barricade. Scale bars: 100 μm (a), 10 μm (b,c), and 50 μm (d). CP, cortical plate; F, interhemispheric fissure; LV, lateral ventricle; PF, perforating fibers; Se, subependymal zone; SP, septal plate.

in the chick optic system is similar to that in our preparations of the mouse forebrain in that the material in both is characterized by a fibrillar core coated with an amorphous substance. Whether the extracellular material is involved in PF fiber outgrowth or cell migration from the subependymal zone is also unknown.

5.2. Embryonic Day 16

On E16 the hemispheres begin to fuse in the caudal septal region, obliterating the caudal part of the ventral interhemispheric fissure (Fig. 6e). It is worth noting that it is in this caudal region of fusion, approximately 200 μm rostral to the lamina terminalis (see Figs. 12a and 13), the glial "sling" forms on E16 (Fig. 6e), and across which the "pioneer" callosal fibers grow on E17 (Silver et al., 1982). On E16, however, only the perforating fibers can be visualized in silver-strained sections projecting ventrally toward the septum within the subplate (Fig. 6b,d).

In the adult mouse the perforating fibers can be found rostral to the fornix for nearly 3 mm, forming a vertical sheet of fibers oriented in a parasagittal plane. Because the corpus callosum is only 250–300 μm long on E17 and subsequently grows rostrally and caudally in a linear fashion (Fig. 7), it is possible to demonstrate the presence of PF by themselves for a few days after E16 by taking coronal sections at points rostral to the front edge of the growing corpus callosum. Using SEM to examine this preparation on E18 vividly demonstrates the PF axons growing in the subplate interface between the cortical plate and the subependymal zone (Fig. 8).

5.3. Embryonic Days 17 and 18

On E17 the pioneer callosal axons cross the midline by migrating on the dorsal surface of the **glial sling** (Silver et al., 1982) (Figs. 9e,f and 10). The sling is a 200–250 μm-long structure composed of flattened primitive astrocytes and their processes (Fig. 12), dorsal to the partially fused septum and whose caudal end is situated approximately 200 μm rostral to lamina terminalis. Rostral to the front end of the sling, however, the hemispheres remain unfused (Fig. 9a). Silver et al. (1982) proposed that the sling serves the specific function of guiding the first wave of callosal axons across the midline at the point where the hemispheres begin to fuse. However, since the corpus callosum will grow to 10 times its E17 length by the end of the first postnatal week, i.e., 200 μm to 2 mm (Fig. 7), the initial sling cannot possibly guide the remainder of the rostral callosum. What does?

At the same time or slightly before the first callosal axons cross in the caudal region, subependymal cells rostral to the sling (i.e., before the rostral callosal fibers have arrived) appear to migrate medially from the subependymal zone to join the ventral edge of the cortical plate and form a thick cellular structure that walls off the septum (Fig. 9a,c; see also Figs. 4d and 6a,c,e). Unlike the sling, this structure, which we have designated the **corticoseptal**

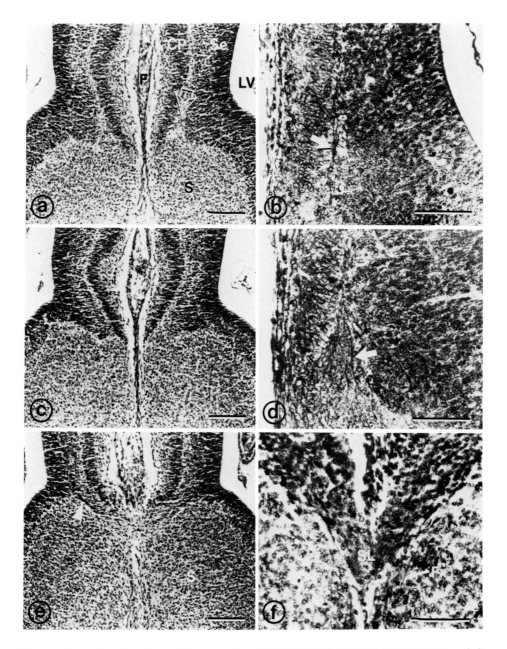

Figure 6. Coronal sections in an E16 mouse taken (a) 430, (b) 510, (c) 340, (d) 450, (e) 250, and (f) 210 μm rostral to lamina terminalis. (a,c,e) Hematoxylin; (b,d,f) silver stain. Note the mesocortical subplate in (a) (open white arrow) and the PF within shown in the matching silver stain sections in (b) and (d) (white arrows). Caudally there is a confined zone of midline fusion (e). Scale bars: 100 μm (a,c,e), and 50 μm (b,d,f). LV, lateral ventricle.

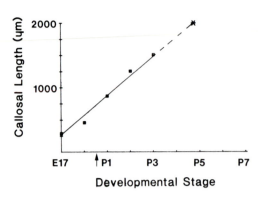

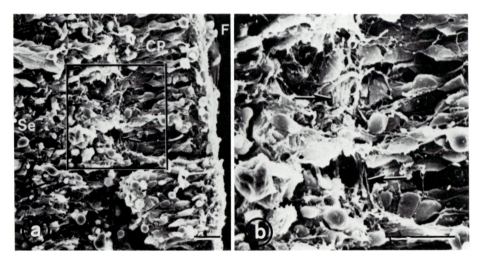

Figure 7. Graphic representation of the rostrocaudal linear growth of the corpus callosum in the mouse as measured at the cerebral midline from E17 to P5. Three animals were measured on each embryonic postnatal day shown. Since our collection did not contain P5 animals, the final point (P5—*) was an extrapolation based on the data of Wahlsten (1984).

barricade, does not span the midline. The perforating fibers remain intact as they traverse this structure (Figs. 9b,d). The corticoseptal barricade, then, is the unfused and thickened rostral extension of the glial sling (Fig. 9a,c); the consequence of its forming a dense cellular structure at the dorsal border of the septum (after the PF have developed but before the arrival of callosal fibers) has led us to propose that the barricade may be responsible for separating the pathways of the two intersecting fiber tracts (Hankin and Silver, 1984).

As the succession of rostral callosal growth occurs, the callosal fibers remain dorsal to the barricade and cross to the opposite hemisphere at that level in a limited zone of midline fusion. However, the two barricades themselves never fuse. In addition, the rostral growing edge of the callosum coincides with the rostral point of fusion of the hemispheres. Therefore, during the development of the corpus callosum, the callosal fibers actually perforate the "perforating fibers," deforming them medially as they push through the PF curtain (Fig. 9c,d). This process also results in the isolation of the ventral tip of the cortical

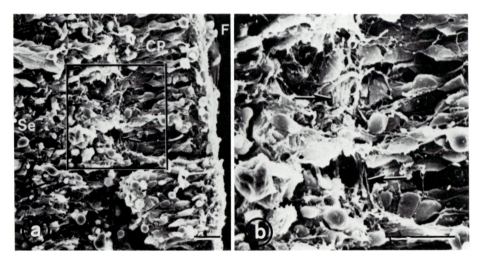

Figure 8. (a) Scanning electron micrograph of the perforating fibers (arrows) discretely located within the mesocortical subplate. (b) High magnification of the box shown in (a). Scale bar: 10 μm. F, interhemispheric fissure; CP, cortical plate; Se, subependymal zone.

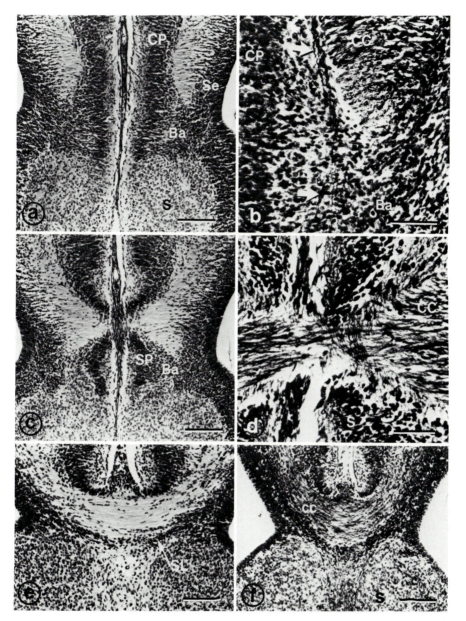

Figure 9. Coronal sections in an E17 mouse taken (a) 550, (b) 600, (c) 510, (d) 550, (e) 230, and (f) 290 μm rostral to lamina terminalis. (a,c,e) Hematoxylin; (b,d,f) silver stain. Caudally (e,f), the pioneer callosal fibers have crossed the hemispheric midline by migrating on the dorsal surface of the sling. Rostrally (a–d) the callosal fibers are just crossing in a zone of pial–glial fusion (c,d) and further forward still they have not yet reached the cerebral midline (a,b). The thickened corticoseptal barricade (a,c), which is the rostral unfused extension of the glial sling (e), walls off the septum. Perforating fibers (PF), seen immediately subjacent to the cortical plate (arrows in b), pass across the callosum and through the barricade into the septum (d). Scale bars: 100 μm (a,c,e,f) and 50 μm (b,d). Ba, barricade; CC, corpus callosum; CG, cingulate cortex; CP, cortical plate; S, septum; Se, subependymal zone; SP, septal plate.

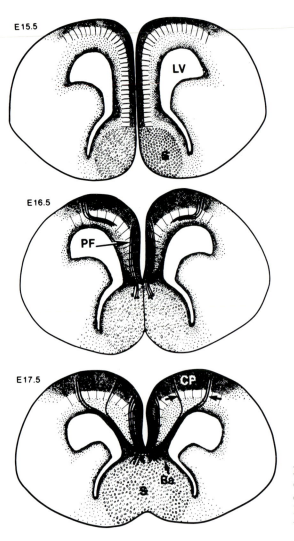

Figure 10. Schematic summary of the developmental events in the ontogeny of the perforating fibers and corpus callosum. CP, cortical plate; LV, lateral ventricle; PF, perforating fibers.

plate within the septum (Fig. 9c). We have called this stranded portion of the cortical plate the **septal plate,** which to our knowledge has not been previously described. It is possible that a remnant of the septal plate is present ventral to the rostral corpus callosum in the adult as the anterior hippocampal continuation. This process of callosal growth is followed throughout the ontogeny of the commissure. The morphogenetic events are summarized in Figure 10.

5.4. Physical Nature and Developmental Fate of the Corticoseptal Barricade

In order to understand more clearly the surface topography of the sling and barricade, the profiles of these structures were traced from light micrographs, and a three-dimensional model was reconstructed with the aid of computer

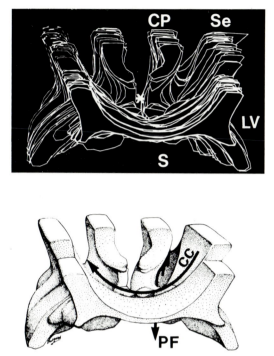

Figure 11. Computer graphic reconstruction (a) and artist's drawing (b) of the sling/barricade junction on E17. The reconstruction was generated from 32 serial 10 μm sections. The sling, seen in the foreground, is much thinner than the more rostral barricade and forms a complete glial "bridge" at the cerebral midline in a special zone of fusion (see Figs. 6e and 9c). The presumptive area of hemispheric fusion is indicated (*). Note the pioneer callosal fibers and the more rostral PF; the latter will be "perforated" by the later crossing callosal fibers. CC, corpus callosum; CP, cortical plate; LV, lateral ventricle; PF, perforating fibers; S, septum; Se, subependymal zone.

graphics. (For a complete description of the methodology used, see Silver *et al.*, 1982.) The reconstruction (Fig. 11) emphasizes that (1) there is a great difference in thickness between the sling and the barricade, (2) the sling is fused across the midline while the barricade is not, and (3) although continuous rostrocaudally, the dorsal surface of this glial structure is found at different levels in the two regions. To reiterate, the pioneer callosal fibers cross the midline by migrating on the dorsal surface of the sling (caudally), while the later-arriving callosal fibers cross dorsal to the barricade (rostrally) through a progressively expanding region of midline hemispheric fusion.

What does the microarchitecture of the dorsal surface of sling and barricade look like? In order to answer this question, a unique dissection was performed. Briefly, E18 mouse brains were dissected free of the cranial vault and fixed in 2% paraformaldehyde/0.5% glutaraldehyde and then dehydrated in ethanol up to 80%. At this level of dehydration, the overlying cortex was peeled away from the dorsal aspect of the corpus callosum. The callosum was then gently lifted away from the underlying septum, thereby permitting SEM analysis of the underside of the callosum and the top of the septum. What is readily evident and very striking in this preparation (Fig. 12) is that the cellullar structure on the underside of the corpus callosum is actually a densely packed "surface" composed of flattened cells and their processes. TEM analysis and immunohistochemical studies (Silver and Mason, 1984) with astrocyte-specific antibodies (anti-GFAP and anti-vimentin) demonstrate that these cells are neuroglia. The cellular elements in the surface do not exhibit any particular geometrical alignment other than to form a dense mat. However, we have also

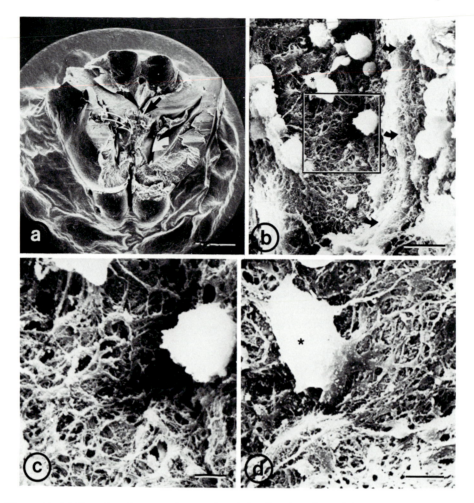

Figure 12. (a) Scanning electron micrographs of the dorsal surface of the sling/barricade, visualized rostral to the lamina terminalis (black arrow in a), between the lateral ventricles (solid white arrow in a). (b–d) A variety of high magnification views of the region indicated in a (open arrow). The sling/barricade is composed of flattened glial cells (*), and dense cell processes that radiate in every direction and intertwine with other processes. Some elements of this structure can also be seen adhering to the undersurface of the reflected callosum (arrows, b). Scale bars: 500 μm (a), 20 μm (b), and 2 μm (c,d).

recently discovered a highly organized system of GFAP$^+$ radial glial cells embedded within the sling and barricade (Silver et al., 1985).

In the adult state, there is no evidence in light and electron micrographs of a dense cellular structure between the corpus callosum and the underlying septum. This suggests that the sling and barricade are transient, present only in the late embryo and early postnate. What is the fate of the cells in this transient structure? Although we cannot elaborate on this point here, counts of pyknotic sites in the sling and barricade regions indicate that the glial scaffold degenerates in a caudal to rostral sequential fashion coincident with, and immediately following, commissurization (Fig. 13).

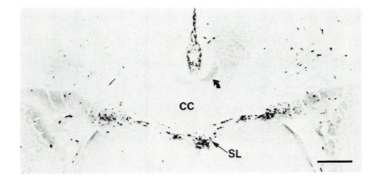

Figure 13. Coronal section from a P3 animal that was injected with horseradish peroxidase (HRP) on P2. Note the selective uptake of peroxidase label by the sling cells and the lack thereof in the remainder of the subependymal and ependymal cells lining the ventricles. Sling cells at this stage are known to be pyknotic. A small number of cells are also labeled within the white matter. In these vibratome preparations, in which ethanol dehydration was not used to process the tissue, the cavum septum underneath the sling is absent. This suggests that it is a potential space that is differentially affected by dehydrants. Scale bar: 100 μm. CC, corpus callosum; SL, sling cells.

How, then, are the corpus callosum and PF guided during development, and do the sling and barricade play a role? Data from our laboratory suggest that the callosal fibers have a high affinity for the subependymal surface, which is also the likely source of cells for the sling and barricade.

5.5. Experimental Manipulations of the Precallosal Environment

The hypothesis that the corticoseptal barricade directs the rostral callosum to make its hairpin turn at the top of the septum and to perforate the "perforating" fibers leads one to make certain predictions. For example, it might be expected that if the barricade could be eliminated or prevented from forming during callosal development, the corticoseptal projection (i.e., PF) should increase in size as the unimpeded callosal axons enter the septum. However, we have not been able to do this, either surgically or chemically.

Although attempts to remove the barricade have not proved successful, we have been able to disrupt the fusion process in the precallosal environment by lesioning the midline with a glass needle on E16 (Fig. 14). This procedure severs the initial interhemispheric contact made by the newly formed sling and all rostral fusion between the medial hemisphere walls. The striking result of preparations in which at least one hemisphere remained undisturbed is that the barricade, although somewhat thinner, still forms with a relatively normal morphology (Fig. 14a,b). It is interesting that the barricade does form even in the absence of midline fusion. Perhaps it does so because it normally forms rostral to the initial point of hemisphere fusion and may therefore be independent of the fusion process. Are the intersecting fiber projections affected? Two days after lesioning the midline, the PF still project to the septum (not visible in Fig. 14a,b due to the plane of focus), but the callosum does not have a normal

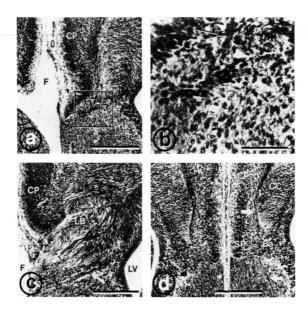

Figure 14. Coronal silver-stained sections from an animal whose midline (sling) was lesioned on E16 and sacrificed on E18 (a–c). (b) High magnification of the boxed region shown in (a). Note the relatively normal morphology of the cortical structures except for the callosal fibers, which whirl into longitudinally oriented neuromas and neither cross the interhemispheric fissure nor enter the septum. Cells in the barricade region still project toward the midline from the ventricle (white arrows). (c) Same is true of the remaining sling cells more caudally (white arrow). The PF, which projected just before lesioning, still project ventrally through the barricade and into the septum (black arrow in b). (d) Normal E18 mouse. Scale bars: 50 μm (a,c,d), and 20 μm (b). Ba, barricade; CC, corpus callosum; CP, cortical plate; F, interhemispheric fissure; LB, longitudinal bundle; LV, lateral ventricle; SP, septal plate.

projection. Although callosal fibers do project toward the midline, they are unable to cross to the opposite hemisphere, apparently because the glial sling was cut and the fusion process disrupted at the time the lesion was made. Instead, the callosal fibers whirl into large longitudinally oriented neuromas, which expand considerably dorsal to, but no farther ventrally than, the dorsal surface of the barricade (Fig. 14c). These neuromas (the "longitudinal bundles" described by Probst, 1901) have also been observed in the genetically acallosal BALB/cCF mouse as well as in acallosal humans (Warkany, 1971).

Callosal axons deprived of their contralateral targets by lesioning do not cross the midline. Why don't the callosal axons enter the septum? Two possibilities come to mind: (1) the callosal fibers may not recognize the septum as an appropriate synaptic target, or (2) the fibers may be physically blocked from entering the septum by the barricade.

Although the barricade cannot be removed experimentally during the early callosal growth period, one can manipulate the callosal fibers after the sling and barricade have degenerated. In this regard, one experimental approach we have pursued was to lesion the precallosal environment on E16, as described

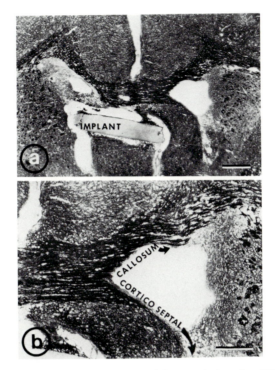

Figure 15. Coronal silver-stained section from a P9 animal that was lesioned on E16, and received an implant (off-center), positioned between the lateral ventricles, on P5. Although many callosal fibers were induced to regrow into the cortex (dorsally), a substantial portion of the commissure follows an ectopic corticoseptal trajectory. It is noteworthy that the ectopic fibers project smoothly to and migrate selectively along the subependymal zone in the septal region. Scale bars: 100 μm (a), and 50 μm (b). (From J. Silver and M. Y. Ogawa, 1983, Postnatally-induced formation of the corpus callosum in acallosal mice on glia-coated cellulose bridge, *Science*, Vol. 220, pp. 1067–1069. Copyright 1983 by the American Association for the Advancement of Science.)

above, but then on postnatal day 5 (P5) to insert an appropriately shaped piece of nitrocellulose as a prosthesis for the sling and barricade (Silver and Ogawa, 1983). The animals were sacrificed on P9. After the implants have been inserted, they become coated with glial cells that have migrated from the subependymal zone; if properly inserted, the implant will span the lateral ventricles (Fig. 16c). The surgical lesion will yield an acallosal subject with prominent Probst bundles. After the implant has been introduced to the animal, the callosal fibers in such neuromas can grow across the midline and enter appropriate cortical areas. In some cases, however, the implant does not completely span the ventricles, i.e., it is off-center (Figs. 15 and 16d). In these animals the callosal fibers leave one Probst bundle and cross the midline, but upon reaching the lateral limits of the implant the callosal fibers appear to travel in two directions, with some fibers migrating dorsally over the lateral ventricles on the subependymal zone as they do in the normal animal and others diverging smoothly from the normal bundle and migrating in the subependymal zone

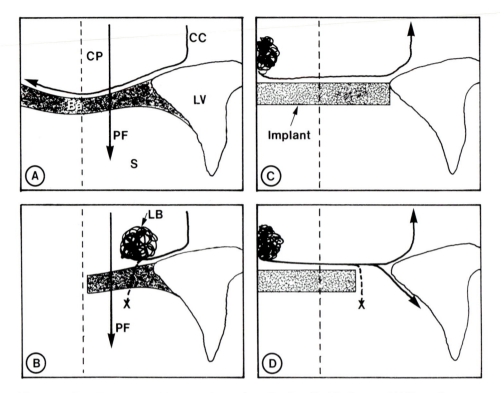

Figure 16. Drawings summarizing experimental results described in the text. (A) Normal arrangement, with a schematic sling/barricade. (B) Lesion experiment with the resultant neuroma. (C) If an implant is placed so that it spans the lateral ventricles, callosal fibers can be induced to leave the whorl and cross to the opposite cortex. (D) If the implant is off-center, the callosal fibers can grow both dorsally, into the cortex, and ventrally, into the septum. The dashed vertical line represents the cerebral midline. Ba, barricade; CC, corpus callosum; LB, longitudinal bundle; LV, lateral ventricle; PF, perforating fibers; S, septum.

adjacent to the septum. Since the barricade in this region degenerated at an earlier stage, there is apparently nothing to prevent the formation of this aberrant projection. This demonstates that the callosal fibers can enter the septum if given the opportunity and suggests that they normally do not do so because the dense barricade, when present, only promotes or permits growth across rather than through its upper surface. During normal development, then, the callosal fibers do not form neuromas, because fusion of the hemispheres at the midline provides a continuous pathway into the opposite hemisphere. The results from the lesion and implant experiments are summarized in Figure 16. These results, in conjunction with the observation that, in fixed tissue, barricade (subependymal) cells associate preferentially with the underside of the corpus callosum (not shown), suggest that there may be general affinity between the subependymal pathway and the callosal fibers. If this is the case, this should affect the way in which the callosum is generated and should be reflected in the pattern of fiber addition during callosal growth.

6. Callosal Fiber Topography and the Subependymal Pathway

The growth of callosal axons on the cell body-dense sling, barricade, and subependymal cells (primitive glia) of the lateral ventricles has led us to hypothesize that the subependymal cells may serve as a general guidance structure, or "pathway," for the axons of the corpus callosum. One would therefore predict that, during development, callosal growth cones would be found primarily in association with these cells rather than with cells of, for example, the cortical plate. To test this prediction, two lines of investigation have been followed. First, discrete HRP injections were made into embryonic cortex to determine topography of the early callosal fibers. These patterns were then compared with the adult pattern found after similarly placed HRP injections. In this way, we hoped to distinguish between two possible mechanisms for the generation of the callosal projection: (1) callosal fibers are generated simultaneously from many different cortical regions and subsequently migrate medially over the lateral ventricles with their growth cones randomly distributed, or (2) the callosal fibers are generated in a spatiotemporal sequence and progressively later-generated fibers "undercut" their predecessors as a result of their continuing association with subependymal cells. The second approach to this problem was to examine the position of callosal growth cones with respect to the subependymal cells, using TEM.

In the adult mouse, HRP injections to medial, dorsomedial, and lateral cortex led to a stereotyped pattern of fiber labeling. Injections to *dorsomedial* cortex labeled callosal fibers in the *dorsal* half of the corpus callosum (Fig. 17a), whereas injections to far *lateral* cortical areas consistently labeled fibers in the *ventral* part (Fig. 17b). Using standard tetramethyl benzidine (TMB) procedures (Mesulam, 1978), the observed retrograde labeling pattern of cortical *neurons* was consistent with that described by Ivy and Killackey (1981), in which the cell bodies resided primarily in two of the six cortical layers (III and V) (Fig. 17).

The most medial and one of the most meager of the regions contributing fibers to the corpus callosum in the adult is the cingulate cortex (Caviness, 1975; Yorke and Caviness, 1975), specifically Brodmann's areas 24, 29b, and 29c. When injections were made to these areas, a small number of fibers were labeled in the most dorsal aspect of the callosal projection (Fig. 18a,b), and labeled cell bodies were seen in superficial layer II (Fig. 18c). The dorsal and ventral fiber labeling patterns lose some of their integrity when the callosal fibers reach the midline where, for example, medial/dorsal fibers can be seen to wander into the ventral region (Fig. 20). The same can be seen for fibers from lateral cortex (see Fig. 17b).

Intrauterine injections of embryos made during early stages of callosal

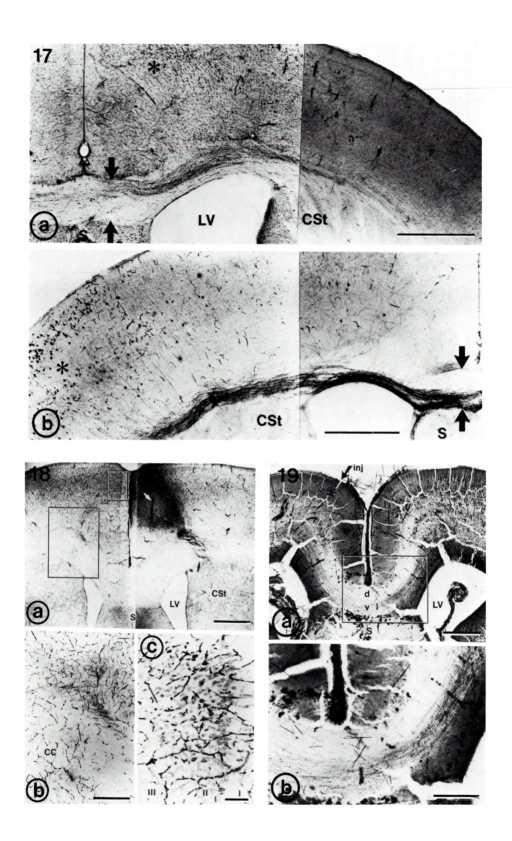

growth (E17–18), however, produced a markedly different fiber pattern from that seen in the adult. When comparable dorsomedial injections were made, the labeled fibers were seen adjacent to the subependymal zone, i.e., in a *ventral* position within the callosal projection (Fig. 19). This finding is in direct contrast with the adult pattern discussed above (compare Figs. 17a and 18 with 19), in which callosal fibers from dorsomedial cortex occupied a *dorsal* position in the corpus callosum. During development, therefore, the dorsomedial fibers must change their topographic position in order to attain the adult pattern. What is the mechanism by which this change occurs? When HRP injections are made to far lateral cortical areas in E17–18 embryos, labeled fibers are also seen in the ventral portion of the callosum, but do not reach as far across the midline. This suggests that the axons of lateral cortical neurons are generated simultaneously with medial ones but that axons from both regions preferentially grow near the subependymal zone. Thus, because of the spatial separation of medial and lateral callosal cells, the axons of the lateral population reach the midline slightly later and may undercut the more medial fibers and grow along the subependymal pathway.

Where do the fibers that occupy the dorsal half of the corpus callosum at these early stages originate? Are these fibers the "pioneer" population of the callosal projection? A likely probability is that these axons originate from cells in the cingulate cortex. However, injections to cingulate cortex in adults label only a small number of fibers in the dorsalmost layer of the callosum (see Fig. 18c) and in numbers much lower than the proportion evident from that region in E17–18 embryos. This situation would then lead to an interesting consequence

← ───

Figure 17. Coronal sections from adult mice injected with horseradish peroxidase (HRP) into (a) dorsomedial and (b) lateral cortex. Callosal fibers from dorsomedial cortical areas occupy the dorsal half of the commissure, whereas the lateral components are found in the ventral part. The asterisk (*) indicates the area corresponding to the injection site in the opposite hemisphere. Scale bars: 500 μm. CSt, corticostriatal fibers; LV, lateral ventricle; S, septum.

Figure 18. (a) Coronal sections from an adult mouse injected with HRP in the cingulate cortex (white arrow). (b) Higher magnification of the large box in (a), demonstrating a small number of cingulate callosal fibers (relative to the size of the injection) in the most dorsal aspect of the callosal projection. (c) Higher magnification of the smaller box in (a) demonstrating retrogradely labeled cell bodies in cortical layers II and III (arrows). Scale bars: 250 μm (a), 100 μm (b), and 25 μm (c). CC, corpus callosum.

Figure 19. Coronal section from an E18 mouse that had received a dorsomedial HRP injection (comparable with Fig. 17a). (a) The position of the injection site is indicated by the arrow but does not represent the actual site. Because the rostral callosal fibers do not cross directly from side to side but rather travel caudally in the forceps minor, the actual injection site is found in a more rostral section. At the actual crossing point (shown here) the fibers occupy a ventral (v) position, in contrast to their dorsal (d) adult location. Note also the labeled sling cells just below the callosal fibers (see Fig. 13). (b) Higher magnification of the box in (a). It is interesting to note the large unlabeled cingulate projection, already located dorsally, which probably represents the pioneer callosal fiber population. This early topographic arrangement suggests that the later-generated callosal fibers from lateral cortical areas undercut their predecessors and migrate preferentially near the subependymal zone, thereby displacing medial fibers into their dorsal adult location. Scale bars: 100 μm (a) and 50 μm (b). S, septum.

for the early cingulate contribution to the corpus callosum. Since the cingulate cortex makes a minimal contribution to the adult callosum, the embryonic cingulate contingent must be retracted or eliminated at some point after their initial projection (see also Innocenti *et al.,* 1977; Ivy *et al.,* 1979). Perhaps related to this observation is the fact that neuronal cell death in the superficial cortical layers of medial cortex has been noted in neonatal hamsters (Finlay and Slattery, 1983). Cingulate efferents do remain, since injections (areas 24 and 29b) in adults label a substantial projection to the ipsilateral corpus striatum (see Fig. 18a). It is not clear, however, when this cingulostriate projection is generated.

7. Conclusions

The intersection of the corpus callosum and its perforating fibers during the ontogeny of the mouse serve to illustrate the potential role of the environment in the guidance of fiber tracts. Our embryological observations have demonstrated that two important events are involved in establishing these fiber tracts in an intersecting pattern. First, the perforating fibers are generated before the callosal fibers. This is important because the problem of "sorting out" two different sets of simultaneously projecting fibers at the point of intersection is avoided by separating them temporally. Second, because of morphological sculpting at their crossroads, each fiber tract encounters a different configuration of the non-neuronal environment.

Until recently, few researchers have studied the role of the physical environment on axon guidance *in vivo*. The discovery of glial "channels" (Silver and Robb, 1979; Singer *et al.,* 1979), the glial sling (Silver *et al.,* 1982) and "barricade," the chiasmatic glial "knot" (Silver, 1984), and the auditory "funnel" (Carney and Silver, 1983) represent the first *in vivo* observations that the physical structures along the pathways of some CNS axon tracts are organized in such a way as to suggest that the substrate may play a role in guiding axons. Why have these guidance structures not been appreciated before? One probable reason is that a likely characteristic of many guidance structures is that they are transient and will therefore be totally missed when examining adult material or in fact any developmental stage other than the one when the guidance structure is present and functional.

Our data have led us to propose a mechanism by which one example of intersecting fibers may be guided during development. On the basis of the observations that (1) fiber tracts are temporally separated, (2) fiber tracts are confronted with different configurations of the enrvironment, and (3) fibers can associate with a specific cell surface (i.e., sling, barricade, and subependymal zone), we propose that populations of fibers may be guided by physical cues in, as well as chemical affinities for, the glial substratum. This mechanism can be contrasted with those proposed by Sperry and Singer, wherein individual fibers are guided by exquisitely *specific* chemical cues along their pathway—a sort of chemically "flavored" line exhibited for every individual axon or set of axons. Since we have found that callosal fibers can grow readily along portions of the

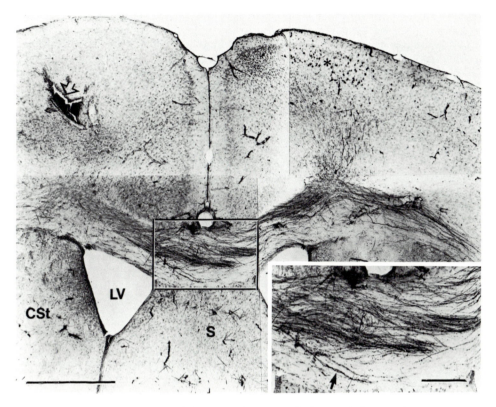

Figure 20. Coronal section of an adult mouse that received a horseradish peroxidase (HRP) injection in the dorsomedial cortex (open arrow). The box, shown at a higher magnification in the inset, dramatically demonstrates the disorder of the callosal fibers at the cerebral midline and between the lateral ventricles. Some dorsal fibers even wander into the extreme ventral part of the commissure before returning to the main dorsal fiber population once reaching the lateral ventricle (arrow in inset). Scale bar: 500 μm (100 μm, inset). CSt, corticostriatal fibers; LV, lateral ventricle; S, septum.

subependymal zone that they do not normally encounter (i.e., that leading into the septum), we suggest that the callosal fibers have a preferential affinity for the entire subependymal surface. In order to account for such plasticities of pathway choice, we propose that the adhesive association that occurs between axon and glial pathway is a "step down" from its level of specificity proposed in the chemoaffinity or blueprint hypotheses. Consequently, in order to generate stereotyped axonal growth routes, we suggest that the physical features in the non-neuronal environment may also play an important role in directing axons as well as in constraining them in critical regions where they might encounter confounding sources of guidance information.

References

Attardi, D. G., and Sperry, R. W., 1963, Preferential selection of central pathways by regenerating optic fibers, *Exp. Neurol.* **7**:46–64.

Balfour, F. M., 1876, On the development of the spinal nerves in elasmobranch fishes, *Philos. Trans. R. Soc. Lond. B* **166**:175–195.

Bentley, D., and Caudy, M., 1983, Pioneer axons lose directed growth after selective killing of guidepost cells, *Nature (Lond.)* **304**:62–65.

Berlot, J., and Goodman, C. S., 1984, Guidance of peripheral pioneer neurons in the grasshopper: Adhesive hierarchy of epithelial and neuronal surfaces, *Science* **223**:493–496.

Bethe, A., 1901, Über die Regeneration peripherischen Nerven, *26 Wandervers. des Südwestdeutsch. Neurol Gesellschaft*, 8, June.

Bidder, F. H., and Küpffer, C., 1857, *Untersuchungen über die Textur des Ruckenmarks und die Entwicklung seiner Formelemente*, Breitkopf and Härtl, Leipzig.

Bunt, S. M., and Horder, T. J., 1983, Evidence for an orderly arrangement of optic axons within the optic nerves of the major nonmammlian vertebrate classes, *J. Comp. Neurol.* **213**:4–14.

Campenot, R. B., 1977, Local control of neurite development by nerve growth factor, *Proc. Natl. Acad. Sci. USA* **74**:4516–4519.

Carney, P. R., and Silver, J., 1983, Studies on cell migration and axon guidance in the developing distal auditory system of the mouse, *J. Comp. Neurol.* **215**:359–369.

Caviness, V. S., 1975, Architectonic map of neocortex of the normal mouse, *J. Comp. Neurol.* **164**:247–264.

Collins, F., and Dawson, A., 1983, An effect of nerve growth factor on parasympathetic neurite outgrowth, *Proc. Natl. Acad. Sci. USA* **80**:2091–2094.

Collins, F., and Lee, M. R., 1984, The spatial control of ganglionic neurite growth by the substrate associated material from conditioned medium: An experimental model of haptotaxis, *J. Neurosci.* **4**:2823–2829.

Constantine-Paton, M., 1983, Trajectories of axons in ectopic VIIIth nerves, *Dev. Biol.* **97**:239–244.

Constantine-Paton, M., and Law, M. I., 1978, Eye-specific termination bands in tecta of three-eyed frogs, *Science* **202**:639–641.

Constantine-Paton, M., Pitts, E. C., and Reh, T. A., 1983, The relationship between retinal axon ingrowth, terminal morphology, and terminal patterning in the optic tectum of the frog, *J. Comp. Neurol.* **218**:297–313.

Cook, J. E., and Horder, T. J., 1977, The multiple factors determining retinotopic order in the growth of optic fibers into the optic tectum, *Philos. Trans. R. Soc. Lond. B* **278**:261–276.

Ebendal, T., and Jacobson, C.-D., 1977, Tissue explants affecting extension and orientation of axons in cultured chick embryo ganglia, *Exp. Cell Res.* **105**:379–387.

Elliot Smith, G., 1897, The fornix superior, *J. Anat.* **31**:80–94.

Fawcett, J. W., and Gaze, R. M., 1981, The organization of regenerating axons in the Xenopus optic nerve, *Brain Res.* **229**:487–490.

Finlay, B. L., and Slattery, M., 1983, Local differences in the amount of early cell death in neocortex predict adult local specializations, *Science* **219**:1349–1351.

Fraser, S. E., and Hunt, R. K., 1980, Retinotectal specificity: Models and experiments in search of a mapping function, *Annu. Rev. Neurosci.* **3**:319–352.

Fujisawa, H., Tani, N., Watanabe, K., and Ibata, Y., 1982, Branching of regenerating retinal axons and preferential selection of appropriate branches for specific neuronal connection in the newt, *Dev. Biol.* **90**:43–57.

Gaze, R. M., and Sharma, S. C., 1970, Axial differences in the reinnervation of the goldfish optic tectum by regenerating optic nerve fibers, *Exp. Cell Res.* **10**:171–181.

Globus, J. H., and Kuhlenbeck, H., 1944, The subependymal plate (matrix) and its relation to brain tumors of the ependymal type, *J. Comp. Neurol.* **3**:1–35.

Goodman, C. S., Raper, J. A., Ho, R., and Chang, S., 1982, Pathfinding by neuronal growth cones in grasshopper embryos, in: *Cytochemical Methods in Neuroanatomy* (V. Chan-Palay and S. L. Palay, eds.), pp. 461–494, Alan R. Liss, New York.

Gundersen, R. W., and Barrett, J. N., 1979, Neuronal chemotaxis: Chick dorsal-root axons turn toward high concentrations of nerve growth factor, *Science* **206**:1079–1080.

Hankin, M. H., 1984, Mechanisms of axonal guidance: The problem of intersecting fiber systems, Doctoral Thesis, Case Western Reserve University, Cleveland, Ohio.

Hankin, M. H., and Silver, J., 1984, The role of the environment in guiding intersecting axon tracts in the mammalian telencephalon, *Soc. Neurosci. Abs.* **10**:372.

Harrison, R. G., 1907, Experiments in transplanting limbs and their bearing upon the development of nerves, *J. Exp. Zool.* **4:**239–281.

Harrison, R. G., 1910, The outgrowth of the nerve fiber as a mode of protoplasmic movement, *J. Exp. Zool.* **9:**787–848.

Harrison, R. G., 1914, The reaction of embryonic cells to solid structures, *J. Exp. Zool.* **17:**521–544.

Hensen, V., 1864, Über die Entwicklung des Gewebes und der Nerven im Schwanze der Froschlarve, *Virchows Arch. [Cell Pathol.]* **31:**51.

Hibbard, W. J., 1959, Central integration of developing nerve tracts from supernumerary grafted eyes and brain in the frog, *J. Exp. Zool.* **141:**323–341.

His, W., 1879, Über die Anfange des peripherischen Nerven Systemes, *Arch. Anat. Physiol. (Anat. Abt.)* 455–482.

His, W., 1887, Die Entwicklung der ersten Nervenbahnen beim menschlichen Embryo. Übersichliche Darstellung, *Arch. Anat. Physiol. (Anat. Abt.)* 368.

His, W., 1890, Die Neuroblasten und deren Entstehung im embryonalen Mark, *Abhandl. Math.-Phys. Cl. Sachs. Gessellsch. Wissensch.* **15:**311.

Horder, T. J., 1971, Retention, by fish optic nerve fibers regenerating to new terminal sites in the tectum, of "chemospecific" affinity for their original sites, *J. Physiol. (Lond.)* **216:**53P.

Horder, T. J., and Martin, K. A. C., 1978, Morphogenetics as an alternative to chemospecificity in the formation of nerve connections, in: *Cell–Cell Recognition* (A. S. Curtis, ed.), pp. 275–358, Cambridge University Press, New York.

Innocenti, G. M., Fiore, L., and Caminiti, R., 1977, Exuberant projections into the corpus callosum from the visual cortex of newborn cats, *Neurosci. Lett.* **4:**237–242.

Ivy, G. D., and Killackey, H. P., 1981, The ontogeny of the distribution of callosal projection neurons in the rat parietal cortex, *J. Comp. Neurol.* **195:**367–389.

Ivy, G. D., Akers, R. M., and Killackey, H. P., 1979, Differential distribution of callosal projection neurons in the neonatal and adult rat, *Brain Res.* **173:**532–537.

Kappers, C. U. A., Huber, G. C., and Crosby, C. C., 1936, *The Comparative Anatomy of the Nervous System of Vertebrates Including Man*, Vol. II, Macmillan New York.

Katz, M. J., and Lasek, R. J., 1979, Substrate pathways which guide growing axons in Xenopus embryos, *J. Comp. Neurol.* **183:**817–832.

Katz, M. J., and Lasek, R. J., 1981, Substrate pathways demonstrated by transplanted Mauthner axons, *J. Comp. Neurol.* **195:**627–641.

Keshishian, H., and Bentley, D., 1983a, Embryogenesis of peripheral nerve pathways in grasshopper legs. I. The initial nerve pathway to the CNS, *Dev. Biol.* **96:**89–102.

Keshishian, H., and Bentley, D., 1983b, Embryogenesis of peripheral nerve pathways in grasshopper legs. II. The major nerve routes, *Dev. Biol.* **96:**103–115.

Keshishian, H., and Bentley, D., 1983c, Embryogenesis of peripheral nerve pathways in grasshopper legs. III. Development without pioneer neurons, *Dev. Biol.* **96:**116–124.

Kölliker, A., 1894, Über den Fornix longus von Forel und die Riechstrahlungen im Gehirn des Kaninchens, *Verhandl. Anat. Gesellsch.* **8:**45.

Kostovic̀, I., and Molliver, M. E., 1974, A new interpretation of the laminar development of cerebral cortex: Synaptogenesis in different layers of neopallium in the human fetus, *Anat. Rec.* **97:**395.

Krayanek, S., 1980, Structure and orientation of extracellular matrix in developing chick optic tectum, *Anat. Rec.* **97:**95–109.

Kromer, L. F., Bjorklund, A., and Stenevi, U, 1981, Regeneration of septohippocampal pathways in adult rats in promoted by using embryonic hippocampal implants as bridges, *Brain Res.* **210:**173–200.

Lance-Jones, C., and Landmesser, L., 1981a, Pathway selection by chick lumbosacral motoneurons during normal development, *Proc. R. Soc. Lond. B* **260:**1–18.

Lance-Jones, C., and Landmesser, L., 1981b, Pathway selection by embryonic chick motoneurons in an experimentally altered environment, *Proc. R. Soc. Lond. B* **260:**19–52.

Letourneau, P. C., 1975, Cell-to-substratum adhesion and guidance of axonal elongation, *Dev. Biol.* **44:**92–101.

Letourneau, P. C., 1978, Chemotactic response of nerve fiber elongation to nerve growth factor, *Dev. Biol.* **66:**183–196.

Levi-Montalcini, R., 1952, Effects of mouse tumor transplantation on the nervous system, *Ann. N.Y. Acad. Sci.* **55**:330–343.

Levi-Montalcini, R., 1962, Analysis of specific nerve growth factor and of its antiserum, *Sci. Rep. Ist. Super. Sanita* **2**:245–368.

Levi-Montalcini, R., 1982, Developmental neurobiology and the natural history of nerve growth factor, *Annu. Rev. Neurosci.* **5**:341–362.

Levine, R., 1983, Neuronal plasticity in the optic tectum of amphibians, in: *Comparative Neurology of the Optic Tectum* (H. Vanegas, ed.), pp. 495–545, Plenum Press, New York.

Lumsden, A. G. S., and Davies, A. M., 1983, Earliest sensory nerve fibers are guided to peripheral targets by attractants other than nerve growth factor, *Nature (Lond.)* **306**:786–788.

Menesini-Chen, M. G., Chen, J. S., and Levi-Montalcini, R., 1978, Sympathetic nerve fibers ingrowth in the central nervous system of neonatal rodent upon intracerebral NGF injections, *Arch. Ital. Biol.* **116**:53–84.

Mesulam, M.-M., 1978, Tetramethyl benzidine for horseradish peroxidase neurohistochemistry: A non-carcinogenic blue reaction-product with superior sensitivity for visualizing neural afferents and efferents, *J. Histochem. Cytochem.* **26**:106–117.

Meyer, R. L., 1979, Retinotectal projection in goldfish to an inappropriate region with a reversal in polarity, *Science* **205**:819–821.

Meyer, R. L., 1980, Mapping for normal and regenerating retinotectal projection of goldfish with autoradiographic methods, *J. Comp. Neurol.* **189**:273–289.

Nakanishi, S., 1983, Extracellular matrix during laminar pattern formation of neocortex in normal and reeler mice, *Dev. Biol.* **95**:305–316.

Nardi, J. B., 1983, Neuronal pathfinding in developing wings of the moth Manduca sexta, *Dev. Biol.* **95**:163–174.

Noble, M., Seang, J. F., and Cogan, J., 1984, Glia are a unique substrate for in vitro growth of CNS neurones, *J. Neurosci.* **4**:1892–1903.

Privat, A., and Leblond, C. P., 1972, The subependymal layer and neighboring region in the brain of the young rat, *J. Comp. Neurol.* **146**:277–302.

Probst, M., 1901, Über den Bau des balkenlosen Grosshirns, sowie uber Mikrogyrie und Heterotopie der grauen Substanz, *Arch. Psychiatry* **34**:709–786.

Rager, G., 1980, Specificity of nerve connections by unspecific mechanisms, *Trends Neurosci.* **3**:43–44.

Ramon y Cajal, S., 1908, Terminacion periferica del nervio acustico de los aves, *Trab. Inst. Cajal Invest. Biol.* **6**:161–176.

Ramon y Cajal, S., 1928, *Degeneration and Regeneration of the Nervous System*, Vols. I and II, repr. 1959, Hafner, New York.

Ramon y Cajal, S., 1937, *Recollections of My Life* (E. H. Cragie, tr.), American Philosophical Society, Philadelphia.

Ramon y Cajal, S., 1960, *Studies on Vertebrate Neurogenesis* (L. Guth, tr.), Charles C Thomas, Springfield, Illinois.

Reh, T. A., Pitts, E., and Constantine-Paton, M., 1983, The organization of the fibers in the optic nerve of normal and tectum-less Rana pipiens, *J. Comp. Neurol.* **218**:282–296.

Roberts, A., and Taylor, J. S. H., 1982, A scanning electron microscope study of the development of a peripheral neurite network, *J. Embryol. Exp. Morphol.* **69**:237–250.

Rutishauser, U., Gall, W. E., and Edelman, G. M., 1978, Adhesion among neural cells of the chick embryo. IV. Role of the cell surface molecule N-CAM in the formation of neurite bundles in cultures of spinal ganglia, *J. Cell Biol.* **79**:382–393.

Schwann, T., 1839, *Mikroskopische Untersuchungen über die Übereinstimmung in der Struktur und den Wachstum der Thiere und Pflantzen*, Sander, Berlin.

Shoukimas, G. M., and Hinds, J. W., 1978, The development of the cerebral cortex in the embryonic mouse. An electron microscopic serial section analysis, *J. Comp. Neurol.* **179**:295–330.

Silver, J., 1984, Studies on the factors that govern directionality of anoxal growth in the embryonic optic nerve and at the chiasm of mice, *J. Comp. Neurol.* **223**:238–251.

Silver, J., and Mason, C. A., 1984, Postnatally induced regeneration of the corpus callosum in acallosal mice, in: *The Eric K. Fernstrom Symposium on Transplantation in the Mammalian CNS*, Lund, Sweden, June, 1984.

Silver, J., and Ogawa, M., 1983, Postnatally induced formation of the corpus callosum in acallosal mice on glia-coated cellulose bridges, *Science* **220**:1067–1069.

Silver, J., and Robb, R. M., 1979, Studies on the development of the eye cup and optic nerve in normal mice and in mutants with congenital optic nerve aplasia, *Dev. Biol.* **68**:175–190.

Silver, J., and Rutishauser, U., 1984, Guidance of optic axons by a preformed adhesive pathway on neuroepithelial endfeet, *Soc. Neurosci. Abs.* **10**:372.

Silver, J., and Sapiro, J., 1981, Axonal guidance during development of the optic nerve: The role of pigmented epithelia and other extrinsic factors, *J. Comp. Neurol.* **202**:521–538.

Silver, J., and Sidman, R. L., 1980, A mechanism for the guidance and topographic patterning of retinal ganglion cell axons, *J. Comp. Neurol.* **189**:101–111.

Silver, J., Lorenz, S. E., Wahlsten, D., and Coughlin, J., 1982, Axonal guidance during development of the great cerebral commissures: Descriptive and experimental studies in vivo on the role of preformed glial pathways. *J. Comp. Neurol.* **210**:10–29.

Silver, J., Smith, G. M., Miller, R. H., and Levitt, P. R., 1985, The immature astrocyte: Its role during normal CNS axon tract development and its ability to reduce scar formation and promote axonal regeneration when transplanted into the brain of adults, *Soc. Neuro Sci. Abs.* **11**:334.

Singer, M., Nordlander, R. H., and Egar, M., 1979, Axonal guidance during embryogenesis and regeneration in the spinal cord of the newt: The Blueprint hypothesis of neuronal pathway patterning, *J. Comp. Neurol.* **185**:1–21.

Smart, I. M. H., 1961, The subependymal layer of the mouse brain and its cell production as shown by radioautography after thymidine-H³ injection, *J. Comp. Neurol.* **116**:325–347.

Speidel, C. C., 1932, Studies of living nerves. I. The movements of individual sheath cells and nerve sprouts correlated with the process of myelin-sheath formation in amphibian larvae, *J. Exp. Zool.* **61**:279–331.

Speidel, C. C., 1933, Studies of living nerves. II. Activities of ameboid growth cones, sheath cells, and myelin segments, as revealed by prolonged observation of individual nerve fibers in frog tadpoles, *Am. J. Anat.* **52**:1–79.

Sperry, R. W., 1944, Optic nerve regeneration with return of vision in anurans, *J. Neurophysiol.* **7**:57–69.

Sperry, R. W., 1963, Chemoaffinity in the orderly growth of nerve fiber patterns and connections, *Proc. Natl. Acad. Sci. USA* **50**:703–710.

Stuermer, C. A. O., and Easter, S. S., 1984, A comparison of the normal and regenerating retinotectal pathways of goldfish, *J. Comp. Neurol.* **223**:57–76.

Taghert, P. H., Bastiani, M. J., Ho, R. K., and Goodman, C. S., 1982, Guidance of pioneer growth cones: Filopodial contacts and coupling revealed with an antibody to Lucifer Yellow, *Dev. Biol.* **94**:391–399.

Thanos, S., Bonhoeffer, F., and Rutishauser, U., 1984, Fiber–fiber interactions and tectal cues influence the development of the chick retinotectal projection, *Proc. Natl. Acad. Sci. USA* **81**:1906–1910.

Thiery, J.-P., Brackenbury, R., Rutishauser, U., and Edelman, G., 1977, Adhesion among neural cells of the chick embryo. II. Purification and characterization of a cell adhesion molecule from neural retina, *J. Cell Biol.* **252**:6841–6845.

Thiery, J.-P., Duband, J.-L., Rutishauser, U., and Edelman, G., 1982, Cell adhesion molecules in early chicken embryogenesis, *Proc. Natl. Acad. Sci. USA* **79**:6737–6741.

Udin, S., 1978, Permanent disorganization of the regenerating optic tract in the frog, *Exp. Neurol.* **58**:455–470.

Van Ness, J., Maxwell, I. H., and Hahn, W. E., 1979, Complex population of non-polyadenylated messenger RNA in mouse brain, *Cell* **18**:1341–1349.

von Büngner, O., 1891, Über die Regeneration und Denervationvorgänge am Nerven nach Verletzungen, *Beitr. Pathol. Anat. Allg. Pathol.* **10**:321–393.

Wahlsten, D., 1984, Growth of the mouse corpus callosum, *Dev. Brain Res.* **15**:59–67.

Warkany, J., 1971, *Cogenital Malformations*, pp. 252–254, Year Book Medical, Chicago.

Weiss, P. A., 1934, In vitro experiments on the factors determining the course of the outgrowing nerve fiber, *J. Exp. Zool.* **68**:393–448.

Weiss, P. A., 1955, Nervous system (neurogenesis), in: *Analysis of Development* (B. H. Willier, P. A. Weiss, and V. Hamburger, eds.), pp. 346–401, W. B. Saunders, Philadelphia.

Weiss, P. A., and Taylor, A. C., 1944, Further experimental evidence against "neurotropism" in nerve regeneration, *J. Exp. Zool.* **95**:233–257.

Wessells, N. K., Letourneau, P. C., Nuttall, R. P., Luduena-Anderson, M., and Geiduschek, J. M., 1980, Responses to cell contacts between growth cones, neurites, and ganglionic non-neuronal cells, *J. Neurocytol.* **9**:647–664.

Whitelaw, V., and Hollyday, M., 1983a, Thigh and calf discrimination in the motor innervation of the chick hindlimb following deletions of limb segments, *J. Neurosci.* **3**:1199–1215.

Whitelaw, V., and Hollyday, M., 1983b, Postition-dependent motor innervation of the chick hindlimb following serial and parallel duplications of limb segments, *J. Neurosci.* **3**:1216–1225.

Whitelaw, V., and Hollyday, M., 1983c, Neural pathway constraints in the motor innervation of the chick hindlimb following dorsoventral rotations of distal limb segments, *J. Neurosci.* **3**:1226–1233.

Wigglesworth, V. B., 1964, *The Life of Insects*, New American Library, New York.

Yoon, M., 1972, Reversibility of the reorganization of retinotectal projection in goldfish, *Exp. Neurol.* **35**:565–577.

Yoon, M., 1975, Readjustment of retinotectal projection following reimplantation of a rotated or innervated tectal tissue in adult goldfish, *J. Physiol. (Lond.)* **252**:137–158.

Yorke, C. H., and Caviness, V. S., 1975, Interhemispheric neocortical connections of the corpus callosum in the normal mouse: A study based on anterograde and retrograde methods, *J. Comp. Neurol.* **164**:233–246.

Chapter 16

Formation of the Vertebrate Neuromuscular Junction

FRANCES MOODY-CORBETT

1. Introduction

The formation of the neuromuscular junction (NMJ) is an example of establishment of contact between unlike cells—nerve and muscle—that results in functional interaction between them. This chapter focuses on the establishment of contact, the morphogenesis of the NMJ, the acquisition of functional competence, and the regulation of these processes at the cellular level. This chapter is not an exhaustive review of the area; rather, coverage is restricted to the development of the NMJs in three organisms: rat, chick, and the amphibian *Xenopus laevis*. The exact sequence of events leading to the formation of a NMJ will in the end be unique to that junction. However, general principles will apply to the development of all NMJs. The aim of the present study is to describe the general features common to the development of the junctions of these three preparations. The formation of the NMJ, including studies pertaining to the effects of denervation and regeneration, has also been recently reviewed by Bennett (1983).

The adult NMJ has a number of morphological characteristics (reviewed by Bowden and Duchen, 1976, and shown in Fig. 1). The presynaptic nerve terminal has an abundance of small electron-lucent vesicles containing the neurotransmitter acetylcholine (ACh). The vesicles are clustered, at regular intervals, next to the membrane. These sites, which are called **active zones,** are believed to be the sites of transmitter release. In freeze-fracture preparations, the membrane in this region contains a pattern of 10–12-nm particles (Dreyer *et al.*, 1973; Rash and Ellisman, 1974). Immediately opposite the nerve terminal, the muscle membrane appears electron dense or thickened. In fast muscle of frog and rat, the membrane has deep folds. However, some muscles (e.g., the myotomal muscle of *Xenopus*) have shallow folds (MacKay *et al.*, 1960; Kullberg *et al.*, 1977), and in others, such as the anterior latissimus dorsi (ALD) and pos-

FRANCES MOODY-CORBETT • Department of Physiology, Tufts University School of Medicine, Boston, Massachusetts 02111. *Present address:* Division of Basic Medical Sciences, Faculty of Medicine, Memorial University of Newfoundland, Saint John's, Newfoundland A1B 3V6, Canada.

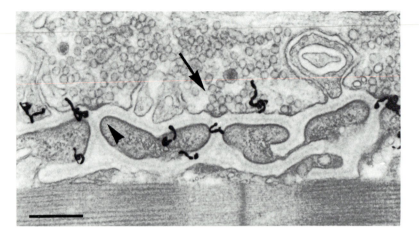

Figure 1. Neuromuscular junction of cutaneous pectoris muscle of the frog. The ACh receptors have been labeled with ^{125}I-α bungarotoxin. The presynaptic terminal contains a large number of synaptic vesicles clustered especially at release sites or active zones (arrow). The synaptic cleft contains basal lamina material that extends into the synaptic fold of the muscle membrane. The postsynaptic membrane appears thickened at the peaks of the folds (arrowhead). Scale bar: 0.5 μm. (Courtesy of J. Matthews-Bellinger.)

terior latissimus dorsi (PLD) of chicken, folds may be absent (Atsumi, 1977; Burden, 1977b). When folds are present, the clefts of the fold are in register with the active zones of the presynaptic terminal. The area of thickened membrane in these cases appears at the peaks and only partway down the fold.

Using ^{125}I-α-bungarotoxin to label ACh receptors, it has been found that the area of thickened membrane corresponds with the site of highest receptor density (Fertuck and Salpeter, 1974, 1976; Matthews-Bellinger and Salpeter, 1978). Beyond the area of the NMJ, the density of ACh receptors drops precipitously (Porter and Barnard, 1975; Fertuck and Salpeter, 1976), which is also evidenced by a drop in ACh sensitivity in the nonjunctional region (Miledi, 1960; Peper and McMahan, 1972; Dreyer and Peper, 1974; Kuffler and Yoshikami, 1975). In freeze-fracture preparations, the ACh receptors are believed to correspond with 8–12 nm particles in the region of thickened membrane (Heuser et al., 1974; Rash and Ellisman, 1974; Peng and Nakajima, 1978). At some adult frog and rat NMJs, a postsynaptic filamentous material, next to the region of thickened plasma membrane, has also been described (Ellisman et al., 1976; Kullberg et al., 1977; Couteaux, 1981; Burden, 1982).

The width of the primary synaptic cleft at any one junction is very uniform and is filled with basal lamina material that is continuous with the external lamina surrounding the muscle membrane (Sanes, 1983). A major component of the basal lamina at the junction is the enzyme **acetylcholinesterase (AChE)**, which hydrolyzes the neurotransmitter ACh (Hall and Kelly, 1971; Betz and Sakmann, 1973; McMahan et al., 1978). Biochemical studies have demonstrated that there is a heavy-molecular-weight form of AChE in innervated muscle (Hall, 1973; Barnard et al., 1984; Vigny et al., 1976; Nicolet and Rieger,

1981; Carson *et al.*, 1979; Massoulie and Bon, 1982; Koenig *et al.*, 1984). In rats, this heavy-molecular-weight form is concentrated at the NMJ (Hall, 1973) and has been referred to as the **synapse-specific form** of the enzyme. However, this form does not appear to be exclusively localized at the junction in other animals (Koenig *et al.*, 1979; Barnard *et al.*, 1984).

Activation of the nerve terminal (e.g., invasion by an action potential) causes the quantal release of ACh (Katz, 1966). The neurotransmitter diffuses across the synaptic cleft and binds to the receptor in the postsynaptic membrane. This binding causes the receptor channels to open and allow passage of cations, particularly Na^+ and K^+. The resultant depolarization, termed an **endplate potential (epp)**, is sufficiently large to initiate a muscle contraction. The epp has a characteristic short time course, which is determined by the presence of AChE in the cleft and the kinetic properties of the individual ACh receptor channels. Cholinesterase, by hydrolyzing ACh, limits the amount of transmitter available for rebinding and thus helps keep the action of the transmitter brief (Katz and Miledi, 1973). Analysis of synaptic currents and ACh-induced noise has shown that the opening of the individual ACh receptor channels in adult muscle is brief (less than 1 msec for rat and frog muscle) and of a relatively high conductance (Katz and Miledi, 1972; Anderson and Stevens, 1973). In the absence of presynaptic action potentials, there is a low level of spontaneous release of neurotransmitter resulting in smaller depolarizations, termed **miniature end-plate potentials (mepps)** (Katz, 1966). As shown in Figure 2, these events occur at a relatively high frequency. Like epps, they have a rapid time course and are the result of the quantal release of ACh.

In developmental studies, the morphological features of the junction that have received considerable attention are the appearance of nerve terminals with an accumulation of synaptic vesicles, the accumulation of a high density of ACh receptors in the muscle membrane, and the accumulation of AChE in the extracellular matrix. Electrophysiological studies have focused on the appearance of mepps and epps during development and on the biophysical properties and alterations in the properties of the individual ACh receptor channels. It is these main features of the developing NMJ that are described below.

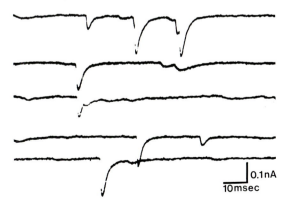

Figure 2. Currents underlying miniature end-plate potentials recorded from the myotomal muscle of mature (stage 49, 12 days) *Xenopus* tadpole. The records were obtained using a single electrode voltage clamp with the membrane potential held at resting level. The events appear slightly distorted because of the limited frequency response (approx. 4 kHz) of the recording device.

0.1nA

10msec

2. Development of the Morphological Characteristics of the Neuromuscular Junction

2.1. Presynaptic Nerve Terminal

The presence of a nerve terminal is best characterized by the appearance of synaptic vesicles. In rat, nerve fibers innervating the intercostal muscles, for example, first enter the rib cage region on day (d) 13 of embryogenesis, 1 day before the appearance of muscle fibers and 8–9 days before birth (Dennis *et al.*, 1981). Vesicles have been detected in axon profiles approaching these muscles as early as d16 (Kelly and Zacks, 1969; Zacks, 1973). However, *accumulations* of vesicles were not seen reliably until d18 (Teravainen, 1968; Kelly and Zacks, 1969). These aggregations are taken as the first appearance of presynaptic specializations.

In chick, muscle differentiates very early in embryogenesis. For example, rudimentary muscle groups are present as early as d3–4, approximately 17–18 days before hatching (Wake, 1964; Giacobini, 1972). However, ultrastructural examination at these early times is not available. The earliest appearance of vesicle accumulations and active zones that has been described is between d8 and d11 in ALD and PLD muscles (Bennett and Pettigrew, 1974; Sisto-Daneo and Filogama, 1975; Atsumi, 1977; Jacob and Lentz, 1979) and at d13 for NMJ of limb muscle (Hirano, 1967a).

In the amphibian *Xenopus laevis*, the first muscle group to form is the myotomal musculature at stage 14 (16 hr after fertilization), just as the neural tube is closing (Nieuwkoop and Faber, 1967). Although nerve fibers are seen in the vicinity of some myotomes as early as stage 19 (21 hr) and at the ends of virtually all myotomes by stage 24 (26 hr) (Chow and Cohen, 1983), nerve terminals with synaptic vesicles have not been seen until stages 25–27 (28–31 hr). By stage 27, some terminals have accumulations of vesicles adjacent to the membrane resembling active zones (Kullberg *et al.*, 1977).

In vivo, the appearance of growth cones in close association with developing muscle cells has been described for both chick (Jacob and Lentz, 1979) and *Xenopus* (Kullberg *et al.*, 1977). These protuberances, as well as containing tubular structures and large irregular vesicles, sometimes contain synapticlike vesicles. In chick, they have been seen as nerve fibers approach the latissimus dorsi muscle groups at d10 of embryogenesis and are sometimes associated with patches of muscle membrane rich in ACh receptors (Jacob and Lentz, 1979). In *Xenopus*, growth cones have been seen between stages 21 and 24 (23–26 hr), which precedes the appearance of ultrastructural synaptic specializations, although, as described below, functional contacts had already been made (Kullberg *et al.*, 1977).

Development of synapses between nerve cells of spinal cord and muscle cells have also been studied *in vitro*. In culture, embryonic muscle cells from either rat or chick differentiate and fuse to form multinucleated myotubes with well-developed contractile apparatus (Fischbach *et al.*, 1973). *Xenopus* myotomal muscle cells also differentiate and develop contractile apparatus in

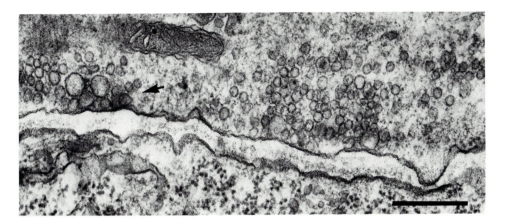

Figure 3. Neuromuscular contact in a 1-day *Xenopus* culture. The nerve fiber contains clusters of 50 nm vesicles, and its plasma membrane displays a patch of increased density (arrow), which may represent a developing active zone. Basal lamina is present within the cleft and the sarcolemma has an increased density and a ridged outline. Portions of basal lamina exhibit a dense central band. Scale bar: 0.5 μm. (From Weldon and Cohen, 1979.)

culture, but the cells remain mononucleated (Cohen, 1981). Nerve fibers from dissociated cells or explants of spinal cord approach and course over muscle cells in co-culture and are capable of releasing neurotransmitter (Fischbach, 1972; Anderson *et al.*, 1979; Nakajima *et al.*, 1980; Kidokoro, 1980; Kidokoro *et al.*, 1980). In rat and *Xenopus* cultures, initially there are a number of structures resembling neuronal growth cones and containing synapticlike vesicles sparsely distributed in the cytoplasm, which are in close proximity with the muscle membrane (Peng *et al.*, 1979; Weldon and Cohen, 1979; Nakajima *et al.*, 1980). In these cultures, the nerve terminal specializations rapidly differentiate and, within 24 hr of co-culture, clusters of vesicles are seen close to the membrane, as shown in Figure 3. In most cases, these sites correspond with sites of postsynaptic specializations, such as a dense region of membrane and basal lamina material; however, on occasion, the presynaptic terminal occurs at a site of unspecialized membrane. Structures resembling active zones in the presynaptic terminal are seen at this time in *Xenopus* cultures (Peng *et al.*, 1979; Weldon and Cohen, 1979). However, in rat cultures, active zones are not apparent for several more days (Nakajima *et al.*, 1980). In chick co-cultures, synaptic vesicles clustered near a thickened presynaptic membrane occur at well-developed nerve–muscle contacts (Frank and Fischbach, 1979).

2.2. Synaptic Cleft

The proximity of the nerve ending with the muscle membrane varies considerably during early stages of development. Regions of very close contact (10 nm) have been identified in d16 rat (Kelly and Zacks, 1969) and at nerve–muscle contacts in *Xenopus* cultures (Weldon and Cohen, 1979). In chick co-cultures, direct electrical coupling has been observed between nerve and muscle (Fischbach, 1972), and gap junctionlike structures have been seen in *Xenopus* co-cultures (Peng *et al.*, 1980). No information is available as to what function these close associations might perform.

By the time a synaptic cleft of uniform width is present, basal lamina material is also seen. Recent studies by McMahan *et al.* (1979) have indicated that the basal lamina may play an important role in synapse formation. Following degeneration, if nerve and muscle are permitted to regenerate in the presence of the original basal lamina sheath, nerve terminals reinnervate the original synaptic site in the absence of muscle regeneration (Marshall *et al.*, 1977); conversely, muscle cells develop postsynaptic specializations at precisely these sites in the absence of nerve regeneration (Burden *et al.*, 1979). However, by the markers currently available, basal lamina material appears after other specializations have already begun developing. Using conventional electron microscopy, basal lamina is present during development of rat intercostal muscles around d17–18 (Teravainen, 1968; Kelly and Zacks, 1969; Zacks, 1973), of chick ALD and PLD muscles between days 8 and 11 (Sisto-Daneo and Filogamo, 1975; Atsumi, 1977; Jacob and Lentz, 1979), of chick limb muscle on d13 (Hirano, 1967b), and of *Xenopus* myotomal muscle between stages 21–24 (25–26 hr) (Kullberg *et al.*, 1977). As described below, antibodies to synaptic basal lamina have also been used to detect basal lamina during development of junctions in culture and at ectopic synapses.

Basal lamina material also develops at sites of nerve–muscle contact *in vitro*. In rat co-cultures, filamentous or amorphous material occurs within 20 hours of plating, but well developed basal lamina at the cleft was not seen until after 4 days of co-culture (Nakajima *et al.*, 1980). In *Xenopus* cultures, however, sites of nerve–muscle contact develop pre- and postsynaptic specializations within 24 hr of co-culture, including basal lamina material (see Fig. 3; Weldon and Cohen, 1979; Peng *et al.*, 1979). Although basal laminalike material is seen this early in culture, it appears more frequently in older cultures, suggesting some period of contact occurs before its accumulation at these sites (Weldon and Cohen, 1979; Nakajima *et al.*, 1980). In chick primary cultures, the first appearance of basal lamina at sites of nerve–muscle contact has not been examined; however, it is present at well-developed contacts (Frank and Fischbach, 1979).

Another approach to examining the appearance of synaptic basal lamina has been to prepare antibodies to antigenic determinants of the basal lamina. This has been done for rat (Sanes and Hall, 1979) and *Xenopus* (Anderson and Fambrough, 1983), and the appearance of the antigenic sites has been examined to a limited extent during development. Two antigenic determinants described by Sanes and Hall (1979) have been found to co-localize with ACh receptor patches on rat myotubes in culture (Sanes and Lawrence, 1983) and at ectopic synapses in denervated rat soleus (Weinberg *et al.*, 1981). The latter investigators found that these antigens develop at ectopic synapses at approximately the same time as synaptic AChE, which is a few days after ACh receptors have begun aggregating. Anderson and Fambrough (1983) obtained an antibody di-

611

rected against a proteoglycan component of the basal lamina of *Xenopus* myotomal muscle (Fig. 4). The antigen is found at adult NMJs of *Xenopus* muscle and co-localizes with ACh receptors at sites of nerve–muscle contact in culture and at patches of ACh receptors on non-innervated muscle cells (Anderson and Fambrough, 1983). It has not been determined when during normal development *in vivo* these antigenic sites arise or what role they may play in development.

The most distinguishing feature of the basal lamina at the NMJ is the presence of a high density of AChE. The appearance of this enzyme signals a change in the functional development of the junction (see Section 3.2). Its histochemical appearance has been used as a marker for the existence of the junction during development.

In rat muscle, AChE first appears widespread over the muscle fiber by d14–15, and patches of high AChE activity at the NMJ first become apparent on d16–17 (Bennett and Pettigrew, 1974; Bevan and Steinbach, 1977; Ziskind-Conhaim and Dennis, 1981). As described earlier, a synapse-specific form of the enzyme is present in rat muscle. During development it can be detected on d14, but—like the histochemical reaction product—it is not found localized exclusively at the junction (Vigny *etal.*, 1976; Koenig *et al.*, 1984; Inestrosa *et al.*, 1984). By birth, there are still relatively high levels of this form in extrajunctional regions (Sketelj and Brzin, 1980; Koenig and Reiger, 1981; Koenig *et al.*, 1984), and the adult distribution of high levels at the junction and negligible levels in extrajunctional regions is not attained for 4–5 weeks postnatal (Koenig and Reiger, 1981).

In chick limb muscle, AChE patches are apparent at sites of the developing junction by d7 (Hirano, 1967a) but become more intense and resemble adult patterns by d12–13 (Wake, 1964, 1976; Filogama and Gabella, 1967; Hirano, 1967a). In ALD and PLD, AChE patches appear as early as d11 (Bennett and Pettigrew, 1974) but are more reliably seen on d13–14 (Gordon *et al.*, 1974; Sisto-Daneo and Filogama, 1975). There is a heavy-molecular-weight form of AChE that has been detected as early as d7 in thigh muscle (Kato *et al.*, 1980). However, unlike the situation in rat, this form does not appear to become synapse-specific in chick muscle (Koenig *et al.*, 1979; Barnard *et al.*, 1984).

In *Xenopus,* myotomal muscle AChE staining was seen at some sites of innervation as early as stage 22 (24 hr) and was always detected beyond stage 24 (26 hr) (Kullberg *et al.*, 1980). The appearance of a synapse-specific form of AChE has not been examined in *Xenopus in vivo.*

In an attempt to determine what regulates the appearance of AChE during development, studies of the effects of blocking synaptic or muscle activity during embryogenesis have been undertaken. Rat intercostal muscles maintained in organ culture will only localize AChE if spinal cord is included in the explant (Ziskind-Conhaim and Bennett, 1982). In chick muscle, blocking activity starting at d10 of embryogenesis results in loss or retardation of the appearance of AChE (Giacobini *et al.*, 1973; Gordon *et al.*, 1974; Giacobini-Robecchi *et al.*, 1975; Oppenheim *et al.*, 1978; Betz *et al.*, 1980), suggesting that synaptic or muscle activity is involved. In *Xenopus* embryos, on the other

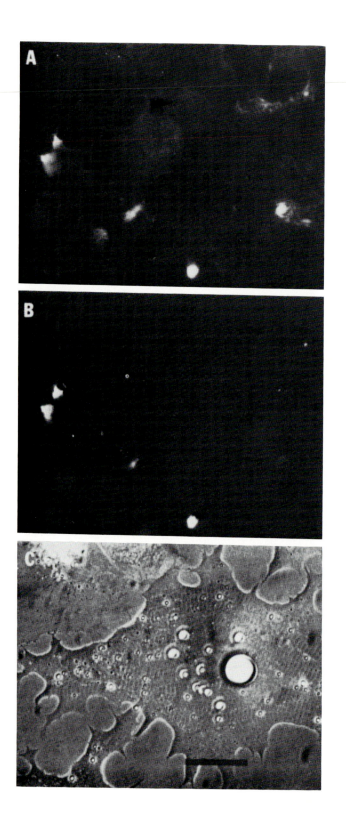

hand, blocking muscle activity from the time of first innervation did not affect the accumulation of AChE at the junction, as measured histochemically or by examining the time course of synaptic currents (Cohen et al., 1984). In the latter study, low-level synaptic activity, which was not blocked, may have played a role in AChE accumulation. However, cell culture studies support the interpretation that activity is not important for cholinesterase development in Xenopus.

AChE develops in mixed cultures containing both nerve and muscle cells. In rat co-cultures of spinal cord cells with muscle, the synapse-specific form of AChE develops, and surface patches of cholinesterase are seen (Vigny and Koenig, 1978; Koenig et al., 1984). In chick cultures, AChE becomes localized at sites of nerve–muscle contact (Frank and Fischbach, 1979; Rubin et al., 1979). As in vivo, cholinesterase (including the heavy molecular weight form) is not detected if synaptic activity is blocked (Rubin et al., 1980). In Xenopus cultures, AChE also becomes localized at sites of nerve–muscle contact (Fig. 5) and—as in vivo—it appears in the absence of muscle activity (Moody-Corbett et al., 1982).

The appearance of cholinesterase has also been examined in muscle cultures grown in the absence of nerve. In these cultures, rat muscle only develops the synapse-specific form of AChE if the muscle had already been exposed to nerve in vivo. As in very early embryonic muscle in vivo, the AChE remains widely distributed on the surface rather than localized in surface patches (Vigny and Koenig, 1978; Reiger et al., 1980; Koenig et al., 1979, 1984). Furthermore, blockage of muscle contractile activity by tetrodotoxin will eliminate the appearance of the synapse-specific form of AChE in these cultures (Koenig et al., 1979). In chick muscle cultures, AChE is seen widely distributed over the muscle surface, or, infrequently, it is localized into patches (Fischbach, 1972; Wilson et al., 1973; Fischbach et al., 1979). The heavy molecular-weight form of AChE is not present in these cultures (Rotundo and Fambrough, 1979; Cisson et al., 1981). However, under certain culture conditions involving nerve extract or the elimination of horse serum, it is present (Kato et al., 1980; Klymkowsky et al., 1983; Popiella et al., 1984). Chick muscle cells show spontaneous fibrillations in culture (Fischbach et al., 1974), and this activity (as in vivo) may play a role in the appearance of AChE in culture (Klymkowsky et al., 1983). In Xenopus muscle cultures, discrete patches of AChE appear, and these sites overlap with a number of other postsynaptic specializations as shown in Figures 6 and 7 (patches of basal lamina, thickened plasma membrane, and

---

Figure 4. Coordinate organization of basal lamina proteoglycan and ACh receptors on living embryonic Xenopus muscle cell developing in a nerve-free culture. The cell was stained with fluorescein-labeled monoclonal antibody (2AC2) and rhodamine-labeled α-bungarotoxin. The proteoglycan stain (A) is organized into diffuse background separating several discrete intensely stained plaques, viewed here on the flat surface facing the collagen substratum. Many such plaques also show dense accumulations of ACh receptors (B). In fact, all ACh receptor clusters are associated with corresponding regions of dense proteoglycan staining. Scale bar: 30 μm. (From Anderson and Fambrough, 1983, reproduced from the Journal of Cell Biology, 1983, Vol. 97, pp. 1396–1411 by copyright permission of the Rockefeller University Press.)

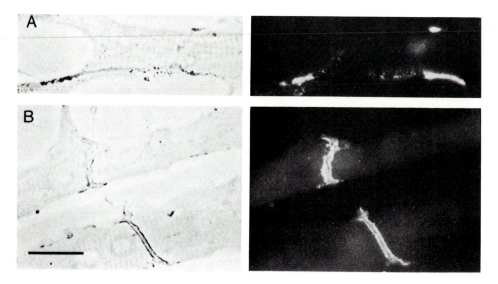

Figure 5. *Xenopus* nerve–muscle cultures, showing correspondence in the location of cholinesterase stain (left) and ACh receptor stain (right) at sites of nerve–muscle contact. (A) Contact on the upper surface of a muscle cell, 2 days after adding nerve to a 2-day-old culture. (B) Contact on the lower surface of two muscle cells, 2 days after adding nerve to a 3-day-old culture. Scale bar: 20 μm. (From Moody–Corbett *et al.*, 1982, from the *Journal of Neurocytology*, Vol. 11, p. 385, published by Chapman and Hall.)

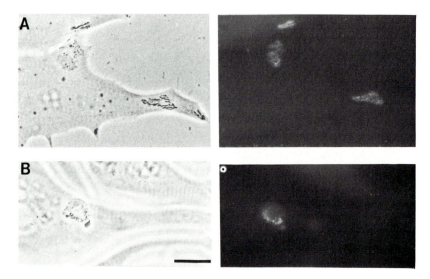

Figure 6. *Xenopus* muscle cells grown in culture in the absence of nerve showing the correspondence between patches of cholinesterase reaction product (left) and patches of ACh receptors labeled with rhodamine-labeled α-bungarotoxin (right). (A) The focus is on the lower surface of the muscle cells, next to the collagen substrate of the culture dish. (B) The focus is on the upper surface. As shown in most cases, there is a high degree of overlap between the two types of patches. Cholinesterase histochemistry is by the method of Karnovsky (1964). Scale bar: 20 μm.

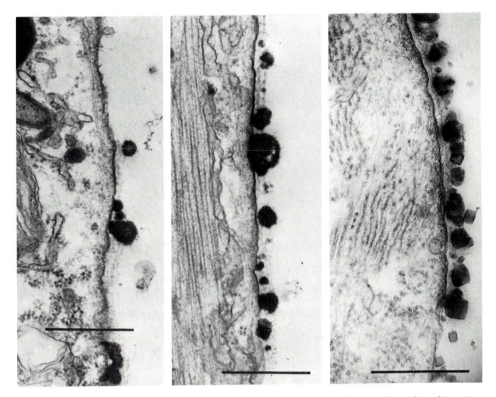

Figure 7. *Xenopus* muscle cells grown in culture in the absence of nerve. Granules of reaction product (depicting the presence of cholinesterase) are associated with regions of sarcolemmal specializations. In each example, the granules are present at sites where there is extracellular (basal lamina) material. Note also the increased electron density of the sarcolemma and subjacent sarcoplasm at these sites. In some cases, the granules appear adherent to the extracellular material rather than to the plasma membrane. From 5-day cultures. Scale bars: 0.5 μm. Cholinesterase staining is by the method of Karnovsky (1964). (From Weldon *et al.*, 1981.)

discrete patches of ACh receptors) (Moody-Corbett and Cohen, 1981; Weldon *et al.*, 1981). In these cultures, a heavy molecular weight form of AChE has been detected (Klymkoswky *et al.*, 1983). *Xenopus* muscle cells are not spontaneously active in culture (Anderson *et al.*, 1977). Therefore, these findings are consistent with the *in vivo* data showing that the appearance of AChE does not require muscle activity (Moody-Corbett and Cohen, 1981).

In rats, the developmental appearance of AChE has also been studied at ectopic synapses of denervated muscle (Guth *et al.*, 1966; Weinberg and Hall, 1979; Lomo and Slater, 1980). As at the original end plate, the synapse-specific form of AChE becomes localized at the ectopic synapse (Weinberg and Hall, 1979; Weinberg *et al.*, 1981). Early denervation of the ectopic site results in a loss of AChE from this site, which can be prevented to some extent by chronic stimulation of the muscle (Lomo and Slater, 1980). Taken together with the findings from cell culture it would appear that, in rat: (1) nerve exposure is required at some point to initiate the localization of AChE and (2) muscle or synaptic ability is required for its maintenance.

From the studies reported in this section, the localization of AChE at the NMJ appears to differ in the three systems presented. In rat muscle, AChE does not develop unless the muscle has been previously exposed to nerve, and activity is important for its localization into patches. In chick muscle, clusters of AChE form only very infrequently in the absence of nerve, and activity is necessary for localization at sites of nerve–muscle contact *in vivo* and *in vitro*. In contrast to the situation in rat and chick muscle, AChE on *Xenopus* muscle develops into discrete patches in the absence of nerve or prior exposure to nerve, and AChE will accumulate at the developing NMJ in the absence of activity. Why differences occur in these three systems is not known; they may represent inherent differences in these animal groups.

2.3. Postsynaptic Membrane

The ultrastructural specializations of the postsynaptic membrane, such as a thickened plasma membrane and cytoplasmic filamentous material, are difficult to discern early in development without some other more obvious marker. In rat intercostal muscle, a postsynaptic membrane that is thickened or electron dense is first described at d17–18, which is the same time that the presynaptic terminal has vesicles and the synaptic cleft has a uniform width and contains basal lamina material (Teravainen, 1968; Kelly and Zacks, 1969; Zacks, 1973). It is also at this time that the postsynaptic membrane shows a furrowed appearance, presumably representing an early stage in the development of synaptic folds, which will not become fully developed until the second to third postnatal week in the intercostal muscles (Teravainen, 1968; Kelly and Zacks, 1969; Zacks, 1973) and even later (>3 weeks) in limb muscle (Juntunen, 1974; Korneliussen and Jansen, 1976).

In chick ALD and PLD muscles, a thickened postsynaptic membrane is apparent on d8 (Atsumi, 1977) and in limb muscle by d13 (Hirano, 1967a). As in rat, the presynaptic nerve terminal is already differentiated at these times. Synaptic folds, although not as deep or as prominent in the adult chicken as in the rat, are believed to begin appearing in limb muscle at d16 (Hirano, 1967a). A thickened postsynaptic membrane in myotomal muscle of *Xenopus* is noted by stage 27 (31.25 hr), a few hours after nerve terminal specializations first appear (Kullberg *et al.*, 1977). In this muscle, innervation occurs at the ends of the muscle cells in the region of the myotendinous junction. In the adult, the myotendinous region has characteristic deep folds and the NMJ has shallow folds or ridges (Kullberg *et al.*, 1977). When the synaptic folds first begin appearing has not been well documented, although elevations and furrows in the muscle membrane in the synaptic region are seen at stage 27 (Kullberg *et al.*, 1977).

A major characteristic of the postsynaptic membrane is the accumulation of a high density of ACh receptors. In the diaphragm muscle of rat, ACh receptors first appear in low density over the entire muscle surface at d15 (Bevan and Steinbach, 1977). The following day, a central cluster of receptors appears at the presumptive site of innervation in the diaphragm as well as in the intercostal and sternomastoid muscles (Bevan and Steinbach, 1977; Steinbach,

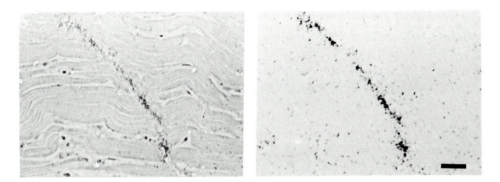

Figure 8. Autoradiograph using ^{125}I-α-bungarotoxin (right) and corresponding phase contrast (left) micrographs of myotomal muscle cells of stage 49 (12-day) *Xenopus*. The grain density is highest at the ends of the muscle cells, which are the site of innervation on these cells. The grain density is also substantial along the myotomes. Scale bar: 20 μm. (Courtesy of I. Chow.)

1981; Ziskind-Conhaim and Bennett, 1982). The background density of ACh receptors remains detectable for several more days, and the adult extrajunctional density is not acquired until 1 week postnatal (Diamond and Miledi, 1962; Bevan and Steinbach, 1977).

In chick limb muscle, ACh receptors can be detected by d4–5 (Giacobini *et al.*, 1973), and in latissimus dorsi muscle mass, receptor clusters are seen as early as d4 (Betz *et al.*, 1980). The reduction in background ACh receptor numbers in the latissimus dorsi muscles begins sometime after embryonic d14, and the adult level is not attained until after hatching (Burden, 1977a). On *Xenopus* myotomal muscle, ACh receptors first become detected by stage 19 (21 hr), and accumulations at the ends of the myotome (site of innervation) occur 3–4 hr later (stage 22) (Blackshaw and Warner, 1976; Chow and Cohen, 1983). The density of ACh receptors in the extrajunctional membrane begins to decline after stage 36 (50 hr). However, as shown in Figure 8, it is still detectable in mature myotomal muscle of these tadpoles (stage 49, 12d) (Chow and Cohen, 1983). In all three organisms under consideration, the accumulation of ACh receptors at the site of innervation occurs before a reduction in the background levels of receptors (Bevan and Steinbach, 1977; Burden, 1977a; Chow and Cohen, 1983).

The formation of receptor clusters during development is not dependent on synaptic or muscle activity. In rat intercostal muscle, a central cluster of ACh receptors appears (as do several randomly dispersed clusters elsewhere on the fibers) after blockage of muscle activity (Ziskind-Conhaim and Bennett, 1982). Complete destruction or removal of the nerve results in the appearance of several receptor clusters occurring randomly over the surface (Braithwaite and Harris, 1979; Ziskind-Conhaim and Dennis, 1981; Ziskind-Conhaim and Bennett, 1982). In ALD and PLD muscles of chicks paralyzed from d10 to d13 of embryogenesis, the ACh receptor clusters at the junction appeared normally (Betz *et al.*, 1980). Paralysis of *Xenopus* embryos at the time of innervation did not prevent the appearance of ACh receptor clusters at the presumptive site of innervation, although removal of spinal cords within the first day of innervation resulted in a reduction in receptor clusters at these sites and an increase in clusters of receptors elsewhere (Chow, 1980).

Formation of receptor clusters has also been studied in culture. In rat (Axelrod *et al.*, 1976), chick (Fischbach, 1972; Vogel *et al.*, 1972; Sytkowski *et al.*, 1973), and *Xenopus* (Anderson *et al.*, 1977) cultures, muscle cells develop regions of high receptor density in the absence of nerve cells. These patches of high ACh receptor density develop on chick and *Xenopus* muscle regardless of whether the muscle has been previously exposed to nerve *in vivo* (Bekoff and Betz, 1976; Moody-Corbett and Cohen, 1981). In rat cultures, ACh receptor patches are sometimes associated with basal lamina material, thickened membrane, and cytoplasmic fuzzy material (Salpeter *et al.*, 1982). These specializations have not been seen at sites of ACh receptor patches on chick myotubes in culture (Hourani *et al.*, 1974; Vogel and Daniels, 1976). On *Xenopus* muscle cells, the receptor patches are co-localized with discrete patches of cholinesterase (Fig. 6) (Moody-Corbett and Cohen, 1981), which are in turn associated with patches of basal lamina material and thickened membrane (Weldon *et al.*, 1981) (Fig. 7). In both chick (Hourani *et al.*, 1974; Yee *et al.*, 1978; S. A. Cohen and Pumplin, 1979) and *Xenopus* (Peng and Nakajima, 1978; Peng *et al.*, 1980) muscle cultures, patches of 10 nm particles have been seen in freeze-fracture replicas of the membrane. The particles presumably represent ACh receptor molecules and, in chick, have been co-localized with the receptor patches. In addition to discrete patches of ACh receptors, muscle cells in culture also have lower levels of receptors randomly distributed over their surface (Vogel *et al.*, 1972; Sytkowski *et al.*, 1973; Fischbach and Cohen, 1973; Anderson *et al.*, 1977; Axelrod *et al.*, 1978).

The patches of high receptor density on chick and *Xenopus* muscle cells in culture are very stable, remaining in one position from a few hours to several days (Frank and Fischbach, 1979; Moody-Corbett and Cohen, 1982a). In rat muscle, it has been further demonstrated that the mobility of ACh receptors within a patch is very low, whereas the diffuse receptors elsewhere on the surface are more mobile (Axelrod *et al.*, 1976).

Given that muscle cells develop patches of high ACh receptor density in the absence of nerve and that clustering of receptors occurs very early during development *in vivo*, it was of interest to determine whether nerve fibers preferentially seek out these sites of high receptor density. Frank and Fishbach (1979) followed ACh sensitivity on chick muscle cells contacted by spinal cord neurites. In these cultures, rather than seeking out sites of high receptor density, the neurites *induced* sites of high ACh sensitivity, although patches of high sensitivity already existed close by. This induction of receptors also occurred in the presence of curare (Rubin *et al.*, 1980), supporting the *in vivo* results that activity is not necessary for the accumulation of ACh receptors at sites of nerve–muscle contact.

In *Xenopus* cultures, muscle cells contacted by nerves have a very different pattern of receptor distribution than noncontacted muscle cells (Anderson *et al.*, 1977). In the former case, ACh receptors are often seen in long streaks at sites of nerve contact (Fig. 5), whereas noncontacted muscle cells have small oval patches of receptors distributed over their surface (Fig. 6). By following ACh receptors that had been prelabeled with fluorescent α-bungarotoxin, Anderson and Cohen (1977) demonstrated that the accumulation of receptors at sites of nerve–muscle contact arise, at least in part, from a redistribution of receptors on the surface of the muscle. This redistribution is also coincident with a loss of receptor patches elsewhere on the surface (Anderson and Cohen, 1977; Moody-Corbett and Cohen, 1982a). In *Xenopus*, as in chick cultures, the accumulation of ACh receptors at sites of nerve–muscle contact is not dependent on activity (Anderson and Cohen, 1977; Moody-Corbett and Cohen, 1982a). Similar experiments have not been performed with rat muscle and spinal cord cells, and it may be that sites of nerve–muscle contact in these cultures do not have a localization of ACh receptors (Kidokoro, 1980).

A number of aspects of the nerve–muscle interaction may be important in localizing ACh receptors. For example, there is evidence that direct surface contact may be involved (Jones and Vrbova, 1974; Jones and Vyskocil, 1975). In culture, although ACh receptor patches form on both the upper surface free of contact with a solid surface and on the lower surface next to the culture dish, the patches appear to have preferential locations. Bloch and Geiger (1980) have found that ACh receptor patches on rat myotubes are often located at sites of close contact with the culture dish (also see Axelrod, 1981). Sites of muscle contact that are strongly adherent to the culture dish have also been found to be sites of high ACh receptor density both in rat and in *Xenopus* muscle in culture (see Fig. 9) (Bloch and Geiger, 1980; Moody-Corbett and Cohen, 1982b). Peng et al. (1981) have further found that polylysine beads are able to aggregate receptors on *Xenopus* muscle cells in culture, suggesting that the polylysine molecules themselves or the positive charge of the beads are important in this phenomenon. It is interesting in this case that the beads also cause receptor patches elsewhere on the muscle surface to disappear.

That the nerve provides a specific influence on ACh receptor localization on muscle cells other than just surface contact is demonstrated by the fact that only specific neuronal cell types can induce these changes. In both chick and *Xenopus* cultures, it has been shown that cholinergic neurons can form synaptic connections and cause ACh receptors to accumulate at the site of contact, but noncholinergic neurons cannot (Hooisma et al., 1975; Betz, 1976; Obata, 1977; Fischbach et al., 1979; M. W. Cohen and Weldon, 1980; Kidokoro et al., 1980). Although it is possible that ACh is important in this behavior, other factors isolated from nerve tissue have also been implicated. In cultures of rat muscle, a high-molecular-weight extract from brain is capable of causing an increase in the number of ACh receptor clusters (Salpeter et al., 1982). Scnaffner and Daniels (1982) also reported that a high-molecular-weight extract from a variety of nervous tissues, including spinal cord, is capable of increasing the number of ACh receptor clusters on rat myotubes. In both cases, the increase in ACh receptor clusters was a result of a redistribution of receptors already present on the membrane. It is interesting to note that the extract used by

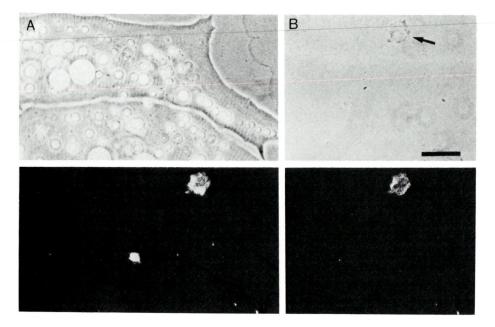

Figure 9. *Xenopus* muscle cells grown in culture in the absence of nerve. Phase-contrast and corresponding fluorescence views of portions of two muscle cells in a 5-day-old culture before (A) and after (B) treatment with 0.5 mg/ml dibucaine. Each muscle cell had an ACh receptor patch (stained using rhodamine α-bungarotoxin) on its lower surface. After dibucaine treatment, the cells rounded up (not shown) and further agitation caused the cells to tear away from the dish. Only a small fragment (arrow) of one of the cells remained attached and this fragment contained one of the ACh receptor patches. Scale bar: 20 μm. (From Moody-Corbett and Cohen, 1982b.)

Salpeter *et al.* (1982) caused an increase in the receptor density within these clusters to a density approximating that found on postnatal NMJs. However, to date there is no evidence that neurites of rat spinal cord in direct contact with rat myotubes in culture are able to induce a local accumulation of ACh receptor (see Kidokoro, 1980).

In chick cultures a low-molecular-weight component of chick brain and spinal cord is capable of increasing the number of ACh receptors by increasing the rate of incorporation of receptors (Jessel *et al.*, 1979; Fischbach and Jessel, 1979; Buc-Caron *et al.*, 1983). This material is also able to increase both the number and size of ACh receptor clusters on these cells, due in part to a redistribution of ACh receptors present on the membrane before extract treatment (Jessel *et al.*, 1979; Buc-Caron *et al.*, 1983). Markelonis *et al.* (1982) have found a much higher-molecular-weight component of sciatic nerve, **sciatin,** which is very similar to transferrin (Oh and Markelonis, 1982), that enhances ACh receptor incorporation and ACh receptor clustering on chick myotubes in culture. The role of these factors during the formation of ACh receptor aggregates *in vivo* has yet to be examined. It will be important to determine whether any of these factors is present in significant amounts in the growing nerve terminal to alter the local accumulation of ACh receptors.

3. Development of the Functional Neuromuscular Junction

3.1. Release of Neurotransmitter

During the very early stages of development of these three animals, it has been possible to detect signs of synaptic transmission. Miniature end-plate potentials can be recorded from the intercostal muscles of 14d rat embryos and from the diaphragm of 15d embryos (Bennett and Pettigrew, 1974; Braithwaite and Harris, 1979; Dennis et al., 1981; Ziskind-Conhaim and Dennis, 1981). In chick latissimus dorsi muscles, mepps and epps can be recorded as early as d9 (Bennett and Pettigrew, 1974). However, there is evidence for synaptic activity earlier than this in chicken. For instance, body movements correlated with spinal cord activity could be seen as early as d4 of embryogenesis (Ripley and Provine, 1972), and spinal cord plus somite grafts taken as early as d3 show twitching (Alcerno, 1965). Stimulation of the spinal roots resulted in contraction of the limb muscle masses as early as d5, which can be blocked by curare (Landmesser and Morris, 1975). The first signs of synaptic activity in *Xenopus* myotomal muscle occurs at stage 21–22 (22.5–24 hr), which can be blocked by curare (Blackshaw and Warner, 1976; Kullberg et al., 1977).

In all three systems, the amplitude of the mepps showed considerable variation, including the occurrence of high-amplitude events (Bennett and Pettigrew, 1974; Kullberg et al., 1977; Dennis et al., 1981). These higher-amplitude events are in part due to the high-input resistance of the cells (Dennis et al., 1981) and to depolarization of the presynaptic terminal, as evidenced by the fact that they can be eliminated by tetrodotoxin (Kullberg et al., 1977; Dennis et al., 1981). In rat and chick muscle, the frequency of mepps is low at the earliest times recorded and increases gradually to adult values (Diamond and Miledi, 1962; Bennett and Pettigrew, 1974; Dennis et al., 1981). In *Xenopus* muscle, the frequency is relatively high, even at the early stages (Kullberg et al., 1977). The quantal content during early synapse formation has only been examined at the rat intercostal muscles. Dennis et al. (1981) found that as early as d15 the quantal content was similar to that determined 1 week postnatal.

Synaptic activity can also be detected very early during development of nerve–muscle associations in culture. Within 105 min of co-culturing rat nerve and muscle cells, mepps could be recorded (Kidokoro, 1980). Cohen (1980) was able to record a postsynaptic depolarization after just 30 min of nerve–muscle contact in chick cultures, and Kidokoro and Yeh (1982) measured epps only minutes after the nerve growth cones advanced on myotomal muscle cells in *Xenopus* cultures. In all three culture systems, this early synaptic activity could be blocked by curare. As *in vivo*, the mepp frequency is low initially and then increases with days in culture (Kano and Shimada, 1971; Fischbach, 1972; Obata, 1977; Kidokoro, 1980; Kidokoro et al., 1980; Nakajima et al., 1980). A further similarity was the variability in the mepp amplitudes, with a number of high-amplitude events. Quantal content measured in both chick (Fischbach, 1972) and *Xenopus* (Kidokoro et al., 1980) cultures was found to be low.

In culture, it has been possible to look more closely at the association between synapse specializations and synaptic activity. For example, in rat co-cultures of spinal cord cells and muscle, mepps could be recorded at a low frequency at a time when only a few vesicles were seen presynaptically, no basal lamina material was present, and the postsynaptic membrane showed no specializations (Nakajima *et al.*, 1980). Although ACh receptor density at sites of nerve–muscle contact has not been directly measured in these cultures, it would appear from examining the properties of the mepps that there is not a significant accumulation of receptors at these junctions (Kidokoro and Patrick, 1978; Kidokoro, 1980).

In chick co-cultures, mepps and epps are commonly recorded at sites of high ACh sensitivity (S. A. Cohen and Fischbach, 1977; Fischbach *et al.*, 1979); however, using signal averaging, small depolarizations have been detected at sites at which there was not a high ACh sensitivity (S. A. Cohen, 1980). As in cultures of rat tissue, these results indicate that an aggregation of ACh receptors is not necessary for transmitter release. Furthermore, it has recently become possible to record transmitter release from growth cones that are not in contact with muscle cells. In these experiments, an excised patch of muscle membrane rich in ACh receptors is placed near the growing nerve tip. Single-channel events activated by ACh released from the nerve can then be recorded. Using these patch electrodes, Hume *et al.* (1983) detected ACh release from growth cones and cell bodies of chick ciliary ganglion cells. [These cells also form synaptic connections with muscle in culture (see Hooisma *et al.*, 1975; Betz, 1976).] Thus, direct contact with the postsynaptic membrane is not necessary for the release of transmitter.

In *Xenopus* cultures, it has been possible to measure the appearance of mepps correlated with the accumulation of ACh receptors as determined by binding with fluorescently labeled α-bungarotoxin (Anderson *et al.*, 1979; Kidokoro *et al.*, 1980). In these co-cultures, mepps can be detected before receptor clustering, and as receptor accumulation along the path of nerve–muscle contact increases, there is an increase in mepp frequency and amplitude (Kidokoro *et al.*, 1980). Using the above-described technique of recording single ACh receptor channel events from an excised patch of membrane, Young and Poo (1983) found that growth cones of *Xenopus* spinal cord cells in culture release a substance that resembles ACh in its activation of channels. Therefore, in *Xenopus*—as in chick—direct contact of the nerve–muscle membranes is not necessary for the release of transmitter. These results raise the possibility that during normal development, the exploring growth cone secretes ACh, ensuring that the earliest contacts with a receptive membrane will be effective in producing a response.

3.2. Activity of Acetylcholinesterase

The high density of AChE present at the adult NMJ limits the amount of available ACh by hydrolyzing the neurotransmitter. This reduces the likelihood of repeated bindings of ACh to its receptor, thereby keeping the duration of the synaptic event brief (Katz and Miledi, 1973). Histochemically, AChE becomes apparent sometime after synaptic activity has been initiated (see Section 2.2). Recordings of synaptic potentials indicate that a change in the time course of these potentials occurs with age, suggesting a change in either the presence of AChE or in the underlying ACh receptor channel kinetics, or both. This section presents physiological evidence for a developmental change in AChE, and the following section describes the developmental change in receptor channel kinetics.

An early study by Diamond and Miledi (1962) first demonstrated that the rise times and decays of mepps recorded from rat diaphragm muscle before birth were longer than those recorded after birth. These findings were confirmed by Bennett and Pettigrew (1974), who found that despite the fast time course of some mepps even at d16, corresponding with a time at which AChE first appears histochemically, it was not until 4 weeks postnatal that all the recorded mepps were fast. In chick ALD and PLD muscle, these investigators were able to record synaptic potentials with a fast time course as early as d9, slightly before the first histochemical evidence of AChE in these muscles. Potentials recorded from *Xenopus* myotomal muscle initially have long time courses, although there is a considerable amount of variability (Kullberg *et al.*, 1977, 1980). By stage 34 (45 hr), the mepps are substantially shorter, and by stage 39 (57 hr), the recorded mepps have time courses characteristic of adult muscle. In order to determine the extent to which these changes are due to the presence of AChE, Kullberg *et al.* (1980) examined the effects of the cholinesterase inhibitor neostigmine on the time course of mepps and epps at different stages of development. No significant effects were seen at stages 24–26 (26–30 hr), but by stage 32–34 (40–45 hr) neostigmine prolonged the time course of the potentials, indicating the development of AChE.

The appearance of AChE at sites of nerve–muscle contact has also been studied by examining the change in the time course of synaptic potentials in culture. In rat cultures, there was no physiological evidence for the appearance of AChE at sites of contact, as was also the case for the accumulation of ACh receptors (see Section 2.3). Histochemically, AChE has been demonstrated in these cultures, but there is no evidence that this cholinesterase is localized at the path of nerve–muscle contact (see Section 2.2). In chick co-cultures, Rubin *et al.* (1979) have demonstrated that the histochemical appearance of AChE occurs concurrently with a shortening of epps. By combining histochemistry with physiology, these workers found that most synapses with fast synaptic potentials had AChE staining at the site of innervation, whereas none of the synapses with a slow time course showed cholinesterase staining. Furthermore, only the fast synaptic potentials were affected by the AChE inhibitor methanesulfonyl fluoride. As noted in Section 2.2, the histochemical appearance of AChE in these cultures could be prevented by growing the cells in curare (Rubin *et al.*, 1980). Consistent with this result is the finding that the time course of the synaptic potentials in these cultures remained slow as well

(Rubin et al., 1980). In Xenopus co-cultures, although AChE has been demonstrated histochemically at sites of nerve–muscle contact (see Section 2.2), the time course of mepps or epps was not examined at the same time.

3.3. Properties of the Acetylcholine Receptor

ACh receptors undergo a change in metabolic stability during development. In adult muscle of rat and chicken, degradation rates for ACh receptors at the junction have been calculated to have a very slow half-life ($t1/2 > 120$ hr) compared with receptors that appear in the extrajunctional membrane after denervation ($t1/2 = 7-30$ hr) (see Fambrough, 1979; Pumplin and Fambrough, 1982). However, early in development of rat ectopic synapses (Reiness and Weinberg, 1981) and chick muscle during normal development (Burden, 1977a,b; Betz et al., 1980), all ACh receptors have fast degradation rates. In rat muscles (diaphragm and soleus), the degradation rate of receptors at the junction becomes slow by birth and in the first postnatal week, whereas the extrajunctional receptors are still being rapidly degraded (Steinbach et al., 1979; Michler and Sakmann, 1980; Reiness and Weinberg, 1981). In chick muscle, ACh receptors of ALD, PLD, and breast muscle are still rapidly degraded after hatching, and the degradation rate of the junctional receptors does not reach adult levels until 5 weeks posthatch (Burden, 1977a,b; Betz et al., 1980). The degradation of ACh receptors on the myotomal muscle of Xenopus has also been reported to become slower during development after stages 26–27 (29–31 hr) (Frair and Cohen, 1981). Therefore, in all three organisms, metabolic stabilization of the ACh receptors occurs after receptor clustering at the NMJ has occurred.

To what extent the nerve may play a role in this change is not clear. In cultures of embryonic muscle, the metabolic rate of the receptors does not change (Devreotes and Fambrough, 1975; Brehm et al., 1983b). In denervated muscle, newly inserted ACh receptors (even at the NMJ) appear to have rapid degradation rates (Brett et al., 1982; Loring and Salpeter, 1982). Recently, it has been reported that a small population of ACh receptors at the NMJ of normally innervated rat muscle also have fast degradation rates (Stanley and Drachman, 1983). These results have suggested that the nerve provides a factor that metabolically stabilizes ACh receptors, which initially have fast degradation rates (Stanley and Drachman, 1983). The results of the developmental studies would indicate that this factor is not immediately available on nerve contact but requires a period of time to become established.

There is also a change in the biophysical properties of the ACh receptor channels during development. In rat muscles (diaphragm, soleus, omohyoideus), the open time of the receptor channels at the NMJ at birth is four times adult values (Fischbach and Schuetze, 1980; Michler and Sakmann, 1980). In the second postnatal week, there begins a shift to faster open times, which is complete by the third week (Fischbach and Schuetze, 1980; Michler and Sakmann, 1980). Recently, a similar change has been reported for ACh receptor channels at ectopic synapses of soleus muscle (Brenner and Sakmann, 1983) as well as the nonjunctional membrane of innervated limb muscle (Schuetze and Vicini, 1983). However, channel conversion can be blocked on rat limb muscle by denervation, indicating a role of innervation in this process (Schuetze and Vicini, 1983). In contrast to the situation in rat, no change was seen in the open time of ACh receptor channels on chick muscle (ALD, PLD, and breast muscles) from d17 of embryogenesis to 14 weeks posthatch (Schuetze, 1980; Harvey and van Helden, 1981), although there was a gradual increase in the conductance of the receptor channels from both ALD and PLD muscles (Harvey and van Helden, 1981). In *Xenopus* myotomal muscle, the kinetics of the ACh receptor channels early in embryogenesis (stage 24, 26 hr) (like those in rat muscle) are slow, and within 30 hr they have become fast (Kullberg *et al.*, 1981). This change in ACh receptor channel kinetics occurs in the receptors clustered at the junction as well as those in the nonjunctional regions of this muscle (Kullberg *et al.*, 1981; Brehm *et al.*, 1984a).

The properties of ACh receptor channels have also been examined in cell culture. Using fluctuation analysis, Fischbach and Schuetze (1980) did not detect a change in the kinetics of the ACh receptor channels on rat myotubes grown in the absence of nerve. More recently, however, recording of single ACh-activated channels has demonstrated that both high-conductance channels with short open times and low-conductance channels with longer open times exist on rat muscle cells in culture (Hamill and Sakmann, 1981). Although the relative proportions of these two classes of receptors have not been examined in these cultures, the results of Fischbach and Schuetze (1980) suggest that the predominant class at all ages has slow kinetics. As *in vivo*, the ACh receptor channels on chick myotubes in culture exhibit slow kinetics regardless of whether receptors are aggregated or muscle cells are innervated (Schuetze *et al.*, 1978). In contrast to these findings, a shift is seen in the kinetics of ACh receptor channels on *Xenopus* muscle cells grown in culture. Even in the absence of nerve, fluctuation analysis reveals a predominance of kinetically slow processes initially and later in culture life a predominance of fast events (Brehm *et al.*, 1982). In fact, addition of neural tube cells causes the kinetics to become slow again (Brehm *et al.*, 1982). The change in ACh receptor channel properties in *Xenopus* cultures has been further demonstrated by recording single-channel events (Brehm *et al.*, 1983a; Moody-Corbett *et al.*, 1983). Two functionally distinct ACh receptor channel populations are present at all ages examined in these cultures (Fig. 10). However, initially most channels have a low-conductance and long-channel open time, whereas in older cultures the majority of channels have a high-conductance and short-channel open time.

Whether channel conversion is due to a modification of one channel type to another or to a difference in the relative proportions of two genetically distinct channels has not been directly determined. Based on a comparison of ACh receptor incorporation and degradation rates on rat muscle with the change in channel properties, Michler and Sakmann (1980) have suggested that channel conversion is not due to a simple replacement of one channel type by another. However, recent studies examining single ACh receptor channels on

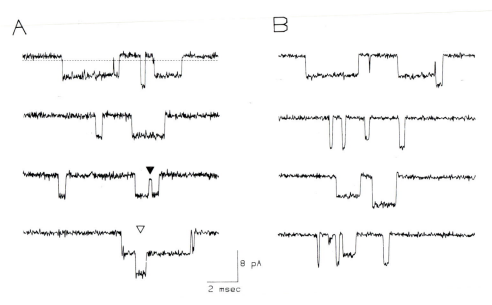

Figure 10. ACh receptor channel currents elicited by 0.2 μM ACh from "cell-attached" patches of 1-day-old (left) and 6-day-old (right) cultured *Xenopus* muscle cells. In the top trace (A), the threshold for event discrimination is indicated by a dashed line, computed to be 2.5 times the standard deviation of baseline fluctuations. The data were digitized at 50-μsec intervals and were redisplayed in analogue fashion on a Hewlett–Packard monitor. The pipette potential was +40 mV, and the intracellular potential was approximately −85 mV in both recordings. This results in an estimated patch potential of −125 mV. An example of an opening to a subconductance state is indicated by the filled triangle, and the coincident opening by two low conductance channels is indicated by the open triangle. (From Brehm *et al.*, 1984*b*.)

Xenopus muscle cells lend support to the view that the change in kinetics seen during development is due to a change in the proportion of two distinct receptor channel classes (Brehm *et al.*, 1982, 1983*a*, 1984*b*).

4. Summary

The sequence of events leading to the formation of the NMJ based on the data presented in this chapter from rat, chick, and *Xenopus* muscle can be divided into three developmental stages, as shown in Table I.

The essential components of the NMJ are acquired early. Acetylcholine is present and can be released from the growing nerve. Acetylcholine receptors are present in the muscle membrane and are functional even at the earliest times. These components of the junction—ACh release and functional ACh receptors—can develop independently of each other; i.e., cell culture studies have shown that nerve cells are capable of releasing ACh before their growing tips have come into contact with the postsynaptic muscle membrane. Conversely, muscle cells grown without nerve synthesize and incorporate in their membranes functional ACh receptors. This situation ensures that functional

Table I. Acquisition of Synaptic Characteristics Grouped into Three
Developmental Stages

	Stage I	Stage II	Stage III
	Transmission AChR[a] clustering	AChE clustering Basal lamina Vesicle clustering Membrane thickening	AChR metabolic stability AChR channel conversion Postsynaptic folds
Rat	d14–16[b]	d16–18	d19–3 wk postnatal
Chick	d4–9[b]	d8–13	d14–5 wk posthatch
Xenopus	21–24 hr[c]	22–31 hr	After 31 hr

[a]Acetylcholine receptor.
[b]Embryonic days.
[c]Time postfertilization.

contacts can occur even at the earliest times. Local accumulation of ACh receptors is also detected at the earliest times of junction formation. Although cell culture studies have demonstrated that receptors can aggregate in the absence of nerve, it would appear that the nerve plays an important role in directing where the highest density of receptors will be localized.

Acetylcholinesterase, identified both histochemically and electrophysiologically, occurs at the presumptive NMJ shortly after synaptic transmission and receptor clustering have begun, suggesting that these events may play a role in localizing cholinesterase. Although the studies on rat and chick muscle support this view, development of AChE on *Xenopus* muscle does not require prior exposure to nerve or muscle activity.

The ultrastructural features characteristic of the adult NMJ also do not become apparent until after synaptic transmission and receptor clustering have been seen. However, detection of small regions of specialization could be easily overlooked at the ultrastructural level, particularly if the tissue has not been serially sectioned. The young tissue is more fragile (Gordon *et al.*, 1974) and may be more susceptible to mechanical damage or alterations from the fixation procedures (Kullberg *et al.*, 1977). For these reasons, results pertaining to when the ultrastructural specializations occur are difficult to interpret and must await identification of these structures by other means.

A number of other changes occur at the NMJ late in development: (1) ACh receptors become metabolically more stable, (2) there is a conversion in the kinetics of the ACh receptor channel, and (3) junctional folds become apparent. The extent to which these changes occur varies among the different organisms discussed. Metabolically stable ACh receptors are present a few days after transmission begins in the rat, but in chick muscle they are not seen for several weeks. The change in ACh receptor channel kinetics seen in rat and *Xenopus* muscle is not apparent in chick muscle. In some cases, the postsynaptic membrane develops extensive folds, whereas in other muscle no folds are present. As the acquisition of these final characteristics is more unique to the individual systems or organisms, understanding how they are regulated will require more careful examination of changes within individual junctions. Clearly, the acquisition of the final characteristics of the NMJ is a complex process that may reflect different degrees of regulation from muscle to muscle or species to species.

ACKNOWLEDGMENTS. The author thanks the Muscular Dystrophy Association of Canada for postdoctoral research support and Dr. P. Brehm, Dr. D. Corbett, and Dr. J. Matthews–Bellinger for helpful comments on the manuscript.

References

Alcerno, B. B., 1965, The nature of the earliest spontaneous activity of the chick embryo, *J. Embryol. Exp. Morphol.* **13**:255–266.

Anderson, M. J., and Cohen, M. W., 1977, Nerve-induced and spontaneous redistribution of acetylcholine receptors on cultured muscle cells, *J. Physiol. (Lond.)* **268**:757–773.

Anderson, M. J., and Fambrough, D. M., 1983, Aggregates of acetylcholine receptors are associated with plaques of a basal lamina heparan sulfate proteoglycan on the surface of skeletal muscle fibers, *J. Cell Biol.* **97**:1396–1411.

Anderson, C. R., and Stevens, C. F., 1973, Voltage clamp analysis of acetylcholine produced endplate current fluctuations at frog neuromuscular junction, *J. Physiol. (Lond.)* **235**:655–691.

Anderson, M. J., Kidokoro, Y., and Gruener, R., 1979, Correlation between acetylcholine receptor localization and spontaneous synaptic potentials in cultures of nerve and muscle, *Brain Res.* **166**:185–190.

Anderson, M. J., Cohen, M. W., and Zorychta, E., 1977, Effects of innervation on the distribution of acetylcholine receptors on cultured amphibian muscle cells, *J. Physiol. (Lond.)* **268**:731–756.

Atsumi, S., 1977, Development of neuromuscular junctions of fast and slow muscles in the chick embryo: A light and electron microscope study, *J. Neurocytol.* **6**:691–709.

Axelrod, D., 1981, Cell–substrate contacts illuminated by total internal reflection fluorescence, *J. Cell Biol.* **89**:141–145.

Axelrod, D., Ravdin, P., Koppel, D. E., Schlessinger, J., Webb, W. W., Elson, E. L., and Podleski, T. R., 1976, Lateral motion of fluorescently labelled acetylcholine receptors in membranes of developing muscle cells, *Proc. Natl. Acad. Sci. USA* **73**:4594–4598.

Axelrod, D., Ravdin, P., and Podleski, T. R., 1978, Control of acetylcholine receptor mobility and distribution in cultured muscle membranes, *Biochim. Biophys. Acta* **511**:23–38.

Barnard, E. A., Lai, J., and Pizzey, J., 1984, Synaptic and extrasynaptic forms of acetylcholinesterase in skeletal muscles: Variation with fiber type and functional considerations, in: *Neuromuscular Diseases* (G. Serratrice, D. Cros, C. Desnuelle, J.-L. Gastaut, J.-F. Pellisier, J. Pouget, and A. Schiano, eds.), pp. 455–463, Raven Press, New York.

Bekoff, A., and Betz, W. H., 1976, Acetylcholine hotspots: Development on myotubes cultured from aneural limb buds, *Science* **193**:915–917.

Bennett, M. R., 1983, Development of neuromuscular synapses, *Physiol. Rev.* **63**:915–1048.

Bennett, M. R., and Pettigrew, A. G., 1974, The formation of synapses in striated muscle during development, *J. Physiol. (Lond.)* **241**:515–545.

Betz, W., 1976, The formation of synapses between chick embryo skeletal muscle and ciliary ganglia grown *in vitro*, *J. Physiol. (Lond.)* **254**:63–73.

Betz, W., and Sakmann, B., 1973, Effects of proteolytic enzymes on function and structure of frog neuromuscular junction, *J. Physiol. (Lond.)* **230**:673–688.

Betz, H., Bourgeois, J.-P., and Changeux, J.-P., 1980, Evolution of cholinergic proteins in developing slow and fast skeletal muscle in chick embryo, *J. Physiol. (Lond.)* **302**:197–218.

Bevan, S., and Steinbach, J. H., 1977, The distribution of alpha-bungarotoxin in binding sites on mammalian skeletal muscle developing *in vivo*, *J. Physiol. (Lond.)* **267**:195–213.

Blackshaw, S., and Warner, A., 1976, Onset of acetylcholine sensitivity and endplate activity in developing myotome muscles of *Xenopus*, *Nature (Lond.)* **262**:217–218.

Bloch, R., and Geiger, B., 1980, Localization of acetylcholine receptor clusters in areas of cell–substrate contact in cultures of rat myotubes, *Cell* **211**:25–35.

Bowden, R., and Duchen, L. W., 1976, The anatomy and pathology of the neuromuscular junction, in: *Neuromuscular Junction* (E. Zaimis, ed.), pp. 1–92, Springer-Verlag, New York.

Braithwaite, A. W., and Harris, A. J., 1979, Neural influence on acetylcholine receptor clusters in embryonic development of skeletal muscle, *Nature (Lond.)* **279**:549–551.

Brehm, P., Steinbach, J. H., and Kidokoro, Y., 1982, Channel open time of acetylcholine receptors on *Xenopus* muscle cells in dissociated cell cultures, *Dev. Biol.* **91**:93–102.

Brehm, P., Moody-Corbett, F., and Kidokoro, Y., 1983a, Developmental alterations in acetylcholine receptor channel properties proceed in the absence of innervation, *Neurosci. Abs.* **9**:1180.

Brehm, P., Yeh, E., Patrick, J., and Kidokoro, Y., 1983b, Metabolism of acetylcholine receptors on embryonic amphibian muscle, *J. Neurosci.* **3**:101–107.

Brehm, P., Kullberg, R., and Moody-Corbett, F., 1984a, Properties of non-junctional acetylcholine receptor channels on innervated muscle of *Xenopus laevis*, *J. Physiol. (Lond.)* **350**:631–648.

Brehm, P., Kidokoro, Y., and Moody-Corbett, F., 1984b, Acetylcholine receptor channel properties during development of *Xenopus* muscle cells in culture, *J. Physiol. (Lond.)* **357**:203–217.

Brenner, H., and Sakmann, B., 1983, Neurotrophic control of channel properties at neuromuscular synapses of rat muscle, *J. Physiol. (Lond.)* **337**:159–171.

Brett, R., Younkin, S. G., Knoieczkowski, and Slugg, R. M., 1982, Accelerated degradation of junctional acetylcholine receptor-alpha-bungarotoxin complexes in denervated rat diaphragm, *Brain Res.* **233**:133–142.

Buc-Caron, M.-H., Nystrom, P., and Fischbach, G. D., 1983, Induction of acetylcholine receptor synthesis and aggregation: Partial purification of low molecular weight activity, *Dev. Biol.* **95**:378–386.

Burden, S. J., 1977a, Development of the neuromusuclar junction in the chick embryo: The number, distribution, and stability of acetylcholine receptors, *Dev. Biol.* **57**:317–329.

Burden, S. J., 1977b, Acetylcholine receptors at the neuromuscular junction: Developmental change in receptor turnover, *Dev. Biol.* **61**:79–85.

Burden, S. J., 1982, Identification of an intracellular postsynaptic antigen at the frog neuromuscular junction, *J. Cell Biol.* **94**:521–530.

Burden, S. J., Sargent, P. B., and McMahan, U. J., 1979, Acetylcholine receptors in regenerating muscle accumulate at original synaptic sites in the absence of the nerve, *J. Cell Biol.* **82**:412–425.

Carson, S., Bon, S., Vigny, M., Massoulie, J., and Fardeau, M., 1979, Distribution of acetyl-cholinesterase molecular forms in neural and non-neural sections of human muscle, *FEBS Lett.* **97**:348–352.

Chow, I., 1980, Distribution of acetylcholine receptors in the myotomes of *Xenopus laevis* during normal development, Ph.D. thesis, McGill University, Montreal, Canada.

Chow, I., and Cohen, M. W., 1983, Developmental changes in the distribution of acetylcholine receptors in the myotomes of *Xenopus laevis*, *J. Physiol. (Lond.)* **339**:553–571.

Cisson, C. M., McQuarrie, C. H., Sketelj, J., McNamee, M. G., and Wilson, B. W., 1981, Molecular forms of acetylcholinesterase in chick embryonic fast muscle: Developmental changes and effects of DFP treatment, *Dev. Neurosci.* **4**:157.

Cohen, M. W., 1981, Development of an amphibian neuromuscular junction *in vivo* and in culture, *J. Exp. Biol.* **89**:43–56.

Cohen, M. W., and Weldon, P. R., 1980, Localization of acetylcholine receptors and synaptic ultrastructure at nerve-muscle contacts in culture: Dependence on nerve type, *J. Cell Biol.* **86**:388–401.

Cohen, M. W., Greschner, M., and Tucci, M., 1984, *In vivo* development of cholinesterase at an amphibian neuromuscular junction in the absence of motor activity, *J. Physiol. (Lond.)* **348**:57–66.

Cohen, S. A., 1980, Early nerve–muscle synapses *in vitro* release transmitter over postsynaptic membrane having low acetylcholine sensitivity, *Proc. Natl. Acad. Sci. USA* **77**:644–648.

Cohen, S. A., and Fischbach, G. D., 1977, Clusters of acetylcholine receptors located at identified nerve-muscle synapses *in vitro*, *Dev. Biol.* **59**:24–38.

Cohen, S. A., and Pumplin, D. W., 1979, Clusters of intramembranous particles associated with binding sites for alpha-bungarotoxin in cultured chick myotubes, *J. Cell Biol.* **82**:494–516.

Couteaux, R., 1981, Structure of the subsynaptic sarcoplasm in the interfolds of the frog neuromuscular junction, *J. Neurocytol.* **10**:947–962.

Dennis, M. J., Ziskind-Conhaim, L., and Harris, A. J., 1981, Development of neuromuscular junctions in rat embryos, *Dev. Biol.* **81**:266–279.

Devreotes, P. N., and Fambrough, D. M., 1975, Acetylcholine receptor turnover in membranes of developing muscle fibers, *J. Cell Biol.* **65**:335–358.

Diamond, J., and Miledi, R., 1962, A study of fetal and new-born rat muscle fibers, *J. Physiol. (Lond.)* **162**:393–408.

Dreyer, F., and Peper, K., 1974, The acetylcholine sensitivity of the neuromuscular junction of the frog, *Pflugers Arch.* **398**:273–286.

Dreyer, F., Peper, K., Akert, K., Sandri, C., and Moor, H., 1973, Ultrastructure of the "active zone" in the frog neuromuscular junction, *Brain Res.* **62**:373–380.

Ellisman, M. H., Rash, J. E., Staehlin, L. A., and Porter, K. R., 1976, Studies of excitable membranes. II. A comparison of specializations at neuromuscular junctions and non-junctional sarcolemmac of mammalian fast and slow twitch muscle fibers, *J. Cell Biol.* **68**:752–774.

Fambrough, D. M., 1979, Control of acetylcholine receptors in skeletal muscle, *Physiol. Rev.* **59**:165–227.

Fertuck, H. C., and Salpeter, M. M., 1974, Localization of acetylcholine receptors by ^{125}I-alpha-bungarotoxin binding at mouse motor endplates, *Proc. Natl. Acad. Sci. USA* **71**:1376–1378.

Fertuck, H. C., and Salpeter, M. M., 1976, Quantitation of junctional and extrajunctional acetylcholine receptors by electron microscope autoradiography after ^{125}I-alpha-bungarotoxin binding at mouse neuromuscular junctions, *J. Cell Biol.* **69**:144–158.

Filogama, G., and Gabella, G., 1967, The development of neuromuscular correlations in vertebrates, *Arch. Biol.* **78**:9–60.

Fischbach, G. D., 1972, Synapse formation between dissociated nerve and muscle cells in low density cell cultures, *Dev. Biol.* **28**:407–429.

Fischbach, G. D., and Cohen, S. A., 1973, The distribution of acetylcholine sensitivity over uninnervated and innervated muscle fibers grown in cell culture, *Dev. Biol.* **31**:147–162.

Fischbach, G. D., and Jessel, T. L., 1979, Induction of acetylcholine receptors and receptor cluster in embryonic myotubes, in: *Ontogenesis and Functional Mechanisms of Peripheral Synapses*, INSERM Symposium 13 (J. Taxi, ed.), pp. 301–311, Elsevier/North-Holland, Amsterdam.

Fischbach, G. D., and Schuetze, S. M., 1980, A post-natal decrease in acetylcholine channel open time at rat end-plates, *J. Physiol. (Lond.)* **303**:125–137.

Fischbach, G. D., Fambrough, D., and Nelson, P. G., 1973, A discussion of neuron and muscle cell cultures, *Fed. Proc.* **32**:1636–1642.

Fischbach, G. D., Cohen, S. A., and Henkart, M. P., 1974, Some observations on trophic interaction between neurons and muscle fibers in cell culture, *Ann. N.Y. Acad. Sci.* **228**:35–46.

Fischbach, G. D., Frank, E., Jessel, T. M., Rubin, L. L., and Schuetze, S. M., 1979, Accumulation of acetylcholine receptors and cholinesterase at newly formed nerve–muscle synapses, *Pharmacol. Rev.* **30**:411–428.

Frair, P., and Cohen, M. W., 1981, Degradation of acetylcholine receptors in innervated myotomes of *Xenopus laevis* maintained in organ culture, *Neurosci. Abs.* **7**:838.

Frank, E., and Fischbach, G. D., 1979, Early events in neuromuscular junction formation *in vitro*. Induction of acetylcholine receptor clusters in the postsynaptic membrane and morphology of newly formed synapses, *J. Cell Biol.* **83**:143–158.

Giacobini, G., 1972, Embryonic and postnatal development of choline acetyltransferase activity in muscles and sciatic nerve of the chick, *J. Neurochem.* **19**:1401–1403.

Giacobini, G., Filogama, G., Weber, M., Boquet, P., and Changeux, J.-P., 1973, Effects of a snake alpha-neurotoxin on the development of innervated skeletal muscles in chick embryo, *Proc. Natl. Acad. Sci. USA* **70**:1708–1712.

Giacobini-Robecchi, M. G., Giacobini, G., Filogama, G., and Changeux, J.-P., 1975, Effects of the type A toxin from *Clostridium botulinum* on the development of skeletal muscles and their innervation in chick embryo, *Brain Res.* **83**:107–121.

Gordon, T., Perry, R., Taffery, A. R., and Vrbova, G., 1974, Possible mechanisms determining synapse formation in developing skeletal muscle of the chick, *Cell Tissue Res.* **155**:13–25.

Guth, L., Zalewski, A. A., and Brown, W. C., 1966, Qualitative changes in cholinesterase activity of denervated soleplates following implantation of nerve into muscle, *Exp. Neurol.* **16**:136–147.

Hall, Z. W., 1973, Multiple forms of acetylcholinesterase and their distribution in endplate and non-endplate regions of rat diaphragm muscle, *J. Neurobiol.* **4**:343–361.

Hall, Z. W., and Kelly, R. B., 1971, Enzymatic detachment of endplate acetylcholinesterase form muscle, *Nature New Biol.* **232**:62–63.

Hamill, O. P., and Sakmann, B., 1981, Multiple conductance states of single acetylcholine receptor channels in embryonic muscle cells, *Nature (Lond.)* **294**:462–464.

Harvey, A. L., and van Helden, D., 1981, Acetylcholine receptors in singly and multiply innervated skeletal muscle fibers of the chicken during development, *J. Physiol. (Lond.)* **317**:397–411.

Heuser, J. E., Reese, T. S., and Landis, D. M. D., 1974, Functional changes in frog neuromuscular junctions studied with freeze-fracture, *J. Neurocytol.* **3**:109–131.

Hirano, H., 1967a, Ultrastructural study on the morphogenesis of the neuromuscular junction in the skeletal muscle of the chick, *Z. Zellforsch.* **79**:198–208.

Hirano, H., 1967b, A histochemical study of the cholinesterase activity in the neuromuscular junction in developing chick skeletal muscles, *Arch. Histol. Jpn.* **28**:89–101.

Hooisma, J., Slaaf, D. W., Meeter, E., and Stevens, W. F., 1975, The innervation of chick striated muscle fibers by the chick ciliary ganglion in tissue culture, *Brain Res.* **85**:79–85.

Hourani, B. T., Tourain, B. F., Henkart, M. P., Carter, R. L., Marchesi, V. T., and Fischbach, G. D., 1974, Acetylcholine receptors of cultured muscle cells demonstrated with ferritin-alpha-bungarotoxin conjugates, *J. Cell Sci.* **16**:473–479.

Hume, R. I., Role, L. W., and Fischbach, G. D., 1983, Acetylcholine release from growth cones detected with patches of acetylcholine receptor rich membranes, *Nature (Lond.)* **305**:632–633.

Inestrosa, N., Ziskind-Conhaim, L., and Hall, Z. W., 1984, Acetylcholinesterase in developing muscle fibers, in: *Neuromuscular Diseases* (G. Serratrice, D. Cros, C. Desnuelle, J.-L. Gastaut, J.-F. Pellisier, J. Pouget, and A. Schiano, eds.), pp. 437–441, Raven Press, New York.

Jacob, M., and Lentz, T. L., 1979, Localization of acetylcholine receptor by means of horseradish peroxidase-alpha-bungarotoxin during formation and development of the neuromuscular junction in the chick embryo, *J. Cell Biol.* **82**:195–211.

Jessel, T. M., Siegel, R. E., and Fischbach, G. D., 1979, Induction of acetylcholine receptors on cultured skeletal muscle by a factor extracted from brain and spinal cord, *Proc. Natl. Acad. Sci. USA* **76**:5397–5401.

Jones, R., and Vrbova, G., 1974, Two factors responsible for the development of denervation hypersensitivity, *J. Physiol. (Lond.)* **236**:517–538.

Jones, R., and Vyskocil, F., 1975, An electrophysiological examination of the changes in skeletal muscle fibers in response to degenerating nerve tissue, *Brain Res.* **88**:309–317.

Juntunen, J., 1974, Induction of the postsynaptic membrane, *Med. Biol.* **52**:164–169.

Kano, M., and Shimada, Y., 1971, Innervation and acetylcholine sensitivity of skeletal muscle cells differentiated *in vitro* from chick embryo, *J. Cell Physiol.* **78**:233–248.

Kao, I., and Drachman, D. B., 1977, Myasthenic immunoglobulin accelerates acetylcholine receptor degradation, *Science* **196**;527–529.

Karnovsky, M. J., 1964, The localization of cholinesterase activity in rat cardiac muscle by electron microscopy, *J. Cell Biol.* **23**:217–232.

Kato, A. C., Vrachliotis, A., Fulpius, B., and Dunant, Y., 1980, Molecular forms of acetylcholinesterase in chick muscle and ciliary ganglion: Embryonic tissues and cultured cells, *Dev. Biol.* **76**:222–228.

Katz, B., 1966, *Nerve, Muscle, and Synapse*, McGraw-Hill, New York.

Katz, B., and Miledi, R., 1972, The statistical nature of the acetylcholine potential and its molecular components, *J. Physiol. (Lond.)* **224**:665–699.

Katz, B., and Miledi, R., 1973, The binding of ACh to receptors and its removal from the synaptic cleft, *J. Physiol. (Lond.)* **231**:549–574.

Kelly, A. M., and Zacks, S. I., 1969, The fine structure of motor endplate morphogenesis, *J. Cell Biol.* **42**:154–169.

Kidokoro, Y., 1980, Developmental changes of spontaneous synaptic potential properties in the rat neuromuscular contact formed in culture, *Dev. Biol.* **78**:231–241.

Kidokoro, Y., and Patrick, J., 1978, Correlation between miniature endplate potential amplitudes and acetylcholine receptor densities in the neuromuscular contact formed *in vitro*, *Brain Res.* **142**:368–373.

Kidokoro, Y., and Yeh, E., 1982, Initial synaptic transmission at the growth cone in *Xenopus* nerve–muscle cultures, *Proc. Natl. Acad. Sci. USA* **79**:6727–6731.

Kidokoro, Y., Anderson, M. J., and Gruener, R., 1980, Changes in synaptic potential properties during acetylcholine receptor accumulation and neurospecific interactions in *Xenopus* nerve–muscle cell culture, *Dev. Biol.* **78**:464–485.

Klymkowsky, M. W., Lappin, R. I., and Rubin, L. L., 1983, Biosynthesis and extracellular transport of acetylcholinesterase in primary muscle cultures, *Neurosci. Abs.* **13(1)**:344.

Koenig, J., and Rieger, F., 1981, Biochemical stability of the acetylcholinesterase molecular forms after cytochemical staining: Postnatal focalization of 16S acetylcholinesterase in rat muscle, *Dev. Neurosci.* **4**:249–257.

Koenig, J., Bournaud, R., and Rieger, F., 1979, Acetylcholinesterase and formation in striated muscle, in: *Ontogenesis and Functional Mechanisms of Peripheral Synapses* (J. Taxi, ed.), pp. 313–326, Elsevier/North-Holland, Amsterdam.

Koenig, J., deLaporte, S., Massoulie, J., and Vigny, M., 1984, Regulation of the acetylcholinesterase molecular forms in nerve-muscle cell cultures, in: *Neuromuscular Diseases* (G. Serratrice, D. Cros, C. Desnuelle, J.-L. Gastaut, J.-F. Pellisier, J. Pouget, and A. Schiano, eds.), pp. 443–446, Raven Press, New York.

Korneliussen, H., and Jansen, J. K. S., 1976, Morphological aspects of the elimination of polyneuronal innervation of skeletal muscle fibers in newborn rats, *J. Neurocytol.* **5**:591–604.

Kuffler, S. W., and Yoshikami, D., 1975, The distribution of acetylcholine sensitivity at the postsynaptic membrane of vertebrate skeletal twitch muscles: Iontophoretic mapping in the micron range, *J. Physiol. (Lond.)* **244**:703–730.

Kullberg, R. W., Lentz, T. L., and Cohen, M. W., 1977, Development of the myotomal neuromuscular junction in *Xenopus laevis*: An electrophysiological and fine-structural study, *Dev. Biol.* **60**:101–120.

Kullberg, R. W., Mickelberg, F. S., and Cohen, M. W., 1980, Contribution of cholinesterase to development decreases in the time course of synaptic potentials at an amphibian neuromuscular junction, *Dev. Biol.* **75**:255–267.

Kullberg, R. W., Brehm, P., and Steinbach, J. H., 1981, Nonjunctional acetylcholine receptor channel open time decreases during development of *Xenopus* muscle, *Nature (Lond.)* **289**:411–413.

Landmesser, L., and Morris, D. G., 1975, The development of functional innervation in the hindlimb of the chick embryo, *J. Physiol. (Lond.)* **249**:301–326.

Lomo, T., and Slater, C. R., 1980, Control of junctional acetylcholinesterase by neural and muscular influence in the rat, *J. Physiol. (Lond.)* **303**:191–202.

Loring, R. H., and Salpeter, M. M., 1980, Denervation increases turnover rate of junctional acetylcholine receptors, *Proc. Natl. Acad. Sci. USA* **77**:2293–2297.

MacKay, B., Muir, A. R., and Peters, A., 1960, Observations on the terminal innervation of segmental muscle fibers in amphibia, *Acta Anat.* **40**:1–12.

Markelonis, G. J., Oh, T. H., Eldefrawi, M. E., and Guth, L., 1982, Sciatin: A myotrophic protein increases the number of acetylcholine receptors and receptor clusters in cultured skeletal muscle, *Dev. Biol.* **89**:353–361.

Marshall, L. M., Sanes, J. R., and McMahan, U. J., 1977, Reinnervation of original synaptic sites on muscle fiber basement after disruption of the muscle cells, *Proc. Natl. Acad. Sci. USA* **74**:3073–3077.

Massoulie, J., and Bon, S., 1982, The molecular forms of cholinesterase and acetylcholinesterase in vertebrates, *Annu. Rev. Neurosci.* **5**:57–106.

Matthews-Bellinger, J., and Salpeter, M., 1978, Distribution of acetylcholine receptors at frog neuromuscular junctions with a discussion of some physiological implications, *J. Physiol. (Lond.)* **279**:197–213.

McMahan, U. J., Sanes, J. R., and Marshall, L. M., 1978, Cholinesterase is associated with the basal lamina at the neuromuscular junction, *Nature (Lond.)* **271**:172–174.

McMahan, U. J., Sargent, P. P., Rubin, L. L., and Burden, B. J., 1979, Factors that influence the organization of acetylcholine receptors in regenerating muscle are associated with the basal lamina at the neuromuscular junction, in: *Ontogenesis and Functional Mechanisms of Peripheral Synapses* (J. Taxi, ed.), pp. 345–354, Elsevier/North-Holland, Amsterdam.

Michler, A., and Sakmann, B., 1980, Receptor stability and channel conversion in the subsynaptic membrane of the developing mammalian neuromuscular junction, *Dev. Biol.* **80**:1–17.

Miledi, R., 1960, Junctional and extrajunctional acetylcholine receptors in skeletal muscle fibers, *J. Physiol. (Lond.)* **151**:24–30.

Moody-Corbett, F., and Cohen, M. W., 1981, Localization of cholinesterase at sites of high acetylcholine receptor density on embryonic amphibian muscle cells cultured without nerve, *J. Neurosci.* **1**:596–605.

Moody-Corbett, F., and Cohen, M. W., 1982a, Influence of nerve on the formation and survival of acetylcholine receptor and cholinesterase patches on embryonic *Xenopus* muscle cells in culture, *J. Neurosci.* **2**:633–646.

Moody-Corbett, F., and Cohen, M. W., 1982b, Increased adhesiveness at sites of high acetylcholine receptor density on embryonic amphibian muscle cells cultured without nerve, *J. Embryol. Exp. Morphol.* **72**:53–69.

Moody-Corbett, F., Weldon, P. R., and Cohen, M. W., 1982, Cholinesterase localization at sites of nerve contact on embryonic amphibian muscle cells in culture, *J. Neurocytol.* **11**:381–394.

Moody-Corbett, F., Brehm, P., and Kullberg, R. W., 1983, Functional properties of non-junctional acetylcholine receptors on developing muscle, *Can. Fed. Biol. Soc.* PA236.

Nakajima, Y., Kidokoro, Y., and Klier, F. G., 1980, Development of functional neuromuscular junctions *in vitro*: An ultra-structural and physiological study, *Dev. Biol.* **77**:52–72.

Nicolet, M., and Reiger, F., 1981, Formes moléculaires de l'acetylcholinesterase du muscle squelettique de Grenouille: Effets de l'enervation, *C.R. Soc. Biol.* **175**:316–322.

Nieuwkoop, P. D., and Faber, J., 1967, *Normal Table of Xenopus laevis (Daudin)*, 2nd ed., North-Holland, Amsterdam.

Obata, K., 1977, Development of neuromuscular transmission in culture with a variety of neurons and in the presence of cholinergic substances and tetrodotoxin, *Brain Res.* **119**:141–153.

Oh, T. H., and Markelonis, G. J., 1982, Chicken serum transferrin duplicates the myotrophic effects of sciatin on cultured muscle cells, *J. Neurosci. Res.* **8**:535–545.

Oppenheim, R. W., Pittman, R., Gray, M., and Maderdrut, J. L., 1978, Embryonic behavior, hatching and neuromuscular development in the chick following a transient reduction of spontaneous motility and sensory input by neuromuscular blocking agents, *J. Comp. Neurol.* **179**:619–640.

Peng, B., and Nakajima, Y., 1978, Membrane particle aggregates in innervated and non-innervated cultures of *Xenopus* embryonic muscle cells, *Proc. Natl. Acad. Sci. USA* **75**:500–504.

Peng, B., Bridgman, P. C., Nakajima, S., Greenberg, A., and Nakajima, Y., 1979, A fast development of presynaptic function and structure of the neuromusucular junction in *Xenopus* tissue culture, *Brain Res.* **167**:379–384.

Peng, B., Nakajima, Y., and Bridgman, P. C., 1980, Development of the post-synaptic membrane in *Xenopus* neuromuscular cultures observed by freeze-fracture and thin-section electron microscopy, *Brain Res.* **196**:11–31.

Peng, B., Cheng, P.-C., and Luther, P. W., 1981, Formation of ACh receptor clusters induced by positively charged latex beads, *Nature (Lond.)* **292**:831–834.

Peper, K., and McMahan, U. J., 1972, Distribution of acetylcholine receptors in the vicinity of nerve terminals on skeletal muscle of the frog, *Proc. R. Soc. Lond. B.* **181**:431–440.

Popiella, H., Beach, R. L., and Festoff, B. W., 1984, Developmental appearance of acetylcholinesterase molecular forms in cultured primary chick muscle cells, in: *Neuromuscular Diseases* (G. Serratrice, D. Cros, C. Desnuelle, J.-L. Gastaut, J.-F. Pellisier, J. Pouget, and A. Schiano, eds.), pp. 447–450, Raven Press, New York.

Porter, C. W., and Barnard, E. A., 1975, The density of cholinergic receptors at the endplate postsynaptic membrane: Ultrastructural studies in two mammalian species, *J. Membrane Biol.* **20**:31–49.

Pumplin, D. W., and Fambrough, D. M., 1982, Turnover of acetylcholine receptors in skeletal muscle, *Annu. Rev. Physiol.* **44**:319–335.

Rash, J. E., and Ellisman, M. H., 1974, Studies of excitable membranes. I. Macromolecular specializations of the neuromuscular junction and the nonjunctional sarcolemma, *J. Cell Biol.* **63:**567–586.

Reiger, F., Koenig, J., and Vigny, M., 1980, Spontaneous contractile activity and the presence of the 16S form of acetylcholinesterase in rat muscle cells in culture: Reversible suppressive action of tetrodotoxin, *Dev. Biol.* **76:**358–365.

Reiness, C. G., and Weinberg, C. B., 1981, Metabolic stabilization of acetylcholine receptors at newly formed neuromuscular junction in rat, *Dev. Biol.* **84:**247–254.

Ripley, K. L., and Provine, R. R., 1972, Neural correlates of embryonic motility in the chick, *Brain Res.* **45:**127–134.

Rotundo, R. L., and Fambrough, D. M., 1979, Molecular forms of chicken embryo acetylcholinesterase *in vitro* and *in vivo*, *J. Biol. Chem.* **254:**4790–4799.

Rubin, L. L., Schuetze, S. M., and Fischbach, G. D., 1979, Accumulation of acetylcholinesterase at newly formed nerve-muscle synapses, *Dev. Biol.* **69:**46–58.

Rubin, L. L., Schuetze, S. M., Weill, C. L., and Fischbach, G. D., 1980, Regulation of acetylcholinesterase appearance at neuromuscular junctions *in vitro*, *Nature (Lond.)* **283:**264–267.

Salpeter, M. M., Spanton, S., Holley, K., and Podleski, T. R., 1982, Brain extract causes acetylcholine receptor redistribution which mimics some early events at developing neuromuscular junctions, *J. Cell Biol.* **93:**417–425.

Sanes, J. R., 1983, Roles of extracellular matrix in neural development, *Annu. Rev. Physiol.* **45:**581–600.

Sanes, J. R., and Hall, Z. W., 1979, Antibodies that bind specifically to synaptic sites on muscle fiber basal lamina, *J. Cell Biol.* **83:**357–370.

Sanes, J. R., and Lawrence, J. C., 1983, Activity-dependent accumulation of basal lamina by cultured rat myotubes, *Dev. Biol.* **97:**123–136.

Sanes, J. R., Marshall, L. M., and McMahan, U. J., 1978, Reinnervation of muscle fiber basal lamina after removal of myofibers, *J. Cell Biol.* **78:**176–198.

Schaffner, A. E., and Daniels, M. P., 1982, Conditioned medium from cultures of embryonic neurons contains a high molecular weight factor which induces acetylcholine receptor aggregation on cultured myotubes, *J. Neurosci.* **2:**623–632.

Schuetze, S. M., 1980, The acetylcholine channel open time in chick muscle is not decreased following innervation, *J. Physiol. (Lond.)* **303:**111–124.

Schuetze, S. M., and Vicini, S., 1983, Denervation blocks the normal postnatal decrease in rat endplate channel open time, *Neurosci. Abs.* **9:**1108.

Schuetze, S. M., Frank, E., and Fischbach, G. D., 1978, Channel open time and metabolic stability of synaptic and extrasynaptic acetylcholine receptors on cultured chick myotubes, *Proc. Natl. Acad. Sci. USA* **75:**520–523.

Sisto-Daneo, L., and Filogama, G., 1975, Differentiation of synaptic area in "slow" and "fast" muscle fibers, *J. Submicrosc. Cytol.* **7:**121–131.

Sketelj, J., and Brzin, M., 1980, 16S acetylcholinesterase in endplate-free regions of developing rat diaphragm, *Neurochem. Res.* **5:**653–658.

Stanley, E. F., and Drachman, B., 1983, Rapid degradation of "new" acetylcholine receptors at neuromuscular junction, *Science* **222:**67–69.

Steinbach, J. H., 1981, Developmental changes in acetylcholine receptor aggregates at rat skeletal neuromuscular junctions, *Dev. Biol.* **84:**267–276.

Steinbach, J. H., Merlie, J., Heinemann, S., and Bloch, R., 1979, Degradation of junctional and extrajunctional acetylcholine receptors by developing rat skeletal muscle, *Proc. Natl. Acad. Sci. USA* **76:**3547–3551.

Sytkowski, A. J., Vogel, Z., and Nirenberg, M. W., 1973, Development of acetylcholine receptor clusters on cultured muscle cells, *Proc. Natl. Acad. Sci. USA* **70:**270–274.

Teravainen, H., 1968, Development of the myoneural junction in the rat, *Z. Zellerforsch.* **87:**249–265.

Vigny, M., and Koenig, J., 1978, Neural induction of the 16S acetylcholinesterase in muscle cell culture, *Nature (Lond.)* **271:**75–77.

Vigny, M., Koenig, J., and Reiger, F., 1976, The motor end-plate specific form of acetylcholinesterase: Appearance during embryogenesis and reinnervation of rat muscle, *J. Neurochem.* **27:**1347–1353.

Vogel, Z., and Daniels, M. P., 1976, Ultrastructure of acetylcholine receptor on cultured muscle fibers, *J. Cell Biol.* **69:**501–507.

Vogel, Z., Sytkowski, A. J., and Nirenberg, M. W., 1972, Acetylcholine receptors of muscle grown *in vitro, Proc. Natl. Acad. Sci. USA* **69:**3180–3184.

Wake, K., 1964, Motor endplates in developing chick embryo skeletal muscle: histological structure and histochemical localization of cholinesterase activity, *Arch. Histol. Jpn.* **25:**23–41.

Wake, K., 1976, Formation of myoneural and myotendinous junctions in the chick embryo, *Cell Tissue Res.* **173:**383–400.

Weinberg, C. B., and Hall, Z. W., 1979, Junctional form of acetylcholinesterase restored at nerve-free endplates, *Dev. Biol.* **68:**631–635.

Weinberg, C. B., Reiness, C. G., and Hall, Z. W., 1981, Topographical segregation of old and new acetylcholine receptors at developing ectopic endplates in adult rat muscle, *J. Cell Biol.* **88:**215–218.

Weldon, P. R., and Cohen, M. W., 1979, Development of synaptic ultrastructure at neuromuscular contacts in an amphibian cell culture system, *J. Neurocytol.* **8:**239–259.

Weldon, P. R., Moody-Corbett, F., and Cohen, M. W., 1981, Ultrastructure of sites of cholinesterase activity on amphibian embryonic muscle cells cultured without nerve, *Dev. Biol.* **84:**341–350.

Wilson, B. W., Neilberg, P. S., Walker, C. R., Linkhart, T A., and Fry, D. M., 1973, Production and release of acetylcholinesterase by cultured chick embryo muscle, *Dev. Biol.* **33:**285–299.

Yee, A., Fischbach, G. D., and Karnovsky, M. J., 1978, Clusters of intramembranous particles on cultured myotubes at sites that are sensitive to acetylcholine, *Proc. Natl. Acad. Sci. USA* **75:**3004–3008.

Young, S., and Poo, M.-M., 1983, Spontaneous release of substances from growth cones of embryonic neurones, *Nature (Lond.)* **305:**634–636.

Zacks, S. I., 1973, *The Motor Endplate*, Krieger, New York.

Ziskind-Conhaim, L., and Bennett, J. I., 1982, The effects of electrical inactivity and denervation on the distribution of acetylcholine receptors in developing rat muscle, *Dev. Biol.* **90:**185–197.

Ziskind-Conhaim, L., and Dennis, M. J., 1981, Development of rat neuromuscular junctions in organ culture, *Dev. Biol.* **85:**243–251.

Index

Acetylcholine, 605, 619, 622, 626
  noise, 607; *see also* Fluctuation analysis
  receptor, 606, 607, 608, 611, 614, 615, 616–621, 622, 623, 624–626
    channel kinetics of, 623, 624–626
    metabolic stability of, 624, 626
  release of, 622, 626
  sensitivity, 606, 622
Acetylcholinesterase, 606–607, 611–616, 618, 622–624, 626
  in neural crest cells, 481
  synapse specific form, 606–607, 611, 612, 615
N-Acetyl-D-galactosamine, 220, 225
Acridine orange, as pH indicator, 73
Actin
  as microfilament component, 4, 5–6
  NBD-phallacidin as a stain for, 69, 70, 81
  in phragmoplast, 49–50
  in sea urchin egg microvilli, 61, 63, 80
  in somite formation, 558–559
  in starfish oocyte cortex, 68–70, 81–84
  treadmilling of, 5–6, 8
  see also Microfilaments, Stress fibers
Actin-binding protein, 7
α-Actinin, 11, 12, 14
  in mouse embryo, 336
Active zones of neuromuscular junctions, 605, 608, 609
Adhesion
  and axon guidance, 574–575
  and epithelial shape, 128–130
  intercellular: *see* Intercellular adhesion
  role of lectins, 145–184
  to substrate, 300, 374, 379, 406–410, 422–424, 439, 442–445, 527, 529–530
Adhesive gradients, 443–445, 525–526

Adhesiveness, of neural crest cells, 518–519
Adiantum, cell wall assembly in, 41
Agglutination, to detect lectin activity, 148–150, 173
Aggregation, of sponge cells, 167–172
Aggregation factors, 168–172
  for sea urchin cells, 224–227
Aggregation receptor, 168–171
Alkaline phosphatase
  as germ cell marker, 436, 437, 440–442, 444
  activity, in sea urchin embryos, 204
Allium, stomatal formation in, 36–39, 48
Ambystoma, pigment pattern of, 532
Ambystoma embryos
  blastocoel fluid of, 304
  epithelial shape changes in, 131
  fate map of, 256–258
  gastrulation of, 257–260, 265, 281–283, 285, 287, 297, 309
  neural crest morphogenesis in, 486–489, 501–502, 505, 519–520, 527
  somite formation in, 556–557
9-Aminoacridine, as pH indicator, 73
Amphibian embryo
  gastrulation of, 241–314
  neurulation of, 126, 129
  primordial germ cells of, 434–435, 436–440, 442–443, 445–446
  somite formation in, 556–557
Amphibians: *see Ambystoma*, Frog, *Pleurodeles, Taricha, Triturus, Xenopus*
amputated mutant, 557–558
Anchorage of cells, 20–21
Anguilla, lectins of, 156
Animal hemisphere, 245–247
Anthocidaris, lectins of, 154
Antiaggregation receptor, 168–171

637